International
REVIEW OF
Neurobiology
Volume 79

International
REVIEW OF
Neurobiology

Volume 79

The Neurobiology of Multiple Sclerosis

EDITED BY

ALIREZA MINAGAR
Louisiana State University Health Sciences Center
Shreveport, Louisiana, USA

ELSEVIER

AMSTERDAM • BOSTON • HEIDELBERG • LONDON
NEW YORK • OXFORD • PARIS • SAN DIEGO
SAN FRANCISCO • SINGAPORE • SYDNEY • TOKYO
Academic Press is an imprint of Elsevier

Academic Press is an imprint of Elsevier
525 B Street, Suite 1900, San Diego, California 92101-4495, USA
84 Theobald's Road, London WC1X 8RR, UK

This book is printed on acid-free paper. ♾

For information on all Academic Press publications visit our Web site at www.books.elsevier.com

ISBN-13: 978-0-12-373736-6
ISBN-10: 0-12-373736-2

Printed and bound by CPI Group (UK) Ltd, Croydon, CR0 4YY

Transferred to Digital Print 2011

CONTENTS

The Role of CD4 T Cells in the Pathogenesis of Multiple Sclerosis

TANUJA CHITNIS

The CD8 T Cell in Multiple Sclerosis: Suppressor Cell or Mediator of Neuropathology?

AARON J. JOHNSON, GEORGETTE L. SUIDAN, JEREMIAH McDOLE, AND ISTVAN PIRKO

Immunopathogenesis of Multiple Sclerosis

SMRITI M. AGRAWAL AND V. WEE YONG

Molecular Mimicry in Multiple Sclerosis

JANE E. LIBBEY, LORI L. MCCOY, AND ROBERT S. FUJINAMI

Molecular "Negativity" May Underlie Multiple Sclerosis: Role of the Myelin Basic Protein Family in the Pathogenesis of MS

ABDIWAHAB A. MUSSE AND GEORGE HARAUZ

Microchimerism and Stem Cell Transplantation in Multiple Sclerosis

BEHROUZ NIKBIN, MANDANA MOHYEDDIN BONAB, AND FATEMEH TALEBIAN

The Insulin-Like Growth Factor System in Multiple Sclerosis

DANIEL CHESIK, NADINE WILCZAK, AND JACQUES DE KEYSER

Cell-Derived Microparticles and Exosomes in Neuroinflammatory Disorders

LAWRENCE L. HORSTMAN, WENCHE JY, ALIREZA MINAGAR, CARLOS J. BIDOT, JOAQUIN J. JIMENEZ, J. STEVEN ALEXANDER, AND YEON S. AHN

Multiple Sclerosis in Children: Clinical, Diagnostic, and Therapeutic Aspects

KEVIN ROSTÁSY

Migraine in Multiple Sclerosis

DEBRA G. ELLIOTT

Multiple Sclerosis as a Painful Disease

MEGHAN KENNER, UMA MENON, AND DEBRA G. ELLIOTT

Multiple Sclerosis and Behavior

JAMES B. PINKSTON, ANITA KABLINGER, AND NADEJDA ALEKSEEVA

Cerebrospinal Fluid Analysis in Multiple Sclerosis

FRANCISCO A. LUQUE AND STEPHEN L. JAFFE

Multiple Sclerosis in Isfahan, Iran

MOHAMMAD SAADATNIA, MASOUD ETEMADIFAR, AND AMIR HADI MAGHZI

Gender Issues in Multiple Sclerosis

ROBERT N. SCHWENDIMANN AND NADEJDA ALEKSEEVA

Differential Diagnosis of Multiple Sclerosis

HALIM FADIL, ROGER E. KELLEY, AND EDUARDO C. GONZALEZ-TOLEDO

Prognostic Factors in Multiple Sclerosis

ROBERTO BERGAMASCHI

Neuroimaging in Multiple Sclerosis

ROBERT ZIVADINOV AND JENNIFER L. COX

Detection of Cortical Lesions is Dependent on Choice of Slice Thickness in Patients with Multiple Sclerosis

ONDREJ DOLEZAL, MICHAEL G. DWYER, DANA HORAKOVA, EVA HAVRDOVA, ALIREZA MINAGAR, SRIVATS BALACHANDRAN, NIELS BERGSLAND, ZDENEK SEIDL, MANUELA VANECKOVA, DAVID FRITZ, JAN KRASENSKY, AND ROBERT ZIVADINOV

The Role of Quantitative Neuroimaging Indices in the Differentiation of Ischemia From Demyelination: An Analytical Study With Case Presentation

ROMY HOQUE, CHRISTINA LEDBETTER, EDUARDO C. GONZALEZ-TOLEDO, VIVEK MISRA, UMA MENON, MEGHAN KENNER, ALEJANDRO A. RABINSTEIN, ROGER E. KELLEY, ROBERT ZIVADINOV, AND ALIREZA MINAGAR

HLA-DRB1*1501, -DQB1*0301, -DQB1*0302, -DQB1 *0602, and -DQB1*0603 Alleles are Associated With More Severe Disease Outcome on MRI in Patients With Multiple Sclerosis

Robert Zivadinov, Laura Uxa, Alessio Bratina, Antonio Bosco, Bhooma Srinivasaraghavan, Alireza Minagar, Maja Ukmar, Su yen Benedetto, and Marino Zorzon

Glatiramer Acetate: Mechanisms of Action in Multiple Sclerosis

Tjalf Ziemssen and Wiebke Schrempf

Evolving Therapies for Multiple Sclerosis

Elena Korniychuk, John M. Dempster, Eileen O'Connor, J. Steven Alexander, Roger E. Kelley, Meghan Kenner, Uma Menon, Vivek Misra, Romy Hoque, Eduardo C. Gonzalez-Toledo, Robert N. Schwendimann, Stacy Smith, and Alireza Minagar

Remyelination in Multiple Sclerosis

Divya M. Chari

Trigeminal Neuralgia: A Modern-Day Review

Kelly Hunt and Ravish Patwardhan

Optic Neuritis and the Neuro-Ophthalmology of Multiple Sclerosis

Paramjit Kaur and Jeffrey L. Bennett

Neuromyelitis Optica: New Findings on Pathogenesis

Dean M. Wingerchuk

CONTRIBUTORS

Numbers in parentheses indicate the pages on which the authors' contributions begin.

Smriti M. Agrawal (99), Department of Clinical Neuroscience, University of Calgary, Calgary, Alberta T2N 4N1, Canada

Yeon S. Ahn (227), Wallace H. Coulter Platelet Laboratory, Department of Medicine, University of Miami, Miami, Florida 33136, USA

Nadejda Alekseeva (323, 377), Department of Psychiatry, Louisiana State University Medical Center, Shreveport, Louisiana 71103, USA; Department of Neurology and Department of Psychiatry, Louisiana State University Health Sciences Center, Shreveport, Louisiana 71103, USA

J. Steven Alexander (1, 227, 571), Department of Cellular and Molecular Physiology, Louisiana State University Health Sciences Center, Shreveport, Louisiana 71103, USA

Srivats Balachandran (475), Department of Neurology, State University of New York at Buffalo, Buffalo Neuroimaging Analysis Center, The Jacobs Neurological Institute, Buffalo, New York 14203, USA

Su yen Benedetto (521), Transfusion Center, Trieste, Italy

Jeffrey L. Bennett (633), Department of Neurology and Department of Ophthalmology, University of Colorado at Denver and Health Sciences Center, Denver, Colorado 80262, USA

Roberto Bergamaschi (423), Department of Clinical Neurology, Multiple Sclerosis Center, Neurological Institute, "C. Mondino," Via Mondino 2, Pavia 27100, Italy

Niels Bergsland (475), Department of Neurology, State University of New York at Buffalo, Buffalo Neuroimaging Analysis Center, The Jacobs Neurological Institute, Buffalo, New York 14203, USA

Carlos J. Bidot (227), Wallace H. Coulter Platelet Laboratory, Department of Medicine, University of Miami, Miami, Florida 33136, USA

Mandana Mohyeddin Bonab (13, 173), Department of Immunology, Immunogenetic Research Center, College of Medicine, Tehran University of Medical Sciences, Tehran 14155, Iran; Immunobiology Laboratory, Department of Immunology, College of Medicine, Tehran University of Medical Sciences, Tehran 14155, Iran

Antonio Bosco (521), Department of Clinical Medicine and Neurology, University of Trieste, Trieste, Italy

Alessio Bratina (521), Department of Clinical Medicine and Neurology, University of Trieste, Trieste, Italy

April Carpenter (1), Department of Cellular and Molecular Physiology, Louisiana State University Health Sciences Center, Shreveport, Louisiana 71103, USA

Divya M. Chari (589), Cambridge Centre for Brain Repair and Veterinary Medicine, University of Cambridge, Cambridge CB3OES, United Kingdom; and Institute for Science and Technology in Medicine, Keele University, Staffordshire ST55BG, United Kingdom

Daniel Chesik (203), Department of Neurology, University Medical Center Groningen, Hanzeplein 1, 9713 GZ Groningen, The Netherlands

Tanuja Chitnis (43), Department of Neurology, Brigham and Women's Hospital, Harvard Medical School, Boston, Massachusetts 02115, USA

Jennifer L. Cox (449), Buffalo Neuroimaging Analysis Center, The Jacobs Neurological Institute, State University of New York at Buffalo, School of Medicine and Biomedical Sciences, Buffalo, New York 14203, USA

Jacques De Keyser (203), Department of Neurology, University Medical Center Groningen, Hanzeplein 1, 9713 GZ Groningen, The Netherlands

John M. Dempster (571), Department of Neurology, Louisiana State University Health Sciences Center, Shreveport, Louisiana 71103, USA

Ondrej Dolezal (475), Department of Neurology, State University of New York at Buffalo, Buffalo Neuroimaging Analysis Center, The Jacobs Neurological Institute, Buffalo, New York 14203, USA; Department of Neurology, 1st School of Medicine, Charles University, Prague, Czech Republic

Michael G. Dwyer (475), Department of Neurology, State University of New York at Buffalo, Buffalo Neuroimaging Analysis Center, The Jacobs Neurological Institute, Buffalo, New York 14203, USA

Debra G. Elliott (281, 303), Department of Neurology, Louisiana State University Health Sciences Center, Shreveport, Louisiana 71103, USA

Masoud Etemadifar (357), Department of Neurology, Isfahan University of Medical Sciences, Isfahan 81744, Iran; Iranian Neurological Science Research Center, Alzahara Hospital, Isfahan 81746, Iran; Isfahan Medical Education Research Center, Isfahan University of Medical Sciences, Isfahan 81744, Iran; Isfahan Research Committee of Multiple Sclerosis, Isfahan MS Society, Isfahan 8158844799, Iran; Medical School, Isfahan University of Medical Sciences, Isfahan 81744, Iran

Halim Fadil (393), Department of Neurology, Louisiana State University Health Sciences Center, Shreveport, Louisiana 71103, USA

David Fritz (475), Department of Neurology, State University of New York at Buffalo, Buffalo Neuroimaging Analysis Center, The Jacobs Neurological Institute, Buffalo, New York 14203, USA

Robert S. Fujinami (127), Department of Neurology, University of Utah School of Medicine, Salt Lake City, Utah 84132, USA

Eduardo Gonzalez-Toledo (393, 491, 571), Department of Neurology and Department of Radiology, Louisiana State University Health Sciences Center, Shreveport, Louisiana 71103, USA

George Harauz (149), Department of Molecular and Cellular Biology, Biophysics Interdepartmental Group University of Guelph, Ontario N1G 2W1, Canada

Eva Havrdova (475), Department of Neurology, 1st School of Medicine, Charles University, Prague, Czech Republic

Romy Hoque (491, 571), Department of Neurology, Louisiana State University Health Sciences Center, Shreveport, Louisiana 71103, USA

Dana Horakova (475), Department of Neurology, 1st School of Medicine, Charles University, Prague, Czech Republic

Lawrence L. Horstman (227), Wallace H. Coulter Platelet Laboratory, Department of Medicine, University of Miami, Miami, Florida 33136, USA

Kelly Hunt (621), The Interactive Neuroscience Center, LLC, Shreveport, Louisiana 71115, USA

Stephen L. Jaffe (341), Department of Neurology, Louisiana State University School of Medicine, Shreveport, Louisiana 71103, USA

Joaquin J. Jimenez (227), Wallace H. Coulter Platelet Laboratory, Department of Medicine, University of Miami, Miami, Florida 33136, USA

Aaron J. Johnson (73), Department of Neurology, University of Cincinnati, Cincinnati, Ohio 45267, USA

Wenche Jy (227), Wallace H. Coulter Platelet Laboratory, Department of Medicine, University of Miami, Miami, Florida 33136, USA

Anita Kablinger (323), Department of Psychiatry, Louisiana State University Medical Center, Shreveport, Louisiana 71103, USA

Paramjit Kaur (633), Department of Neurology, University of Colorado at Denver and Health Sciences Center, Denver, Colorado 80262, USA

Roger E. Kelley (393, 491, 571), Department of Neurology, Louisiana State University Health Sciences Center, Shreveport, Louisiana 71103, USA

Meghan Kenner (303, 491, 571), Department of Neurology, Louisiana State University Health Sciences Center, Shreveport, Louisiana 71103, USA

Farideh Khosravi (13), Department of Immunology, Immunogenetic Research Center, College of Medicine, Tehran University of Medical Sciences, Tehran 14155, Iran

Elena Korniychuk (571), Department of Neurology, Louisiana State University Health Sciences Center, Shreveport, Louisiana 71103, USA

Jan Krasensky (475), Department of Radiology, 1st School of Medicine, Charles University, Prague, Czech Republic

Christina Ledbetter (491), Department of Psychiatry, Louisiana State University Health Sciences Center, Shreveport, Louisiana 71103, USA

Jane E. Libbey (127), Department of Neurology, University of Utah School of Medicine, Salt Lake City, Utah 84132, USA

Francisco A. Luque (341), Neurology Service, Overton Brooks VA Medical Center, Shreveport, Louisiana 71101, USA

Amir Hadi Maghzi (357), Isfahan Research Committee of Multiple Sclerosis, Isfahan MS Society, Isfahan 8158844799, Iran; Medical School, Isfahan University of Medical Sciences, Isfahan 81744, Iran

Lori L. McCoy (127), Department of Neurology, University of Utah School of Medicine, Salt Lake City, Utah 84132, USA

Jeremiah McDole (73), University of Cincinnati Neuroscience Program, Vontz Center for Molecular Studies, University of Cincinnati College of Medicine, Cincinnati, Ohio 45267, USA

Uma Menon (303, 491, 571), Department of Neurology, Louisiana State University Health Sciences Center, Shreveport, Louisiana 71103, USA

Alireza Minagar (1, 227, 475, 491, 521, 571), Department of Neurology, Louisiana State University Health Sciences Center, Shreveport, Louisiana 71103, USA

Vivek Misra (491, 571), Department of Neurology, Louisiana State University Health Sciences Center, Shreveport, Louisiana 71103, USA

Abdiwahab A. Musse (149), Department of Molecular and Cellular Biology, Biophysics Interdepartmental Group, University of Guelph, Ontario N1G 2W1, Canada

Behrouz Nikbin (13, 173), Department of Immunology, Immunogenetic Research Center, College of Medicine, Tehran University of Medical Sciences, Tehran 14155, Iran; Immunobiology Laboratory, Department of Immunology, College of Medicine, Tehran University of Medical Sciences, Tehran 14155, Iran

Eileen O'Connor (571), Biogen-Idec, Boston, Massachusetts 02142, USA

Ravish Patwardhan (621), The Interactive Neuroscience Center, LLC, Shreveport, Louisiana 71115, USA

James B. Pinkston (323), Department of Neurology and Department of Psychiatry, Louisiana State University Medical Center, Shreveport, Louisiana 71103, USA

Istvan Pirko (73), Department of Neurology, University of Cincinnati, Cincinnati, Ohio 45267, USA

Alejandro A. Rabinstein (491), Department of Neurology, Mayo Clinic, Rochester, Minesota, USA

Kevin Rostásy (269), Department of Pediatrics, Division of Pediatric Neurology, Medical University of Innsbruck, Innsbruck 6020, Austria

Mohammad Saadatnia (357), Department of Neurology, Isfahan University of Medical Sciences, Isfahan 81744, Iran; Iranian Neurological Science Research Center, Alzahara Hospital, Isfahan 81746, Iran; Isfahan Medical Education Research Center, Isfahan University of Medical Sciences, Isfahan 81744, Iran; Isfahan Research Committee of Multiple Sclerosis, Isfahan MS Society, Isfahan 8158844799, Iran; Medical School, Isfahan University of Medical Sciences, Isfahan 81744, Iran

Wiebke Schrempf (537), Multiple Sclerosis Center Dresden, Neurological University Clinic, Dresden University of Technology, Dresden 01307, Germany

Robert N. Schwendimann (377, 571), Department of Neurology and Department of Psychiatry, Louisiana State University Health Sciences Center, Shreveport, Louisiana 71103, USA

Zdenek Seidl (475), Department of Radiology, 1st School of Medicine, Charles University, Prague, Czech Republic

Stacy Smith (571), Department of Neurology, Louisiana State University Health Sciences Center, Shreveport Louisiana 71103, USA

Bhooma Srinivasaraghavan (521), Department of Neurology, State University of New York at Buffalo, Buffalo Neuroimaging Analysis Center, The Jacobs Neurological Institute, State University at Buffalo, New York 14203, USA

Georgette L. Suidan (73), University of Cincinnati Neuroscience Program, Vontz Center for Molecular Studies, University of Cincinnati College of Medicine, Cincinnati, Ohio 45267, USA

Fatemeh Talebian (13, 173), Department of Immunology, Immunogenetic Research Center, College of Medicine, Tehran University of Medical Sciences, Tehran 14155, Iran; Immunobiology Laboratory, Department of Immunology, College of Medicine, Tehran University of Medical Sciences, Tehran 14155, Iran

Maja Ukmar (521), Department of Clinical, Morphological and Technological Sciences, University of Trieste, Trieste, Italy

Laura Uxa (521), Transfusion Center, Trieste, Italy

Manuela Vaneckova (475), Department of Radiology, 1st School of Medicine, Charles University, Prague, Czech Republic

Nadine Wilczak (203), Department of Neurology, University Medical Center Groningen, Hanzeplein 1, 9713 GZ Groningen, The Netherlands

Dean M. Wingerchuk (665), Mayo Clinic College of Medicine, Scottsdale, Arizona 85259, USA

V. Wee Yong (99), Department of Clinical Neuroscience and Department of Oncology, University of Calgary, Calgary, Alberta T2N 4N1, Canada

Tjalf Ziemssen (537), Multiple Sclerosis Center Dresden, Neurological University Clinic, Dresden University of Technology, Dresden 01307, Germany

Robert Zivadinov (449, 475, 491, 521), Department of Neurology, State University of New York at Buffalo, Buffalo Neuroimaging Analysis Center, The Jacobs Neurological Institute, State University of New York at Buffalo, School of Medicine and Biomedical Sciences, Buffalo, New York 14203, USA

Marino Zorzon (521), Department of Clinical Medicine and Neurology, University of Trieste, Trieste, Italy

PREFACE

Multiple sclerosis (MS) is an immune-mediated neurodegenerative disorder of the human central nervous system (CNS), which usually affects young adults with certain genetic background who are then exposed to certain precipitating environmental antigen(s). Despite major advances of the past two decades in understanding the pathophysiology of MS, and in spite of the introduction of new immunomodulatory and immunosuppressive agents, which may slow down disease progression and delay the onset of disability, the "cause" and the "cure" for MS remain elusive. What traditionally has been viewed and researched as a demyelinating disease is a more widespread destructive process than was initially conceived and affects several resident cells of the CNS. In addition, what has been previously termed as "MS" is not a single syndrome but a conglomeration of various syndromes with unlike pathogenesis and dissimilar responses to available treatments. Although we have learned much more about the most common form of MS (relapsing-remitting MS), the molecular pathogenesis of clinical relapses and remissions remains only partially understood.

This volume of *International Review of Neurobiology* focuses on MS and related disorders. The volume can be divided into various sections with the main emphasis on MS pathogenesis, clinical features and epidemiology, neuroimaging, and treatment. These papers are followed by three other chapters on MS-related disorders.

The first section of this volume contains 10 chapters on pathogenesis and immunopathogenesis of MS. First, Minagar *et al.* (2007) reviews the interactions of leukocytes and cerebral endothelial cells in the pathogenesis of MS. This chapter is followed by three other chapters discussing the role of B cells (Nikbin *et al.*, 2007a), CD4 T cells (Chitnis, 2007), and CD8 cells (Johnson *et al.*, 2007) in the pathogenesis of MS, with emphasis on the most recent findings on these cell groups. This section then continues with an excellent review of immunopathogenesis of MS by Agrawal and Yong (2007), which provides readers with a detailed analysis of pathogenic mechanisms. The next five chapters provide readers with detailed reviews of some of the most intensely debated aspects of MS. Libbey *et al.* (2007) discusses the concept of molecular mimicry in pathogenesis of MS and present a detailed discussion about all proposed agents with molecular similarities to self-antigens and their possible involvement in the initiation of the inflammatory cascade. Next, Musse and Harauz (2007) present an excellent discussion on the role of the myelin basic protein family in pathogenesis of MS. The last two chapters of this section review look at the mechanisms in the world of MS.

Nikbin *et al.* (2007b) describe the role of microchimerism in immune-mediated diseases in general, but MS in particular. The authors raise new questions about the possible role of this fascinating phenomenon in the development of, and also protection against, MS. This interesting chapter also probes the intensely discussed issue of stem cell therapy in MS. This remarkable chapter is followed by an interesting chapter on the involvement of insulin-like growth factors in MS, providing our avid readers with the latest findings in this field (Chesik *et al.*, 2007). The last chapter delves into another fascinating subject: microparticles, exosomes, and their role in neuropathologic syndromes (Horstman *et al.*, 2007). The last decade has witnessed remarkable advances concerning the role of these newly detected participants in the inflammatory cosmos; and many scientists around the globe are now involved in exploring their nature and function.

The second section of this issue is devoted to the clinical features and epidemiology of MS. Elliot and colleagues provide the readers with two superb chapters, one on migraine and MS (Elliott, 2007) and the other on the various pain syndromes occurring in the MS population (Kenner *et al.*, 2007). Generally, MS is not considered a "painful disease"; however, this is an erroneous view that disregards the daily pain and agony experienced by MS patients. The authors of these two chapters review the commonality of migraine and other painful syndromes in MS patients. This section then continues with two outstanding reviews. The first involves the neuropsychiatric manifestations of MS (Pinkston *et al.*, 2007), and the next focuses on the role of cerebrospinal fluid analysis in diagnosing MS and excluding other disorders which can mimic MS (Luque and Jaffe, 2007). The next chapter in this section discusses the clinical features and therapeutic options for MS in very young patients (Rotasy, 2007). MS in children is not a well-described entity, and only during the last few years, there has been a surge of interest in this significant issue. The following chapter of this second section describes the epidemiology and clinical features of MS in Isfahan, Iran (Saadatnia *et al.*, 2007). The literature on MS in Middle Eastern countries, particularly Iran, is limited. The authors provide us with a new window into the clinical behavior of MS in Iranian patients, including the significant features that set them apart from the disease process, which one observes in the Western Hemisphere. The last three chapters in this section discuss differential diagnosis, prognostic factors, and gender issues in MS. Fadil *et al.* (2007) provide readers with an extensive review of other disorders which comprise the differential diagnosis of MS, potentially assisting our younger colleagues in setting MS apart from MS imitators. This useful chapter is followed by another significant chapter concerning the prognosis of MS. MS patients often make inquiry about their prognosis, that is, "what lies ahead of them." Bergamaschi (2007) has compiled a comprehensive chapter that, in great detail, examines the prognostic factors in MS. This section ends with another thorough review, the role of gender in the development of MS. Schwendimann and Alekseeva (2007) attempt to provide some answers to the fundamental questions: Why MS is more frequent among female than male individuals? The authors examine the available literature on this significant topic and provide readers with interesting and thought-provoking concepts.

The third section of this volume concentrates on the various aspects of neuroimaging of MS. At first, Zivadinov and Cox (2007) present an ample review of the latest findings in the field of MS neuroimaging. This chapter sets the stage for the next two chapters. The first discusses newest MRI techniques for detection of cortical lesions (Dolezal *et al.*, 2007), and the next discusses differentiation of ischemic lesions from demyelinating lesions (Hoque *et al.*, 2007). Finally, this section contains a chapter correlating various HLA class II markers with MRI-measured disease severity (Zivadinov *et al.*, 2007).

The treatment section contains two significant chapters on the mechanisms of action of glatiramer acetate, followed by a chapter of emerging treatments for MS. Ziemssen and Schrempf (2007) present their detailed chapter delineating the immunomodulatory mechanisms of glatiramer acetate. Then, Korniychuk *et al.* (2007) provide readers with an up-to-date report on the upcoming therapies for MS. This interesting chapter presents the latest findings on the therapeutic as well as adverse effects of monoclonal antibodies, which are currently under investigation for treatment of MS. The final chapter of this section is provided by Chari (2007) who reviews the concept of remyelination in MS and presents the latest knowledge on this key issue to our readers.

The final section of this issue of the *International Review of Neurobiology* concentrates on three other disorders that often occur in MS or are in its differential diagnosis. First, Hunt and Patwardhan (2007) review the subject of trigeminal neuralgia with emphasis on its relevance to MS, since significant number of MS patients initially present with trigeminal neuralgia. Then, Kaur and Bennett (2007) review optic neuritis and other neuro-ophthalmologic presentation in MS. This excellent chapter provides clinicians with the latest findings in the field, assisting them to interpret these important visual symptoms and signs in the MS population. The issue finishes with another remarkable chapter that concerns one of the editor's favorite disorders: neuromyelitis optica (Wingerchuk, 2007). Despite major advances in our understanding of the pathogenesis of neuromyelitis optica, the cause and treatment of this unique and devastating syndrome still remain elusive. In this chapter, Wingerchuk, a world authority, presents the latest developments regarding the pathogenesis of this disorder.

In summary, the contributors to this issue of *International Review of Neurobiology* have attempted to extend the readers' knowledge of the various aspects of MS and MS-related syndromes. Each chapter addresses a critical issue in the study of MS, with the hope of stimulating further research into the pathogenesis of this elusive disease.

Acknowledgments

I would like to acknowledge and commend the great work and effort of all the contributors to this volume of the "International Review of Neurobiology," who have provided me with their superior chapters. Without their effort and dedication, this issue would have never become a reality. I also would like to thank Dr. Ronald Bradley, Mr. Rogue Schindler, Ms. Cindy Minor, Ms. Stacy J. Smith, Ms. Brenda Shivers, and Mr. Prasenjit Bakshi for all their effort and time expended on this issue.

References

Agrawal, S. M., and Yong, V. W. (2007). Immunopathogenesis of multiple sclerosis. *Int. Rev. Neurobiol.* **79,** 99–126.

Bergamaschi, R. (2007). Prognostic factors in multiple sclerosis. *Int. Rev. Neurobiol.* **79,** 423–447.

Chari, D. M. (2007). Remyelination in multiple sclerosis. *Int. Rev. Neurobiol.* 589–620.

Chesik, D., Wilczak, N., and DeKeyser, J. (2007). The insulin-like growth factor system in multiple sclerosis. *Int. Rev. Neurobiol.* **79,** 203–226.

Chitnis, T. (2007). The role of CD4 T cells in the pathogenesis of multiple sclerosis. *Int. Rev. Neurobiol.* **79,** 43–72.

Dolezal, O., Dwyer, M. G., Horakova, D., Havrodava, E., Minagar, A., Balachandran, S., Bergsland, N., Seidl, Z., Vaneckova, M., Fritz, D., Krasensky, J., and Zivadinov, R. (2007). Detection of cortical lesions is dependent on choice of slice thickness in patients with multiple sclerosis. *Int. Rev. Neurobiol.* **79,** 475–489.

Elliott, D. G. (2007). Migraine in multiple sclerosis. *Int. Rev. Neurobiol.* **79,** 279–300.

Fadil, H., Kelley, R. E., and Gonzalez-Toledo, E. (2007). Differential diagnosis of multiple sclerosis. *Int. Rev. Neurobiol.* **79,** 391–420.

Horstman, L. L., Jy, W., Minagar, A., Bidot, C. J., Jimenez, J. J., Alexander, J. S., and Ahn, Y. S. (2007). Cell-derived microparticles and exosomes in neuroinflammatory disorders. *Int. Rev. Neurobiol.* **79,** 227–268.

Hoque, R., Ledbetter, C., Gonzalez-Toledo, E., Misra, V., Menon, U., Kenner, M., Zivadinov, R., Rabinstein, A. A., Kelley, R. E., and Minagar, A. (2007). The role of quantitative neuroimaging indices in the differentiation of ischemia from demyelination: An analytical study with case presentation. *Int. Rev. Neurobiol.* **79,** 491–519.

Hunt, K., and Patwardhan, R. (2007). Trigeminal neuralgia: A modern-day review. *Int. Rev. Neurobiol.* **79,** 621–631.

Luque, F., and Jaffe, S. L. (2007). Cerebrsopinal fluid analysis in multiple sclerosis. *Int. Rev. Neurobiol.* **79,** 339–354.

Johnson, A. J., Suidan, G. L., McDole, J., and Pirko, I. (2007). The CD8 T cell in multiple sclerosis: Suppressor cell or mediator of neuropathology. *Int. Rev. Neurobiol.* **79,** 73–97.

Kaur, P., and Bennett, J. L. (2007). Optic neuritis and the neuro-ophthalmology of multiple sclerosis. *Int. Rev. Neurobiol.* **79,** 633–663.

Kenner, M., Menon, U., and Elliott, D. G. (2007). Multiple sclerosis as a painful disease. *Int. Rev. Neurobiol.* **79,** 303–321.

Korniychuk, E., Dempster, J. M., O'Connor, E., Alexander, J. S., Kelley, R. E., Kenner, M., Menon, U., Misra, V., Hoque, R., Gonzalez-Toledo, E., Schwendimann, R. N., and Minagar, A. (2007). Evolving therapies for multiple sclerosis. *Int. Rev. Neurobiol.* **79,** 571–588.

Libbey, J. E., McCoy, L. L., and Fujinami, R. S. (2007). Molecular mimicry in multiple sclerosis. *Int. Rev. Neurobiol.* **79,** 127–147.

Minagar, A., Carpenter, A., and Alexander, J. S. (2007). The destructive alliance: Interactions of leukocytes, cerebral endothelial cells, and the immune cascade in pathogenesis of multiple sclerosis. *Int. Rev. Neurobiol.* **79,** 1–11.

Musse, A. A., and Harauz, G. (2007). Molecular "negativity" may underlie multiple sclerosis: Role of myelin basic protein family in the pathogenesis of MS. *Int. Rev. Neurobiol.* **79,** 149–172.

Nikbin, B., Bonab, M. M., Khosravi, F., and Talebian, F. (2007a). Role of B cells in pathogenesis of multiple sclerosis. *Int. Rev. Neurobiol.* **79,** 13–42.

Nikbin, B., Bonab, M. M., and Talebian, F. (2007b). Microchimerism and stem cell tansplantation in multiple sclerosis. *Int. Rev. Neurobiol.* **79,** 173–202.

Pinkston, J. B., Kablinger, A., and Alekseeva, N. (2007). Multiple sclerosis and behavior. *Int. Rev. Neurobiol.* **79,** 321–338.

Rotasy, K. (2007). Multiple sclerosis in children: Clinical, diagnostic, and therapeutic aspects. *Int. Rev. Neurobiol.* **79,** 267–277.

Saadatnia, M., Etemadifar, M., and Maghzi, A. H. (2007). Multiple sclerosis in Isfahan, Iran. *Int. Rev. Neurobiol.* **79,** 357–375.

Schwendimann, R. N., and Alekseeva, N. (2007). Gender issues in multiple sclerosis. *Int. Rev. Neurobiol.* **79,** 375–390.

Ziemssen, T., and Schrempf, W. (2007). Glatiramer acetate: Mechanisms of action in multiple sclerosis. *Int. Rev. Neurobiol.* **79,** 537–570.

Zivadinov, R., and Cox, J. L. (2007). Neuroimaging in multiple sclerosis. *Int. Rev. Neurobiol.* **79,** 449–474.

Zivadinov, R., Uxa, L., Bratina, A., Bosco, A., Srinivasaraghavan, B., Minagar, A., Ukmar, M., Benedetto, S., and Zorzon, M. (2007). HLA-DRB1*1501, -DQB1*0301, -DQB1*0302, -DQB1*0602, and -DQB1*0603 alleles are associated with more severe disease outcome on MRI in patients with multiple sclerosis. *Int. Rev. Neurobiol.* **79,** 521–535.

Wingerchuk, D. M. (2007). Neuromyelitis optica: New findings on pathogenesis. *Int. Rev. Neurobiol.* **79,** 665–688.

THE DESTRUCTIVE ALLIANCE: INTERACTIONS OF LEUKOCYTES, CEREBRAL ENDOTHELIAL CELLS, AND THE IMMUNE CASCADE IN PATHOGENESIS OF MULTIPLE SCLEROSIS

Alireza Minagar,* April Carpenter,[†] and J. Steven Alexander[†]

*Department of Neurology, Louisiana State University Health Sciences Center
Shreveport, Louisiana 71103, USA
[†]Department of Cellular and Molecular Physiology
Louisiana State University Health Sciences Center
Shreveport, Louisiana 71103, USA

Multiple sclerosis (MS) is an inflammatory disease of the human central nervous system (CNS) which develops predominantly in young adults with certain predisposing genetic characteristics, often following exposure to initiating environmental insult(s) including viral infections. The causes of MS remain elusive and no entire cure is in sight. However, it is well known that interactions between the immune system and the CNS play a central role in MS pathogenesis. Patients with MS generate CD4+ autoreactive T cells that at some point differentiate to Th1 phenotype cells, which are the major players in maintaining a continuous destructive immune response against brain and spinal cord antigens. Other significant participants in MS pathogenesis involved in the destruction of the target tissue are cerebral endothelial cells, CD8+ T cells, B cells, complement, autoantibodies, cytokines, and chemokines. The presence and interactions of all these participants further complicate the pathogenesis of MS, and make finding a cure for MS challenging. This chapter looks at the roles of these factors in the development of MS.

INTERNATIONAL REVIEW OF
NEUROBIOLOGY, VOL. 79
DOI: 10.1016/S0074-7742(07)79001-3

0074-7742/07 $35.00

I. Introduction

Multiple sclerosis (MS) is an immune-mediated neurodegenerative disorder which affects the human central nervous system (CNS) (Frohman *et al.*, 2006; Noseworthy *et al.*, 2000). In the United States alone, more than 350,000 individuals suffer from MS, and it is the most common neurological cause of disability in young adults. Generally, MS is not lethal, but imposes devastating neurological and psychiatric limits on patients. It is important to realize that MS is not a single disease but a conglomerate of different neurological syndromes with different pathological bases and hence dissimilar responses to therapeutic intervention. Relapsing-remitting MS (RRMS), the most common form of MS, affects females twice as often as males (Keegan and Noseworthy, 2002). Many patients with RRMS, with or without treatment, will later develop secondary progressive MS (SPMS). SPMS is characterized as a neurodegenerative syndrome with less clinical relapses, less response to immunomodulatory treatments, and persistent progression of the disease with accumulation of disability. Primary progressive MS (PPMS) is manifested by an insidious onset and steady disease progression. Patients with PPMS show the least responsiveness to therapy. Neuropathological studies have shown that each of these different forms of MS show different patterns of morphological and histological alterations in the CNS (Lucchinetti *et al.*, 2000). For example, RRMS is pathologically characterized by the predominant presence of inflammatory cells, while the neuropathology of PPMS mainly consists of less severe inflammation, more prolonged T-cell infiltration, more pronounced oligodendrocyte loss, and an ongoing low level of axonal damage in PPMS (Bruck *et al.*, 2002). The factors responsible for this diverse underlying neuropathology remain largely unknown but most likely reflect compound interactions between individual genetics and environmental antigenic "triggers," all of which provoke immune responses and/or increased levels of susceptibility to the inflammatory cascade. In addition to these initiating factors, chronic inflammation also alters and impairs mechanisms related to repair and restitution in the CNS (Frohman *et al.*, 2006).

Currently, there are two main views about disease onset in MS: central versus peripheral. Some neuropathologists and neuroscientists believe that MS begins within the CNS and then propagates to the peripheral immune system (Barnett and Prineas, 2004). A neuropathology study by Barnett and Prineas (2004) of patients with RRMS strongly supports this concept. The investigators reported clinical and pathological findings in 12 patients with relapsing and remitting MS, who died during or shortly after the onset of a relapse. On the basis of their observations, neuropathological changes not previously associated with the formation of new symptomatic lesions were observed in seven cases, and included extensive oligodendrocyte apoptosis. Microglia were activated in the myelinated tissue which contained few or no lymphocytes or myelin-associated phagocytes. The findings of this study raise the possibility that MS may be initiated within the CNS and then move to the periphery.

On the other hand, the majority of neuroscientists believe that MS is triggered only after an individual's exposure to certain environmental factors like viral agents (Steinman and Zamvil, 2006). This initial viral exposure/infection activates CD4+ T cells against CNS tissue antigens. These imprinted cells eventually gain access to the CNS microenvironment, which in turn perpetuates ongoing cycles of neuroinflammation and neurodegeneration (Lovett-Racke and Racke, 2006). Most neuroimmunologists favor the peripheral onset model of MS based on composition of cellular infiltrates of the CSF and MS lesions, and data obtained from experimental allergic (autoimmune) encephalomyelitis (EAE). In EAE and perhaps in MS, CD4+ T cells somehow become sensitized against myelin basic proteins (MBPs), and eventually cross the endothelial barrier of the blood–brain barrier (BBB). This concept of MS pathogenesis is further supported by the fact that certain class II HLA molecules act as the antigen-presenting molecules to the activated pathogenic CD4+ T cells (Kort *et al.*, 2006).

The dramatic increase in research into MS pathogenesis during the last two decades has expanded our knowledge of the involvement of other cellular/molecular elements in pathogenesis of this elusive and complicated neurological disease. In recent years, we have learned more about the various roles of B cells, CD8+ T cells, cerebral endothelial cells, and various pro- and anti-inflammatory cytokines in the development and maintenance of the continuous pathology of MS. The roles of B cells, CD4+ T cells, and CD8+ T cells have been described in detail in separate chapters in this volume. Therefore, we will only focus on the role of cerebral endothelial cells and their interactions with activated leukocytes in pathogenesis of MS.

II. Role of Activated Cerebral Endothelial Cells in Pathogenesis of MS

Cerebral endothelial cells create the anatomic and physiological barrier of the BBB and play an essential role in forming the demyelinating lesions of MS. Under normal circumstances, CNS endothelial cells prevent transendothelial escape of various blood-borne molecules and the migration of leukocytes into the CNS. Indeed, the combined endothelial barrier and the supporting astrocytes keep the CNS out of reach of the peripheral immune system, homeostatically maintaining a stable environment for neurons to function. Cerebral endothelial cells possess several unique characteristics which enable them to protect the CNS. Cerebral endothelial cells are metabolically highly active, lack fenestrations, create a complex network of tight junctions that adheres between adjacent cells, and block paracellular movement of cells and molecules into the CNS compartment (Abbott *et al.*, 2006; Minagar and Alexander, 2003). Cerebral endothelial cells contain selective transport mechanisms which mediate transport of nutrients into and toxic metabolites out of the CNS (Abbott *et al.*, 2006). Under normal circumstances, cerebral endothelial cells do not express class II HLA molecules and cannot act as antigen-presenting cells. In addition, CNS does not have a lymphatic system, a feature which makes CNS more immunoprivileged (Hatterer *et al.*, 2006).

However, during neuroinflammatory disorders, such as MS, the inflammatory cells (especially MBP reactive CD4+ T cells), macrophages, and B cells readily gain access to and initiate a continuous cascade of destruction within the CNS.

III. Potential Role of Endothelial Microparticles in Pathogenesis of MS

It is known that under normal circumstances, endothelial cells continuously shed small membranous vesicles called "endothelial microparticles" (EMP) $<1\ \mu m$, consisting mainly of membrane phospholipids and proteins. Released EMP also carry the same adhesion molecules as their parent cells and have been extensively studied in the context of various pathological conditions including MS (Jimenez *et al.*, 2005). During pathogenesis of active MS *in vivo* and on exposure of cerebral endothelial cells to elevated serum levels of proinflammatory cytokines, such as TNF-α and IFN-γ, high numbers of EMP are released into peripheral circulation (Jy *et al.*, 2004). We reported elevated plasma of EMP carrying CD31/PECAM-1 (CD31 + EMP) during MS relapses (Minagar *et al.*, 2001). We observed the elevation of CD51/endoglin + EMP during both relapse and remission. Elevated plasma levels of CD31 + EMP was associated with presence of contrast-enhancing lesions on brain MRI scans. In another set of experiments, Jy *et al.* (2004) showed that EMP formed bonds with monocytes and created EMP-monocyte conjugates. The plasma levels of these CD54/ICAM-1 + EMP-monocyte conjugates were elevated in MS patients with active disease. The investigators also assessed the role of EMP-monocyte conjugates in transendothelial migration of monocytes and observed that the formation of such complexes facilitated transendothelial migration of monocytes. The investigators concluded that EMP could enhance inflammation and increase transendothelial migration of monocytes in MS by binding to and activating monocytes through CD54/ICAM-1. Further research into role of EMP in MS pathogenesis revealed that beta-interferons (IFN-β1a and IFN-β1b) were able to reduce the amount of EMP released by MS patients, both *in vitro* (Jimenez *et al.*, 2005) and *in vivo* (Sheremata *et al.*, 2006).

IV. Interactions Among Endothelial Cells and Activated Leukocytes in Pathogenesis of MS

Transendothelial migration of activated leukocytes in pathogenesis of MS is a dynamic and complicated process which remains marginally understood. Passage of the activated leukocytes through the junctions between adjacent endothelial cells involves interactions among passing cells and junctional components.

The classic model of leukocyte migration through endothelium consists of a number of key steps: rolling and tethering, interactions of activated leukocytes with adhesion molecules expressed by stimulated endothelium, which leads to adhesion, and migration. Tethering and rolling captures the activated leukocytes and is a crucial step in MS pathogenesis. Initially, leukocytes come in contact with the underlying endothelium, a process which is mediated by adhesion molecules of the selectin family and their corresponding carbohydrate ligands. The rolling process slows leukocytes, letting them contact the endothelial surface sampling for the presence of platelet-activating factors and chemokines. Suppressing rolling behavior inhibits leukocyte adhesion, but inhibiting firm adhesion does not block rolling.

The selectins are type 1 transmembrane glycoproteins which attach to sialylated carbohydrate motifs in a Ca^{2+}-dependent manner. Three different types of selectins are identified: L-, P-, and E-selectins and all three groups of selectin genes are located on chromosome 1. These selectins share a lectin-like domain at the N-terminus followed by an epidermal growth factor (EGF)-like domain. L-selectins (CD62L) is expressed by neutrophils, myeloid, naive T cells, memory, and activated T cells. P-selectins (CD62P) are expressed constitutively by platelets (in their α-granules), and later on the activated platelet surface. Similarly, endothelial Weibel–Palade bodies store P-selectin which is rapidly mobilized to the cell surface following activation. E-selectin (CD62E) is expressed exclusively by endothelial cells. However, E- and P-selectins and PGSL-1 are not involved in pathogenesis of EAE and the entry of immune cells to the CNS and thus may play a relatively minor role in human MS (Engelhardt *et al.*, 2005; Osmers *et al.*, 2005).

Firm adhesion of the activated leukocyte precedes transendothelial migration. Under normal circumstances, the complexion of the resting cerebral endothelium differs significantly from activated cells and has low selectin levels, but constitutively expresses ICAM-1 and ICAM-2. Studies on animals with EAE demonstrated that the expression of intercellular adhesion molecule-1 (ICAM-1) and vascular cell adhesion molecule-1 (VCAM-1) are dramatically upregulated by inflamed cerebral endothelium (Baron *et al.*, 1993; Cross *et al.*, 1990; Steffen *et al.*, 1994). The activated leukocytes within the perivenular cuffs of MS lesions express ligands for ICAM-1 and VCAM-1 (e.g., LFA-1 and $\alpha4\beta1$ integrins), suggesting these are involved in the efflux of these cells (Engelhardt *et al.*, 1998). Upregulated VCAM-1 and ICAM-1 expression play a significant role in the pathogenesis of EAE and MS since studies have demonstrated adhesion of encephalitogenic CD4+ T cells and monocytes to mouse or human cerebral endothelial cells via $\alpha 4\beta 1$ integrin-VCAM-1- and LFA-1-ICAM-1-mediated interactions (Floris *et al.*, 2002; Greenwood *et al.*, 1995; Laschinger and Engelhardt, 2000; Seguin *et al.*, 2003).

Mucosal addressin cell adhesion molecule-1 (MAdCAM-1) is another adhesion molecule which may be intimately involved in the pathogenesis of EAE (Engelhardt *et al.*, 2001) and human MS. The ligand for MAdCAM-1 is $\alpha 4\beta 7$ (Briskin *et al.*, 1993) and under inflamed conditions, epithelial cells of the choroid plexus express ICAM-1, VCAM-1, and MAdCAM-1 (Engelhardt *et al.*, 2001). The precise role of MAdCAM-1 in MS remains unknown, but MAdCAM-1 expression is known to be induced by cytokines, TNF-α, in particular (Sasaki *et al.*, 2004, 2005; Sikorski *et al.*, 1993) in an oxidant-dependent, NO-inhibitable manner (Oshima *et al.*, 2001). MAdCAM-1 is implicated in MS since it is abundantly expressed within the chronically inflamed cerebrum in human MS and in MS models (Kanwar *et al.*, 2000; Kumpfel *et al.*, 2002), and importantly antibodies against MAdCAM-1 block disease progression in EAE when given early in its course (Kanwar *et al.*, 2000). The humanized anti-$\alpha 4$ and anti-$\alpha 7$ integrin monoclonal antibody, "*Tysabri*," the most recent treatment for MS (Miller *et al.*, 2003) and Crohn's disease (Ghosh *et al.*, 2003), targets both VCAM-1- and MAdCAM-1-dependent adhesion mechanisms (Kumpfel *et al.*, 2002).

During EAE pathogenesis, two members of $\beta 2$ integrin family LFA-1 and Mac-1 and their ligands ICAM-1 and ICAM-2 have also been implicated in transendothelial migration of T cells (Laschinger *et al.*, 2002). The role of $\beta 2$ integrins in the extravasation of encephalitogenic T cells into the CNS is also supported by the fact that ICAM-1 and ICAM-2 endothelial knockouts exhibit significantly decreased T-cell migration (Lyck *et al.*, 2003), but it is not clear which of the different structural motifs in ICAM-1 are involved in the adhesion and migration of autoreactive CD4+ T cells in MS (Greenwood *et al.*, 2003; Lyck *et al.*, 2003).

Platelet-endothelial cell adhesion molecule (PECAM-1/CD31) is a 130-kDa glycoprotein member of the immunoglobulin family supporting extravasation of activated leukocytes into the CNS in MS. In MS pathogenesis, both the soluble (Losy *et al.*, 1999) and insoluble (microparticulate) forms of PECAM-1 (Minagar *et al.*, 2001) have been implicated. PECAM-1′s role in extravasation is complex, and may both inhibit and facilitate leukocyte emigration. While Qing *et al.* (2001), using T-cell receptor transgenic mice, showed that intravenous injection of anti-PECAM-1 antibody or PECAM-Ig chimera diminished migration of α/β TCR$^+$ V$\beta 3^+$ Mac1$^-$ cells into the CNS, Graesser *et al.* (2002) demonstrated that PECAM-1 KO mice in the EAE model of MS showed an increased extravasation of leukocytes across the BBB.

Other molecules implicated in leukocyte migration in MS pathophysiology may include CD99. CD99 is a 32-kD dense type I transmembrane protein expressed by hematopoietic cells, leukocytes, and endothelial cells. CD99 is densely *O*-glycosylated and its concentration at cell junctions may suggest that it supports transjunctional movement. For example, Schenkel *et al.* (2002) demonstrated that anti-CD99 lowered monocytes migration across the endothelial cell layers by more than 90%. The junctional adhesion molecules or JAMs are also expressed at endothelial tight junctions, and may support monocytes and neutrophils extravasation in MS as well (Del Maschio *et al.*, 1999; Martin-Padura *et al.*, 1998).

V. MS and Endothelial Tight Junctions

The molecular basis of the final step in trafficking of antigen-specific CD4+ T cells into the CNS—migration beyond the endothelium—remains largely unknown. The two mechanisms for transendothelial leukocyte migration include paracellular and transcellular migration. A major unanswered question is whether activated leukocytes, particularly activated CD4+ T cells, migrate paracellulary through circulating endothelial cell (CEC) junctions, or whether they penetrate the BBB, membrane/cytoplasm with no junction involvement (Greenwood *et al.*, 1994; Lossinsky *et al.*, 1989; McMenamin *et al.*, 1992; Raine *et al.*, 1990; Wekerle *et al.*, 1991; Wolburg, 2005). To explore endothelial tight junctions in pathogenesis of MS, Plumb *et al.* (2002) investigated the expression of tight junction (TJ) proteins (occludin and ZO-1) in blood vessels in active MS lesions from eight cases of MS, and in six cases of normal-appearing white matter (NAWM). Blood vessels were imaged by confocal microscopy, which revealed abnormal TJs in vessels with premortem serum protein leakage. The investigators concluded that the existence of abnormal or opened TJs, associated with inflammation, might contribute to BBB leakage in enhancing MRI lesions and may also be involved in subtle leakage in nonenhancing focal and diffuse lesions in NAWM.

Our own observations show that serum from MS patients disturbs tight and adherens junctions [through downregulation of the expression of occludin and vascular endothelium (VE) cadherin] (Minagar *et al.*, 2003) and cytokine-altered BBB junctions may represent an MS-specific mechanism creating a pathway for both protein and cell leakage in MS.

VI. Conclusions

Interactions among many components of the inflammatory cascade: inflamed endothelial cells, activated leukocytes, inflammatory cytokines, and chemokines play central roles in initiating and sustaining the cerebrovascular consequences of the unchecked immune system in MS. Each step of this complex immune cascade, however, represents a potential therapeutic opportunity. The currently generalized and nonspecific immunosuppressive therapy for MS is gradually giving way to more selective approaches, as we improve our understanding of the major mechanisms in MS. Consequently, continued research into the basis of these interactions is needed, which will develop more specific effective therapies with less adverse effects for relieving MS symptomatology and disease progression.

References

Abbott, N. J., Ronnback, L., and Hansson, E. (2006). Astrocyte-endothelial interactions at the blood-brain barrier. *Nat. Rev. Neurosci.* **7,** 41–53.

Barnett, M. H., and Prineas, J. W. (2004). Relapsing and remitting multiple sclerosis: Pathology of the newly forming lesion. *Ann. Neurol.* **55,** 458–468.

Baron, J. L., Madri, J. A., Ruddle, N. H., Hashim, G., and Janeway, C. A., Jr. (1993). Surface expression of alpha 4 integrin by CD4 T cells is required for their entry into brain parenchyma. *J. Exp. Med.* **177,** 57–68.

Briskin, M. J., McEvoy, L. M., and Butcher, E. C. (1993). MAdCAM-1 has homology to immunoglobulin and mucin-like adhesion receptors and to IgA1. *Nature* **363**(6428), 461–464.

Bruck, W., Lucchinetti, C., and Lassmann, H. (2002). The pathology of primary progressive multiple sclerosis. *Mult. Scler.* **8,** 93–97.

Cross, A. H., Cannella, B., Brosnan, C. F., and Raine, C. S. (1990). Homing to central nervous system vasculature by antigen-specific lymphocytes. I. Localization of 14C-labeled cells during acute, chronic, and relapsing experimental allergic encephalomyelitis. *Lab. Invest.* **63,** 162–170.

Del Maschio, A., De Luigi, A., Martin-Padura, I., Brockhaus, M., Bartfai, T., Fruscella, P., Adorini, L., Martino, G., Furlan, R., De Simoni, M. G., and Dejana, E. (1999). Leukocyte recruitment in the cerebrospinal fluid of mice with experimental meningitis is inhibited by an antibody to junctional adhesion molecule (JAM). *J. Exp. Med.* **190,** 1351–1356.

Engelhardt, B., Laschinger, M., Schulz, M., Samulowitz, U., Vestweber, D., and Hoch, G. (1998). The development of experimental autoimmune encephalomyelitis in the mouse requires α4-integrin but not $\alpha 4\beta 7$-integrin. *J. Clin. Invest.* **102,** 2096–2105.

Engelhardt, B., Wolburg-Buchholz, K., and Wolburg, H. (2001). Involvement of the choroid plexus in central nervous system inflammation. *Microsc. Res. Tech.* **52,** 112–129.

Engelhardt, B., Kempe, B., Merfeld-Clauss, S., Laschinger, M., Furie, B., Wild, M. K., and Vestweber, D. (2005). P-selectin glycoprotein ligand 1 is not required for the development of experimental autoimmune encephalomyelitis in SJL and C57BL/6 mice. *J. Immunol.* **175,** 1267–1275.

Floris, S., Ruuls, S. R., Wierinckx, A., van der Pol, S. M., Dopp, E., van der Meide, P. H., Dijkstra, C. D., and De Vries, H. E. (2002). Interferon-beta directly influences monocyte infiltration into the central nervous system. *J. Neuroimmunol.* **127,** 69–79.

Frohman, E. M., Racke, M. K., and Raine, C. S. (2006). Multiple sclerosis—the plaque and its pathogenesis. *N. Engl. J. Med.* **354,** 942–955.

Ghosh, S., Goldin, E., Gordon, F. H., Malchow, H. A., Rask-Madsen, J., Rutgeerts, P., Vyhnalek, P., Zadorova, Z., Palmer, T., Donoghue, S., and Natalizumab Pan-European Study Group (2003). Natalizumab for active Crohn's disease. *N. Engl. J. Med.* **348,** 24–32.

Graesser, D., Solowiej, A., Bruckner, M., Osterweil, E., Juedes, A., David, S., Ruddle, N. H., Engelhardt, B., and Madri, J. A. (2002). Altered vascular permeability and early onset of experimental autoimmune encephalomyelitis in PECAM-1-deficient mice. *J. Clin. Invest.* **109,** 383–392.

Greenwood, J., Howes, R., and Lightman, S. (1994). The blood-retinal barrier in experimental autoimmune uveoretinitis. Leukocyte interactions and functional damage. *Lab. Invest.* **70,** 39–52.

Greenwood, J., Wang, Y., and Calder, V. L. (1995). Lymphocyte adhesion and transendothelial migration in the central nervous system: The role of LFA-1, ICAM-1, VLA-4 and VCAM-1. *Immunology* **86,** 408–415.

Greenwood, J., Amos, C. L., Walters, C. E., Courand, P. O., Lyck, R., Engelhardt, B., and Adamson, P. (2003). Intracellular domain of brain endothelial intercellular adhesion molecule-1 is essential for T lymphocyte-mediated signaling and migration. *J. Immunol.* **171,** 2099–2108.

Hatterer, E., Davoust, N., Didier-Bazes, M., Vuaillat, C., Malcus, C., Belin, M. F., and Nataf, S. (2006). How to drain without lymphatics? Dendritic cells migrate from the cerebrospinal fluid to the B-cell follicles of cervical lymph nodes. *Blood* **107,** 806–812.

Jimenez, J., Jy, W., Mauro, L. M., Horstman, L. L., Ahn, E. R., Ahn, Y. S., and Minagar, A. (2005). Elevated endothelial microparticle-monocyte complexes induced by multiple sclerosis plasma and the inhibitory effects of interferon-beta 1b on release of endothelial microparticles, formation and transendothelial migration of monocyte-endothelial microparticle complexes. *Mult. Scler.* **11,** 310–315.

Jy, W., Minagar, A., Jimenez, J. J., Sheremata, W. A., Mauro, L. M., Horstman, L. L., Bidot, C., and Ahn, Y. S. (2004). Endothelial microparticles (EMP) bind and activate monocytes: Elevated EMP-monocyte conjugates in multiple sclerosis. *Front. Biosci.* **9,** 3137–3144.

Kanwar, J. R., Kanwar, R. K., Wang, D., and Krissansen, G. W. (2000). Prevention of a chronic progressive form of experimental autoimmune encephalomyelitis by an antibody against mucosal addressin cell adhesion molecule-1, given early in the course of disease progression. *Immunol. Cell Biol.* **78,** 641–645.

Keegan, B. M., and Noseworthy, J. H. (2002). Multiple sclerosis. *Annu. Rev. Med.* **53,** 285–302.

Kort, J. J., Kawamura, K., Fugger, L., Weissert, R., and Forsthuber, T. G. (2006). Efficient presentation of myelin oligodendrocyte glycoprotein peptides but not protein by astrocytes from HLA-DR2 and HLA-DR4 transgenic mice. *J. Neuroimmunol.* **173,** 23–34.

Kumpfel, T., Heydari, N., and Hohlfeld, R. (2002). Antegren (natalizumab). A promising new approach to therapy of multiple sclerosis. *Nervenarzt* **73,** 552–555.

Laschinger, M., and Engelhardt, B. (2000). Interaction of α4-integrin with VCAM-1 is involved in adhesion of encephalitogenic T cell blasts to brain endothelium but not in their transendothelial migration *in vitro*. *J. Neuroimmunol.* **102,** 32–43.

Laschinger, M., Vajkoczy, P., and Engelhardt, B. (2002). Encephalitogenic T cells use LFA-1 for transendothelial migration but not during capture and initial adhesion strengthening in healthy spinal cord microvessels *in vivo*. *Eur. J. Immunol.* **32,** 3598–3606.

Lossinsky, A. S., Badmajew, V., Robson, J. A., Moretz, R. C., and Wisniewski, H. M. (1989). Sites of egress of inflammatory cells and horseradish peroxidase transport across the blood-brain barrier in a murine model of chronic relapsing experimental allergic encephalomyelitis. *Acta Neuropathol. (Berl.)* **78,** 359–371.

Losy, J., Niezgoda, A., and Wender, M. (1999). Increased serum levels of soluble PECAM-1 in multiple sclerosis patients with brain gadolinium-enhancing lesions. *J. Neuroimmunol.* **99,** 169–172.

Lovett-Racke, A. E., and Racke, M. K. (2006). Epstein-Barr virus and multiple sclerosis. *Arch. Neurol.* **63,** 810–811.

Lucchinetti, C., Bruck, W., Parisi, J., Scheithauer, B., Rodriguez, M., and Lassmann, H. (2000). Heterogeneity of multiple sclerosis lesions: implications for the pathogenesis of demyelination. *Ann. Neurol.* **47,** 707–717.

Lyck, R., Reiss, Y., Gerwin, N., Greenwood, J., Adamson, P., and Engelhardt, B. (2003). T-cell interaction with ICAM-1/ICAM-2 double-deficient brain endothelium *in vitro*: The cytoplasmic tail of endothelial ICAM-1 is necessary for transendothelial migration of T cells. *Blood* **102,** 3675–3683.

Martin-Padura, I., Lostaglio, S., Schneemann, M., Williams, L., Romano, M., Fruscella, P., Panzeri, C., Stoppacciaro, A., Ruco, L., Villa, A., Simmons, D., and Dejana, E. (1998). Junctional adhesion molecule, a novel member of the immunoglobulin superfamily that distributes at intercellular junctions and modulates monocyte transmigration. *J. Cell Biol.* **142,** 117–127.

McMenamin, P. G., Forrester, J. V., Steptoe, R. J., and Dua, H. S. (1992). Ultrastructural pathology of experimental autoimmune uveitis. Quantitative evidence of activation and possible high endothelial venule-like changes in retinal vascular endothelium. *Lab. Invest.* **67,** 42–55.

Miller, D. H., Khan, O. A., Sheremata, W. A., Blumhardt, L. D., Rice, G. P., Libonati, M. A., Willmer-Hulme, A. J., Dalton, C. M., Miszkiel, K. A., O'Connor, P. W., and International Natalizumab Multiple Sclerosis Trial Group (2003). A controlled trial of natalizumab for relapsing multiple sclerosis. *N. Engl. J. Med.* **348,** 15–23.

Minagar, A., and Alexander, J. S. (2003). Blood-brain barrier disruption in multiple sclerosis. *Mult. Scler.* **9,** 540–549.

Minagar, A., Jy, W., Jimenez, J. J., Sheremata, W. A., Mauro, L. M., Mao, W. W., Horstman, L. L., and Ahn, Y. S. (2001). Elevated plasma endothelial microparticles in multiple sclerosis. *Neurology* **56,** 1319–1324.

Minagar, A., Ostanin, D., Long, A. C., Jennings, M., Kelley, R. E., Sasaki, M., and Alexander, J. S. (2003). Serum from patients with multiple sclerosis downregulates occludin and VE-cadherin expression in cultured endothelial cells. *Mult. Scler.* **9**(3), 235–238.

Noseworthy, J. H., Lucchinetti, C., Rodriguez, M., and Weinshenker, B. G. (2000). Multiple sclerosis. *N. Engl. J. Med.* **343,** 938–952.

Oshima, T., Pavlick, K. P., Laroux, F. S., Verma, S. K., Jordan, P., Grisham, M. B., Williams, L., and Alexander, J. S. (2001). Regulation and distribution of MAdCAM-1 in endothelial cells *in vitro*. *Am. J. Physiol. Cell Physiol.* **281**(4), C1096–C1105.

Osmers, I., Bullard, D. C., and Barnum, S. R. (2005). PSGL-1 is not required for development of experimental autoimmune encephalomyelitis. *J. Neuroimmunol.* **166,** 193–196.

Plumb, J., McQuaid, S., Mirakhur, M., and Kirk, J. (2002). Abnormal endothelial tight junctions in active lesions and normal-appearing white matter in multiple sclerosis. *Brain Pathol.* **12,** 154–169.

Qing, Z., Sandor, M., Radvany, Z., Sewell, D., Falus, A., Potthoff, D., Muller, W. A., and Fabry, Z. (2001). Inhibition of antigen-specific T cell trafficking into the central nervous system via blocking PECAM1/CD31 molecule. *J. Neuropathol. Exp. Neurol.* **60,** 798–807.

Raine, C. S., Cannella, B., Duijvestijn, A. M., and Cross, A. H. (1990). Homing to central nervous system vasculature by antigen-specific lymphocytes. II. Lymphocyte/endothelial cell adhesion during the initial stages of autoimmune demyelination. *Lab. Invest.* **63,** 476–489.

Sasaki, M., Elrod, J. W., Jordan, P., Itoh, M., Joh, T., Minagar, A., and Alexander, J. S. (2004). CYP450 dietary inhibitors attenuate TNF-alpha-stimulated endothelial molecule expression and leukocyte adhesion. *Am. J. Physiol. Cell Physiol.* **286,** 931–939.

Sasaki, M., Jordan, P., Welbourne, T., Minagar, A., Joh, T., Itoh, M., Elrod, J. W., and Alexander, J. S. (2005). Troglitazone, a PPAR-gamma activator prevents endothelial cell adhesion molecule expression and lymphocyte adhesion mediated by TNF-alpha. *BMC Physiol.* **5**(1), 3.

Schenkel, A. R., Mamdouh, Z., Chen, X., Liebman, R. M., and Muller, W. A. (2002). CD99 plays a major role in the migration of monocytes through endothelial junctions. *Nat. Immunol.* **3,** 143–150.

Seguin, R., Biernacki, K., Rotondo, R. L., Prat, A., and Antel, J. P. (2003). Regulation and functional effects of monocyte migration across human brain-derived endothelial cells. *J. Neuropathol. Exp. Neurol.* **62,** 412–419.

Sheremata, W. A., Jy, W., Delgado, S., Minagar, A., McLarty, J., and Ahn, Y. (2006). Interferon-beta1a reduces plasma CD31+ endothelial microparticles (CD31 + EMP) in multiple sclerosis. *J. Neuroinflammation* **3,** 23.

Sikorski, E. E., Hallmann, R., Berg, E. L., and Butcher, E. C. (1993). The Peyer's patch high endothelial receptor for lymphocytes, the mucosal vascular addressin, is induced on a murine endothelial cell line by tumor necrosis factor-alpha and IL-1. *J. Immunol.* **151**(10), 5239–5250.

Steffen, B. J., Butcher, E. C., and Engelhardt, B. (1994). Evidence for involvement of ICAM-1 and VCAM-1 in lymphocyte interaction with endothelium in experimental autoimmune encephalomyelitis in the central nervous system in the SJL/J mouse. *Am. J. Pathol.* **145,** 189–201.

Steinman, L., and Zamvil, S. S. (2006). How to successfully apply animal studies in experimental allergic encephalomyelitis to research on multiple sclerosis. *Ann. Neurol.* **60,** 12–21.

Wekerle, H., Engelhardt, B., Risau, W., and Meyermann, R. (1991). Interaction of T lymphocytes with cerebral endothelial cells *in vitro*. *Brain. Pathol.* **1,** 107–114.

Wolburg, H., Wolburg-Buchholz, K., and Engelhardt, B. (2005). Diapedesis of mononuclear cells across cerebral venules during experimental autoimmune encephalomyelitis leaves tight junctions intact. *Acta Neuropathol. (Berl.)* **109,** 181–190.

ROLE OF B CELLS IN PATHOGENESIS OF MULTIPLE SCLEROSIS

Behrouz Nikbin, Mandana Mohyeddin Bonab, Farideh Khosravi, and Fatemeh Talebian

Department of Immunology, Immunogenetic Research Center, College of Medicine
Tehran University of Medical Sciences, Tehran 14155, Iran

Despite the current limited understanding of the etiology of multiple sclerosis (MS), genetic susceptibility and environmental influences are known driving factors. MS is considered a T-cell-mediated disease given the prevalence of T cells in plaques. Plaque formation is characteristic of this disease attributable to immune mechanisms, triggered by an autoimmune attack aimed at antigens in the myelin sheath or oligodendrocyte proteins. The attack consists of the following:

- B cells
- The complement system
- The respective antigen

The role of the B cells is twofold: first, as autoreactive B cells they produce autoantibodies, secrete cytokines, clonally replicate memory B cells, and long-living plasma cells which serve to advance the diseased state by their constant production of autoantibodies. Second, as antigen-presenting cells they activate the autoreactive T cells. For this reason, the stipulation that T cell is the cornerstone of

INTERNATIONAL REVIEW OF
NEUROBIOLOGY, VOL. 79
DOI: 10.1016/S0074-7742(07)79002-5

MS must be reevaluated. Various studies on pathogenesis of MS have indicated that B cells, as the humoral component of the adaptive immune system, are active participants in pathogenesis and lesion maintenance throughout the disease process. The active role of B cells and autoantibodies makes them an encouraging therapeutic target. Advances in the understanding of B-cell development and activity would allow for an enhanced strategy in the design of autoimmune treatment. For this reason, further investigation is necessary to determine whether depletion of B cells or antibodies may restore immune function.

I. Introduction

The immune system is characterized by its ability to recognize self and tolerate it and to react against nonself. This tolerance toward self is not hereditary. How is it that self-antigens (Ags) do not activate the immune system? Because in the process of T and B lymphocyte growth, development of autoreactive lymphocytes is prevented. The phenomenon of autoreactive lymphocyte inhibition occurs in two phases of development:

- Immature T and B cells in the central organs [thymus and bone marrow (BM)] interact with self-Ag through their surface receptors leading to clonal deletion via apoptosis (Hogquist *et al.*, 2005; Pillai, 2005).
- Autoreactive mature B and T cells, which have escaped from central organs, are conditioned to unresponsiveness or anergy due to the regulatory mechanisms (Lineberry and Fathman, 2006; Melchers and Rolink, 2006) in the periphery.

Peripheral tolerance has a central role in unresponsiveness toward Ags found exclusively in the peripheral organs (Taams and Akbar, 2005). Therefore, once lymphocytes mature and leave the central and peripheral lymphoid organs, the repertoire lacking clones should be able to recognize self-Ags in most tissues. In the event that autoreactive cells gain access to the body and circulate freely, or worse find their target and initiate an immune response against that component, what will result is a phenomenon known as autoimmunity. So essentially, autoimmunity could result from the breakdown of natural processes responsible for maintaining tolerance.

There is a third way of preventing autoimmunity's evasion. For example, the nervous system enjoys an immune privileged status under physiological conditions. The nervous system in effect hides self-tissues behind a tissue barrier impermeable to circulating immune cells (Wekerle, 2006).

The predisposition of an individual to develop an autoimmune disease is based on a number of criteria, including antigen-presenting cell (APC), T- and B-cell dysfunctions, genetic susceptibility, tissue damage, and infectious agents (Ji *et al.*, 1999). One of the complications in defining the underlying mechanisms responsible for autoimmune induction in humans is the inability to diagnose the initial Ag or trigger. Therefore, the specific etiology of most of these diseases remains unanswered. Multiple sclerosis (MS) is one of these and lingers as one of the most intriguing examples because it is wholly unknown etiologically, and almost unrelated to any other organs and affecting only the central nervous system (CNS).

Many studies show (Pettinelli and McFarlin, 1981; Raine *et al.*, 1984) that MS can be transferred from one animal with experimental autoimmune encephalomyelitis (EAE) to another, but only through transfer of activated mature T cells of the diseased animal (it cannot be transferred passively). The disease is originally induced by myelin-related proteins such as myelin basic protein (MBP), proteolipid protein (PLP), or myelin oligodendrocyte glycoprotein (MOG). This feature has allowed recognition of specific T-cell clones against myelin in humans (Kerlero de Rosbo *et al.*, 1993). These T-cell clones isolated both from normal individuals and MS patients seem to play a central role in the initiation of immune cascades and lead to plaque formation (the hallmark of MS pathology). This forms the basis on which MS is largely considered a T-cell-mediated disease; as such, therapeutic strategies have focused on this consideration in their attempts for treatment options (Bansil *et al.*, 1995). Recently, however, this assertion has come into dispute causing a new found focus and interest on the role of B cells in the pathology of MS. This new thesis has yet to emerge to the forefront of studies on the subject. However, an overview of the events leading up to plaque formation would shed light on the argument that redirecting the focus from T cells to B cells as a therapeutic target may in fact be an improved prospect.

According to *in vitro* and *in vivo* studies, this basic outline can be suggested for a possible mechanism of attack (Frohman *et al.*, 2006; Sospedra and Martin, 2005). A diverse disease process stemming from a variety of triggers including environmental factors, infectious agents, inflammation, and autoimmune reactions induces autoreactive CD4+ T cells to be activated in the periphery. The activated CD4+ T cells adhere to the CNS endothelial and by transendothelial migration gain access to the CNS. An inflammatory medium is formed where antibodies (Abs) play an integrated role by identifying and attaching to their autoantigen target. It is important to note that although CD4+ T cells most likely initiate the process, they are almost bystanders in the effector phase. Here cytotoxic cells, soluble factors, plasma cells, macrophages, antibodies, and proteases come into play and cause the real damage. In support of this pathogenetic view, development and growth of MS lesions is directly dependent on other immunological factors such as pathological Abs (Archelos *et al.*, 2000).

In this chapter, we will expound on the more understood basis of MS etiology followed by an outline of production and development of B cells, plus the role which these cells play in the immune response. Next, we will review the contribution of B cells in the mechanisms of MS development. Finally, B cells and the role they might play in the treatment of MS will be presented.

II. Etiology of MS

While the current etiology of MS is unknown, there are some unrefuted hypotheses shedding some light on the development of this disease. Among these are primary or infectious agent induced. The evidence to support the effect of external factors on the development of MS is the existence of geographically high and low prevalent areas. Population studies have shown that migration to high prevalence areas prior to puberty increases the risk of MS development (Kurtzke and Bui, 1980; Kurtzke *et al.*, 1998). Alternatively, the autoimmune response may be triggered through mechanisms known as molecular mimicry (MM) or antibody cross-reactivity. The oldest known example is antistreptococcal Ab and its cross-reaction with a similar self-antigen from heart and kidney tissues. Among examples of MM in CNS tissue are Guillain-Barre syndrome (GBS) with *Campylobacter jejuni* and Human T-lymphotropic virus 1 (HTLV-1)-associated myelopathy (Ang *et al.*, 2004; Levin *et al.*, 2002). Some evidence suggest a pathogen, most probably a viral infection as the triggering factor in MS disease, most likely due to MM leading to an autoimmune attack against myelin membrane in a genetically susceptible individual (Wekerle and Hohlfeld, 2003). Lang *et al.* (2002) showed certain Epstein-Barr virus (EBV) peptides, when presented by major histocompatibility complex (MHC) very closely resemble to the MBP–MHC complex and thus promote T-cell receptor (TCR) cross-reactivity giving rise to autoreactive immune responses.

A. Autoantigen Induced and Susceptible Genes

Certain genetic factors are involved in the etiology of MS. First evidence of genetic association was related to MHC as reported by Jersild *et al.* (1972). Later Olerup and Hillert (1991) reported a correlation between MS and the HLA-DRB1*1501-DQB1*0602 haplotype. Martinez *et al.* (2005) reported that early B-cell factor (EBF1) "an effective transcription factor in B-cell development" has a causative role in MS pathogenesis. B-cell-activating factor for the TNF family (BAFF) is one of the popular factors in B-cell molecular studies. BAFF is a B-cell surface marker which induces expression of the oncogene BCL1 (prosurvival receptor) (Mackay and Tangye, 2004). With abnormal expression of BAFF, B-cell survival rate intensifies and thus autoreactive B cells escape apoptosis and move to the periphery. New studies show signal intensity through B-cell receptor (BCR), CD40R, and BAFF modulates the destiny of the B cell. Variation in signaling magnitude by these molecules could lead to immunological dysregulation and loss of self-tolerance (Noelle and Erickson, 2005). In support of this idea, Krumbholz *et al.* (2005) detected higher BAFF expression in brain tissue of MS patients as well as individuals suffering from other autoimmune dysfunctions.

III. Generation and Development of B Cells

B cells originate from hematopoeitic stem cells (HSC) in the BM (Nagasawa, 2006). Their maturation in the BM is independent of Ag and their progenitors are CD19−CD20− stem cells. From these stem cells, Pro-B, then Pre-B, and eventually mature naive CD20+CD19+CD27−, IgD+, and IgM+ B cells are produced. Mature B cells in the circulation are divided into two types: CD5+ known as B1 and CD5− known as B2. They are all from the same precursor. B1 cells undergo T-cell-independent maturation in the periphery (usually under the skin area), mainly producing IgM against nonprotein Ags. B2 are the classic B cells which undergo two stages of development: (1) Ag dependent and (2) Ag independent, and their development occurs within the lymphoid organs and germinal center (GC). CD5+ cells usually have fetal origin while CD5− cells have adult origin (Brezinschek *et al.*, 1997; Hannet *et al.*, 1992; Kantor, 1991).

According to the clonal selection theory, B cells specific to any Ag are present before Ag encounter. Therefore, it can be stated that information concerning a diverse Ab repertoire is present in the DNA of each individual. Interaction of immature B cells with any Ag (the autoantigens in the BM) leads to death or unresponsiveness in these cells. This phenomenon is necessary for development of tolerance toward local autoantigens in the BM. Once a B cell forms a complete immunoglobulin (Ig) molecule (both heavy and light chain), it will not be able to build another Ig with a different variable (V) region. Mature B cells which complete their education leave the BM and enter secondary lymphoid organs with the resultant phenotype. From there on, the Ag-dependent development begins in the GC of secondary lymphoid organs (Liu *et al.*, 1992). When naive B cells encounter the respective Ag and the necessary costimulatory factors, they undergo Ig isotype switching and turn into active mature B cells. B-cell selection inside GC obeys the Darwinian law based on affinity. Those displaying high affinity to the respective Ag survive, while low-affinity B cells undergo apoptosis. Positively selected B cells in this period, which possess the potential to clonally expand, gradually lose their surface Ig molecules, increase their secretory form of

Abs, and differentiate into early plasmablasts (CD27^{high}, CD40L^{+}), late plasmablasts (CD27+, CD38+, BAFF+), and eventually Ab-producing plasma cells (CD27+) (Avery *et al.*, 2005; Goldsby, 2000).

Some of the activated mature B cells do not begin to release their surface Abs and remain as surface Ab-expressing cells. These are called the "memory B cells" (CD27+, CD38+, BAFF+). These memory B cells remain circulating in the peripheral blood (PB) and lymph nodes for a long time. When these cells are stimulated by their respective Ag, a secondary antibody response ensues.

The members of each B-cell lineage express a specific V region against a single Ag and preserve this specificity during their life span. Each B cell possesses specified Ig on its surface made up of light (L) and heavy (H) chains. Each chain consists of V D J (VHDHJH, VLDLJL) and constant regions (Fig. 1). Each part of VL and VH is made up of three hypervariable regions which bind Ag (Burnett, 1976). As a result, response to Ag causes small variations in the V region known as somatic mutation. These somatic mutations increase the affinity of the Ab for its Ag. Repeated encounter of B cell with Ag leads to affinity maturation. Therefore, Ab produced later possesses higher affinity. B cells with the highest affinity stay alive and differentiate into plasma and memory cells. Subsequent encounter with persistent Ag leads to proliferation of clonal memory B cells and production of antibody-secreting plasma cells with high affinity. This phenomenon is observed

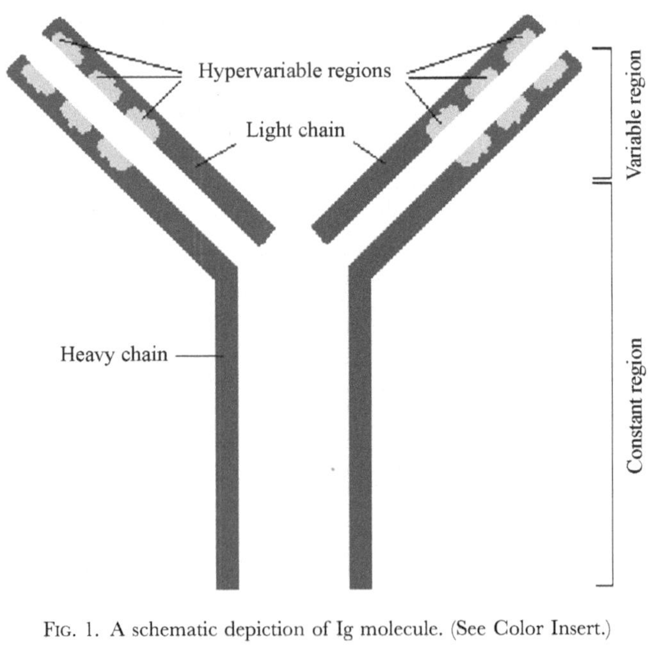

FIG. 1. A schematic depiction of Ig molecule. (See Color Insert.)

in humoral immune responses after vaccination, autoimmunity, and B-cell malignancies. Somatic hypermutation of IgV gene and clonal expansion of B cells are the hallmark of humoral immunity to Ag.

IV. The Role of B Cells in Immune Responses

All adoptive immune responses are triggered in peripheral lymphoid organs (lymph node, spleen, mucosal nodes). In light of recent studies (Chan *et al.*, 1999; Fillatreau *et al.*, 2002; Ji *et al.*, 2002; Lipsky, 2001; Mauri *et al.*, 2003; Saito *et al.*, 2002; Weinstein *et al.*, 2004), the role of B cells in immune responses can be classified as the following:

1. Ig Production
2. Ag Presentation
3. Cytokine production
4. Regulation of lymphoid organogenesis
5. Differentiation of effector T cells
6. Regulation of antigen presenting function of dendritic cell (DC)

Given the fundamental role of B cells in immune responses, it is evident how B-cell dysregulation may lead to production and development of autoimmunity. Therefore, while the role of B cells in MS pathology is yet to be determined, it is possible that B/plasma cells secrete Ab against an unknown Ag which is in fact the primary cause of MS. These cells also secrete Ab against myelin Ag, or they can secrete an anti-idiotypic Ab which could be a cause for the relapsing phase of MS.

V. The Role of B Cells in MS Pathogenesis

Ab and complement deposition and T/B/macrophage cell infiltration to the perivascular region are accompanied by demyelination of axons (Lucchinetti *et al.*, 2000). This is the most common pathological pattern of MS lesions. Due to the presence of autoreactive T cells against various autoantigens, MS has been classified as a predominantly T-cell-induced autoimmune disease. Histopathological (biopsy/autopsy) samples of active MS lesions reveal the complexity of this disease. These lesions show the necessity for the existence of other etiological means for the onset of such an illness (Lucchinetti *et al.*, 2000). In many MS subgroups, B cells and Abs seem important, as well. From an immunological point of view, MS lesions have been classified into four subgroups as listed below (Frohman *et al.*, 2006):

1. Macrophage-mediated demyelination
2. Antibody-mediated demyelination

3. Distal oligodendropathy
4. Demyelination secondary to oligodendrocyte damage

The autoantibodies produced by autoreactive plasmablast and plasma cells play a pivotal role in autoreactive pathogenesis. They are active participants in the pathways leading to myelin damage, complement activation, macrophage recognition phase, axonal loss, and neurodegeneration (Tan, 1991).

Although Abs against many infectious agents and Ags have been found in correlation with MS, the oligoclonal feature of the identified IgG in the cerebrospinal fluid (CSF) remains elusive. In our discussion, we attempt to address concerns arising from the findings provided by recent studies. The concerns include the following:

- Are these Abs disease specific or disease related?
- Are they quantity dependent?
- Are they part of the immune response?

Studies (Baranzini *et al.*, 1999; Colombo *et al.*, 2000; Qin *et al.*, 1998) show B-cell immunity in the CNS of patients is constitutively active (Fig. 2). Such events are observed during early phases of MS pathogenesis and continue through to terminal stages of disease.

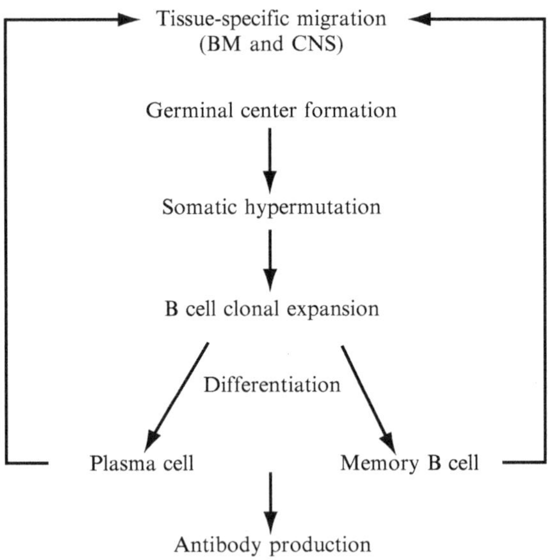

FIG. 2. B-cell immunity in patients' CNS.

A. GC Formation

During CNS inflammatory responses, T cells constitute the greater population of CNS leucocytes, mostly of the memory phenotype (Giunti *et al.*, 2003). However, B-cell population occupies about 25% of CSF infiltrating leukocytes. A perceptible increase in the number of B cells has been associated with a progressive clinical course (Cepok *et al.*, 2001). Although the genetic structure of the Ag against which B cells react remains unknown, molecular analysis has shown altered IgV region genes in CSF (Qin *et al.*, 1998). Such findings suggest the possibility that GC like reactions during autoimmune attacks is against CNS structure as a whole.

Examining the Ig genes of B cells obtained from CSF and lesions of MS patients revealed that existing B cells exhibit an Ag-driven adoptive immune response. This includes: T cell–B cell interaction accompanied by somatic hypermutation, immune memory development, clonal B-cell selection, and proliferation (Correale and de Los Milagros Bassani Molinas, 2002; Cross, 2000).

We still do not know the origin of persistent antibody-secreting cells (ASCs) in the CNS which causes ongoing damage in MS. There are some speculations based on their characteristics as mentioned below (Dorner and Radbruch, 2005):

1. Short-living plasma cells formed from a repertoire of newly activated B cells
2. Long-living plasma cells or memory B cells formed from the repertoire of GC B cells

The differential feature is their longevity, their high expansion potential, and their migration to the BM, where they differentiate to end stage cells. These end stage cells participate in humoral immune responses with a long-lasting nature (Noelle and Erickson, 2005). This dichotomous possibility justifies the nonoverlapping of IgV gene sequences of clonally expanded B cells and plasma cells in patient CNS. Similar differences in IgV region gene sequence have been observed in B and plasma cells isolated from other autoimmune diseases as well (Kim *et al.*, 1999). Molecular analysis of IgV region shows the mutation of these genes (Baranzini *et al.*, 1999; Owens *et al.*, 1998). This demonstrates that GC must be present in the CNS during autoimmune attacks. This was confirmed when a great number of VH gene sequences was found in CSF of MS patients. Such a finding proves that different B cells result from diversification of one or a few progenitor clones (Colombo *et al.*, 2000).

We know that the centroblast is exclusively found in GC. Therefore, the discovery of $CD38^{high}CD77+bcl2-$ centroblasts in the CNS of MS patients as shown by Corcione *et al.* (2004) confirms GC formation in CNS.

To produce secondary lymphoid organs, inducing factors must come into play in order to form secondary GC. Studies highlight the role of cytokines such as lymphotoxin α, CXCL-12, CXCL-13, BAFF in aiding B- and T-cell homing and formation of secondary lymphoid follicles. These cytokines have been reported in the CNS of MS patients (Corcione *et al.*, 2004; Krumbholz *et al.*, 2005; Thangarajh *et al.*, 2004).

In many chronic infection and autoimmune disorders, atypic lymphoid follicles are produced in targeted tissue. It appears that these follicles are responsible for local T-cell priming and B-cell maturation (Hjelmstrom, 2001). Presence of these atypical follicles in the meninges of patients with secondary progressive MS (SPMS) has been reported by Serafini *et al.* (2004). They demonstrated that follicular DC (FDC), proliferating B cells, and plasma cells are present within these follicles and believed that atypical lymphoid structure provides an appropriate microenvironment for B-cell differentiation.

B. Somatic Hypermutation

ASCs present in inflamed CNS produce oligoclonal Ig. Their detection in CNS is an indication of MS. The presence of clonally expanded B cells with mutated genes in the variable region found in brain and CSF of MS patients (Baranzini *et al.*, 1999; Owens *et al.*, 1998) supports affinity maturation of B cells occurring in the CNS. This phenomenon is probably due to one or a few nonself- or self-Ag. Mutation in the IgV region of B cells exposed to hypermutation and selection in GC have been observed (Wagner and Neuberger, 1996). Active B cells can grow as nonmutated plasmablasts outside the GC (Ochsenbein *et al.*, 2000).

Mattson *et al.* (1980, 1982)—using isoelectric focusing—showed that oligoclonal band (OCBs) profile of a single MS patient's plaque do overlap but are not identical. However, the same evaluation on OCB of subacute sclerosing panencephalitis (SSPE) patients taken from various sites of the brain showed an almost identical pattern. It seems the somatic mutation and repertoire shift in Ab which occurs after recurrent Ag encounter must be the means for establishing the various OCB profiles.

Some researchers have amplified heavy chain variable region sequences of the Ig in CSF and lesions of several MS patients. The analysis of their study from VH region has led to a restricted Ig repertoire indicating a targeted humoral response (Baranzini *et al.*, 1999; Burgoon *et al.*, 2004; Owens *et al.*, 1998; Prineas *et al.*, 2001; Smith-Jensen *et al.*, 2000). They also compared the VH sequence found in CNS plaques of MS and brain tissue of SSPE patients, finding no significance in the specificity and level of somatic mutation.

C. B-Cell Clonal Expansion and Ab Secretion

Clonal B-cell expansion in MS CSF and control subjects suffering from different neurological inflammations (viral encephalitis, disseminated encephalitis) has also been observed. However, CSF of some noninflammatory neurological disorders (migraine, epilepsy) does not show B-cell clonal expansion (Colombo *et al.*, 2000;

Owens *et al.*, 2003; Qin and Duquette, 2003; Qin *et al.*, 1998). MS is thought to be a T-cell-mediated disease. However, most patients possess OCB Ab in their CSF which is a sign of humoral immune system involvement. These Abs are generated by B cells in the CNS (Esiri, 1977; Tourtellotte, 1970). Usually, intrathecal Igs are restricted to IgG1 and IgG3 subsets in many patients (Sesboue *et al.*, 1985). IgM and IgA subsets are also present but they are rare and not considered a constant feature (Giles and Wroe, 1990). It has been reported (Burgoon *et al.*, 2003; Grimaldi *et al.*, 1988) that normal individuals have 13% IgG and OCB. OCB can be detected in the CSF of patients suffering from CNS infection by agents such as neurosyphilis, tuberculosis, and fungal or viral meningitis. However, Cole *et al.* (1998) show that 90–95% of MS patients have OCB in their CSF. OCB is important in diagnosing early MS especially in patients with normal magnetic resonance imaging (MRI). MS patients with optic neuritis (an early sign of MS) display 34–72% OCB positivity in their CSF (Filippini *et al.*, 1994). These Igs in CSF are against different myelin Ags such as MOG, myelin-associated glycoprotein (MAG), MBP, PLP (Cross *et al.*, 2001) and against various viruses such as *Measles virus* (*MeV*), Varicella-zoster virus (VZV), and Herpes simplex virus (HSV) (Horikawa *et al.*, 1973). The Ig profile within the CNS and CSF of MS patients is widely heterogeneous and only a very small percentage of these are MS specific.

Studies of advanced MS plaques (Prineas *et al.*, 2001) show that IgG does not interact with the surface of intact myelin sheath, even when encountering antimyelin IgG. We could speculate that antimyelin Ab is a factor involved in myelin damage. Therefore, the determinant Ag might not be an accessible Ag on the surface of intact myelin.

A high percentage of CD5+ B cells have been observed in MS patients (Mix *et al.*, 1990). As mentioned earlier, CD5+ B cells are a subgroup of B cells producing IgM against nonprotein Ag. Lolli *et al.* (1991) reports that IgM is predominantly observed in early MS patients. Spinal antimyelin lipid IgM has also been reported as a predictor of aggressive MS progression (Villar *et al.*, 2005).

Beside Ab production, B cells secrete proinflammatory cytokines such as TNF-α, IL-10, IL-6. IL-6 and IL-10 promote a stimulatory feedback process for B-cell proliferation. This could be a sign for disease persistence. On the other hand, IL-10 and TNF-α activate macrophages directly and modify the function of other immunoregulatory cells (Fillatreau *et al.*, 2002; Weinstein *et al.*, 2004).

D. Memory B-Cell and Plasma Cell Development

Due to the chronic nature of MS, the role of secondary immune responses is more pronounced. Among these secondary responses, memory B cells and long-living plasma cells have a greater role in MS pathogenesis. Memory B cells with long life are clones of B cells that preserve the adoptive Ab response to a certain

Ag. In persistent Ag stimulation, memory B-cell clones expand and exhibit a specific humoral response during the life of an organism. Such Ag stimulation is necessary to maintain the presence of the respective Ab with high titer. Therefore, B-cell clones need Ag stimulation for their maintenance (Lassmann *et al.*, 1998). In the absence of Ag, they are expelled from the host body. The role of Ag in preserving memory and clonal proliferation is essential, so persistent Ag presence is a key factor in autoimmunity.

Memory B cells lying within the CNS are active because costimulatory ligands on their surface are upregulated and exhibit the ability to secrete proinflammatory cytokines such as IFN-γ. These CNS memory B cells express high levels of chemokine receptors (CCR1, CCR2, CCR4) compared to PB memory B cells. These receptors aid the homing of B cells to the CNS and are a key element in guiding them toward the inflamed CNS site (Corcione *et al.*, 2005).

The quality of B-cell activation directs these cells toward a long-living or short-living fate. If B cells are activated in the absence of Th cells they turn into short-living plasma cells and are present in the extra follicular regions of secondary lymphoid organs, like the spleen (Vos *et al.*, 2000). As a result, these B cells do not have memory phenotypes, and their peak of action is in the fifth day after stimulation (Erickson *et al.*, 2000; Sze *et al.*, 2000). B cells, which are stimulated in the presence of antigens with the assistance of Th cells through their CD40L, develop into long-living plasma cells within the GC (Foy *et al.*, 1996; O'connor *et al.*, 2003). CD40L is an important component because blocking of CD40L leads to inhibition of GC formation (Ahonen *et al.*, 2002; Erickson *et al.*, 2002). GC forms around day 7 inside secondary lymphoid organs, after naive B cells interact with antigen through Th. Within GC, active B cells are called centeroblasts. After a few cell divisions, somatic hypermutation and isotype switching, their BCR evolves. Centrocytes, daughter cells to centroblasts, compete with one another in order to capture the respective antigen based on affinity maturation. The centrocyte(s) that captures the antigen successfully (having higher affinity to the antigen) is granted the survival signal and leaves the GC to continue its development as long-living plasma cells and memory B cells (Gray *et al.*, 1996).

Post-GC B cells, expressing CD138 and CD44, give rise to two types of precursors: CD138+CD44− and CD138−CD44+ both are being long-living PC. Short-living PC precursors are CD138+CD44+ (O'Connor *et al.*, 2002). CD138−CD44− are a very rare population and largely unknown. It has been demonstrated (Manz *et al.*, 1997) that plasma cells can survive in the BM for long periods (months to years). The reason these cells stay viable and functional in the BM is attributed to the plasma cell survival niche. Studies show (Hauser *et al.*, 2002) IL-5-, IL-6-, TNF-α-, SDF-1-, and CD44-associated signals are survival factors for plasma cells. Most plasma cells undergo apoptosis within the confines of GC (Smith *et al.*, 1996), and the rest move out of the secondary lymphoid organs in search of survival niches elsewhere in the body (Manz *et al.*, 2002). After immunization, less than 5% of their specific antigen-associated plasma cells can be found in the spleen. The other 95% or so are detectable in the BM (Manz *et al.*, 1998). Interaction of these PC precursors with BM stroma can lead to further development or even positive/negative selection in the BM vicinity. This might be effective in creating a specific self-antigen PC repertoire (Hoyer *et al.*, 2004).

Newly generated PCs have a migratory phenotype because they express CX-chemokine-receptor 3 (CXCR3) and CXCR4. These cells migrate to inflamed tissue via CX-chemokine-receptor ligand 9 (CXCL9), 10, and 11 (CXCR3 ligands). CXCL12, expressed in the BM, interacts with CXCR4 and allows migration to BM. These receptors are expressed on PC for 2 weeks, after which they begin to decline (Moser *et al.*, 2006). Therefore, local PCs lose their migratory potential after this period. CXCL12 is not only a chemokine but also it represents one of the survival signals presiding in the respective niches. So the PCs that enter these areas must compete with previous PCs for attaining access to the survival niche. The successful competent PC is the one with higher mobility. PCs which cannot secure a place in the survival niche are doomed to death (Odendahl *et al.*, 2005).

Other factors such as IL-7, stem cell factor (SCF), and FLT3 ligand have a role in growth and development of B cells. These factors must be produced by certain cells at certain periods to promote B-cell development at the desired time. This means B-cell development is certainly dependent on specific niches. Reticullar $CXCL12^{hi}$ cells surrounding BM sinuses have the greatest interaction with B cells (Lichtman, 1981). However, the role of other cells such as osteoblasts and IL-7-secreting cells in producing this microenvironment must not be ignored. Long-living plasma cells, produced by Ag stimulation outside the BM, will home back to these specified survival niches within the BM via CXCR4–CXCL12 interaction (Tokoyoda *et al.*, 2004). Expression of CXCR4 and its migration toward CXCL12 (SDF-1) is upregulated constantly at the period of differentiation, even in the absence of T-cell signals (Muehlinghaus *et al.*, 2005). CXCL12 must have a variety of roles in this microenvironment (Egawa *et al.*, 2001), all of which promote:

1. Home PC to specific BM niches
2. Increase adherence of cells to the niche (possibly via integrin activation)
3. Support intraniche survival of the cells

Hoyer *et al.* (2004) demonstrated that NZB/W mice (SLE animal model) have a ten fold greater number of survival niches encompassing long-living plasma cells in their spleen compared to normal mice. However, other reports show this number is almost equal in the BM (Cassese *et al.*, 2001). Therefore, long-living PCs are effective in MS pathogenesis due to their constitutive Ab secretion. This gains even greater attention when their resistance to immunosuppressive therapy is noted. Hoyer *et al.* (2004) examined NZB/W mice and showed that antiproliferative immunosuppresive therapy affects only the short-living plasma cell population. The long-living cells survive and continue to secrete autoAb within the diseased body.

Odendahl *et al.* (2005) preformed an interesting study on plasma cells and showed that 6–8 days after vaccination, along with vaccine-specific PC entrance to PB, polyspecific PCs are also mobilized and join their ranks. These cells have a heterogeneous phenotype and do not produce vaccine-specific Ab. Using HLA-DR staining, they showed that half of the PB plasmablasts are newly formed due to their high level of HLA-DR expression (Arce *et al.*, 2004). Therefore, all the cells expressing high levels of HLA-DR are vaccine-specific Ab-producing plasmablasts, and those with low levels of HLA-DR are mature resident plasmablasts with various specificities unrelated to the vaccine.

It appears that with the entrance of each new Ag and the production of their Ag-specific memory B cells, previously formed memory B cells can be detected alongside them. Perhaps the older memory B cells are being reactivated for their specificity and functionality while adding a new memory B-cell specificity to their repertoire. This phenomenon assists the recruitment of an active humoral immune response in normal individuals, but perhaps in predisposed or sensitive individuals it could induce autoreactive B-cell activation. This may explain the existence of various autoAb in (MS) patients.

Memory B cells play an important role in the chronicity of autoimmune dysregulation. Compared to naive B cells, memory B cells have a much lower activation threshold. Conversely, certain memory B cells express CXCR5 on their surface which recruits B cells to sites of inflammation. Long-living memory plasma cells constitutively produce Abs. Not only are Abs secreted by long-living plasma cells, but also these cells have a specific role in the immunopathogenesis of the disease. These cells are resistant to immunosuppressive treatments available at this time. They constantly secrete Abs into their surrounding environment independent of stimulation. Long-living plasma cells persist in Ab secretion even after immunosuppressive therapy and during remission of disease states (Hoyer *et al.*, 2004).

Hela-Felicitas and Reske (2005) performed a follow-up study on PRMS patients within 22 months (mean). Their results show the importance of activated B cells on determining the outcome of progress and development of disease. They have demonstrated that intrathecal antibody production remains stable over time once it has been successfully established. In addition, an Ab-specific expansion occurs in some patients, leading to extensive B-cell activation correlated with clinical worsening of disease.

Compared to other inflammatory neurological disorders, an increase of CD19+, CD138+ plasma cells and CD138+, CD19+, CD27+, HLA-DR+ short-living plasmablasts in CSF of MS patients has been reported. It has also been shown that the presence of plasmablasts has a significant relationship with CNS inflammation (Cepok *et al.*, 2005; Corcione *et al.*, 2004). CNS memory B cells and plasma cells can be produced in three ways:

1. When naive B cells encounter a local Ag, they become activated B cells, creating GC in the CNS. This microenvironment gives rise to long-living plasma cells.

2. When memory PB cells enter the CNS, they home to GC, differentiating into long-living plasma cells.
3. When PB-activated B cells find their way into the CNS, they differentiate into short-living PC independent of GC maturation stage.

E. The Role of Antibody

In humoral immunity-mediated demyelination, Ab and clonal B-cell infiltration is necessary. These B-cell clones secrete Abs or autoAbs responsible for myelin degeneration. Abs are attached to their corresponding Ags or myelin surface through the Fab region. The Fcs, which are free, can activate degenerative mechanisms (Fig. 3) through two pathways:

1. Attachment of C1q to Fc; this starts a cascade of complement component cleavage and activity until it reaches C7–9 forming the membrane attack complex (MAC). This complex penetrates the cell membrane and causes pore formation thus rupturing of the cell membrane.
2. Attachment of Fc to FcR on macrophage, neutrophil, or Tc cells. It is presumed that this complex attacks the myelin and begins the destructive process through antibody-dependent cell cytotoxicity (ADCC).

The other products of these two pathways, cytokines or chemokines, recruit inflammatory cells and factors to the mircroenvironment which in turn propagate and enhance ongoing damage. A historical analysis of 135 active MS lesions from 83 autopsies and biopsies has shown that the most prominent pathological abnormality is demyelination accompanied by codeposition of IgG and complement activity (Lucchinetti *et al.*, 2000). Complete activation of the complement system especially when involving C9 during MS attacks is mostly observed in very advanced stages of the disease, and is directly and significantly related to neurological disability index (Sellebjerg *et al.*, 1998). There are two means by which complement could penetrate and be present in the CNS:

1. Leakage into CNS through BBB breakdown
2. Synthesis by astrocytes, microglia, and macrophages infiltrated to the lesion area with a considerable cytokine profile exhibiting a proinflammatory character (Morgan and Gasque, 1996)

Studies (Adams, 1975; Gay and Esiri, 1991; Storch *et al.*, 1998) have demonstrated that IgG and C1q proteins are present within the plaque, while C3 protein is mostly observed in the vicinity of the demyelinating plaque, and C3d spotted on the surface of most macrophages in active lesion areas.

Studies (Barrington *et al.*, 2002; Qin *et al.*, 2000) indicate that immune complex (IC) (Ab + soluble Ag) affects the severity and quality of memory B-cell response. Some suggest (Tangye *et al.*, 2003; Wu *et al.*, 2000) that Ag–Ab complex

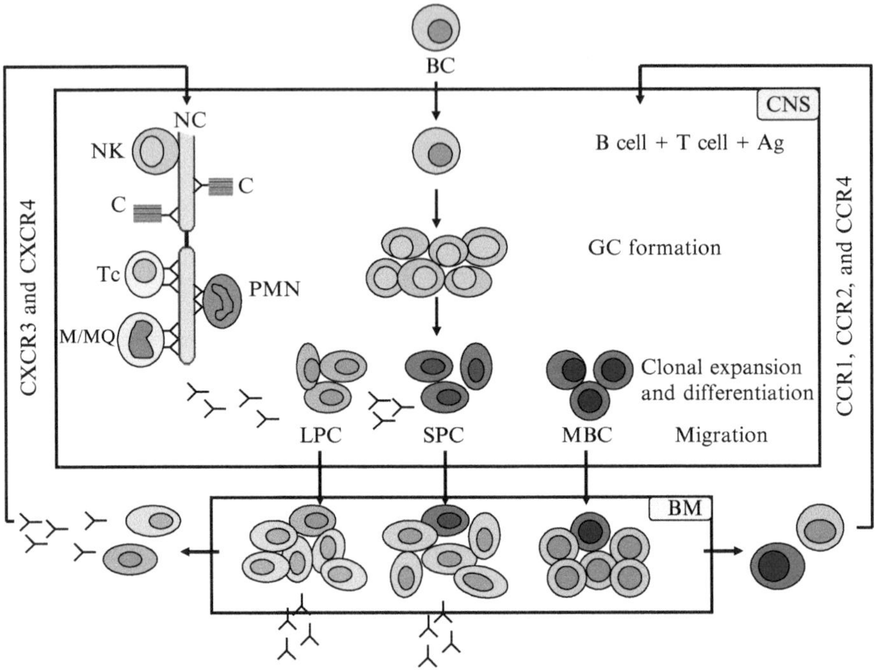

FIG. 3. B-cell responses in CNS. When active BCs enter CNS they form GC, undergo clonal expansion, and differentiation into short-living PC (SPC), long-living PC (LPC), and MBC. Plasma cells secrete autoAbs/Abs (⅄) inside CNS which initiate a damage process on nerve cells (NCs) using complement (C) system and effector cells such as M/MQ, and so on. These cells can also migrate and home to other sites mainly the BM. In the BM, they undergo further maturation. When recalled, these cells plus the BM repertoire of PCs and MBCs can migrate to CNS via appropriate chemokines. This is a cycle in the CNS of MS patient and evidence of constitutive Abs presence in that area. NK, natural killer; Tc, T cytotoxic; M/MQ, monocyte macrophages; PMN, polymorphonuclear cells; Th, T helper; Ag, antigen. (See Color Insert.)

formation using FCR and complement receptor (CR) facilitates the presence and affinity response of memory B cells.

F. Tissue-Specific Migration

Migration of cells between various sites in the body is a natural phenomenon. This ensures homeostasis in the body. Without cell migration, growth and development of cells, interaction among cells, and production of suitable responses could not occur. If the body's regulatory processes perform abnormally under certain circumstances, the normal functions of the cell support this dysregulatory phenomenon. Chemokines participate in the migratory pathways and enhance the autoreactive component migration to the inflamed area.

Ag-specific B cells are able to enter all parts of a normal human brain; however, this amount is infinitesimal (Anthony *et al.*, 2003). Human B cells constitutively express adhesion molecules such as very late activation antigen-4 (VLA-4) and LFA-1 on their surface. Their corresponding ligands, vascular cellular adhesion molecule-1 (VCAM-1) and ICAM-1, are upregulated on BBB endothelium by chemokines such as MCP-1 and IL-8 (Alter *et al.*, 2003). So anytime the above chemokines are secreted into the environment for any reason (inflammation or infection), B cells can easily cross the BBB. Scientists have noted that human B cells cross the endothelium faster than their T-cell counterparts (Anthony *et al.*, 2003).

However, during inflammation or the duration of this demyelinating reaction, MCP-1 and IL-8 chemokines interact with their corresponding receptors (CXCR1, CXCR2, CCR2) on the surface of B cells, facilitating B-cell entry into the CNS (Alter *et al.*, 2003). It has been reported (Koch *et al.*, 1981) that 80% of serum Ig is produced by long-living PC residing in the BM. VLA-4, LFA-1, and CD44, which interact with BM stroma, cause long-living PC accumulation in this area. Except for BM, PC can be found in inflamed tissue. Perhaps a number of PC are produced *in situ*; however, most of the population migrate to the respective tissue (Cassese *et al.*, 2001). It seems that migration is not dependent on the antigen. However, T cells are potentially significant because CXCR3 expression is dependent on IFN-γ. This causes plasmablast migration to inflamed sites and release of their Ab in an attempt to eliminate the pathogen. Therefore, CXCR3 expression is dependent on T-cell activation (Muehlinghaus *et al.*, 2005). It has been shown (Egawa *et al.*, 2001) that CXCL12, a late phase BM PC homing chemokine, has a critical role in early B-cell development (Pre-pro-B-cell stage). CXCL12 is also effective in other developmental processes like neurogenesis and angiogenesis (Ma *et al.*, 1998). IP-10 or CXCL-10 is one of the inflammatory chemokines that could be detected in the inflamed tissue (Hauser *et al.*, 2002). IP-10 receptor, CXCR-3, is expressed on plasmablast surfaces. Moser *et al.* (2006) has reported that CXCR-3 and its ligand, IP-10, play an important role in attraction of these cells to the site of inflammation. They showed the accumulation of autoreactive plasma cells in spleen and inflamed tissues of NZB/W mice. They believe that such plasma cells come into the vicinity of the diseased tissue long before onset of the illness itself. Therefore, these cells can play an inductive role in immunopathology of the disease.

G. Antigens in MS

The presence of an Ab indicates the existence of an Ag. When we speak of an Ab, we are in fact referring to the Ag which induces its production. In MS, a number of Ags have been implicated in disease pathogenesis. Plaque formation is a consequence of immune reaction and adoptive mechanisms. Once triggered, an autoimmune attack is launched against myelin sheath Ags. As we know, EAE can be induced in mammals through brain tissue or myelin-specific protein injection (Wekerle, 1998). These myelin proteins include MOG, MBP, MAG, and PLP. Here, we present some of the more prevalent Ags.

1. *Myelin Oligodendrocyte Glycoprotein*

MOG is the strongest immunogen molecule in the myelin sheath composition. This protein is very small and is a component of the outer lamella of the myelin sheath (Brunner *et al.*, 1989). MOG is the most prominent target of demyleinating Abs in EAE (Lebar *et al.*, 1986; Linington and Lassmann, 1987). Anti-MOG Abs are more abundantly detected in MS patients compared to other neurological diseases (Xiao *et al.*, 1991). However, this amount is insignificant (Karni *et al.*, 1999; Reindl *et al.*, 1999). It seems that anti-MOG Abs have a secondary role in demyelination, but are not the causative agents of disease onset. However, in B-cell-deficient mice, EAE induction via MOG caused a disease with little severity (Svensson *et al.*, 2002). The discrepancy among the evidence does not allow definite conclusions on the role MOG plays in MS.

2. *Myelin Basic Protein*

MBP is the most studied autoantigen in MS. This protein is present in oligodendrocyte cytoplasmic membranes and constitutes 30% of myelin (Cruz *et al.*, 1987). A rise of Ab titer against MBP proteins or MBP peptides in MS patients is related to disease activity (Warren *et al.*, 1994). Anti-MBP Ab can be found in other neurological disorders such as optic neuritis or aseptic meningoencephalitis, a feature common in CNS dysfunction (Soderstrom *et al.*, 1993). The presence of a heterogeneous and diverse Ab repertoire against MBP and the considerable number of anti-myelin IgG in MS patients lends further support to the chronic nature of MS.

3. *2′3′ Cyclic Nucleotide 3′ Phosphodiesterase (CNP1)*

CNP1 is present on the surface of oligodendrocytes. It is one of the autoantigens against which MS patients synthesize IgM. Testing the antigenicity and reactivity of these antigens, the serum of 52 of 70 MS patients (74%) reacted with CNP1 (Walsh and Murray, 1998). Only 3 of 33 (~1%) patients suffering from other neurological disorders reacted to this antigen. Although the main oligoclonal response in CSF of MS patients consists of IgG, CNP1-reacting IgG (Link and Laurenzi, 1979) has not been isolated.

4. *Proteolipid Protein*

PLP is the other component of myelin. About 2% of MS patients carry anti-PLP Ab in their CSF, while patients with other neurological diseases have an anti-PLP Ab detection rate of about 11% (Warren and Catz, 1994; Warren *et al.*, 1994).

5. *Myelin-Associated Glycoprotein*

MAG is only detectable with sensitive tests such as radioimmunoassay (RIA) and is only present in the CSF; such finding indicates that MAG is only marginally immunogenic (Moller *et al.*, 1989).

6. *Oligodendrocyte-Specific Protein*

Using ELIZA technique, antibody against oligodendrocyte-specific protein has been detected in the CSF of 80% of RRMS patients, while such antibody has not been reported in SPMS patients (Bronstein *et al.*, 1999).

7. *Other Autoantigens*

Non-CNS autoantigens which have been detected in serum and lesions biopsies of MS patients include α,β-crystalin (Agius *et al.*, 1999), heat-shock protein (hsp) (Birnbaum and Kotilinek, 1997), transadolase (Banki *et al.*, 1994), proteasome (Mayo *et al.*, 2002), and DNA (Williamson *et al.*, 2001).

8. *Viral Ag and Ab*

The association of a variety of infectious agents, including mumps, measles, rubella, HSV, VZV, vaccinia, parainfluenza1, with MS have been examined a number of times (Cremer *et al.*, 1980; Forghani *et al.*, 1978; Norrby, 1978; Salmi *et al.*, 1983). These studies appear to come to the same end point; antibody against viral antigen can be detected in control and patients, but MS patients tend to have antibody relating to more than one virus in their CSF (Salmi *et al.*, 1983), the only viral Ab that seems to have a correlation with MS is measles (Cremer *et al.*, 1980; Salmi *et al.*, 1983). One study suggests that viral infection in MS is a sex-biased issue (Fuller *et al.*, 2005).

VI. The Role of B Cells in Therapy

Studies (Baker *et al.*, 2003; Matsushita *et al.*, 2006; Tedder *et al.*, 2005) indicate a protective role for B cells in certain autoimmune systems. One study (Matsushita *et al.*, 2006) showed that CD19 surface marker deficiency has a critical role in EAE etiology. They showed that CD19 is a fundamental factor for T-cell differentiation and shifts to Th2 phenotype. CD19 acts as an inhibitory molecule in the development of EAE. Tedder *et al.* (2005) demonstrated that the absence of CD19 on B cells increases the severity of disease and delayed recovery. CD19 plays a role in regulation of basal signaling threshold and accelerates BCR signaling.

Monoclonal anti-BLYS Ab suppresses *in vitro* B-cell proliferation and induces B-cell depletion in spleen and lymph nodes of Cynomolgus monkey (Baker *et al.*, 2003). As we mentioned before, upregulation of BAFF also called BLYS is observed in MS patients (Krumbholz *et al.*, 2005). Therefore, anti-BLYS Ab may serve as a potential therapy in MS.

Depletion of B cell is followed by a dramatic decreases in Ab, diminishes IC formation, and a decline in cytokine secretion (IL-10, IL-6, TNF-α). Absence of B cells also means their APC activity is deterred; therefore, costimulatory signals which are needed for clonal T cell and macrophage stimulation are missing.

Rituximab is a synthetic monoclonal antibody against CD20. This monoclonal antibody selectively depletes CD20+ pre- and mature B cells. This depletion involves a variety of mechanisms such as ADCC, complement-mediated lysis, and B-cell apoptosis (Dalakas, 2006; Reff *et al.*, 1994). Stuve *et al.* (2005) reported that the effect of these monoclonal Abs is more pronounced in primary progressive MS (PPMS) patients, when B cells are depleted from PB rather than CSF. To prove the effect of Ab decreases in disease improvement, Keegan *et al.* (2005) showed that after therapeutic plasmapheresis in MS patients, some improvement was observed in disease condition. Therefore, B-cell-targeted therapy can be a powerful tool in autoimmune treatment.

VII. Conclusions

Most immunologic studies on MS have focused on T cells as the main participants in disease induction and maintenance to the point that the roles of other components and cells involved in pathogenesis of MS have gone largely unnoticed. In a complex system such as the immune network, isolated components do not provide a comprehensive reflection of the scope of their activity and their causes. Hence, B cells and their end stage products, the antibodies, must not be forgotten in this cyclic network. The late entry of B cells into the disease arena, that is after the entry of T cells, is likely the reason for the lack of attention given them in the disease process. As discussed above, it is an established fact that T cells are in most cases the initiator. T cells induce the events that lead to an effector phase where lesions are made and then maintained through a dimly understood feedback system. This is most likely attributable to pathological long-living plasma cells which produce their specific Ab irrespective of the presence of a trigger. Therefore, it can be said that B cells and plasma cells enter the network at a later time. So their detection in the site of action could be less pronounced at certain stages.

Another issue worth mentioning when considering the role of B cells in pathogenesis of MS is the constant presence of OCB in their serum. This is an early observation in MS diagnosis. MS patients also have a pronounced B-cell clonal expansion detectable within the CNS and brain lesion vicinity. Besides, in MS pathological classifications, antibody-mediated demyelination of lesions is a common observation (Frohman *et al.*, 2006).

A new focus in therapeutic research in MS is B-cell depletion. The simple idea is that if pathological Ab-producing cells are eliminated, there should be a perceptible improvement in the patient's condition. This option has proven itself an encouraging alternative worthy of consideration.

Abbreviations

Ab, antibody; ADCC, antibody-dependent cell cytotoxicity; Ag, antigen; APC, antigen-presenting cell; ASC, antibody-secreting cell; BAFF, B-cell-activating factor for the TNF family; BBB, blood–brain barrier; BCR, B-cell receptor; BM, bone marrow; CCR, chemokine C-motif receptor; CNP1, 2′3′ cyclic nucleotide 3′ phosphodiesterase; CNS, central nervous system; CR, complement receptor; CSF, cerebrospinal fluid; CXCL, CX-chemokine receptor ligand; CXCR, CX-chemokine receptor; DJ, diversity and junction; EAE, experimental autoimmune encephalomyelitis; EBF-1, early B-cell factor; EBV, Ebstein-Barr virus; Fab, fragment of antigen binding; Fc, fragment of crystallization; FDC, follicular dendritic cells; FLT3, Fms-related tyrosine kinase 3; GBS, guillain barre syndrome; GC, germinal center; HSC, hematopoeitic stem cells; HSV, Herpes simplex virus; HTLV-1, Human T-lymphotropic virus; IC, immune complex; ICAM-1, intracellular adhesion molecule-1; IFN-γ, interferon-γ; Ig, immunoglobulin; LFA-1, leukocyte function-associated antigen-1; MAC, membrane attack complex; MHC, major histocompatibility complex; MM, molecular mimicry; MS, multiple sclerosis; MAG, myelin-associated glycoprotein; MBP, myelin basic protein; MOG, myelin oligoclonal glycoprotein; MRI, magnetic resonance imaging; OCB, oligoclonal bands; OSP, oligodendrocyte-specific protein; PB, peripheral blood; PC, plasma cell; PLP, proteolipid protein; PPMS, primary progressive multiple sclerosis; RA, rheumatoid arthritis; SCF, stem cell factor; SPMS, secondary progressive multiple sclerosis; SSPE, subacute sclerosing panencephalitis; TCR, T-cell receptor; Th, helper T cell; TNF-α, tumor necrosis factor-α; VCAM-1, vascular cellular adhesion molecule-1; VLA-4, very late activation antigen-4; VZV, Varicella-zoster virus.

References

Adams, C. V. (1975). The onset and progression of the lesion in multiple sclerosis. *J. Neurol. Sci.* **25,** 165–182.

Agius, M. A., Kirvan, C. A., Schafer, A. L., Gudipati, E., and Zhu, S. (1999). High prevalence of anti-alpha-crystallin antibodies in multiple sclerosis: Correlation with severity and activity of disease. *Acta Neurol. Scand.* **100,** 139–147.

Ahonen, C. L., Ahonen, C., Manning, E., Erickson, L. D., O'Connor, B., Lind, E. F., Pullen, S. S., Kehry, M. R., and Noelle, R. J. (2002). The CD40-TRAF6 axis controls affinity maturation and the generation of long-lived plasma cells. *Nat. Immunol.* **3,** 451–456.

Alter, A., Duddy, M., Hebert, S., Biernacki, K., Prat, A., Antel, J. P., Yong, V. W., Nuttall, R. K., Pennington, C. J., Edwards, D. R., and Bar-Or, A. (2003). Determinants of human B cell migration across brain endothelial cells. *J. Immunol.* **170,** 4497–4505.

Ang, C. W., Jacobs, B. C., and Laman, J. D. (2004). The Guillain-Barre syndrome: A true case of molecular mimicry. *Trends Immunol.* **25,** 61–66.

Anthony, I. C., Crawford, D. H., and Bell, J. E. (2003). B lymphocytes in the normal brain: Contrasts with HIV-associated lymphoid infiltrates and lymphomas. *Brain* **126,** 1058–1067.

Arce, S., Luger, E., Muehlinghaus, G., Cassese, G., Hauser, A., Horst, A., Lehnert, K., Odendahl, M., Honemann, D., Heller, K. D., Kleinschmidt, H., Dorner, T., *et al.* (2004). CD38 low IgG-secreting cells are precursors of various CD38 high-expressing plasma cell populations. *J. Leukoc. Biol.* **75,** 1022–1028.

Archelos, J. J., Storch, M. K., and Hartung, H. P. (2000). The role of B cells and autoantibodies in multiple sclerosis. *Ann. Neurol.* **47,** 694–706.

Avery, D. T., Ellyard, J. I., Mackay, F., Corcoran, L. M., Hodgkin, P. D., and Tangye, S. G. (2005). Increased expression of CD27 on activated human memory B cells correlates with their commitment to the plasma cell lineage. *J. Immunol.* **174**(7), 4034–4042.

Baker, K. P., Edwards, B. M., Main, S. H., Choi, G. H., Wager, R. E., Halpern, W. G., Lappin, P. B., Riccobene, T., Abramian, D., Sekut, L., Sturm, B., Poortman, C., *et al.* (2003). Generation and characterization of LymphoStat-B, a human monoclonal antibody that antagonizes the bio-activities of B lymphocyte stimulator. *Arthritis Rheum.* **48,** 3253–3265.

Banki, K., Colombo, E., Sia, F., Halladay, D., Mattson, D. H., Tatum, A. H., Massa, P. T., Phillips, P. E., and Perl, A. (1994). Oligodendrocyte-specific expression and autoantigenicity of transaldolase in multiple sclerosis. *J. Exp. Med.* **180**(5), 1649–1663.

Bansil, S., Cook, S. D., and Rohowsky-Kochan, C. (1995). Multiple sclerosis: Immune mechanism and update on current therapies. *Ann. Neurol.* **37**(Suppl. 1), S87–S101.

Baranzini, S. E., Jeong, M. C., Butunoi, C., Murray, R. S., Bernard, C. C. A., and Oksenberg, J. R. (1999). B cell repertoire diversity and clonal expansion in multiple sclerosis brain lesions. *J. Immunol.* **163,** 5133–5144.

Barrington, R. A., Pozdnyakova, O., Zafari, M. R., Benjamin, C. D., and Carroll, M. C. (2002). B lymphocyte memory: Role of stromal cell complement and FcγRIIB receptors. *J. Exp. Med.* **196,** 1189–1199.

Birnbaum, G., and Kotilinek, L. (1997). Heat shock or stress proteins and their role as autoantigens in multiple sclerosis. *Ann. N. Y. Acad. Sci.* **835,** 157–167.

Brezinschek, H. P., Foster, S. J., Brezinschek, R. I., Dorner, T., Domiati-Saad, R., and Lipsky, P. E. (1997). Analysis of the human VH gene repertoire differential effects of selection and somatic hypermutation on human peripheral CD5(+)/IgM+ and CD5(−)/IgM+ B cells. *J. Clin. Invest.* **99,** 2488–2501.

Bronstein, J. M., Lallone, R. L., Seitz, R. S., Ellison, G. W., and Myers, L. W. (1999). A humoral response to oligodendrocyte-specific protein in MS: A potential molecular mimic. *Neurology* **53,** 154–161.

Brunner, C., Lassmann, H., Weahneldt, T. V., Matthieu, J. M., and Linington, C. (1989). Differential ultrastructural localization of myelin basic protein, myelin/oligodendroglial glycoprotein, and 2′,3′-cyclic nucleotide 3′-phospho diesterase in the CNS of adult rat. *J. Neurochem.* **52,** 296–304.

Burgoon, M. P., Hammack, B. N., Owens, G. P., Maybach, A. L., Eikelenboom, M. J., and Gilden, D. H. (2003). Oligoclonal immunoglobulins in cerebrospinal fluid during Varicella-zoster virus (VZV) vasculopathy are directed against VZV. *Ann. Neurol.* **54,** 459–463.

Burgoon, M. P., Gilden, D. H., and Owens, G. P. (2004). B cells in multiple sclerosis. *Front. Biosci.* **9,** 786–796.

Burnet, F. M. (1976). A modification of Jerne's theory of antibody production using the concept of clonal selection. *CA Cancer J. Clin.* **26,** 119–121.

Cassese, G., Lindenau, S., de Boer, B., Arce, S., Hauser, A., Riemekasten, G., Berek, C., Hiepe, F., Radbruch, A., and Manz, R. A. (2001). Inflamed kidneys of NZB/W mice are a major site for the homeostasis of plasma cells. *Eur. J. Immunol.* **31,** 2726–2732.

Cepok, S., Jacobsen, M., Schock, S., Omer, B., Jaekel, S., Boddeker, I., Oertel, W. H., Sommer, N., and Hemmer, B. (2001). Patterns of cerebrospinal fluid pathology correlate with disease progression in multiple sclerosis. *Brain* **124,** 2169–2176.

Cepok, S., Rosche, B., Grummel, V., Vogel, F., Zhou, D., Sayn, J., Sommer, N., Hartung, H. P., and Hemmer, B. (2005). Short-lived plasma blasts are the main B cell effector subset during the course of multiple sclerosis. *Brain* **128,** 1667–1676.

Chan, O. T., Madaio, M. P., and Shlomchik, M. J. (1999). B cells are required for lupus nephritis in the polygenic, Fas-intact MRL model of systemic autoimmunity. *J. Immunol.* **163,** 3592–3596.

Cole, S. R., Beck, R. W., Moke, P. S., Kaufman, D. I., and Tourtellotte, W. W. (1998). The predictive value of CSF oligoclonal banding for MS 5 years after optic neuritis. Optic Neuritis Study Group. *Neurology* **51,** 885–887.

Colombo, M., Dono, M., Gazzola, P., Roncella, S., Valetto, A., Chiorazzi, N., Mancardi, G. L., and Ferrarini, M. (2000). Accumulation of clonally related B lymphocytes in the cerebrospinal fluid of multiple sclerosis patients. *J. Immunol.* **164,** 2782–2789.

Corcione, A., Casazza, S., Ferretti, E., Giunti, D., Zappia, E., Pistorio, A., Gambini, C., Mancardi, L., Uccelli, A., and Pistoia, V. (2004). Recapitulation of B cell differentiation in the central nervous system of patients with multiple sclerosis. *Proc. Natl. Acad. Sci. USA* **101,** 11064–11069.

Corcione, A., Aloisi, F., Serafini, B., Capello, E., Mancardi, G. L., Pistoia, V., and Uccelli, A. (2005). B-cell differentiation in the CNS of patients with multiple sclerosis. *Autoimmun. Rev.* **4,** 549–554.

Correale, J., and de Los Milagros Bassani Molinas, M. (2002). Oligoclonal bands and antibody responses in multiple sclerosis. *J. Neurol.* **249,** 375–389.

Cremer, N. E., Johnson, K. P., Fein, G., and Likosky, W. H. (1980). Comprehensive viral immunology of multiple sclerosis. II. Analysis of serum and CSF antibodies by standard serologic methods. *Arch. Neurol.* **37,** 610–615.

Cross, A. H. (2000). The return of the B-cell. *Neurology* **54,** 1214–1215.

Cross, A. H., Trotter, J. L., and Lyons, J. A. (2001). B-cells and antibodies in CNS demyelinating disease. *J. Neuroimmunol.* **112,** 1–14.

Cruz, M., Olsson, T., Ernerudh, J., Hojeberg, B., and Link, H. (1987). Immunoblot detection of oligoclonal anti-myelin basic protein IgG antibodies in cerebrospinal fluid in multiple sclerosis. *Neurology* **37,** 1515–1519.

Dalakas, M. C. (2006). B cells in the pathophysiology of autoimmune neurological disorders: A credible therapeutic target. *Pharmacol. Ther.* **112,** 57–70.

Dorner, T., and Radbruch, A. (2005). Selecting B cells and plasma cells to memory. *J. Exp. Med.* **201,** 497–499.

Egawa, T., Kawabata, K., Kawamoto, H., Amada, K., Okamoto, R., Fujii, N., Kishimoto, T., Katsura, Y., and Nagasawa, T. (2001). The earliest stages of B cell development require a chemokine stromal cellderived factor/pre-B cell growth-stimulating factor. *Immunity* **15,** 323–334.

Erickson, L. D., Vogel, L. A., Cascalho, M., Wong, J., Wabl, M., Durell, B. G., and Noelle, R. J. (2000). B cell immunopoiesis: Visualizing the impact of CD40 engagement on the course of T cell-independent immune responses in an Ig transgenic system. *Eur. J. Immunol.* **30,** 3121–3131.

Erickson, L. D., Durell, B. G., Vogel, L. A., O'Connor, B. P., Cascalho, M., Yasui, T., Kikutani, H., and Noelle, R. J. (2002). Short-circuiting long-lived humoral immunity by the heightened engagement of CD40. *J. Clin. Invest.* **109,** 613–620.

Esiri, M. M. (1977). Immunoglobulin containing cells in multiple sclerosis plaques. *Lancet* **2,** 478.

Filippini, G., Comi, G. C., Cosi, V., Bevilacqua, L., Ferrarini, M., Martinelli, V., Bergamaschi, R., Filippi, M., Citterio, A., D'Incerti, L., Campi, A., Sberna, M, *et al.* (1994). Sensitivities and predictive values of paraclinical tests for diagnosing multiple sclerosis. *J. Neurol.* **241,** 132–137.

Fillatreau, S., Sweenie, C. H., McGeachy, M. J., Gray, D., and Anderton, S. M. (2002). B cells regulate autoimmunity by provision of IL-10. *Nat. Immunol.* **3,** 944–950.

Forghani, B., Cremer, N. E., Johnson, K. P., Ginsberg, A. H., and Likosky, W. H. (1978). Viral antibodies in cerebrospinal fluid of multiple sclerosis and control patients: Comparison between radioimmunoassay and conventional techniques. *J. Clin. Microbiol.* **7,** 63–69.

Foy, T., Aruffo, A., Bajorath, J., Buhlmann, J. E., and Noelle, R. J. (1996). Immune regulation by CD40 and its ligand GP39. *Annu. Rev. Immunol.* **14,** 591–617.

Frohman, E. M., Racke, M. K., and Raine, C. S. (2006). Multiple sclerosis—the plaque and its pathogenesis. *N. Engl. J. Med.* **354,** 942–955.

Fuller, A. C., Kang, B., Kang, H. K., Yahikozowa, H., Dal Canto, M. C., and Kim, B. S. (2005). Gender bias in Theiler's virus-induced demyelinating disease correlates with the level of antiviral immune responses. *J. Immunol.* **175,** 3955–3963.

Gay, D., and Esiri, M. (1991). Blood–brain barrier damage in acute multiple sclerosis plaques. An immunocytological study. *Brain* **114,** 557–572.

Giles, P. D., and Wroe, S. J. (1990). Cerebrospinal fluid oligoclonal IgM in multiple sclerosis: Analytical problems and clinical limitations. *Ann. Clin. Biochem.* **27,** 199–207.

Giunti, D., Borsellino, G., Benelli, R., Marchese, M., Capello, E., Valle, M. T., Pedemonte, E., Noonan, D., Albini, A., Bernardi, G., Mancardi, G. L., Battistini, L., *et al.* (2003). Phenotypic and functional analysis of T cells homing into the CSF of subjects with inflammatory diseases of the CNS. *J. Leukoc. Biol.* **73,** 584–590.

Goldsby, R. A., Kindt, T. J., and Osborne, B. A. (2000). Kuby Immunology, 4th edn. WH Freeman and Company. New York.

Gray, D., Siepmann, K., van Essen, D., Poudrier, J., Wykes, M., Jainandunsing, S., Bergthorsdottir, S., and Dullforce, P. (1996). B–T lymphocyte interactions in the generation and survival of memory cells. *Immunol. Rev.* **150,** 45–61.

Grimaldi, L. M., Roos, R. P., Devare, S. G., Casey, J. M., Maruo, Y., Hamada, T., and Tashiro, K. (1988). HTLV-I-associated myelopathy: Oligoclonal immunoglobulin G bands contain anti-HTLV-I p24 antibody. *Ann. Neurol.* **24,** 727–731.

Hannet, I., Erkeller-Yuksel, F., Lydyard, P., Deneys, V., and DeBruyere, M. (1992). Developmental and maturational changes in human blood lymphocyte subpopulations. *Immunol. Today* **13,** 215.

Hauser, A. E., Debes, G. F., Arce, S., Cassese, G., Hamann, A., Radbruch, A., and Manz, R. A. (2002). Chemotactic responsiveness toward ligands for CXCR3 and CXCR4 is regulated on plasma blasts during the time course of a memory immune response. *J. Immunol.* **169,** 1277–1282.

Hela-Felicitas, P., and Reske, D. (2005). Expansion of antibody reactivity in the cerebrospinal fluid of multiple sclerosis patients—follow-up and clinical implications. *Cerebrospinal Fluid Res.* **2,** 3.

Hjelmstrom, P. (2001). Lymphoid neogenesis: *De novo* formation of lymphoid tissue in chronic inflammation through expression of homing chemokines. *J. Leukoc. Biol.* **69,** 331–339.

Hogquist, K. A., Baldwin, T. A., and Jameson, S. C. (2005). Central tolerance: Learning self-control in the thymus. *Nat. Rev. Immunol.* **5,** 772–782.

Horikawa, Y., Tsubaki, T., and Nakajima, M. (1973). Rubella antibody in multiple sclerosis. *Lancet* **1,** 996–997.

Hoyer, B. F., Moser, K., Hauser, A. E., Peddinghaus, A., Voigt, C., Eilat, D., Radbruch, A., Hiepe, F., and Manz, F. (2004). Short-lived plasmablasts and long-lived plasma cells contribute to chronic humoral autoimmunity in NZB/W mice. *J. Exp. Med.* **199,** 1577–1584.

Jersild, C., Svejgaard, A., and Fog, T. (1972). HL-A antigens and multiple sclerosis. *Lancet* **1,** 1240–1241.

Ji, H., Korganow, A. S., Mangialaio, S., Hoglund, P., Andre, I., Luhder, F., Gonzalez, A., Poirot, L., Benoist, C., and Mathis, D. (1999). Different modes of pathogenesis in T-cell dependent autoimmunity: Clues from two TCR transgenic systems. *Immunol. Rev.* **169,** 139–146.

Ji, H., Ohmura, K., Mahmood, U., Lee, D. M., Hofhuis, F. M., Boackle, S. A., Takahashi, K., Holers, V. M., Walport, M., Gerard, C., Ezekowitz, A., Carroll, M. C., *et al.* (2002). Arthritis critically dependent on innate immune system players. *Immunity* **16,** 157–168.

Karni, A., Bakimer-Kleiner, R., Abramsky, O., and Ben-Nun, A. (1999). Elevated levels of antibody to myelin oligodendrocyte glycoprotein is not specific for patients with multiple sclerosis. *Arch. Neurol.* **56,** 311–315.

Keegan, M., Konig, F., McClelland, R., Bruck, W., Morales, Y., Bitsch, A., Panitch, H., Lassmann, H., Weinshenker, B., Rodriguez, M., Parisi, J., and Lucchinetti, C. F. (2005). Relation between humoral pathological changes in multiple sclerosis and response to therapeutic plasma exchange. *Lancet* **366,** 526–527.

Kerlero de Rosbo, N., Milo, R., Lees, M. B., Burger, D., Bernard, C. C., and Ben-Nun, A. (1993). Reactivity to myelin antigens in multiple sclerosis. Peripheral blood lymphocytes respond predominantly to myelin oligodendrocyte glycoprotein. *J. Clin. Invest.* **92,** 2602–2608.

Kim, H. J., Krenn, V., Steinhauser, G., and Berek, C. (1999). Plasma cell development in synovial germinal centers in patients with rheumatoid and reactive arthritis. *J. Immunol.* **162,** 3053–3062.

Kantor, A. (1991). A new nomenclature for B cells. *Immunol. Today* **12,** 388–391.

Koch, G., Osmond, D. G., Julius, M. H., and Benner, R. (1981). The mechanism of thymus-dependent antibody formation in bone marrow. *J. Immunol.* **126,** 1447–1451.

Krumbholz, M., Theil, D., Derfuss, T., Rosenwald, A., Schrader, F., Monoranu, C. M., Kalled, S. L., Hess, D. M., Serafini, B., Aloisi, F., Wekerle, H., Hohlfeld, R., *et al.* (2005). BAFF is produced by astrocytes and up-regulated in multiple sclerosis lesions and primary central nervous system lymphoma. *J. Exp. Med.* **201,** 195–200.

Kurtzke, J. F., and Bui-Quoc-Huong, J. F. (1980). Multiple sclerosis in a migrant population: 2. Half-orientals immigrating in childhood. *Ann. Neurol.* **8**(3), 256–260.

Kurtzke, J. F., Delasnerie-Laupretre, N., and Wallin, M. T. (1998). Multiple sclerosis in North African migrants to France. *Acta Neurol. Scand.* **98,** 302–309.

Lang, H. L. E., Jacobsen, H., Ikemizu, S., Andersson, C., Harlos, K., Madsen, L., Hjorth, P., Sondergaard, L., Svejgaard, A., Wucherpfenning, K., Stuart, D. I., Bell, J. I., *et al.* (2002). A functional and structural basis for TCR cross-reactivity in multiple sclerosis. *Nat. Immunol.* **3,** 940–943.

Lassmann, H., Raine, C. S., Antel, J., and Prineas, J. W. (1998). Immunopathology of multiple sclerosis: Report on an international meeting held at the Institute of Neurology of the University of Vienna. *J. Neuroimmunol.* **86,** 213–217.

Lebar, R., Lubetzki, C., Vincent, C., Lombrail, P., and Boutry, J. M. (1986). The M2 autoantigen of central nervous system myelin, a glycoprotein present in oligodendrocyte membrane. *Clin. Exp. Immunol.* **66,** 423–434.

Levin, M. C., Lee, S. M., Morcos, Y., Brady, J., and Stuart, J. (2002). Cross-reactivity between immunodominant human T lymphotropic virus type I tax and neurons: Implications for molecular mimicry. *J. Infect. Dis.* **186,** 1514–1517.

Lichtman, M. A. (1981). The ultrastructure of the hemopoietic environment of the marrow: A review. *Exp. Hematol.* **9,** 391–410.

Lineberry, N., and Fathman, C. G. (2006). T cell anergy: Where it's LAT. *Immunity* **24,** 501–503.

Linington, C., and Lassmann, H. (1987). Antibody responses in chronic relapsing experimental allergic encephalomyelitis: Correlation of serum demyelinating activity with antibody titer to the myelin/oligodendrocyte glycoprotein (MOG). *J. Neuroimmunol.* **17,** 61–69.

Link, H., and Laurenzi, M. A. (1979). Immunoglobulin class and light chain type of oligoclonal bands in CSF in multiple sclerosis determined by agarose gel electrophoresis and immunofixation. *Ann. Neurol.* **6,** 107–110.

Lipsky, P. E. (2001). Systemic lupus erythematosus: An autoimmune disease of B cell hyperactivity. *Nat. Immunol.* **2,** 764–766.

Liu, Y. J., Johnson, G. D., Gordon, J., and MacLennan, I. C. (1992). Germinal centres in T-cell-dependent antibody responses. *Immunol. Today* **13,** 17–21.

Lolli, F., Siracusa, G., Amato, M. P., Fratiglioni, L., Dal Pozzo, G., Galli, E., and Amaducci, L. (1991). Intrathecal synthesis of free immunoglobulin light chains and IgM in initial multiple sclerosis. *Acta Neurol. Scand.* **83,** 239–243.

Lucchinetti, C., Bruck, W., Parisi, J., Scheithauer, B., Rodriguez, M., and Lassmann, H. (2000). Heterogeneity of multiple sclerosis lesions: Implications for the pathogenesis of demyelination. *Ann. Neurol.* **47,** 707–717.

Ma, Q., Jones, D., Borghesani, P. R., Segal, R. A., Nagasawa, T., Kishimoto, T., Bronson, R. T., and Springer, T. A. (1998). Impaired B-lymphopoiesis, myelopoiesis, and derailed cerebellar neuron migration in CXCR4- and SDF-1-deficient mice. *Proc. Natl. Acad. Sci. USA* **95,** 9448–9453.

Mackay, F., and Tangye, S. G. (2004). The role of the BAFF/APRIL system in B cell homeostasis and lymphoid cancers. *Curr. Opin. Pharmacol.* **4,** 347–354.

Manz, R. A., Thiel, A., and Radbruch, A. (1997). Lifetime of plasma cells in the bone marrow. *Nature* **388,** 133–134.

Manz, R. A., Lohning, M., Cassese, G., Thiel, A., and Radbruch, A. (1998). Survival of long-lived plasma cells is independent of antigen. *Int. Immunol.* **10,** 1703–1711.

Manz, R. A., Arce, S., Cassese, G., Hauser, A. E., Hiepe, F., and Radbruch, A. (2002). Humoral immunity and long-lived plasma cells. *Curr. Opin. Immunol.* **14,** 517–521.

Martinez, A., Mas, A., de las Heras, V., Arroyo, R., Fernandez-Arquero, M., de la Concha, E. G., and Urcelay, E. (2005). Early B-cell factor gene association with multiple sclerosis in the Spanish population. *BMC Neurol.* **5,** 19.

Matsushita, T., Fujimoto, M., Hasegawa, M., Komura, K., Takehara, K., Tedder, T. F., and Sato, S. (2006). Inhibitory role of CD19 in the progression of experimental autoimmune encephalomyelitis by regulating cytokine response. *Am. J. Pathol.* **168,** 812–821.

Mattson, D. H., Roos, R. P., and Arnason, B. G. (1980). Isoelectric focusing of IgG eluted from multiple sclerosis and subacute sclerosing panencephalitis brains. *Nature* **287,** 335–337.

Mattson, D. H., Roos, R. P., and Arnason, B. G. (1982). Oligoclonal IgG in multiple sclerosis and subacute sclerosing panencephalitis brains. *J. Neuroimmunol.* **2,** 261–276.

Mauri, C., Gray, D., Mushtaq, N., and Londei, M. (2003). Prevention of arthritis by interleukin 10-producing B cells. *J. Exp. Med.* **197,** 489–501.

Mayo, I., Arribas, J., Villoslada, P., Alvarez DoForno, R., Rodriguez-Vilarino, S., Montalban, X., De Sagarra, M. R., and Castano, J. G. (2002). The proteasome is a major autoantigen in multiple sclerosis. *Brain* **125,** 2658–2667.

Melchers, F., and Rolink, A. R. (2006). B cell tolerance—how to make it and how to break it. *Curr. Top. Microbiol. Immunol.* **305,** 1–23.

Mix, E., Olsson, T., Correale, J., Baig, S., Kostulas, V., Olsson, O., and Link, H. B. (1990). Cells expressing CD5 are increased in cerebrospinal fluid of patients with multiple sclerosis. *Clin. Exp. Immunol.* **79,** 21–27.

Moller, J. R., Johnson, D., Brady, R. O., Tourtellotte, W. W., and Quarles, R. H. (1989). Antibodies to myelin associated glycoprotein (MAG) in the cerebrospinal fluid of multiple sclerosis patients. *J. Neuroimmunol.* **22,** 55–61.

Morgan, B. P., and Gasque, P. (1996). Expression of complement in the brain: Role in health and disease. *Immunol. Today* **17,** 461–466.

Moser, K., Muehlinghaus, G., Manz, R., Mei, H., Voigt, C., Yoshida, T., Dorner, T., Hiepe, F., and Radbruch, A. (2006). Long-lived plasma cells in immunity and immunopathology. *Immunol. Lett.* **103,** 83–85.

Muehlinghaus, G., Cigliano, L., Huehn, S., Peddinghaus, A., Leyendeckers, H., Hauser, A. E., Hiepe, F., Radbruch, A., Arce, S., and Manz, R. A. (2005). Regulation of CXCR3 and CXCR4 expression during terminal differentiation of memory B cells into plasma cells. *Blood* **105,** 3965–3971.

Nagasawa, T. (2006). Microenvironmental niches in the bone marrow required for B-cell development. *Nat. Rev. Immunol.* **6,** 107–116.

Noelle, R. J., and Erickson, L. D. (2005). Determinations of B cell fate in immunity and autoimmunity. *Curr. Dir. Autoimmun.* **8,** 1–24.

Norrby, E. (1978). Viral antibodies in multiple sclerosis. *Prog. Med. Virol.* **24,** 1–39.

O'Connor, B. P., Cascalho, M., and Noelle, R. J. (2002). Short-lived and long-lived bone marrow plasma cells are derived from a novel precursor population. *J. Exp. Med.* **195,** 737–745.

O'Connor, B. P., Gleeson, M. W., Noelle, R. J., and Erickson, L. D. (2003). The rise and fall of long-lived humoral immunity: Terminal differentiation of plasma cells in health and disease. *Immunol. Rev.* **194,** 61–76.

Ochsenbein, A. F., Pinschewer, D. D., Sierro, S., Horvath, E., Hengartner, H., and Zinkernagel, R. M. (2000). Protective long-term antibody memory by antigen-driven and T help-dependent differentiation of long-lived memory B cells to short-lived plasma cells independent of secondary lymphoid organs. *Proc. Natl. Acad. Sci. USA* **97,** 13263–13268.

Odendahl, M., Mei, H., Hoyer, B. F., Jacobi, A. M., Hansen, A., Muehlinghaus, G., Berek, C., Hiepe, F., Manz, R., Radbruch, A., and Dorner, T. (2005). Generation of migratory antigen-specific plasma blasts and mobilization of resident plasma cells in a secondary immune response. *Blood* **105,** 1614–1621.

Olerup, O., and Hillert, J. (1991). HLA class II-associated genetic susceptibility in multiple sclerosis: A critical evaluation. *Tissue Antigens* **38,** 1–15.

Owens, G. P., Kraus, H., Burgoon, M. P., Smith-Jensen, T., Devlin, M. E., and Gilden, D. H. (1998). Restricted use of VH4 germline segments in an acute multiple sclerosis brain. *Ann. Neurol.* **43,** 236–243.

Owens, G. P., Ritchie, A. M., Burgoon, M. P., Williamson, R. A., Corboy, J. R., and Gilden, D. H. (2003). Single-cell repertoire analysis demonstrates that clonal expansion is a prominent feature of the B cell response in multiple sclerosis cerebrospinal fluid. *J. Immunol.* **171,** 2725– 2733.

Pettinelli, C. B., and McFarlin, D. E. (1981). Adoptive transfer of experimental allergic encephalomyelitis in SJL/J mice after *in vitro* activation of lymph node cells by myelin basic protein. Requirement for $1^+ 2^-$ T lymphocytes. *J. Immunol.* **127,** 1420–1423.

Pillai, S. (2005). Two lymphoid roads diverge—but does antigen bade B cells to take the road less traveled? *Immunity* **23,** 242–244.

Prineas, J. W., Kwon, E. E., Cho, E. S., Sharer, L. R., Barnett, M. H., Oleszak, E. L., Hoffman, B., and Morgan, B. P. (2001). Immunopathology of secondary-progressive multiple sclerosis. *Ann. Neurol.* **50,** 646–657.

Qin, D., Wu, J., Vora, K. A., Ravetch, J. V., Szakal, A. K., Manser, T., and Tew, J. G. (2000). Fc gamma receptor IIB on follicular dendritic cells regulates the B cell recall response. *J. Immunol.* **164,** 6268–6275.

Qin, Y., and Duquette, P. (2003). B-cell immunity in MS. *Int. MS J.* **10,** 110–120.

Qin, Y. F., Duquette, P., Zhang, Y. P., Talbot, P., Poole, R., and Antel, J. (1998). Clonal expansion and somatic hypermutation of V-H genes of B cells from cerebrospinal fluid in multiple sclerosis. *J. Clin. Invest.* **102,** 1045–1050.

Raine, C. S., Mokhtarian, F., and Mac Farlin, D. E. (1984). Adoptively transferred chronic relapsing experimental autoimmune encephalomyelitis in mouse: Neuropathologic analysis. *Lab. Invest.* **51,** 534–546.

Reff, M. E., Carner, K., Chambers, K. S., Chinn, P. C., Leonard, J. E., Raab, R., Newman, R. A., Hanna, N., and Anderson, D. R. (1994). Depletion of B cells *in vivo* by a chimeric mouse human monoclonal antibody to CD20. *Blood* **83,** 435–445.

Reindl, M., Linington, C., Brehm, U., Egg, R., Dilitz, E., Deisenhammer, F., Poewe, W., and Berger, T. (1999). Antibodies against the myelin oligodendrocyte glycoprotein and the myelin basic protein in multiple sclerosis and other neurological diseases: A comparative study. *Brain* **122,** 2047–2056.

Saito, E., Fujimoto, M., Hasegawa, M., Komura, K., Hamaguchi, Y., Kaburagi, Y., Nagaoka, T., Takehara, K., Tedder, T. F., and Sato, S. (2002). CD19-dependent B lymphocyte signaling thresholds influence skin fibrosis and autoimmunity in the tight-skin mouse. *J. Clin. Invest.* **109,** 1453–1462.

Salmi, A., Reunanen, M., Ilonen, J., and Panelius, M. (1983). Intrathecal antibody synthesis to virus antigens in multiple sclerosis. *Clin. Exp. Immunol.* **52,** 241–249.

Sellebjerg, F., Jaliashvili, I., Christiansen, M., and Garred, P. (1998). Intrathecal activation of the complement system and disability in multiple sclerosis. *J. Neurol. Sci.* **157,** 168–174.

Serafini, B., Rosicarelli, B., Magliozzi, R., Stigliano, E., and Aloisi, F. (2004). Detection of ectopic B-cell follicles with germinal centers in the meninges of patients with secondary progressive multiple sclerosis. *Brain Pathol.* **14,** 164–174.

Sesboue, R., Daveau, M., Degos, J. D., Martin-Mondiere, C., Goust, J. M., and Schuller, E. (1985). IgG (Gm) allotypes and multiple sclerosis in a French population: Phenotype distribution and quantitative abnormalities in CSF with respect to sex, disease severity, and presence of intrathecal antibodies. *Clin. Immunol. Immunopathol.* **37,** 143–153.

Smith, K. G., Hewitson, T. D., Nossal, G. J., and Tarlinton, D. M. (1996). The phenotype and fate of the antibody-forming cells of the splenic foci. *Eur. J. Immunol.* **26,** 444–448.

Smith-Jensen, T., Burgoon, M. P., Anthony, J., Kraus, H., Gilden, D. H., and Owens, G. P. (2000). Comparison of immunoglobulin G heavy-chain sequences in MS and SSPE brains reveals an antigen-driven response. *Neourology* **54,** 1227–1232.

Soderstrom, M., Link, H., Xu, Z., and Fredriksson, S. (1993). Optic neuritis and multiple sclerosis: Anti-MBP and anti-MBP peptide antibody-secreting cells are accumulated in CSF. *Neurology* **43,** 1215–1222.

Sospedra, M., and Martin, R. (2005). Immunology of multiple sclerosis. *Annu. Rev. Immunol.* **23,** 683–747.

Storch, M. K., Piddlesden, S., Haltia, M., Iivanainen, M., Morgan, P., and Lassmann, H. (1998). Multiple sclerosis: *In situ* evidence for antibody and complement-mediated demyelination. *Ann. Neurol.* **43,** 465–471.

Stuve, O., Cepok, S., Elias, B., Saleh, A., Hartung, H. P., Hemmer, B., and Kieseier, B. C. (2005). Clinical stabilization and effective B-lymphocyte depletion in the cerebrospinal fluid and peripheral blood of a patient with fulminant relapsing-remitting multiple sclerosis. *Arch. Neurol.* **62,** 1620–1623.

Svensson, L., Abdul-Majid, K. B., Bauer, J., Lassmann, H., Harris, R. A., and Holmdahl, R. (2002). A comparative analysis of B cell-mediated myelin oligodendrocyte glycoprotein-experimental autoimmune encephalomyelitis pathogenesis in B cell-deficient mice reveals an effect on demyelination. *Eur. J. Immunol.* **32,** 1939–1946.

Sze, D. M., Toellner, K. M., Garcia de Vinuesa, C., Taylor, D. R., and MacLennan, I. C. (2000). Intrinsic constraint on plasmablast growth and extrinsic limits of plasma cell survival. *J. Exp. Med.* **192,** 813–821.

Taams, L. S., and Akbar, A. N. (2005). Peripheral generation and function of CD4+CD25+ regulatory T cells. *Curr. Top. Microbiol. Immunol.* **293,** 115–131.

Tan, E. M. (1991). Autoantibodies in pathology and cell biology. *Cell* **67,** 841–842.

Tangye, S. G., Avery, D. T., and Hodgkin, P. D. (2003). A division-linked mechanism for the rapid generation of Ig-secreting cells from human memory B cells. *J. Immunol.* **170,** 261–269.

Thangarajh, M., Gomes, A., Masterman, T., Hillert, J., and Hjelmstrom, P. (2004). Expression of B-cell-activating factor of the TNF family (BAFF) and its receptors in multiple sclerosis. *J. Neuroimmunol.* **152,** 183–190.

Tedder, T. F., Poe, J. C., Fujimoto, M., Haas, K. M., and Sato, S. (2005). The CD19–CD21 signal transduction complex of B lymphocytes regulates the balance between health and autoimmune disease: Systemic sclerosis as a model system. *Curr. Dir. Autoimmun.* **8,** 55–90.

Tokoyoda, K., Egawa, T., Sugiyama, T., Choi, B. I., and Nagasawa, T. (2004). Cellular niches controlling B lymphocyte behaviour within bone marrow during development. *Immunity* **20,** 707–718.

Tourtellotte, W. (1970). On cerebrospinal fluid immunoglobulin-G (IgG) quotients in multiple sclerosis and other diseases. A review and a new formula to estimate the amount of IgG synthesized per day by the central nervous system. *J. Neurol. Sci.* **10,** 279–304.

Villar, L. M., Sadaba, M. C., Roldan, E., Masjuan, J., Gonzalez-Porque, P., Villarrubia, N., Espino, M., Garcia-Trujillo, J. A., Bootello, A., and Alvarez-Cermeno, J. C. (2005). Intrathecal synthesis of oligoclonal IgM against myelin lipids predicts an aggressive disease course in MS. *J. Clin. Invest.* **115,** 187–194.

Vos, Q., Lees, A., Wu, Z. Q., Snapper, C. M., and Mond, J. J. (2000). B-cell activation by T-cell-independent type 2 antigens as an integral part of the humoral immune response to pathogenic microorganisms. *Immunol. Rev.* **176,** 154–170.

Wagner, S. D., and Neuberger, M. S. (1996). Somatic hypermutation of immunoglobulin genes. *Annu. Rev. Immunol.* **14,** 441–457.

Walsh, M. J., and Murray, J. M. (1998). Dual implication of 2′,3′-cyclic nucleotide 3′ phosphodiesterase as major autoantigen and C3 complement-binding protein in the pathogenesis of multiple sclerosis. *J. Clin. Invest.* **101,** 1923–1931.

Warren, K. G., and Catz, I. (1994). Relative frequency of autoantibodies to myelin basic protein and proteolipid protein in optic neuritis and multiple sclerosis cerebrospinal fluid. *J. Neurol. Sci.* **121,** 66–73.

Warren, K. G., Catz, I., Johnson, E., and Mielke, B. (1994). Anti-myelin basic protein and anti-proteolipid protein specific forms of multiple sclerosis. *Ann. Neurol.* **35,** 280–289.

Weinstein, E., Peeva, E., Putterman, C., and Diamond, B. (2004). B-cell biology. *Rheum. Dis. Clin. North Am.* **30,** 159–174.

Wekerle, H. (1998). The viral triggering of autoimmune disease. *Nat. Med.* **4,** 770–771.

Wekerle, H. (2006). Breaking ignorance: The case of the brain. *Curr. Top. Microbiol. Immunol.* **305,** 25–50.

Wekerle, H., and Hohlfeld, R. (2003). Molecular mimicry in multiple sclerosis. *N. Engl. J. Med.* **349,** 185–186.

Williamson, R. A., Burgoon, M. P., Owens, G. P., Ghausi, O., Leclerc, E., Firme, L., Carlson, S., Corboy, J., Parren, P. W., Sanna, P. P., Gilden, D. H., and Burton, D. R. (2001). Anti-DNA antibodies are a major component of the intrathecal B cell response in multiple sclerosis. *Proc. Natl. Acad. Sci. USA* **98,** 1793–1798.

Wu, X., Jiang, N., Fang, Y. F., Xu, C., Mao, D., Singh, J., Fu, Y. X., and Molina, H. (2000). Impaired affinity maturation in Cr2−/− mice is rescued by adjuvants without improvement in germinal center development. *J. Immunol.* **165,** 3119–3127.

Xiao, B. G., Linington, C., and Link, H. (1991). Antibodies to myelin-oligodendrocyte glycoprotein in cerebrospinal fluid from patients with multiple sclerosis and controls. *J. Neuroimmunol.* **31,** 91–96.

THE ROLE OF CD4 T CELLS IN THE PATHOGENESIS OF MULTIPLE SCLEROSIS

Tanuja Chitnis

Department of Neurology, Brigham and Women's Hospital
Harvard Medical School, Boston, Massachusetts 02115, USA

T lymphocytes play a central role in the pathogenesis of multiple sclerosis (MS) (Zhang *et al.*, 1992). Both CD4+ and CD8+ T cells have been demonstrated in MS lesions, with CD4+ T cells predominating in acute lesions and CD8+ T cells being observed more frequently in chronic lesions (Raine, 1994). Additionally, T cells are found in all four of the described histopathologic subtypes of MS (Lucchinetti *et al.*, 2000). Activated myelin-reactive CD4+ T cells are present in the blood and cerebrospinal fluid (CSF) of MS patients; in contrast, only nonactivated myelin-reactive T cells are present in the blood of controls (Zhang *et al.*, 1994). The success of several T-cell-targeted therapies in MS reinforces the importance of the role

INTERNATIONAL REVIEW OF
NEUROBIOLOGY, VOL. 79
DOI: 10.1016/S0074-7742(07)79003-7

0074-7742/07 $35.00

of the T cell in MS pathogenesis. Here, we outline basic concepts in CD4+ T-cell immunology and summarize the current understanding of the role of CD4+ T cells in the pathogenesis of MS.

I. Overview of CD4+ T-Cell Immunology

A. The T-Cell Subset of Inflammatory Cells

T cells as well as B cells are critical components of the adaptive immune system. Both T cells and B cells are equipped with specific antigen recognition receptors, the former with the T-cell receptor (TCR), while B cells secrete immunoglobulin (Ig) molecules. T cells originate and differentiate in the thymus. Every T cell that leaves the thymus is conferred with a unique specificity for recognizing antigens through its TCR. The TCR consists of two glycosylated polypeptide chains, the alpha (α) and beta (β) chains, which are linked by disulfide bonds. Each chain consists of variable (V), joining (J), and constant (C) regions closely resembling Ig chains. T cells that recognize self-antigens with high affinity are either deleted or rendered tolerant within the thymus, through a process called central tolerance.

B. Subsets of T Cells

T cells may be divided into two groups on the basis of their expression of either the CD4+ or the CD8+ surface molecules. Functionally, CD4+ T cells are involved in delayed-type hypersensitivity (DTH) responses and also provide help for B-cell differentiation, and hence are termed T helper (Th) cells. In contrast, CD8+ T cells are involved in class I-restricted lysis of antigen-specific targets, and hence are termed cytotoxic T cells. The CD4 molecule binds to a nonpolymorphic site on the major histocompatibility complex (MHC) class II β chain that is expressed by antigen-presenting cells (APCs). In contrast CD8 binds to the α-3 domain of the MHC class I molecule expressed by most cell types. The MHC molecule serves to present antigen to the T cell via the TCR.

C. Activation of T Cells

Signaling through surface molecules by second messengers deliver signals for cell division to the nucleus. The CD3 molecule is part of the TCR complex, although the TCR interacts with the MHC–peptide complex on APCs, signals for the subsequent enactment of T-cell activation and proliferation are delivered by the CD3 antigen. The cytoplasmic tail of the CD3 proteins contains one copy of a sequence motif important for signaling functions, called the immunoreceptor tyrosine-based activation motif (ITAM). Phosphorylation of the ITAM initiates intracellular signaling events. The interaction of MHC–peptide complex with T cells, while necessary, is insufficient for T-cell activation. Additional classes of molecules are involved in T-cell antigen recognition, activation, intracellular signaling, adhesion, and trafficking of T cells to their target organs.

Two signals are required for T lymphocyte activation. According to this "two-signal" model (Bretscher and Cohn, 1970), "signal 1" consists of the interaction of the TCR with antigen, presented by the MHC on the surface of APCs. "Signal 2" consists of the engagement of costimulatory receptors on the T cell by ligands present on the surface of APCs (Alegre *et al.*, 2001; Bretscher, 1999). After contact with specific antigen–MHC complex and adequate costimulatory signals, T cells begin to proliferate, differentiate, and deliver a series of signals, enabling effector functions to other cells such as B cells and NK cells. T cells can thereby orchestrate the immune response.

Costimulatory molecules may deliver either a stimulatory (positive) or an inhibitory (negative) signal for T-cell activation (Brunet *et al.*, 1987). Examples of molecules delivering a positive costimulatory signal for T-cell activation include the B7-CD28 and CD40-CD154 pathways. Examples of molecular pathways delivering a negative signal for T-cell activation include B7-CTLA4 and PD1-PD ligand. The delicate balance between positive and negative regulatory signals can determine the outcome of a specific immune response. Importantly, in the absence of adequate costimulatory signals, T cells can die or become anergic *in vitro*, and thus, fail to initiate an effective immune response *in vivo*. Therefore, manipulation of costimulatory signals represents an important mechanism to inhibit immune-activation.

D. Memory T Cells

On exposure to an antigen, antigen-specific T cells proliferate and differentiate into effector T cells (Sprent and Surh, 2002). The vast majority of effector T cells undergo apoptosis as the immune response progresses, and the few lymphocytes that survive become long-lived memory T cells (Dutton *et al.*, 1998). Memory T cells are specific to the antigen encountered during the primary immune response and react rapidly and vigorously on reencounter with the same antigen. Functionally, in terms of activation requirements, memory T cells can be activated by lower concentrations of anti-CD3 (Byrne *et al.*, 1988), require less costimulation by anti-CD28 (Kuiper *et al.*, 1994), and readily secrete more effector cytokines (Bird *et al.*, 1998; Ehlers and Smith, 1991; Lee *et al.*, 1990) than naive T-cell counterparts, indicating a state of hyperresponsiveness.

E. Migration of T Cells

Molecules primarily involved in cell migration into tissues include chemokines, integrins, selectins, and matrix metalloproteinases (MMPs). Chemokines constitute a large family of chemoattractant peptides that regulate the vast spectrum of leukocyte migration events through interactions with chemokine receptors. The integrin family includes vascular cell adhesion molecule-1 (VCAM-1), intercellular adhesion molecule-1 (ICAM-1), and leukocyte function antigen 3 (LFA-3), CD45, and CD2. The integrin family also mediates T-cell adhesion, facilitates interaction with the APCs, as well mediates adhesion to nonhematopoietic cells such as endothelial cells and guides cell traffic. L-Selectins facilitate the rolling of leucocytes along the surface of endothelial cells and function as a homing receptor to target peripheral lymphoid organs. The MMPs are a family of proteinases, secreted by inflammatory cells, which digest specific components of the extracellular matrix, thereby facilitating lymphocyte entry through basement membranes, including the blood–brain barrier (BBB).

F. T-Cell Cytokine Production

The Th cells play a critical role in the orchestration of the immune response, in part, through the production of cytokines that provide secondary signals to other cells in the immune cascade. Two major types of Th cell responses have been described. Th1 cells produce IL-2, TNF-α, and interferon (IFN)-γ, while Th2 cells produce IL-4, IL-5, IL-10, and IL-13. A Th3 cell that primarily secretes TGF-β has been described in the context of oral tolerance to myelin antigens (Fukaura *et al.*, 1996; Hafler *et al.*, 1997) and in other immune-mediated settings (Minguela *et al.*, 1999). These subsets of T cells interregulate each other's development, with Th1 cytokines suppressing Th2 differentiation and vice versa (Fig. 1).

A subset of T cells that predominantly produces IL-17 has been described (Yao *et al.*, 1995). These cells are believed to represent a distinct subset from IFN-γ-producing Th1 cells, evidenced by the dependence of $Th_{IL\text{-}17}$ cells on IL-16 and TGF-β for differentiation (Bettelli *et al.*, 2006; Mangan *et al.*, 2006; Veldhoen *et al.*, 2006) and IL-23 for expansion (Aggarwal *et al.*, 2003; Langrish *et al.*, 2005), as opposed to Th1 cells which are dependent on IL-12 and IL-2, respectively, for differentiation and expansion. Both Th1 and Th2 cytokines have been shown to suppress the development of Th17 cells (Harrington *et al.*, 2005; Park *et al.*, 2005). An intriguing relationship between the generation of pathogenic Th17 and regulatory CD4+CD25+Foxp3+ cells has recently been demonstrated. TGF-β is critical for the differentiation and generation of CD4+CD25+Foxp3+ regulatory T cells; however, if additionally exposed to IL-6 during culture, these cells differentiate into Th17 cells (Bettelli *et al.*, 2006).

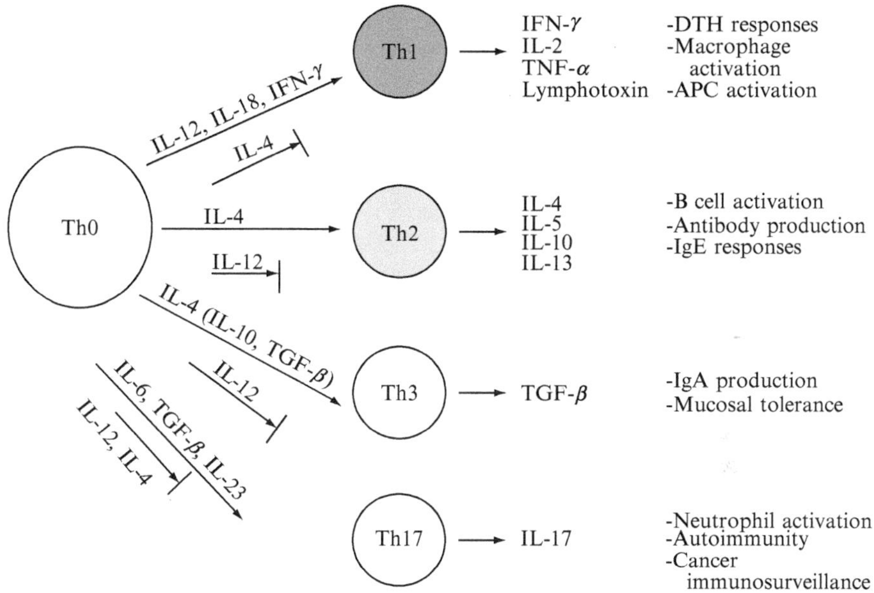

FIG. 1. Factors that influence T helper cell differentiation.

Traditionally, Th cell subsets have been distinguished by their patterns of cytokine production, however identification of distinguishing surface molecule markers has been a major advance in the field. T-cell-, Ig-, and mucin-domain-containing molecules (Tim) represent an important family of molecules which encode cell-surface receptors involved in the regulation of Th1- and Th2-cell-mediated immunity. Tim-3 is specifically expressed on Th1 cells and negatively regulates Th1 responses through interaction with the Tim-3 ligand, galactin-9, also expressed on CD4+ T cells (Monney *et al.*, 2002; Sabatos *et al.*, 2003; Zhu *et al.*, 2005). Tim-2 is expressed on Th2 cells (Chakravarti *et al.*, 2005) and appears to negatively regulate Th2 cell proliferation, although this has not been fully established. Tim-1 is expressed on Th2 cells > Th1 cells and interacts with Tim-4 on APCs to induce T-cell proliferation (Meyers *et al.*, 2005).

G. T-Cell-Signaling Pathways

Intracellular signaling mechanisms provide the link between the binding of the cytokine with its receptor and the effect of the cytokine on cellular function. The Janus kinase and signal transducer and activator of transcription (Jak/STAT) family of transducer/transcription-activating factors plays a critical role in the signaling of many cytokine receptors. Cytokine binding to the specific receptor activates the Jak molecule associated with the receptor, causing phosphorylation of tyrosine residues, and binding of the Jak molecules to its receptor. This facilitates binding of STAT proteins to the phosphorylated receptor, which subsequently dissociates from the receptor, dimerizes, and activates transcription of genes containing specific *cis*-regulatory STAT-binding sequences. Different cytokine receptors are associated with different Jak/STAT proteins. The IL-12 receptor is associated with Jak-2 and STAT3 and STAT4 (Jacobson *et al.*, 1995). The IL-4 receptor is associated with Jak-1–3 and STAT6 (Kaplan *et al.*, 1996; Takeda *et al.*, 1996) (Fig. 1). Mice deficient in STAT6 display a reduction in Th2 cytokine production, decreased IL-4-induced B-cell proliferation and reduced IgE (Kaplan *et al.*, 1996; Takeda *et al.*, 1996). In contrast, STAT4 plays a pivotal role in Th1 immune responses. STAT4 is activated after IL-12 interacts with the IL-12 receptor, inducing transcription of IFN-γ (Jacobson *et al.*, 1995). Mice deficient in STAT4 lack IL-12-induced IFN-γ production and Th1 differentiation (Kaplan *et al.*, 1996; Thierfelder *et al.*, 1996), and display a predominantly Th2 phenotype (Kaplan *et al.*, 1996).

T-bet is a Th1-specific T box transcription factor that directly controls the expression of the hallmark Th1 cytokine, IFN-γ, and IL-12Rβ2 expression, thus facilitating Th1 cell differentiation (Szabo *et al.*, 2000). The transcription factor c-maf enhances IL-4 production and represents an important step in the induction of Th2 cells (Ho *et al.*, 1996). GATA-3 was found to be an STAT6-independent inducer of Th2 differentiation, and is located upstream of c-maf, thus representing a master switch in Th2 development and commitment (Ouyang *et al.*, 2000).

Th17 differentiation is independent of STAT4 and STAT6 signaling (Park *et al.*, 2005); however, it has been demonstrated that STAT3 signaling is activated by both IL-6 and IL-23, and binds to IL-17 gene promoters (Chen *et al.*, 2006). SOCS-3 is a major regulator of IL-23-mediated STAT3 phosphorylation and subsequent Th17 generation (Chen *et al.*, 2006).

H. Regulatory T Cells

Several populations of regulatory or suppressor T cells have been described in humans. These include CD4+CD25+Foxp3 regulatory T cells (Baecher-Allan *et al.*, 2001; Dieckmann *et al.*, 2001; Levings *et al.*, 2001; Stephens *et al.*, 2001; Yagi *et al.*, 2004), CD8+CD28− T cells (Koide and Engleman, 1990), IL-10-producing Th2 cells (Bacchetta *et al.*, 1994), and TGF-β-producing Th3 cells (Kitani *et al.*, 2000; Roncarolo and Levings, 2000). Regulatory T cells suppress T-cell proliferation through a variety of mechanisms, including the production of immunosuppressive cytokines, or through T–T-cell interactions. Several studies have demonstrated that these cells play an important role in the control of the immune response in multiple sclerosis (MS) and that the function of regulatory T cells may be enhanced by immunomodulatory therapies (Crucian *et al.*, 1995; Fukaura *et al.*, 1996; Hafler *et al.*, 1997; Karaszewski *et al.*, 1991; Viglietta *et al.*, 2004).

II. T-Cell Immunologic Studies in MS

CD4+ T cells are believed to play a central role in the pathogenesis of MS. In this section, we summarize the prevailing theory of the pathogenesis of MS, and evidence for the role of CD4+ T cells from both human MS and animal models of disease.

A. Animal Models of MS

Much of the current understanding of the potential mechanistics and role of CD4+ T cells in MS comes from the animal models simulating features of MS. Experimental autoimmune encephalomyelitis (EAE) is an inflammatory central nervous system (CNS) demyelinating disease, and may be induced in several animal types via immunization with myelin proteins or peptides. Disease is primarily mediated by myelin-reactive Th1 cells, which precipitate an inflammatory, demyelinating response within the CNS (Chitnis *et al.*, 2001b). Transfer of myelin basic protein (MBP)-specific T-cell clones restricted to class II (Ia) antigens of the MHC into naive recipient animals causes a similar inflammatory demyelinating disease (Zamvil *et al.*, 1985). EAE reproduces many of the clinical and immunologic aspects of MS, and has been widely used to study the mechanisms of CD4+ T-cell priming and response to myelin components (Bettelli *et al.*, 1998; Chitnis *et al.*, 2001a) as well as to test potential therapies for MS (Aharoni *et al.*, 1999; Yednock *et al.*, 1992).

Theiler's murine encephalomyelitis virus-induced demyelinating disease (TMEV-IDD) model is a virally mediated model of CNS inflammatory demyelination, with some resemblance to MS and is induced by direct CNS infection of the neurotropic TMEV picornavirus, initially resulting in an immune-mediated reaction primarily involving TMEV-specific CD4 and CD8 T cells (Clatch *et al.*, 1986; Rodriguez *et al.*, 1996). However, during the chronic stages of disease, T-cell reactivity to host myelin peptides has been observed, indicating epitope spreading has occurred, causing secondary T-cell responses to myelin breakdown products, and resulting in a disseminated autoimmune response (Miller *et al.*, 1997).

A summary of T-cell immunology related to animal models of MS is beyond the scope of this chapter, and can be found elsewhere in this volume or in alternate sources (Chitnis and Khoury, 2003a,b); however, selected topics in MS that are illuminated by animal studies are discussed.

B. Molecular Mimicry and the Initiation of an Immune Response in MS

The prevailing theory of the etiology of MS is that of "molecular mimicry" whereby CD4+ T cells activated by a foreign antigen cross-react with myelin antigens. Activated myelin-reactive CD4+ T cells are present in the blood and cerebrospinal fluid (CSF) of MS patients; in contrast, only nonactivated myelin-reactive T cells are present in the blood of controls (Zhang *et al.*, 1994). Sequences in MBP has been shown to resemble several viral sequences, and in some cases, cross-reactive T-cell responses have been demonstrated. Although no pathogen has definitely been proven to be the cause of MS, it is conceivable that certain pathogens serve as molecular mimics to CNS components or play a role in the activation of myelin-specific CD4+ T cells. Examples of cross-reactive T cells with MBP antigens include *human herpesvirus* 6 (HHV-6) (Tejada-Simon *et al.*, 2003), staphylococcal enterotoxin antigens (Zhang *et al.*, 1995), coronavirus (Talbot *et al.*, 1996), influenza virus hemagglutinin (Markovic-Plese *et al.*, 2005), and Epstein–Barr virus (EBV) (Lang *et al.*, 2002). Proteolipid protein (PLP) shares common sequences with *Haemophilus influenzae* (Olson *et al.*, 2001), while *Semliki Forest virus* (SFV) peptides mimic epitopes of myelin oligodendrocyte glycoprotein (MOG) (Mokhtarian *et al.*, 1999).

These activated T cells are then thought to migrate to the CNS, where they undergo reactivation in response to nascent myelin antigens. The reactivation of T cells heralds an inflammatory response within the CNS, resulting in more tissue damage and release of secondary antigens. Subsequent T-cell reactivity to secondary antigens is termed "epitope spreading." Evidence of epitope spreading has been demonstrated in animal models of MS (McMahon *et al.*, 2005; Vanderlugt *et al.*, 1998), and may play an important role in the pathogenesis of the human disease.

Although this is the most widely accepted paradigm of MS pathogenesis, and is supported by evidence from studies discussed in this chapter, there still remain many unanswered questions, which include the identity of the initiating foreign cross-reactive antigen(s), the identity of the initiating self-antigen directed T-cell response, and the nature of epitope spreading within the CNS. Moreover, the paucity of CD4+ T cells in certain pathological subtypes of MS questions whether alternate mechanisms may predominate in subsets of this heterogeneous disease.

C. CD4+ T Cells in the Peripheral Immune System of MS Patients

Because of the focus on myelin proteins and, in particular, MBP as a potential autoantigen in MS (Allegretta *et al.*, 1990; Chou *et al.*, 1992; Zhang *et al.*, 1994), considerable interest has developed in the role of T-cell responses to MBP. Moreover, MS disease-associated MHC class II allele, DRB1*1501 has been shown to be effective in presenting MBP peptide to T-cell clones isolated from MS patients (Wucherpfennig *et al.*, 1995, 1997). Activated myelin-reactive CD4+ T cells are present in the blood and CSF of MS patients; in contrast, only nonactivated myelin-reactive T cells are present in the blood of controls (Zhang *et al.*, 1994). Furthermore, MBP-reactive T cells isolated from the CSF of MS patients display increased expression of the IL-2 receptor (Zhang *et al.*, 1994), consistent with a previously activated or memory phenotype. In MS patients, but not in healthy controls, these cells can be activated in the absence of CD28-B7 costimulation, thus implying that they have been previously activated *in vivo* (Markovic-Plese *et al.*, 2001; Scholz *et al.*, 1998). MBP-reactive T cells from MS patients were found to be less responsive to CTLA-4 blockade compared to those from healthy controls (Oliveira *et al.*, 2003), signifying that in MS patients these T cells are not subject to the normal regulatory mechanisms. Although the contribution of MBP-reactive T cells to the pathogenesis of MS is currently unknown, their differential phenotype and costimulatory requirements indicate a memory and potentially dysregulated cell population.

T-cell reactivity to other myelin proteins and peptides in MS has been explored. Studies examining PLP T-cell responses demonstrated T-cell proliferation to certain epitopes (Pelfrey *et al.*, 1994), with some differential reactivity when compared to controls (Markovic-Plese *et al.*, 1995; Zhang *et al.*, 1994). T-cell responses to recombinant MOG appeared to be similar in MS patients and healthy controls (Diaz-Villoslada *et al.*, 1999), although other studies demonstrated increased reactivity (Kerlero de Rosbo *et al.*, 1993) or altered T-cell properties (Van der Aa *et al.*, 2003a) in MS patients. Other studies have examined T-cell responses to myelin oligodendrocyte basic protein (MOBP) (Holz *et al.*, 2000) or 2′,3′-cyclic nucleotide 3′-phosphodiesterase (CNPase) (Muraro *et al.*, 2002), with some reactivity demonstrated in T-cell lines isolated from select MS patients. T-cell reactivity to other CNS antigens has not been fully explored due to the technical difficulties with performing and interpreting the results of such assays. Many studies currently employ strategies to expand T cells using nonspecific methods or mixtures of myelin peptides.

Studies in patients with postinfectious encephalomyelitis or acute disseminated encephalomyelitis (ADEM) have consistently found robust T-cell reactivity to myelin peptides in both the blood and CSF (Hafler *et al.*, 1987; Hemachudha *et al.*, 1988; Pohl-Koppe *et al.*, 1998), and suggest an intriguing relationship in the pathophysiology of ADEM and MS.

D. TCR Repertoires in MS

Studies examining TCR repertoire in MS patients have demonstrated a bias for use of β chain variable region (Vβ) 5.2 and 5.3 (Kotzin *et al.*, 1991; Lozeron *et al.*, 1998; Oksenberg *et al.*, 1993), and this has led to the exploration of TCR Vβ 5.2/5.3-targeted therapies. TCR vaccines employing TCR Vβ 5.2 peptides are thought to exert their effects by enhancing the function of regulatory T-cell populations recognizing TCR determinants (Vandenbark, 2005; Vandenbark *et al.*, 1996), and are currently undergoing pilot studies in MS. Phase II trials using antibodies specifically targeting the Vβ 5.2/5.3 sequence of the TCR have shown some success and no significant adverse effects (Killestein *et al.*, 2002; Olsson *et al.*, 2002). However, other groups (Gran *et al.*, 1998; Lozeron *et al.*, 1998; Musette *et al.*, 1996) have found predominant usage of other Vβ chains, indicative of potential limitations with TCR Vβ 5.2-targeted therapies. Studies in twins demonstrated similar selection of TCR Vα chains in concordant twins in response to MBP, when compared to discordant twins or controls, suggesting a genetic basis for the evolution of self-antigen T-cell responses (Utz *et al.*, 1993). MS patients treated with autologous hematopoietic stem cell transplantation demonstrated an increase in naive compared to memory CD4+ T cells, with increased TCR diversity indicative of broader clonal phenotypes 2 years following therapy (Muraro *et al.*, 2005). A separate study ascertained similar findings and found that MBP-reactive T cells demonstrated broader epitope recognition following reconstitution (Sun *et al.*, 2004).

E. T Cells in MS Lesions

Pathologically, MS lesions are characterized by perivascular infiltrates of CD4+ and CD8+ T cells and macrophages (Prineas and Wright, 1978; Traugott *et al.*, 1983b). CD4+ and CD8+ T cells and macrophages are also found toward the periphery of the lesion and in the normal appearing white matter (Traugott *et al.*, 1983a). CD4+ T cells were shown to predominate in acute lesions, while CD8+ T cells were observed more frequently in chronic lesions (Raine, 1994). Similar densities of CD3+ T cells have been demonstrated in all four of the histopathologic subtypes of MS, although type III and IV lesions are additionally characterized by prominent oligodendrocyte degeneration (Lucchinetti *et al.*, 1996). Attempts to isolate T-cell clones from the brains of MS patients failed to show either MBP or PLP reactivity (Hafler *et al.*, 1987). TCR analysis from MS lesions demonstrated a broad TCR Vα and Vβ repertoires in active lesions, while fewer TCR V genes were detected in chronic plaques and control samples (Wucherpfennig *et al.*, 1992). Other studies demonstrated restricted TCR specificities, with rearranged Vβ 5.2 genes found in the brains of all patients who were HLA DRB1*1501, DQA1*0102, DQB1*0602, and DPB1*0401 positive, suggesting that MHC class II genotype may play a role in VDJ rearrangements in MS lesions (Oksenberg *et al.*, 1993). T cells in parenchymal MS lesions lacked CCR7, indicating a differentiation of central-memory T cells into effector memory cells presumably on restimulation by antigen within the CNS (Kivisakk *et al.*, 2004).

In summary, multiple studies have demonstrated the presence of CD4+ T cells in MS lesions, arguing for a central role in MS pathogenesis. The lack of consensus regarding T-cell specificities suggest a heterogeneity in T-cell responses at the time of analysis, which may be a result of epitope spreading in chronic disease.

F. T-Cell Activation and Costimulation

Costimulatory pathways deliver a positive or negative signal for T-cell activation, and thus represent an important control step in the immune response. The CD28/CTLA-4-B7-1/2 family of costimulatory molecules represents an important step in the activation of CD4+ T cells. Lesions in the CNS of patients with MS were found to be exclusively associated with the expression of B7-1 in perivenular lymphocytes, while B7-2 was expressed on macrophages in both MS and in other neurological diseases (Windhagen *et al.*, 1995). Peripheral blood mononuclear cells (PBMCs), isolated from MS patients, showed increased expression of B7-1 on both CD4+ and CD8+ cell patients with rapidly progressive disease, compared to those with stable disease, or normal controls (Mena and Rohowsky-Kochan, 1999). In a separate study, B7-1 expression localized to B cells was found to be increased during MS relapses, and treatment with IFN-β 1b reduced the number of B7-1-expressing B cells but increased the number of B7-2 monocytes (Genc *et al.*, 1997). In total, these observations suggest that the B7-CD28-CTLA4 pathway is activated in MS, and that B7-1, in particular, may play an important role in regulating disease activity.

Genetic polymorphisms of costimulatory molecules may contribute to disease susceptibility. Three CTLA-4 gene polymorphisms were found in MS patients, but not in healthy controls (Ligers *et al.*, 1999), however no association was found with disease course or severity (Masterman *et al.*, 2002). In an Olmstead County study, two polymorphisms were associated with the presence of MS (Kantarci *et al.*, 2003). The Canadian Collaborative Study found no association of CTLA-4 polymorphisms with the disease course of MS (Dyment *et al.*, 2002). The exon 1 A/G polymorphism was associated with the presence of oligoclonal bands in the CSF (Fukazawa *et al.*, 1999). Thus, dysregulation of CTLA-4 signaling may contribute to susceptibility to MS. A phase I safety study of CTLA4Ig (Repligen-RG2077) as well as a multicenter study of CTLA4Ig (BMS-188667) for MS are ongoing.

Interaction of CD40 on APCs with CD154 on T cells induces APC production of IL-12, a major factor in Th1 cell differentiation (Kelsall *et al.*, 1996). Expression of both CD40 and CD154 were increased in lesions from postmortem MS brains compared with controls, with CD40 found predominantly on macrophages and microglia, while CD154 colocalized with the CD4 T-cell marker (Gerritse *et al.*, 1996). Expression of CD154 was found to be higher in peripheral blood monocytes isolated from SPMS compared with RRMS or healthy controls (Filion *et al.*, 2003; Jensen *et al.*, 2001), and was reduced by IFN-β treatment (Teleshova *et al.*, 2000). PBMCs from SPMS patients produced more IL-12 and IFN-γ when restimulated *in vitro*, compared with healthy controls (Karni *et al.*, 2002). In summary, these studies indicate that the CD40-CD154 pathway is important for the regulation of Th1 cytokine production in MS. Clinical trials with an anti-CD154 antibody (Biogen) in autoimmune disease such as ITP and lupus were terminated because of the occurrence of thromboembolic events. Another formulation of the antibody (IDEC pharmaceuticals) is under investigation. A phase I clinical trial in MS patients was recently performed with good safety data, and therapeutic effects are currently under investigation.

Programmed death-1 (PD-1) is expressed on T cells and is a negative regulator of T-cell activation. PD-1 polymorphism was shown to be a genetic modifier of the progression of MS, and this may relate to PD-1-mediated inhibition of T-cell activation (Kroner *et al.*, 2005).

The Tim family of molecules are important cell-surface markers as well as costimulatory regulators of Th1/Th2 responses. In MS patients, CSF T-cell clones demonstrated reduced levels of Tim-3 and T-bet and secreted higher amounts of IFN-γ than did those from control subjects, indicating that Tim-3 may represent an important regulator of Th1 responses in MS (Koguchi *et al.*, 2006).

In summary, there is significant evidence that costimulatory molecules represent an important step in the control of T-cell activation in MS and are viable therapeutic targets.

G. T-Cell Cytokine Production in MS

In the context of MS, Th1 cytokines are thought to mediate disease, while Th2 cytokines are believed to play a protective role. However, as our understanding of the disease evolves, it is clear that this paradigm is not absolute. Moreover, evidence of a distinct lineage of Th17-producing T cells has led to reevaluation of the role of Th1 cytokines in MS.

Th1 cytokines are predominantly found in the brains of MS patients, while a paucity of Th2 cytokines, in particular TGF-β, and IL-10 was observed (Cannella and Raine, 1995b; Hofman *et al.*, 1986; Woodroofe and Cuzner, 1993). Studies using semiquantitative RT-PCR and immunocytochemistry found increased expression of B7-1 and IL12p40 in acute MS plaques, compared with samples isolated from inflammatory infarcts (Windhagen *et al.*, 1995). IL-6, IFN-γ, and TNF-α were expressed by cells located in the perivascular cuffs, suggesting that in acute MS lesions, inflammatory cells are the most important source of these cytokines (Woodroofe and Cuzner, 1993). Expression of TNF-α was localized to macrophages, microglia, and astrocytes (Cannella and Raine, 1995b;

Hofman *et al.*, 1989; Selmaj *et al.*, 1991) in chronic-active lesions. A separate study found that IL-2 was expressed predominantly in association with perivascular inflammatory cells examining acute MS lesions (Hofman *et al.*, 1986). Contrary to the Th1/Th2 paradigm of MS, high levels of IL-4 were expressed in both acute and chronic-active MS lesions with no obvious correlation to the resolution of the lesion (Cannella and Raine, 1995b). Studies using RNA microarrays in MS brains at autopsy found increased transcripts of genes encoding for IL-6, IL-17, and IFN-γ (Lock *et al.*, 2002), indicating a potential role for both Th1 and Th17 cells in proinflammatory responses in MS.

Several studies have demonstrated enhanced production of the hallmark Th1 cytokine, IFN-γ, from PBMCs restimulated *ex vivo* in MS patients compared to controls (Balashov *et al.*, 1997; Comabella *et al.*, 1998). Clinical attacks correlated with increased IFN-γ production *in vitro* (Beck *et al.*, 1988). In a similar study, IFN-β was found to blunt increased production of IFN-γ during relapse (Becher *et al.*, 1999). TCR-mediated IFN-γ and IL-10 secretions are increased in relapsing-remitting (RR) and secondary progressive (SP) patients, but not in primary progressive disease, suggesting a dysregulation of this signaling pathway in certain MS subtypes (Balashov *et al.*, 2000). Interestingly, SPMS patients also exhibit seasonal variations of IFN-γ production with increased expression in autumn and winter months compared with spring and summer months, which was not observed in normal controls (Balashov *et al.*, 1998). A progressive course of MS was found to be significantly more frequent in carriers of the IFN-γ receptor-2 allele Arg64 (Schrijver *et al.*, 2004). Administration of IFN-γ to MS patients precipitated clinical attacks, confirming the role of IFN-γ as a proinflammatory cytokine in MS (Panitch *et al.*, 1987a,b).

Studies of the prototypic Th2 cytokine IL-4 in MS are limited, IL-4 was expressed in high levels in both acute- and chronic-active MS lesions (Cannella and Raine, 1995a). High frequencies of T-cell clones reactive to MBP- and PLP-expressing IL-4 were found in MS patients compared with untreated patients (Chou *et al.*, 1992). Increased expression of IL-4 secretion by CD3-stimulated PBMCs was demonstrated in SPMS patients treated with cyclophosphamide/methylprednisolone compared with untreated patients (Smith *et al.*, 1997). Thus, the role of IL-4 in the pathogenesis of MS is unclear, however may be associated with responses to therapy.

Studies examining Th17 cell activity in MS found that dendritic cells from MS patients secrete elevated amounts of IL-23 and express increased levels of IL-23p19 mRNA, and are associated with increased T-cell production of IL-17 (Vaknin-Dembinsky *et al.*, 2006). A Japanese study examining cytokine expression found that CSF levels of IL-17, IL-8, and IL-5 were significantly higher in opticospinal-MS patients than in conventional RRMS patients, and may be associated with pronounced neutrophilic infiltrates typically found in the opticospinal disease variant (Ishizu *et al.*, 2005).

H. Regulatory T Cells in MS

Alterations in the function of several populations of regulatory T cells have been demonstrated in MS. Induction of regulatory antigen-specific Th3 cells through oral tolerance with myelin proteins has been described (Fukaura *et al.*, 1996; Hafler *et al.*, 1997). Deficiency in the CD8+CD28− subset of suppressor cells has been demonstrated in MS patients (Crucian *et al.*, 1995). Moreover, increases in β-adrenergic receptor density on CD8+CD28− cells were found in MS compared with controls (Karaszewski *et al.*, 1991). Much attention has focused on the CD4+CD25+ subset of regulatory T cells. Induction of CD4+CD25+ T cells is controlled by the transcription regulator FOXP3 which also serves as a cell-specific marker (Hori *et al.*, 2003). Defects in the effector function of CD4+CD25+ regulatory T cells have been demonstrated in MS patients (Viglietta *et al.*, 2004). Thus, defects in regulatory T-cell populations may facilitate the development and/or progression of autoimmunity.

I. T-Cell Migration in MS

T-cell migration into the CNS in MS is believed to follow the sequence of capture, rolling, activation, adhesion strengthening, and finally transmigration through the BBB. The specifics of these events may vary depending on the region of the CNS, as well as the activation state of T cells and microvasculature.

Two processes appear to be important in T-cell migration into the CNS: (1) migration of T cells from the blood into the CSF with interactions with dendritic elements on the luminal surface of the choroid plexus resulting in immunosurveillance and (2) T-cell migration through inflamed endothelial BBB and interactions with perivascular APCs within the Virchow-Robbins space (Engelhardt and Ransohoff, 2005). Evidence exists for both of these processes in MS, however the relative contribution of each is unclear, and may depend on the stage and subtype of disease.

Adhesion molecule ICAM-1 is expressed on the inflamed BBB in MS, while LFA-1 is expressed on infiltrating T cells, suggesting an important role for this pathway in T-cell migration in MS (Bo *et al.*, 1996). Inhibition of interactions between integrin molecule $\alpha 4\beta 1$ present on the surface of T cells with VCAM-1 present on endothelial cells of the BBB was shown to suppress the development of EAE (Yednock *et al.*, 1992). This led to the development of an $\alpha 4\beta 1$-integrin antibody, natalizumab, which has been effective in reducing MS relapses (Miller *et al.*, 2003). However, clinical trials with a natalizumab led to several cases of progressive multifocal leukoencephalopathy, resulting in the reevaluation of the role of such drugs in MS therapeutics (Kleinschmidt-Demasters and Tyler, 2005; Langer-Gould *et al.*, 2005; Van Assche *et al.*, 2005).

CXCR3 is postulated to be the major chemokine involved in the trafficking of T cells from the blood to the CSF for immunosurveillance in both normals and MS patients (Kivisakk *et al.*, 2002). Studies of chemokine expression on peripheral blood T cells in MS patients found a positive correlation between CXCR3 expression on blood T cells (Eikelenboom *et al.*, 2002) and CSF T cells (Sindern *et al.*, 2002) on MRI measures of disease activity in MS patients.

T-cell migration is an important step in the pathogenesis of MS, as confirmed by the success of $\alpha4\beta1$-integrin antibody therapy (natalizumab). However, effective blockade of CNS immunosurveillance has produced profound adverse effects, as evidenced by the development of PML in three treated patients. The risk of infection and emergence of tumors due to blockade of immunosurveillance must be balanced with therapeutic effect when utilizing such powerful therapeutics.

J. T-Cell Interaction with Axons and Neurons

Much attention has focused on the presence of axonal damage in MS plaques as a substrate for chronic progressive disease (Trapp *et al.*, 1998, 1999). Although mediators of axonal damage include cytokines, complement, antibody, and nitric oxide, T cells may play a significant role both in neurodegeneration and in neuroprotection. The presence of CD8+ T cells, but not CD4+ T cells, in the MS lesion correlates with axonal damage (Bitsch *et al.*, 2000). In the TMEV model of MS, demyelination but no axonal damage was found in the CNS of MHC class I-deficient mice after infection with Theiler's virus (Rivera-Quinones *et al.*, 1998), suggesting that axonal damage is class I mediated. These mice also displayed a deficiency of CD8-positive cells in their CNS lesions. Similarly, *in vitro*, T cell-mediated neurotoxicity was dependent on IFN-γ-induced expression of MHC class I (Medana *et al.*, 2001). Collectively, these studies suggest that class I expression in the CNS, particularly on neurons, may enhance a cytotoxic CD8+ T-cell response.

In vitro coculture of human fetal neurons with OKT3-activated CD4+ or CD8+ T cells has been found to produce apoptosis of neurons (Giuliani *et al.*, 2003). This process required physical contact of the cells, as demonstrated by transwell experiments, and was not dependent on MHC I. Protection could be conferred by blocking CD40 on both T cells and neurons, and FasL on neurons. Attention has focused on the regulatory role that neurons may play on T cells. In a Lewis rat model of EAE, motoneurons were demonstrated to engulf T lymphocytes through a process consistent with emperipolesis (Smith *et al.*, 2000). Neuronal production of TGF-β has been shown to play a significant role in the induction of CD4+CD25+Foxp3+ regulatory T cells in a murine EAE model (Liu *et al.*, 2006). These studies have important implications for understanding the T cell-neuronal interactions in MS.

Although MBP-reactive T cells have been implicated in the pathogenesis of MS (Zhang *et al.*, 1992, 1994), an intriguing study has shown that passive transfer of MBP-reactive T cells resulted in protection of the retinal ganglion cell body after optic nerve crush injury *in vivo* (Moalem *et al.*, 1999). Interestingly, transfer of T cells specific for other antigens was not protective. Furthermore, expression of mRNA for several nerve growth factors, including brain-derived neurotrophic factor (BDNF), was upregulated after antigen-activation of these T cells, while the injured nerve expressed mRNA for nerve growth factor receptors (Moalem *et al.*, 2000). Two major types of Th cell responses have been described, based on cytokine secretion. Th1 cells produce the cytokines IFN-γ and TNF-α, while Th2 cells produce IL-4, IL-5, and IL-10. EAE has been associated with a Th1 response, while Th2 cytokines are generally protective. In support of the above findings, a separate study demonstrated that the presence of myelin-reactive Th2 cells conferred neuroprotection in organotypic hippocampal slice cultures (Wolf *et al.*, 2002). These findings have several implications for MS. First, myelin-reactive T cells may be neuroprotective under certain conditions including trauma, suggesting that the MS brain may be more susceptible to inflammation-induced damage. Second, the local production of neurotrophic factors by inflammatory cells may be neuroprotective. However, in contrast to these findings, other studies have shown that MBP-TCR transgenic mice sustained more CNS damage and inflammation following traumatic spinal cord injury, compared with wild-type controls (Jones *et al.*, 2002). T-cell infiltrates were localized to areas of demyelination and axonal loss, suggesting that autoreactive T cells can trigger autoimmune demyelination in the setting of trauma under certain circumstances. Therefore, the therapeutic potential of neuroprotective T lymphocytes should be entertained only with caution.

Glatiramer acetate (GA) is an established treatment for RRMS. Its mechanism of action relies, in part, on the migration of Th2 cells into the CNS (Aharoni *et al.*, 2000, 2002), where they presumably downregulate local inflammatory responses. Interestingly, these T cells have been shown to produce BDNF in the EAE model (Aharoni *et al.*, 2003), and GA-reactive T cells harvested from MS patients treated with GA produce BDNF on restimulation *in vitro* (Ziemssen *et al.*, 2002). However, the role of BDNF in neuroprotection in MS or EAE has not been established. Thus, the role of this and other neurotrophic factors requires further exploration.

III. T-Cell-Targeted Therapies in MS

The success of several T cell-targeted therapies in MS reinforces the importance of the role of the T cell in MS pathogenesis. Of the six approved therapies for MS, the effects of GA and natalizumab can be directly related to modulation of T-cell function. The β-interferons (IFN-β1a and IFN-β1b) modulate some T-cell functions including T-cell migration and Th1 cytokine production (Yong, 2002), while mitoxantrone is a cytotoxic agent that nonspecifically abrogates T- and B-cell proliferation.

An altered peptide ligand (APL) may be defined as "any peptide that serves as a receptor ligand in which substitutions of a single or multiple amino acids lead to changes in the functional outcome of receptor signaling" (Bielekova and Martin, 2001). APLs have most commonly been used as TCR ligands to alter T-cell responses to presumed immunogenic or target antigens presumably resulting in immune suppression or immune deviation as well as induction of a regulatory T-cell population reactive to the APL itself, which then serves to downregulate the inflammatory disease process through bystander suppression.

Glatiramer acetate (GA; Copolymer-1; Copaxone®) is an FDA approved therapy for the treatment of RRMS. GA is an APL that was originally developed to mimic MBP. It is composed of a random sequence of the amino acids *g*lutamic acid, *l*ysine, *a*lanine, and *t*yrosine present in a specific molar ratio (0.14:0.34:0.43:0.09). Copaxone is administered by daily subcutaneous injection, and in a phase III clinical trial was found to reduce relapse frequency by 29%, as well as decrease the incidence of new gadolinium enhancing lesions on MRI (Ge *et al.*, 2000; Johnson *et al.*, 1995, 2001). Despite its crude resemblance to MBP, evidence from several studies showing GA stimulates several nonmyelin antigen T-cell lines suggests that GA acts as a "universal" or degenerate T-cell antigen (Duda *et al.*, 2000). GA has been shown to inhibit responses to MBP-specific T-cell lines *in vitro* (Racke *et al.*, 1992), and *in vivo* treatment with GA induces a hyporesponsiveness to this antigen (Schmied *et al.*, 2003). Interestingly, GA-reactive T-cell lines isolated from both treated patients as well as untreated controls were found to cross-react with a variety of peptides, suggesting degenerate antigenicity. In addition, Th2 cytokine deviation was noted in GA-reactive T-cell lines (Duda *et al.*, 2000). In the EAE model, Th2-producing GA-reactive T cells were shown to accumulate in the CNS and attenuate disease (Aharoni *et al.*, 2000). Thus, the principal mechanism of action of Copaxone may be the induction of Th2 responses, which exert bystander suppression of inflammation within the CNS.

APLs targeting MBP have been widely studied in MS because of the interest in this potential autoantigen. An APL to MBP_{87-99} peptide was shown to be effective in ameliorating disease in the EAE model (Karin *et al.*, 1994). An initial phase I clinical trial tested four doses of an APL to MBP_{83-99} administered subcutaneously for 4 weeks demonstrated no safety concerns (Bielekova and Martin, 2001). Two phase II trials using MBP_{83-99} APLs were initiated: A small NIH-based trial tested the highest dose of APL CGP77116 (50 mg) administered weekly for 9 months. Three of eight patients developed atypical MS exacerbations, characterized by a high gadolinium-enhancing lesion load, tumefactive-type lesion, or a flaccid paralysis with inflammatory involvement of the peripheral nervous system (Bielekova *et al.*, 2000). 2/3 of exacerbations, correlated with enhanced reactivity to MBP (Bielekova *et al.*, 2000). A second larger multicenter study testing three doses of APL NBI-5788 (5, 20, or 50 mg) versus placebo in 144 total patients was terminated because of the occurrence of APL-induced systemic hypersensitivity reactions in 9% of enrolled patients (Kappos *et al.*, 2000). In both studies, enhanced *ex vivo* T-cell responses to APL were observed following treatment. In patients who developed hypersensitivity reactions, enhanced Th2 responses to the APL could be demonstrated (Kappos *et al.*, 2000). Further phase II studies investigating the effects of low-dose APL (NBI-5788) are currently undergoing safety evaluation.

T-cell vaccination strategies attempt to eliminate pathogenic T cells through the enhancement of regulatory immune responses to autoreactive T cells. This approach requires the isolation of autoreactive T-cell clones from the individual patient's blood or CSF, and subcutaneous reinjection in the form of an immunizing vaccine. Pilot trials of T-cell vaccination with autologous MBP-specific T cells from peripheral blood, in 28 RRMS and 26 SPMS patients, demonstrated a modest reduction in posttreatment relapse rate (Medaer *et al.*, 1995; Zhang *et al.*, 2002). In this study, the frequency of gadolinium-enhancing lesions was largely unchanged posttreatment. A second pilot trial using autologous MBP and MOG-reactive T-cell vaccines in 20 RRMS nonresponders demonstrated a significant reduction in relapse rate ($p = 0.026$) as well as gadolinium enhancing and T2 lesion load (Achiron *et al.*, 2004). In both studies, no serious adverse events were noted. In a small study utilizing myelin-reactive CD4+ T cells derived from autologous CSF, no adverse effects were observed in any of the five treated patients (Van der Aa *et al.*, 2003b). Further larger phase II trials using T-cell vaccination are planned.

TCR vaccination strategies target TCR sequences believed to be critical in the immunopathogenesis of MS. In MS, Vβ 5.2/5.3+ has been identified as a dominant TCR variable region sequence involved in MBP T-cell reactivity (Kotzin *et al.*, 1991; Lozeron *et al.*, 1998; Oksenberg *et al.*, 1993). TCR vaccines are thought to exert their effects by enhancing the function of regulatory T-cell populations recognizing TCR determinants (Vandenbark, 2005; Vandenbark *et al.*, 1996). TCR peptides derived from the Vβ 5.2 region of the TCR have been used as a vaccine in MS patients. In a double-blind pilot study, 23 patients were treated with weekly to monthly injections of the peptide. All patients carried the HLADRB1*1501 allele. Enhanced T-cell responses to the immunizing peptide correlated with clinical improvement (Bourdette *et al.*, 1994; Vandenbark *et al.*, 1996). T-cell responses to MBP trended downward in responders. No major adverse events were observed in treated patients. ATM-027 is an antibody specifically targeting the Vβ 5.2/5.3 sequence of TCR. Results from a multicenter phase II study in 47 MS patients treated with a run-in regimen of ATM-027 monthly for 6 months, showed no significant reduction in new gadolinium MRI lesions posttreatment despite a significant reduction in Vβ 5.2/5.3+ T cells (Killestein *et al.*, 2002; Olsson *et al.*, 2002). MS relapses occurred in three treated patients, however no other adverse events directly related to drug were observed. These negative results suggest that there is considerable variability in Vβ profiling in individual MS patients. An alternative explanation is that by the time the disease presents clinically, epitope spreading has occurred, negating the use of a single Vβ-depleting agent.

CD4 is a cell-surface marker of Th cells. In a randomized phase II double-blind trial, an anti-CD4 antibody (cM-T412) was administered intravenously to 35 RRMS and SPMS patients (van Oosten *et al.*, 1997). Administration of the antibody resulted in a rapid and sustained reduction in circulating CD4+ T cells. Infusion-related side effects including nausea, fever, and tachycardia were limited to 24-h postinfusion. After 9 months, treated patients demonstrated an approximately 40% reduction in relapse rate compared to placebo controls, however there was no significant change in the number of gadolinium-enhancing lesions on MRI. Although anti-CD4 therapy was effective in reducing relapse rate, lack of efficacy on primary MRI measures has led to questions regarding the effectiveness in MS.

The Jak/STAT family of transducer/transcription-activating factors plays a critical role in the signaling of many cytokine receptors. 3-Hydroxy-3-methylglutaryl coenzyme A (HMG-CoA) reductase inhibitors such as atorvastatin has been shown to exert its protective effects in EAE through the induction of STAT6 phosphorylation and secretion of Th2 cytokines, with concomitant inhibition of STAT4 phosphorylation and secretion of Th1 cytokines (Youssef *et al.*, 2002). A pilot clinical trial of simvastatin in MS showed positive results and a larger trial of Atorvastatin is under way.

Inhibition of molecular pathways involved in T-cell migration has been effective in reducing MS relapses (Miller *et al.*, 2003), however clinical trials with an α4-integrin antibody led to several cases of progressive multifocal leukoencephalopathy, resulting in the reevaluation of the role of such drugs in MS therapeutics (Kleinschmidt-Demasters and Tyler, 2005; Langer-Gould *et al.*, 2005; Van Assche *et al.*, 2005).

IV. Conclusions

In conclusion, we have summarized the evidence for the central role of CD4+ T cells in the pathogenesis of MS, which represents the work of many teams of investigators over many years. Although much progress has been made in the field, which has led to important therapeutic advances, several questions remain unanswered, including the nature of the initiating T-cell responses and the mechanisms of propagation of the disease within the CNS. The field of T-cell biology remains an integral part of the MS question, and further advances will undoubtedly lead to improved treatment strategies for patients.

References

Achiron, A., Lavie, G., Kishner, I., Stern, Y., Sarova-Pinhas, I., Ben-Aharon, T., Barak, Y., Raz, H., Lavie, M., Barliya, T., Faibel, M., Cohen, I. R., and Mandel, M. (2004). T cell vaccination in multiple sclerosis relapsing-remitting nonresponders patients. *Clin. Immunol.* **113,** 155–160.

Aggarwal, S., Ghilardi, N., Xie, M. H., de Sauvage, F. J., and Gurney, A. L. (2003). Interleukin-23 promotes a distinct CD4 T cell activation state characterized by the production of interleukin-17. *J. Biol. Chem.* **278,** 1910–1914.

Aharoni, R., Teitelbaum, D., Arnon, R., and Sela, M. (1999). Copolymer 1 acts against the immunodominant epitope 82–100 of myelin basic protein by T cell receptor antagonism in addition to major histocompatibility complex blocking. *Proc. Natl. Acad. Sci. USA* **96,** 634–639.

Aharoni, R., Teitelbaum, D., Leitner, O., Meshorer, A., Sela, M., and Arnon, R. (2000). Specific Th2 cells accumulate in the central nervous system of mice protected against experimental autoimmune encephalomyelitis by copolymer 1. *Proc. Natl. Acad. Sci. USA* **97,** 11472–11477.

Aharoni, R., Meshorer, A., Sela, M., and Arnon, R. (2002). Oral treatment of mice with copolymer 1 (glatiramer acetate) results in the accumulation of specific Th2 cells in the central nervous system. *J. Neuroimmunol.* **126,** 58–68.

Aharoni, R., Kayhan, B., Eilam, R., Sela, M., and Arnon, R. (2003). Glatiramer acetate-specific T cells in the brain express T helper 2/3 cytokines and brain-derived neurotrophic factor *in situ*. *Proc. Natl. Acad. Sci. USA* **100**(24), 14157–14162.

Alegre, M. L., Frauwirth, K. A., and Thompson, C. B. (2001). T-cell regulation by CD28 and CTLA-4. *Nat. Rev. Immunol.* **1,** 220–228.

Allegretta, M., Nicklas, J. A., Sriram, S., and Albertini, R. J. (1990). T cells responsive to myelin basic protein in patients with multiple sclerosis. *Science* **247,** 718–721.

Bacchetta, R., Bigler, M., Touraine, J. L., Parkman, R., Tovo, P. A., Abrams, J., de Waal Malefyt, R., de Vries, J. E., and Roncarolo, M. G. (1994). High levels of interleukin 10 production *in vivo* are associated with tolerance in SCID patients transplanted with HLA mismatched hematopoietic stem cells. *J. Exp. Med.* **179,** 493–502.

Baecher-Allan, C., Brown, J. A., Freeman, G. J., and Hafler, D. A. (2001). CD4+ $CD25^{high}$ regulatory cells in human peripheral blood. *J. Immunol.* **167,** 1245–1253.

Balashov, K. E., Smith, D. R., Khoury, S. J., Hafler, D. A., and Weiner, H. L. (1997). Increased interleukin 12 production in progressive multiple sclerosis: Induction by activated CD4+ T cells via CD40 ligand. *Proc. Natl. Acad. Sci. USA* **94,** 599–603.

Balashov, K. E., Olek, M. J., Smith, D. R., Khoury, S. J., and Weiner, H. L. (1998). Seasonal variation of interferon-gamma production in progressive multiple sclerosis. *Ann. Neurol.* **44,** 824–828.

Balashov, K. E., Comabella, M., Ohashi, T., Khoury, S. J., and Weiner, H. L. (2000). Defective regulation of IFNγ and IL-12 by endogenous IL-10 in progressive MS. *Neurology* **55,** 192–198.

Becher, B., Giacomini, P. S., Pelletier, D., McCrea, E., Prat, A., and Antel, J. P. (1999). Interferon-gamma secretion by peripheral blood T-cell subsets in multiple sclerosis: Correlation with disease phase and interferon-beta therapy. *Ann. Neurol.* **45,** 247–250.

Beck, J., Rondot, P., Catinot, L., Falcoff, E., Kirchner, H., and Wietzerbin, J. (1988). Increased production of interferon gamma and tumor necrosis factor precedes clinical manifestation in multiple sclerosis: Do cytokines trigger off exacerbations? *Acta Neurol. Scand.* **78,** 318–323.

Bettelli, E., Das, M. P., Howard, E. D., Weiner, H. L., Sobel, R. A., and Kuchroo, V. K. (1998). IL-10 is critical in the regulation of autoimmune encephalomyelitis as demonstrated by studies of IL-10- and IL-4-deficient and transgenic mice. *J. Immunol.* **161,** 3299–3306.

Bettelli, E., Carrier, Y., Gao, W., Korn, T., Strom, T. B., Oukka, M., Weiner, H. L., and Kuchroo, V. K. (2006). Reciprocal developmental pathways for the generation of pathogenic effector TH17 and regulatory T cells. *Nature* **441,** 235–238.

Bielekova, B., and Martin, R. (2001). Antigen-specific immunomodulation via altered peptide ligands. *J. Mol. Med.* **79,** 552–565.

Bielekova, B., Goodwin, B., Richert, N., Cortese, I., Kondo, T., Afshar, G., Gran, B., Eaton, J., Antel, J., Frank, J. A., McFarland, H. F., and Martin, R. (2000). Encephalitogenic potential of the myelin basic protein peptide (amino acids 83–99) in multiple sclerosis: Results of a phase II clinical trial with an altered peptide ligand. *Nat. Med.* **6,** 1167–1175.

Bird, J. J., Brown, D. R., Mullen, A. C., Moskowitz, N. H., Mahowald, M. A., Sider, J. R., Gajewski, T. F., Wang, C. R., and Reiner, S. L. (1998). Helper T cell differentiation is controlled by the cell cycle. *Immunity* **9,** 229–237.

Bitsch, A., Schuchardt, J., Bunkowski, S., Kuhlmann, T., and Bruck, W. (2000). Acute axonal injury in multiple sclerosis. Correlation with demyelination and inflammation. *Brain* **123,** 1174–1183.

Bo, L., Peterson, J. W., Mork, S., Hoffman, P. A., Gallatin, W. M., Ransohoff, R. M., and Trapp, B. D. (1996). Distribution of immunoglobulin superfamily members ICAM-1, -2, -3, and the beta 2 integrin LFA-1 in multiple sclerosis lesions. *J. Neuropathol. Exp. Neurol.* **55,** 1060–1072.

Bourdette, D. N., Whitham, R. H., Chou, Y. K., Morrison, W. J., Atherton, J., Kenny, C., Liefeld, D., Hashim, G. A., Offner, H., and Vandenbark, A. A. (1994). Immunity to TCR peptides in multiple sclerosis. I. Successful immunization of patients with synthetic Vβ 5.2 and Vβ 6.1 CDR2 peptides. *J. Immunol.* **152,** 2510–2519.

Bretscher, P., and Cohn, M. (1970). A theory of self-nonself discrimination. *Science* **169,** 1042–1049.

Bretscher, P. A. (1999). A two-step, two-signal model for the primary activation of precursor helper T cells. *Proc. Natl. Acad. Sci. USA* **96,** 185–190.

Brunet, J. F., Denizot, F., Luciani, M. F., Roux-Dosseto, M., Suzan, M., Mattei, M. G., and Golstein, P. (1987). A new member of the immunoglobulin superfamily—CTLA-4. *Nature* **328,** 267–270.

Byrne, J. A., Butler, J. L., and Cooper, M. D. (1988). Differential activation requirements for virgin and memory T cells. *J. Immunol.* **141,** 3249–3257.

Cannella, B., and Raine, C. S. (1995). The adhesion molecule and cytokine profile of multiple sclerosis lesions. *Ann. Neurol.* **37,** 424–435.

Chakravarti, S., Sabatos, C. A., Xiao, S., Illes, Z., Cha, E. K., Sobel, R. A., Zheng, X. X., Strom, T. B., and Kuchroo, V. K. (2005). Tim-2 regulates T helper type 2 responses and autoimmunity. *J. Exp. Med.* **202,** 437–444.

Chen, Z., Laurence, A., Kanno, Y., Pacher-Zavisin, M., Zhu, B. M., Tato, C., Yoshimura, A., Hennighausen, L., and O'Shea, J. J. (2006). Selective regulatory function of Socs3 in the formation of IL-17-secreting T cells. *Proc. Natl. Acad. Sci. USA* **103,** 8137–8142.

Chitnis, T., and Khoury, S. J. (2003a). Cytokine shifts and tolerance in experimental autoimmune encephalomyelitis. *Immunol. Res.* **28,** 223–239.

Chitnis, T., and Khoury, S. J. (2003b). Role of costimulatory pathways in the pathogenesis of multiple sclerosis and experimental autoimmune encephalomyelitis. *J. Allergy Clin. Immunol.* **112,** 837–849; quiz 850.

Chitnis, T., Najafian, N., Abdallah, K. A., Dong, V., Yagita, H., Sayegh, M. H., and Khoury, S. J. (2001a). CD28-independent induction of experimental autoimmune encephalomyelitis. *J. Clin. Invest.* **107,** 575–583.

Chitnis, T., Najafian, N., Benou, C., Salama, A. D., Grusby, M. J., Sayegh, M. H., and Khoury, S. J. (2001b). Effect of targeted disruption of STAT4 and STAT6 on the induction of experimental autoimmune encephalomyelitis. *J. Clin. Invest.* **108,** 739–747.

Chou, Y. K., Bourdette, D. N., Offner, H., Whitham, R., Wang, R. Y., Hashim, G. A., and Vandenbark, A. A. (1992). Frequency of T cells specific for myelin basic protein and myelin proteolipid protein in blood and cerebrospinal fluid in multiple sclerosis. *J. Neuroimmunol.* **38,** 105–113.

Clatch, R. J., Lipton, H. L., and Miller, S. D. (1986). Characterization of Theiler's murine encephalomyelitis virus (TMEV)-specific delayed-type hypersensitivity responses in TMEV-induced demyelinating disease: Correlation with clinical signs. *J. Immunol.* **136,** 920–927.

Comabella, M., Balashov, K., Issazadeh, S., Smith, D., Weiner, H. L., and Khoury, S. J. (1998). Elevated interleukin-12 in progressive multiple sclerosis correlates with disease activity and is normalized by pulse cyclophosphamide therapy. *J. Clin. Invest.* **102,** 671–678.

Crucian, B., Dunne, P., Friedman, H., Ragsdale, R., Pross, S., and Widen, R. (1995). Alterations in levels of CD28−/CD8+ suppressor cell precursor and CD45RO+/CD4+ memory T lymphocytes in the peripheral blood of multiple sclerosis patients. *Clin. Diagn. Lab. Immunol.* **2,** 249–252.

Diaz-Villoslada, P., Shih, A., Shao, L., Genain, C. P., and Hauser, S. L. (1999). Autoreactivity to myelin antigens: Myelin/oligodendrocyte glycoprotein is a prevalent autoantigen. *J. Neuroimmunol.* **99,** 36–43.

Dieckmann, D., Plottner, H., Berchtold, S., Berger, T., and Schuler, G. (2001). *Ex vivo* isolation and characterization of CD4(+)CD25(+) T cells with regulatory properties from human blood. *J. Exp. Med.* **193,** 1303–1310.

Duda, P. W., Schmied, M. C., Cook, S. L., Krieger, J. I., and Hafler, D. A. (2000). Glatiramer acetate (Copaxone) induces degenerate, Th2-polarized immune responses in patients with multiple sclerosis. *J. Clin. Invest.* **105,** 967–976.

Dutton, R. W., Bradley, L. M., and Swain, S. L. (1998). T cell memory. *Annu. Rev. Immunol.* **16,** 201–223.

Dyment, D. A., Steckley, J. L., Willer, C. J., Armstrong, H., Sadovnick, A. D., Risch, N., and Ebers, G. C. (2002). No evidence to support CTLA-4 as a susceptibility gene in MS families: The Canadian Collaborative Study. *J. Neuroimmunol.* **123,** 193–198.

Ehlers, S., and Smith, K. A. (1991). Differentiation of T cell lymphokine gene expression: The *in vitro* acquisition of T cell memory. *J. Exp. Med.* **173,** 25–36.

Eikelenboom, M. J., Killestein, J., Izeboud, T., Kalkers, N. F., van Lier, R. A., Barkhof, F., Uitdehaag, B. M., and Polman, C. H. (2002). Chemokine receptor expression on T cells is related to new lesion development in multiple sclerosis. *J. Neuroimmunol.* **133,** 225–232.

Engelhardt, B., and Ransohoff, R. M. (2005). The ins and outs of T-lymphocyte trafficking to the CNS: Anatomical sites and molecular mechanisms. *Trends Immunol.* **26,** 485–495.

Filion, L. G., Matusevicius, D., Graziani-Bowering, G. M., Kumar, A., and Freedman, M. S. (2003). Monocyte-derived IL12, CD86 (B7-2) and CD40L expression in relapsing and progressive multiple sclerosis. *Clin. Immunol.* **106,** 127–138.

Fukaura, H., Kent, S. C., Pietrusewicz, M. J., Khoury, S. J., Weiner, H. L., and Hafler, D. A. (1996). Induction of circulating myelin basic protein and proteolipid protein-specific transforming growth factor-beta1-secreting Th3 T cells by oral administration of myelin in multiple sclerosis patients. *J. Clin. Invest.* **98,** 70–77.

Fukazawa, T., Yanagawa, T., Kikuchi, S., Yabe, I., Sasaki, H., Hamada, T., Miyasaka, K., Gomi, K., and Tashiro, K. (1999). CTLA-4 gene polymorphism may modulate disease in Japanese multiple sclerosis patients. *J. Neurol. Sci.* **171,** 49–55.

Ge, Y., Grossman, R. I., Udupa, J. K., Fulton, J., Constantinescu, C. S., Gonzales-Scarano, F., Babb, J. S., Mannon, L. J., Kolson, D. L., and Cohen, J. A. (2000). Glatiramer acetate (Copaxone) treatment in relapsing-remitting MS: Quantitative MR assessment. *Neurology* **54,** 813–817.

Genc, K., Dona, D. L., and Reder, A. T. (1997). Increased CD80(+) B cells in active multiple sclerosis and reversal by interferon beta-1b therapy. *J. Clin. Invest.* **99,** 2664–2671.

Gerritse, K., Laman, J. D., Noelle, R. J., Aruffo, A., Ledbetter, J. A., Boersma, W. J., and Claassen, E. (1996). CD40-CD40 ligand interactions in experimental allergic encephalomyelitis and multiple sclerosis. *Proc. Natl. Acad. Sci. USA* **93,** 2499–2504.

Giuliani, F., Goodyer, C. G., Antel, J. P., and Yong, V. W. (2003). Vulnerability of human neurons to T cell-mediated cytotoxicity. *J. Immunol.* **171,** 368–379.

Gran, B., Gestri, D., Sottini, A., Quiros Roldan, E., Bettinardi, A., Signorini, S., Primi, D., Ballerini, C., Taiuti, R., Amaducci, L., and Massacesi, L. (1998). Detection of skewed T-cell receptor V-beta gene usage in the peripheral blood of patients with multiple sclerosis. *J. Neuroimmunol.* **85,** 22–32.

Hafler, D. A., Benjamin, D. S., Burks, J., and Weiner, H. L. (1987). Myelin basic protein and proteolipid protein reactivity of brain- and cerebrospinal fluid-derived T cell clones in multiple sclerosis and postinfectious encephalomyelitis. *J. Immunol.* **139,** 68–72.

Hafler, D. A., Kent, S. C., Pietrusewicz, M. J., Khoury, S. J., Weiner, H. L., and Fukaura, H. (1997). Oral administration of myelin induces antigen-specific TGF-beta 1 secreting T cells in patients with multiple sclerosis. *Ann. N. Y. Acad. Sci.* **835,** 120–231.

Harrington, L. E., Hatton, R. D., Mangan, P. R., Turner, H., Murphy, T. L., Murphy, K. M., and Weaver, C. T. (2005). Interleukin 17-producing CD4+ effector T cells develop via a lineage distinct from the T helper type 1 and 2 lineages. *Nat. Immunol.* **6,** 1123–1132.

Hemachudha, T., Griffin, D. E., Johnson, R. T., and Giffels, J. J. (1988). Immunologic studies of patients with chronic encephalitis induced by post-exposure Semple rabies vaccine. *Neurology* **38,** 42–44.

Ho, I. C., Hodge, M. R., Rooney, J. W., and Glimcher, L. H. (1996). The proto-oncogene c-maf is responsible for tissue-specific expression of interleukin-4. *Cell* **85,** 973–983.

Hofman, F. M., von Hanwehr, R. I., Dinarello, C. A., Mizel, S. B., Hinton, D., and Merrill, J. E. (1986). Immunoregulatory molecules and IL 2 receptors identified in multiple sclerosis brain. *J. Immunol.* **136,** 3239–3245.

Hofman, F. M., Hinton, D. R., Johnson, K., and Merrill, J. E. (1989). Tumor necrosis factor identified in multiple sclerosis brain. *J. Exp. Med.* **170,** 607–612.

Holz, A., Bielekova, B., Martin, R., and Oldstone, M. B. (2000). Myelin-associated oligodendrocytic basic protein: Identification of an encephalitogenic epitope and association with multiple sclerosis. *J. Immunol.* **164,** 1103–1109.

Hori, S., Nomura, T., and Sakaguchi, S. (2003). Control of regulatory T cell development by the transcription factor Foxp3. *Science* **299,** 1057–1061.

Ishizu, T., Osoegawa, M., Mei, F. J., Kikuchi, H., Tanaka, M., Takakura, Y., Minohara, M., Murai, H., Mihara, F., Taniwaki, T., and Kira, J. (2005). Intrathecal activation of the IL-17/IL-8 axis in opticospinal multiple sclerosis. *Brain* **128,** 988–1002.

Jacobson, N. G., Szabo, S. J., Weber-Nordt, R. M., Zhong, Z., Schreiber, R. D., Darnell, J. E., Jr., and Murphy, K. M. (1995). Interleukin 12 signaling in T helper type 1 (Th1) cells involves tyrosine phosphorylation of signal transducer and activator of transcription (Stat)3 and Stat4. *J. Exp. Med.* **181,** 1755–1762.

Jensen, J., Krakauer, M., and Sellebjerg, F. (2001). Increased T cell expression of CD154 (CD40-ligand) in multiple sclerosis. *Eur. J. Neurol.* **8,** 321–328.

Johnson, K. P., Brooks, B. R., Cohen, J. A., Ford, C. C., Goldstein, J., Lisak, R. P., Myers, L. W., Panitch, H. S., Rose, J. W., and Schiffer, R. B. (1995). Copolymer 1 reduces relapse rate and improves disability in relapsing-remitting multiple sclerosis: Results of a phase III multicenter, double-blind placebo-controlled trial. The Copolymer 1 Multiple Sclerosis Study Group. *Neurology* **45,** 1268–1276.

Johnson, K. P., Brooks, B. R., Cohen, J. A., Ford, C. C., Goldstein, J., Lisak, R. P., Myers, L. W., Panitch, H. S., Rose, J. W., Schiffer, R. B., Vollmer, T., Weiner, L. P., *et al.* (2001). Copolymer 1 reduces relapse rate and improves disability in relapsing-remitting multiple sclerosis: Results of a phase III multicenter, double-blind, placebo-controlled trial. 1995. *Neurology* **57,** S16–S24.

Jones, T. B., Basso, D. M., Sodhi, A., Pan, J. Z., Hart, R. P., MacCallum, R. C., Lee, S., Whitacre, C. C., and Popovich, P. G. (2002). Pathological CNS autoimmune disease triggered by traumatic spinal cord injury: Implications for autoimmune vaccine therapy. *J. Neurosci.* **22,** 2690–2700.

Kantarci, O. H., Hebrink, D. D., Achenbach, S. J., Atkinson, E. J., Waliszewska, A., Buckle, G., McMurray, C. T., de Andrade, M., Hafler, D. A., and Weinshenker, B. G. (2003). CTLA4 is associated with susceptibility to multiple sclerosis. *J. Neuroimmunol.* **134,** 133–141.

Kaplan, M. H., Sun, Y. L., Hoey, T., and Grusby, M. J. (1996). Impaired IL-12 responses and enhanced development of Th2 cells in Stat4-deficient mice. *Nature* **382,** 174–177.

Kappos, L., Comi, G., Panitch, H., Oger, J., Antel, J., Conlon, P., and Steinman, L. (2000). Induction of a non-encephalitogenic type 2 T helper-cell autoimmune response in multiple sclerosis after administration of an altered peptide ligand in a placebo-controlled, randomized phase II trial. The Altered Peptide Ligand in Relapsing MS Study Group. *Nat. Med.* **6,** 1176–1182.

Karaszewski, J. W., Reder, A. T., Anlar, B., Kim, W. C., and Arnason, B. G. (1991). Increased lymphocyte beta-adrenergic receptor density in progressive multiple sclerosis is specific for the CD8+, CD28− suppressor cell. *Ann. Neurol.* **30,** 42–47.

Karin, N., Mitchell, D. J., Brocke, S., Ling, N., and Steinman, L. (1994). Reversal of experimental autoimmune encephalomyelitis by a soluble peptide variant of a myelin basic protein epitope: T cell receptor antagonism and reduction of interferon gamma and tumor necrosis factor alpha production. *J. Exp. Med.* **180,** 2227–2237.

Karni, A., Koldzic, D. N., Bharanidharan, P., Khoury, S. J., and Weiner, H. L. (2002). IL-18 is linked to raised IFN-gamma in multiple sclerosis and is induced by activated CD4+ T cells via CD40-CD40 ligand interactions. *J. Neuroimmunol.* **125,** 134–140.

Kelsall, B. L., Stuber, E., Neurath, M., and Strober, W. (1996). Interleukin-12 production by dendritic cells. The role of CD40-CD40L interactions in Th1 T-cell responses. *Ann. N. Y. Acad. Sci.* **795,** 116–126.

Kerlero de Rosbo, N., Milo, R., Lees, M. B., Burger, D., Bernard, C. C., and Ben-Nun, A. (1993). Reactivity to myelin antigens in multiple sclerosis. Peripheral blood lymphocytes respond predominantly to myelin oligodendrocyte glycoprotein. *J. Clin. Invest.* **92,** 2602–2608.

Killestein, J., Olsson, T., Wallstrom, E., Svenningsson, A., Khademi, M., Blumhardt, L. D., Fagius, J., Hillert, J., Landtblom, A. M., Edenius, C., Arfors, L., Barkhof, F., *et al.* (2002). Antibody-mediated suppression of Vβ5.2/5.3(+) T cells in multiple sclerosis: Results from an MRI-monitored phase II clinical trial. *Ann. Neurol.* **51,** 467–474.

Kitani, A., Chua, K., Nakamura, K., and Strober, W. (2000). Activated self-MHC-reactive T cells have the cytokine phenotype of Th3/T regulatory cell 1 T cells. *J. Immunol.* **165,** 691–702.

Kivisakk, P., Trebst, C., Liu, Z., Tucky, B. H., Sorensen, T. L., Rudick, R. A., Mack, M., and Ransohoff, R. M. (2002). T-cells in the cerebrospinal fluid express a similar repertoire of inflammatory chemokine receptors in the absence or presence of CNS inflammation: Implications for CNS trafficking. *Clin. Exp. Immunol.* **129,** 510–518.

Kivisakk, P., Mahad, D. J., Callahan, M. K., Sikora, K., Trebst, C., Tucky, B., Wujek, J., Ravid, R., Staugaitis, S. M., Lassmann, H., and Ransohoff, R. M. (2004). Expression of CCR7 in multiple sclerosis: Implications for CNS immunity. *Ann. Neurol.* **55,** 627–638.

Kleinschmidt-Demasters, B. K., and Tyler, K. L. (2005). Progressive multifocal leukoencephalopathy complicating treatment with natalizumab and interferon beta-1a for multiple sclerosis. *N. Engl. J. Med.* **353**(4), 369–374.

Koguchi, K., Anderson, D. E., Yang, L., O'Connor, K. C., Kuchroo, V. K., and Hafler, D. A. (2006). Dysregulated T cell expression of TIM3 in multiple sclerosis. *J. Exp. Med.* **203,** 1413–1418.

Koide, J., and Engleman, E. G. (1990). Differences in surface phenotype and mechanism of action between alloantigen-specific CD8+ cytotoxic and suppressor T cell clones. *J. Immunol.* **144,** 32–40.

Kotzin, B. L., Karuturi, S., Chou, Y. K., Lafferty, J., Forrester, J. M., Better, M., Nedwin, G. E., Offner, H., and Vandenbark, A. A. (1991). Preferential T-cell receptor beta-chain variable gene use in myelin basic protein-reactive T-cell clones from patients with multiple sclerosis. *Proc. Natl. Acad. Sci. USA* **88,** 9161–9165.

Kroner, A., Mehling, M., Hemmer, B., Rieckmann, P., Toyka, K. V., Maurer, M., and Wiendl, H. (2005). A PD-1 polymorphism is associated with disease progression in multiple sclerosis. *Ann. Neurol.* **58,** 50–57.

Kuiper, H., Brouwer, M., de Boer, M., Parren, P., and van Lier, R. A. (1994). Differences in responsiveness to CD3 stimulation between naive and memory CD4+ T cells cannot be overcome by CD28 costimulation. *Eur. J. Immunol.* **24,** 1956–1960.

Lang, H. L., Jacobsen, H., Ikemizu, S., Andersson, C., Harlos, K., Madsen, L., Hjorth, P., Sondergaard, L., Svejgaard, A., Wucherpfennig, K., Stuart, D. I., Bell, J. I., *et al.* (2002). A functional and structural basis for TCR cross-reactivity in multiple sclerosis. *Nat. Immunol.* **3,** 940–943.

Langer-Gould, A., Atlas, S. W., Bollen, A. W., and Pelletier, D. (2005). Progressive multifocal leukoencephalopathy in a patient treated with natalizumab. *N. Engl. J. Med.* **353**(4), 375–381.

Langrish, C. L., Chen, Y., Blumenschein, W. M., Mattson, J., Basham, B., Sedgwick, J. D., McClanahan, T., Kastelein, R. A., and Cua, D. J. (2005). IL-23 drives a pathogenic T cell population that induces autoimmune inflammation. *J. Exp. Med.* **201,** 233–240.

Lee, W. T., Yin, X. M., and Vitetta, E. S. (1990). Functional and ontogenetic analysis of murine CD45Rhi and CD45Rlo CD4+ T cells. *J. Immunol.* **144,** 3288–3295.

Levings, M. K., Sangregorio, R., and Roncarolo, M. G. (2001). Human CD25+ CD4+ T regulatory cells suppress naive and memory T cell proliferation and can be expanded *in vitro* without loss of function. *J. Exp. Med.* **193,** 1295–1302.

Ligers, A., Xu, C., Saarinen, S., Hillert, J., and Olerup, O. (1999). The CTLA-4 gene is associated with multiple sclerosis. *J. Neuroimmunol.* **97,** 182–190.

Liu, Y., Teige, I., Birnir, B., and Issazadeh-Navikas, S. (2006). Neuron-mediated generation of regulatory T cells from encephalitogenic T cells suppresses EAE. *Nat. Med.* **12,** 518–525.

Lock, C., Hermans, G., Pedotti, R., Brendolan, A., Schadt, E., Garren, H., Langer-Gould, A., Strober, S., Cannella, B., Allard, J., Klonowski, P., Austin, A., *et al.* (2002). Gene-microarray analysis of multiple sclerosis lesions yields new targets validated in autoimmune encephalomyelitis. *Nat. Med.* **8,** 500–508.

Lozeron, P., Chabas, D., Duprey, B., Lyon-Caen, O., and Liblau, R. (1998). T cell receptor Vβ 5 and Vβ 17 clonal diversity in cerebrospinal fluid and peripheral blood lymphocytes of multiple sclerosis patients. *Mult. Scler.* **4,** 154–161.

Lucchinetti, C., Bruck, W., Parisi, J., Scheithauer, B., Rodriguez, M., and Lassmann, H. (2000). Heterogeneity of multiple sclerosis lesions: Implications for the pathogenesis of demyelination. *Ann. Neurol.* **47,** 707–717.

Lucchinetti, C. F., Bruck, W., Rodriguez, M., and Lassmann, H. (1996). Distinct patterns of multiple sclerosis pathology indicates heterogeneity on pathogenesis. *Brain Pathol.* **6,** 259–274.

Mangan, P. R., Harrington, L. E., O'Quinn, D. B., Helms, W. S., Bullard, D. C., Elson, C. O., Hatton, R. D., Wahl, S. M., Schoeb, T. R., and Weaver, C. T. (2006). Transforming growth factor-beta induces development of the T(H)17 lineage. *Nature* **441,** 231–234.

Markovic-Plese, S., Fukaura, H., Zhang, J., al-Sabbagh, A., Southwood, S., Sette, A., Kuchroo, V. K., and Hafler, D. A. (1995). T cell recognition of immunodominant and cryptic proteolipid protein epitopes in humans. *J. Immunol.* **155,** 982–992.

Markovic-Plese, S., Cortese, I., Wandinger, K. P., McFarland, H. F., and Martin, R. (2001). CD4+ CD28-costimulation-independent T cells in multiple sclerosis. *J. Clin. Invest.* **108,** 1185–1194.

Markovic-Plese, S., Hemmer, B., Zhao, Y., Simon, R., Pinilla, C., and Martin, R. (2005). High level of cross-reactivity in influenza virus hemagglutinin-specific CD4+ T-cell response: Implications for the initiation of autoimmune response in multiple sclerosis. *J. Neuroimmunol.* **169,** 31–38.

Masterman, T., Ligers, A., Zhang, Z., Hellgren, D., Salter, H., Anvret, M., and Hillert, J. (2002). CTLA4 dimorphisms and the multiple sclerosis phenotype. *J. Neuroimmunol.* **131,** 208–212.

McMahon, E. J., Bailey, S. L., Castenada, C. V., Waldner, H., and Miller, S. D. (2005). Epitope spreading initiates in the CNS in two mouse models of multiple sclerosis. *Nat. Med.* **11,** 335–339.

Medaer, R., Stinissen, P., Truyen, L., Raus, J., and Zhang, J. (1995). Depletion of myelin-basic-protein autoreactive T cells by T-cell vaccination: Pilot trial in multiple sclerosis. *Lancet* **346,** 807–808.

Medana, I., Martinic, M. A., Wekerle, H., and Neumann, H. (2001). Transection of major histocompatibility complex class I-induced neurites by cytotoxic T lymphocytes. *Am. J. Pathol.* **159,** 809–815.

Mena, E., and Rohowsky-Kochan, C. (1999). Expression of costimulatory molecules on peripheral blood mononuclear cells in multiple sclerosis. *Acta Neurol. Scand.* **100,** 92–96.

Meyers, J. H., Chakravarti, S., Schlesinger, D., Illes, Z., Waldner, H., Umetsu, S. E., Kenny, J., Zheng, X. X., Umetsu, D. T., DeKruyff, R. H., Strom, T. B., and Kuchroo, V. K. (2005). TIM-4 is the ligand for TIM-1, and the TIM-1-TIM-4 interaction regulates T cell proliferation. *Nat. Immunol.* **6,** 455–464.

Miller, D. H., Khan, O. A., Sheremata, W. A., Blumhardt, L. D., Rice, G. P., Libonati, M. A., Willmer-Hulme, A. J., Dalton, C. M., Miszkiel, K. A., and O'Connor, P. W. (2003). A controlled trial of natalizumab for relapsing multiple sclerosis. *N. Engl. J. Med.* **348,** 15–23.

Miller, S. D., Vanderlugt, C. L., Begolka, W. S., Pao, W., Yauch, R. L., Neville, K. L., Katz-Levy, Y., Carrizosa, A., and Kim, B. S. (1997). Persistent infection with Theiler's virus leads to CNS autoimmunity via epitope spreading. *Nat. Med.* **3,** 1133–1136.

Minguela, A., Torio, A., Marin, L., Sanchez-Bueno, F., Garcia-Alonso, A. M., Ontanon, J., Parrilla, P., and Alvarez-Lopez, M. R. (1999). Implication of Th1, Th2, and Th3 cytokines in liver graft acceptance. *Transplant. Proc.* **31,** 519–520.

Moalem, G., Leibowitz-Amit, R., Yoles, E., Mor, F., Cohen, I. R., and Schwartz, M. (1999). Autoimmune T cells protect neurons from secondary degeneration after central nervous system axotomy. *Nat. Med.* **5,** 49–55.

Moalem, G., Gdalyahu, A., Shani, Y., Otten, U., Lazarovici, P., Cohen, I. R., and Schwartz, M. (2000). Production of neurotrophins by activated T cells: Implications for neuroprotective autoimmunity. *J. Autoimmun.* **15,** 331–345.

Mokhtarian, F., Zhang, Z., Shi, Y., Gonzales, E., and Sobel, R. A. (1999). Molecular mimicry between a viral peptide and a myelin oligodendrocyte glycoprotein peptide induces autoimmune demyelinating disease in mice. *J. Neuroimmunol.* **95,** 43–54.

Monney, L., Sabatos, C. A., Gaglia, J. L., Ryu, A., Waldner, H., Chernova, T., Manning, S., Greenfield, E. A., Coyle, A. J., Sobel, R. A., Freeman, G. J., and Kuchroo, V. K. (2002). Th1-specific cell surface protein Tim-3 regulates macrophage activation and severity of an autoimmune disease. *Nature* **415,** 536–541.

Muraro, P. A., Kalbus, M., Afshar, G., McFarland, H. F., and Martin, R. (2002). T cell response to 2′,3′-cyclic nucleotide 3′-phosphodiesterase (CNPase) in multiple sclerosis patients. *J. Neuroimmunol.* **130,** 233–242.

Muraro, P. A., Douek, D. C., Packer, A., Chung, K., Guenaga, F. J., Cassiani-Ingoni, R., Campbell, C., Memon, S., Nagle, J. W., Hakim, F. T., Gress, R. E., McFarland, H. F., *et al.* (2005). Thymic output generates a new and diverse TCR repertoire after autologous stem cell transplantation in multiple sclerosis patients. *J. Exp. Med.* **201,** 805–816.

Musette, P., Bequet, D., Delarbre, C., Gachelin, G., Kourilsky, P., and Dormont, D. (1996). Expansion of a recurrent Vβ 5.3+ T-cell population in newly diagnosed and untreated HLA-DR2 multiple sclerosis patients. *Proc. Natl. Acad. Sci. USA* **93,** 12461–12466.

Oksenberg, J. R., Panzara, M. A., Begovich, A. B., Mitchell, D., Erlich, H. A., Murray, R. S., Shimonkevitz, R., Sherritt, M., Rothbard, J., Bernard, C. C., and Steinman, L. (1993). Selection for T-cell receptor V beta-D beta-J beta gene rearrangements with specificity for a myelin basic protein peptide in brain lesions of multiple sclerosis. *Nature* **362,** 68–70.

Oliveira, E. M., Bar-Or, A., Waliszewska, A. I., Cai, G., Anderson, D. E., Krieger, J. I., and Hafler, D. A. (2003). CTLA-4 dysregulation in the activation of myelin basic protein reactive T cells may distinguish patients with multiple sclerosis from healthy controls. *J. Autoimmun.* **20,** 71–81.

Olson, J. K., Croxford, J. L., Calenoff, M. A., Dal Canto, M. C., and Miller, S. D. (2001). A virus-induced molecular mimicry model of multiple sclerosis. *J. Clin. Invest.* **108,** 311–318.

Olsson, T., Edenius, C., Ferm, M., Samuelson, P., Torrang, A., Wallstrom, E., Khademi, M., Andersson, M., and Arfors, L. (2002). Depletion of Vβ 5.2/5.3 T cells with a humanized antibody in patients with multiple sclerosis. *Eur. J. Neurol.* **9,** 153–164.

Ouyang, W., Lohning, M., Gao, Z., Assenmacher, M., Ranganath, S., Radbruch, A., and Murphy, K. M. (2000). Stat6-independent GATA-3 autoactivation directs IL-4-independent Th2 development and commitment. *Immunity* **12,** 27–37.

Panitch, H. S., Hirsch, R. L., Haley, A. S., and Johnson, K. P. (1987a). Exacerbations of multiple sclerosis in patients treated with gamma interferon. *Lancet* **1,** 893–895.

Panitch, H. S., Hirsch, R. L., Schindler, J., and Johnson, K. P. (1987b). Treatment of multiple sclerosis with gamma interferon: Exacerbations associated with activation of the immune system. *Neurology* **37,** 1097–1102.

Park, H., Li, Z., Yang, X. O., Chang, S. H., Nurieva, R., Wang, Y. H., Wang, Y., Hood, L., Zhu, Z., Tian, Q., and Dong, C. (2005). A distinct lineage of CD4 T cells regulates tissue inflammation by producing interleukin 17. *Nat. Immunol.* **6,** 1133–1141.

Pelfrey, C. M., Trotter, J. L., Tranquill, L. R., and McFarland, H. F. (1994). Identification of a second T cell epitope of human proteolipid protein (residues 89–106) recognized by proliferative and cytolytic CD4+ T cells from multiple sclerosis patients. *J. Neuroimmunol.* **53,** 153–161.

Pohl-Koppe, A., Burchett, S. K., Thiele, E. A., and Hafler, D. A. (1998). Myelin basic protein reactive Th2 T cells are found in acute disseminated encephalomyelitis. *J. Neuroimmunol.* **91,** 19–27.

Prineas, J. W., and Wright, R. G. (1978). Macrophages, lymphocytes, and plasma cells in the perivascular compartment in chronic multiple sclerosis. *Lab. Invest.* **38,** 409–421.

Racke, M. K., Martin, R., McFarland, H., and Fritz, R. B. (1992). Copolymer-1-induced inhibition of antigen-specific T cell activation: Interference with antigen presentation. *J. Neuroimmunol.* **37,** 75–84.

Raine, C. S. (1994). The Dale E. McFarlin Memorial Lecture: The immunology of the multiple sclerosis lesion. *Ann. Neurol.* **36**(Suppl.), S61–S72.

Rivera-Quinones, C., McGavern, D., Schmelzer, J. D., Hunter, S. F., Low, P. A., and Rodriguez, M. (1998). Absence of neurological deficits following extensive demyelination in a class I-deficient murine model of multiple sclerosis. *Nat. Med.* **4,** 187–193.

Rodriguez, M., Pavelko, K. D., Njenga, M. K., Logan, W. C., and Wettstein, P. J. (1996). The balance between persistent virus infection and immune cells determines demyelination. *J. Immunol.* **157,** 5699–5709.

Roncarolo, M. G., and Levings, M. K. (2000). The role of different subsets of T regulatory cells in controlling autoimmunity. *Curr. Opin. Immunol.* **12,** 676–683.

Sabatos, C. A., Chakravarti, S., Cha, E., Schubart, A., Sanchez-Fueyo, A., Zheng, X. X., Coyle, A. J., Strom, T. B., Freeman, G. J., and Kuchroo, V. K. (2003). Interaction of Tim-3 and Tim-3 ligand regulates T helper type 1 responses and induction of peripheral tolerance. *Nat. Immunol.* **4,** 1102–1110.

Schmied, M., Duda, P. W., Krieger, J. I., Trollmo, C., and Hafler, D. A. (2003). *In vitro* evidence that subcutaneous administration of glatiramer acetate induces hyporesponsive T cells in patients with multiple sclerosis. *Clin. Immunol.* **106,** 163–174.

Scholz, C., Patton, K. T., Anderson, D. E., Freeman, G. J., and Hafler, D. A. (1998). Expansion of autoreactive T cells in multiple sclerosis is independent of exogenous B7 costimulation. *J. Immunol.* **160,** 1532–1538.

Schrijver, H. M., Hooper-Van Veen, T., Van Belzen, M. J., Crusius, J. B., Pena, A. S., Barkhof, F., Polman, C. H., and Uitdehaag, B. M. (2004). Polymorphisms in the genes encoding interferon-gamma and interferon-gamma receptors in multiple sclerosis. *Eur. J. Immunogenet.* **31,** 133–140.

Selmaj, K., Raine, C. S., Cannella, B., and Brosnan, C. F. (1991). Identification of lymphotoxin and tumor necrosis factor in multiple sclerosis lesions. *J. Clin. Invest.* **87,** 949–954.

Sindern, E., Patzold, T., Ossege, L. M., Gisevius, A., and Malin, J. P. (2002). Expression of chemokine receptor CXCR3 on cerebrospinal fluid T-cells is related to active MRI lesion appearance in patients with relapsing-remitting multiple sclerosis. *J. Neuroimmunol.* **131,** 186–190.

Smith, D. R., Balashov, K. E., Hafler, D. A., Khoury, S. J., and Weiner, H. L. (1997). Immune deviation following pulse cyclophosphamide/methylprednisolone treatment of multiple sclerosis: Increased interleukin-4 production and associated eosinophilia. *Ann. Neurol.* **42,** 313–318.

Smith, T., Groom, A., Zhu, B., and Turski, L. (2000). Autoimmune encephalomyelitis ameliorated by AMPA antagonists. *Nat. Med.* **6,** 62–66.

Sprent, J., and Surh, C. D. (2002). T cell memory. *Annu. Rev. Immunol.* **20,** 551–579.

Stephens, L. A., Mottet, C., Mason, D., and Powrie, F. (2001). Human CD4+ CD25+ thymocytes and peripheral T cells have immune suppressive activity *in vitro*. *Eur. J. Immunol.* **31,** 1247–1254.

Sun, W., Popat, U., Hutton, G., Zang, Y. C., Krance, R., Carrum, G., Land, G. A., Heslop, H., Brenner, M., and Zhang, J. Z. (2004). Characteristics of T-cell receptor repertoire and myelin-reactive T cells reconstituted from autologous haematopoietic stem-cell grafts in multiple sclerosis. *Brain* **127,** 996–1008.

Szabo, S. J., Kim, S. T., Costa, G. L., Zhang, X., Fathman, C. G., and Glimcher, L. H. (2000). A novel transcription factor, T-bet, directs Th1 lineage commitment. *Cell* **100,** 655–669.

Takeda, K., Kamanaka, M., Tanaka, T., Kishimoto, T., and Akira, S. (1996). Impaired IL-13-mediated functions of macrophages in STAT6-deficient mice. *J. Immunol.* **157,** 3220–3222.

Talbot, P. J., Paquette, J. S., Ciurli, C., Antel, J. P., and Ouellet, F. (1996). Myelin basic protein and human coronavirus 229E cross-reactive T cells in multiple sclerosis. *Ann. Neurol.* **39,** 233–240.

Tejada-Simon, M. V., Zang, Y. C., Hong, J., Rivera, V. M., and Zhang, J. Z. (2003). Cross-reactivity with myelin basic protein and human herpesvirus-6 in multiple sclerosis. *Ann. Neurol.* **53,** 189–197.

Teleshova, N., Bao, W., Kivisakk, P., Ozenci, V., Mustafa, M., and Link, H. (2000). Elevated CD40 ligand expressing blood T-cell levels in multiple sclerosis are reversed by interferon-beta treatment. *Scand. J. Immunol.* **51,** 312–320.

Thierfelder, W. E., van Deursen, J. M., Yamamoto, K., Tripp, R. A., Sarawar, S. R., Carson, R. T., Sangster, M. Y., Vignali, D. A., Doherty, P. C., Grosveld, G. C., and Ihle, J. N. (1996). Requirement for Stat4 in interleukin-12-mediated responses of natural killer and T cells. *Nature* **382,** 171–174.

Trapp, B. D., Peterson, J., Ransohoff, R. M., Rudick, R., Mork, S., and Bo, L. (1998). Axonal transection in the lesions of multiple sclerosis. *N. Engl. J. Med.* **338,** 278–285.

Trapp, B. D., Ransohoff, R., and Rudick, R. (1999). Axonal pathology in multiple sclerosis: Relationship to neurologic disability. *Curr. Opin. Neurol.* **12,** 295–302.

Traugott, U., Reinherz, E. L., and Raine, C. S. (1983a). Multiple sclerosis. Distribution of T cells, T cell subsets and Ia-positive macrophages in lesions of different ages. *J. Neuroimmunol.* **4,** 201–221.

Traugott, U., Reinherz, E. L., and Raine, C. S. (1983b). Multiple sclerosis: Distribution of T cell subsets within active chronic lesions. *Science* **219,** 308–310.

Utz, U., Biddison, W. E., McFarland, H. F., McFarlin, D. E., Flerlage, M., and Martin, R. (1993). Skewed T-cell receptor repertoire in genetically identical twins correlates with multiple sclerosis. *Nature* **364,** 243–247.

Vaknin-Dembinsky, A., Balashov, K., and Weiner, H. L. (2006). IL-23 is increased in dendritic cells in multiple sclerosis and down-regulation of IL-23 by antisense oligos increases dendritic cell IL-10 production. *J. Immunol.* **176,** 7768–7774.

van Oosten, B. W., Lai, M., Hodgkinson, S., Barkhof, F., Miller, D. H., Moseley, I. F., Thompson, A. J., Rudge, P., McDougall, A., McLeod, J. G., Ader, H. J., and Polman, C. H. (1997). Treatment of multiple sclerosis with the monoclonal anti-CD4 antibody cM-T412: Results of a randomized, double-blind, placebo-controlled, MR-monitored phase II trial. *Neurology* **49,** 351–357.

Van Assche, G., Van Ranst, M., Sciot, R., Dubois, B., Vermeire, S., Noman, M., Verbeeck, J., Geboes, K., Robberecht, W., and Rutgeerts, P. (2005). Progressive multifocal leukoencephalopathy after natalizumab therapy for Crohn's disease. *N. Engl. J. Med.* **353,** 362–368.

Van der Aa, A., Hellings, N., Bernard, C. C., Raus, J., and Stinissen, P. (2003a). Functional properties of myelin oligodendrocyte glycoprotein-reactive T cells in multiple sclerosis patients and controls. *J. Neuroimmunol.* **137,** 164–176.

Van der Aa, A., Hellings, N., Medaer, R., Gelin, G., Palmers, Y., Raus, J., and Stinissen, P. (2003b). T cell vaccination in multiple sclerosis patients with autologous CSF-derived activated T cells: Results from a pilot study. *Clin. Exp. Immunol.* **131,** 155–168.

Vandenbark, A. A. (2005). TCR peptide vaccination in multiple sclerosis: Boosting a deficient natural regulatory network that may involve TCR-specific CD4+ CD25+ Treg cells. *Curr. Drug Targets Inflamm. Allergy* **4,** 217–229.

Vandenbark, A. A., Chou, Y. K., Whitham, R., Mass, M., Buenafe, A., Liefeld, D., Kavanagh, D., Cooper, S., Hashim, G. A., and Offner, H. (1996). Treatment of multiple sclerosis with T-cell receptor peptides: Results of a double-blind pilot trial. *Nat. Med.* **2,** 1109–1115.

Vanderlugt, C. L., Begolka, W. S., Neville, K. L., Katz-Levy, Y., Howard, L. M., Eagar, T. N., Bluestone, J. A., and Miller, S. D. (1998). The functional significance of epitope spreading and its regulation by co-stimulatory molecules. *Immunol. Rev.* **164,** 63–72.

Veldhoen, M., Hocking, R. J., Atkins, C. J., Locksley, R. M., and Stockinger, B. (2006). TGFβ in the context of an inflammatory cytokine milieu supports *de novo* differentiation of IL-17-producing T cells. *Immunity* **24,** 179–189.

Viglietta, V., Baecher-Allan, C., Weiner, H. L., and Hafler, D. A. (2004). Loss of functional suppression by CD4+ CD25+ regulatory T cells in patients with multiple sclerosis. *J. Exp. Med.* **199,** 971–979.

Windhagen, A., Newcombe, J., Dangond, F., Strand, C., Woodroofe, M. N., Cuzner, M. L., and Hafler, D. A. (1995). Expression of costimulatory molecules B7–1 (CD80), B7–2 (CD86), and interleukin 12 cytokine in multiple sclerosis lesions. *J. Exp. Med.* **182,** 1985–1996.

Wolf, S. A., Fisher, J., Bechmann, I., Steiner, B., Kwidzinski, E., and Nitsch, R. (2002). Neuroprotection by T-cells depends on their subtype and activation state. *J. Neuroimmunol.* **133,** 72–80.

Woodroofe, M. N., and Cuzner, M. L. (1993). Cytokine mRNA expression in inflammatory multiple sclerosis lesions: Detection by non-radioactive *in situ* hybridization. *Cytokine* **5,** 583–588.

Wucherpfennig, K. W., Newcombe, J., Li, H., Keddy, C., Cuzner, M. L., and Hafler, D. A. (1992). T cell receptor V alpha-V beta repertoire and cytokine gene expression in active multiple sclerosis lesions. *J. Exp. Med.* **175,** 993–1002.

Wucherpfennig, K. W., Hafler, D. A., and Strominger, J. L. (1995). Structure of human T-cell receptors specific for an immunodominant myelin basic protein peptide: Positioning of T-cell receptors on HLA-DR2/peptide complexes. *Proc. Natl. Acad. Sci. USA* **92,** 8896–8900.

Wucherpfennig, K. W., Catz, I., Hausmann, S., Strominger, J. L., Steinman, L., and Warren, K. G. (1997). Recognition of the immunodominant myelin basic protein peptide by autoantibodies and HLA-DR2-restricted T cell clones from multiple sclerosis patients. Identity of key contact residues in the B-cell and T-cell epitopes. *J. Clin. Invest.* **100,** 1114–1122.

Yagi, H., Nomura, T., Nakamura, K., Yamazaki, S., Kitawaki, T., Hori, S., Maeda, M., Onodera, M., Uchiyama, T., Fujii, S., and Sakaguchi, S. (2004). Crucial role of FOXP3 in the development and function of human CD25+ CD4+ regulatory T cells. *Int. Immunol.* **16,** 1643–1656.

Yao, Z., Painter, S. L., Fanslow, W. C., Ulrich, D., Macduff, B. M., Spriggs, M. K., and Armitage, R. J. (1995). Human IL-17: A novel cytokine derived from T cells. *J. Immunol.* **155,** 5483–5486.

Yednock, T. A., Cannon, C., Fritz, L. C., Sanchez-Madrid, F., Steinman, L., and Karin, N. (1992). Prevention of experimental autoimmune encephalomyelitis by antibodies against alpha 4 beta 1 integrin. *Nature* **356,** 63–66.

Yong, V. W. (2002). Differential mechanisms of action of interferon-beta and glatiramer aetate in MS. *Neurology* **59,** 802–808.

Youssef, S., Stuve, O., Patarroyo, J. C., Ruiz, P. J., Radosevich, J. L., Hur, E. M., Bravo, M., Mitchell, D. J., Sobel, R. A., Steinman, L., and Zamvil, S. S. (2002). The HMG-CoA reductase inhibitor, atorvastatin, promotes a Th2 bias and reverses paralysis in central nervous system autoimmune disease. *Nature* **420,** 78–84.

Zamvil, S., Nelson, P., Trotter, J., Mitchell, D., Knobler, R., Fritz, R., and Steinman, L. (1985). T-cell clones specific for myelin basic protein induce chronic relapsing paralysis and demyelination. *Nature* **317,** 355–358.

Zhang, J., Weiner, H. L., and Hafler, D. A. (1992). Autoreactive T cells in multiple sclerosis. *Int. Rev. Immunol.* **9,** 183–201.

Zhang, J., Markovic-Plese, S., Lacet, B., Raus, J., Weiner, H. L., and Hafler, D. A. (1994). Increased frequency of interleukin 2-responsive T cells specific for myelin basic protein and proteolipid protein in peripheral blood and cerebrospinal fluid of patients with multiple sclerosis. *J. Exp. Med.* **179,** 973–984.

Zhang, J., Vandevyver, C., Stinissen, P., Mertens, N., van den Berg-Loonen, E., and Raus, J. (1995). Activation and clonal expansion of human myelin basic protein-reactive T cells by bacterial superantigens. *J. Autoimmun.* **8,** 615–632.

Zhang, J. Z., Rivera, V. M., Tejada-Simon, M. V., Yang, D., Hong, J., Li, S., Haykal, H., Killian, J., and Zang, Y. C. (2002). T cell vaccination in multiple sclerosis: Results of a preliminary study. *J. Neurol.* **249,** 212–218.

Zhu, C., Anderson, A. C., Schubart, A., Xiong, H., Imitola, J., Khoury, S. J., Zheng, X. X., Strom, T. B., and Kuchroo, V. K. (2005). The Tim-3 ligand galectin-9 negatively regulates T helper type 1 immunity. *Nat. Immunol.* **6,** 1245–1252.

Ziemssen, T., Kumpfel, T., Klinkert, W. E., Neuhaus, O., and Hohlfeld, R. (2002). Glatiramer acetate-specific T-helper 1- and 2-type cell lines produce BDNF: Implications for multiple sclerosis therapy. Brain-derived neurotrophic factor. *Brain* **125,** 2381–2391.

THE CD8 T CELL IN MULTIPLE SCLEROSIS: SUPPRESSOR CELL OR MEDIATOR OF NEUROPATHOLOGY?

Aaron J. Johnson,* Georgette L. Suidan,† Jeremiah McDole,† and Istvan Pirko*

*Department of Neurology, University of Cincinnati, Cincinnati, Ohio 45267, USA
†University of Cincinnati Neuroscience Program, Vontz Center for Molecular Studies
University of Cincinnati College of Medicine, Cincinnati, Ohio 45267, USA

Multiple sclerosis (MS) is the most common human demyelinating disease of the central nervous system. It is universally accepted that the immune system plays a major role in the pathogenesis of MS. For decades, CD4 T cells have been considered the predominant mediator of neuropathology in MS. This perception was largely due to the similarity between MS and CD4 T-cell-driven experimental allergic encephalomyelitis, the most commonly studied murine model of MS. Over the last decade, several new observations in MS research imply an emerging role for CD8 T cells in neuropathogenesis. In certain experimental autoimmune encephalomyelitis (EAE) models, CD8 T cells are considered suppressors of pathology, whereas in other EAE models, neuropathology can be exacerbated by adoptive transfer of CD8 T cells. Studies using the Theiler's murine encephalomyelitis virus (TMEV) model have demonstrated preservation of motor function and axonal integrity in animals deficient in CD8 T cells or their effector molecules. CD8 T cells have also been demonstrated to be important regulators of blood–brain barrier permeability. There is also an emerging role for CD8 T cells in human MS. Human genetic studies reveal an important role for HLA class I molecules in MS susceptibility. In addition, neuropathologic studies demonstrate that CD8 T cells are the most numerous inflammatory infiltrate in MS lesions at all stages of lesion

INTERNATIONAL REVIEW OF
NEUROBIOLOGY, VOL. 79
DOI: 10.1016/S0074-7742(07)79004-9

development. CD8 T cells are also capable of damaging neurons and axons *in vitro*. In this chapter, we discuss the neuropathologic, genetic, and experimental evidence for a critical role of CD8 T cells in the pathogenesis of MS and its most frequently studied animal models. We also highlight important new avenues for future research.

I. Introduction

Multiple sclerosis (MS) is the most common chronic demyelinating disease of the central nervous system (CNS) and is the leading cause of disability in young adults in the western world. Clinically, the course of MS most usually follows a relapsing-remitting pattern, with the majority of cases later converting to a progressive disease. Some cases are never associated with a relapsing-remitting course, but are progressive from onset (Kantarci and Weinshenker, 2005). MS is a complex disease, with environmental influences and genetics playing important roles in disease onset (Noseworthy *et al.*, 2000). MS is an inflammatory disease, where the main pathogenic role is attributed to the immune system. Using the broad definition of autoimmunity, in that disease is caused by an immune response that is harmful to the host, MS can be considered an autoimmune disease. However, unlike in classic autoimmune diseases, where the antigen is known and can be identified via blood tests, a strong antigen in MS has not been identified to date. Several weak myelin-related antigens are known, but none of them has been proven to be the driving force of the immune response. A strong antigen was identified as a serum marker for Devic's disease (Wingerchuk *et al.*, 2006) also known as neuromyelitis optica. The presence of NMO-IgG, an antibody specific for the aquaporin-4 water channel, provides a sensitive and specific blood test for Devic's disease (Lennon *et al.*, 2004). It is intriguing to believe that MS may also once become amenable to easy serum tests to support a more accurate diagnosis. However, unlike MS, Devic's disease shows many similarities to classic autoimmune conditions (Wingerchuk, 2004; Wingerchuk *et al.*, 1999). Devic's disease is known to be associated with "autoimmune overlap syndromes": several autoantibodies may be positive in Devic's cases, including antinuclear antibody (ANA), extractable nuclear antigens (ENA), and others. In MS, the coexistence of other autoimmune disease is considered rare. Devic's also shows a more prominent female predominance compared to MS. All these data suggest that MS may have a different immune-mediated mechanism than classic autoimmune diseases.

Presence of the macrophages and CD4 T cells within MS lesions have been known for decades (Markovic-Plese and McFarland, 2001; Weiner, 2004). However, newly emerging data demonstrates an important role for CD8 T cells in the disease process (Lassmann and Ransohoff, 2004). Carefully conducted neuropathologic studies clearly outline that CD8 T cells predominate in all lesions, regardless of stage or MS subtype. In some MS lesions, CD8 T cells outnumber CD4 T cells 10 to 1. In this chapter, we review the potential protective and pathogenic qualities of these CD8 T cells based on clinical observations and animal models, and we discuss strategies for the future to clearly define the role of CD8 T cells and potential therapeutic avenues.

II. Genetic Association Between Class I Genes and Predisposition to MS

Human class I alleles have been identified that confer both protection and susceptibility to MS. HLA-A*0201, the most common class I allele in the human population, confers protection against MS, reducing the risk of acquiring the disease by approximately 50% (Fogdell-Hahn *et al.*, 2000; Harbo *et al.*, 2004). Meanwhile, the less prevalent HLA*A301 class I allele doubles the risk of developing MS (Fogdell-Hahn *et al.*, 2000; Harbo *et al.*, 2004). The presence of these class I alleles work in combination with class II alleles that are known to be render susceptibility to MS. The class II allele HLA-DR2, which normally triples the risk of acquiring MS, works in synergy with the HLA*A201 and HLA*A301 alleles. Possessing both HLA-DR2 and HLA*A301 increases the risk of developing MS five- to sevenfold. In contrast, HLA-DR2 susceptibility is tempered by the presence of the protective HLA-A*0201 allele, reducing MS susceptibility 1.5-fold (Fogdell-Hahn *et al.*, 2000; Harbo *et al.*, 2004). These genetic studies clearly define important protective and pathogenic roles for class I alleles in the onset of MS. These studies also demonstrate the independence of the effects generated by class I alleles from those elicited by class II. On the basis of this genetic evidence of protection and susceptibility to MS elicited by class I alleles, determining mechanisms by which CD8 T cells provide protection or pathogenesis in MS is emerging as an important avenue of research.

III. Pathological Studies Implicate a Role for CD8 T Cells in MS Lesions

The strongest evidence for a role of CD8 T cells in neuropathology in MS hinges on the results of new pathological studies (Babbe *et al.*, 2000; Bruck and Stadelmann, 2003; Monteiro *et al.*, 1995; Skulina *et al.*, 2004). The presence and numeric dominance of CD8 T cells in MS lesions is a relatively new concept (Babbe *et al.*, 2000; Booss *et al.*, 1983). Unlike for CD4 T cells, antibodies that enabled immunohistochemistry for detection of CD8 molecule on formalin-fixed CNS tissue were not widely available. In addition, for decades it has been well established that CD4 T cells were critical mediators of neuropathology in experimental autoimmune encephalomyelitis (EAE), the most commonly studied mouse model of MS. This combination of events compounded by the relative scarcity of MS patient CNS tissue with active lesions resulted in reduced interest in CD8 T-cell biology in MS lesions.

Even though there is now evidence for clonal expansion of CD8 T cells (Babbe *et al.*, 2000; Monteiro *et al.*, 1995; Skulina *et al.*, 2004), the immunodominant epitopes recognized by these CD8 T cells are yet to be defined. However, some insight into the nature of the CD8 T-cell response was demonstrated by Babbe *et al.* (2000). Using single cell PCR, this study demonstrated that CD8 T cells were clonally expanded. This strongly suggests that these CD8 T cells had T-cell receptor specificity for discrete epitopes. This level of CD8 T-cell clonal expansion for immunodominant epitopes in the CNS is similar to that observed in mouse models (Butz and Bevan, 1998; Johnson *et al.*, 1999, 2001; Murali-Krishna *et al.*, 1998).

If we suspect that CD8 T cells may play a role as mediators of CNS injury, it must be determined whether class-I HLA molecules are expressed by CNS cells. It has now been universally accepted that all CNS cells, including oligodendrocytes, microglia, astrocytes, and neurons, express class I HLA molecules under inflammatory conditions (Hoftberger *et al.*, 2004; Neumann *et al.*, 1995). These cell types are therefore potential targets for CD8 T-cell-mediated cytotoxicity through effector molecules. Parallel studies *in vitro* have confirmed, in principle, that CD8 T cells can recognize and damage each of these cell types (Hoftberger *et al.*, 2004; Neumann *et al.*, 1995).

Microscopy studies also lend support to a pathogenic role of CD8 T cells in MS lesions. Bitsch *et al.* found a statistically significant correlation between the presence of CD8 T cells and axonal damage in MS lesions (Bitsch *et al.*, 2000; Kuhlmann *et al.*, 2002). The capacity of CD8 T cells to transect axons in culture has also been demonstrated (Medana *et al.*, 2001; Neumann *et al.*, 2002). Subsequent experiments demonstrated that Fas ligand was the effector molecule that promoted cytotoxic damage to the neurons (Neumann *et al.*, 1995, 2002). In addition, studies of CD8 T cells interacting with CNS tissue slices demonstrated a role for Fas ligand in cell-mediated cytotoxicity against neurons (Medana *et al.*, 2000). Interestingly, in both of these studies, perforin did not appear to have a major role in CD8 T-cell-mediated neuronal damage. It appears that CD8 T cells do not kill neurons through a perforin-dependent process, which is consistent with mouse models in which CD8 T cells utilize perforin to control HSV infection of sensory neurons without initiating cytotoxicity (Khanna *et al.*, 2003; Liu *et al.*, 2000). Neurons therefore appear to be relatively resistant to perforin-mediated cytotoxicity.

Determining the protective and pathogenic attributes of the CD8 T cells that infiltrate MS lesions is further complicated by the heterogeneity of the neuropathology of MS patients. On the basis of discrete differences in lesion pathology in tissue banks acquired from biopsies and autopsies, the early formation of MS lesions has now been subdivided into four separate patterns (Lucchinetti *et al.*, 2000, 2004):

Type I (macrophage mediated)—Perivenous distribution of lesions. Lesions are characterized by radial expansion and sharp margins. Inflammatory infiltrate is composed of T cells and macrophages. Activated macrophages and microglia are associated with degenerating myelin.
Type II (antibody mediated)—Lesions are similar to type I. In addition, antibody and activated complement are deposited in areas of active demyelination.
Type III (distal oligodendrogliopathy)—Inflammation is characterized by presence of T cells and macrophages. Lesions have small vessel vasculitis with endothelial cell damage and microvessel thrombosis. Lesions also have degeneration of distal oligodendrocyte processes, which appear to be followed by oligodendrocyte apoptosis and demyelination.
Type IV (primary oligodendrocyte damage with secondary demyelination)—Lesions are similar to type I. In addition, there is oligodendrocyte degeneration in a small rim of periplaque white matter.

CD4 and CD8 T cells are present in each of these MS lesion patterns (Lucchinetti *et al.*, 2000, 2004). However, it is clear that depending on lesion pattern, T-cell infiltration into MS lesions is accompanied by additional inflammatory mediators that ultimately result in different pathology. Determining the role of various immune cells including CD8 T cells in these patterns will be an important subject for future research.

IV. What Have We Learned About CD8 T Cells from Peripheral Blood of MS Patients?

It is becoming more established that CD8 T cells are present in MS lesions. However, with the current state of technology and poor availability of MS patients' CNS tissue with active lesions, it remains difficult to determine the epitopes recognized by CNS infiltrating CD8 T cells. In the meantime, there have been numerous reports of atypical CD8 T-cell activity in the peripheral blood of MS patients. Studies in peripheral blood may therefore provide some insight into CD8 T-cell activity in the CNS, although peripheral and CNS immune cells are known to be different.

There have been studies of cytokine expression among CD8 T cells in peripheral blood of MS patients during different phases of disease (Killestein *et al.*, 2001, 2003; Sepulcre *et al.*, 2005). Among these studies, Sepulcre *et al.* (2005) showed a statistically higher expression of interferon gamma (IFN-γ) by CD4 and CD8

T cells isolated from relapsing-remitting MS (RRMS) patients. This study demonstrated that there were more CD8 T cells that could be nonspecifically induced to express IFN-γ on PMA/ionomycin in MS patients than in healthy controls. Also in this study, the magnitude of IFN-γ expressing CD8 T cells correlated with disability of MS patients. A similar study performed by Killestein *et al.* (2001) also looked at the capacity of peripheral blood CD4 and CD8 T cells isolated from MS patients to express cytokine following culture with PMA/Ionomycin. This study determined that the presence of activated cytokine expressing CD8 T cells in the peripheral blood was predictive of changes in T1-weighted magnetic resonance imaging (MRI) lesion load (Killestein *et al.*, 2003). This study determined that activation of CD8 T cells in the peripheral blood correlated with increased clinical disease and with MRI markers of the disease.

To enhance studies of CD8 T cells in MS, defining the immunodominant antigens will be of paramount importance. Most studies focused predominantly on peptides that bind the HLA-A2 class I molecule due to its high frequency in the human population. Cytotoxic T-cell lines have also been generated from CD8 T cells isolated from MS patients. These CD8 lines reacted against myelin basic protein (MBP), proteolipid protein (PLP), myelin-associated glycoprotein (MAG) (Tsuchida *et al.*, 1994), and more recently, transaldolase peptide (Niland *et al.*, 2005). HLA-A2:$MBP_{110-118}$ peptide-specific CD8 T-cell lines were able to kill oligodendrocytes *in vitro* (Jurewicz *et al.*, 1998). Zhang *et al.* demonstrated that MBP transfected HLA-A2 expressing target cells could be killed by CD8 T-cell lines generated from MS patients (Zang *et al.*, 2004; Zhang *et al.*, 2001). In these studies, it was determined that CD8 T cells specific for CNS epitopes were easier to expand if they were acquired from MS patients rather than healthy controls. Mixtures of CNS-derived peptides have also been reported to be more readily recognized by CD8 T cells isolated from MS patients than normal healthy controls (Crawford *et al.*, 2004).

The above studies demonstrate that CD8 T cells can become activated when T-cell receptor engages CNS antigens presented to them by the HLA-A2 molecule. However, several questions remain regarding this work. The effect of cultured CD8 T cells toward the antigen(s) of interest may result in erroneous interpretation of data: even for brief periods of time, CD8 T-cell lines can be markedly skewed toward epitopes that were not initially recognized. This taken into account with the observation that the T-cell receptor is a lot more promiscuous than previously thought is also of concern (Huseby *et al.*, 2005; Tallquist *et al.*, 1996). Finally, it would be interesting to determine the role of CD8 T cells that are restricted to a human class I HLA molecule that is known to render susceptibility to MS, such as HLA-A3. It is also possible that the immunodominant epitope that CD8 T cells recognize in MS patients may not be restricted to the HLA-A2 class I and thus far, only minor epitopes have been observed. A study by Crawford *et al.* (2004) avoided these potential pitfalls. They studied the expansion of CD8 T cells toward a mixture of peptides presented by all HLA molecules. In this study, it was determined that RRMS patients had increased CD8 T-cell reactivity toward myelin oligodendrocyte-associated protein (Crawford *et al.*, 2004). However, these studies again showed broad range of CD8 T-cell reactivity toward several CNS antigens among different patients. This broad reactivity among CD8 T cells in the peripheral blood should be reconciled with the observation of epitope immunodominance observed among CD8 T cells in the CNS MS lesion. The observation by Babbe *et al.* (2000) demonstrated that CD8 T cells are clonally expanded in the CNS and are therefore highly specific for particular antigens. Nevertheless, continued work to define the class I epitopes in the CNS using peripheral blood could ultimately define the immunodominant peptides recognized by CD8 T cells. Knowledge of these specific peptides would ultimately enable therapeutic avenues to modify the suppressive and pathogenic attributes of CD8 T cells in human MS (Johnson *et al.*, 2001; Neville *et al.*, 2002). Epitope discovery will therefore continue to be an important avenue of MS research.

In summary, there is clear CD8 T-cell activity, both in MS lesion and in peripheral blood. These CD8 T cells are clonally expanded and colocalize with areas of axonal damage. The significance of these results is yet to be defined. Meanwhile, a model can be put forward describing the potential routes CD8 Tcells can promote axonal damage (Fig. 1). Expression of class I molecules has been observed in all major cell types in the CNS. Therefore, all CNS cell types are candidate targets for neuropathology by CD8 T cells. A direct mechanism of CD8 T-cell-mediated axonal damage involves direct attack of the neuron or axon by CD8 T cells (Fig. 1A).

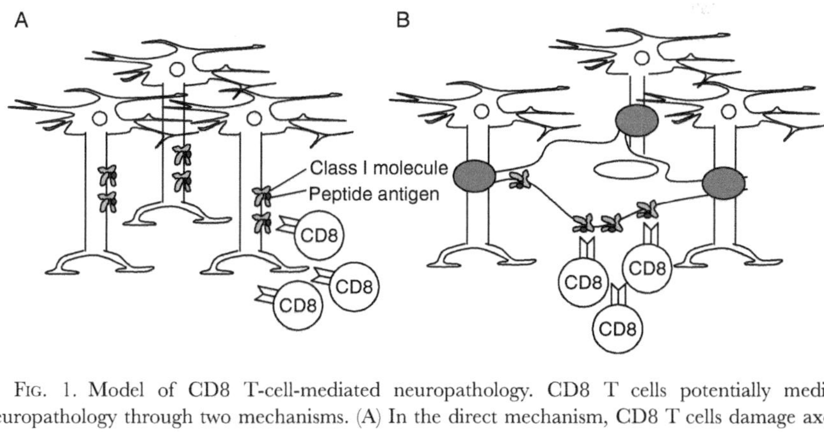

FIG. 1. Model of CD8 T-cell-mediated neuropathology. CD8 T cells potentially mediate neuropathology through two mechanisms. (A) In the direct mechanism, CD8 T cells damage axons through direct recognition of class I peptide epitopes on the surface of axons. (B) In the indirect mechanism, CD8 T cells potentially damage axons indirectly by killing glial cells that present the class I peptide epitope, leaving axons exposed to other forms of inflammation. Shown is one example of an indirect mechanism where an oligodendrocyte presenting class I peptide antigen to CD8 T cells is targeted.

An indirect mechanism entails CD8 T cells damaging cell types that surround the neuron that provides a nurturing environment, including oligodendrocytes and astrocytes (Fig. 1B). Animal models will continue to be important for defining the capacity of CD8 T cells to damage axons through the direct or indirect mechanism.

V. CD8 T Cells as Suppressors of Neuropathology in MS

A subset of CD4 T cells has received considerable attention as a major regulator of the inflammatory immune response (Baecher-Allan and Hafler, 2006). These regulatory T cells (Tregs) have been demonstrated to reduce immune responses in both infectious and autoimmune models (Belkaid and Rouse, 2005). Less characterized is the role of regulatory NK cells and suppressor CD8 T cells in autoimmune disease (French and Yokoyama, 2004; Jiang *et al.*, 1992; Koh *et al.*, 1992; Smeltz *et al.*, 1999). The role of suppressor CD8 T cells in reducing neuropathology has been observed in the EAE model. In 1992, it was determined that CD8 T cells protect against EAE relapses (Jiang *et al.*, 1992). This initial observation was later confirmed through adoptive transfer of CD8 T cells isolated from EAE recovered mice into MBP-immunized recipients. Mice reconstituted with EAE-experienced CD8 T cells were resistant to MBP vaccination-induced EAE (Jiang *et al.*, 1992). These early experiments demonstrated the capacity of CD8 T cells to inhibit the MBP-specific CD4 T-cell response.

Some insight into the mechanism(s) by which suppressor CD8 T cells inhibit immune responses was revealed with the discovery that this cell type is restricted to the Qa-1 nonclassical class I molecule (Jiang *et al.*, 1992). Qa-1 presents endogenous peptides, including signal sequences of other class I molecules, in its groove on the cell surface (Jiang *et al.*, 1992). The working model is that suppressor CD8 T cells recognize this signal peptide (or another peptide) presented on the surface of CD4 T cells (Madakamutil *et al.*, 2003). The CD8 T cells then eliminate the EAE-inducing CD4 T-cell subtype. Support for this model has been reported in studies involving Qa-1-deficient mice. Induction of EAE in Qa-1-deficient mice results in exaggerated secondary but not primary CD4 T-cell responses to foreign and self-peptides. In this process, EAE was exacerbated in Qa-1-deficient animals demonstrating that the suppressor CD8 T cells were necessary to inhibit the autoimmune CD4 T-cell responses in this model (Hu *et al.*, 2004).

Another report of CD8 T-cell-mediated suppression of the EAE has been reported by Najafian *et al.* (2003). In this system, adoptive transfer of CD8 T cells from CD28-deficient mice into CD8-deficient mice conferred these animals resistant to EAE. *In vitro* analysis of this effect determined that CD4 T-cell suppression was dependent on cell to cell contact and the presence of antigen presenting cells

(APCs) (Najafian *et al.*, 2003). The role of APCs in inducing protection from exacerbation of EAE has also been shown to involve CD40L and TGF-β (Faunce *et al.*, 2004; Gilliet *et al.*, 2003). Other mechanisms of CD8 T-cell suppression are mediated through cytokine responses (Filaci *et al.*, 2001; Gilliet and Liu, 2002).

In MS, the role of CD8 T-cell suppressors is more difficult to define. Some insight has been obtained through the study of MS patients that have been prescribed glatiramer acetate, a drug that has been approved for RRMS patients. This drug consists of a degenerative peptide library and is designed to promote anergy in CD4 T-cell responses (Dhib-Jalbut, 2003). However, it has now been shown that CD8 T-cell responses are also significantly reduced following a regimen with this drug, demonstrating that part of the treatment affect may be alteration of CD8 T-cell responses (Biegler *et al.*, 2006; Karandikar *et al.*, 2002). The beneficial effect of glatiramer acetate has been found to correlate with the presence of a subset of CD8 T cells in MS patients. This CD8 T-cell response is capable of killing CD4 T cells *ex vivo* in a glatiramer acetate-dependent manner. This demonstrated that suppression by these CD8 T cells may be mediated through cytolytic processes *in vivo*. The role of HLA-E, the Qa-1 human homologue, was not determined in these experiments.

We are therefore left with two animal model systems describing the potential role of suppressor CD8 T cells in MS. Again, it appears that CD8 T cells could have two potential mechanisms by which they promote suppression of inflammation (Fig. 2).

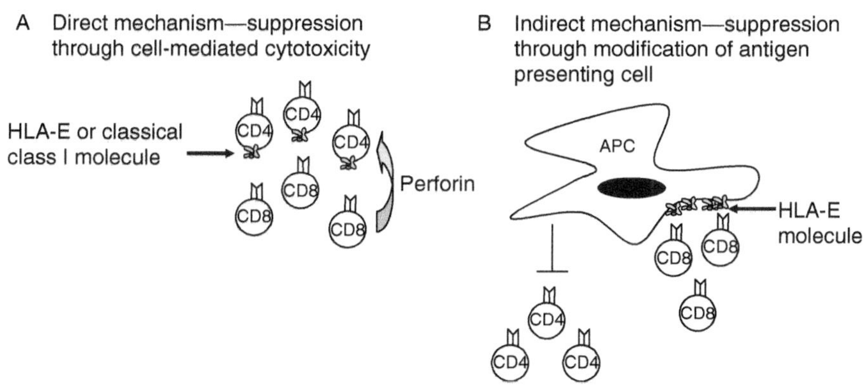

FIG. 2. Model of CD8 T-cell-mediated suppression of MS. CD8 T cells potentially mediate suppression through two mechanisms. (A) In the direct mechanism, CD8 T cells mediate suppression through direct recognition of class I molecules on CD4 T cells. This results in cytotoxicity of the CD4 T cell. (B) In the indirect mechanism, CD8 T cells mediate suppression through interaction with APCs, triggering a licensing effect that mediates CD4 T-cell unresponsiveness.

The above studies involving Qa-1 and the clinical studies on glatiramer acetate both suggest a direct killing of CD4 T cells by CD8 T cells in MS (Fig. 2A). An alternative pathway of CD4 T-cell suppression involves CD8 T-cell interaction with APCs (Fig. 2B). Through this mechanism, the CD8 T cell appears to license the APC to inhibit CD4 T cells, which in turn dampens the inflammatory response.

In summary, CD8 T cells have biological significance in both the MS lesion and peripheral blood. Their role in MS, however, remains unclear, given the obvious inability to perform mechanistic work in MS patients beyond clinical trials. Animal models will therefore continue to be integral in efforts to define mechanisms by which CD8 T cells promote protective or neuropathologic effects in MS.

VI. CD8 T Cells as Mediators of Neuropathology and Motor Deficit in Animal Models of MS

The two major animal models of MS are EAE and Theiler's murine encephalomyelitis virus (TMEV) infection. EAE is generally induced through injection of spinal cord homogenate complete with adjuvant into a mouse. The T cells generated through this vaccination induce neuropathology and paralysis in the vaccinated animal (Steinman and Zamvil, 2005, 2006; Zamvil and Steinman, 1990). Alternatively, through adoptive transfer, T cells from these spinal cord-vaccinated animals can promote disease in a recipient mouse. In contrast, TMEV infection in susceptible strains of mice results in chronic infection of the CNS, inflammation, demyelination, axonal dropout, and motor dysfunction (Nelson *et al.*, 2004; Oleszak *et al.*, 2004). Disease in both of these models has been shown to be dependent on both CD4 and CD8 T-cell responses.

The vast majority of published EAE studies have favored a CD4-mediated mechanism of disease. However, there are now several EAE models in which neuropathology is mediated by CD8 T cells. EAE can be induced by adoptively transferring CD8 T cells from an animal vaccinated with CNS protein into a naive recipient animal. CD8 T-cell epitopes in this model include MBP, MOG, and an assortment of other proteins endogenously expressed by oligodendrocytes (Crawford *et al.*, 2004; Tsuchida *et al.*, 1994). TMEV induces persistent viral infection of oligodendrocytes and glial cells in susceptible strains. Both models are characterized by demyelination and axonal loss accompanied by clinical dysfunction resembling symptoms of MS. Mice with genetic disruption of alleles critical for CD8 T-cell function have been useful to determine the role CD8 T cells in these models. Among the mice used in early studies of CD8 T-cell-mediated pathology in MS models were the β-2-microglobulin (β_2m)-deficient mouse, recombinant activation gene (RAG)-deficient mouse, and the perforin-deficient mouse. These mice exhibit deficiencies in class I molecule assembly, T-cell and B-cell development, and CD8 T-cell effector killing, respectively.

Huseby *et al.* (2001) were the first to develop an EAE model of MS-like demyelination using MBP-specific CD8 T cells. In this model, C3H wild-type, C3H and MBP-deficient C3H shiver mice were immunized with an MBP 79–87 peptide. T cells from the draining lymph nodes were cloned. Six independent MBP-specific CD8 T-cell lines were obtained from the wild-type mice and 10 from the C3H mice. All cell lines were fully capable of killing target cells either expressing MBP or pulsed with MBP 79–87 peptide. Activated CD8 T-cell lines isolated from the wild-type C3H mice were then adoptively transferred into either C3H wild-type or C3H severe combined immunodeficiency (SCID) recipient mice. Both mouse types incurred severe neurological disease including ataxia, spasticity, brisk reflexes, and loss of coordinated movements. CNS damage was consistent with focal cytotoxic insults further implicating CD8 T cells as the perpetuators of the observed destruction.

In another murine model, the adoptive transfer of CD8 T cells specific for MOG peptide 35–55 was able to induce EAE in susceptible mice (Sun *et al.*, 2001). C57BL/6 mice were immunized with MOG 35–55 and their CD8 T cells were harvested. Primed MOG-specific CD8 T cells were adoptively transferred into syngeneic C57BL/6, RAG-deficient, and β_2m-deficient mice. The CD8 T cells induced severe EAE in C57BL/6- and RAG-deficient mice but not in the β_2m-deficient mice. Furthermore, MOG-specific CD8 T cells could be isolated for up to 287 days in the affected mice. Subsequent work by Ford *et al.* has shown that CD8 T cells mediate damage in this EAE model (Ford and Evavold, 2005). C57BL/6 mice were immunized with MOG 35–55 and their CD8 T cells were harvested. Enzyme-linked immunosorbent assay (ELISA) analysis of IFN-γ production by CD8 T cells confirmed their specificity for MOG 35–55. Primed CD8 T cells were transferred into both SCID recipients and naive wild-type C57BL/6 mice. Both strains developed comparable levels of demyelination and motor dysfunction consistent with symptoms of EAE.

In addition to EAE, the role of CD8 T cells in exacerbating CNS disease in the TMEV model of MS has been well established. A study by Rivera-Quinones *et al.* (1998) was the first to implicate the role of CD8 T cells in neurological dysfunction in mice infected with TMEV. This study utilized C57BL/6 and 129/J β_2m-deficient mice for infection. Control mice were of the SJL/J strain, which are susceptible to chronic TMEV infection and display motor abnormalities.

Interestingly, both strains of mice displayed comparable distribution and severity of demyelinating lesions. However, when spontaneous motor activity and hindlimb motor-evoked potentials were measured, the 129/J β_2m-deficient mice were functionally normal. The authors hypothesized that deficiency in CD8 T cells resulted in preservation of axons and preserved motor function. The concept of CD8 T-cell-mediated damage to axons was also demonstrated by Ure and Rodriguez. They compared two susceptible mouse strains with TMEV infection (Ure and Rodriguez, 2002). Both SJL/J- and β_2m-deficient strains of mice develop chronic inflammation and demyelination on intracranial injection of TMEV. However, SJL/J mice develop pronounced motor dysfunction, while β_2m-deficient mice, a strain deficient in CD8 T cells, incur limited motor deficits. On examination, infected mice of each strain shared similar lesion load and expansion, remyelination, and lesion distributions. However, the β_2m-deficient mice demonstrated a highly significant increase in preservation of neuronal and axonal integrity over SJL/J mice. To determine if perforin, one of the effector molecules of CD8 T cells is involved in CNS pathology, Murray analyzed motor ability and demyelination in TMEV infected C57BL/6 perforin-deficient mice (Murray *et al.*, 1998). This study determined that mice deficient in perforin-mediated MHC class I-restricted cytotoxicity incurred demyelination comparable with controls. However, similar to β_2m-deficient mice, C57BL/6 perforin-deficient mice displayed limited motor deficits. Axonal integrity in C57BL/6 perforin-deficient mice was not assessed in this study. Therefore, the question of whether perforin can contribute to axonal damage remains an interesting, yet unexplored avenue of research.

Johnson *et al.* (2001) have developed a murine model for MS that allows direct assessment of how virus-specific CD8 T cells contribute to neurological deficit (Johnson *et al.*, 2001). This model employs the use of chronic infection of IFN-γR-deficient mice with TMEV. Both IFN-γR-deficient and C57BL/6 mouse strains mount a heightened class I-restricted CD8 T-cell response toward the TMEV VP$2_{121-130}$ peptide presented in the context of the D^b class I molecule. However, while C57BL/6 mice clear the TMEV infection by 28 days postinfection, IFN-γR-deficient mice rapidly develop severe paralysis within weeks following infection. This condition is usually fatal within 7–8 weeks. In order to explore the role that the D^b:VP$2_{121-130}$ epitope-specific CD8 T cells play in the motor deficits incurred by the IFN-γR-deficient mice, VP$2_{121-130}$ peptide was injected to these mice 1 day before TMEV infection. By doing so, CD8 T cells specific for this peptide are effectively eliminated from the lymphocyte population. The motor functioning of mice was analyzed 45 days post-TMEV infection using the rotarod assay. As a result of D^b:VP$2_{121-130}$-specific CD8 T-cell elimination, IFN-γR-deficient mice demonstrated significantly preserved motor functioning over their mock-treated counterparts (Fig. 3).

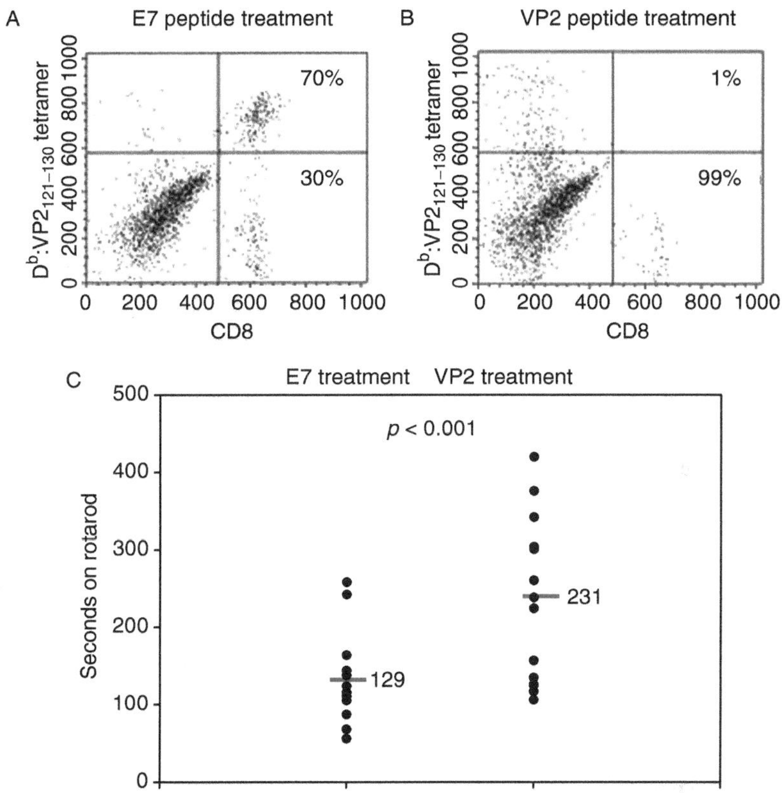

FIG. 3. CD8 T cells contribute to motor dysfunction in TMEV-infected IFN-γR-deficient mice. (A) Intravenous administration of mock control E7 peptide does not inhibit the expansion of D^b:VP2$_{121-130}$ epitope-specific CD8 T cells. (B) Pretreatment and weekly treatment with VP2$_{121-130}$ peptide inhibits the expansion D^b:VP2$_{121-130}$ epitope-specific CD8 T cells. (C) Five week TMEV-infected IFN-γR-deficient mice receiving VP2$_{121-130}$ peptide treatment to inhibit the expansion D^b:VP2$_{121-130}$ epitope-specific CD8 T cells have preservation of motor function.

VII. CD8 T Cells as Potential Regulators of the Blood–Brain Barrier

Inflammation of the CNS has been shown to promote opening of the blood–brain barrier (BBB) through partially understood mechanisms. Unregulated vascular permeability of the BBB is also a feature of many other diseases, including viral hemorrhagic fevers, HIV dementia, shock, and cerebral malaria (Green *et al.*, 2004; Kirk *et al.*, 2003; Medana and Turner, 2006; Minagar and Alexander, 2003; Shacklett *et al.*, 2004; Solomon *et al.*, 2000). There are reports that inflammatory mediators released from CNS infiltrating lymphocytes contribute to opening of the BBB, but very little is known about the mechanisms by which this occurs (Blamire *et al.*, 2000; Chavarria and Alcocer-Varela, 2004; Minagar and Alexander, 2003; Petereit *et al.*, 2003; Stoll *et al.*, 2000).

The BBB has several unique features that enable it to create a solute impermeable barrier (Begley, 2004). On the basis of electron microscopy data, the current model of the BBB cerebral endothelial cells (CECs) form tight junctions with one another, creating a seal between the blood and CNS tissue (Fanning *et al.*, 1999; Itoh *et al.*, 1999; Tsukita *et al.*, 1999; Wolburg and Lippoldt, 2002). Greater than 90% of the abluminal surface of these CECs are in direct contact with astrocytes (Abbott, 2002; Demeuse *et al.*, 2002). Other cell types include microglia, pericytes, and neurons (Begley, 2004). Among these cell types, astrocytes have been shown *in vitro* to promote the opening and maintenance of CEC tight junctions on stimulation (Abbott, 2002; Demeuse *et al.*, 2002). For this reason, it is believed that astrocytes play an important role in maintaining cerebral endothelial cell integrity through direct contact and chemical messengers *in vivo* (Abbott, 2005; Haseloff *et al.*, 2005).

While astrocytes appear to control entry of proteins and cells into the BBB, tight junctions between CECs serve as the actual gate (Fig. 4). Tight junctions are composed of many cytoplasmic and transmembrane proteins. These proteins are linked to an actin-based cytoskeleton allowing for a tight seal (Petty and Lo, 2002). Cytoplasmic proteins collectively called membrane-associated guanylate kinase-like homologue family provide structural support and play an organizational role for CECs (Hawkins and Davis, 2005). Also heavily researched are the transmembrane proteins occludin, claudin, and junctional adhesion molecules (JAMs) shown in Fig. 4. Occludin and claudin appear to form the primary seal of the tight junction, whereas JAMs seem to be involved in monocyte and leukocyte adhesion and transmigration through the BBB (Ballabh *et al.*, 2004; Hawkins and Davis, 2005). The zona-occludin proteins, ZO-1, ZO-2, and ZO-3, have been shown to connect the transmembrane proteins to actin, providing stability for tight junction formation (Kubota *et al.*, 1999). Expression of occludin has been found to be much higher in neural endothelial cells when compared to peripheral endothelial cells, whereas claudins are found in both (Ballabh *et al.*, 2004). Occludin could therefore contribute to structural differences in CNS tight junctions as compared to tight junctions found in peripheral tissue.

Levels of expression of occludin and claudin in rats have been studied during chronic inflammation induced by complete Freund's adjuvant (CFA) leading to BBB permeability. These proteins are far from static under inflammatory conditions. Injection of CFA results in a decrease of occludin expression by 60% while claudin expression increased by over 200% (Brooks *et al.*, 2005). Efforts to link inflammatory mediators to expression of tight junction proteins have revealed that cytokines may play an important role in tight junction stability. *In vitro* studies

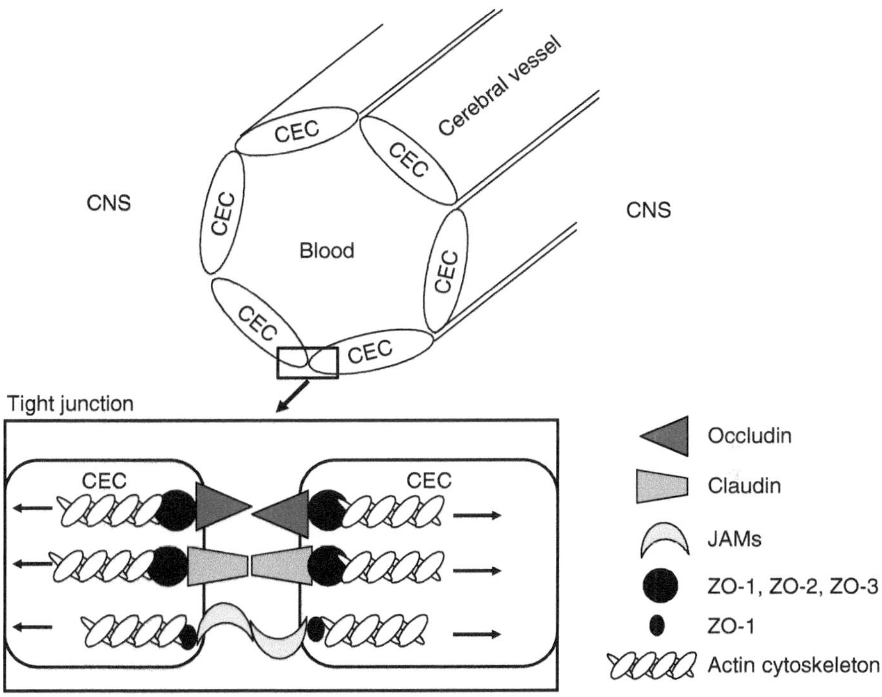

FIG. 4. Molecular model of tight junctions between CECs. Occludins, claudins, and JAMs regulate adhesion of CECs. Occludins and claudins adhere to the zona-occludin complex, which consists of an interaction of ZO-1, ZO-2, and ZO-3 proteins. In contrast, JAMs adhere only to zona-occludin-1 (ZO-1). The zona-occludin proteins in turn adhere to actin cytoskeleton proteins. Opening of tight junctions during inflammation appears to be an orchestrated event of degradation and expression of occludins and claudins, respectively, and contraction of actin cytoskeleton.

using cultured human brain microvessel endothelial cells have demonstrated that TNF-α, IL-1β, IFN-γ, or lipopolysaccharide increase permeability of these cells to horse radish peroxidase, an indicator enzyme, that would not pass through the BBB under normal conditions (Wong *et al.*, 2004). IFN-γ, a cytokine secreted by activated lymphocytes, has also been shown *in vitro* to decrease expression of occludin (Oshima *et al.*, 2001), thus bridging the idea that activated lymphocytes may play a role in tight junction abnormality which may lead to BBB permeability.

There is increasing evidence that CD8 T cells provide an important link between inflammatory disease and vascular permeability. This concept has been supported by numerous studies in viral hemorrhagic fevers. Expansion and global activation among CD8 T cells have been reported in patients with dengue hemorrhagic fever (DHF) (Kurane *et al.*, 1991; Loke *et al.*, 2001; Mongkolsapaya *et al.*, 2003;

Stephens *et al.*, 2002; Zivna *et al.*, 2002). This expansion is followed by large-scale apoptosis among dengue-specific CD8 T cells (Mongkolsapaya *et al.*, 2003). Epidemiological studies have also supported CD8 T-cell involvement in the pathogenesis of DHF. Expression of specific classical class I HLA-A and HLA-B molecules correlates with a positive or negative clinical outcome following dengue infection. For example, ethnic Thais that express HLA-A*0203 are less susceptible to DHF following secondary infection with a different dengue strain than individuals that express HLA-A*0207 (Stephens *et al.*, 2002). A correlation between HLA type and susceptibility to hemorrhagic fever has also been observed in cases of Hantavirus pulmonary syndrome in southwestern regions of the United States (Kilpatrick *et al.*, 2004; Zivna *et al.*, 2002).

A new mouse model of CD8 T-cell-mediated vascular permeability in the CNS using the TMEV model of MS was reported (Johnson *et al.*, 2005). At 7 days post-TMEV infection, there is massive expansion of TMEV-specific CD8 T cells in the CNS. At this time point, 70% of CD8 T cells found in the CNS are specific for the TMEV peptide, $VP2_{121-130}$, presented in the context of the D^b class I molecule. The condition is induced by intravenously injecting this immunodominant virus peptide $VP2_{121-130}$ at 7 days postinfection coinciding with the expansion of D^b:$VP2_{121-130}$ epitope-specific CD8 T cells (Fig. 5). Following induction, glial cells become activated in the CNS as demonstrated by upregulation of glial fibrillary acidic protein (GFAP) by astrocytes and F4/80 by microglia. Phenotypically, C57BL/6 mice experience signs of severe neurological disorder, including rapidly progressive paralysis and ataxia within 24 h of injection with peptide. MRI and histology show extensive vascular leakage throughout both hemispheres leading to microhemorrhages supporting the hypothesis that vascular permeability of the BBB may be mediated by CD8 T cells. The observed pathology is specific to the CNS, suggesting that vascular permeability is limited to the organ that harbors the CD8 T cells that incite this condition (Johnson *et al.*, 2005).

We have developed a working model of CD8 T-cell-mediated vascular permeability (Fig. 6), in which CD8 T cells enter the CNS through tight junctions or by transendothelial migration. CNS-infiltrating CD8 T cells become hyperstimulated, potentially through the overabundance of available antigen expressed by class I molecules. Following this hyperactivation, CD8 T cells promote CNS vascular permeability though direct or indirect activation of glial cell astrocytes. Determining the inflammatory mediators expressed by CD8 T cells that contribute to CNS vascular permeability remains an important area of research. We have determined that there are strain-specific differences in susceptibility to this condition. C57BL/6 mice are very susceptible, whereas the 129/Sv strain is not. This occurs despite the CD8 T-cell response being very similar between these two strains of mice (Johnson *et al.*, 2005). If genetic variation accounts for the differences

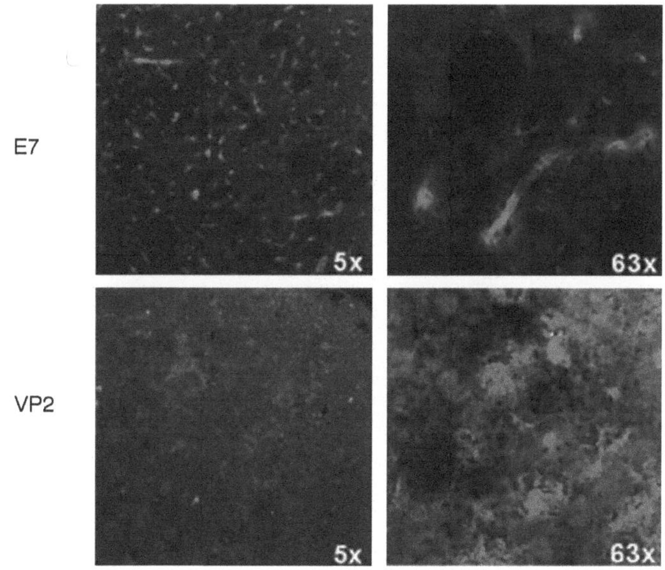

FIG. 5. CD8 T cells as mediators of vascular permeability. Confocal microscopic analysis of CNS vascular leak in the striatum of mice following activation of CD8 T cells with $VP2_{121-130}$ peptide. Mock E7 peptide-treated animals have an absence of vascular permeability (top panels). C57BL/6 mice intravenously administered $VP2_{121-130}$ peptide during the peak of the CNS infiltrating D^b:$VP2_{121-130}$ epitope-specific CD8 T-cell response have extensive leakage of FITC-albumin into CNS tissue (bottom panels). In all sections, FITC-albumin fluoresces green whereas astrocytes expressing GFAP fluoresce red. (See Color Insert.)

observed between these two strains of mice, this model could be useful for mapping the location of powerful genes that greatly influence susceptibility to CD8 T-cell-mediated CNS vascular permeability.

VIII. Future Directions: Define the CD8 T-Cell Epitopes and Exploit Them Therapeutically

Animal models have clearly demonstrated the necessity of T-cell receptor specificity in T-cell homing (Calzascia *et al.*, 2005; Savinov *et al.*, 2003). In order to home to an inflamed organ, T cells need to engage antigen, and the CNS is not any different in this regard (Karman *et al.*, 2004). It is highly unlikely that CD8 T cells present in the CNS lesions of MS patients are simple bystanders not recognizing antigen. The observation that these CD8 T cells are clonally expanded also demonstrates specificity toward a common yet undefined antigen. Determination of this antigen is of critical importance as it will enable specific targeting of epitope-specific CD8 T cells through immunotherapy. There are now

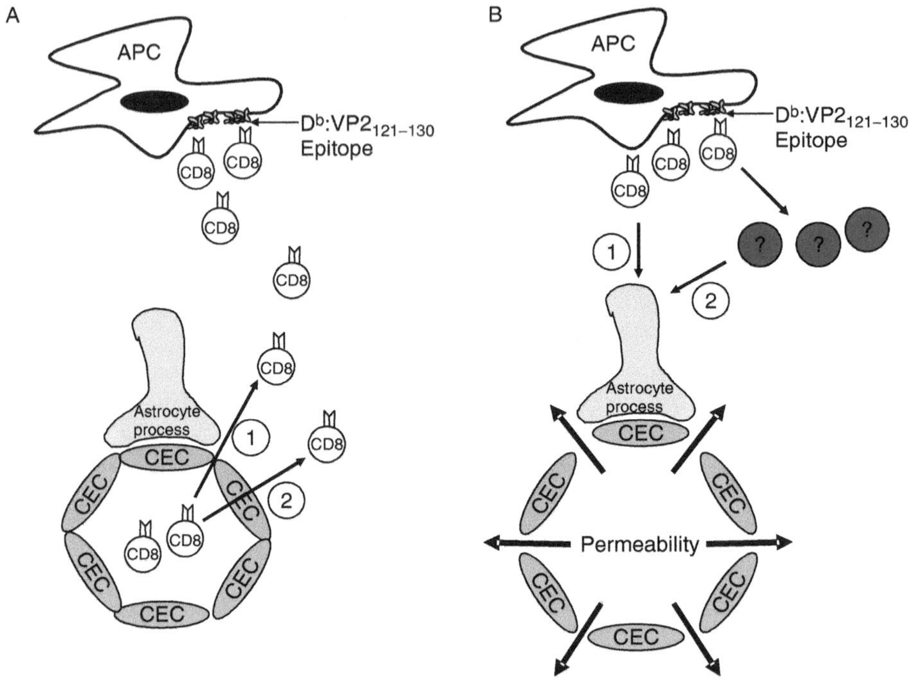

FIG. 6. (A) Migration of CD8 T cells during TMEV infection. D^b:VP$2_{121-130}$-specific CD8 T cells enter the CNS by (1) crossing tight junctions or (2) transendothelial migration to access the VP$2_{121-130}$ peptide that is presented by the APC. (B) CD8 T-cell-mediated CNS vascular permeability following induction of PIFS in C57BL/6 mice. PIFS is induced by intravenous injection of VP$2_{121-130}$ peptide during the peak of the D^b:VP$2_{121-130}$-specific CD8 T-cell response in the CNS. Induction of PIFS in C57BL/6 mice results in heightened activation of expanded CNS infiltrating D^b:VP$2_{121-130}$-specific CD8 T cells interacting with class I expressing APCs. D^b:VP$2_{121-130}$-specific CD8 T cells then (1) directly activate glial cells that line the BBB through undefined inflammatory mediators. Alternatively, (2) D^b:VP$2_{121-130}$-specific CD8 T cells activate glial cells that line the BBB indirectly through inflammatory mediators expressed by blood-derived cells.

numerous reports of antigen-specific inhibition of CD8 T cells (Johnson *et al.*, 2001; Neville *et al.*, 2002).

If we can prove that CD8 T cells contribute to the neuropathologic processes that lead to disability in MS patients, removal of epitope-specific CD8 T cells would deplete the inflammatory component that mediates CNS pathology while retaining protective immunity. This would be a clear advantage over current therapeutic approaches that target all immune cells. Alternatively, should it be determined that CD8 T cells in MS are protective through promoting suppression of immune-mediated pathology, subsequent therapeutic designs could be designed to expand these suppressive effects.

The approach to define CD8 T-cell epitopes will be no small task. However, unlike their CD4 T-cell counterparts, it appears that CD8 T cells engage a far smaller pool of immunodominant peptides during inflammatory conditions (Babbe *et al.*, 2000; Butz and Bevan, 1998; Johnson *et al.*, 1999; Murali-Krishna *et al.*, 1998). In some cases, one immunodominant peptide appears to be recognized by the majority of the CD8 T-cell response (Johnson *et al.*, 1999; Murali-Krishna *et al.*, 1998). Such a situation bodes well for identifying a CD8 T-cell epitope in MS. In the NOD mouse model of diabetes, an immunodominant autoantigen recognized by CD8 T cells was determined using a molecular biology approach utilizing an mRNA tissue library and an immortalized T-cell hybridoma line (Wong *et al.*, 1999). It is probable that a similar molecular approach will be necessary to determine immunodominant CD8 T-cell epitopes among an overwhelming pool of potential target peptides in human MS. We may need to be ready for surprising results. T-cell receptor specificity has been shown to be far more promiscuous than expected (Huseby *et al.*, 2005; Tallquist *et al.*, 1996). In addition, suppressive CD8 T cells have been demonstrated to recognize nonclassical class I HLA molecules. The eventual CD8 T-cell epitope may also prove to be pathogen derived (Giovannoni *et al.*, 2006; Kurtzke and Heltberg, 2001; Moses and Sriram, 2001; Theil *et al.*, 2001). High-throughput systems designed to identify CD8 T-cell epitopes in MS will therefore have to entertain all such possibilities.

IX. Conclusions

An emerging role for CD8 T cells in MS has clearly been outlined. New microscopy data demonstrates that CD8 T cells are a prevalent cell type in MS lesions. How these CD8 T cells contribute to MS is not yet known. Studies using tissue culture and animal models demonstrate both suppressive and pathogenic roles for the CD8 T cell. In particular, CD8 T cells have been demonstrated to have the capacity to transect axons, kill CNS cell types, and promote vascular permeability. All of these forms of pathology are observed in MS lesions.

The difficulty of defining the role of CD8 T cells in MS will be further complicated by the rising acceptance of heterogeneity of disease. MS lesion formation is now classified into four different subtypes. These different types demonstrate additional roles for antibody, complement and CNS susceptibility to inflammation as additional factors in the onset of pathology. For this reason, it will be difficult to dissect any one inflammatory factor as the sole mediator of MS. Nevertheless, a comprehensive effort to define the epitopes utilized by CD8 T cells will be vital to exploit therapeutic avenues designed to modify their role in MS.

References

Abbott, N. J. (2002). Astrocyte-endothelial interactions and blood-brain barrier permeability. *J. Anat.* **200,** 629–638.

Abbott, N. J. (2005). Dynamics of CNS barriers: Evolution, differentiation, and modulation. *Cell. Mol. Neurobiol.* **25,** 5–23.

Babbe, H., Roers, A., Waisman, A., Lassmann, H., Goebels, N., Hohlfeld, R., Friese, M., Schroder, R., Deckert, M., Schmidt, S., Ravid, R., and Rajewsky, K. (2000). Clonal expansions of CD8(+) T cells dominate the T cell infiltrate in active multiple sclerosis lesions as shown by micromanipulation and single cell polymerase chain reaction. *J. Exp. Med.* **192,** 393–404.

Baecher-Allan, C., and Hafler, D. A. (2006). Human regulatory T cells and their role in autoimmune disease. *Immunol. Rev.* **212,** 203–216.

Ballabh, P., Braun, A., and Nedergaard, M. (2004). The blood-brain barrier: An overview: Structure, regulation, and clinical implications. *Neurobiol. Dis.* **16,** 1–13.

Begley, D. J. (2004). Delivery of therapeutic agents to the central nervous system: The problems and the possibilities. *Pharmacol. Ther.* **104,** 29–45.

Belkaid, Y., and Rouse, B. T. (2005). Natural regulatory T cells in infectious disease. *Nat. Immunol.* **6,** 353–360.

Biegler, B. W., Yan, S. X., Ortega, S. B., Tennakoon, D. K., Racke, M. K., and Karandikar, N. J. (2006). Glatiramer acetate (GA) therapy induces a focused, oligoclonal CD8+ T-cell repertoire in multiple sclerosis. *J. Neuroimmunol.* **180**(1–2), 159–171.

Bitsch, A., Schuchardt, J., Bunkowski, S., Kuhlmann, T., and Bruck, W. (2000). Acute axonal injury in multiple sclerosis. Correlation with demyelination and inflammation. *Brain* **123**(Pt. 6), 1174–1183.

Blamire, A. M., Anthony, D. C., Rajagopalan, B., Sibson, N. R., Perry, V. H., and Styles, P. (2000). Interleukin-1beta-induced changes in blood-brain barrier permeability, apparent diffusion coefficient, and cerebral blood volume in the rat brain: A magnetic resonance study. *J. Neurosci.* **20,** 8153–8159.

Booss, J., Esiri, M. M., Tourtellotte, W. W., and Mason, D. Y. (1983). Immunohistological analysis of T lymphocyte subsets in the central nervous system in chronic progressive multiple sclerosis. *J. Neurol. Sci.* **62,** 219–232.

Brooks, T. A., Hawkins, B. T., Huber, J. D., Egleton, R. D., and Davis, T. P. (2005). Chronic inflammatory pain leads to increased blood-brain barrier permeability and tight junction protein alterations. *Am. J. Physiol. Heart Circ. Physiol.* **289,** H738–H743.

Bruck, W., and Stadelmann, C. (2003). Inflammation and degeneration in multiple sclerosis. *Neurol. Sci.* **24**(Suppl. 5), S265–S267.

Butz, E. A., and Bevan, M. J. (1998). Massive expansion of antigen-specific CD8+ T cells during an acute virus infection. *Immunity* **8,** 167–175.

Calzascia, T., Masson, F., Di Berardino-Besson, W., Contassot, E., Wilmotte, R., Aurrand-Lions, M., Ruegg, C., Dietrich, P. Y., and Walker, P. R. (2005). Homing phenotypes of tumor-specific CD8 T cells are predetermined at the tumor site by crosspresenting APCs. *Immunity* **22,** 175–184.

Chavarria, A., and Alcocer-Varela, J. (2004). Is damage in central nervous system due to inflammation? *Autoimmun. Rev.* **3,** 251–260.

Crawford, M. P., Yan, S. X., Ortega, S. B., Mehta, R. S., Hewitt, R. E., Price, D. A., Stastny, P., Douek, D. C., Koup, R. A., Racke, M. K., and Karandikar, N. J. (2004). High prevalence of autoreactive, neuroantigen-specific CD8+ T cells in multiple sclerosis revealed by novel flow cytometric assay. *Blood* **103,** 4222–4231.

Demeuse, P., Kerkhofs, A., Struys-Ponsar, C., Knoops, B., Remacle, C., and van den Bosch de Aguilar, P. (2002). Compartmentalized coculture of rat brain endothelial cells and astrocytes: A syngenic model to study the blood-brain barrier. *J. Neurosci. Methods* **121,** 21–31.

Dhib-Jalbut, S. (2003). Glatiramer acetate (Copaxone) therapy for multiple sclerosis. *Pharmacol. Ther.* **98,** 245–255.

Fanning, A. S., Mitic, L. L., and Anderson, J. M. (1999). Transmembrane proteins in the tight junction barrier. *J. Am. Soc. Nephrol.* **10,** 1337–1345.

Faunce, D. E., Terajewicz, A., and Stein-Streilein, J. (2004). Cutting edge: *In vitro*-generated tolerogenic APC induce CD8+ T regulatory cells that can suppress ongoing experimental autoimmune encephalomyelitis. *J. Immunol.* **172,** 1991–1995.

Filaci, G., Bacilieri, S., Fravega, M., Monetti, M., Contini, P., Ghio, M., Setti, M., Puppo, F., and Indiveri, F. (2001). Impairment of CD8+ T suppressor cell function in patients with active systemic lupus erythematosus. *J. Immunol.* **166,** 6452–6457.

Fogdell-Hahn, A., Ligers, A., Gronning, M., Hillert, J., and Olerup, O. (2000). Multiple sclerosis: A modifying influence of HLA class I genes in an HLA class II associated autoimmune disease. *Tissue Antigens* **55,** 140–148.

Ford, M. L., and Evavold, B. D. (2005). Specificity, magnitude, and kinetics of MOG-specific CD8+ T cell responses during experimental autoimmune encephalomyelitis. *Eur. J. Immunol.* **35,** 76–85.

French, A. R., and Yokoyama, W. M. (2004). Natural killer cells and autoimmunity. *Arthritis Res. Ther.* **6,** 8–14.

Gilliet, M., and Liu, Y. J. (2002). Generation of human CD8 T regulatory cells by CD40 ligand-activated plasmacytoid dendritic cells. *J. Exp. Med.* **195,** 695–704.

Gilliet, M., Soumelis, V., Watanabe, N., Hanabuchi, S., Antonenko, S., de Waal-Malefyt, R., and Liu, Y. J. (2003). Human dendritic cells activated by TSLP and CD40L induce proallergic cytotoxic T cells. *J. Exp. Med.* **197,** 1059–1063.

Giovannoni, G., Cutter, G. R., Lunemann, J., Martin, R., Munz, C., Sriram, S., Steiner, I., Hammerschlag, M. R., and Gaydos, C. A. (2006). Infectious causes of multiple sclerosis. *Lancet Neurol.* **5,** 887–894.

Green, R., Scott, L. K., Minagar, A., and Conrad, S. (2004). Sepsis associated encephalopathy (SAE): A review. *Front. Biosci.* **9,** 1637–1641.

Harbo, H. F., Lie, B. A., Sawcer, S., Celius, E. G., Dai, K. Z., Oturai, A., Hillert, J., Lorentzen, A. R., Laaksonen, M., Myhr, K. M., Ryder, L. P., Fredrikson, S., *et al.* (2004). Genes in the HLA class I region may contribute to the HLA class II-associated genetic susceptibility to multiple sclerosis. *Tissue Antigens* **63,** 237–247.

Haseloff, R. F., Blasig, I. E., Bauer, H. C., and Bauer, H. (2005). In search of the astrocytic factor(s) modulating blood-brain barrier functions in brain capillary endothelial cells *in vitro*. *Cell. Mol. Neurobiol.* **25,** 25–39.

Hawkins, B. T., and Davis, T. P. (2005). The blood-brain barrier/neurovascular unit in health and disease. *Pharmacol. Rev.* **57,** 173–185.

Hoftberger, R., Aboul-Enein, F., Brueck, W., Lucchinetti, C., Rodriguez, M., Schmidbauer, M., Jellinger, K., and Lassmann, H. (2004). Expression of major histocompatibility complex class I molecules on the different cell types in multiple sclerosis lesions. *Brain Pathol.* **14,** 43–50.

Hu, D., Ikizawa, K., Lu, L., Sanchirico, M. E., Shinohara, M. L., and Cantor, H. (2004). Analysis of regulatory CD8 T cells in Qa-1-deficient mice. *Nat. Immunol.* **5,** 516–523.

Huseby, E. S., Liggitt, D., Brabb, T., Schnabel, B., Ohlen, C., and Goverman, J. (2001). A pathogenic role for myelin-specific CD8(+) T cells in a model for multiple sclerosis. *J. Exp. Med.* **194,** 669–676.

Huseby, E. S., White, J., Crawford, F., Vass, T., Becker, D., Pinilla, C., Marrack, P., and Kappler, J. W. (2005). How the T cell repertoire becomes peptide and MHC specific. *Cell* **122,** 247–260.

Itoh, M., Furuse, M., Morita, K., Kubota, K., Saitou, M., and Tsukita, S. (1999). Direct binding of three tight junction-associated MAGUKs, ZO-1, ZO-2, and ZO-3, with the COOH termini of claudins. *J. Cell Biol.* **147,** 1351–1363.

Jiang, H., Zhang, S. I., and Pernis, B. (1992). Role of CD8+ T cells in murine experimental allergic encephalomyelitis. *Science* **256,** 1213–1215.

Johnson, A. J., Njenga, M. K., Hansen, M. J., Kuhns, S. T., Chen, L., Rodriguez, M., and Pease, L. R. (1999). Prevalent class I-restricted T-cell response to the Theiler's virus epitope Db: VP2121–130 in the absence of endogenous CD4 help, tumor necrosis factor alpha, gamma interferon, perforin, or costimulation through CD28. *J. Virol.* **73,** 3702–3708.

Johnson, A. J., Upshaw, J., Pavelko, K. D., Rodriguez, M., and Pease, L. R. (2001). Preservation of motor function by inhibition of CD8+ virus peptide-specific T cells in Theiler's virus infection. *FASEB J.* **15,** 2760–2762.

Johnson, A. J., Mendez-Fernandez, Y., Moyer, A. M., Sloma, C. R., Pirko, I., Block, M. S., Rodriguez, M., and Pease, L. R. (2005). Antigen-specific CD8+ T cells mediate a peptide-induced fatal syndrome. *J. Immunol.* **174,** 6854–6862.

Jurewicz, A., Biddison, W. E., and Antel, J. P. (1998). MHC class I-restricted lysis of human oligodendrocytes by myelin basic protein peptide-specific CD8 T lymphocytes. *J. Immunol.* **160,** 3056–3059.

Kantarci, O. H., and Weinshenker, B. G. (2005). Natural history of multiple sclerosis. *Neurol. Clin.* **23,** 17–38.

Karandikar, N. J., Crawford, M. P., Yan, X., Ratts, R. B., Brenchley, J. M., Ambrozak, D. R., Lovett-Racke, A. E., Frohman, E. M., Stastny, P., Douek, D. C., Koup, R. A., and Racke, M. K. (2002). Glatiramer acetate (Copaxone) therapy induces CD8(+) T cell responses in patients with multiple sclerosis. *J. Clin. Invest.* **109,** 641–649.

Karman, J., Ling, C., Sandor, M., and Fabry, Z. (2004). Initiation of immune responses in brain is promoted by local dendritic cells. *J. Immunol.* **173,** 2353–2361.

Khanna, K. M., Bonneau, R. H., Kinchington, P. R., and Hendricks, R. L. (2003). Herpes simplex virus-specific memory CD8+ T cells are selectively activated and retained in latently infected sensory ganglia. *Immunity* **18,** 593–603.

Killestein, J., Den Drijver, B. F., Van der Graaff, W. L., Uitdehaag, B. M., Polman, C. H., and Van Lier, R. A. (2001). Intracellular cytokine profile in T-cell subsets of multiple sclerosis patients: Different features in primary progressive disease. *Mult. Scler.* **7,** 145–150.

Killestein, J., Eikelenboom, M. J., Izeboud, T., Kalkers, N. F., Ader, H. J., Barkhof, F., Van Lier, R. A., Uitdehaag, B. M., and Polman, C. H. (2003). Cytokine producing CD8+ T cells are correlated to MRI features of tissue destruction in MS. *J. Neuroimmunol.* **142,** 141–148.

Kilpatrick, E. D., Terajima, M., Koster, F. T., Catalina, M. D., Cruz, J., and Ennis, F. A. (2004). Role of specific CD8+ T cells in the severity of a fulminant zoonotic viral hemorrhagic fever, hantavirus pulmonary syndrome. *J. Immunol.* **172,** 3297–3304.

Kirk, J., Plumb, J., Mirakhur, M., and McQuaid, S. (2003). Tight junctional abnormality in multiple sclerosis white matter affects all calibres of vessel and is associated with blood-brain barrier leakage and active demyelination. *J. Pathol.* **201,** 319–327.

Koh, D. R., Fung-Leung, W. P., Ho, A., Gray, D., Acha-Orbea, H., and Mak, T. W. (1992). Less mortality but more relapses in experimental allergic encephalomyelitis in CD8−/− mice. *Science* **256,** 1210–1213.

Kubota, K., Furuse, M., Sasaki, H., Sonoda, N., Fujita, K., Nagafuchi, A., and Tsukita, S. (1999). Ca(2+)-independent cell-adhesion activity of claudins, a family of integral membrane proteins localized at tight junctions. *Curr. Biol.* **9,** 1035–1038.

Kuhlmann, T., Lingfeld, G., Bitsch, A., Schuchardt, J., and Bruck, W. (2002). Acute axonal damage in multiple sclerosis is most extensive in early disease stages and decreases over time. *Brain* **125,** 2202–2212.

Kurane, I., Innis, B. L., Nimmannitya, S., Nisalak, A., Meager, A., Janus, J., and Ennis, F. A. (1991). Activation of T lymphocytes in dengue virus infections. High levels of soluble interleukin 2 receptor, soluble CD4, soluble CD8, interleukin 2, and interferon-gamma in sera of children with dengue. *J. Clin. Invest.* **88,** 1473–1480.

Kurtzke, J. F., and Heltberg, A. (2001). Multiple sclerosis in the Faroe islands: An epitome. *J. Clin. Epidemiol.* **54,** 1–22.

Lassmann, H., and Ransohoff, R. M. (2004). The CD4-Th1 model for multiple sclerosis: A critical [correction of crucial] re-appraisal. *Trends Immunol.* **25,** 132–137.

Lennon, V. A., Wingerchuk, D. M., Kryzer, T. J., Pittock, S. J., Lucchinetti, C. F., Fujihara, K., Nakashima, I., and Weinshenker, B. G. (2004). A serum autoantibody marker of neuromyelitis optica: Distinction from multiple sclerosis. *Lancet* **364,** 2106–2112.

Liu, T., Khanna, K. M., Chen, X., Fink, D. J., and Hendricks, R. L. (2000). CD8(+) T cells can block herpes simplex virus type 1 (HSV-1) reactivation from latency in sensory neurons. *J. Exp. Med.* **191,** 1459–1466.

Loke, H., Bethell, D. B., Phuong, C. X., Dung, M., Schneider, J., White, N. J., Day, N. P., Farrar, J., and Hill, A. V. (2001). Strong HLA class I—restricted T cell responses in dengue hemorrhagic fever: A double-edged sword? *J. Infect. Dis.* **184,** 1369–1373.

Lucchinetti, C., Bruck, W., Parisi, J., Scheithauer, B., Rodriguez, M., and Lassmann, H. (2000). Heterogeneity of multiple sclerosis lesions: Implications for the pathogenesis of demyelination. *Ann. Neurol.* **47,** 707–717.

Lucchinetti, C. F., Bruck, W., and Lassmann, H. (2004). Evidence for pathogenic heterogeneity in multiple sclerosis. *Ann. Neurol.* **56,** 308.

Madakamutil, L. T., Maricic, I., Sercarz, E., and Kumar, V. (2003). Regulatory T cells control autoimmunity *in vivo* by inducing apoptotic depletion of activated pathogenic lymphocytes. *J. Immunol.* **170,** 2985–2992.

Markovic-Plese, S., and McFarland, H. F. (2001). Immunopathogenesis of the multiple sclerosis lesion. *Curr. Neurol. Neurosci. Rep.* **1,** 257–262.

Medana, I., Martinic, M. A., Wekerle, H., and Neumann, H. (2001). Transection of major histocompatibility complex class I-induced neurites by cytotoxic T lymphocytes. *Am. J. Pathol.* **159,** 809–815.

Medana, I. M., and Turner, G. D. (2006). Human cerebral malaria and the blood-brain barrier. *Int. J. Parasitol.* **36,** 555–568.

Medana, I. M., Gallimore, A., Oxenius, A., Martinic, M. M., Wekerle, H., and Neumann, H. (2000). MHC class I-restricted killing of neurons by virus-specific CD8+ T lymphocytes is effected through the Fas/FasL, but not the perforin pathway. *Eur. J. Immunol.* **30,** 3623–3633.

Minagar, A., and Alexander, J. S. (2003). Blood-brain barrier disruption in multiple sclerosis. *Mult. Scler.* **9,** 540–549.

Mongkolsapaya, J., Dejnirattisai, W., Xu, X. N., Vasanawathana, S., Tangthawornchaikul, N., Chairunsri, A., Sawasdivorn, S., Duangchinda, T., Dong, T., Rowland-Jones, S., Yenchitsomanus, P. T., McMichael, A., *et al.* (2003). Original antigenic sin and apoptosis in the pathogenesis of dengue hemorrhagic fever. *Nat. Med.* **9,** 921–927.

Monteiro, J., Hingorani, R., Pergolizzi, R., Apatoff, B., and Gregersen, P. K. (1995). Clonal dominance of CD8+ T-cell in multiple sclerosis. *Ann. N Y Acad. Sci.* **756,** 310–312.

Moses, H., Jr., and Sriram, S. (2001). An infectious basis for multiple sclerosis: Perspectives on the role of Chlamydia pneumoniae and other agents. *BioDrugs* **15,** 199–206.

Murali-Krishna, K., Altman, J. D., Suresh, M., Sourdive, D. J., Zajac, A. J., Miller, J. D., Slansky, J., and Ahmed, R. (1998). Counting antigen-specific CD8 T cells: A reevaluation of bystander activation during viral infection. *Immunity* **8,** 177–187.

Murray, P. D., McGavern, D. B., Lin, X., Njenga, M. K., Leibowitz, J., Pease, L. R., and Rodriguez, M. (1998). Perforin-dependent neurologic injury in a viral model of multiple sclerosis. *J. Neurosci.* **18,** 7306–7314.

Najafian, N., Chitnis, T., Salama, A. D., Zhu, B., Benou, C., Yuan, X., Clarkson, M. R., Sayegh, M. H., and Khoury, S. J. (2003). Regulatory functions of CD8+CD28− T cells in an autoimmune disease model. *J. Clin. Invest.* **112,** 1037–1048.

Nelson, A. L., Bieber, A. J., and Rodriguez, M. (2004). Contrasting murine models of MS. *Int. MS J.* **11,** 95–99.

Neumann, H., Cavalie, A., Jenne, D. E., and Wekerle, H. (1995). Induction of MHC class I genes in neurons. *Science* **269,** 549–552.

Neumann, H., Medana, I. M., Bauer, J., and Lassmann, H. (2002). Cytotoxic T lymphocytes in autoimmune and degenerative CNS diseases. *Trends Neurosci.* **25,** 313–319.

Neville, K. L., Padilla, J., and Miller, S. D. (2002). Myelin-specific tolerance attenuates the progression of a virus-induced demyelinating disease: Implications for the treatment of MS. *J. Neuroimmunol.* **123,** 18–29.

Niland, B., Banki, K., Biddison, W. E., and Perl, A. (2005). CD8+ T cell-mediated HLA-A*0201-restricted cytotoxicity to transaldolase peptide 168–176 in patients with multiple sclerosis. *J. Immunol.* **175,** 8365–8378.

Noseworthy, J. H., Lucchinetti, C., Rodriguez, M., and Weinshenker, B. G. (2000). Multiple sclerosis. *N. Engl. J. Med.* **343,** 938–952.

Oleszak, E. L., Chang, J. R., Friedman, H., Katsetos, C. D., and Platsoucas, C. D. (2004). Theiler's virus infection: A model for multiple sclerosis. *Clin. Microbiol. Rev.* **17,** 174–207.

Oshima, T., Laroux, F. S., Coe, L. L., Morise, Z., Kawachi, S., Bauer, P., Grisham, M. B., Specian, R. D., Carter, P., Jennings, S., Granger, D. N., Joh, T., *et al.* (2001). Interferon-gamma and interleukin-10 reciprocally regulate endothelial junction integrity and barrier function. *Microvasc. Res.* **61,** 130–143.

Petereit, H. F., Pukrop, R., Fazekas, F., Bamborschke, S. U., Ropele, S., Kolmel, H. W., Merkelbach, S., Japp, G., Jongen, P. J., Hartung, H. P., and Hommes, O. R. (2003). Low interleukin-10 production is associated with higher disability and MRI lesion load in secondary progressive multiple sclerosis. *J. Neurol. Sci.* **206,** 209–214.

Petty, M. A., and Lo, E. H. (2002). Junctional complexes of the blood-brain barrier: Permeability changes in neuroinflammation. *Prog. Neurobiol.* **68,** 311–323.

Rivera-Quinones, C., McGavern, D., Schmelzer, J. D., Hunter, S. F., Low, P. A., and Rodriguez, M. (1998). Absence of neurological deficits following extensive demyelination in a class I-deficient murine model of multiple sclerosis. *Nat. Med.* **4,** 187–193.

Savinov, A. Y., Wong, F. S., Stonebraker, A. C., and Chervonsky, A. V. (2003). Presentation of antigen by endothelial cells and chemoattraction are required for homing of insulin-specific CD8+ T cells. *J. Exp. Med.* **197,** 643–656.

Sepulcre, J., Sanchez-Ibarrola, A., Moreno, C., and de Castro, P. (2005). Association between peripheral IFN-gamma producing CD8+ T-cells and disability score in relapsing-remitting multiple sclerosis. *Cytokine* **32,** 111–116.

Shacklett, B. L., Cox, C. A., Wilkens, D. T., Karl Karlsson, R., Nilsson, A., Nixon, D. F., and Price, R. W. (2004). Increased adhesion molecule and chemokine receptor expression on CD8+ T cells trafficking to cerebrospinal fluid in HIV-1 infection. *J. Infect. Dis.* **189,** 2202–2212.

Skulina, C., Schmidt, S., Dornmair, K., Babbe, H., Roers, A., Rajewsky, K., Wekerle, H., Hohlfeld, R., and Goebels, N. (2004). Multiple sclerosis: Brain-infiltrating CD8+ T cells persist as clonal expansions in the cerebrospinal fluid and blood. *Proc. Natl. Acad. Sci. USA* **101,** 2428–2433.

Smeltz, R. B., Wolf, N. A., and Swanborg, R. H. (1999). Inhibition of autoimmune T cell responses in the DA rat by bone marrow-derived NK cells *in vitro*: Implications for autoimmunity. *J. Immunol.* **163,** 1390–1397.

Solomon, T., Dung, N. M., Vaughn, D. W., Kneen, R., Thao, L. T., Raengsakulrach, B., Loan, H. T., Day, N. P., Farrar, J., Myint, K. S., Warrell, M. J., James, W. S., *et al.* (2000). Neurological manifestations of dengue infection. *Lancet* **355,** 1053–1059.

Steinman, L., and Zamvil, S. S. (2005). Virtues and pitfalls of EAE for the development of therapies for multiple sclerosis. *Trends Immunol.* **26,** 565–571.

Steinman, L., and Zamvil, S. S. (2006). How to successfully apply animal studies in experimental allergic encephalomyelitis to research on multiple sclerosis. *Ann. Neurol.* **60,** 12–21.

Stephens, H. A., Klaythong, R., Sirikong, M., Vaughn, D. W., Green, S., Kalayanarooj, S., Endy, T. P., Libraty, D. H., Nisalak, A., Innis, B. L., Rothman, A. L., Ennis, F. A., *et al.* (2002). HLA-A and -B allele associations with secondary dengue virus infections correlate with disease severity and the infecting viral serotype in ethnic Thais. *Tissue Antigens* **60,** 309–318.

Stoll, G., Jander, S., and Schroeter, M. (2000). Cytokines in CNS disorders: Neurotoxicity versus neuroprotection. *J. Neural. Transm. Suppl.* **59,** 81–89.

Sun, D., Whitaker, J. N., Huang, Z., Liu, D., Coleclough, C., Wekerle, H., and Raine, C. S. (2001). Myelin antigen-specific CD8+ T cells are encephalitogenic and produce severe disease in C57BL/6 mice. *J. Immunol.* **166,** 7579–7587.

Tallquist, M. D., Yun, T. J., and Pease, L. R. (1996). A single T cell receptor recognizes structurally distinct MHC/peptide complexes with high specificity. *J. Exp. Med.* **184,** 1017–1026.

Theil, D. J., Tsunoda, I., Rodriguez, F., Whitton, J. L., and Fujinami, R. S. (2001). Viruses can silently prime for and trigger central nervous system autoimmune disease. *J. Neurovirol.* **7,** 220–227.

Tsuchida, T., Parker, K. C., Turner, R. V., McFarland, H. F., Coligan, J. E., and Biddison, W. E. (1994). Autoreactive CD8+ T-cell responses to human myelin protein-derived peptides. *Proc. Natl. Acad. Sci. USA* **91,** 10859–10863.

Tsukita, S., Furuse, M., and Itoh, M. (1999). Structural and signalling molecules come together at tight junctions. *Curr. Opin. Cell Biol.* **11,** 628–633.

Ure, D. R., and Rodriguez, M. (2002). Preservation of neurologic function during inflammatory demyelination correlates with axon sparing in a mouse model of multiple sclerosis. *Neuroscience* **111,** 399–411.

Weiner, H. L. (2004). Multiple sclerosis is an inflammatory T-cell-mediated autoimmune disease. *Arch. Neurol.* **61,** 1613–1615.

Wingerchuk, D. M. (2004). Neuromyelitis optica: Current concepts. *Front. Biosci.* **9,** 834–840.

Wingerchuk, D. M., Hogancamp, W. F., O'Brien, P. C., and Weinshenker, B. G. (1999). The clinical course of neuromyelitis optica (Devic's syndrome). *Neurology* **53,** 1107–1114.

Wingerchuk, D. M., Lennon, V. A., Pittock, S. J., Lucchinetti, C. F., and Weinshenker, B. G. (2006). Revised diagnostic criteria for neuromyelitis optica. *Neurology* **66,** 1485–1489.

Wolburg, H., and Lippoldt, A. (2002). Tight junctions of the blood-brain barrier: Development, composition and regulation. *Vascul. Pharmacol.* **38,** 323–337.

Wong, D., Dorovini-Zis, K., and Vincent, S. R. (2004). Cytokines, nitric oxide, and cGMP modulate the permeability of an *in vitro* model of the human blood-brain barrier. *Exp. Neurol.* **190,** 446–455.

Wong, F. S., Karttunen, J., Dumont, C., Wen, L., Visintin, I., Pilip, I. M., Shastri, N., Pamer, E. G., and Janeway, C. A., Jr. (1999). Identification of an MHC class I-restricted autoantigen in type 1 diabetes by screening an organ-specific cDNA library. *Nat. Med.* **5,** 1026–1031.

Zamvil, S. S., and Steinman, L. (1990). The T lymphocyte in experimental allergic encephalomyelitis. *Annu. Rev. Immunol.* **8,** 579–621.

Zang, Y. C., Li, S., Rivera, V. M., Hong, J., Robinson, R. R., Breitbach, W. T., Killian, J., and Zhang, J. Z. (2004). Increased CD8+ cytotoxic T cell responses to myelin basic protein in multiple sclerosis. *J. Immunol.* **172,** 5120–5127.

Zhang, D., Beresford, P. J., Greenberg, A. H., and Lieberman, J. (2001). Granzymes A and B directly cleave lamins and disrupt the nuclear lamina during granule-mediated cytolysis. *Proc. Natl. Acad. Sci. USA* **98,** 5746–5751.

Zivna, I., Green, S., Vaughn, D. W., Kalayanarooj, S., Stephens, H. A., Chandanayingyong, D., Nisalak, A., Ennis, F. A., and Rothman, A. L. (2002). T cell responses to an HLA-B*07-restricted epitope on the dengue NS3 protein correlate with disease severity. *J. Immunol.* **168,** 5959–5965.

IMMUNOPATHOGENESIS OF MULTIPLE SCLEROSIS

Smriti M. Agrawal* and V. Wee Yong*,†

*Department of Clinical Neuroscience, University of Calgary
Calgary, Alberta T2N 4N1, Canada
†Department of Oncology, University of Calgary, Calgary, Alberta T2N 4N1, Canada

Multiple sclerosis (MS) is considered an immune-mediated disorder in which immune cells cross the blood–brain barrier to enter the central nervous system (CNS) wherein they augment the neuropathology of the disease. This chapter discusses the role of various immune cell types that contribute to the development and progression of MS. Specifically, the role of T cells, antigen-presenting cells, and components of the innate immune system such as macrophages, B cells, and the complement system are discussed. The involvement of CNS-specific cells such as microglia, astrocytes, and neurons in MS are discussed and the immunosuppressive role of regulatory T cells is considered. We introduce the involvement of chemokines and matrix metalloproteinases which helps recruit immune cells into the CNS in MS.

INTERNATIONAL REVIEW OF
NEUROBIOLOGY, VOL. 79
DOI: 10.1016/S0074-7742(07)79005-0

0074-7742/07 $35.00

Although the causes of MS are unknown, various factors such as genetic influences, environmental effects, and involvement of infectious agents as potential contributors to MS immune dysfunctions are also considered. With this background, we discuss the mechanisms of the immunomodulators that are used to treat MS.

I. Introduction

Multiple sclerosis (MS), an inflammatory disorder of the central nervous system (CNS) in humans, is characterized by myelin loss, gliosis, and varying degrees of axonal and oligodendrocyte pathology. MS usually begins in early adulthood, and paralysis, sensory disturbances, lack of coordination, and visual impairment are its common features. The disease often starts with an "attack" or "flare" that lasts from a few days to weeks; this is followed by remission that lasts from a few months to years. Disease course is not the only factor that differs among MS patients, since morphological alterations in CNS anatomy detected by magnetic resonance imaging (MRI) and histopathologic evaluations also vary significantly. Although the causes of MS are unknown, disease mechanisms are often inferred from the effector phase of an animal model of MS, experimental autoimmune encephalomyelitis (EAE). Briefly, EAE commences when leukocytes (mainly T cells and monocyte/macrophages) receive cues to tether to and roll along the activated endothelial cell layer of venules in the CNS. Leukocytes utilize cell adhesion molecules and proteases to adhere to and migrate through the endothelial cell layer, and then across the two underlying basement membrane barriers to enter the CNS parenchyma. Within the CNS, leukocytes encounter CNS resident antigen-presenting cells (APCs, e.g., microglia) to become restimulated and in turn destroy myelin and axonal processes, resulting in the production of remnant epitopes of degraded myelin (which augment a secondary wave of restimulation) and finally in the formation of demyelinated plaques. MS therapies focus on targeting various aspects of this complicated process in CNS destruction. In this chapter, we discuss the immune abnormalities in MS and we provide information on how immune modulators induce disease course.

II. MS as an Inflammatory Disease of the CNS

Although the cause and pathogenesis of MS are unknown, the current theories favor MS as an autoimmune inflammatory disorder of the CNS, wherein autoreactive T lymphocytes that recognize CNS-specific proteins result in inflammation and demyelination. On the basis of cellular infiltrates detected in brain and cerebrospinal fluid (CSF) from patients and data from rodent models such as EAE, MS is largely considered to be a CD4+ T helper 1 (Th1)-mediated inflammatory disease (LaSalle *et al.*, 1991), although CD4+ Th17 effector T cells are postulated to have more important roles (Bettelli *et al.*, 2006; Iwakura and Ishigame, 2006; Veldhoen and Stockinger, 2006). In the EAE model, it has been shown that the injection of myelin components into susceptible animals leads to a CD4+-mediated autoimmune disease that shares similarities with MS (Martin *et al.*, 1992); EAE can also be induced by the adoptive transfer of encephalitogenic CD4+ T cells into a naive animal (Martin *et al.*, 1992; Pettinelli and McFarlin, 1981; Zamvil and Steinman, 1990). As will be discussed later, CD8+ T cells and other leukocyte subsets are also represented in MS lesions where they are thought to play important pathogenic roles.

In MS and EAE CNS vessels, proinflammatory cytokines such as tumor necrosis factor-α (TNF-α) induce the expression of endothelial cell surface adhesion molecules such as VCAM-1 and LFA-1. Together with chemokine attractants, of which potential candidates are CCL19 and CCL21 (Alt *et al.*, 2002), this facilitates leukocyte tethering and rolling along the endothelial cell surface. Via $\alpha 4\beta 1$ integrin receptors (clinically known as VLA-4) on their surface, T cells bind to VCAM-1 on the endothelial cell surface, allowing them to attach to and finally transmigrate through the endothelial cell layer (Bar-Or, 2005; Engelhardt, 2006; Fig. 1). Once through the endothelial cell layer, the migrating leukocytes traverse the two underlying basement membrane barriers by binding specifically to laminin subtypes which constitute the basement membranes (Sixt *et al.*, 2001) and by producing metalloproteases such as the matrix metalloproteinases (MMPs) to specifically cleave components of the basement membrane barriers (Agrawal *et al.*, 2006), resulting in their entrance into the CNS parenchyma. Immunohistochemical analysis of both MS and EAE tissues have found that leukocytes accumulate in the Virchow–Robin space to form perivascular cuffs (Prineas, 1975; van Horssen *et al.*, 2005) before they enter the CNS parenchyma to result in a massive local inflammation wherein they act to degrade myelin protein components and eventually destroy the underlying axons, resulting in the formation of lesions or plaques within the CNS.

On entry to the CNS compartment, activated Th1 cells must be restimulated, as otherwise they die or leave the CNS. Restimulation, perhaps via an autoantigen or a microbial antigen presented on microglia or resident dendritic cells, causes clonal expansion of the Th1 cells. Th1 cells release an array of proinflammatory cytokines, including interleukin-1 (IL-1), interferon-γ (IFN-γ), and TNF-α leading to monocytes and microglia activation. Activated monocytes and macrophages can also release a variety of neurotoxic elements, including nitric oxide, oxygen free-radicals, and MMPs, all of which contribute to the destruction

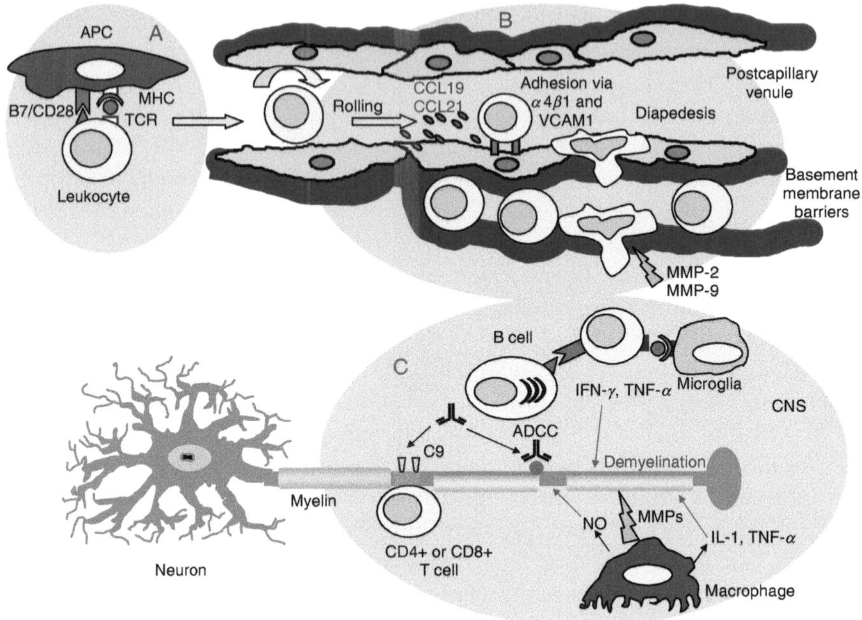

FIG. 1. Leukocytes in MS: From activation to demyelination. Leukocytes undergo three major steps before they bring about demyelination of CNS neurons. (A) *Activation*: leukocytes are presented self-antigens in peripheral organs by specific antigen presenting cells, allowing them to clonally expand and become activated. (B) *Adhesion and transmigration*: within a postcapillary venule of the CNS, activated leukocytes encounter chemokines secreted by activated endothelial cells. Leukocytes respond to the chemokine signal, they roll along and tether to the endothelial cell, and eventually migrate through the endothelial layer and the underlying basement membrane barriers to enter the CNS. (C) *Reactivation and demyelination*: within the CNS, leukocytes are represented the self-antigens, where they produce proinflammatory cytokines, MMPs, activate B cells, macrophage/microglia, and the complement system to eventually lead to demyelination of surround neurons. (See Color Insert.)

of the myelin sheath and axons (Dhib-Jalbut, 2002). MRI scans utilizing gadolinium are used to identify and quantify such lesions which are indicative of blood–brain barrier breach and disease extent.

The role of CD4+ T cells in MS is supported by many parallels with EAE, but it is also supported indirectly by the result that certain HLA class II molecules represent the strongest genetic risk factor for MS, presumably via their role as antigen-presenting molecules to pathogenic CD4+ T cells (Bar-Or, 2005). Also, in a trial of an altered peptide ligand of myelin basic protein (MBP), some MS patients were worsened and this was correlated with the massive expansion of MBP-reactive CD4+ T cells (Bielekova *et al.*, 2000).

III. Genetic and Environmental Aspects of MS and Their Impact on Immune Functions

A. Genetic Influences in MS

Although the causes of MS are unknown, at present it is thought to be contributed by the influences of both genes and the environment (Fig. 2). The disease is reported to predominantly affect northern Europeans, and there is a higher age-adjusted risk for siblings (3%; full siblings, the risk for half siblings is less), parents (2%), and children (2%) than for second- and third-degree relatives (Compston and Coles, 2002). Concordance in monozygotic twins is around 35%, and is higher in children with both parents affected (20%) as opposed to offsprings of single-affected couples (2%). Conversely, the risk is not increased either for individuals adopted into a family with an affected individual or in the nonbiological relatives of adoptees who themselves develop MS.

MS does not result from gene mutations or aberrations; rather, it seems to be contributed by the effect of normal polymorphisms in a number of genes. These polymorphisms act independently or through epistasis, and each can exert a small

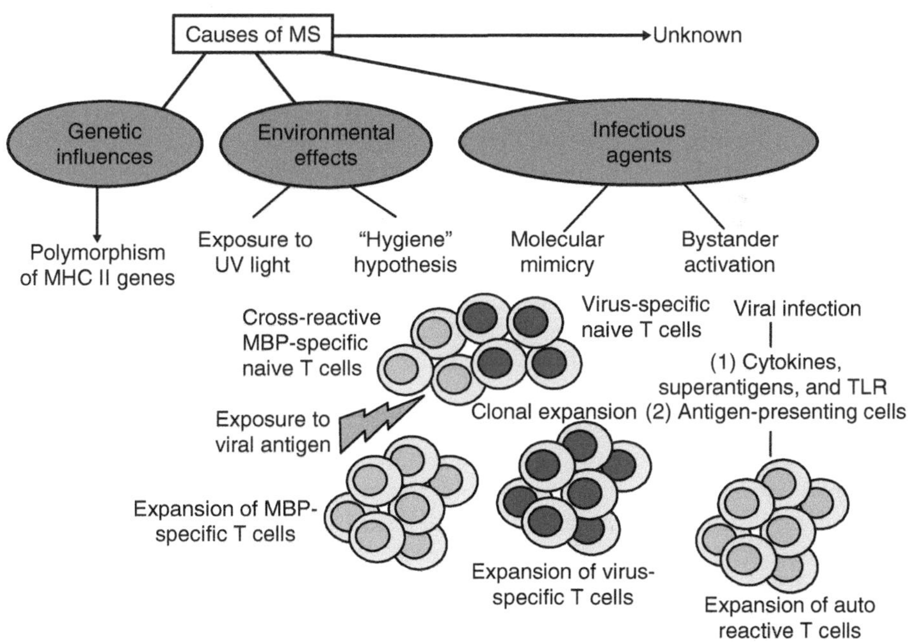

FIG. 2. Various factors influence the onset of MS. Genetic influences, environmental effects, and infectious agents all contribute to the causes of MS. Infectious agents act via two mechanisms: molecular mimicry and bystander effect. (See Color Insert.)

contributory effect on some as yet undefined structure or physiological function (Compston and Coles, 2002; Motsinger *et al.*, 2007; Willer *et al.*, 2006). Extensive searches and results from population studies have yielded few secure candidate regions for susceptibility genes. An association between a few class II MHC alleles and the gene for TNF-α are encoded within the same linkage group. A specifically different association is seen in Mediterranean populations not susceptible to MS. Although no major susceptibility gene has yet been identified, several promising chromosomal linkages are for the time being linked with MS—at 1p, 6p, 10p, 17q, and 19q.13 (reviewed in Brassat *et al.*, 2006).

B. Environmental Effects in MS

Despite the genetic evidence, the lack of complete concurrence for MS in identical twins suggests a role for nongenetic factors in the disease (reviewed in Ponsonby *et al.*, 2005b). An environmental influence is suggested by migration studies. In this regard, regions to which northern European descendents migrated have high prevalence rates of MS. The rates among Caucasians outside Europe are about half of those in many parts of northern Europe. Also, if one migrated from an area of high prevalence of MS to an area of low prevalence before age 15–16, the lower risk of the new area was acquired; migration after 15–16 did not change the risk (Kurtzke, 1983). Various factors have been taken into consideration to explain this MS prevalence pattern. Sunlight exposure depending on the latitude is thought to be one causative factor. UV radiation may exert its effects either by influencing immunoregulatory cells or by the biosynthesis of vitamin D (Hayes, 2000), which has many immunoregulatory properties (van Etten and Mathieu, 2005). As well, melatonin secretion depends on sunlight exposure, and the lack of sunlight could induce an excess of melatonin which enhances Th1 response (Kidd, 2003; Ponsonby *et al.*, 2005a).

Other factors contributing to increase in prevalence rates of MS include socioeconomic development related to industrialization, urban living, pollution, occupational exposures to solvents, changes in diet, breastfeeding, and smoking habits. Finally, a delayed exposure to or overall reduction in childhood infections in developed countries, the "hygiene hypothesis," suggests a possible skewed immune responsiveness and increased susceptibility to develop autoimmune reactions/diseases (Th1-mediated) and allergy (Th2-mediated) in later life. This is hypothesized to be relevant in MS prevalence; however, it is difficult to prove (Sospedra and Martin, 2005).

C. Infectious Agents as Triggers of MS

Viral and bacterial infections are logical candidates as infectious triggers of MS. Almost 100% of transgenic mice expressing a T-cell receptor that is specific for an encephalitogenic peptide of MBP develop EAE when the transgenic mice are housed under non-pathogen-free conditions, whereas the same animals housed in a specific-pathogen-free facility remained disease free (Goverman *et al.*, 1993). Current data suggest that MS could be induced and/or exacerbated by many different microbial infections, and the responsible agents are most likely ubiquitous pathogens that are highly prevalent in the general population. Two main mechanisms have been proposed to explain how infections could induce MS (Fig. 2). The first is *molecular mimicry*, wherein autoreactive T cells are activated by cross-reactivity between self-antigens and foreign agents. For example, during thymic selection, T cells may recognize self-antigens along with foreign antigens resulting in a positive selection of this population of T cells and their export to the periphery. Cross-reactivity of these "potentially" self-reactive T cells with foreign antigens can lead to their activation during infection and migration across the blood–brain barrier; if these cells recognize antigens expressed in the brain (such as MBP), tissue damage and potentially an autoimmune disease like MS may result (Wucherpfennig *et al.*, 1997).

A second mechanism to explain how infection could induce MS is that of *bystander activation*, which assumes that autoreactive cells become activated due to nonspecific inflammatory events that occur during viral or bacterial infections. This mechanism is classified into two catogories, the first category encompasses T-cell receptor independent bystander cell activation of T cells via inflammatory cytokines, superantigens, and activation of Toll-like receptors, and the second encompasses the role of APCs which are activated directly by host antigens and the adjuvant effect of infectious agents (Murali-Krishna *et al.*, 1998; Tough and Sprent, 1996; Waldner *et al.*, 2004). Pender (2003) suggested a role for autoreactive B cells in MS, where autoreactive B cells infected by a virus (such as EBV) begin to proliferate and transform into B cells resistant to apoptosis, thereby triggering costimulation of autoreactive T cells or preventing these cells from undergoing activation-induced cell death. The result is the persistence and accumulation of autoreactive T cells in the inflamed area.

IV. Immune Components of MS

A. CD4+ T Cells

In the early reports on MS studies, one of the most striking features was the involvement of MBP-specific CD4+ T cells in MS patients versus healthy controls (Burns *et al.*, 1983). The observation that such autoreactive T cells from the normal T-cell cohort of Lewis rats could induce EAE (Schluesener and Wekerle, 1985) suggested to investigators that their equivalent in humans might be relevant for MS (Fig. 3). Since then CD4+ T cells have been a subject of comprehensive research in MS.

A Adaptive immunity

CD4+ T cells- Th1 — ↑ *Proinflammatory cytokines*
Th2 — ↑ *Antiinflammatory cytokines*
Th17 — ↑ *IL-17*

CD8+ T cells — *Target MHC/expressing CNS resident cells*

CD4+ CD25+ Foxp3+ Tregs — *Immunosupressive activity,* ↓ *disease progress*

Bystander immune cells—Macrophage — ↑ *MMP, iNOS*
Dendritic cells — *Antigen presentation, T-cell restimulation*
Neutrophils — ↑ *MMP, iNOS*
B cells — ↑ *Antibody, complement activation*

B Innate immunity

Mast cells — *Release histamine and tryptase, activates MMP cascade*

Toll-like receptors — *Break tolerance, induce APC reactivity to self-antigens*

Complement — *Binds to myelin,* ↑ *Demyelination*

C CNS contribution

Microglia — *Express costimulatory molecules,* ↑ *proinflammatory cytokines*

Astrocytes — *Express costimulatory molecules,* ↑ *proinflammatory cytokines*

Neurons — ↑ *Proinflammatory cytokines*

FIG. 3. Cellular contributions to MS. Components of both the adaptive and innate immunities contribute to MS exacerbation, with the exception of Tregs which are immunosuppressive. CNS-specific cells also play active roles in promoting the disease.

B. Th1 Verus Th2 CD4+ T Cells

On activation during a normal immune response, T cells undergo different developmental pathways wherein they acquire specialized properties and effector functions. Th cells are traditionally thought to differentiate into Th1- and Th2-cell subsets. The Th1 cells are necessary in the clearance of intracellular pathogens and Th2 cells are essential in the removal of extracellular organisms (Bottomly, 1988; Mosmann and Coffman, 1989). A subset of IL-17-producing, IL-6-dependent T (Th17) cells distinct from Th1 or Th2 cells have been described to have a crucial role in autoimmune tissue injury (Harrington *et al.*, 2005; Langrish *et al.*, 2005; Park *et al.*, 2005).

How T cells differentiate is heavily dependent on their microenvironment during the initial activation stage. Cytokines profoundly influence T-cell differentiation; for example, the presence of IL-12 or IL-23 can preferentially generate Th1 cells that subsequently secrete IFN-γ and TNF-α (Mosmann and Coffman, 1989; Murphy *et al.*, 2003). Th2 cells develop in the absence of IL-12 but in the presence of IL-4, subsequently producing cytokines such as IL-4, IL-5, IL-13, and possibly IL-10. Normally, a balance between the Th1 and Th2 responses regulates disease states. For example, a Th2 response may contribute to asthma, whereas a Th1 response is thought to result in autoimmune diseases. The inflammatory states in MS and EAE are attributed to Th1 cell types, wherein amounts of proinflammatory cytokines such as TNF-α, IFN-γ, and IL-12 prevail. During a relapse in MS, Th2 cells predominate and anti-inflammatory cytokines such as IL-5, IL-13, and IL-10 are produced. A shift from Th1 to Th2, T-cell response is protective of autoimmune diseases.

C. CD8+ T CELLS

CD8+ cytotoxic T cells are likely involved in MS (Fig. 3). Huseby *et al.* (2001) have shown that when encephalitogenic $MBP_{(79-87)}$-specific CD8+ cytotoxic T cells from C3H mice are transferred into wild-type mice, they produce a disease which resembles MS more closely in terms of ataxia and spasticity than that observed with CD4+ T-cell-mediated EAE models. CD8+ T cells are more prevalent in CNS, CSF, and blood of MS subjects (Skulina *et al.*, 2004) than CD4+ T cells. They are also thought to be much better suited to mediate CNS damage, since CNS resident cells (neurons, oligodendrocytes, astrocytes) express MHC class I and therefore are preferentially recognized by CD8+ T cells (Jurewicz *et al.*, 1998; Medana *et al.*, 2000) rather than MHC class II-restricted CD4+ T cells. However, it is recognized that both classes of T cells can injure neurons through soluble mediators (Wang *et al.*, 2006) or through non-MHC-dependent cell–cell contact (Giuliani *et al.*, 2003).

D. BYSTANDER IMMUNE CELLS

Aside from T cells, blood-borne bystander cells such as macrophages, dendritic cells, and neutrophils have been shown to play a role in both MS and EAE (Fig. 3). Agrawal *et al.* (2006) showed that when circulating macrophages/monocytes were specifically depleted with clodronate liposomes in mice, EAE onset was delayed. FACS data from this study demonstrated that when disease eventually occurred, this corresponded with repopulation of circulating macrophage/monocyte populations. These findings are consistent with previous studies (Tran *et al.*, 1998) which showed that mice lacking in macrophage/monocyte populations were resistant to EAE onset. Toft-Hansen *et al.* (2004) used RT-PCR to assess the expression of metalloproteinases in various cell populations to implicate their roles in EAE. In their reports, among various cell types isolated from the CNS of mice with EAE, macrophages were identified as a major source of MMP-12 and tissue inhibitor of metalloproteinases (TIMP)-1. Neutrophils were a major source of MMP-8; a related metalloproteinase, ADAM-12 (a disintegrin and metalloproteinase-12), was expressed primarily by T cells. Microglia expressed MMP-15, the levels of which were downregulated in EAE.

The observations that immunoglobulins (Igs) accumulate in the CSF of MS patients, and correlation studies that show increases in CSF Igs with worsening of disease, have suggested a role for both B cells and Igs in MS (Baranzini *et al.*, 1999; Walsh and Tourtellotte, 1986). Normally, B cells cannot cross the blood–brain barrier. However, once an inflammation has begun, B cells, Igs, complement, and other serum proteins enter the CNS. A reported increase in Igs in CSF but not in serum of MS patients has suggested a local CNS production of antibodies (Kabat *et al.*, 1950). Various roles for B cells and Igs have been suggested in MS pathology: B cells can function as APCs for T cells and participate in CNS-specific restimulation of encephalitogenic T cells, they can provide costimulatory signals to T cells and recruit T cells to the CNS (Sun *et al.*, 1991), and B-cell-induced myelin-specific antibodies may cause myelin destruction within the CNS. Antibodies in the CNS may not only function in myelin destruction but are also suggested to influence the cytokine profile (Th1 to Th2), resulting in either increased inflammation or protection and repair (Saoudi *et al.*, 1995). Consistent with roles of B cells in MS is the observation that antibody deposition and complement activation are commonly observed in lesions in MS (Lucchinetti *et al.*, 2000) and that autoantibodies to MBP can be found on degenerating myelin (Genain *et al.*, 1999). The role of antibodies in MS is complicated, however, since some studies have suggested that antibodies can promote the remyelination process in animal models to result in remission (Rodriguez and Lennon, 1990).

V. Contribution of Components of Innate Immunity in MS

Although the innate immune system's main role is self-protection and maintenance of homeostasis, some innate immune mechanisms can result in destructive immunity. Various components of the innate immune system have been shown to play roles in MS and EAE, a few of which are discussed here. The first example is the family of Toll-like receptors (TLR) on cells such as microglia/macrophages and dendritic cells, which in normal situations function as sentinels by recognizing preprogrammed pathogen-associated molecular patterns to generate proinflammatory signals which then initiate an adaptive immune response. Also, under normal conditions, antigen-APCs remain in their resting state and induce tolerance in autoreactive T cells, whereas in the presence of foreign antigens, these APCs are activated and may convert tolerized autoreactive T cells into effector cells. In MS, stimulation of TLR may break tolerance and render lymph node-specific APCs reactive against self-antigen to which they were previously unresponsive, thereby creating self-reactive effector T cells (Prinz *et al.*, 2006).

Mast cells are another example of the innate immune response playing a role in MS. Normally, mast cells are activated during allergic reactions when surface IgE receptors cross-link. Mast cells respond by releasing several molecules (histamine and tryptase) that destroy the allergen. Mast cells have been detected in low numbers in normal CNS tissue; however, their function there has not yet been elucidated. In MS, mast cells are thought to be attracted to areas of lesion via chemokines such as RANTES and by self-antigens (possibly remnant epitopes of destroyed myelin) where they release histamine and tryptase which activate metalloprotease pathways, potentially contributing to the opening of the BBB and the facilitation of leukocytes influx into the CNS (Sospedra and Martin, 2005; Ziemssen and Ziemssen, 2005). Both neutrophils and macrophages are capable of generating enzymes such as inducible nitric oxide synthase (iNOS) which catalyzes the production of nitric oxide (NO), a short-lived, highly bioactive-free radical, produced in high amounts in normal bacterial infections. High iNOS levels have been found to be associated with MS lesions (Bo *et al.*, 1994) and NO is thought to augment microglial cytotoxic actions and bring about the destruction of myelin and oligodendrocytes in MS.

Other components, normally involved in innate immunity, have also been implicated in MS pathogenesis. For example, complement binding to myelin has been reported as a source of demyelination, and astrocytes may be a major CNS source for complements (Vanguri and Shin, 1986). Also complement activation has been shown to result in oligodendrocyte lysis and macrophage chemoattraction in MS (Johns and Bernard, 1997; Piddlesden and Morgan, 1993). Both NKT and γ/δ T cells have also been reported to participate in MS pathology (Gausling *et al.*, 2001; Triebel and Hercend, 1989).

VI. CNS-Specific Cellular Contribution in MS

For the large part, the contribution of the affected organ, the CNS, in mediating neuroinflammation in MS has widely understudied. Pathological and imaging studies, as well as research of the molecular aspects of the disease in EAE and MS, now provide ample evidence that CNS-specific factors are important. The kinetics of expression of costimulatory molecules and cytokines on astrocytes and neurons in the CNS of mice during EAE relapses have suggested strongly that the CNS resident cells are not passive bystanders in the disease but play active roles in EAE remission. Studies by Issazadeh *et al.* (1998) have shown that CD28 and CTLA-4 molecules, the two known ligands for the T-cell costimulatory molecules B7–1 and B7–2, have distinct expression patterns in the CNS: CD28/B7–2 expression increased in macrophages/microglia as well as astrocytes with increasing clinical severity of EAE. CTLA-4, on the other hand, was expressed by substantially fewer cells during the effector phase of disease and peaked during remission.

The expression of CD40 and CD40L in the CNS was increased during clinical attacks investigated in EAE. The expression of IL-12, IFN-γ, and TNF-α correlated with disease activity and severity, while TGF-β was upregulated in neurons during the recovery and remission phase. This study demonstrates that the kinetics of the *in vivo* expression of costimulatory molecules, their ligands, and cytokines in an autoimmune disease model influence or are modulated by remissions and relapses. Aside from CNS cellular involvement (Fig. 3), receptors for glutamate, a neurotransmitter in the CNS, are reported to be upregulated within MS lesions (Bolton and Paul, 2006).

VII. Regulatory Mechanisms in MS

The activation of myelin reactive T cells is also seen in other conditions such as spinal cord injury (Jones *et al.*, 2005). Why do these conditions not give rise to MS? Here, it is important to note that while we have discussed mechanisms by which T cells become activated (e.g., through molecular mimicry), there are also other mechanisms that prevent the excessive activation of T cells. These regulatory mechanisms include CD4+CD25+Foxp3+ regulatory T (Treg) cells which are TGF-β dependent (reviewed in Fehervari and Sakaguchi, 2004). Bettelli *et al.* (2006) showed that a balance exists in the generation of pathogenic T cells (Th17) which induce autoimmunity and regulatory (Foxp3) T cells that inhibit autoimmune tissue injury, suggesting an important role for Tregs in the determination of the fate of the autoimmune disease.

Tregs are T cells with an immunosuppressive function. Several different types of Tregs, including naturally arising CD25+CD4+ Tregs, IL-10-secreting Tr1 cells, TGF-β-secreting Th3 cells, Qa-1-restricted CD8+ T cells, CD8+CD28− T cells, CD8+CD122+ T cells, γ/δ T cells, and NKT cells have been reported to date (Baecher-Allan and Hafler, 2006; Lu *et al.*, 2006; Roncarolo *et al.*, 2006; Sakaguchi *et al.*, 2006; Shevach, 2006). Some Treg populations are naturally generated as functionally distinct populations, while others are adaptively induced from naive T cells as a consequence of antigen exposure (reviewed in Sakaguchi, 2006).

Tregs can actively suppress an immune response and are thereby thought to be capable of inhibiting ongoing pathogenic autoimmunity, allowing self-tolerance to be restored in patients suffering from autoimmune diseases (Tang and Bluestone, 2006). Tregs can act through dendritic cells to limit autoreactive T-cell activation, thereby preventing T-cell differentiation and activation (Baecher-Allan and Hafler, 2006). By limiting the supply of activated pathogenic cells, Tregs prevent or slow down the progression of autoimmune diseases (Nishikawa *et al.*, 2006). However, this protective mechanism appears insufficient in autoimmune patients, most likely due to a shortage of circulating Tregs and/or the development and accumulation of Treg-resistant pathogenic T cells over the disease course (Baecher-Allan and Hafler, 2006; Tang and Bluestone, 2006).

Viglietta *et al.* (2004) were the first to show an altered function of CD4+ CD25hi Tregs in patients with MS; they reported a significant decrease in the effector function of CD4+CD25hi regulatory T cells from peripheral blood of patients with MS as compared with healthy donors. They also demonstrated using *in vitro* cloning experiments that the cloning frequency of CD4+CD25hi Tregs was significantly reduced in MS patients as compared with normal controls, suggesting important function and possible future therapeutic roles for Tregs in MS (reviewed in Hafler *et al.*, 2005).

VIII. Chemokines in MS

Normal physiological trafficking and positioning of immune cells is controlled by lymphoid chemokines during hematopoiesis and immunosurveillance. However, during inflammation, lymphoid cells express inflammatory (or induced) chemokines to recruit inflammatory cells into these sites (Moser and Loetscher, 2001). Together with their receptors, chemokines play important roles in MS and EAE (reviewed in Babcock and Owens, 2003) at various levels. On EAE induction, CCL2 and CXCL10 levels are upregulated in the liver and various secondary lymphatic organs (Glabinski *et al.*, 1995) prior to appearance of clinical symptoms, suggesting a chemokine-dependent systemic immune activation. A study has suggested that chemokine signals (CXCL12) from glial cells from within the CNS are responsible for the migration of activated T cells from secondary lymphatic organs to the CNS (Flugel *et al.*, 2001); however, how these chemokines exit the blood–brain barrier is still unknown.

In MS, chemokine levels are altered in the blood and CXCR3 and CCR5 expressions are reported to be increased in circulating T cells (Calabresi *et al.*, 1999; Strunk *et al.*, 2000); these T cells are shown to secrete large amounts of proinflammatory cytokines. In the CSF, levels of CCL5 (RANTES) and CXCL10 (IP-10) are elevated, and CCL2 (MCP-1) levels are significantly reduced. These changes in chemokine levels correlate with the presence of CNS inflammation and gadolinium-enhanced lesions (detected by MRI), although the source of these chemokines in the CSF is yet to be identified. Similar to blood, CCR5 and CXCR3 are also detected in the CSF, while CXCR3 expression probably facilitates the entry of T cells into the CSF, and CXCL10 (IP-10) mediates their retention in the inflamed CNS.

At the level of the postcapillary venules within the CNS, chemokines have been implicated in leukocyte extravasation processes. A gradient of CCL19 and CCL21 across the blood vessel endothelium not only acts as homing signals for CCR7-expressing T cells but also strengthens the adhesion of extravasating T lymphocytes to endothelial cells (Alt *et al.*, 2002; Opdenakker *et al.*, 2003) by a proposed increase in the avidity of $\alpha4\beta1$-VCAM-1-mediated interactions (Alt *et al.*, 2002). Once in the CNS parenchyma, T cells interact with astrocytes to produce CCL2, which attracts peripheral/perivascular macrophages (expressing CCR2) to enter the CNS parenchyma.

The chemokine profile becomes complicated on disease establishment, when polymorphonuclear neutrophils (PMN)-derived CCL5 (RANTES), CCL3 (macrophage inflammatory protein-1 CCL2 (monocyte chemoatractant protein-1) CCR5, CXCR4, CXCL1, CXCL2, and CXCL3, and astroglial-derived CCL2, CXCL10, CCL3, CCL4, and CCL5 are all detected in the CNS and thought to augment subsequent leukocyte recruitment into the CNS (reviewed in Babcock and Owens, 2003; Opdenakker *et al.*, 2003). A more detailed knowledge about regulatory pathways is needed to detangle the complex cytokine and chemokine responses in EAE *in vivo*.

IX. Role for Metalloproteinases in MS

A role for various MMPs has been investigated in MS brain, CSF, and serum and also in rodent EAE, particularly in the degenerative processes of the disease. In EAE, the majority of the studies to date have focused on investigation of mRNA levels, which give an indication of the MMPs that are likely to be involved and their cellular source, but do not reflect MMP activity. One such comprehensive study of murine EAE, induced by adoptive transfer into SJL mice, has revealed increased mRNA levels for several metalloproteinases and their inhibitors, including MMP-8, -10, -12, ADAM-12, and TIMP-1, and downregulation of MMP-15 (Toft-Hansen *et al.*, 2004). Other studies on murine models of EAE have reported upregulation of mRNA levels for MMP-8, MMP-9 (Nygardas and Hinkkanen, 2002), and MMP-14 (Teesalu *et al.*, 2001), and of MMP-2 in encephalitogenic T cells (Graesser *et al.*, 1998).

Studies in a rat adoptive transfer model of EAE have shown strong upregulation of mRNA levels for MMP-9 and MMP-7, peaking at maximal disease severity, but unaltered levels for MMP-2, MMP-3, MMP-11 and MMP-13 (Clements *et al.*, 1997b; Kieseier *et al.*, 1998). The collective MMP data is clearly heterogeneous, probably due to the use of different species, different modes of disease induction (active vs adoptive transfer), and different stages of EAE examined, but nevertheless, cumulatively support a role for MMPs in EAE. This is substantiated by independent studies on MMP substrates, which have resulted in the identification of molecules relevant to inflammatory disease: for example, MMP-9 cleaves MBP (Gijbels *et al.*, 1993; Proost *et al.*, 1993) and the CNS proteoglycan, neural glia antigen 2 (NG2) (Larsen *et al.*, 2003); several chemokines are cleaved by MMP-9 (McQuibban *et al.*, 2002; Overall *et al.*, 2002) and TNF-α is cleaved by ADAM-17. MMPs have also been shown to regulate the inflammatory response by influencing chemotactic gradients and inducing cytokine production (D'haese *et al.*, 2000; Opdenakker *et al.*, 2001; Van den Steen *et al.*, 2000).

In support of a pathogenic role of MMPs in EAE, synthetic inhibitors of metalloproteinases have been shown to attenuate EAE disease course: minocycline, probably because of its Zn^{2+}-chelating properties, reduces production of MMP-9 and is efficacious against both mild and severe EAE in mice (Brundula *et al.*, 2002). The antioxidant α lipoic acid, which reduces MMP-9 activity, is found to reduce inflammation, demyelination, and axonal loss in the spinal cords of EAE-induced mice (Marracci *et al.*, 2002); in clinical trials, α lipoic acid was not only well tolerated by patients but also reduced MMP-9 and sICAM-1 levels indicating its beneficial role in MS treatment (Yadav *et al.*, 2005). The hydroxamate inhibitors of MMPs such as BB1101 (Clements *et al.*, 1997a), Ro31–9790 (Hewson *et al.*, 1995), and GM-6001 (Gijbels *et al.*, 1994) all reduce EAE clinical scores but do not protect against disease induction.

Although various MMPs and TIMPs have been genetically altered in mice (reviewed in Coussens *et al.*, 2001; Flannery, 2006), EAE has been investigated in only few null mice, specifically that of MMP-2, MMP-9, and MMP-12. Young MMP-9 null mice are less susceptible to EAE induction than wild-type mice (Dubois *et al.*, 1999); however, this difference is no longer apparent in older mice, which show normal disease onset and minimal reduction in disease severity.

MMP-2 null mice have been reported to be more susceptible to EAE induction than wild-type controls, due to a compensatory upregulation of MMP-9 activity (Esparza *et al.*, 2004). However, these studies show an unusually late onset of EAE symptoms in the wild-type controls and data have shown that EAE onset and progression in MMP-2 null mice are similar to that observed in WT mice (Agrawal *et al.*, 2006). On the other hand, MMP-12 null mice show increased disease severity and a significantly higher Th1 (INF-γ; T-bet) to Th2 (IL-5; GATA-3) cytokine and transcriptional regulator ratio (Weaver *et al.*, 2005), suggesting a protective role for MMP-12 in EAE but also a role in the T effector cell polarization.

MMPs, implicated in the degradation of BM components as well as myelin components in MS, are detectable in the spinal fluid of MS patients, and MMP-9 immunoreactivity is present in endothelial cells, pericytes, macrophages, and astrocytes of MS lesions (Bar-Or *et al.*, 2003). Myelin-specific T-cell clones derived from MS patients also produce MMP-9 on activation with antigen, and the presence of MMP-9 in the perivascular infiltrate demonstrated by immunohistochemistry is thought to be associated with disruption of the underlying basement membranes and is critical in the opening of the blood–brain barrier.

X. MS Therapies and Their Effect on Immune Abnormalities in MS

Given the above discussion, therapeutic agents for MS are being designed and tested based on their capability of affecting the immune system. The following section briefly describes the main modes of action attributed to immunomodulators currently used to treat MS patients and Fig. 4 summarizes the mode of action of each of the following.

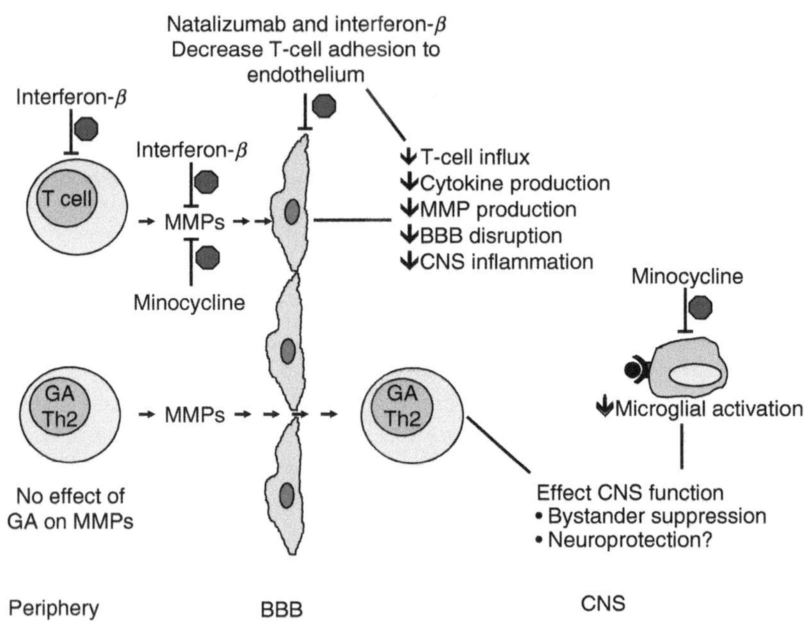

FIG. 4. Immunomodulators of MS. Shown here are the mechanisms and sites by which various immunomodulators exert their positive effects in MS. While Natalizumab and Interferon-β act mainly at the level of the blood–brain barrier, preventing immune cells from entering the CNS, minocycline influences both cellular entry into the CNS and cellular activation within the CNS. Glatiramer acetate (GA) has no effect at the level of the blood–brain barrier, but it generates GA-specific Th2 cells that render neuroprotection and suppress bystander activation within the CNS. (See Color Insert.)

A. Natalizumab

Monoclonal antibodies against $\alpha 4$ integrin abrogate the development of EAE in mice, consistent with this integrin being important in the trafficking of T cells into the CNS by interacting with its ligand on endothelial cells, vascular cell adhesion molecule-1 (Tubridy *et al.*, 1999; Yednock *et al.*, 1992). Natalizumab, one such monoclonal antibody, has been administered intravenously in the treatment of MS (blockage of integrin $\alpha 4\beta 1$) and Crohn's disease (blockage of integrin $\alpha 4\beta 1$) (von Andrian and Engelhardt, 2003). Administration in phase II and phase III trials in MS have produced significant benefits with results showing both a marked reduction in the formation of new lesions and a significant reduction in the risk of exacerbations within 6 months after the initiation of therapy (Miller *et al.*, 2003; Polman *et al.*, 2006). In the CSF of MS patients treated with Natalizumab, there is a prolonged reduction in the number of T lymphocytes (Stuve *et al.*, 2006), further supporting the mechanism of Natalizumab in reducing T-cell trafficking in humans.

Unfortunately, the success of Natalizumab had been curtailed when three MS patients administered the drug developed progressive multifocal leukoencephalopathy (PML) (Kleinschmidt-DeMasters and Tyler, 2005; Langer-Gould *et al.*, 2005); two of these died from the PML. Natalizumab was voluntarily withdrawn from the market after the PML incidents (Sheremata *et al.*, 2005), but it has now been reintroduced to the North American market where its use is monitored by strict guidelines (http://www.mssociety.ca/en/research/DMT061004.htm). Despite the PML setbacks, it is widely believed that targeting leukocyte trafficking using Natalizumab may still be a promising mode of treatment in MS therapy (Li *et al.*, 2006).

B. Interferon-β

Type I interferons (α and β) were discovered about 50 years ago to have antiviral properties. IFN-α (or "leukocyte IFN") is clinically used as an antiviral agent, whereas IFN-β (fibroblast IFN) results in anti-inflammatory effects and constitutes a cytokine used in the treatment of MS. Different forms of human recombinant IFN-β have been approved for treatment of relapsing-remitting MS because they reduce relapse rate, relapse severity, and the progression of disability. Type II IFN ("immune IFN" or IFN-γ) is produced by activated T cells and NK cells and has mainly proinflammatory actions.

IFN-β acts partly by reducing the levels and activity of proinflammatory cytokines, and inducing the expression of anti-inflammatory cytokines (Yong, 2002). In addition, *in vitro* studies show that IFN-β effectively decreased T-cell migration through reduction of MMP-9 levels (Stuve *et al.*, 1996); as

T-cell infiltration damages the blood–brain barrier, the decrease of T-cell migration by IFN-β can help account for the clinical finding that IFN-β rapidly reduces gadolinium-enhancing MRI activity and inflammation (Yong, 2002). To extend on this, it has been noted that MMP-9 activity in dendritic cells, which are shown to initiate an inflammatory response in MS, is diminished by IFN-β treatment in culture (Bartholome *et al.*, 2001).

Other *in vitro* studies have shown that IFN-β decreased blood–brain barrier cerebral endothelial cell permeability, accompanied by minor changes in immunostaining for tight-junction-associated proteins in the brain endothelial cell monolayers (Kraus *et al.*, 2004b). *In vivo*, the serum concentrations of the soluble forms of ICAM-1 and VCAM-1 are elevated after 3 months of IFN-β treatment in MS patients compared to the untreated situation (Kraus *et al.*, 2004a); the soluble forms of these adhesion molecules are thought to bind integrins on leukocytes before the latter have access to their ligands on endothelial cells. These findings support a stabilization effect of IFN-β on the integrity of the blood–brain barrier.

Another means by which the IFN-βs affect the blood–brain barrier is by their impact on MMPs, which, as noted earlier, are produced by leukocytes and contribute to the disruption of the blood–brain barrier. Clinical studies have supported the possibility that IFN-β regulates MMP levels in patients with MS. Boz *et al.* (2006) obtained blood and CSF samples from 14 patients with relapsing-remitting MS before and 6 months after IFN-β therapy. They report that after IFN-β treatment, serum MMP-9 as well as MMP-9:TIMP-1 ratio were significantly decreased from pretreatment levels. Furthermore, there was an increase in CSF levels of TIMP-1. Trojano *et al.* (1999) examined serum MMP-9 levels over a 24-month period in patients on IFN-β and they reported that a significant decrease was found by 3 months of treatment, the first time point evaluated, and that this was maintained for 18 months; patients with neutralizing antibodies to IFN-β tended to have higher MMP-9 levels. Notably, Galboiz *et al.* (2002) showed that IFN-β treatment was associated with a significant suppression of MMP-9 and -7 transcript levels in peripheral blood leukocytes from MS subjects. In another study (Boz *et al.*, 2006), 21 relapsing-remitting patients underwent simultaneous blood withdrawals and MRI examinations before and at different times during 48 weeks of IFN-β therapy. The serum MMP-9:TIMP-1 ratio, as well as the numbers of gadolinium-enhancing lesions, were found to be decreased during treatment. Moreover, the serum MMP-9:TIMP-1 ratio was a good predictor of the number of gadolinium-enhancing lesions. In yet another study, Karabudak *et al.* (2004) reported that while MMP-9 levels did not change during a 1-year period with IFN-β therapy, levels of TIMP-1 increased. Overall, these studies support the possibility that a major mechanism of IFN-β in MS is the decrease of MMP levels and activity, or the increase in levels of physiological inhibitors of MMPs such that the transmigratory capacity of leukocytes into the CNS is reduced.

C. Glatiramer Acetate

Glatiramer acetate (GA, Copaxone©), a random mixture of four amino acids (Ala, Lys, Glu, and Tyr) suppresses EAE (Arnon *et al.*, 1996) and has long-term efficacy and tolerability in the treatment of relapsing-remitting MS (Ford *et al.*, 2006). Different immunomechanisms (Farina *et al.*, 2005; Yong, 2002) have been suggested to contribute to the beneficial effects of GA, including its ability to affect the antigen-binding groove of MHC class II molecules, thereby potentially competing with other epitopes (e.g., autoreactive ones) for antigen binding. Other activites of GA include the generation of Th2 cytokine reactivity with anti-inflammatory effects (Aharoni *et al.*, 2000; Neuhaus *et al.*, 2000), the increase in activity of CD4+CD25+ regulatory cells (Hong *et al.*, 2005), the modulation of macrophage activity (Weber *et al.*, 2004), and the increase in CD8+ cells with suppressor activity (Karandikar *et al.*, 2002) (summarized in Fig. 4). Finally, the GA-reactive Th2 cells may also enter into the CNS to confer neuroprotection (Yong, 2002).

D. Minocycline

The semisynthetic tetracycline derivative, minocycline, has multiple actions including that of the inhibition of MMP activity (Yong *et al.*, 2004). In mice afflicted with EAE, minocycline has been described to reduce the expression and activity of MMP-9 in T cells and the administration of minocycline to EAE-afflicted animals has been shown to alleviate disease severity and neuropathology (Brundula *et al.*, 2002). Results from the animal studies have encouraged a clinical trial with the use of minocycline in 10 patients with relapsing-remitting MS. With gadolinium (gad)-enhancing MRI activity as a marker of clinical activity, patients beginning the trial showed reduced activity within the first 2 months of treatment with minocycline (Metz *et al.*, 2004). Serum samples from the patients in the minocycline trial before and at different times following minocycline treatment have demonstrated that MMP-9 activity in the serum of these individuals following minocycline treatment is reduced compared to pre-treatment values and is maintained for up to 18 months of therapy, the latest time point examined in this immune study (Zabad *et al.*, 2007). Besides the effect of minocycline at the level of the blood–brain barrier, other mechanisms that may contribute to its efficacy include the inhibition of microglia and T-cell activity, the reduction of neural cell apoptosis, the inhibition of glutamate cytotoxicity, and others (reviewed in Yong *et al.*, 2004) (summarized in Fig. 4).

XI. Conclusions

Vast amounts of work in the past few decades have contributed to a better understanding of MS immunopathogenesis. While some concepts on MS have become clearer, others remain elusive, revealing its extremely complicated nature. Our current standing is that MS is not only a CD4+ T-cell-mediated autoimmune disease but that it is a complex inflammatory disease involving many leukocyte subsets of innate and adaptive immunity, and influenced by both genetic background and environmental factors. Additional factors other than the immune system, such as the involvement of neural cells, also play active roles in MS. Despite this current knowledge, the heterogeneity among different patients and the unpredictable response to various modes of treatment only make the understanding of MS more challenging. Nonetheless, this encourages research to investigate the disease more thoroughly, working through the complexities to eventually unravel better-suited treatment strategies than the ones currently in use.

References

Agrawal, S., Anderson, P., Durbeej, M., van Rooijen, N., Ivars, F., Opdenakker, G., and Sorokin, L. (2006). Dystroglycan is selectively cleaved at the parenchymal basement membrane at sites of leukocyte extravasation in experimental autoimmune encephalomyelitis. *J. Exp. Med.* **203,** 1007–1019.

Aharoni, R., Teitelbaum, D., Leitner, O., Meshorer, A., Sela, M., and Arnon, R. (2000). Specific Th2 cells accumulate in the central nervous system of mice protected against experimental autoimmune encephalomyelitis by copolymer 1. *Proc. Natl. Acad. Sci. USA* **97,** 11472–11477.

Alt, C., Laschinger, M., and Engelhardt, B. (2002). Functional expression of the lymphoid chemokines CCL19 (ELC) and CCL 21 (SLC) at the blood-brain barrier suggests their involvement in G-protein-dependent lymphocyte recruitment into the central nervous system during experimental autoimmune encephalomyelitis. *Eur. J. Immunol.* **32,** 2133–2144.

Arnon, R., Sela, M., and Teitelbaum, D. (1996). New insights into the mechanism of action of copolymer 1 in experimental allergic encephalomyelitis and multiple sclerosis. *J. Neurol.* **243,** S8–S13.

Babcock, A., and Owens, T. (2003). Chemokines in experimental autoimmune encephalomyelitis and multiple sclerosis. *Adv. Exp. Med. Biol.* **520,** 120–132.

Baecher-Allan, C., and Hafler, D. A. (2006). Human regulatory T cells and their role in autoimmune disease. *Immunol. Rev.* **212,** 203–216.

Baranzini, S. E., Jeong, M. C., Butunoi, C., Murray, R. S., Bernard, C. C., and Oksenberg, J. R. (1999). B cell repertoire diversity and clonal expansion in multiple sclerosis brain lesions. *J. Immunol.* **163,** 5133–5144.

Bar-Or, A. (2005). Immunology of multiple sclerosis. *Neurol. Clin.* **23,** 149–175.

Bar-Or, A., Nuttall, R. K., Duddy, M., Alter, A., Kim, H. J., Ifergan, I., Pennington, C. J., Bourgoin, P., Edwards, D. R., and Yong, V. W. (2003). Analyses of all matrix metalloproteinase members in leukocytes emphasize monocytes as major inflammatory mediators in multiple sclerosis. *Brain* **126,** 2738–2749.

Bartholome, E. J., Van Aelst, I., Koyen, E., Kiss, R., Willems, F., Goldman, M., and Opdenakker, G. (2001). Human monocyte-derived dendritic cells produce bioactive gelatinase B: Inhibition by IFN-beta. *J. Interferon Cytokine Res.* **21,** 495–501.

Bettelli, E., Carrier, Y., Gao, W., Korn, T., Strom, T. B., Oukka, M., Weiner, H. L., and Kuchroo, V. K. (2006). Reciprocal developmental pathways for the generation of pathogenic effector TH17 and regulatory T cells. *Nature* **441,** 235–238.

Bielekova, B., Goodwin, B., Richert, N., Cortese, I., Kondo, T., Afshar, G., Gran, B., Eaton, J., Antel, J., Frank, J. A., McFarland, H. F., and Martin, R. (2000). Encephalitogenic potential of the myelin basic protein peptide (amino acids 83–99) in multiple sclerosis: Results of a phase II clinical trial with an altered peptide ligand. *Nat. Med.* **6,** 1167–1175.

Bo, L., Dawson, T. M., Wesselingh, S., Mork, S., Choi, S., Kong, P. A., Hanley, D., and Trapp, B. D. (1994). Induction of nitric oxide synthase in demyelinating regions of multiple sclerosis brains. *Ann. Neurol.* **36,** 778–786.

Bolton, C., and Paul, C. (2006). Glutamate receptors in neuroinflammatory demyelinating disease. *Mediators Inflamm.* **2006,** 1–12.

Bottomly, K. (1988). A functional dichotomy in CD4+ T lymphocytes. *Immunol. Today* **9,** 268–274.

Boz, C., Ozmenoglu, M., Velioglu, S., Kilinc, K., Orem, A., Alioglu, Z., and Altunayoglu, V. (2006). Matrix metalloproteinase-9 (MMP-9) and tissue inhibitor of matrix metalloproteinase (TIMP-1) in patients with relapsing-remitting multiple sclerosis treated with interferon beta. *Clin. Neurol. Neurosurg.* **108,** 124–128.

Brassat, D., Motsinger, A. A., Caillier, S. J., Erlich, H. A., Walker, K., Steiner, L. L., Cree, B. A., Barcellos, L. F., Pericak-Vance, M. A., Schmidt, S., Gregory, S., Hauser, S. L., *et al.* (2006). Multifactor dimensionality reduction reveals gene-gene interactions associated with multiple sclerosis susceptibility in African Americans *Genes Immun.* **7,** 310–315.

Brundula, V., Rewcastle, N. B., Metz, L. M., Bernard, C. C., and Yong, V. W. (2002). Targeting leukocyte MMPs and transmigration: Minocycline as a potential therapy for multiple sclerosis. *Brain* **125,** 1297–1308.

Burns, J., Rosenzweig, A., Zweiman, B., and Lisak, R. P. (1983). Isolation of myelin basic protein-reactive T-cell lines from normal human blood. *Cell. Immunol.* **81,** 435–440.

Calabresi, P. A., Martin, R., and Jacobson, S. (1999). Chemokines in chronic progressive neurological diseases: HTLV-1 associated myelopathy and multiple sclerosis. *J. Neurovirol.* **5,** 102–108.

Clements, J. M., Cossins, J. A., Wells, G. M., Corkill, D. J., Helfrich, K., Wood, L. M., Pigott, R., Stabler, G., Ward, G. A., Gearing, A. J., and Miller, K. M. (1997a). Matrix metalloproteinase expression during experimental autoimmune encephalomyelitis and effects of a combined matrix metalloproteinase and tumour necrosis factor-alpha inhibitor. *J. Neuroimmunol.* **74,** 85–94.

Clements, J. M., Cossins, J. A., Wells, G. M. A., Corkill, D. J., Helfrich, K., Wood, L. M., Pigott, R., Stabler, G., Ward, G. A., Gearing, A. J. H., and Miller, K. M. (1997b). Matrix metalloproteinase expression during experimental autoimmune encephalomyelitis and effects of a combined matrix metalloproteinase and tumour necrosis factor-α inhibitor. *J. Neuroimmunol.* **74,** 85–94.

Compston, A., and Coles, A. (2002). Multiple sclerosis. *Lancet* **359,** 1221–1231.

Coussens, L. M., Shapiro, S. D., Soloway, P. D., and Werb, Z. (2001). Models for gain-of-function and loss-of-function of MMPs. Transgenic and gene targeted mice. *Methods Mol. Biol.* **151,** 149–179.

D'haese, A., Wuyts, A., Dillen, C., Dubois, B., Billiau, A., Heremans, H., Van Damme, J., Arnold, B., and Opdenakker, G. (2000). *In vivo* neutrophil recruitment by granulocyte chemotactic protein-2 is assisted by gelatinase B/MMP-9 in the mouse. *J. Interferon Cytokine Res.* **20,** 667–674.

Dhib-Jalbut, S. (2002). Mechanisms of action of interferons and glatiramer acetate in multiple sclerosis. *Neurology* **58,** S3–S9.

Dubois, B., Masure, S., Hurtenbach, U., Paemen, L., Heremans, H., van den Oord, J., Sciot, R., Meinhardt, T., Hammerling, G., Opdenakker, G., and Arnold, B. (1999). Resistance of young gelatinase B-deficient mice to EAE and necrotizing tail lesions. *J. Clin. Invest.* **104,** 1507–1515.

Engelhardt, B. (2006). Molecular mechanisms involved in T cell migration across the blood-brain barrier. *J. Neural Trans.* **113,** 477–485.

Esparza, J., Kruse, M., Lee, J., Michaud, M., and Madri, J. A. (2004). MMP-2 null mice exhibit an early onset and severe EAE due to an increase in MMP-9 expression and activity. *FASEB J.* **18,** 1682–1691.

Farina, C., Weber, M. S., Meinl, E., Wekerle, H., and Hohlfeld, R. (2005). Glatiramer acetate in multiple sclerosis: Update on potential mechanisms of action. *Lancet Neurol.* **4,** 567–575.

Fehervari, Z., and Sakaguchi, S. (2004). Control of Foxp3+CD25+CD4+ regulatory cell activation and function by dendritic cells. *Int. Immunol.* **16,** 1769–1780.

Flannery, C. R. (2006). MMPs and ADAMTSs: Functional studies. *Front. Biosci.* **11,** 544–569.

Flugel, A., Berkowicz, T., Ritter, T., Labeur, M., Jenne, D. E., Li, Z., Ellwart, J. W., Willem, M., Lassmann, H., and Wekerle, H. (2001). Migratory activity and functional changes of green fluorescent effector cells before and during experimental autoimmune encephalomyelitis. *Immunity* **14,** 547–560.

Ford, C. C., Johnson, K. P., Lisak, R. P., Panitch, H. S., Shifronis, G., and Wolinsky, J. S. (2006). A prospective open-label study of glatiramer acetate: Over a decade of continuous use in multiple sclerosis patients. *Mult. Scler.* **12,** 309–320.

Galboiz, Y., Shapiro, S., Lahat, N., and Miller, A. (2002). Modulation of monocytes matrix metalloproteinase-2, MT1-MMP and TIMP-2 by interferon-gamma and -beta: Implications to multiple sclerosis. *J. Neuroimmunol.* **131,** 191–200.

Gausling, R., Trollmo, C., and Hafler, D. A. (2001). Decreases in interleukin-4 secretion by invariant CD4(−)CD8(−)V alpha 24J alpha Q T cells in peripheral blood of patients with relapsing-remitting multiple sclerosis. *Clin. Immunol.* **98,** 11–17.

Genain, C. P., Cannella, B., Hauser, S. L., and Raine, C. S. (1999). Identification of autoantibodies associated with myelin damage in multiple sclerosis. *Nat. Med.* **5,** 170–175.

Gijbels, K., Proost, P., Masure, S., Carton, H., Billiau, A., and Opdenakker, G. (1993). Gelatinase B is present in the cerebrospinal fluid during experimental autoimmune encephalomyelitis and cleaves myelin basic protein. *J. Neurosci. Res.* **36,** 432–440.

Gijbels, K., Galardy, R. E., and Steinman, L. (1994). Reversal of experimental autoimmune encephalomyelitis with a hydroxamate inhibitor of MMP. *J. Clin. Invest.* **94,** 2177–2182.

Giuliani, F., Goodyer, C. G., Antel, J. P., and Yong, V. W. (2003). Vulnerability of human neurons to T cell-mediated cytotoxicity. *J. Immunol.* **171,** 368–379.

Glabinski, A. R., Tani, M., Tuohy, V. K., Tuthill, R. J., and Ransohoff, R. M. (1995). Central nervous system chemokine mRNA accumulation follows initial leukocyte entry at the onset of acute murine experimental autoimmune encephalomyelitis. *Brain Behav. Immun.* **9,** 315–330.

Goverman, J., Woods, A., Larson, L., Weiner, L. P., Hood, L., and Zaller, D. M. (1993). Transgenic mice that express a myelin basic protein-specific T cell receptor develop spontaneous autoimmunity. *Cell* **72,** 551–560.

Graesser, D., Mahooti, S., Haas, T., Davis, S., Clark, R. B., and Madri, J. A. (1998). The interrelationship of $\alpha 4$ integrin and matrix metalloproteinase-2 in the apthogenesis of experimental autoimmune encephalomyelitis. *Lab. Invest.* **78,** 1445–1458.

Hafler, D. A., Slavik, J. M., Anderson, D. E., O'Connor, K. C., De Jager, P., and Baecher-Allan, C. (2005). Multiple sclerosis. *Immunol. Rev.* **204,** 208–231.

Harrington, L. E., Hatton, R. D., Mangan, P. R., Turner, H., Murphy, T. L., Murphy, K. M., and Weaver, C. T. (2005). Interleukin 17-producing CD4+ effector T cells develop via a lineage distinct from the T helper type 1 and 2 lineages. *Nat. Immunol.* **6,** 1123–1132.

Hayes, C. E. (2000). Vitamin D: A natural inhibitor of multiple sclerosis. *Proc. Nutr. Soc.* **59,** 531–535.

Hewson, A. K., Smith, T., Leonard, J. P., and Cuzner, M. L. (1995). Suppression of experimental allergic encephalomyelitis in the Lewis rat by the MMP inhibitor Ro31–9790. *Inflamm. Res.* **44,** 345–349.

Hong, J., Li, N., Zhang, X., Zheng, B., and Zhang, J. Z. (2005). Induction of CD4+ CD25+ regulatory T cells by copolymer-I through activation of transcription factor Foxp3. *Proc. Natl. Acad. Sci. USA* **102,** 6449–6454.

Huseby, E. S., Liggitt, D., Brabb, T., Schnabel, B., Ohlen, C., and Goverman, J. (2001). A pathogenic role for myelin-specific CD8(+) T cells in a model for multiple sclerosis. *J. Exp. Med.* **194,** 669–676.

Issazadeh, S., Navikas, V., Schaub, M., Sayegh, M., and Khoury, S. (1998). Kinetics of expression of costimulatory molecules and their ligands in murine relapsing experimental autoimmune encephalomyelitis *in vivo*. *J. Immunol.* **161,** 1104–1112.

Iwakura, Y., and Ishigame, H. (2006). The IL-23/IL-17 axis in inflammation. *J. Clin. Invest.* **116,** 1218–1222.

Johns, T. G., and Bernard, C. C. (1997). Binding of complement component Clq to myelin oligodendrocyte glycoprotein: A novel mechanism for regulating CNS inflammation. *Mol. Immunol.* **34,** 33–38.

Jones, T. B., Hart, R. P., and Popovich, P. G. (2005). Molecular control of physiological and pathological T-cell recruitment after mouse spinal cord injury. *J. Neurosci.* **25,** 6576–6583.

Jurewicz, A., Biddison, W. E., and Antel, J. P. (1998). MHC class I-restricted lysis of human oligodendrocytes by myelin basic protein peptide-specific CD8 T lymphocytes. *J. Immunol.* **160,** 3056–3059.

Kabat, E. A., and Freedman, D. A. (1950). A study of the crystalline albumin, gamma globulin and total protein in the cerebrospinal fluid of 100 cases of multiple sclerosis and in other diseases. *Am. J. Med. Sci.* **219,** 55–64.

Karabudak, R., Kurne, A., Guc, D., Sengelen, M., Canpinar, H., and Kansu, E. (2004). Effect of interferon beta-1a on serum matrix metalloproteinase-9 (MMP-9) and tissue inhibitor of matrix metalloproteinase (TIMP-1) in relapsing remitting multiple sclerosis patients. One year follow-up results. *J. Neurol.* **251,** 279–283.

Karandikar, N. J., Crawford, M. P., Yan, X., Ratts, R. B., Brenchley, J. M., Ambrozak, D. R., Lovett-Racke, A. E., Frohman, E. M., Stastny, P., Douek, D. C., Koup, R. A., and Racke, M. K. (2002). Glatiramer acetate (Copaxone) therapy induces CD8(+) T cell responses in patients with multiple sclerosis. *J. Clin. Invest.* **109,** 641–649.

Kidd, P. (2003). Th1/Th2 balance: The hypothesis, its limitations, and implications for health and disease. *Altern. Med. Rev.* **8,** 223–246.

Kieseier, B. C., Kiefer, R., Clements, J. M., Miller, K., Wells, G. M., Schweitzer, T., Gearing, A. J., and Hartung, H. P. (1998). Matrix metalloproteinase-9 and -7 are regulated in experimental autoimmune encephalomyelitis. *Brain* **121,** 159–166.

Kleinschmidt-DeMasters, B. K., and Tyler, K. L. (2005). Progressive multifocal leukoencephalopathy complicating treatment with natalizumab and interferon beta-1a for multiple sclerosis. *N. Engl. J. Med.* **353,** 369–374.

Kraus, J., Bauer, R., Chatzimanolis, N., Engelhardt, B., Tofighi, J., Bregenzer, T., Kuehne, B. S., Stolz, E., Blaes, F., Morgen, K., Traupe, H., Kaps, M., *et al.* (2004a). Interferon-beta 1b leads to a short-term increase of soluble but long-term stabilisation of cell surface bound adhesion molecules in multiple sclerosis. *J. Neurol.* **251,** 464–472.

Kraus, J., Ling, A. K., Hamm, S., Voigt, K., Oschmann, P., and Engelhardt, B. (2004b). Interferon-beta stabilizes barrier characteristics of brain endothelial cells *in vitro*. *Ann. Neurol.* **56,** 192–205.

Kurtzke, J. F. (1983). Rating neurologic impairment in multiple sclerosis: An expanded disability status scale (EDSS). *Neurology* **33,** 1444–1452.

Langer-Gould, A., Atlas, S. W., Green, A. J., Bollen, A. W., and Pelletier, D. (2005). Progressive multifocal leukoencephalopathy in a patient treated with natalizumab. *N. Engl. J. Med.* **353,** 375–381.

Langrish, C. L., Chen, Y., Blumenschein, W. M., Mattson, J., Basham, B., Sedgwick, J. D., McClanahan, T., Kastelein, R. A., and Cua, D. J. (2005). IL-23 drives a pathogenic T cell population that induces autoimmune inflammation. *J. Exp. Med.* **201,** 233–240.

Larsen, P. H., Wells, J. E., Stallcup, W. B., Opdenakker, G., and Yong, V. W. (2003). MMP-9 facilitates remyelination in part by processing the inhibitory NG2 proteoglycan. *J. Neurosci.* **23,** 11127–11135.

LaSalle, J. M., Ota, K., and Hafler, D. A. (1991). Presentation of autoantigen by human T cells. *J. Immunol.* **147,** 774–780.

Li, Y. Y., Perez, H. D., and Zollner, T. M. (2006). Fatalities in natalizumab treatment—a 'no go' for leukocyte recirculation approaches? *Expert Opin. Ther. Targets* **10,** 489–499.

Lu, L., Werneck, M. B., and Cantor, H. (2006). The immunoregulatory effects of Qa-1. *Immunol. Rev.* **212,** 51–59.

Lucchinetti, C., Bruck, W., Parisi, J., Scheithauer, B., Rodriguez, M., and Lassman, H. (2000). Heterogeneity of multiple sclerosis lesions: Implications for the pathogenesis of demyelination. *Ann. Neurol.* **47,** 707–717.

Marracci, G. H., Jones, R. E., McKeon, G. P., and Bourdette, D. N. (2002). Alpha lipoic acid inhibits T cell migration into the spinal cord and suppresses and treats experimental autoimmune encephalomyelitis. *J. Neuroimmunol.* **131,** 104–114.

Martin, R., McFarland, H. F., and McFarlin, D. E. (1992). Immunological aspects of demyelinating diseases. *Annu. Rev. Immunol.* **10,** 153–187.

McQuibban, G. A., Gong, J. H., Wong, J. P., Wallace, J. L., Clark-Lewis, I., and Overall, C. M. (2002). Matrix metalloproteinase processing of monocyte chemoattractant proteins generates CC chemokine receptor antagonists with anti-inflammatory properties *in vivo*. *Blood* **100,** 1160–1167.

Medana, I. M., Gallimore, A., Oxenius, A., Martinic, M. M., Wekerle, H., and Neumann, H. (2000). MHC class I-restricted killing of neurons by virus-specific CD8+ T lymphocytes is effected through the Fas/FasL, but not the perforin pathway. *Eur. J. Immunol.* **30,** 3623–3633.

Metz, L. M., Zhang, Y., Yeung, M., Patry, D. G., Bell, R. B., Stoian, C. A., Yong, V. W., Patten, S. B., Duquette, P., Antel, J. P., and Mitchell, J. R. (2004). Minocycline reduces gadolinium-enhancing magnetic resonance imaging lesions in multiple sclerosis. *Ann. Neurol.* **55,** 756.

Miller, D. H., Khan, O. A., Sheremata, W. A., Blumhardt, L. D., Rice, G. P., Libonati, M. A., Willmer-Hulme, A. J., Dalton, C. M., Miszkiel, K. A., and O'Connor, P. W. (2003). A controlled trial of natalizumab for relapsing multiple sclerosis. *N. Engl. J. Med.* **348,** 15–23.

Moser, B., and Loetscher, P. (2001). Lymphocyte traffic control by chemokines. *Nat. Immunol.* **2,** 123–128.

Mosmann, T. R., and Coffman, R. L. (1989). TH1 and TH2 cells: Different patterns of lymphokine secretion lead to different functional properties. *Annu. Rev. Immunol.* **7,** 145–173.

Motsinger, A. A., Brassat, D., Caillier, S. J., Erlich, H. A., Walker, K., Steiner, L. L., Barcellos, L. F., Pericak-Vance, M. A., Schmidt, S., Gregory, S., Hauser, S. L., Haines, J. L., *et al.* (2007). Complex gene-gene interactions in multiple sclerosis: A multifactorial approach reveals associations with inflammatory genes. *Neurogenetics* **8,** 11–20.

Murali-Krishna, K., Altman, J. D., Suresh, M., Sourdive, D. J., Zajac, A. J., Miller, J. D., Slansky, J., and Ahmed, R. (1998). Counting antigen-specific CD8 T cells: A reevaluation of bystander activation during viral infection. *Immunity* **8,** 177–187.

Murphy, C. A., Langrish, C. L., Chen, Y., Blumenschein, W., McClanahan, T., Kastelein, R. A., Sedgwick, J. D., and Cua, D. J. (2003). Divergent pro- and antiinflammatory roles for IL-23 and IL-12 in joint autoimmune inflammation. *J. Exp. Med.* **198,** 1951–1957.

Neuhaus, O., Farina, C., Yassouridis, A., Wiendl, H., Then Bergh, F., Dose, T., Wekerle, H., and Hohlfeld, R. (2000). Multiple sclerosis: Comparison of copolymer-1-reactive T cell lines from treated and untreated subjects reveals cytokine shift from T helper 1 to T helper 2 cells. *Proc. Natl. Acad. Sci. USA* **97,** 7452–7457.

Nishikawa, H., Qian, F., Tsuji, T., Ritter, G., Old, L. J., Gnjatic, S., and Odunsi, K. (2006). Influence of CD4+ CD25+ regulatory T cells on low/high-avidity CD4+ T cells following peptide vaccination. *J. Immunol.* **176,** 6340–6346.

Nygardas, P., and Hinkkanen, A. (2002). Up-regulation of MMP-8 and MMP-9 activity in the BALB/c mouse spinal cord correlates with the severity of EAE. *Clin. Exp. Immunol.* **128,** 245–254.

Opdenakker, G., Van den Steen, P. E., and Van Damme, J. (2001). Gelatinase B: A tuner and amplifier of immune functions. *Trends Immunol.* **22,** 571–579.

Opdenakker, G., Nelissen, I., and Van Damme, J. (2003). Functional roles and therapeutic targeting of gelatinase B and chemokines in multiple sclerosis. *Lancet Neurol.* **2,** 747–756.

Overall, C. M., McQuibban, G. A., and Clark-Lewis, I. (2002). Discovery of chemokine substrates for matrix metalloproteinases by exosite scanning: A new tool for degradomics. *Biol. Chem.* **383,** 1059–1066.

Park, H., Li, Z., Yang, X. O., Chang, S. H., Nurieva, R., Wang, Y. H., Wang, Y., Hood, L., Zhu, Z., Tian, Q., and Dong, C. (2005). A distinct lineage of CD4 T cells regulates tissue inflammation by producing interleukin 17. *Nat. Immunol.* **6,** 1133–1141.

Pender, M. P. (2003). Infection of autoreactive B lymphocytes with EBV, causing chronic autoimmune diseases. *Trends Immunol.* **24,** 584–588.

Pettinelli, C. B., and McFarlin, D. E. (1981). Adoptive transfer of experimental allergic encephalomyelitis in SJL/J mice after *in vitro* activation of lymph node cells by myelin basic protein: Requirement for Lyt 1+ 2-T lymphocytes. *J. Immunol.* **127,** 1420–1423.

Piddlesden, S. J., and Morgan, B. P. (1993). Killing of rat glial cells by complement: Deficiency of the rat analogue of CD59 is the cause of oligodendrocyte susceptibility to lysis. *J. Neuroimmunol.* **48,** 169–175.

Polman, C. H., O'Connor, P. W., Havrdova, E., Hutchinson, M., Kappos, L., Miller, D. H., Phillips, J. T., Lublin, F. D., Giovannoni, G., Wajgt, A., Toal, M., Lynn, F., *et al.* (2006). A randomized, placebo-controlled trial of natalizumab for relapsing multiple sclerosis. *N. Engl. J. Med.* **354,** 899–910.

Ponsonby, A. L., Lucas, R. M., and van der Mei, I. A. (2005a). UVR, vitamin D and three autoimmune diseases—multiple sclerosis, type 1 diabetes, rheumatoid arthritis. *Photochem. Photobiol.* **81,** 1267–1275.

Ponsonby, A. L., van der Mei, I., Dwyer, T., Blizzard, L., Taylor, B., and Kemp, A. (2005b). Birth order, infection in early life, and multiple sclerosis. *Lancet Neurol.* **4,** 793–794. Author reply 795.

Prineas, J. (1975). Pathology of the early lesion in multiple sclerosis. *Hum. Pathol.* **6,** 531–554.

Prinz, M., Garbe, F., Schmidt, H., Mildner, A., Gutcher, I., Wolter, K., Piesche, M., Schroers, R., Weiss, E., Kirschning, C. J., Rochford, C. D., Bruck, W., *et al.* (2006). Innate immunity mediated by TLR9 modulates pathogenicity in an animal model of multiple sclerosis. *J. Clin. Invest.* **116,** 456–464.

Proost, P., Van Damme, J., and Opdenakker, G. (1993). Leukocyte gelatinase B cleavage releases encephalitogens from human myelin basic protein. *Biochem. Biophys. Res. Commun.* **192,** 1175–1181.

Rodriguez, M., and Lennon, V. A. (1990). Immunoglobulins promote remyelination in the central nervous system. *Ann. Neurol.* **27,** 12–17.

Roncarolo, M. G., Gregori, S., Battaglia, M., Bacchetta, R., Fleischhauer, K., and Levings, M. K. (2006). Interleukin-10-secreting type 1 regulatory T cells in rodents and humans. *Immunol. Rev.* **212,** 28–50.

Sakaguchi, S. (2006). Regulatory T cells. *Springer Semin. Immunopathol.* **28,** 1–2.

Sakaguchi, S., Ono, M., Setoguchi, R., Yagi, H., Hori, S., Fehervari, Z., Shimizu, J., Takahashi, T., and Nomura, T. (2006). Foxp3+CD25+CD4+ natural regulatory T cells in dominant self-tolerance and autoimmune disease. *Immunol. Rev.* **212,** 8–27.

Saoudi, A., Simmonds, S., Huitinga, I., and Mason, D. (1995). Prevention of experimental allergic encephalomyelitis in rats by targeting autoantigen to B cells: Evidence that the protective mechanism depends on changes in the cytokine response and migratory properties of the autoantigen-specific T cells. *J. Exp. Med.* **182,** 335–344.

Schluesener, H. J., and Wekerle, H. (1985). Autoaggressive T lymphocyte lines recognizing the encephalitogenic region of myelin basic protein: *In vitro* selection from unprimed rat T lymphocyte populations. *J. Immunol.* **135,** 3128–3133.

Sheremata, W. A., Minagar, A., Alexander, J. S., and Vollmer, T. (2005). The role of alpha-4 integrin in the aetiology of multiple sclerosis: Current knowledge and therapeutic implications. *CNS Drugs* **19,** 909–922.

Shevach, E. M. (2006). From vanilla to 28 flavors: Multiple varieties of T regulatory cells. *Immunity* **25,** 195–201.

Sixt, M., Engelhardt, B., Pausch, F., Hallmann, R., Wendler, O., and Sorokin, L. M. (2001). Endothelial cell laminin isoforms, laminins 8 and 10, play decisive roles in T cell recruitment across the blood-brain barrier in experimental autoimmune encephalomyelitis. *J. Cell Biol.* **153,** 933–946.

Skulina, C., Schmidt, S., Dornmair, K., Babbe, H., Roers, A., Rajewsky, K., Wekerle, H., Hohlfeld, R., and Goebels, N. (2004). Multiple sclerosis: Brain-infiltrating CD8+ T cells persist as clonal expansions in the cerebrospinal fluid and blood. *Proc. Natl. Acad. Sci. USA* **101,** 2428–2433.

Sospedra, M., and Martin, R. (2005). Immunology of multiple sclerosis. *Annu. Rev. Immunol.* **23,** 683–747.

Strunk, T., Bubel, S., Mascher, B., Schlenke, P., Kirchner, H., and Wandinger, K. P. (2000). Increased numbers of CCR5+ interferon-gamma- and tumor necrosis factor-alpha-secreting T lymphocytes in multiple sclerosis patients. *Ann. Neurol.* **47,** 269–273.

Stuve, O., Dooley, N. P., Uhm, J. H., Antel, J. P., Francis, G. S., Williams, G., and Yong, V. W. (1996). Interferon beta-1b decreases the migration of T lymphocytes *in vitro*: Effects on matrix metalloproteinase-9. *Ann. Neurol.* **40,** 853–863.

Stuve, O., Marra, C. M., Jerome, K. R., Cook, L., Cravens, P. D., Cepok, S., Frohman, E. M., Phillips, J. T., Arendt, G., Hemmer, B., Monson, N. L., and Racke, M. K. (2006). Immune surveillance in multiple sclerosis patients treated with natalizumab. *Ann. Neurol.* **59,** 743–747.

Sun, J., Link, H., Olsson, T., Xiao, B. G., Andersson, G., Ekre, H. P., Linington, C., and Diener, P. (1991). T and B cell responses to myelin-oligodendrocyte glycoprotein in multiple sclerosis. *J. Immunol.* **146,** 1490–1495.

Tang, Q., and Bluestone, J. A. (2006). Regulatory T-cell physiology and application to treat autoimmunity. *Immunol. Rev.* **212,** 217–237.

Teesalu, T., Hinkkanen, A. E., and Vaheri, A. (2001). Coordinated induction of extracellular proteolysis systems during EAE in mice. *Am. J. Pathol.* **159,** 2227–2237.

Toft-Hansen, H., Nuttall, R. K., Edwards, D. R., and Owens, T. (2004). Key metalloproteinases are expressed by specific cell types in experimental autoimmune encephalomyelitis. *J. Immunol.* **173,** 5209–5218.

Tough, D. F., and Sprent, J. (1996). Viruses and T cell turnover: Evidence for bystander proliferation. *Immunol. Rev.* **150,** 129–142.

Tran, E. H., Hoekstra, K., Rooijen, N. v., Dijkstra, C. D., and Owens, T. (1998). Immune evasion of the central nervous system parenchyma and experimental allergic encephalomyelitis, but not leukocyte extravasation from blood, are prevented in macrophage-depleted mice. *J. Immunol.* **161,** 3767–3775.

Triebel, F., and Hercend, T. (1989). Subpopulations of human peripheral T gamma delta lymphocytes. *Immunol. Today* **10,** 186–188.

Trojano, M., Avolio, C., Liuzzi, G. M., Ruggieri, M., Defazio, G., Liguori, M., Santacroce, M. P., Paolicelli, D., Giuliani, F., Riccio, P., and Livrea, P. (1999). Changes of serum sICAM-1 and MMP-9 induced by rIFNbeta-1b treatment in relapsing-remitting MS. *Neurology* **53,** 1402–1408.

Tubridy, N., Behan, P. O., Capildeo, R., Chaudhuri, A., Forbes, R., Hawkins, C. P., Hughes, R. A., Palace, J., Sharrack, B., Swingler, R., Young, C., Moseley, I. F., *et al.* (1999). The effect of anti-alpha4 integrin antibody on brain lesion activity in MS. The UK Antegren Study Group. *Neurology* **53,** 466–472.

Van den Steen, P. E., Proost, P., Wuyts, A., Van Damme, J., and Opdenakker, G. (2000). Neutrophil gelatinase B potentiates interleukin-8 tenfold by aminoterminal processing, whereas it degrades CTAP-III, PF-4, and GRO-alpha and leaves RANTES and MCP-2 intact. *Blood* **96,** 2673–2681.

van Etten, E., and Mathieu, C. (2005). Immunoregulation by 1,25-dihydroxyvitamin D3: Basic concepts. *J. Steroid Biochem. Mol. Biol.* **97,** 93–101.

van Horssen, J., Bo, L., Vos, C. M., Virtanen, I., and de Vries, H. E. (2005). Basement membrane proteins in multiple sclerosis-associated inflammatory cuffs: Potential role in influx and transport of leukocytes. *J. Neuropathol. Exp. Neurol.* **64,** 722–729.

Vanguri, P., and Shin, M. L. (1986). Activation of complement by myelin: Identification of C1-binding proteins of human myelin from central nervous tissue. *J. Neurochem.* **46,** 1535–1541.

Veldhoen, M., and Stockinger, B. (2006). TGFbeta1, a 'Jack of all trades': The link with pro-inflammatory IL-17-producing T cells. *Trends Immunol.* **27,** 358–361.

Viglietta, V., Baecher-Allan, C., Weiner, H. L., and Hafler, D. A. (2004). Loss of functional suppression by CD4+ CD25+ regulatory T cells in patients with multiple sclerosis. *J. Exp. Med.* **199,** 971–979.

von Andrian, U. H., and Engelhardt, B. (2003). Alpha4 integrins as therapeutic targets in autoimmune disease. *N. Engl. J. Med.* **348,** 68–72.

Waldner, H., Collins, M., and Kuchroo, V. K. (2004). Activation of antigen-presenting cells by microbial products breaks self tolerance and induces autoimmune disease. *J. Clin. Invest.* **113,** 990–997.

Walsh, M. J., and Tourtellotte, W. W. (1986). Temporal invariance and clonal uniformity of brain and cerebrospinal IgG, IgA, and IgM in multiple sclerosis. *J. Exp. Med.* **163,** 41–53.

Wang, T., Allie, R., Conant, K., Haughey, N., Turchan-Chelowo, J., Hahn, K., Rosen, A., Steiner, J., Keswani, S., Jones, M., Calabresi, P. A., and Nath, A. (2006). Granzyme B mediates neurotoxicity through a G-protein-coupled receptor. *FASEB J.* **20,** 1209–1211.

Weaver, A., Goncalves da Silva, A., Nuttall, R. K., Edwards, D. R., Shapiro, S. D., Rivest, S., and Yong, V. W. (2005). An elevated matrix metalloproteinase (MMP) in an animal model of multiple sclerosis is protective by affecting Th1/Th2 polarization. *FASEB J.* **19,** 1668–1670.

Weber, M. S., Starck, M., Wagenpfeil, S., Meinl, E., Hohlfeld, R., and Farina, C. (2004). Multiple sclerosis: Glatiramer acetate inhibits monocyte reactivity *in vitro* and *in vivo*. *Brain* **127,** 1370–1378.

Willer, C. J., Herrera, B. M., Morrison, K. M., Sadovnick, A. D., and Ebers, G. C. (2006). Association between microchimerism and multiple sclerosis in Canadian twins. *J. Neuroimmunol.* **179,** 145–151.

Wucherpfennig, K. W., Catz, I., Hausmann, S., Strominger, J. L., Steinman, L., and Warren, K. G. (1997). Recognition of the immunodominant myelin basic protein peptide by autoantibodies and HLA-DR2-restricted T cell clones from multiple sclerosis patients. Identity of key contact residues in the B-cell and T-cell epitopes. *J. Clin. Invest.* **100,** 1114–1122.

Yadav, V., Marracci, G., Lovera, J., Woodward, W., Bogardus, K., Marquardt, W., Shinto, L., Morris, C., and Bourdette, D. (2005). Lipoic acid in multiple sclerosis: A pilot study. *Mult. Scler.* **11,** 159–165.

Yednock, T. A., Cannon, C., Fritz, L. C., Sanchez-Madrid, F., Steinman, L., and Karin, N. (1992). Prevention of experimental autoimmune encephalomyelitis by antibodies against $\alpha4\beta1$ integrin. *Nature* **356,** 63–66.

Yong, V. W. (2002). Differential mechanisms of action of interferon-beta and glatiramer acetate in MS. *Neurology* **59,** 802–808.

Yong, V. W., Wells, J., Giuliani, F., Casha, S., Power, C., and Metz, L. M. (2004). The promise of minocycline in neurology. *Lancet Neurol.* **3,** 744–751.

Zabad, R. K., M. L., Todoruk, T. R., Zhang, Y., Mitchell, J. R., Yeung, M., Patry, D. G., Bell, R. M., and Yong, V. W. (2007). The clinical response to minocycline in MS is accompanied by immune changes. *Mult. Scler.*

Zamvil, S. S., and Steinman, L. (1990). The T lymphocyte in experimental allergic encephalomyelitis. *Annu. Rev. Immunol.* **8,** 579–621.

Ziemssen, T., and Ziemssen, F. (2005). The role of the humoral immune system in multiple sclerosis (MS) and its animal model experimental autoimmune encephalomyelitis (EAE). *Autoimmun. Rev.* **4,** 460–467.

MOLECULAR MIMICRY IN MULTIPLE SCLEROSIS

Jane E. Libbey, Lori L. McCoy, and Robert S. Fujinami

Department of Neurology, University of Utah School of Medicine, Salt Lake City
Utah 84132, USA

One of the most common demyelinating central nervous system (CNS) diseases in humans is multiple sclerosis (MS). The disease can be very debilitating with vision loss, motor and sensory disturbances, and cognitive impairment. The clinical course may present as a relapsing-remitting disease course, a progressive disease course, or a combination thereof. The etiology of MS is unknown. Though many viruses have been shown to be associated with MS, no one virus has ever been demonstrated to be the cause of MS. In addition, MS is thought to have an autoimmune component. Molecular mimicry is one hypothesis put forth which could reconcile the diverse pathology and etiology of MS. Molecular mimicry occurs when peptides from pathogens share sequence or structural similarities with self-antigens. Infection with various pathogens, each with its individual molecular mimic to a CNS antigen, may explain the inability of investigators to link one specific virus to MS. Molecular mimicry may be mediated through human leukocyte antigen class I- and class II-restricted T cells and antibodies, which may explain the diversity in phenotype. Aspects of molecular mimicry will be discussed in relation to each of these immune system components. Examples of various molecular mimics will be discussed with a particular focus on the CNS and MS. Molecular mimicry alone may not be able to induce disease; priming of the immune system by infection with a pathogen that carries a molecular mimic to self may have to be followed by a later nonspecific immunologic challenge in order for disease to be initiated. Recent research into this priming and triggering of disease will be discussed in relation to an animal model for MS.

INTERNATIONAL REVIEW OF
NEUROBIOLOGY, VOL. 79
DOI: 10.1016/S0074-7742(07)79006-2

0074-7742/07 $35.00

I. Introduction

Multiple sclerosis (MS) is the most common demyelinating disease in humans. MS has prevalence rates between 50 and 100 per 100,000 Caucasians; other ethnic groups have somewhat lower prevalence rates and women are more afflicted than men by a 2:1 ratio (Kurtzke, 1997). The inflammatory demyelinating lesions characteristic of MS are limited to the central nervous system (CNS) (Dejong, 1970). In most instances, MS patients have oligoclonal immunoglobulin (Ig)G bands in the cerebral spinal fluid (CSF), and a mild mononuclear pleocytosis may also be present. Clinical features of the disease include vision loss, motor and sensory disturbances, and cognitive impairment. The clinical course of MS can include relapses and remissions and may be progressive in nature. MS has been proposed to be mediated by autoreactive CNS-specific $CD4^+$ T cells (Markovic-Plese and McFarland, 2001; Noseworthy *et al.*, 2000). However, significant numbers of $CD8^+$ T cells are found in MS lesions (Hayashi *et al.*, 1988; Sobel, 1989) and are likely to be involved in pathogenesis.

There have been many studies which attribute the neuroinflammation and neurodegeneration in MS to $CD8^+$ T cells, $CD4^+$ T cells, or antibody recognition of self. In this chapter, we summarize some salient points on each of these components of the immune system as they relate to molecular mimicry and disease. However, first we will provide background information and define molecular mimicry in relation to other key concepts.

Molecular mimicry was first hypothesized to be a potential mechanism for the initiation of autoimmune disease in the early 1980s. It is said to occur when peptides from viral (Fujinami *et al.*, 1983) or bacterial (Zabriskie, 1986) proteins share sequence or structural similarities with self-peptides. This similarity could stimulate autoreactive immune responses causing the production of self-reactive antibodies. Fujinami, Oldstone, and colleagues (Fujinami and Oldstone, 1985; Fujinami *et al.*, 1983) initially showed that a cross-reactive epitope between the *Hepatitis B virus* polymerase and the encephalitogenic epitope of myelin basic protein (MBP) for the rabbit could induce an experimental autoimmune encephalomyelitis (EAE)-like disease when used as an immunogen. For EAE to occur, this cross-reaction had to occur at the level of induction/activation of autoreactive $CD4^+$ T cells. This was the first demonstration of autoimmune disease induced by a viral peptide (Fujinami and Oldstone, 1985).

In the past, peptide similarities had been identified by computer searches for similar amino acid sequences; however, based on more recent studies, molecular mimicry has also been shown to occur with incomplete sequence matching provided the major histocompatibility complex (MHC) and T-cell receptor (TCR) contact motifs are preserved (Lang *et al.*, 2002). Additionally, peptide elution studies have found that MHC molecules can potentially bind hundreds of different peptides (Hunt *et al.*, 1992). The work by Lang *et al.* (2002) suggested that molecular mimicry occurs rather frequently having shown that different peptides bound to class II molecules can lead to cross-reactivity by the same TCR provided the complexes have similar charge distribution and overall shape. The flexibility of TCR recognition plays a major role in forming the T-cell repertoire through thymic selection, and is highly important in protecting the host against potential pathogen-derived antigens that are more wide ranging than the limited number of memory T cells (Casrouge *et al.*, 2000). TCR degeneracy raises the potential for cross-reactivity between pathogen-derived and self-antigens.

The various forms of molecular mimicry are illustrated in Fig. 1. There are many molecular mimics that have been identified in MS studies, both viral and bacterial, as shown in Table I, which supports the hypothesis that molecular mimicry frequently occurs.

The Welsh and Selin laboratories have investigated molecular mimicry, or the sharing of immunologic cross-reactive epitopes, between viruses, as opposed to between a pathogen and its host, using such viruses as *Lymphocytic choriomeningitis virus* (LCMV), *Pichinde virus* (PV), *Vaccinia virus* (VV), and murine cytomegalovirus (MCMV) (Brehm *et al.*, 2002; Chen *et al.*, 2001; Welsh *et al.*, 2000). They coined the term "heterologous immunity" to describe the protection provided by one

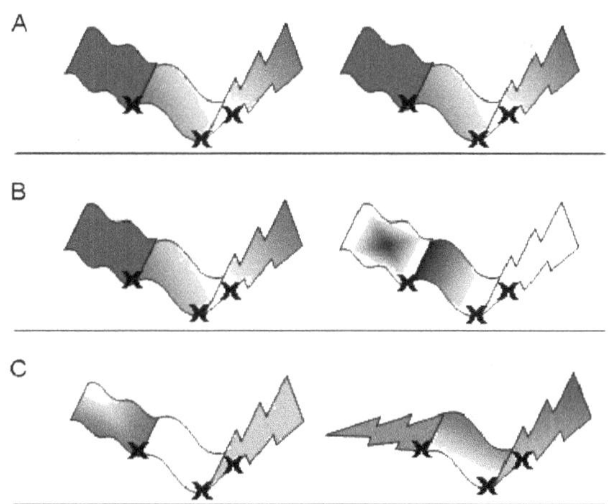

FIG. 1. Molecular mimicry with and without sequence homology. The two peptides shown in (A) share sequence homology as well as T-cell receptor (TCR) contact sites (marked as an "x"). The peptide sequences in (B) have the same overall shape and TCR contact residues, but do not share sequence homology. In (C), the sequences share TCR contact sites only; they do not share sequence homology and the overall shape is slightly different. Molecular mimicry may still occur due to the core region of TCR recognition; however, this recognition may be lower affinity.

TABLE I
MOLECULAR MIMICS IDENTIFIED IN MULTIPLE SCLEROSIS STUDIES

Molecular mimic	Infectious agent
Arginine-enriched domains of CNS proteins	Torque Teno Virus (Sospedra *et al.*, 2005)
Myelin basic protein	*Acinetobacter calcoaceticus* 4-carboxymuconolactone decarboxylase[a]
	Chlamydia pneumoniae (Lenz *et al.*, 2001)
	Epstein-Barr virus (Lang *et al.*, 2002)[b]
	Human coronavirus 229E[c]
	Human hepatitis B virus polymerase (Fujinami *et al.*, 1983)
	Human herpesvirus 6[d] (Tejada-Simon *et al.*, 2003)
	Maedi Visna virus[e]
	Pseudomonas aeruginosa γ-carboxymucono lactone decarboxylase[a]
Myelin oligodendrocyte glycoprotein	Butyrophilin[f]
	Rubella[g]
Myelin proteolipid protein	*Haemophilus influenzae* serine protease IV (Croxford *et al.*, 2005)

[a]Hughes *et al.* (2003).
[b]Ufret-Vincenty *et al.* (1998).
[c]Boucher *et al.* (2001).
[d]Cirone *et al.* (2002).
[e]Davies *et al.* (1996).
[f]Stefferl *et al.* (2000).
[g]Besson Duvanel *et al.* (2001).

infection against a different viral infection at a later time (Chen *et al.*, 2001). This protection to the second infection was mediated by CD8$^+$ T cells and the production of interferon (IFN)-γ (Brehm *et al.*, 2002). They also found that for CD8$^+$ T cells, cross-reactivity impacts T-cell kinetics and the hierarchy of T cells responding to epitopes encoded by the infecting virus, which changes T-cell responsiveness (Chen *et al.*, 2001; Selin *et al.*, 1998). Foreign proteins generally have several epitopes capable of being presented by MHC molecules; however, the cellular immune response is usually focused toward a narrow subset of these epitopes. This narrow response is known as immunodominance. When dominant epitopes have been deleted from a pathogenic organism, other epitopes emerge as dominant (Yewdell and Bennink, 1999). This phenomenon of immunodominance can be affected by prior infection with an organism encoding an epitope that generates T cells that cross-react with a second pathogen (Brehm *et al.*, 2002). The skewing of T-cell immunodominance may be beneficial or harmful to the host. Additional examples of cross-reactivity resulting in altered immune response include: primary infection with influenza virus followed by Epstein-Barr virus (EBV), *Human papillomavirus* and *Human coronavirus*, and that which occurs between different strains of *Dengue virus* (Kim *et al.*, 2005). For example, dengue virus serotypes encode subdominant epitopes that share sequence homology between them. When a host with prior immunity to one of these serotypes is infected with another viral serotype, the weakly shared epitope dominates the immune response resulting in better recognition of the first serotype. This is due to memory T cells specific for the shared epitope proliferating and expanding, giving the subdominant epitope a distinct advantage (Mongkolsapaya *et al.*, 2003). Unfortunately, not all cases of heterologous immunity are as predictable. For instance, previous infection with LCMV has been shown to protect mice against VV infection resulting in reduced viral titers and a change in the T-cell-dependent pathology (Chen *et al.*, 2001; Selin *et al.*, 1998). In contrast, prior VV infection has little effect on LCMV immunity. It has been reported that VV infection of LCMV-immune mice results in expansion of MHC class I-restricted $CD8^+$ T cells with different epitopes dominating the response (Chen *et al.*, 2001; Kim *et al.*, 2002, 2005; Selin *et al.*, 1998). Following VV infection, LCMV-immune mice had a reduction in the frequency of LCMV-specific $CD8^+$ T memory cells to some epitopes while others remained constant or increased (Kim *et al.*, 2005).

Heterologous immunity and molecular mimicry may be able to explain the low concordance rates for MS in monozygotic twins. With a 30% concordance rate for this group, it is clear that MS has a genetic component but that environmental factors also play a role in disease pathogenesis (Ebers *et al.*, 1986; Mumford *et al.*, 1994; Sadovnick *et al.*, 1993). Perhaps a series of infections that generate immune reactivity to otherwise subdominant epitopes mimicking self-proteins are required in a genetically predisposed individual in order for disease to occur. This response or T-cell reactivity would be different for identical twins since their TCRs are rearranged randomly.

An alternate hypothesis currently being explored states that environmental factors in fetal or neonatal stages of life cause genetically susceptible individuals to be primed for autoimmune disease. The déjà vu theory takes this step further stating that there is an initial (fetal or neonatal) viral infection that persists and primes the individual for autoimmune disease provided a subsequent infection shares T-cell epitopes with the persisting virus (Merkler *et al.*, 2006). In support of this, a study by Rothwell *et al.* (1996) of patients with type-1 diabetes found that disease prevalence was higher in patients born in the spring and early summer than during the winter months. However, this hypothesis is still controversial as studies from Slovenia (Ursic-Bratina *et al.*, 2001) and Israel (Laron *et al.*, 2005) had similar results with higher prevalence rates for type-1 diabetes in individuals born in the spring and early summer, but an analysis of data from Germany showed opposite results with fewer type-1 diabetes cases born between April and September (Neu *et al.*, 2000).

II. Class I Molecules and Mimicry

Thus far, MHC class II-restricted $CD4^+$ T-cell responses to self have been more extensively investigated in MS patients and animal models than the MHC class I-restricted $CD8^+$ T-cell response. However, clinical and experimental research indicates that $CD8^+$ T-cell reactivity could be involved in the pathology of MS (Wekerle and Lassmann, 2005). Cytotoxic T lymphocytes (CTLs) could destroy oligodendrocytes leading to demyelination. *In vitro* experiments have shown that oligodendrocytes can be induced by IFN-γ to express MHC class I antigen but not class II antigen (Grenier *et al.*, 1989). In both MS and Theiler's murine encephalomyelitis virus (TMEV) infection, apoptosis of oligodendrocytes has been observed (Dowling *et al.*, 1997; Tsunoda and Fujinami, 1996, 1999; Tsunoda *et al.*, 1997). TMEV belongs to the family *Picornaviridae*. Infection with TMEV causes an extensive demyelinating disease in the CNS of persistently infected mice (Tsunoda and Fujinami, 1996, 1999). We monitored cytotoxic T-cell responses in SJL/J mice, which are susceptible to TMEV-induced demyelinating disease. We found that spleen cells and T-cell clones from TMEV-infected mice were highly cytotoxic to uninfected syngeneic cells, but not to allogeneic cells following stimulation with TMEV-infected antigen-presenting cells. These autoreactive cells were found to be $CD3^+CD8^+$ by flow cytometry and antibody-blocking cytotoxicity assays (Tsunoda *et al.*, 2005). Double chamber ^{51}Cr release assays showed that direct cell-to-cell contact was required for lysis. It was also found that cell killing was mediated by the Fas-FasL pathway (Tsunoda *et al.*, 2002). Intracerebral adoptive transfer of activated TMEV-induced autoreactive cells and T-cell clones into naive mice resulted in degenerating lesions in the spinal cord, suggesting that these MHC class I-restricted $CD8^+$ T cells could play an effector role in CNS pathology in this model (Tsunoda *et al.*, 2005). We went on to show that these CTLs could be efficiently induced by a recombinant VV encoding the TMEV viral capsid proteins (Tsunoda *et al.*, 2006). We hypothesized that the autoreactive CTLs recognize TMEV capsid proteins that share molecular mimicry with CNS antigen. We are in the process of narrowing down the TMEV capsid epitope and characterizing the mouse CNS antigen that is the target of these CTLs.

Previous studies on memory $CD8^+$ T cells suggested that these cells were a resting, nondividing population lying in wait until reexposure to the antigen that resulted in their initial differentiation (Jamieson and Ahmed, 1989). Other studies suggest that memory $CD8^+$ T cells continuously undergo a low level of homeostatic division for a long period of time following antigen clearance (Kos and Mullbacher, 1993; Tough and Sprent, 1994; Zimmerman *et al.*, 1996). This homeostatic division can be augmented by type I IFN, interleukin (IL)-15, and IFN inducers such as poly (I:C) as well as viral infection (Tough *et al.*, 1996;

Zhang *et al.*, 1998). One study used mice whose immune system had been reconstituted with labeled splenocytes from LCMV-immune mice and examined the fate of these cells following poly (I:C) treatment or infection with LCMV, PV, and VV (Kim *et al.*, 2002). Previous reports from this group indicated that LCMV-immune mice were partially resistant to PV and VV and that LCMV-immune splenocytes could provide resistance to both PV and VV following adoptive transfer into naive animals; however, this resistance was lost if $CD8^+$ T cells were depleted (Selin *et al.*, 1998). In the Kim *et al.* (2002) study, it was found that the heterologous viruses (PV and VV) had the capacity to induce several cycles of proliferative expansion of memory $CD8^+$ T cells which altered the antigen hierarchy of T cells specific to earlier pathogens. LCMV infection resulted in a greater than 300-fold increase in the number of $CD8^+$ T cells with a majority of cells being LCMV specific. In contrast, poly (I:C) treatment only resulted in limited cell division without a net increase in cell number or changes in the hierarchy. This phenomenon is likely due to poly (I:C) induction of apoptosis, which offsets cell division (Kim *et al.*, 2002). Therefore, it was concluded that memory cell division with a net increase in number is characteristic of antigen-specific stimulation while division without an increase in cell number is a characteristic of cytokine-induced bystander stimulation. Interestingly, it was also found that CTLs generated against either LCMV or PV were not capable of lysing cells infected with the heterologous virus, which indicated little cross-reactivity between the viruses (Brehm *et al.*, 2002). However, CTLs from PV-infected LCMV-immune mice were able to lyse cells infected with either virus (Selin *et al.*, 1994). It was found that PV encodes a subdominant epitope that shares six of eight amino acids with an LCMV epitope (Brehm *et al.*, 2002). The fate of memory cells was assessed following PV infection of mice reconstituted with LCMV-immune cells (Kim *et al.*, 2005). Results indicate heterologous infection caused an increase in the number of T cells specific for the subdominant PV epitope. In contrast, there were considerable differences between individual LCMV-immune mice in terms of what epitope-specific T cells were stimulated by VV infection. Even though these responses were unpredictable, patterns of epitope recognition did emerge, with three epitopes more likely to be dominant. These differences in epitope-specific response are thought to be due to private specificities of the TCR repertoire generated in each mouse following random rearrangement of the variable (V), diversity (D), and joining (J) TCR gene segments (Kim *et al.*, 2005). Studies examining the composition, maintenance, and alteration of the T-cell memory pool have important implications for MS, and may help to explain the low concordance rates seen in identical twins.

A novel transgenic mouse model was created by Oldstone and colleagues (Evans *et al.*, 1996) to examine whether molecular mimicry to a protein expressed in the CNS could lead to autoimmune disease. In this model, mice were generated that express nucleoprotein (NP) or glycoprotein from LCMV under the control of the MBP promoter as self in oligodendrocytes (Evans *et al.*, 1996). These mice were infected with the Armstrong strain of LCMV and monitored for clinical signs of disease. It was found that LCMV infected tissues in the periphery but not the CNS, and that a vigorous $CD8^+$ CTL response cleared virus from all tissues by 7–14 days postinfection. At 3-weeks postinfection, the transgenic mice had 150–300 $CD8^+$ T cells per sagittal section of brain while nontransgenic control mice had less than 30 lymphocytes per section. Perivascular cuffing was seen in the transgenic mice and lymphocytes were seen in the parenchyma, brain stem, and spinal cord. Interestingly, brains of the transgenic mice infected with LCMV were found to contain these elevated levels of $CD8^+$ T-cell infiltrates 1 year after infection. At 3-months postinfection, a clustering of $CD4^+$ and $CD8^+$ T cells was found predominately in the white matter of the CNS including the corpus callosum, internal capsule, fimbria hippocampus, brain stem, and spinal cord. Clinical signs, including ruffled fur, weight loss, and balance difficulties, were also evident in 75% of the transgenic mice at 3-months postinfection. This study also investigated the effect of a second infection with LCMV or VV encoding the NP transgene on the transgenic mice 6 weeks after the initial LCMV infection. Stimulation of the memory response by these viruses resulted in enhanced pathology including an increase in the ratio of $CD8^+$ to $CD4^+$ T lymphocytes and a greater likelihood that these infiltrates would be localized to the white matter as shown by immunohistochemical staining (Evans *et al.*, 1996). This enhanced disease pathology, characterized by demyelination and motor dysfunction, following secondary viral infection is similar to human CNS autoimmune diseases, and these results suggest that infection by a virus that shares immune determinants with a protein expressed in oligodendrocytes can induce a chronic inflammatory disease of the CNS.

To further validate the role viruses may play in the induction of autoimmune disease, a molecular mimicry model utilizing an H-$2D^b$-restricted immunodominant epitope of LCMV, $NP_{396–404}$ was used to identify structurally homologous murine self-proteins (Hudrisier *et al.*, 2001). The MHC class I-restricted $CD8^+$ T-cell response against the LCMV NP protein has been shown to be directed against this immunodominant $NP_{396–404}$ epitope. Typically in an LCMV-infected mouse that has cleared the virus, there is activation and proliferation of $CD8^+$ T cells which stay at a high level throughout the animal's life despite the fact that there is no detectable virus or viral antigens. These CTLs represent a potential source of autoreactivity, especially if their cytolytic activity remains functional and intact. Six nonameric sequences from endogenous proteins which shared structural and functional homology with LCMV $NP_{396–404}$ were identified through a database search. Five of these peptides were shown to have high H-$2D^b$-binding affinity and shared the main TCR contact site with LCMV $NP_{396–404}$. Three of these five peptides also shared an auxiliary TCR contact residue and one of these, tumor necrosis factor (TNF) receptor $I_{302–310}$, also shared a third residue with the

LCMV peptide resulting in a marked functional similarity between the self and viral epitopes. Despite the presence of TCR contact residues, none of the epitopes identified were able to stimulate antiviral CTLs in cytotoxicity assays. However, these peptides did behave as antagonists of lysis by LCMV-specific CTLs which indicated that they were capable of interacting with the TCR, though the affinity was relatively low. Low affinity recognition of self-MHC promoted thymocyte survival, and cells cultured with peptides were maintained for more than 2 months, suggesting these epitopes were in fact low-affinity molecular mimics. Importantly, these self-peptides were shown to allow LCMV-specific and potentially autoreactive $CD8^+$ T cells to maintain their cytolytic function in the absence of viral antigens over a period of months (Hudrisier *et al.*, 2001).

III. Class II Molecules and Mimicry

In the 1930s Rivers and colleagues (Rivers and Schwentker, 1935; Rivers *et al.*, 1933) published reports on induction of encephalomyelitis using CNS homogenates; and throughout the 1950s and 1960s research continued on myelin proteins in EAE and MS. In the 1940s it was shown that injection of myelin proteins with adjuvant into naive animals could cause relapsing-remitting, acute, or chronic encephalomyelitis (Freund *et al.*, 1947; Kabat *et al.*, 1946, 1947; Kopeloff and Kopeloff, 1947; Morgan, 1946, 1947; Morrison, 1947). Early studies also determined that EAE could be transferred by adoptive transfer of $CD4^+$ T cells (Pettinelli and McFarlin, 1981; Zamvil and Steinman, 1990).

MHC class II was found to have a genetic link with a variety of autoimmune diseases including MS (Jersild *et al.*, 1973a,b). MHC class II molecules present self-peptides to $CD4^+$ T cells; therefore, $CD4^+$ T cells are thought to play a major role in the initiation and/or progression of many autoimmune diseases. Several animal models of MS and human studies using blood and CSF from MS patients implicate $CD4^+$ T cells, mimicry, and various microbes such as *Chlamydia pneumoniae*, influenza virus, Torque Teno virus (TTV), EBV, and *Human herpesvirus 6* (HHV-6) (Croxford *et al.*, 2005; Lang *et al.*, 2002; Markovic-Plese *et al.*, 2005; Sospedra *et al.*, 2005; Sriram *et al.*, 1999; Tejada-Simon *et al.*, 2003).

C. pneumoniae has been associated with MS (Sriram *et al.*, 1999). In one study, CSF from 17 patients with relapsing-remitting MS, 20 patients with progressive MS, and 27 patients with other neurological diseases was tested for *C. pneumoniae* (Sriram *et al.*, 1999). Bacterial DNA was detected in 97% of these MS patients compared to only 18% of patients with other neurological diseases. They were also able to isolate *C. pneumoniae* from the CSF of MS patients in addition to identifying anti-*C. pneumoniae* antibodies. This work inspired Lenz *et al.* (2001) to identify a protein from *C. pneumoniae*, Cnp0483, with similarity to MBP. Cnp0483, an MBP_{68-86} homologue peptide, was used to sensitize Lewis rats. Animals began showing signs of disease approximately 12-days postsensitization. Pathologically, perivascular cuffing and mononuclear cell infiltrates were found in the spinal cords of Cpn0483-immunized rats. Additionally, splenocytes from Cpn0483-injected mice that had been cultured with peptide caused disease when they were transferred into naive animals. Interestingly, the Cpn0483 peptide shares only seven amino acids with MBP_{68-86}. This small degree of similarity happens to constitute a structural motif that permits interaction of the peptide with MHC class II gene products (Lenz *et al.*, 2001).

Influenza virus is an example of a virus infection where there is flexibility in TCR recognition and the degree of sequence and structural similarity necessary for cross-reaction. One group derived a $CD4^+$ T-cell clone from an MS patient's peripheral blood mononuclear cells (PBMCs) during an acute respiratory infection with influenza-A (Markovic-Plese *et al.*, 2005). They determined that this T-cell clone, GP5F11, reacted with the immunodominant influenza hemagglutinin (Flu-HA) epitope 306–318. GP5F11 was found to be a high avidity pathogen-specific T-cell clone that had demonstrated cross-reactivity against 14 Flu-HA variants, 11 viral, 15 human, and 3 myelin-derived peptides. Of the 11 viral mimics, it was found that equine influenza-A HA peptide had greater than 70 times higher stimulatory potential than the human influenza peptide, suggesting that there is a high potential for cross-recognition even for a T-cell clone with a stringent (high avidity) TCR response. Of 12 potentially stimulatory myelin-derived peptides, 2 myelin oligodendrocyte glycoprotein (MOG) peptides and one 2′,3′-cyclic nucleotide 3′-phosphodiesterase (CNPase)-derived peptide were found to be extremely potent stimulators of the T-cell clone. Interestingly, the CNPase-derived peptide had no sequence similarity to the Flu-HA peptide (Markovic-Plese *et al.*, 2005). This broad range of cross-reactivity of a single TCR would assure protection against much more than the original infecting influenza virus with the potential side effect of autoimmune responses against self-antigens having structural or sequence similarity to pathogenic peptides.

Another series of experiments found that one $CD4^+$ T-cell clone (MN19) isolated from the CSF of an MS patient during an exacerbation was stimulated by arginine-enriched protein domains from common viruses as well as the nonpathogenic TTV (Sospedra *et al.*, 2005). The T-cell clones used in this experiment were found to be expanded during exacerbations of MS and reduced during periods of remission. Biometric analysis was used to predict stimulatory peptides for the T-cell clones and then peptides were selected from human infectious agents. Many peptides were identified using this method; however, following normalization only two related DNA viruses were selected. These viruses were TTV and TTV-like mini virus, both of which are recognized as ubiquitous, nonpathogenic viruses in the human population. There were several stimulatory peptides identified from TTV, and surprisingly, the majority were mapped to a 74-amino acid sequence in the N-terminal region of open reading frame 1. This N-terminal region is enriched in positively charged amino acids, mainly arginine. Other stimulatory peptides were also identified for the T-cell clone from other human viruses, including adenovirus and papillomavirus. The adenovirus peptide was from the pVII protein which is enriched in arginine and performs a histone-like function. The papillomavirus peptides were found in the minor capsid protein L2, which is enriched in basic amino acids. Human stimulatory peptides from the human CNS proteins adrenergic receptor α-1B, α-1A, and α-2C, arginine-rich protein, and dopamine D2 receptor were also identified for the MN19 T-cell clone. Interestingly, arginine-rich domains are frequent in viruses as well as eukaryotes and prokaryotes (Sospedra *et al.*, 2005). This recognition of an arginine-enriched conserved domain may be involved in inducing and perpetuating autoimmune responses against arginine-rich self-antigens.

EBV has long been associated with MS and viral reactivation is linked to disease activity (Bray *et al.*, 1983; Larsen *et al.*, 1985). EBV is a large DNA virus that causes mononucleosis in humans and has been shown to persist in B cells. Infection with EBV late in life has been correlated with an increased risk of developing MS. Lang *et al.* (2002) used a $CD4^+$ T-cell clone isolated from an MS patient with a relapsing-remitting disease course to examine the TCR contact surfaces of a cross-reactive TCR that recognizes both the MBP_{85-99} peptide and an EBV DNA polymerase peptide ($EBV_{627-641}$). They found that MBP_{85-99} was recognized in the context of DRB1*1501 and $EBV_{627-641}$ was recognized in the context of DRB5*0101 based on relative binding affinities (Lang *et al.*, 2002). DRB1*1501, DRB5*0101, as well as DQB1*0602 are MHC class II alleles of the DR2 haplotype that have been shown to be risk factors for MS. Transgenic mouse studies determined that spleen cells from TCR-DRB1*1501 double-transgenic mice only responded in the form of proliferation and cytokine production to the MBP_{85-99} peptide, while spleen cells from TCR-DRB5*0101 double-transgenic mice only responded to the $EBV_{627-641}$ peptide. Spleen cells from triple-transgenic mice with TCR-DRB1*1501-DRB5*0101 responded to both the MBP_{85-99} and $EBV_{627-641}$ peptides. These transgenic mice recognized the MBP and EBV peptides in the context of two different MHC class II molecules. Next, the crystal structures of both of these MHC-peptide complexes were determined and compared. It was found that the TCR contact surfaces of the DRB5*0101-$EBV_{627-641}$ and DRB1*1501-MBP_{85-99} complexes were structurally very similar. The nature of MHC class II binding to peptide, with the peptide chain held in a highly conserved extended conformation, is thought to allow for structural mimicry in TCR recognition (Lang *et al.*, 2002).

Clinical studies extending the *in vitro* work of Lang *et al.* (2002) on EBV have found that $CD4^+$ T cells present in the CSF of an MS patient cross-react with $EBV_{627-641}$ and the immunodominant peptide MBP_{85-99} (Holmøy *et al.*, 2004). The presence of $EBV_{627-641}$-specific $CD4^+$ T cells in the CSF proves that

EBV-specific T cells can gain access to the intrathecal compartment and suggests that they could target MBP in the CNS (Holmøy *et al.*, 2004).

Another virus that encodes proteins with peptide mimics to human brain proteins is HHV-6, the agent responsible for infantile exanthem. The virus has a tropism for $CD4^+$ T cells and can reactivate from its latent state following immunosuppression (Braun *et al.*, 1997; Salahuddin *et al.*, 1986). Challoner *et al.* (1995) found that oligodendrocytes from MS patients expressed HHV-6 virion proteins. A protein found in both the A and B variants of HHV-6, known as U24, has amino acid sequence similarity with MBP. Two 13-mer peptides corresponding to residues 1–13 of the HHV-6 U24 protein and MBP_{93-105} were synthesized and used to determine whether T cells would cross-react with the two peptides or were specific for only one of the sequences (Tejada-Simon *et al.*, 2003). T-cell cultures from PBMCs of MS patients were incubated with MBP_{93-105} or HHV-6 $U24_{1-13}$ for 1 week, tested for specificity, and characterized as MBP or HHV-6 reactive or cross-reactive. The frequency of T cells reactive to both HHV-6 and MBP peptides was determined to be significantly higher in MS patients than in the control group (Tejada-Simon *et al.*, 2003). The cytokine profiles of the peptide-specific and cross-reactive T-cell lines generated from MS patients were found to display a Th1 phenotype, predominately producing IFN-γ and TNF-α, but not IL-4 and IL-10, regardless of peptide specificity. These cells were shown to be $CD4^+$ by flow cytometry.

IV. Antibody and Mimicry

The presence of oligoclonal antibodies in the CSF of patients with MS is a consistent immunologic marker of the disease (Link, 1978; Link and Kostulas, 1983). Two or more oligoclonal bands consisting of IgG are routinely seen by isoelectric focusing followed by immunoblotting of CSF in up to 95% of MS patients (Link and Huang, 2006). Each band represents antibody, though identification of the antigen that these antibodies bind has been difficult in part because of the low amount of antibodies contained in the CSF. Some of the antigens to which the oligoclonal IgG antibodies have been found to be specific include MBP, HHV-6, measles, rubella, Varicella-zoster virus, and Herpes simplex virus 1 (HSV-1) (Cruz *et al.*, 1987; Derfuss *et al.*, 2005; Reiber *et al.*, 1998).

One method that has been used as an alternative source of antigens to study the specificity of antibodies is phage-displayed random peptide libraries (RPL) (Cortese *et al.*, 1996; Dunn, 1996; Smith, 1991). The peptides in these libraries are referred to as mimotopes because it is not required for them to have sequence similarity, but rather they mimic binding properties or conformation of natural epitopes. There have been reports of mimotopes from RPL selection that were recognized by antibodies from the CSF of MS patients (Cortese *et al.*, 1996, 1998; Dybwad *et al.*, 1997). Interestingly, some of these antibodies were detected in sera from normal individuals as well as MS patients, suggesting that the mimotopes mimic common antigens (Cortese *et al.*, 2001). One mimotope family, MS17, was chosen because of its reactivity with CSF antibodies and high frequency of reactivity with the sera of MS patients. Two CSF samples from MS patients that recognize MS17 were tested for reactivity in an enzyme-linked immunosorbent assay (ELISA) to neurotropic viruses, including EBV, measles, mumps, rubella, HSV-1, and cytomegalovirus (Cortese *et al.*, 2001). Both CSF samples tested positive against measles and HSV-1; however, only the reactivity to HSV-1 was competitively inhibited by MS17 phage. In addition, whole cell extract from an HSV-1-infected cell line preincubated with CSF from the MS patients completely abolished recognition of the phage (Cortese *et al.*, 2001). A consensus sequence profile was then derived from multiple alignments of all HSV-1 isolates, and this was used to search for homology between eight members of the MS17 family and HSV-1 protein sequences. The highest score from this profile came from the N-terminal region of the HSV-1 envelope glycoprotein B (gB). The MS17 mimotope corresponds to the sequence of HSV-1 gB between amino acid positions 474–482. Antibodies against MS17 were raised in rabbits, and extensive characterization of these antibodies demonstrated that the MS17 phage is both an antigenic and immunogenic mimic of the HSV-1 gB epitope. Using the anti-MS17 antibodies in place of the CSF samples, the antibodies were found to specifically interact with an as yet uncharacterized protein present in protein extract from the brain of an MS patient (Cortese *et al.*, 2001). These studies represent an important step in identification of the natural antigens that antibodies from MS patients recognize as well as providing an example of molecular mimicry between a viral and CNS epitope that is mediated through antibody. The RPL technique has the potential to aid in explaining the pathogenesis of autoimmune diseases.

Monoclonal antibodies generated against various virus-specific proteins have been shown to cross-react with host cell components, thus serving as additional examples of molecular mimicry at the level of antibodies. Studies have shown that monoclonal antibodies generated against viral proteins from measles, HSV-1, and VV react with host cell components (Dales *et al.*, 1983; Fujinami *et al.*, 1983); common determinants existed between viruses and cellular- or tissue-specific elements (Srinivasappa *et al.*, 1986). In another report, a monoclonal antibody, H8, to the TMEV Daniels (DA) strain reacted with both TMEV viral protein-1 and lipid-like moieties including galactocerebroside, a major component of myelin (Fujinami *et al.*, 1988). In mouse brain cultures, cells that bound the TMEV monoclonal antibody, H8, were also labeled with antibody to MBP (Yamada *et al.*, 1990). This double staining indicated that the H8 antibody recognized epitopes on oligodendrocytes. When the antibody was injected into mice with EAE, H8 caused an increase in the area of demyelination within spinal cords. Furthermore, a competition ELISA for galactocerebroside and TMEV found that sera from mice contained antibody with the same specificity as H8. These results indicate that the immune response which generates antibodies specific for TMEV that cross-react with myelin and oligodendrocytes could contribute to demyelination through an antibody-mediated process (Yamada *et al.*, 1990).

There are several other examples of molecular mimics at the level of antibodies in autoimmune diseases other than MS. They include Guillain-Barre syndrome in which antibody cross-reactivity between peripheral nerve and ganglioside has been shown (Ogino *et al.*, 1995; Yu *et al.*, 2006), and rheumatic heart disease in which antistreptococcal antibodies cross-react with *N*-acetyl-β-D-glucosamine, myosin, and other self-epitopes (Guilherme *et al.*, 2006; Mertens *et al.*, 2000). In addition, autoantibody development is seen in systemic lupus erythematosus (Poole *et al.*, 2006). EBV has been associated with lupus through serological and DNA studies and antibodies specific for Epstein-Barr nuclear antigen-1 (EBNA-1) cross-react with lupus-associated autoantigens (Poole *et al.*, 2006). Alternately, infection with human T-lymphotropic virus type-1 (HTLV-1) can cause HTLV-1-associated myelopathy/tropical spastic paraparesis (HAM/TSP), an immune-mediated disease of the CNS (Lee *et al.*, 2006). Molecular mimicry has been found in antibodies that cross-react with heterogeneous nuclear ribonucleoprotein A1 (hnRNP A1), a protein found in CNS neurons, and the HTLV-1 tax protein. Two core epitopes within the C-terminal region of hnRNP A1, both functionally important regions, were found to react with antibodies purified from HAM/TSP patients. Additionally, monoclonal antibodies raised against HTLV-1 tax protein also reacted with the two core epitopes from hnRNP A1 (Lee *et al.*, 2006).

V. Discussion

Potentially, infection with viruses that have molecular mimicry with CNS antigens can prime autoreactive immune cells specific for the CNS in hosts. Our laboratory has explored whether a virus having molecular mimicry with self-CNS antigens could induce an autoimmune disease in mice which had previously been inducible only by injection of CNS antigen in complete Freund's adjuvant (CFA) (Theil *et al.*, 2001). Three recombinant VVs were constructed: encoding myelin proteolipid protein (VV_{PLP}), encoding myelin-associated glycoprotein (VV_{MAG}), and encoding glial fibrillary acidic protein (VV_{GFAP}). Mice infected with VV_{PLP}, VV_{MAG}, or VV_{GFAP} showed no clinical or histological signs of CNS disease. This suggests that molecular mimicry alone cannot result in high enough numbers or activation of CNS-specific autoimmune cells for induction of CNS disease. Thus, 5 weeks after the first infection, when VV was cleared, we challenged mice nonspecifically with CFA to activate CNS-specific autoimmune cells (bystander activation).

Clinically, some mice showed paralysis in the tail, similar to EAE, following the CFA challenge. At 1-month post-CFA challenge, mice were sacrificed and CNS tissues were examined for pathological changes. All mice (15/15) were found to have inflammatory lesions with $CD3^+$ T cells in the CNS (Theil *et al.*, 2001). As a negative control, mice were infected with VV_{SC11}, a recombinant VV that encodes β-galactosidase, which has no known molecular mimicry with CNS antigens; following CFA challenge, no inflammatory changes were seen in the CNS (Theil *et al.*, 2001). These data indicate that infections with viruses encoding molecular mimics can substantially prime animals for autoimmune disease and at a later time a nonspecific immunologic challenge could initiate the disease.

As reviewed by Trinchieri (1995), MCMV infection in mice causes a significant burst of IL-12, which promotes the development of Th1 cells. Therefore, we have tested whether VV_{PLP}-sensitized mice developed clinical disease following a second, unrelated viral challenge with MCMV. Preliminary findings suggest this to be the case. Five out of nine mice primed with VV_{PLP} and challenged with MCMV were found to have meningitis and perivascular cuffing in the CNS. In contrast, mice primed with VV_{PLP} and challenged with the wild-type strain of VV were found to have no obvious lesions. In another experiment, mice primed with VV_{PLP} and challenged with MCMV showed marked weight loss and had righting reflex disturbances, compared with mice injected with phosphate-buffered saline (PBS) or VV_{SC11} followed by MCMV challenge (Ikuo Tsunoda, Jane E. Libbey, and Robert S.Fujinami, unpublished data). Thus far, the definitive experiments to determine whether disease in this model is caused by MHC class I or class II cells have not been completed. However, these experiments are important in that they provide a working model mirroring what may be occurring in human patients with MS.

Molecular mimicry has been found in many autoimmune diseases on a variety of levels. There is significant evidence that autoreactive $CD4^+$ and $CD8^+$ T cells as well as autoantibodies can contribute to disease progression in various animal models of MS and other autoimmune diseases. However, it has yet to be shown that a single molecular mimic is responsible for initiation of disease. None of the autoimmune diseases described here are simple cause-and-effect diseases and as such the mechanisms responsible for their manifestation are likely to be a combination of molecular mimicry, bystander activation, epitope spreading, and heterologous immunity.

Acknowledgments

We wish to thank Ikuo Tsunoda, MD, Ph.D., for many helpful discussions and Ms. Kathleen Borick for her excellent preparation of the chapter. This work was supported by NIH grant AI0581501.

References

Besson Duvanel, C., Honegger, P., and Matthieu, J. M. (2001). Antibodies directed against rubella virus induce demyelination in aggregating rat brain cell cultures. *J. Neurosci. Res.* **65,** 446–454.

Boucher, A., Denis, F., Duquette, P., and Talbot, P. J. (2001). Generation from multiple sclerosis patients of long-term T-cell clones that are activated by both human coronavirus and myelin antigens. *Adv. Exp. Med. Biol.* **494,** 355–362.

Braun, D. K., Dominguez, G., and Pellett, P. E. (1997). Human herpesvirus 6. *Clin. Microbiol. Rev.* **10,** 521–567.

Bray, P. F., Bloomer, L. C., Salmon, V. C., Bagley, M. H., and Larsen, P. D. (1983). Epstein-Barr virus infection and antibody synthesis in patients with multiple sclerosis. *Arch. Neurol.* **40,** 406–408.

Brehm, M. A., Pinto, A. K., Daniels, K. A., Schneck, J. P., Welsh, R. M., and Selin, L. K. (2002). T cell immunodominance and maintenance of memory regulated by unexpectedly cross-reactive pathogens. *Nat. Immunol.* **3,** 627–634.

Casrouge, A., Beaudoing, E., Dalle, S., Pannetier, C., Kanellopoulos, J., and Kourilsky, P. (2000). Size estimate of the alpha beta TCR repertoire of naive mouse splenocytes. *J. Immunol.* **164,** 5782–5787.

Challoner, P. B., Smith, K. T., Parker, J. D., MacLeod, D. L., Coulter, S. N., Rose, T. M., Schultz, E. R., Bennett, J. L., Garber, R. L., Chang, M., Schad, P. A., Stewart, P. M., *et al.* (1995). Plaque-associated expression of human herpesvirus 6 in multiple sclerosis. *Proc. Natl. Acad. Sci. USA* **92,** 7440–7444.

Chen, H. D., Fraire, A. E., Joris, I., Brehm, M. A., Welsh, R. M., and Selin, L. K. (2001). Memory $CD8^+$ T cells in heterologous antiviral immunity and immunopathology in the lung. *Nat. Immunol.* **2,** 1067–1076.

Cirone, M., Cuomo, L., Zompetta, C., Ruggieri, S., Frati, L., Faggioni, A., and Ragona, G. (2002). Human herpesvirus 6 and multiple sclerosis: A study of T cell cross-reactivity to viral and myelin basic protein antigens. *J. Med. Virol.* **68,** 268–272.

Cortese, I., Tafi, R., Grimaldi, L. M., Martino, G., Nicosia, A., and Cortese, R. (1996). Identification of peptides specific for cerebrospinal fluid antibodies in multiple sclerosis by using phage libraries. *Proc. Natl. Acad. Sci. USA* **93,** 11063–11067.

Cortese, I., Capone, S., Tafi, R., Grimaldi, L. M., Nicosia, A., and Cortese, R. (1998). Identification of peptides binding to IgG in the CSF of multiple sclerosis patients. *Mult. Scler.* **4,** 31–36.

Cortese, I., Capone, S., Luchetti, S., Cortese, R., and Nicosia, A. (2001). Cross-reactive phage-displayed mimotopes lead to the discovery of mimicry between HSV-1 and a brain-specific protein. *J. Neuroimmunol.* **113,** 119–128.

Croxford, J. L., Anger, H. A., and Miller, S. D. (2005). Viral delivery of an epitope from *Haemophilus influenzae* induces central nervous system autoimmune disease by molecular mimicry. *J. Immunol.* **174,** 907–917.

Cruz, M., Olsson, T., Ernerudh, J., Hojeberg, B., and Link, H. (1987). Immunoblot detection of oligoclonal anti-myelin basic protein IgG antibodies in cerebrospinal fluid in multiple sclerosis. *Neurology* **37,** 1515–1519.

Dales, S., Fujinami, R. S., and Oldstone, M. B. A. (1983). Infection with vaccinia favors the selection of hybridomas synthesizing autoantibodies against intermediate filaments, one of them cross-reacting with the virus hemagglutinin. *J. Immunol.* **131,** 1546–1553.

Davies, J. M., Watt, N. J., Torsteinsdottir, S., and Carnegie, P. R. (1996). Mimicry of a 21.5 kDa myelin basic protein peptide by a Maedi Visna virus polymerase peptide does not contribute to the pathogenesis of encephalitis in sheep. *Vet. Immunol. Immunopathol.* **55,** 127–139.

Dejong, R. N. (1970). Multiple sclerosis. History, definition and general considerations. *In* "Multiple Sclerosis and Other Demyelinating Diseases" (P. J. Vinken and G. W. Bruyn, Eds.), pp. 45–62. North-Holland Publishing Company, Amsterdam.

Derfuss, T., Hohlfeld, R., and Meinl, E. (2005). Intrathecal antibody (IgG) production against human herpesvirus type 6 occurs in about 20% of multiple sclerosis patients and might be linked to a polyspecific B-cell response. *J. Neurol.* **252,** 968–971.

Dowling, P., Husar, W., Menonna, J., Donnenfeld, H., Cook, S., and Sidhu, M. (1997). Cell death and birth in multiple sclerosis brain. *J. Neurol. Sci.* **149,** 1–11.

Dunn, I. S. (1996). Phage display of proteins. *Curr. Opin. Biotechnol.* **7,** 547–553.

Dybwad, A., Førre, Ø., and Sioud, M. (1997). Probing for cerebrospinal fluid antibody specificities by a panel of random peptide libraries. *Autoimmunity* **25,** 85–89.

Ebers, G. C., Bulman, D. E., Sadovnick, A. D., Paty, D. W., Warren, S., Hader, W., Murray, T. J., Seland, T. P., Duquette, P., Grey, T., Nelson, R., Nicolle, M., *et al.* (1986). A population-based study of multiple sclerosis in twins. *N. Engl. J. Med.* **315,** 1638–1642.

Evans, C. F., Horwitz, M. S., Hobbs, M. V., and Oldstone, M. B. A. (1996). Viral infection of transgenic mice expressing a viral protein in oligodendrocytes leads to chronic central nervous system autoimmune disease. *J. Exp. Med.* **184,** 2371–2384.

Freund, J., Stern, E. R., and Pisani, T. M. (1947). Isoallergic encephalomyelitis and radiculitis in guinea pigs after one injection of brain and mycobacteria in water-in-oil emulsion. *J. Immunol.* **57,** 179–193.

Fujinami, R. S., and Oldstone, M. B. A. (1985). Amino acid homology between the encephalitogenic site of myelin basic protein and virus: Mechanism for autoimmunity. *Science* **230,** 1043–1045.

Fujinami, R. S., Oldstone, M. B. A., Wroblewska, Z., Frankel, M. E., and Koprowski, H. (1983). Molecular mimicry in virus infection: Crossreaction of measles virus phosphoprotein or of herpes simplex virus protein with human intermediate filaments. *Proc. Natl. Acad. Sci. USA* **80,** 2346–2350.

Fujinami, R. S., Zurbriggen, A., and Powell, H. C. (1988). Monoclonal antibody defines determinant between Theiler's virus and lipid-like structures. *J. Neuroimmunol.* **20,** 25–32.

Grenier, Y., Ruijs, T. C. G., Robitaille, Y., Olivier, A., and Antel, J. P. (1989). Immunohistochemical studies of adult human glial cells. *J. Neuroimmunol.* **21,** 103–115.

Guilherme, L., Kalil, J., and Cunningham, M. (2006). Molecular mimicry in the autoimmune pathogenesis of rheumatic heart disease. *Autoimmunity* **39,** 31–39.

Hayashi, T., Morimoto, C., Burks, J. S., Kerr, C., and Hauser, S. L. (1988). Dual-label immunocytochemistry of the active multiple sclerosis lesion: Major histocompatibility complex and activation antigens. *Ann. Neurol.* **24,** 523–531.

Holmøy, T., Kvale, E. Ø., and Vartdal, F. (2004). Cerebrospinal fluid $CD4^+$ T cells from a multiple sclerosis patient cross-recognize Epstein-Barr virus and myelin basic protein. *J. Neurovirol.* **10,** 278–283.

Hudrisier, D., Riond, J., Burlet-Schiltz, O., von Herrath, M. G., Lewicki, H., Monsarrat, B., Oldstone, M. B. A., and Gairin, J. E. (2001). Structural and functional identification of major histocompatibility complex class I-restricted self-peptides as naturally occurring molecular mimics of viral antigens. Possible role in $CD8^+$ T cell-mediated, virus-induced autoimmune disease. *J. Biol. Chem.* **276,** 19396–19403.

Hughes, L. E., Smith, P. A., Bonell, S., Natt, R. S., Wilson, C., Rashid, T., Amor, S., Thompson, E. J., Croker, J., and Ebringer, A. (2003). Cross-reactivity between related sequences found in Acinetobacter sp., *Pseudomonas aeruginosa,* myelin basic protein and myelin oligodendrocyte glycoprotein in multiple sclerosis. *J. Neuroimmunol.* **144,** 105–115.

Hunt, D. F., Henderson, R. A., Shabanowitz, J., Sakaguchi, K., Michel, H., Sevilir, N., Cox, A. L., Appella, E., and Engelhard, V. H. (1992). Characterization of peptides bound to the class I MHC molecule HLA-A2.1 by mass spectrometry. *Science* **255,** 1261–1263.

Jamieson, B. D., and Ahmed, R. (1989). T cell memory. Long-term persistence of virus-specific cytotoxic T cells. *J. Exp. Med.* **169,** 1993–2005.

Jersild, C., Fog, T., Hansen, G. S., Thomsen, M., Svejgaard, A., and Dupont, B. (1973a). Histocompatibility determinants in multiple sclerosis, with special reference to clinical course. *Lancet* **2,** 1221–1225.

Jersild, C., Svejgaard, A., Fog, T., and Ammitzboll, T. (1973b). HL-A antigens and diseases. I. Multiple sclerosis. *Tissue Antigens* **3,** 243–250.

Kabat, E. A., Wolf, A., and Bezer, A. E. (1946). Rapid production of acute disseminated encephalomyelitis in Rhesus monkeys by injection of brain tissue with adjuvants. *Science* **104,** 362–363.

Kabat, E. A., Wolf, A., and Bezer, A. E. (1947). The rapid production of acute disseminated encephalomyelitis in Rhesus monkeys by infection of heterologous and homologous brain tissue with adjuvants. *J. Exp. Med.* **85,** 117–128.

Kim, S.-K., Brehm, M. A., Welsh, R. M., and Selin, L. K. (2002). Dynamics of memory T cell proliferation under conditions of heterologous immunity and bystander stimulation. *J. Immunol.* **169,** 90–98.

Kim, S.-K., Cornberg, M., Wang, X. Z., Chen, H. D., Selin, L. K., and Welsh, R. M. (2005). Private specificities of CD8 T cell responses control patterns of heterologous immunity. *J. Exp. Med.* **201,** 523–533.

Kopeloff, L. M., and Kopeloff, N. (1947). Neurologic manifestations in laboratory animals produced by organ (adjuvant) emulsions. *J. Immunol.* **57,** 229–237.

Kos, F. J., and Mullbacher, A. (1993). Transfer of $CD8^+$ T cells into SCID mice and activation of memory virus-specific cytotoxic T cells. *Scand. J. Immunol.* **37,** 523–528.

Kurtzke, J. F. (1997). The epidemiology of multiple sclerosis. *In* "Multiple Sclerosis: Clinical and Pathogenetic Basis" (C. S. Raine, H. F. McFarland, and W. W. Tourtellotte, Eds.), pp. 91–140. Chapman & Hall, London.

Lang, H. L. E., Jacobsen, H., Ikemizu, S., Andersson, C., Harlos, K., Madsen, L., Hjorth, P., Sondergaard, L., Svejgaard, A., Wucherpfennig, K., Stuart, D. I., Bell, J. I., *et al.* (2002). A functional and structural basis for TCR cross-reactivity in multiple sclerosis. *Nat. Immunol.* **3,** 940–943.

Laron, Z., Lewy, H., Wilderman, I., Casu, A., Willis, J., Redondo, M. J., Libman, I., White, N., and Craig, M. (2005). Seasonality of month of birth of children and adolescents with type 1 diabetes mellitus in homogenous and heterogeneous populations. *Isr. Med. Assoc. J.* **7,** 381–384.

Larsen, P. D., Bloomer, L. C., and Bray, P. F. (1985). Epstein-Barr nuclear antigen and viral capsid antigen antibody titers in multiple sclerosis. *Neurology* **35,** 435–438.

Lee, S. M., Dunnavant, F. D., Jang, H., Zunt, J., and Levin, M. C. (2006). Autoantibodies that recognize functional domains of hnRNPA1 implicate molecular mimicry in the pathogenesis of neurological disease. *Neurosci. Lett.* **401,** 188–193.

Lenz, D. C., Lu, L., Conant, S. B., Wolf, N. A., Gérard, H. C., Whittum-Hudson, J. A., Hudson, A. P., and Swanborg, R. H. (2001). A *Chlamydia pneumoniae*-specific peptide induces experimental autoimmune encephalomyelitis in rats. *J. Immunol.* **167,** 1803–1808.

Link, H. (1978). The immune response in human demyelinating diseases. *Adv. Exp. Med. Biol.* **100,** 529–543.

Link, H., and Huang, Y. M. (2006). Oligoclonal bands in multiple sclerosis cerebrospinal fluid: An update on methodology and clinical usefulness. *J. Neuroimmunol.* **180,** 17–28[Epub ahead of print].

Link, H., and Kostulas, V. (1983). Utility of isoelectric focusing of cerebrospinal fluid and serum on agarose evaluated for neurological patients. *Clin. Chem.* **29,** 810–815.

Markovic-Plese, S., and McFarland, H. F. (2001). Immunopathogenesis of the multiple sclerosis lesion. *Curr. Neurol. Neurosci. Rep.* **1,** 257–262.

Markovic-Plese, S., Hemmer, B., Zhao, Y., Simon, R., Pinilla, C., and Martin, R. (2005). High level of cross-reactivity in influenza virus hemagglutinin-specific $CD4^+$ T-cell response: Implications for the initiation of autoimmune response in multiple sclerosis. *J. Neuroimmunol.* **169,** 31–38.
Merkler, D., Horvath, E., Bruck, W., Zinkernagel, R. M., Del la Torre, J. C., and Pinschewer, D. D. (2006). "Viral déjà vu" elicits organ-specific immune disease independent of reactivity to self. *J. Clin. Invest.* **116,** 1254–1263.
Mertens, N. M. J., Galvin, J. E., Adderson, E. E., and Cunningham, M. W. (2000). Molecular analysis of cross-reactive anti-myosin/anti-streptococcal mouse monoclonal antibodies. *Mol. Immunol.* **37,** 901–913.
Mongkolsapaya, J., Dejnirattisai, W., Xu, X. N., Vasanawathana, S., Tangthawornchaikul, N., Chairunsri, A., Sawasdivorn, S., Duangchinda, T., Dong, T., Rowland-Jones, S., Yenchitsomanus, P. T., McMichael, A., *et al.* (2003). Original antigenic sin and apoptosis in the pathogenesis of dengue hemorrhagic fever. *Nat. Med.* **9,** 921–927.
Morgan, I. M. (1946). Allergic encephalomyelitis in monkeys in response to injection of normal monkey cord. *J. Bacteriol.* **51,** 614–615.
Morgan, I. M. (1947). Allergic encephalomyelitis in monkeys in response to injection of normal monkey nervous tissue. *J. Exp. Med.* **85,** 131–140.
Morrison, L. R. (1947). Disseminated encephalomyelitis experimentally produced by the use of homologous antigen. *Arch. Neurol. Psychiatry* **58,** 391–416.
Mumford, C. J., Wood, N. W., Kellar-Wood, H., Thorpe, J. W., Miller, D. H., and Compston, D. A. (1994). The British Isles survey of multiple sclerosis in twins. *Neurology* **44,** 11–15.
Neu, A., Kehrer, M., Ashkenazi, I., and Laron, Z. (2000). Seasonality of birth in children (0–14 years) with diabetes mellitus type 1 in Baden-Wuerttemberg, Germany. *J. Pediatr. Endocrinol. Metab.* **13,** 1081–1085.
Noseworthy, J. H., Lucchinetti, C., Rodriguez, M., and Weinshenker, B. G. (2000). Multiple sclerosis. *N. Engl. J. Med.* **343,** 938–952.
Ogino, M., Orazio, N., and Latov, N. (1995). IgG anti-GM1 antibodies from patients with acute motor neuropathy are predominantly of the IgG1 and IgG3 subclasses. *J. Neuroimmunol.* **58,** 77–80.
Pettinelli, C. B., and McFarlin, D. E. (1981). Adoptive transfer of experimental allergic encephalomyelitis in SJL/J mice after *in vitro* activation of lymph node cells by myelin basic protein: Requirement for Lyt 1+2− T lymphocytes. *J. Immunol.* **127,** 1420–1423.
Poole, B. D., Scofield, R. H., Harley, J. B., and James, J. A. (2006). Epstein-Barr virus and molecular mimicry in systemic lupus erythematosus. *Autoimmunity* **30,** 63–70.
Reiber, H., Ungefehr, S., and Jacobi, C. (1998). The intrathecal, polyspecific and oligoclonal immune response in multiple sclerosis. *Mult. Scler.* **4,** 111–117.
Rivers, T. M., and Schwentker, F. F. (1935). Encephalomyelitis accompanied by myelin destruction experimentally produced in monkeys. *J. Exp. Med.* **61,** 689–702.
Rivers, T. M., Sprunt, D. H., and Berry, G. P. (1933). Observations on attempts to produce acute disseminated encephalomyelitis in monkeys. *J. Exp. Med.* **58,** 39–53.
Rothwell, P. M., Staines, A., Smail, P., Wadsworth, E., and McKinney, P. (1996). Seasonality of birth of patients with childhood diabetes in Britain. *BMJ* **312,** 1456–1457.
Sadovnick, A. D., Armstrong, H., Rice, G. P. A., Bulman, D., Hashimoto, L., Paty, D. W., Hashimoto, S. A., Warren, S., Hader, W., Murray, T. J., Seland, T. P., Metz, L., *et al.* (1993). A population-based study of multiple sclerosis in twins: Update. *Ann. Neurol.* **33,** 281–285.
Salahuddin, S. Z., Ablashi, D. V., Markham, P. D., Josephs, S. F., Sturzenegger, S., Kaplan, M., Halligan, G., Biberfeld, P., Wong-Staal, F., Kramarsky, B., and Gallo, R. C. (1986). Isolation of a new virus, HBLV, in patients with lymphoproliferative disorders. *Science* **234,** 596–601.
Selin, L. K., Nahill, S. R., and Welsh, R. M. (1994). Cross-reactivities in memory cytotoxic T lymphocyte recognition of heterologous viruses. *J. Exp. Med.* **179,** 1933–1943.

Selin, L. K., Varga, S. M., Wong, I. C., and Welsh, R. M. (1998). Protective heterologous antiviral immunity and enhanced immunopathogenesis mediated by memory T cell populations. *J. Exp. Med.* **188,** 1705–1715.

Smith, G. P. (1991). Surface presentation of protein epitopes using bacteriophage expression systems. *Curr. Opin. Biotechnol.* **2,** 668–673.

Sobel, R. A. (1989). T-lymphocyte subsets in the multiple sclerosis lesion. *Res. Immunol.* **140,** 208–211.

Sospedra, M., Zhao, Y., Muraro, P. A., Hamashin, C., de Villiers, E. M., Pinilla, C., zur Hausen, H., and Martin, R. (2005). Recognition of conserved amino acid motifs of common viruses and its role in autoimmunity. *PLoS. Pathog.* **1,** 335–348.

Srinivasappa, J., Saegusa, J., Prabhakar, B. S., Gentry, M. K., Buchmeier, M. J., Wiktor, T. J., Koprowski, H., Oldstone, M. B., and Notkins, A. L. (1986). Molecular mimicry: Frequency of reactivity of monoclonal antiviral antibodies with normal tissues. *J. Virol.* **57,** 397–401.

Sriram, S., Stratton, C. W., Yao, S., Tharp, A., Ding, L., Bannan, J. D., and Mitchell, W. M. (1999). *Chlamydia pneumoniae* infection of the central nervous system in multiple sclerosis. *Ann. Neurol.* **46,** 6–14.

Stefferl, A., Schubart, A., Storch, M., Amini, A., Mather, I., Lassmann, H., and Linington, C. (2000). Butyrophilin, a milk protein, modulates the encephalitogenic T cell response to myelin oligodendrocyte glycoprotein in experimental autoimmune encephalomyelitis. *J. Immunol.* **165,** 2859–2865.

Tejada-Simon, M. V., Zang, Y. C. Q., Hong, J., Rivera, V. M., and Zhang, J. Z. (2003). Cross-reactivity with myelin basic protein and human herpesvirus-6 in multiple sclerosis. *Ann. Neurol.* **53,** 189–197.

Theil, D. J., Tsunoda, I., Rodriguez, F., Whitton, J. L., and Fujinami, R. S. (2001). Viruses can silently prime for and trigger central nervous system autoimmune disease. *J. Neurovirol.* **7,** 220–227.

Tough, D. F., and Sprent, J. (1994). Turnover of naive- and memory-phenotype T cells. *J. Exp. Med.* **179,** 1127–1135.

Tough, D. F., Borrow, P., and Sprent, J. (1996). Induction of bystander T cell proliferation by viruses and type I interferon *in vivo*. *Science* **272,** 1947–1950.

Trinchieri, G. (1995). Interleukin-12: A proinflammatory cytokine with immunoregulatory functions that bridge innate resistance and antigen-specific adaptive immunity. *Annu. Rev. Immunol.* **13,** 251–276.

Tsunoda, I., and Fujinami, R. S. (1996). Two models for multiple sclerosis: Experimental allergic encephalomyelitis and Theiler's murine encephalomyelitis virus. *J. Neuropathol. Exp. Neurol.* **55,** 673–686.

Tsunoda, I., and Fujinami, R. S. (1999). Theiler's murine encephalomyelitis virus. *In* "Persistent Viral Infections" (R. Ahmed and I. Chen, Eds.), pp. 517–536. John Wiley & Sons Ltd., Chichester, West Sussex, England.

Tsunoda, I., Kurtz, C. I. B., and Fujinami, R. S. (1997). Apoptosis in acute and chronic central nervous system disease induced by Theiler's murine encephalomyelitis virus. *Virology* **228,** 388–393.

Tsunoda, I., Kuang, L.-Q., and Fujinami, R. S. (2002). Induction of autoreactive $CD8^+$ cytotoxic cells during Theiler's murine encephalomyelitis virus infection: Implications for autoimmunity. *J. Virol.* **76,** 12834–12844.

Tsunoda, I., Kuang, L.-Q., Kobayashi-Warren, M., and Fujinami, R. S. (2005). Central nervous system pathology caused by autoreactive $CD8^+$ T cell clones following virus infection. *J. Virol.* **79,** 14640–14646.

Tsunoda, I., Libbey, J. E., Kobayashi-Warren, M., and Fujinami, R. S. (2006). IFN-γ production and astrocyte recognition by autoreactive T cells induced by Theiler's virus infection: Role of viral strains and capsid proteins. *J. Neuroimmunol.* **172,** 85–93.

Ufret-Vincenty, R. L., Quigley, L., Tresser, N., Pak, S. H., Gado, A., Hausmann, S., Wucherpfennig, K. W., and Brocke, S. (1998). *In vivo* survival of viral antigen-specific T cells that induce experimental autoimmune encephalomyelitis. *J. Exp. Med.* **188,** 1725–1738.

Ursic-Bratina, N., Battelino, T., Krzisnik, C., Laron-Kenet, T., Ashkenazi, I., and Laron, Z. (2001). Seasonality of birth in children (0–14 years) with type 1 diabetes mellitus in Slovenia. *J. Pediatr. Endocrinol. Metab.* **14,** 47–52.

Wekerle, H., and Lassmann, H. (2005). The immunology of inflammatory demyelinating disease. *In* "McAlpine's Multiple Sclerosis" (A. Compston, C. Confavreux, H. Lassmann, I. McDonald, D. Miller, J. Noseworthy, K. Smith, and H. Wekerle, Eds.), 4th ed., pp. 491–555. Churchill Livingstone Elsevier, London.

Welsh, R. M., McNally, J. M., Brehm, M. A., and Selin, L. K. (2000). Consequences of cross-reactive and bystander CTL responses during viral infections. *Virology* **270,** 4–8.

Yamada, M., Zurbriggen, A., and Fujinami, R. S. (1990). Monoclonal antibody to Theiler's murine encephalomyelitis virus defines a determinant on myelin and oligodendrocytes, and augments demyelination in experimental allergic encephalomyelitis. *J. Exp. Med.* **171,** 1893–1907.

Yewdell, J. W., and Bennink, J. R. (1999). Immunodominance in major histocompatibility complex class I-restricted T lymphocyte responses. *Annu. Rev. Immunol.* **17,** 51–88.

Yu, R. K., Usuki, S., and Ariga, T. (2006). Ganglioside molecular mimicry and its pathological roles in Guillain-Barre syndrome and related diseases. *Infect. Immun.* **74,** 6517–6527 [Epub ahead of print].

Zabriskie, J. B. (1986). Rheumatic fever: A model for the pathological consequences of microbial- host mimicry. *Clin. Exp. Rheumatol.* **4,** 65–73.

Zamvil, S. S., and Steinman, L. (1990). The T lymphocyte in experimental allergic encephalomyelitis. *Annu. Rev. Immunol.* **8,** 579–621.

Zhang, X., Sun, S., Hwang, I., Tough, D. F., and Sprent, J. (1998). Potent and selective stimulation of memory-phenotype $CD8^+$ T cells *in vivo* by IL-15. *Immunity* **8,** 591–599.

Zimmerman, C., Brduscha-Riem, K., Blaser, C., Zinkernagel, R. M., and Pircher, H. (1996). Visualization, characterization, and turnover of $CD8^+$ memory T cells in virus-infected hosts. *J. Exp. Med.* **183,** 1367–1375.

MOLECULAR "NEGATIVITY" MAY UNDERLIE MULTIPLE SCLEROSIS: ROLE OF THE MYELIN BASIC PROTEIN FAMILY IN THE PATHOGENESIS OF MS

Abdiwahab A. Musse and George Harauz

Department of Molecular and Cellular Biology, Biophysics Interdepartmental Group
University of Guelph, Ontario N1G 2W1, Canada

Myelin basic protein (MBP) binds to negatively charged lipids on the cytosolic surface of oligodendrocyte membranes and is responsible for adhesion of these surfaces in the multilayered myelin sheath. The pattern of extensive posttranslational modifications of MBP is dynamic during normal central nervous system development and during myelin degeneration in multiple sclerosis (MS), affecting its interactions with the myelin membranes and other proteins. In particular, the degree of deimination (or citrullination) of MBP is correlated with the severity of MS, and may represent a primary defect that precedes neurodegeneration due to autoimmune attack. That MBP deimination also affects topological accessibility of an otherwise partially buried immunodominant epitope of the protein indicates that this modification may play a major role in the autoimmune pathogenesis of the disease. In this chapter, we describe the structural and functional consequences of MBP deimination in healthy and diseased myelin.

I. Multiple Sclerosis

The neurodegenerative human disease, multiple sclerosis (MS), is characterized by the active destruction of the insulating myelin sheath around the axons,

INTERNATIONAL REVIEW OF
NEUROBIOLOGY, VOL. 79
DOI: 10.1016/S0074-7742(07)79007-4

0074-7742/07 $35.00

leaving lesions which comprise demyelinated axon fibers (Lassmann, 2004). The infiltration and/or accumulation of resident macrophages, activated astrocytes and microglia, as well as lymphocytic cells, responding to myelin antigens, are also common features of MS lesions. Persistent demyelination of axons impairs the signal conduction of these fibers (loss of saltatory conduction) and contributes immensely to the neurological disabilities suffered by patients with MS, resulting eventually in axonal degeneration (Ferguson *et al.*, 1997; Trapp *et al.*, 1998). Although the etiology of MS still remains elusive, growing evidence suggests that the pathogenesis of the disease is a consequence of multiple genetic, immunologic, and environmental factors (Compston, 2004; Ibrahim and Gold, 2005; Lutton *et al.*, 2004; Oksenberg and Barcellos, 2005; Sospedra and Martin, 2005).

The disease is highly variable (Lassmann *et al.*, 2001; Lucchinetti *et al.*, 1999), but the most common manifestation is chronic and relapse-remitting in which the myelin sheath is repaired to a certain extent to enable a moderate clinical remission (Lublin, 2004). The loss of the myelin sheath in the central nervous system (CNS) of patients with MS is also accompanied by an inadequate repair capacity and failure to remyelinate denuded axons after persistent demyelination episodes due to the resurgence of early developmental signaling pathways (Franklin, 2002; John *et al.*, 2002). However, the exact details of why the capacity for remyelination is lost in CNS as the disease progresses remain elusive. Even more importantly, the triggers of the destruction of the myelin sheath are still a mystery. Consequently, the design for therapeutic interventions for MS remains a challenging goal over a century after the disease was first described.

Extensive investigations have focused on the components of the myelin sheath as both the target and the possible initiators of the autoimmune aspect of the disease. Of these candidate autoantigens, myelin basic protein (MBP), the second most abundant protein in myelin, has been widely studied (Boggs, 2006; Campagnoni and Campagnoni, 2004; Campagnoni and Skoff, 2001; Harauz *et al.*, 2004) and its importance to the immunopathogenesis of MS is extensively reviewed elsewhere (Lutton *et al.*, 2004; Sospedra and Martin, 2005; Steinman *et al.*, 1995). MBP and its peptides are highly encephalitogenic and immunization of MBP in rodents and primates induces experimental autoimmune encephalomyelitis (EAE), used widely as a model for MS. The mechanisms by which the initial MBP antigens are released, and by which targeted immune responses are initiated in MS patients, remain unidentified. It has been argued that immune responses to viral- and/or bacterial-derived antigens with significant structural similarities (molecular mimicry) to MBP peptides or other myelin antigens may trigger the initiation of autoimmunity in MS patients (Wucherpfennig and Strominger, 1995). While certain epidemiological data do support the involvement in MS of exogenous agents such as viral infections (Kurtzke, 1993), molecular mimicry has never been proven. Therefore, a more direct involvement of self-derived autoantigens in the autoimmune pathogenesis of MS cannot be ruled out. In this chapter, we will examine the role by which deimination of MBP plays a role in demyelination of axons, as well as the early course of the autoimmune pathogenesis of MS.

II. MBP: A Membrane Velcro in CNS Myelin

Myelin is the fatty material that surrounds axons in tightly packed multilamellae in both the central and peripheral nervous systems, and facilitates rapid conduction of nerve impulses by saltatory conduction (Baumann and Pham-Dinh, 2001; Trapp and Kidd, 2004). One of the major protein components of CNS myelin is MBP, constituting roughly 30% of the total protein by weight. The MBP family arises from the genetic unit called *g*enes of the *oli*godendrocyte *li*neage (Golli), via use of several different transcription sites (tss1, tss2, and tss3) and alternative splicing of the 10 (human) or 11 (murine) exons (Campagnoni and Campagnoni, 2004; Campagnoni *et al.*, 1993; Pribyl *et al.*, 1993). The transcription of this gene complex is regulated by a combinatorial network of control sequences (Farhadi and Peterson, 2006; Farhadi *et al.*, 2003; Givogri *et al.*, 2001; Landry *et al.*, 1996). The so-called "classic" 18.5-kDa MBP isoform is one of the major protein components of adult human and bovine CNS myelin (14 kDa in rats and mice). In this chapter, the term "MBP" shall be used to mean primarily the 18.5-kDa isoform (Boggs, 2006; Harauz *et al.*, 2004).

The primary role of MBP in the CNS is generally considered to be the maintenance of compaction of the myelin sheath (Fig. 1). In electron micrographs of sectioned CNS myelin, MBP is localized in the major dense lines of the spiral multilamellar structure of the compact myelin. These electron dense lines are formed by the fusion of the cytoplasmic leaflets of the myelin membrane (Fig. 1). In this structure, MBP serves as an adhesive molecule that maintains the tight multilamellar packing of the sheath (Brady *et al.*, 1981b; Sedzik *et al.*, 1984; Trapp and Kidd, 2004). In addition, the *in vitro* reconstitution of MBP with unilamellar vesicles containing anionic lipids results in the formation of multilamellar structures similar to those seen in myelin sheaths (Brady *et al.*, 1981a,b; Cristofolini *et al.*, 2005; Deber *et al.*, 1986; Riccio *et al.*, 2000). The importance of MBP to the proper formation of compact myelin sheaths in the CNS has been demonstrated by light and electron microscopy analyses of myelin from the dysmyelinating mutant mouse, *shiverer* (Nave, 1994). These animals, which carry an autosomal recessive trait, have a deletion of the last five exons of the gene-encoding MBP and show no detectable MBP in their CNS. Morphologically, the *shiverer* mice are characterized by the nearly total absence of myelin around the axons of the affected animals. The extent of myelination is greatly reduced and, where myelin is observed, it shows aberrant morphology and a periodicity typified by the absence of the major dense lines.

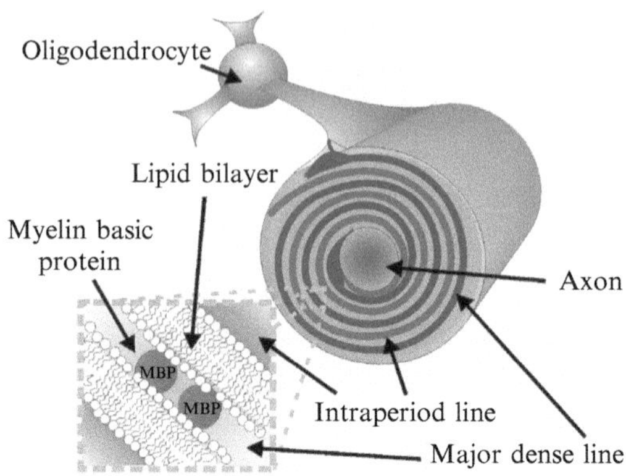

FIG. 1. Illustration of myelinated axons. The plasma membrane processes of oligodendrocytes form tightly packed multilamellae ensheathing axons of the CNS to facilitate the rapid conduction of nerve impulses by saltatory conduction. Here, MBP is localized in the major dense lines of the spiral multilamellar structure of the compact myelin. These electron dense lines are formed by the fusion of the cytoplasmic leaflets of the myelin membrane, whereby MBP serves as an adhesive molecule that maintains the tight multilamellar packing of the sheath. Figure courtesy of Dr. Noboru Ishiyama.

III. MBP: A Conformationally Adaptable and Multifunctional Protein

Structurally, MBP is a member of the emerging class of "intrinsically disordered" proteins (IDPs), which have also been referred to as "intrinsically unstructured," "natively unstructured," or "conformationally adaptable" (Dunker *et al.*, 2002; Fink, 2005; Receveur-Bréchot *et al.*, 2006; Uversky *et al.*, 2000; Wright and Dyson, 1999). Roughly, a third of all eukaryotic proteins fall into this category, carrying out critically important functions in molecular recognition, molecular assembly, protein modification, and entropic chain activities (Dunker *et al.*, 2002; Receveur-Bréchot *et al.*, 2006; Tompa, 2002, 2005; Tompa *et al.*, 2005; Ward *et al.*, 2004). The lack of defined structure of these types of proteins is largely due to their high net charge coupled with a low mean hydrophobicity, thus maximizing the effect of intramolecular electrostatic repulsion and minimizing the contribution of a hydrophobic collapse (Tompa, 2002). In this context, MBP has an extremely high net positive charge, with $pI > 10$, and minimum mean hydrophobicity. In aqueous solution, MBP, as shown by various spectroscopic studies, is an extended protein with significant conformational plasticity and minimal ordered secondary structure (Bates and Harauz, 2003; Bates *et al.*, 2000, 2003; Farès *et al.*, 2006; Gow and Smith, 1989; Haas *et al.*, 2004; Polverini *et al.*, 1999). Association of this protein with lipids affords the protein-increased α-helical and β-sheet secondary structure (Bates and Harauz, 2003; Bates *et al.*, 2000; Farès *et al.*, 2006; Musse *et al.*, 2006; Polverini *et al.*, 1999).

From a physicochemical standpoint, the extreme net positive charge of MBP is critical to how this protein discharges its role in myelin, allowing it to interact with negatively charged lipids through a plethora of synergistic interactions with the cytoplasmic membrane (Cristofolini *et al.*, 2005; Hu *et al.*, 2004) during compact myelin formation. As shown in model myelin membranes, using force measurements and atomic force microscopy, the synergistic MBP–lipid interactions rely on a balance between the opposite charges of MBP and the lipid bilayer in order to assemble the proper multilamellar structure seen in myelin sheaths (Hu *et al.*, 2004). Small changes to this balance could result in significant changes in myelin adhesion or stability. Consequently, alteration of the cationicity of MBP may represent a regulatory mechanism for normal myelin assembly, or a degradative mechanism in MS. In this regard, isolated MBP shows extensive posttranslational modifications with varying degrees of deimination, phosphorylation, deamidation, methylation, and N-terminal acylation (Boulias *et al.*, 1995; Costentino *et al.*, 1994; Harauz *et al.*, 2004; Kim *et al.*, 2003; Moscarello *et al.*, 1992). These varied posttranslational modifications of MBP are interdependent and alter its conformation and interactions with the myelin membrane and other proteins (as reviewed in Boggs, 2006; Harauz *et al.*, 2004). The reasons for this extensive modification are unknown. It has been proposed (Yang, 2005) that "multisite modification on a protein constitutes a complex regulatory program that resembles a 'dynamic molecular barcode' and transduces information to and from signaling pathways." This latter paradigm may potentially be applied to MBP if we regard this protein family as having physiological roles beyond simple structural stabilization of the myelin sheath.

The aforementioned modifications of MBP give rise to multiple charge variants denoted as C1–C8, when separated on carboxymethyl (CM52) cellulose resin. The C1 component represents the least modified and the most cationic component with net charge +19 at neutral pH. This component is the most abundant form in healthy adult humans (Moscarello *et al.*, 1994) and is the most effective lipid-aggregating charge variant of MBP (Brady *et al.*, 1985; Wood and Moscarello, 1989). The C2 charge isomer of MBP arises as a consequence of deamidation of Gln/Asn, giving this component a net loss of one unit of positive charge. The remaining components, known as C3–C5, differ by the successive additional loss of one unit of positive charge as a result of phosphorylations of seryl and threonyl residues. As shown by both liquid X-ray diffraction experiments as well as lipid aggregation studies, the effectiveness of these charge isomers in inducing multilamellar membrane structures is greatly reduced as a function of their overall net positive charge (i.e., C1 > C2 > C3 > C4 > C5 > C8) (Brady *et al.*, 1985; Wood and Moscarello, 1989). Similar qualitative results with respect to the membrane competency of these charge isomers as a function of their net positive charge have also been obtained from a molecular-level membrane adhesion study using optical waveguide spectrometry (Shanshiashvili *et al.*, 2003).

IV. MBP Deimination and MS

The most severe posttranslational modification of MBP, in terms of its reduction in overall cationicity, is the extensive arginine deimination of the protein, which gives rise to the C8 charge isomer. This posttranslational modification results in reduction in net positive charge of the protein by one unit for each arginyl residue that is converted to citrulline. In mammalian cells, the conversion of protein-contained arginyl residues to citrulline is catalyzed by the Ca^{2+}-dependent enzyme peptidylarginine deiminase (PAD, EC 3.5.3.15), of which the known isoforms are PAD1, PAD2, PAD3, PAD4/PAD5, and PAD6, identified mainly in mammals (Balandraud *et al.*, 2005; Vossenaar *et al.*, 2003). These proteins differ in terms of their tissue and cellular distributions. The PAD2 isotype has the broadest tissue distribution and is found in myelinating cells of the CNS and in myelin, where one of its target substrates is MBP.

As far as we know, peptidylarginine deimination is an irreversible reaction, although it has been speculated that the appropriate "aminotransferase" enzymes may exist and are yet to be discovered (Bannister and Kouzarides, 2005). Arginine is a large, positively charged amino acid that will contribute significantly to a protein's structural and functional properties by its electrostatic nature (Woods, 2004; Woods and Ferre, 2005). In contrast, citrulline is polar and uncharged. Therefore, interconversion of arginyl residues in proteins to uncharged citrulline is expected to disrupt existing intramolecular interactions involving the modified arginine through ionic pairing with acidic residues. Such intramolecular interactions may be of critical importance to the structure of the protein, and their loss causes protein unfolding, making it more susceptible to degradation. In addition, deimination of arginyl residues in proteins could cause loss of intermolecular interactions of the target protein with other ligands, leading to loss of function. Therefore, the conversion of arginine to uncharged citrulline by PAD has profound consequences on a protein's structure and function (van Venrooij and Pruijn, 2000; Vossenaar *et al.*, 2003). As shown in *in vitro* studies, extensive deimination of trichohyalin and filaggrin results in pronounced structural loss and protein unfolding (Tarcsa *et al.*, 1996). Citrullination of vimentin at the N-terminal head domain has also been shown to result in the disassembly of vimentin intermediate filaments (Inagaki *et al.*, 1989). Excellent reviews of the structural and functional effects of PAD protein deimination can be found in (György *et al.*, 2006; van Venrooij and Pruijn, 2000; Vossenaar *et al.*, 2003).

Citrulline was first identified in MBP in 1971 (Finch *et al.*, 1971). In general, the C8 charge isomer of MBP contains citrulline at six conserved site (residues 25, 31, 122, 130, 159, and 170 using human 18.5-kDa MBP sequence numbering) as assessed by amino acid analysis (Wood and Moscarello, 1989), resulting in a net positive of only +13 compared to the +19 of the C1 charge isomer. As shown by a more recent proteomics analysis by mass spectrometry, the citrulline content of the C8 isomer may also include additional sites, albeit with lower frequencies (Kim *et al.*, 2003). Studies from our and other laboratories have observed deiminated MBP to be structurally less-ordered and more susceptible to proteolysis under solution conditions (Bates *et al.*, 2002, 2003; Cao *et al.*, 1999; D'Souza and Moscarello, 2006; Pritzker *et al.*, 2000a,b). Furthermore, deiminated MBP is extremely less effective, compared to the highly cationic C1 charge isomer, in organizing lipids into a multilamellar structure, owing to its reduced ability to cause adhesion of apposed membranes (Bates *et al.*, 2002; Beniac *et al.*, 2000; Boggs *et al.*, 1997, 1999; Shanshiashvili *et al.*, 2003; Wood and Moscarello, 1989). In addition, when myelin is fractionated by discontinuous sucrose gradient centrifugation, the highly deiminated C8 charge isomer is found in the loose myelin fractions that contain fewer lamellae (Cruz and Moscarello, 1985). By comparison, the highly cationic C1 isomer is found in the compact myelin fractions with extensive multilamellae (Cruz and Moscarello, 1985). Subsequent immunoelectron microscopical studies have shown epitopes of C8 to be present in the (extracellular) intraperiod line of myelin, whereas C1 is found only in the (cytoplasmic) major dense line (McLaurin *et al.*, 1993). These observations have led to the conclusion that the extent of MBP deimination may have a profound impact on the stability of myelin sheaths in the CNS. Consequently, increased deimination of MBP, *in vivo*, may cause loss in adhesion of the cytoplasmic leaflets of the myelin membrane, and loosened myelin compaction.

At the molecular level, the severity of MS correlates strongly with the degree of arginine loss of MBP due to deimination (Moscarello *et al.*, 1994; Wood *et al.*, 1996). For instance, in individuals afflicted with chronic MS, nearly 50% of isolated MBP molecules are deiminated compared to only 20% in normal brain (Moscarello *et al.*, 1994; Wood *et al.*, 1996). In comparison, in the rare case of acute MS known as Marburg's syndrome, this proportion is increased to over 90%, indicating a correlation between the physical, structural characteristics of MBP, the degree of myelin degeneration, and the severity of MS (Wood and Moscarello, 1984). In the case of Marburg's syndrome, which represents an extreme situation, almost all arginyl residues in MBP are deiminated. Addition of this highly deiminated MBP to anionic lipid vesicles caused fragmentation similar to myelin vacuoles seen during myelin breakdown in MS patients (Boggs *et al.*, 1999), suggesting a perhaps pathogenic role of this charge isomer not only in loss of myelin compaction but also myelin breakdown.

Additional support for the link between reduced MBP cationicity and myelin stability has been gleaned from a transgenic mouse model for demyelination, ND4. The ND4 mouse line was generated by the incorporation of 70 copies of the cDNA for the PLP variant DM20 into the genome of CD1 mice (Mastronardi *et al.*, 1993; Simons-Johnson *et al.*, 1995). The mechanism of demyelination in the mice containing the transgene appears to involve increased fatty acid acylation that affects proper insertion of PLP and DM20 into the myelin bilayer (Barrese *et al.*, 1998), but which also has other consequences at the molecular level (Mastronardi *et al.*, 1996a,c). In these animals, increases in PAD activity and PAD mRNA levels have been observed prior to an increase in the amount of deiminated MBP, and the concomitant appearance of any clinical and pathological signs, preceding both by about 1 month (Mastronardi *et al.*, 1996a,c; Moscarello *et al.*, 2002). The resulting increase in deiminated MBP correlated well with the onset of demyelination in these animals. The membrane-bound fraction of PAD2 was preponderant compared to the soluble enzyme (Pritzker *et al.*, 1999). Thus, it has been suggested that the upregulation of membrane-bound PAD2 activity represents an important early molecular change in demyelinating diseases, thereby precipitating the destabilization of myelin in MS brain white matter (Mastronardi and Moscarello, 2005; Moscarello *et al.*, 2002).

V. MBP Deimination and Autoimmunity in MS

The foregoing correlation between isolated MBP deimination and MS implies that chemical modifications to MBP *in vivo* may contribute directly to structural instability of myelin. The cornerstone of this argument is, as shown by studies *in vitro*, that reduction in net positive charge of the protein alters its physicochemical and lipid-organizing properties that might be operative in myelin *in vivo*. Of critical importance is whether the reduction in positive charge of MBP, as seen in MS, could contribute toward abnormal membrane-bound conformational state of the protein that may serve as a substrate for autoimmunity in MS. In this context, it is important to determine the structural and functional changes effected on MBP by deimination, in order to understand the progression of the disease, and to be able to devise effective strategies for therapeutic management.

To this end, we have exploited the charge differences of two recombinant isomers, *rmC1* and *rmC8*, of murine MBP corresponding to the most cationic, C1, and least cationic, C8, variants of the protein, respectively (Bates *et al.*, 2000, 2002). Using site-directed spin-labeling (SDSL) and electron paramagnetic resonance (EPR) spectroscopy, we have shown that this deimination mimic results in the central and C-terminal regions of the protein being significantly less strongly associated with myelin-like membranes (Bates *et al.*, 2003; Fig. 2A). Such observations have led us to suspect that in addition to myelin compaction loss, deimination of MBP may also contribute to autoantigenic presentation of this protein, *in vivo*, during the early stages of MS.

We have performed subsequent detailed SDSL/EPR structural analyses of the membrane-associated conformational state of the primary immunodominant

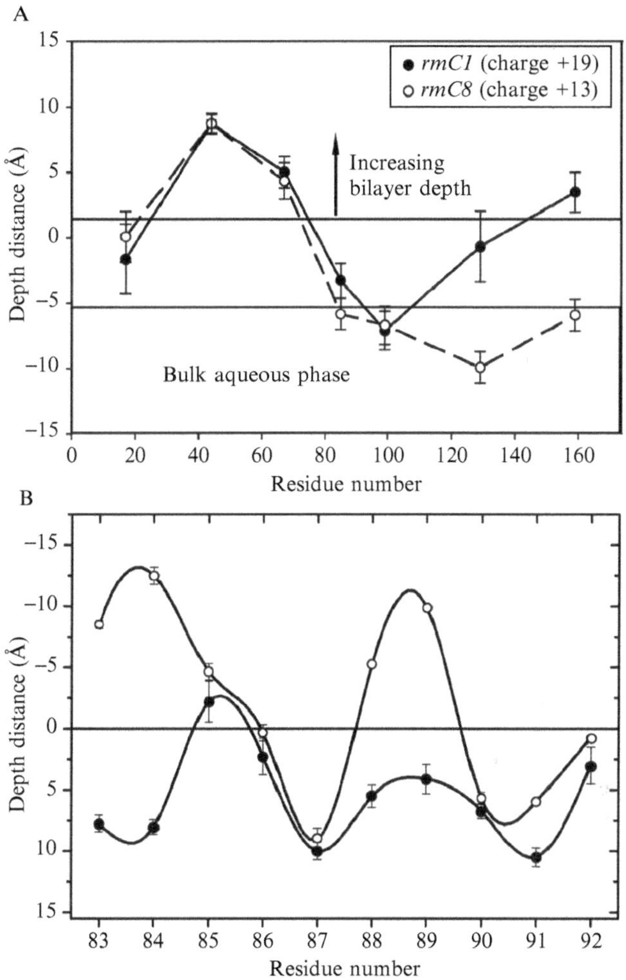

FIG. 2. Membrane depth measurement of spin-labeled residues of normal (*rmC1*) and disease-associated (*rmC8*) MBP charge isomers. (A) Global survey of membrane depths of the two isomers reveal that the membrane-bound topologies of the two charge isomers differ in disposition at the central and C-terminal regions. The disease-associated charge isomer, *rmC8*, shows significant surface exposure at the C-terminus and, to a lesser degree, central regions of the protein (adapted from Bates *et al.*, 2003). (B) Membrane depth measurements of the primary immunodominant epitope of MBP in normal (*rmC1*) and disease-associated (*rmC8*) charge isomers. In the normal myelin-associated charge isomer of MBP (*rmC1*), this entire epitope (V83–T92, murine 18.5-kDa numbering) forms a highly membrane-sequestered amphipathic α-helix, with a 9° tilt with respect to the plane of the bilayer. In contrast, in the disease-associated and less cationic charge isomer (*rmC8*), this epitope is found to be highly surface exposed and to form a shorter α-helix than in *rmC1* (adapted from Bates *et al.*, 2004; Musse *et al.*, 2006).

epitope (V83–T92, murine sequence numbering) of these two recombinant isomers (Bates *et al.*, 2004; Musse *et al.*, 2006; Fig. 2B). This segment of the protein corresponds to the core elements of the minimum epitope for T-cell recognition of human MBP (residues P85–P99, human sequence numbering) with the highest affinity for the major histocompatibility complex (MHC) class II haplotype, HLA-DR, believed to be associated with increased susceptibility to MS (Lang *et al.*, 2002; Ota *et al.*, 1990; Valli *et al.*, 1993; Wucherpfennig *et al.*, 1994). The T and B cells found in the CNS of MS patients show consistent specificity for this epitope, and antibodies against this epitope are found in both the brain and spinal fluid of patients with MS, emphasizing both humoral and cellular immune response to this peptide in the pathogenesis of the disease (Martino *et al.*, 1991; Wucherpfennig *et al.*, 1997). Detailed requirements for the binding of this epitope to HLA class II molecules and the autoimmune T-cell receptor (TCR) were provided by the X-ray crystallographic analysis of the structure of this peptide complexed with the two MHC class II isotypes, HLA-DRa and HLA-DRb, and TCR (Hahn *et al.*, 2005; Li *et al.*, 2000, 2005; Smith *et al.*, 1998). The linear polypeptide D82-ENPVVHFFKNIVTPR-T98 (human numbering) has shown significant clinical efficacy as an immunotherapeutic agent in a phase II clinical trial, especially in patients with HLA haplotypes DR2 and/or DR4 (Warren *et al.*, 2006), suggesting great promise for antigen-specific therapy for this disease (Steinman, 2006).

We have shown in the membrane-bound state of the highly cationic *rmC1* that the V83–T92 epitope forms an amphipathic α-helix situated semiparallel to the membrane surface, with an approximately 9° helical tilt toward the N-terminus (Bates *et al.*, 2004). The estimated membrane depth penetration was up to 10 Å, indicating that this epitope would be largely sequestered within the leaflet of a normally compacted myelin sheath (Fig. 2B). By contrast, the primary immunodominant epitope of the less cationic recombinant murine MBP isomer, *rmC8*, forms a more highly surface-exposed and shorter amphipathic α-helix than in the unmodified form, *rmC1*, analogous to the most cationic and abundant isomer of MBP in normal myelin (Musse *et al.*, 2006; Fig. 2B). Whereas the entire V83–T92 sequence of the membrane-bound *rmC1* isomer was mapped to an α-helix, the observed EPR parameters for the *rmC8* mutants showed only a short α-helix consisting of residues H85–T92 (Musse *et al.*, 2006). We have confirmed the topological difference of this epitope in these two isomers of MBP by performing proteolytic digestion of the membrane-bound proteins using cathepsin D (Fig. 3B). Cathepsin D was used because of its unique cut sites in MBP—F42–F43 and F86–F87—of which the latter site overlaps the probed epitope (Fig. 3B). Under these conditions, cathepsin D digested membrane-bound *rmC8* three-fold faster than *rmC1*, and cleavage at F86–F87 occurred much more readily in *rmC8* than in *rmC1* (Musse *et al.*, 2006; Fig. 3B).

VI. Does Deimination of MBP Precipitate the Autoimmune Response in MS?

Our observations of the surface exposure of this primary immunodominant epitope of the disease-associated MBP isomer are further supported by immunohistochemical analyses of demyelinating lesions in MS, using a unique antibody [encephalitogenic peptide (EP) antiserum] that recognizes the synthetic peptide QDENPVV of MBP (Matsuo *et al.*, 1997). This peptide corresponds to the N-terminal region of the above-described primary immunodominant epitope of MBP, and is highly conserved among all species (Fig. 3A). Immunohistochemical staining of brain sections from MS patients with this antibody showed unique labeling of demyelinating regions and abnormal processes of oligodendrocytes (Matsuo *et al.*, 1997). Such staining could not be seen when the antibody was used against normal human, rat, or guinea pig brain. Consistent with our report for the increased accessibility of the targeted epitope in disease-associated MBP isomers, Matsuo *et al.* (1997) argued that the antibody only recognized this region on its becoming accessible in demyelinating areas of MS brains. On the basis of these observations, we argue that the reduction in net positive of MBP by deimination, as seen in MS, contributes not only to loss of myelin compaction and stability but also to possible autoantigen presentation, *in vivo*, by causing the surface exposure of highly encephalitogenic epitopes of this protein that may initiate or sustain the onslaught of autoimmune destruction of myelin in MS patients.

It appears that in MS, the QDENPVV epitope, recognized by their unique antibody, is in a conformation that renders it exposed in actively demyelinating fibers and abnormally appearing OL cell bodies at the margins of plaques within proximity to highly activated microglia and myelin-laden macrophages (Matsuo *et al.*, 1997). While this study did not provide a mechanism by which this region of the protein could be exposed under disease conditions, our recent structural investigations of this region of the protein in its normal and disease-associated charge isomers provide a plausible mechanism by which established biochemical changes within myelin and myelin-producing cells, in MS, could be attributed to the abnormal exposure of this highly encephalitogenic epitope of MBP. Taken together, these observations suggest that myelin degeneration—whatever its origin may be—could lead to conformationally accessible epitopes of MBP that can be processed in MS by activated microglia and/or macrophages for the purpose of MHC class II-restricted antigen presentation to T cells, which can either initiate or sustain autoimmune reaction in the CNS.

It should also be noted that the unmasking of the epitope recognized by the EP antiserum may not only be specific to MS but could also be ascribed to other neurological diseases and diverse brain insults. Among the neurological diseases thus far characterized to have abnormal exposure of this epitope of MBP, as in

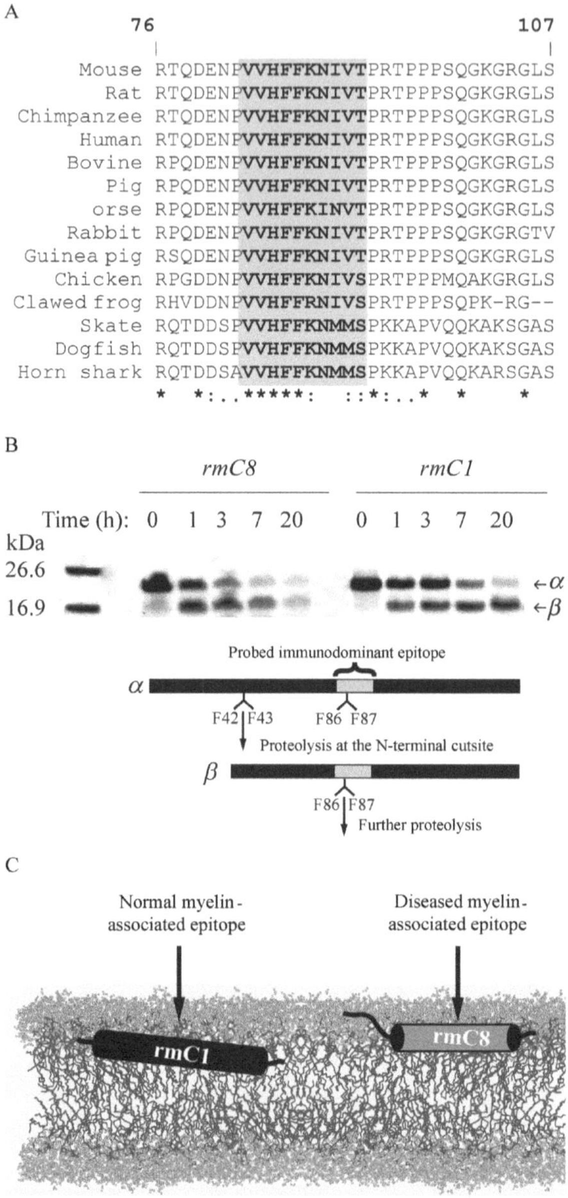

FIG. 3. Topology of the primary immunodominant epitope of MBP in normal (*rmC1*) and disease-associated (*rmC8*) charge isomers. (A) Comparison of amino acid sequences of the primary immunodominant epitope from various species, highlighting the high degree of sequence identity of this region of the protein. Symbols mean that residues in that column are (*) identical in all sequences, (:) substitutions are conservative, and (.) substitutions are semiconservative. (B) Cathepsin

MS, there is Multiple System Atrophy. This disease is characterized by the degenerative formation of cytoplasmic inclusions in OLs, causing widespread abnormalities both in white and in gray matters of the CNS (Matsuo *et al.*, 1998). Equally, the abnormal exposure of this epitope has been observed after ischemic damage to the CNS. In addition, glutamate receptor-mediated excitoxic injury to isolated spinal cord white matter has been shown to result in targeted calcium-dependent excitoxic damage to myelin and glia in which the epitope recognized by the EP antiserum was extensively exposed (Li and Stys, 2000).

Clearly, whether the degeneration of myelin occurs as consequence of an inflammatory process to the CNS in MS, or whether such degenerative processes in myelin and myelin-producing cells occur prior to any inflammatory process, represents the proverbial "chicken and egg" conundrum. For the former, the overwhelming EAE studies, as a model for MS, provide compelling evidence that inflammatory responses induced by myelin-specific antigens can lead to the destruction of myelin in the CNS much like in MS. Such inflammatory responses have also been shown to include hyperdeimination of CNS proteins, including MBP, in areas of highest inflammation, indicating a link between inflammation in the CNS, protein hyperdeimination, and myelin destruction (Raijmakers *et al.*, 2005). However, studies involving an animal model (ND4 mice) for demyelinating disease in which the temporal relationships between the onset of the neurological disease and the appearance of deiminated MBP were explored, showed a "precursor–product" type relationship (Mastronardi *et al.*, 1996a,c; Moscarello *et al.*, 2002). As alluded to in early discussions, the ND4 model was generated through genetic lesions by incorporating 70 copies of the myelin proteolipid protein DM20. As the transgenic mice were made on a CD1 background, the spontaneous demyelination injuries of these animals are devoid of inflammatory processes, and are solely the result of abnormal metabolic changes in myelin and myelin-producing cells. The apparent disconnect, vis-à-vis the origins of protein deimination, demyelination, and inflammation, suggested by the above observations only illustrates the complexity and the interdependency of demyelination and inflammatory responses to the CNS in MS.

D digestion of membrane-bound *rmC8* and *rmC1*. Cleavage of the intact protein (denoted α, $M_r \approx$ 19.5 kDa) is expected to release a combination of five bands depending on the cut site (F42–F43 and/or F86–F87). Single cleavage at F42–F43 releases the β peptide ($M_r \approx$ 14.4 kDa), which can be subsequently digested at F86–F87 (the probed epitope). (C) Schematic topology of the primary immunodominant epitope of MBP in normal and disease-associated charge isomers (on the basis of the data in Figs. 2B and 3B). This schematic illustrates the dependence on the overall net positive charge of the protein of the membrane disposition of the primary immunodominant epitope of MBP, segment V83–T92 (murine 18.5-kDa numbering). Reduction in global net positive charge of MBP by deimination reduces the overall helicity of this epitope and renders it highly surface-exposed and susceptible to proteolysis (adapted from Musse *et al.*, 2006).

VII. PAD and Protein Deimination

PADs are highly regulated proteins both at the translational and enzymatic levels. These enzymes are calcium-dependent, yet the concentration of Ca^{2+} ions (100 nM) in the cytoplasm of cells at rest is not sufficient to activate them. As shown in *in vitro* studies, the concentration of Ca^{2+} ions for half-maximum activity of the two most widely studied PAD family members, PAD2 and PAD4 (0.65 and 0.75 mM, respectively) (Nakayama-Hamada *et al.*, 2005) is nearly 10,000-fold higher than that of normal cells at rest. As such, activation of these enzymes, *in vivo*, requires the influx through the plasma membrane, or efflux from intracellular stores, of Ca^{2+} ions into the cytoplasm. This process is also ubiquitously regulated in all cell types (as reviewed in Berridge, 2004; Berridge *et al.*, 1998, 2000; Bootman *et al.*, 2001). The origin of the stimuli required to evoke this increase in cytosolic Ca^{2+} concentrations could be associated with normal cellular processes, as the case is for the many fundamental cellular processes that are regulated by Ca^{2+}ions. Similarly, increased Ca^{2+} concentrations in the cytoplasm could also be the result of the manifestation of disease states or the consequences of environmental insults. The Ca^{2+} influx in oligodendrocytes caused by response to an external "signal" is a well-studied phenomenon (Dyer and Benjamins, 1990, 1991; Micu *et al.*, 2006; Paz Soldan *et al.*, 2003; Tzeng *et al.*, 1995; Waxman, 1992).

MBP is not the only CNS protein to be deiminated. Ursell *et al.* (1995) have described an unusual 60-kDa dynamin–MBP fusion protein containing citrulline. The function of this protein is unknown, but it was suggested to be involved in intracellular transport. Nor is deimination representative only of MS—it is a response to various other stresses. In the CNS or CNS-derived cells, there is increased protein deimination (particularly of MBP but also of glial fibrillary acidic protein, GFAP, and vimentin, a cytoskeletal protein) in *jimpy* (dysmyelinating mutant due to PLP single-point mutation) mice (Fannon and Moscarello, 1990), in cultured rat oligodendrocytes (Akiyama *et al.*, 1999; see discussion in Vossenaar *et al.*, 2003), in mice deficient in Fyn tyrosine kinase (Seiwa *et al.*, 2000), in human-cultured astrocytes subjected to hypoxia (Sambandam *et al.*, 2004), in spinal cords of mice with MOG-induced EAE (Nicholas *et al.*, 2005; Raijmakers *et al.*, 2005) (contrasting with PLP-induced EAE in Mastronardi *et al.*, 1996b), and in brains of patients with Alzheimer's disease (Ishigami *et al.*, 2005) (contrasting with an earlier study in Moscarello *et al.*, 1994). GFAP has also been suggested to be present in multiple citrullinated isoforms in MS (Nicholas *et al.*, 2004).

Increased deimination of histones by PAD4 has been reported to be part of the molecular hallmark of MS and of an animal model of demyelination much like MBP deimination by PAD2 (Mastronardi *et al.*, 2006). Histones are the substrate of the nuclear PAD4 isoform. The nuclear translocation of this enzyme has been ascribed to involve elevated levels of tumor necrosis factor-α (TNF-α) in normal-appearing white matter. A functional haplotype of the PAD4 gene has been associated with rheumatoid arthritis in Japanese and Korean populations (Kang *et al.*, 2006; Suzuki *et al.*, 2003), although similar studies failed to show any association in European populations (Barton *et al.*, 2004; Caponi *et al.*, 2005; Martinez *et al.*, 2005). Similarly, a study of MS patients showed no familial-based association between PAD4 gene haplotypes or single nucleotide polymorphism in a French population (Tommasi *et al.*, 2006). Though the exact details of the involvement of PAD4 and histone deimination in MS are not well understood, it is thought that the decreased net positive charge of histones, after deimination, may interfere with DNA chromatin structuring and result in apoptosis of oligodendrocytes. In fact, overexpression and activation of PAD4 in hematopoietic cells were shown to induce cell cycle arrest, suggesting that a similar mechanism may be at work in MS as well (Liu *et al.*, 2006). As such, the link between the activation of PAD2 and PAD4 in MS, as per the abnormal deimination of their cellular substrates (MBP and histones, respectively), may represent a progressive degenerative process, of unknown origin, in oligodendrocytes. In this regard, the manifestation of this degenerative process may include, but is not limited to, increased intercellular Ca^{2+}, subsequent activation of PADs (isoforms 2 and 4), increased deimination of MBP and histones, loss of myelin and chromatin structuring, cell cycle arrest and apoptosis, exposure and release of encephalitogenic epitopes of myelin antigens, initiation and sustenance of inflammation in CNS, loss of cellular repair, and finally axonal degeneration.

VIII. Conclusions

The pathogenic roles of the MBP family in MS have been the subject of extensive investigations. Given the highly encephalitogenic nature of its peptides, the presence of autoantibodies and autoreactive T cells with specificity for it in the brain lesions of MS patients, and the critical role it plays in myelin structure and maintenance, the MBP protein family has been long considered as one of the candidate autoantigens for autoimmune pathogenesis of MS. In addition to its autoantigenicity, isolated MBPs from normal-appearing white matter in MS show apparent posttranslational modifications that can be correlated with the severity of the disease. One of the most critical modifications is the conversion of positively charged arginyl residues to uncharged citrullines (deimination) by the enzyme PAD2. Deimination of MBP undermines, among other things, the ability of this protein to organize lipids into a multilamellar structure, *in vitro*, and has been argued to contribute toward instability and loss of myelin compaction in MS. Moreover, we have shown recently through structural analysis that such modification renders an otherwise partially sequestered immunodominant epitope (V86–T95, human 18.5-kDa MBP numbering) of this protein highly surface exposed, consistent with earlier findings which showed the unusual accessibility of this epitope, *in vivo*, in degenerating myelin of MS lesions in proximity to activated microglia and myelin-laden macrophages. Proteolytic processing of such a surface-exposed epitope by myelin-associated proteases, and/or phagocytosis of the damaged myelin by immune-derived cells, could result in the release of this encephalitogenic epitope from the CNS, and its processing for antigen presentation to the peripheral immune system. As such, we argue that apparent deimination of MBP may, in addition to reducing the degree of compaction of myelin, also participate in the autoimmune pathogenesis of the disease by revealing otherwise inaccessible antigens of myelin to the surveillance of immune cells. Such antigens could serve as initiators or sustainers of the autoimmune onslaught in the CNS of MS patients. How the topologies of other known epitopes of this MBP are affected by the deimination of the protein is presently under investigation.

Given that the enzyme responsible for the deimination of MBP, PAD2, is highly regulated, both transcriptionally and functionally, the disease-associated hyperdeimination of MBP in MS indicates inherent changes within the myelin-producing cells of the CNS, the oligodendrocytes. The origin of these changes is presently unclear, and our ongoing search for the nature of these changes will undoubtedly advance our understandings of the disease and result in more effective therapeutic interventions. Such therapeutic interventions would be expected to go beyond the mere modulations of the inflammatory process of the presently available treatments.

Abbreviations

C1–C8, MBP charge components (charge isomers) 1–8; CNS, central nervous system; EAE, experimental autoimmune encephalomyelitis; EP, encephalitogenic peptide; EPR, electron paramagnetic resonance; DM20, truncated isoform of proteolipid protein; GFAP, glial fibrillary acidic protein; Golli, genes of the oligodendrocyte lineage; MBP, myelin basic protein (hMBP, human; rmMBP, recombinant murine); MHC, major histocompatibility complex; MOG, myelin/oligodendrocyte glycoprotein; MS, multiple sclerosis; ND4, mouse model for demyelination with 70 copies of DM20; PAD, peptidylarginine deiminase, EC 3.5.3.15; PLP, proteolipid protein; *rmC1*, *rmC8*, recombinant murine analogues of natural C1, C8 charge isomers; SDSL, site-directed spin-labeling; TNF, tumor necrosis factor.

Acknowledgments

Our work has been supported over the years by grants to GH from the Natural Sciences and Engineering Research Council of Canada, the Multiple Sclerosis Society of Canada (MSSC), and the Canadian Institutes for Health Research. AAM is the recipient of an MSSC Doctoral Studentship.

We are grateful to many colleagues, past and present, but particularly Dr. Ian Bates (formerly of our group), and to Drs. Mario Moscarello, Fabrizio Mastronardi, and Joan Boggs (Hospital for Sick Children, Toronto).

References

Akiyama, K., Sakurai, Y., Asou, H., and Senshu, T. (1999). Localization of peptidylarginine deiminase type II in a stage-specific immature oligodendrocyte from rat cerebral hemisphere. *Neurosci. Lett.* **274,** 53–55.

Balandraud, N., Gouret, P., Danchin, E. G., Blanc, M., Zinn, D., Roudier, J., and Pontarotti, P. (2005). A rigorous method for multigenic families' functional annotation: The peptidyl arginine deiminase (PADs) proteins family example. *BMC Genomics* **6,** 153.

Bannister, A. J., and Kouzarides, T. (2005). Reversing histone methylation. *Nature* **436,** 1103–1106.

Barrese, N., Mak, B., Fisher, L., and Moscarello, M. A. (1998). Mechanism of demyelination in DM20 transgenic mice involves increased fatty acylation. *J. Neurosci. Res.* **53,** 143–152.

Barton, A., Bowes, J., Eyre, S., Spreckley, K., Hinks, A., John, S., and Worthington, J. (2004). A functional haplotype of the PADI4 gene associated with rheumatoid arthritis in a Japanese population is not associated in a United Kingdom population. *Arthritis Rheum.* **50,** 1117–1121.

Bates, I. R., and Harauz, G. (2003). Molecular dynamics exposes alpha-helices in myelin basic protein. *J. Mol. Model.* (Online) **9,** 290–297.

Bates, I. R., Matharu, P., Ishiyama, N., Rochon, D., Wood, D. D., Polverini, E., Moscarello, M. A., Viner, N. J., and Harauz, G. (2000). Characterization of a recombinant murine 18.5-kDa myelin basic protein. *Protein Exp. Purif.* **20,** 285–299.

Bates, I. R., Libich, D. S., Wood, D. D., Moscarello, M. A., and Harauz, G. (2002). An Arg/Lys → Gln mutant of recombinant murine myelin basic protein as a mimic of the deiminated form implicated in multiple sclerosis. *Protein Exp. Purif.* **25,** 330–341.

Bates, I. R., Boggs, J. M., Feix, J. B., and Harauz, G. (2003). Membrane-anchoring and charge effects in the interaction of myelin basic protein with lipid bilayers studied by site-directed spin labeling. *J. Biol. Chem.* **278,** 29041–29047.

Bates, I. R., Feix, J. B., Boggs, J. M., and Harauz, G. (2004). An immunodominant epitope of myelin basic protein is an amphipathic alpha-helix. *J. Biol. Chem.* **279,** 5757–5764.

Baumann, N., and Pham-Dinh, D. (2001). Biology of oligodendrocyte and myelin in the mammalian central nervous system. *Physiol. Rev.* **81,** 871–927.

Beniac, D. R., Wood, D. D., Palaniyar, N., Ottensmeyer, F. P., Moscarello, M. A., and Harauz, G. (2000). Cryoelectron microscopy of protein-lipid complexes of human myelin basic protein charge isomers differing in degree of citrullination. *J. Struct. Biol.* **129,** 80–95.

Berridge, M. J. (2004). Calcium signal transduction and cellular control mechanisms. *Biochim. Biophys. Acta* **1742,** 3–7.

Berridge, M. J., Bootman, M. D., and Lipp, P. (1998). Calcium-a life and death signal. *Nature* **395,** 645–648.

Berridge, M. J., Lipp, P., and Bootman, M. D. (2000). The versatility and universality of calcium signalling. *Nat. Rev. Mol. Cell Biol.* **1,** 11–21.

Boggs, J. M. (2006). Myelin basic protein: A multifunctional protein. *Cell. Mol. Life Sci.* **63**(17), 1945–1961.

Boggs, J. M., Yip, P. M., Rangaraj, G., and Jo, E. (1997). Effect of posttranslational modifications to myelin basic protein on its ability to aggregate acidic lipid vesicles. *Biochemistry* **36,** 5065–5071.

Boggs, J. M., Rangaraj, G., Koshy, K. M., Ackerley, C., Wood, D. D., and Moscarello, M. A. (1999). Highly deiminated isoform of myelin basic protein from multiple sclerosis brain causes fragmentation of lipid vesicles. *J. Neurosci. Res.* **57,** 529–535.

Bootman, M. D., Lipp, P., and Berridge, M. J. (2001). The organisation and functions of local Ca(2+) signals. *J. Cell Sci.* **114,** 2213–2222.

Boulias, C., Pang, H., Mastronardi, F., and Moscarello, M. A. (1995). The isolation and characterization of four myelin basic proteins from the unbound fraction during CM52 chromatography. *Arch. Biochem. Biophys.* **322,** 174–182.

Brady, G. W., Fein, D. B., Wood, D. D., and Moscarello, M. A. (1981a). The interaction of basic proteins from normal and multiple sclerosis myelin with phosphatidylglycerol vesicles. *FEBS Lett.* **125,** 159–160.

Brady, G. W., Murthy, N. S., Fein, D. B., Wood, D. D., and Moscarello, M. A. (1981b). The effect of basic myelin protein on multilayer membrane formation. *Biophys. J.* **34,** 345–350.

Brady, G. W., Fein, D. B., Wood, D. D., and Moscarello, M. A. (1985). The role of charge microheterogeneity of human myelin basic protein in the formation of phosphatidylglycerol multilayers. *Biochem. Biophys. Res. Commun.* **126,** 1161–1165.

Campagnoni, A. T., and Campagnoni, C. (2004). Myelin basic protein gene. *In* "Myelin Biology and Disorders" (R. A. Lazzarini, J. W. Griffin, H. Lassman, K.-A. Nave, R. H. Miller, and B. D. Trapp, Eds.), Vol. 1, pp. 387–4000. Elsevier Academic Press, San Diego.

Campagnoni, A. T., and Skoff, R. P. (2001). The pathobiology of myelin mutants reveal novel biological functions of the MBP and PLP genes. *Brain Pathol.* **11,** 74–91.

Campagnoni, A. T., Pribyl, T. M., Campagnoni, C. W., Kampf, K., Amur-Umarjee, S., Landry, C. F., Handley, V. W., Newman, S. L., Garbay, B., and Kitamura, K. (1993). Structure and developmental regulation of *Golli-mbp*, a 105-kilobase gene that encompasses the myelin basic protein gene and is expressed in cells in the oligodendrocyte lineage in the brain. *J. Biol. Chem.* **268,** 4930–4938.

Cao, L., Goodin, R., Wood, D., Moscarello, M. A., and Whitaker, J. N. (1999). Rapid release and unusual stability of immunodominant peptide 45–89 from citrullinated myelin basic protein. *Biochemistry* **38,** 6157–6163.

Caponi, L., Petit-Teixeira, E., Sebbag, M., Bongiorni, F., Moscato, S., Pratesi, F., Pierlot, C., Osorio, J., Chapuy-Regaud, S., Guerrin, M., Cornelis, F., Serre, G., *et al.* (2005). A family based study shows no association between rheumatoid arthritis and the PADI4 gene in a white French population. *Ann. Rheum. Dis.* **64,** 587–593.

Compston, A. (2004). Genetic susceptibility and epidemiology. *In* "Myelin Biology and Disorders" (R. A. Lazzarini, J. W. Griffin, H. Lassman, K.-A. Nave, R. H. Miller, and B. D. Trapp, Eds.), Vol. 2, pp. 701–731. Elsevier Academic Press, San Diego.

Costentino, M., Pritzker, L., Boulias, C., and Moscarello, M. A. (1994). Acylation of myelin basic protein peptide 1–21 with alkyl carboxylates 2–10 carbons long affects secondary structure and posttranslational modification. *Biochemistry* **33,** 4155–4162.

Cristofolini, L., Fontana, M. P., Serra, F., Fasano, A., Riccio, P., and Konovalov, O. (2005). Microstructural analysis of the effects of incorporation of myelin basic protein in phospholipid layers. *Eur. Biophys. J.* **34,** 1041–1048.

Cruz, T. F., and Moscarello, M. A. (1985). Characterization of myelin fractions from human brain white matter. *J. Neurochem.* **44,** 1411–1418.

D'Souza, C. A., and Moscarello, M. A. (2006). Differences in susceptibility of MBP charge isomers to digestion by stromelysin-1 (MMP-3) and release of an immunodominant epitope. *Neurochem. Res.* **31**(8), 1045–1054.

Deber, C. M., Hughes, D. W., Fraser, P. E., Pawagi, A. B., and Moscarello, M. A. (1986). Binding of human normal and multiple sclerosis-derived myelin basic protein to phospholipid vesicles: Effects on membrane head group and bilayer regions. *Arch. Biochem. Biophys.* **245,** 455–463.

Dunker, A. K., Brown, C. J., Lawson, J. D., Iakoucheva, L. M., and Obradovic, Z. (2002). Intrinsic disorder and protein function. *Biochemistry* **41,** 6573–6582.

Dyer, C. A., and Benjamins, J. A. (1990). Glycolipids and transmembrane signaling: Antibodies to galactocerebroside cause an influx of calcium in oligodendrocytes. *J. Cell Biol.* **111,** 625–633.

Dyer, C. A., and Benjamins, J. A. (1991). Galactocerebroside and sulfatide independently mediate Ca2+ responses in oligodendrocytes. *J. Neurosci. Res.* **30,** 699–711.

Fannon, A. M., and Moscarello, M. A. (1990). Myelin basic protein is affected by reduced synthesis of myelin proteolipid protein in the jimpy mouse. *Biochem. J.* **268,** 105–110.

Farès, C., Libich, D. S., and Harauz, G. (2006). Solution NMR structure of an immunodominant epitope of myelin basic protein. *FEBS J.* **273,** 601–614.

Farhadi, H. F., and Peterson, A. C. (2006). The myelin basic protein gene: A prototype for combinatorial mammalian transcriptional regulation. *Adv. Neurol.* **98,** 65–76.

Farhadi, H. F., Lepage, P., Forghani, R., Friedman, H. C., Orfali, W., Jasmin, L., Miller, W., Hudson, T. J., and Peterson, A. C. (2003). A combinatorial network of evolutionarily conserved myelin basic protein regulatory sequences confers distinct glial-specific phenotypes. *J. Neurosci.* **23,** 10214–10223.

Ferguson, B., Matyszak, M. K., Esiri, M. M., and Perry, V. H. (1997). Axonal damage in acute multiple sclerosis lesions. *Brain* **120,** 393–399.

Finch, P. R., Wood, D. D., and Moscarello, M. A. (1971). The presence of citrulline in a myelin protein fraction. *FEBS Lett.* **15,** 145–148.

Fink, A. L. (2005). Natively unfolded proteins. *Curr. Opin. Struct. Biol.* **15,** 35–41.

Franklin, R. J. (2002). Why does remyelination fail in multiple sclerosis? *Nat. Rev. Neurosci.* **3,** 705–714.

Givogri, M. I., Bongarzone, E. R., Schonmann, V., and Campagnoni, A. T. (2001). Expression and regulation of golli products of myelin basic protein gene during *in vitro* development of oligodendrocytes. *J. Neurosci. Res.* **66,** 679–690.

Gow, A., and Smith, R. (1989). The thermodynamically stable state of myelin basic protein in aqueous solution is a flexible coil. *Biochem. J.* **257,** 535–540.

György, B., Tóth, E., Tarcsa, E., Falus, A., and Buzás, E. I. (2006). Citrullination: A posttranslational modification in health and disease. *Int. J. Biochem. Cell Biol.* **38,** 1662–1677.

Haas, H., Oliveira, C. L., Torriani, I. L., Polverini, E., Fasano, A., Carlone, G., Cavatorta, P., and Riccio, P. (2004). Small angle x-ray scattering from lipid-bound myelin basic protein in solution. *Biophys. J.* **86,** 455–460.

Hahn, M., Nicholson, M. J., Pyrdol, J., and Wucherpfennig, K. W. (2005). Unconventional topology of self peptide-major histocompatibility complex binding by a human autoimmune T cell receptor. *Nat. Immunol.* **6,** 490–496.

Harauz, G., Ishiyama, N., Hill, C. M. D., Bates, I. R., Libich, D. S., and Farès, C. (2004). Myelin basic protein-diverse conformational states of an intrinsically unstructured protein and its roles in myelin assembly and multiple sclerosis. *Micron* **35,** 503–542.

Hu, Y., Doudevski, I., Wood, D. D., Moscarello, M. A., Husted, C., Genain, C., Zasadzinski, J. A., and Israelachvili, J. (2004). Synergistic interactions of lipids and myelin basic protein. *Proc. Natl. Acad. Sci. USA* **101,** 13466–13471.

Ibrahim, S. M., and Gold, R. (2005). Genomics, proteomics, metabolomics: What is in a word for multiple sclerosis? *Curr. Opin. Neurol.* **18,** 231–235.

Inagaki, M., Takahara, H., Nishi, Y., Sugawara, K., and Sato, C. (1989). Ca2+-dependent deimination-induced disassembly of intermediate filaments involves specific modification of the amino-terminal head domain. *J. Biol. Chem.* **264,** 18119–18127.

Ishigami, A., Ohsawa, T., Hiratsuka, M., Taguchi, H., Kobayashi, S., Saito, Y., Murayama, S., Asaga, H., Toda, T., Kimura, N., and Maruyama, N. (2005). Abnormal accumulation of citrullinated proteins catalyzed by peptidylarginine deiminase in hippocampal extracts from patients with Alzheimer's disease. *J. Neurosci. Res.* **80,** 120–128.

John, G. R., Shankar, S. L., Shafit-Zagardo, B., Massimi, A., Lee, S. C., Raine, C. S., and Brosnan, C. F. (2002). Multiple sclerosis: Re-expression of a developmental pathway that restricts oligodendrocyte maturation. *Nat. Med.* **8,** 1115–1121.

Kang, C. P., Lee, H. S., Ju, H., Cho, H., Kang, C., and Bae, S. C. (2006). A functional haplotype of the PADI4 gene associated with increased rheumatoid arthritis susceptibility in Koreans. *Arthritis Rheum.* **54,** 90–96.

Kim, J. K., Mastronardi, F. G., Wood, D. D., Lubman, D. M., Zand, R., and Moscarello, M. A. (2003). Multiple sclerosis: An important role for post-translational modifications of myelin basic protein in pathogenesis. *Mol. Cell. Proteomics* **2,** 453–462.

Kurtzke, J. F. (1993). Epidemiologic evidence for multiple sclerosis as an infection. *Clin. Microbiol. Rev.* **6,** 382–427.

Landry, C. F., Ellison, J. A., Pribyl, T. M., Campagnoni, C., Kampf, K., and Campagnoni, A. T. (1996). Myelin basic protein gene expression in neurons: Developmental and regional changes in protein targeting within neuronal nuclei, cell bodies, and processes. *J. Neurosci.* **16,** 2452–2462.

Lang, H. L., Jacobsen, H., Ikemizu, S., Andersson, C., Harlos, K., Madsen, L., Hjorth, P., Sondergaard, L., Svejgaard, A., Wucherpfennig, K., Stuart, D. I., Bell, J. I., *et al.* (2002). A functional and structural basis for TCR cross-reactivity in multiple sclerosis. *Nat. Immunol.* **3,** 940–943.

Lassmann, H. (2004). Cellular damage and repair in multiple sclerosis. *In* "Myelin Biology and Disorders" (R. W. Lazzarini, J. W. Griffin, H. Lassman, K.-A. Nave, R. H. Miller, and B. D. Trapp, Eds.), Vol. 2, pp. 733–762. Elsevier Academic Press, San Diego.

Lassmann, H., Bruck, W., and Lucchinetti, C. (2001). Heterogeneity of multiple sclerosis pathogenesis: Implications for diagnosis and therapy. *Trends Mol. Med.* **7,** 115–121.

Li, S., and Stys, P. K. (2000). Mechanisms of ionotropic glutamate receptor-mediated excitotoxicity in isolated spinal cord white matter. *J. Neurosci.* **20,** 1190–1198.

Li, Y., Li, H., Martin, R., and Mariuzza, R. A. (2000). Structural basis for the binding of an immunodominant peptide from myelin basic protein in different registers by two HLA-DR2 proteins. *J. Mol. Biol.* **304,** 177–188.

Li, Y., Huang, Y., Lue, J., Quandt, J. A., Martin, R., and Mariuzza, R. A. (2005). Structure of a human autoimmune TCR bound to a myelin basic protein self-peptide and a multiple sclerosis-associated MHC class II molecule. *EMBO J.* **24,** 2968–2979.

Liu, G. Y., Liao, Y. F., Chang, W. H., Liu, C. C., Hsieh, M. C., Hsu, P. C., Tsay, G. J., and Hung, H. C. (2006). Overexpression of peptidylarginine deiminase IV features in apoptosis of haematopoietic cells. *Apoptosis* **11,** 183–196.

Lublin, F. D. (2004). Multiple sclerosis classification and overview. *In* "Myelin Biology and Disorders" (R. A. Lazzarini, J. W. Griffin, H. Lassman, K.-A. Nave, R. H. Miller, and B. D. Trapp, Eds.), Vol. 2, pp. 691–699. Elsevier Academic Press, San Diego.

Lucchinetti, C., Bruck, W., Parisi, J., Scheithauer, B., Rodriguez, M., and Lassmann, H. (1999). A quantitative analysis of oligodendrocytes in multiple sclerosis lesions. A study of 113 cases. *Brain* **122,** 2279–2295.

Lutton, J. D., Winston, R., and Rodman, T. C. (2004). Multiple sclerosis: Etiological mechanisms and future directions. *Exp. Biol. Med.* (Maywood) **229,** 12–20.

Martinez, A., Valdivia, A., Pascual-Salcedo, D., Lamas, J. R., Fernandez-Arquero, M., Balsa, A., Fernandez-Gutierrez, B., de la Concha, E. G., and Urcelay, E. (2005). PADI4 polymorphisms are not associated with rheumatoid arthritis in the Spanish population. (Oxf.) *Rheumatology* **44,** 1263–1266.

Martino, G., Olsson, T., Fredrikson, S., Hojeberg, B., Kostulas, V., Grimaldi, L. M., and Link, H. (1991). Cells producing antibodies specific for myelin basic protein region 70–89 are predominant in cerebrospinal fluid from patients with multiple sclerosis. *Eur. J. Immunol.* **21,** 2971–2976.

Mastronardi, F. G., and Moscarello, M. A. (2005). Molecules affecting myelin stability: A novel hypothesis regarding the pathogenesis of multiple sclerosis. *J. Neurosci. Res.* **80,** 301–308.

Mastronardi, F. G., Ackerley, C. A., Arsenault, L., Roots, B. I., and Moscarello, M. A. (1993). Demyelination in a transgenic mouse: A model for multiple sclerosis. *J. Neurosci. Res.* **36,** 315–324.

Mastronardi, F. G., Ackerley, C. A., Roots, B. I., and Moscarello, M. A. (1996a). Loss of myelin basic protein cationicity in DM20 transgenic mice is dosage dependent. *J. Neurosci. Res.* **44,** 301–307.

Mastronardi, F. G., al Sabbagh, A., Nelson, P. A., Rego, J., Roots, B. I., and Moscarello, M. A. (1996b). Myelin basic protein in experimental allergic encephalomyelitis is not affected at the posttranslational level: Implications for demyelinating disease. *J. Neurosci. Res.* **44,** 344–349.

Mastronardi, F. G., Mak, B., Ackerley, C. A., Roots, B. I., and Moscarello, M. A. (1996c). Modifications of myelin basic protein in DM20 transgenic mice are similar to those in myelin basic protein from multiple sclerosis. *J. Clin. Invest.* **97,** 349–358.

Mastronardi, F. G., Wood, D. D., Mei, J., Raijmakers, R., Tseveleki, V., Dosch, H.-M., Probert, L., Casaccia-Bonnefil, P., and Moscarello, M. A. (2006). Increased citrullination of histone H3 in MS brain and animal models for demyelination: A role for TNF-induced PAD4 translocation. *J. Neurosci.* **26**(44), 11387–11396.

Matsuo, A., Lee, G. C., Terai, K., Takami, K., Hickey, W. F., McGeer, E. G., and McGeer, P. L. (1997). Unmasking of an unusual myelin basic protein epitope during the process of myelin degeneration in humans: A potential mechanism for the generation of autoantigens. *Am. J. Pathol.* **150,** 1253–1266.

Matsuo, A., Akiguchi, I., Lee, G. C., McGeer, E. G., McGeer, P. L., and Kimura, J. (1998). Myelin degeneration in multiple system atrophy detected by unique antibodies. *Am. J. Pathol.* **153,** 735–744.

McLaurin, J., Ackerley, C. A., and Moscarello, M. A. (1993). Localization of basic proteins in human myelin. *J. Neurosci. Res.* **35,** 618–628.

Micu, I., Jiang, Q., Coderre, E., Ridsdale, A., Zhang, L., Woulfe, J., Yin, X., Trapp, B. D., McRory, J. E., Rehak, R., Zamponi, G. W., Wang, W., *et al.* (2006). NMDA receptors mediate calcium accumulation in myelin during chemical ischaemia. *Nature* **439,** 988–992.

Moscarello, M. A., Pang, H., Pace-Asciak, C. R., and Wood, D. D. (1992). The N terminus of human myelin basic protein consists of C2, C4, C6, and C8 alkyl carboxylic acids. *J. Biol. Chem.* **267,** 9779–9782.

Moscarello, M. A., Wood, D. D., Ackerley, C., and Boulias, C. (1994). Myelin in multiple sclerosis is developmentally immature. *J. Clin. Invest.* **94,** 146–154.

Moscarello, M. A., Pritzker, L., Mastronardi, F. G., and Wood, D. D. (2002). Peptidylarginine deiminase: A candidate factor in demyelinating disease. *J. Neurochem.* **81,** 335–343.

Musse, A. A., Boggs, J. M., and Harauz, G. (2006). Deimination of membrane-bound myelin basic protein in multiple sclerosis exposes an immunodominant epitope. *Proc. Natl. Acad. Sci. USA* **103,** 4422–4427.

Nakayama-Hamada, M., Suzuki, A., Kubota, K., Takazawa, T., Ohsaka, M., Kawaida, R., Ono, M., Kasuya, A., Furukawa, H., Yamada, R., and Yamamoto, K. (2005). Comparison of enzymatic properties between hPADI2 and hPADI4. *Biochem. Biophys. Res. Commun.* **327,** 192–200.

Nave, K. A. (1994). Neurological mouse mutants and the genes of myelin. *J. Neurosci. Res.* **38,** 607–612.

Nicholas, A. P., Sambandam, T., Echols, J. D., and Tourtellotte, W. W. (2004). Increased citrullinated glial fibrillary acidic protein in secondary progressive multiple sclerosis. *J. Comp. Neurol.* **473,** 128–136.

Nicholas, A. P., Sambandam, T., Echols, J. D., and Barnum, S. R. (2005). Expression of citrullinated proteins in murine experimental autoimmune encephalomyelitis. *J. Comp. Neurol.* **486,** 254–266.

Oksenberg, J. R., and Barcellos, L. F. (2005). Multiple sclerosis genetics: Leaving no stone unturned. *Genes Immun.* **6,** 375–387.

Ota, K., Matsui, M., Milford, E. L., Mackin, G. A., Weiner, H. L., and Hafler, D. A. (1990). T-cell recognition of an immunodominant myelin basic protein epitope in multiple sclerosis. *Nature* **346,** 183–187.

Paz Soldan, M. M., Warrington, A. E., Bieber, A. J., Ciric, B., Van, K.V, Pease, L. R., and Rodriguez, M. (2003). Remyelination-promoting antibodies activate distinct Ca2+ influx pathways in astrocytes and oligodendrocytes: Relationship to the mechanism of myelin repair. *Mol. Cell Neurosci.* **22,** 14–24.

Polverini, E., Fasano, A., Zito, F., Riccio, P., and Cavatorta, P. (1999). Conformation of bovine myelin basic protein purified with bound lipids. *Eur. Biophys. J.* **28,** 351–355.

Pribyl, T. M., Campagnoni, C. W., Kampf, K., Kashima, T., Handley, V. W., McMahon, J., and Campagnoni, A. T. (1993). The human myelin basic protein gene is included within a 179-kilobase transcription unit: Expression in the immune and central nervous systems. *Proc. Natl. Acad. Sci. USA* **90,** 10695–10699.

Pritzker, L. B., Nguyen, T. A., and Moscarello, M. A. (1999). The developmental expression and activity of peptidylarginine deiminase in the mouse. *Neurosci. Lett.* **266,** 161–164.

Pritzker, L. B., Joshi, S., Gowan, J. J., Harauz, G., and Moscarello, M. A. (2000a). Deimination of myelin basic protein. 1. Effect of deimination of arginyl residues of myelin basic protein on its structure and susceptibility to digestion by cathepsin D. *Biochemistry* **39,** 5374–5381.

Pritzker, L. B., Joshi, S., Harauz, G., and Moscarello, M. A. (2000b). Deimination of myelin basic protein. 2. Effect of methylation of MBP on its deimination by peptidylarginine deiminase. *Biochemistry* **39,** 5382–5388.

Raijmakers, R., Vogelzangs, J., Croxford, J. L., Wesseling, P., van Venrooij, W. J., and Pruijn, G. J. (2005). Citrullination of central nervous system proteins during the development of experimental autoimmune encephalomyelitis. *J. Comp. Neurol.* **486,** 243–253.

Receveur-Bréchot, V., Bourhis, J. M., Uversky, V. N., Canard, B., and Longhi, S. (2006). Assessing protein disorder and induced folding. *Proteins* **62,** 24–45.

Riccio, P., Fasano, A., Borenshtein, N., Bleve-Zacheo, T., and Kirschner, D. A. (2000). Multilamellar packing of myelin modeled by lipid-bound MBP. *J. Neurosci. Res.* **59,** 513–521.

Sambandam, T., Belousova, M., Accaviti-Loper, M. A., Blanquicett, C., Guercello, V., Raijmakers, R., and Nicholas, A. P. (2004). Increased peptidylarginine deiminase type II in hypoxic astrocytes. *Biochem. Biophys. Res. Commun.* **325,** 1324–1329.

Sedzik, J., Blaurock, A. E., and Höchli, M. (1984). Lipid/myelin basic protein multilayers. A model for the cytoplasmic space in central nervous system myelin. *J. Mol. Biol.* **174,** 385–409.

Seiwa, C., Sugiyama, I., Yagi, T., Iguchi, T., and Asou, H. (2000). Fyn tyrosine kinase participates in the compact myelin sheath formation in the central nervous system. *Neurosci. Res.* **37,** 21–31.

Shanshiashvili, L. V., Suknidze, N. C., Machaidze, G. G., Mikeladze, D. G., and Ramsden, J. J. (2003). Adhesion and clustering of charge isomers of myelin basic protein at model myelin membranes. *Arch. Biochem. Biophys.* **419,** 170–177.

Simons-Johnson, R., Roder, J. C., and Riordan, J. R. (1995). Over-expression of the DM-20 myelin proteolipid causes central nervous system demyelination in transgenic mice. *J. Neurochem.* **64,** 967–976.

Smith, K. J., Pyrdol, J., Gauthier, L., Wiley, D. C., and Wucherpfennig, K. W. (1998). Crystal structure of HLA-DR2 (DRA*0101, DRB1*1501) complexed with a peptide from human myelin basic protein. *J. Exp. Med.* **188,** 1511–1520.

Sospedra, M., and Martin, R. (2005). Immunology of multiple sclerosis. *Annu. Rev. Immunol.* **23,** 683–747.

Steinman, L. (2006). The coming of age for antigen-specific therapy for multiple sclerosis. *Eur. J. Neurol.* **13,** 793–794.

Steinman, L., Waisman, A., and Altmann, D. (1995). Major T-cell responses in multiple sclerosis. *Mol. Med. Today* **1,** 79–83.

Suzuki, A., Yamada, R., Chang, X., Tokuhiro, S., Sawada, T., Suzuki, M., Nagasaki, M., Nakayama-Hamada, M., Kawaida, R., Ono, M., Ohtsuki, M., Furukawa, H., *et al.* (2003). Functional haplotypes of PADI4, encoding citrullinating enzyme peptidylarginine deiminase 4, are associated with rheumatoid arthritis. *Nat. Genet.* **34,** 395–402.

Tarcsa, E., Marekov, L. N., Mei, G., Melino, G., Lee, S. C., and Steinert, P. M. (1996). Protein unfolding by peptidylarginine deiminase. Substrate specificity and structural relationships of the natural substrates trichohyalin and filaggrin. *J. Biol. Chem.* **271,** 30709–30716.

Tommasi, C., Petit-Teixeira, E., Cournu-Rebeix, I., Caponi, L., Pierlot, C., Fontaine, B., Cornelis, F., and Migliorini, P. (2006). PADI4 gene in multiple sclerosis: A family-based association study. *J. Neuroimmunol.* **177**(1–2), 142–145.

Tompa, P. (2002). Intrinsically unstructured proteins. *Trends Biochem. Sci.* **27,** 527–533.

Tompa, P. (2005). The interplay between structure and function in intrinsically unstructured proteins. *FEBS Lett.* **579,** 3346–3354.

Tompa, P., Szasz, C., and Buday, L. (2005). Structural disorder throws new light on moonlighting. *Trends Biochem. Sci.* **30,** 484–489.

Trapp, B. D., and Kidd, G. J. (2004). Structure of the myelinated axon. *In* "Myelin Biology and Disorders" (R. A. Lazzarini, J. W. Griffin, H. Lassman, K.-A. Nave, R. H. Miller, and B. D. Trapp, Eds.), Vol. 1, pp. 3–27. Elsevier Academic Press, San Diego.

Trapp, B. D., Peterson, J., Ransohoff, R. M., Rudick, R., Mork, S., and Bo, L. (1998). Axonal transection in the lesions of multiple sclerosis. *N. Engl. J. Med.* **338,** 278–285.

Tzeng, S. F., Deibler, G. E., and DeVries, G. H. (1995). Exogenous myelin basic protein promotes oligodendrocyte death via increased calcium influx. *J. Neurosci. Res.* **42,** 768–774.

Ursell, M. R., McLaurin, J., Wood, D. D., Ackerley, C. A., and Moscarello, M. A. (1995). Localization and partial characterization of a 60 kDa citrulline-containing transport form of myelin basic protein from MO3-13 cells and human white matter. *J. Neurosci. Res.* **42,** 41–53.

Uversky, V. N., Gillespie, J. R., and Fink, A. L. (2000). Why are "natively unfolded" proteins unstructured under physiologic conditions? *Proteins* **41,** 415–427.

Valli, A., Sette, A., Kappos, L., Oseroff, C., Sidney, J., Miescher, G., Hochberger, M., Albert, E. D., and Adorini, L. (1993). Binding of myelin basic protein peptides to human histocompatibility leukocyte antigen class II molecules and their recognition by T cells from multiple sclerosis patients. *J. Clin. Invest.* **91,** 616–628.

van Venrooij, W. J., and Pruijn, G. J. (2000). Citrullination: A small change for a protein with great consequences for rheumatoid arthritis. *Arthritis Res.* **2,** 249–251.

Vossenaar, E. R., Zendman, A. J., van Venrooij, W. J., and Pruijn, G. J. (2003). PAD, a growing family of citrullinating enzymes: Genes, features and involvement in disease. *Bioessays* **25,** 1106–1118.

Ward, J. J., Sodhi, J. S., McGuffin, L. J., Buxton, B. F., and Jones, D. T. (2004). Prediction and functional analysis of native disorder in proteins from the three kingdoms of life. *J. Mol. Biol.* **337,** 635–645.

Warren, K. G., Catz, I., Ferenczi, L. Z., and Krantz, M. J. (2006). Intravenous synthetic peptide MBP8298 delayed disease progression in an HLA class II-defined cohort of patients with progressive multiple sclerosis: Results of a 24-month double-blind placebo-controlled clinical trial and 5 years of follow-up treatment. *Eur. J. Neurol.* **13,** 887–895.

Waxman, S. G. (1992). Demyelination in spinal cord injury and multiple sclerosis: What can we do to enhance functional recovery? *J. Neurotrauma* **9,** S105–S117.

Wood, D. D., and Moscarello, M. A. (1984). Is the myelin membrane abnormal in multiple sclerosis? *J. Membr. Biol.* **79,** 195–201.

Wood, D. D., and Moscarello, M. A. (1989). The isolation, characterization, and lipid-aggregating properties of a citrulline containing myelin basic protein. *J. Biol. Chem.* **264,** 5121–5127.

Wood, D. D., Bilbao, J. M., O'Connors, P., and Moscarello, M. A. (1996). Acute multiple sclerosis (Marburg type) is associated with developmentally immature myelin basic protein. *Ann. Neurol.* **40,** 18–24.

Woods, A. S. (2004). The mighty arginine, the stable quaternary amines, the powerful aromatics, and the aggressive phosphate: Their role in the noncovalent minuet. *J. Proteome Res.* **3,** 478–484.

Woods, A. S., and Ferre, S. (2005). Amazing stability of the arginine-phosphate electrostatic interaction. *J. Proteome Res.* **4,** 1397–1402.

Wright, P. E., and Dyson, H. J. (1999). Intrinsically unstructured proteins: Re-assessing the protein structure-function paradigm. *J. Mol. Biol.* **293,** 321–331.

Wucherpfennig, K. W., and Strominger, J. L. (1995). Molecular mimicry in T cell-mediated autoimmunity: Viral peptides activate human T cell clones specific for myelin basic protein. *Cell* **80,** 695–705.

Wucherpfennig, K. W., Sette, A., Southwood, S., Oseroff, C., Matsui, M., Strominger, J. L., and Hafler, D. A. (1994). Structural requirements for binding of an immunodominant myelin basic protein peptide to DR2 isotypes and for its recognition by human T cell clones. *J. Exp. Med.* **179,** 279–290.

Wucherpfennig, K. W., Catz, I., Hausmann, S., Strominger, J. L., Steinman, L., and Warren, K. G. (1997). Recognition of the immunodominant myelin basic protein peptide by autoantibodies and HLA-DR2-restricted T cell clones from multiple sclerosis patients. Identity of key contact residues in the B-cell and T-cell epitopes. *J. Clin. Invest.* **100,** 1114–1122.

Yang, X. J. (2005). Multisite protein modification and intramolecular signaling. *Oncogene* **24,** 1653–1662.

MICROCHIMERISM AND STEM CELL TRANSPLANTATION IN MULTIPLE SCLEROSIS

Behrouz Nikbin, Mandana Mohyeddin Bonab, and Fatemeh Talebian

Immunogenetic Research Center, Department of Immunology
College of Medicine, Tehran University of Medical Sciences
Tehran 14155, Iran

Scientific advances have demonstrated that autoreactive cells are a component of the healthy immune repertoire. If we define autoimmunity as an active induction of autoreaction, the solution should be an active induction of self-tolerance, and may indicate the direction to explore the future therapies.

Microchimerism (MC) refers to the presence of a limited number of nonhost cells in the body of an individual. These cells can enter via blood transfusion and organ transplantation or naturally through pregnancy. Chimeric cells engraft in the host body, develop, proliferate, and are accepted by the immune system as self.

INTERNATIONAL REVIEW OF
NEUROBIOLOGY, VOL. 79
DOI: 10.1016/S0074-7742(07)79008-6

0074-7742/07 $35.00

These include stem cells that enter the maternal body during fetal stages. These stem cells are also postulated to be helpful reservoirs in protecting the host body.

MC has been considered a risk factor in autoimmune disease induction. However, today we know it is a natural phenomenon. MC can be considered a natural model of successful transplantation, the earliest engrafting cells being fetal mesenchymal stem cells (MSCs). MSCs have two notable features. They have an immunosuppressive quality when encountering the adoptive immune system and they display repair-inducing potential within damaged tissues.

For the fetus, MC appears to be an effective factor in maternal tolerance induction toward the fetal graft and for the mother; these novel fetal cells might be useful in disease conditions occurring after pregnancy.

Hematopoietic stem cell transplantation has become an accepted treatment option for both malignant and nonmalignant diseases and this unique procedure is now being investigated as a potential therapy for multiple sclerosis (MS). Due to the dichotomous properties of MSC, suppressing aggressive immune dysfunction while promoting damaged tissue repair, they may be appropriate therapy for MS.

I. Introduction

We have come a long way since the days when Burnet (1957) presented his view of autoimmunity. After the theory of the "forbidden clone" was presented, therapy was directed toward "elimination" of the autoreactive cells (Coutinho, 1989). Today, scientific advances in immunology indicate that autoreactive cells are not harmful components by themselves. In fact, these cells are a component of the healthy immune repertoire (Moalem *et al.*, 1999). Deletion of a component in the *systemic network* of the body cannot be a "cure", or a long-term effective therapeutic option. This is particularly true when one deals with chronic autoimmune disorders such as multiple sclerosis (MS).

Dynamic issues demand dynamic solutions. That is the underlying reason why passive therapies have not met our medical demands (Hohlfeld and Wekerle, 2004) in finding a curative solution for MS and similar diseases. We must ask ourselves that: If passive, deletional methods are unreliable, on what framework must we design our therapeutic theories? The most intriguing answer to this question seems to be *counteraction.* So, if we define autoimmunity as an active induction of autoreactions (or *im*proper response to self-antigens), the solution should be sought in active induction of self-tolerance (or reestablishing proper *un*responsiveness) in that individual. And true enough, as we look at the evolution of therapy over the past decades, we see a trend with chemotherapeutic options being replaced by biotechnological solutions, and chemically synthesized polymers being modified into biologic-based complexes. This is the wave of the future. Human solutions have always been replications of the work of nature, or lessons learned by observing what happens in the world around them. That is how the Wright Brothers learned how to fly, and how Pasteur made his great breakthrough in microbiology.

The roots of therapy too seem to lie in the works of nature itself. It seems the climax lies in the realization that the most effective way to fight nature's work (dysregulation) could in fact be achieved by mimicking nature's own solutions (conditional modulatory elements)!

In this chapter, we shall try to look at the science surrounding us from a variety of outlooks. This may not increase the scope of our knowledge, but it will probably give new insights applicable to research and therapy.

In this chapter, at first a review of microchimerism (MC) and then scientific advances offering new avenues in pursuit of functional MC will be presented. In short, MC shall be presented in a different light. This view will help pursue a new line of research that could eventually lead to a greater understanding of the physiology and immunobiology of chimerism and offer us novel tools and modulations applicable to prognostic and therapeutic science.

Later the *in vivo* use of stem cell as a novel source of therapy will be pointed out. This could be an exciting therapeutic option to consider for MS.

II. Microchimerism

MC has become a familiar term in the past few years. Many groups all over the globe, specializing in a diverse array of basic and medical sciences, have turned their attention to MC and its possible role in disease or repair processes. There is also global effort by various scientists to further explore the mechanism of action of MC in the host body.

Simply put MC is the presence of foreign or nonhost cells in the body. By definition, these are nonself cells that live, differentiate, and persist in the host body.

A. Portal of Entry

Microchimeric cells enter the host body through various pathways. The most well-known pathway for microchimeric cells to gain access to a host body is organ transplant [solid, bone marrow (BM), stem cell]. For many decades, scientists have been debating over the interpretation of this phenomenon. Establishment and persistence of engrafted donor cell in the recipient body is a sign of transplantation success (Bettens *et al.*, 2005). The result is development of tolerance toward foreign organ and creation of chimeric state. How long this chimeric state will last, whether these cells will induce or be induced to create a chronic complication in the long run, or will these two genetically distinct cell types live peacefully in one body to the end of the host's life is the essence of the ongoing discussion. These unanswered questions have stimulated many scientific minds.

MC is induced by different means. Our focus in this chapter is the chimeric state produced by fetomaternal contact or placenta-induced MC. Since the 1970s, it became certain that some of the fetal cells can cross the placenta and then the question was raised as what happens to these fetal cells in the maternal body. Today we know that all these cells are not necessarily destroyed in the recipient body. However, we do not have a clear idea that what happens to these cells which survive.

B. Origin of the Hypothesis

The first published manuscripts on fetal MC (Bianchi *et al.*, 1996) viewed fetal MC as a source of noninvasive prenatal diagnosis (recovering fetal cells from maternal blood and analyzing these cells for fetal genetic abnormalities early in pregnancy). Then, Nelson (1998) presented a hypothesis regarding MC as a possible risk factor for autoimmune disease. Some of her reasons included:

1. Autoimmune disease occurs more frequently in women.
2. The onset of most immune-mediated diseases is during child-bearing age.
3. Pregnancy can lead to MC because fetal cells cross the placenta and enter the maternal body.

Chronic graft versus host disease (GVHD) after some transplantations shares mechanisms and features with certain autoimmune disorders.

Many groups applied this logic to their studies of autoimmune disorders, in pursuit of a strong foundation for a link between MC, pregnancy, and autoimmunity. To this date, although speculations go on and the idea has not been refuted, its link has not been confirmed either. The results encompass a wide range, from no detection of MC to over 50% detection or more, and from almost negligible quantity in those detected to detection of a considerable number of MC cells. And yet, after three decades of research the question(s) remains: "What are these cells doing in a foreign body? Do they have a set role? Are they harmful or helpful? When do they play an active role in the body? What is that active role? What defines their role and responsibility? What factor(s) triggers it all? . . .."

C. A New View

Despite the fact that during the last decades no definite conclusion regarding MC has been drawn, we have learned more about this interesting phenomenon. A review of the available findings indicates that:

1. MC is a natural phenomenon that can affect any host. The persistence of the microchimeric cells is not confined or even more pronounced in individuals with any diseases. Evidence shows microchimeric state is created in individuals without any apparent possibility of receiving foreign cells like women with no history of pregnancy and no blood transfusion and so on.

2. We know that fetal cells entering the body are almost exclusively stem cells. Some call these cells a "fountain of youth" entering the mother's body. Campagnoli *et al.* (2001) claim that most of the persistent microchimeric cells entering mother's body during the first trimester of pregnancy are the mesenchymal stem cells (MSCs). This is incidentally the time period when most immune-mediated diseases of the affected mother enter a natural remission (Abramsky *et al.*, 1984; Spector and Da Silva, 1992; Wilder, 1998). This raises the possibility that they could be very helpful in repair mechanisms, as speculated and investigated by some studies (Artlett, 2005). These cells have been found in sites where various damaged tissue are experiencing degrees of repair. Tan *et al.* (2005) demonstrate (in animal experience) that these cells enter the brain and localize to the site of injury "in response to host cues." They observed these fetal cells expressing a variety of cell markers, including neural cells. Johnson *et al.* (2002) analyzed a liver biopsy specimen from a woman with hepatitis C who stopped treatment (against medical advice) but, despite this, the patient did well clinically and her disease abated. Her liver specimen contained thousands of male cells detected by dual-color fluorescence *in situ* hybridization (FISH). Patient had never received a blood transfusion and was not a twin. Follow-up studies using DNA polymorphism analyses indicated that the probable source of the male cells in her liver was a pregnancy that she had terminated 17–19 years earlier. In this case, the male cells in the liver were morphologically indistinguishable from surrounding liver tissue, which suggests that they were hepatocytes (Johnson *et al.*, 2002). Similar studies (Khosrotehrani *et al.*, 2003, 2004) have examined the phenotypes of fetal microchimeric cells from patients who have high numbers of such cells by combining *in situ* hybridization to detect the fetal cells and immunolabeling to identify their phenotype. They found that different number of fetal cells (4–60%) expresses the marker of the invaded tissues. Most of the other fetal microchimeric cells in these tissues express CD45, the common leukocyte antigen, indicating a likely hematopoietic origin. In all cases, the morphology of the fetal cells suggests that they have differentiated. These results suggest that fetal cells, possibly hematopoietic in origin, home to the site of injury and adopt the maternal local tissue phenotype.

3. Animal studies (Evans *et al.*, 1999; Khosrotehrani *et al.*, 2005; Loubiere *et al.*, 2006) show that many of these MC fetal cells in the host body proceed to populate the lymph organs—although they are dispersed within the host body and travel to many organs. This most probably indicates that they eventually differentiate into immune cells, either T cell, B cell, or dendritic cells (DCs). Considering these immune cells have the responsibility of defending the body, maybe the MC's helpful role should be looked into in more detail in the future.

D. Dual-Role Performance

We must not forget that nature always has two sides for every story, so one cannot ignore the possibility that MC cells could turn pathological under certain conditions. Some considered that the existing evidence pointed toward an "apparent dichotomous role of donor cell chimerism in the processes of organ rejection and acceptance" (Starzl and Lakkis, 2006). From the beginning of the twenty-first century, medical literature was peppered with experiments, opinions, and successful or hopeful signs of the positive effect of MC on graft acceptance, whether from solid organ or BM transplants (Umeda *et al.*, 2003). And the year 1996 witnessed a redirection of interest toward MC and pregnancy with a paper reporting long-term fetal cell survival, up to 27 years to be exact (Bianchi *et al.*, 1996). However, if these cells are presumably involved in pathogenesis of immune-mediated disorders and are capable to home systemically all over the host body, how can one claim that in the normal host these cells are prevented from functioning? Why these cells cannot be part of the effector cell population performing like "self" cells?

E. An Interesting Observation

Silman (1992) demonstrated that the average onset of autoimmunity in women without pregnancy was much earlier than those who had become pregnant (26 vs 44 years of age). A study in 2002 (Artlett *et al.*, 2002) observed that there was a "statistically significant delayed age of onset for systemic sclerosis (SSc) with increasing gravity and parity." The authors also claimed that nonpregnant women have a more severe course with more diseased progression. Now if one combines this date with the high levels of fetal MC detected in affected tissues of women with scleroderma, then one may speculate that the novel stem cell pool in the mother's body is the cause of such delay. These cells have been protecting the mother and serving as a fresh, more competent "soldier population" in her body, fending off illness for as long as possible and thus giving the host a better chance to compensate for the period of pregnancy during which the mother protected the fetus.

The role of MC in pathogenesis of immune-mediated neurological disorders has not been studied. Only one study by Willer *et al.* (2002) has pointed toward MC as a potential mechanism in pathogenesis of MS and by combining various characteristics of MS, the authors recommend further studies to further explore the role of MC in pathogenesis of MS.

F. Lessons Learned

Animal studies of fetomaternal MC have (Khosrotehrani *et al.*, 2005; Wang *et al.*, 2004) further explored this phenomenon. Almost all of these studies show acceptance of fetal cells and successful engraftment not only in hematopoietic tissues but also in different organs. These studies have strongly suggested that microchimeric stem cells harbored in the body have yet undiscovered repair potential. Some investigators imply that immunoablation followed by stem cell rescue could be of potential therapeutic benefit. From our viewpoint, this is exactly what happens in the body normally. The pregnant mother is being conditioned to the "immunoablated or immunosuppressed" state, so it can tolerate and accept the "graft," or fetus. Then "stem cells"—new, unused, and full of potential—enter this conditioned host or recipient, home to different organs, and stay there until being called on. We are copying this act of nature as a possible therapeutic solution for hopeless disease states, why it is not possible that nature's purpose here is beneficial as well?

Using allogeneic male mice heterozygous for green fluorescent protein expression, Imaizumi *et al.* (2002) performed a study on mice with experimental autoimmune thyroiditis (EAT). The male mice were mated with females and after pregnancy, thyroid digests in the female mice showed the cell population which included fetal cells of T-cell and DC lineage. Such finding indicated that the inflamed thyroid gland was capable of accumulating fetal cells. On the basis of this observation, one must ask if these cells were "nonself" why they would stay in the body of immunocompetent hosts for such long period. It does not appear that these cells are hostile and based on the available evidence one may raise the possibility that these cells are "protective" rather than "destructive."

MC has been implicated repeatedly in hyporesponsiveness of the fetal/maternal immune system toward the fetal allograft as well as the longevity of organ transplants (Ichinohe *et al.*, 2005). Some scientists have taken a more positive position on this phenomenon in the wake of its seeming helpfulness in haploidentical graft acceptance (Artlett, 2002; Ichinohe *et al.*, 2002; Umeda *et al.*, 2003). This phenomenon may indeed cease many disease states from developing in the maternal body before they present clinically. Why one should not consider this hypothesis in reverse, assuming that clinical disease was stronger than the combined effort of the body's defense system and microchimeric fetal stem cells to combat?

By 2000, the concept that fetal cells with stem cell potential cross the placenta had become subject of intense debate. The report by O'Donoghue *et al.* (2003) further supported this concept. The authors found a rare population of MSCs in the maternal body—MSCs with fetal markers—and indicated that these cells were not routinely detected postnatally. However, the authors raised the possibility that this could have been due to their engraftment in maternal tissues early on. This concept deserves more investigation. Nowadays, our knowledge of MSCs has expanded and later in this chapter we will review their immunosuppressive or immunomodulatory potentials. In addition, one study on posttransplantation engraftment in 2004 (Villaron *et al.*, 2004) demonstrated the presence of MSC in peripheral blood (PB) after allogeneic transplantation and also showed their ability to engraft in BM. Another study focused on engraftment potential of amniotic and chorionic cells (Bailo *et al.*, 2004). The authors concluded that these cells have MSCs-like properties, so they "do not induce neither allogeneic nor xenogeneic lymphocyte proliferation responses and they are able to actively suppress lymphocyte responsiveness." Such pool of cells may serve as "an advantageous source of progenitor cells with potential applications in a variety of cell therapy and transplantation-based procedures."

Taking the view that fetal cells enter maternal circulation as stem cells, a study which was performed on organ specimens supported this concept (Khosrotehrani *et al.*, 2004). Namely, Y+ cells were detected in epithelial, hematopoietic, and hepatic cells bearing respective markers cytokeratin, CD45, and heppar-1, respectively. These stem cells entering the body are thought to be MSC (O'Donoghue *et al.*, 2004) accounting for their low level of detection. MSC enter the maternal circulation in the first trimester of pregnancy and rapidly engraft in the marrow where they reside and possibly differentiate at a later time. This is interesting because this rare population of MSC is noted for differentiating into all nonhematopoietic tissue lineages.

It is exciting to consider that we may be able to develop targeted therapies after understanding the mechanisms underlying MSC. The questions that need to be answered include: (1) Under which conditions the semiallogenic fetus is tolerated? (2) What modifications does maternal body undergo? and (3) Where, when, and why do microchimeric cells enter the circulation or act as active cells? Lambert *et al.* (2005), who had originally reported the idea of targeted therapy development, published a paper showing microchimeric detection in women with no history of male pregnancy. They emphasized that there were various sources for the induction of MC (Yan *et al.*, 2005) and left the issue of "mechanisms of MC" to further investigations.

G. Our Experience

Recently we completed a study on MC in mothers with MS (Talebian *et al.*, 2005). Compared to other immune-mediated disease, MS affects individuals at a younger age and many female patients with MS develop the relapses after delivery. Postpartum period has been regarded as a critical stage in female MS patients when the natural immunosuppressive effects of pregnancy have been resolved and these patients develop devastating relapses with severe neurological deficits. On the basis of the facts that MC does exist in normal individuals and fetal cells exhibit stem cell characteristics, we tested the hypothesis that higher number of microchimeric cells in MS patients' peripheral circulation can serve as an indicator of disease severity. This hypothesis was based on the observation that inflamed tissue recruit immune and stem cells to the site of injury. So, MS patients should have higher numbers of stem cells in their peripheral circulation and these cells are recruited to the inflamed site. As these cells normally reside in different organs and are not freely roaming the PB, their lower detection rate in normal individuals or stable MS patients would be acceptable.

However, when we performed the tests, our results gave no indication of such relationship between MC and MS. The frequency of MC in patients and control subjects was equal. We did, however, observe that the patients with MC had significantly higher blood titers of microchimeric cells. The limitations of our study included lack of long-term follow-up of the patients and the small size of our subjects; therefore, it was not possible to determine whether the higher blood titer of microchimeric cells in these patients correlated with disease severity or whether it could serve as a prognostic marker to predict the onset of a relapse in the patients. More extensive studies involving more subjects and longer periods of follow-up are necessary to provide better explanations for our preliminary findings. However, what may be concluded from our limited study is (1) the presence of microchimeric cells in the PB may indicate greater rate of migration of these cells into areas with ongoing inflammation; therefore, microchimeric cells are being recruited to sites of active inflammation or ongoing repair; (2) MS is a chronic immune-mediated disorder and it is possible that microchimeric cells, as instigating agents of the first MS relapse, have been present in the peripheral circulation of MS patients long before the initial relapse.

H. MC and MSCs: Unforeseen Potentials

The relationship between MC and MSCs is another underexplored area of research in the field of MC. As we mentioned before, the earliest fetal cells noted to enter the maternal body is in the first trimester and they engraft to various organs with alacrity. This cell population has been characterized as mostly fetal MSCs. The significant unanswered question is whether the presence or activity of these cells can be responsible for improvement observed in MS patients during pregnancy. True, this improvement has been attributed to hormonal changes and T-cell shift from a Th1 profile to a Th2 profile. However, with the new emerging evidence, MSC could well be a plausible addition to the array of factors establishing the suppressed immune state a pregnant women experiences.

Study of microchimeric cells in the context of organ transplantation may assist us to explain the acceptance of the grafted nonself tissue by the recipient, an essential step in research into the nature of MC which eventually translates into development of new therapies for immune-mediated diseases. Tan *et al.* (2005) found the microchimeric fetal cells in maternal mouse brain and such interesting finding has opened new research avenues into potential role of these cells in CNS repair. MC may be viewed as a natural model of successful transplantation. Therefore, further studies of the mechanisms of successful entry of the microchimeric cells into mother's circulation will provide us with necessary knowledge to employ them as more effective therapies for immune-mediated diseases such as MS.

III. Therapy

MS is an immune-mediated chronic disabling disorder with a complex pathogenesis. What we know is that MS is a conglomerate of different neurological disorders with various clinical courses and dissimilar pathologies. The heterogeneity of MS pathology makes it difficult to devise any definite cure(s) for these patients. Traditionally, the CNS has been regarded as an "inner sanctum," an immunoprivileged area with no lymphatic system, no circulating immune cells, and no expression of HLA antigens by its residential cells. However, further research into the interactions between the CNS and immune system has demonstrated that normally a small number of immune cells do cross the blood–brain barrier (BBB) and enter the CNS (Compston *et al.*, 1999). Lymphocytes can cross the BBB in a state of high activation, but resting T cells do not. Despite such significant findings, the mechanisms of leukocyte entry to the CNS remain largely unknown. In fact, it remains unclear how and when in the pathogenesis of MS, the inflammatory cells enter the CNS and initiate the destructive cascade of neuroinflammation and neurodegeneration.

Current therapies for management of MS include corticosteroids, IFN-β1a, and IFN-β1b, glatiramer acetate (GA), natalizumab, and mitoxantrone. A number of investigational treatments for MS include those which block transendothelial migration of activated T lymphocytes, agents which deplete B cells, immunotherapies such as anti-cytokine agents and T-cell vaccines, and stem cell therapy with hematopoietic stem cells (HSCs) and MSCs. Of all these investigational therapies for MS, we will focus on only those which involve use of stem cells.

IV. Stem Cells

All cells of a living organism originate from a primary pool of cells in the embryo. These embryonic stem cells are unspecialized pluripotent cells, which possess regeneration potential and differentiate into all cell types: ectodermal, mesodermal, and endodermal (Oh and Choo, 2006; Semb, 2005; Vats *et al.*, 2002).

Adult (or tissue-derived) stem cells (ASC) are localized in different tissues of the adult body (Krabbe *et al.*, 2005). They have the potential to self-renew indefinitely and generally replenish the tissue they occupy. However, their differentiation potential is not limited to their organ of residence. As turnover of cells in different tissues is variable, the amount of stem cell and the microenvironment they are placed in within different tissues of an organism vary. This can be better understood by an example from two extreme tissues, the blood forming tissue (with the greatest turnover) and the CNS (with almost no turnover). These stem cells stay in their appointed microenvironment and only divide in response to stimuli recruiting them to maintain the population and integrity of the concerned tissue (Vats *et al.*, 2002). Previous studies (Cat and My, 2003; Vescovi *et al.*, 2002) have raised the concept of "stem cell plasticity": the possibility that the stem cells from one organ or tissue transdifferentiate into stem cells of a completely different tissue. This transdifferentiation potential has been demonstrated repeatedly in the past decade, ranging from BM turning into muscle cells or neural stem cells into hematopoietic cells (Kraus and Kirker-Head, 2006; Orlic, 2005; Zemchikhina and Golubeva, 2005). This is due to the modulatory microenvironment where these multipotential cells are placed. For example, HSCs, which are the precursors of blood cells, can also generate liver, heart, or skeletal muscle cells (Ferrari *et al.*, 1998; Mendez-Ferrer *et al.*, 2006; Mezey *et al.*, 2000; Van Haaften and Thebaud, 2006). However, this concept is not fully established, which means that the extent of stem cell plasticity is not completely understood. For instance, MSCs are one of the most versatile stem cell populations. And yet many are still speculative regarding the potential of MSC to fully and stably transdifferentiate into neural stem cells (or any cell type for that matter) and consider this concept as a possible artifact (Krabbe *et al.*, 2005).

Therefore stem cells, regardless of their source, exhibit two unique characteristics. They regenerate for unlimited periods and can differentiate into functionally specific cell types. However, the type of cells into which they differentiate depends on their physiological or experimental microenvironment.

BM is the oldest and most renowned source of stem cells in the adult body. Within the BM, three types of stem cells have been identified. To date, the best characterized population is the HSCs that generate all blood cells (red blood cells, white blood cells, and platelets) (Shizuru *et al.*, 2005). The second population is endothelial precursor cells. Three decades earlier, Friedenstein *et al.* (1970) characterized the third population within the BM, which is known as stromal cells. Because of their multiple mesenchymal lineage differentiation potential, they have been designated the term MSC.

V. Hematopoietic Stem Cells

A. Their Characterization

HSCs are recognized by their specific cell surface marker: CD34+. Using this marker, HSCs were tracked *in vivo* for better functional understanding. One of the methods used to study HSC for its stem cell properties was by subsequent passaging of a marked population of HSC into a number of lethally irradiated mice. When these cells reconstituted hemopoiesis in serial animals and retained their multipotentiality, not only were they confirmed as stem cells but also a great amount of information was complied regarding these cells. Today this reconstitution is being performed in patients.

B. HSC Transplantation

HSC transplantation (HSCT) is the first therapy designed to ameliorate or possibly cure hematological malignancies. In this form of therapy, after intensive immunosuppression eradicates the "dysfunctional" immune system and the new population of HSC which enters the recipient should be able to reconstitute the immune and hematopoietic systems. HSCT use has been broadened to include immune-mediated diseases such as MS (Van Gelder and van Bekkum,1996). When applying HSCT to MS patients, the reconstructed immune system should be tolerant of the CNS antigens. HSCT was first applied to MS patients in 1997 (Fassas *et al.*, 1997). Then, 15 patients with progressive MS underwent autologous HSCT (AHSCT). The results of this study confirmed the feasibility of this therapeutic method for patients with severe MS. Blanco *et al.* (2005) evaluated the results of AHSCT in 250 patients with severe MS. The authors concluded that AHSCT was a safe therapy for patients with severe MS and AHSCT was more effective than allogeneic HSCT. Burt *et al.* (1998) observed a reduction in CD4+ T cells accompanied by an increase in the number of CD8+ T cells up to 12 months after AHSCT. Saccardi *et al.* (2005) performed 19 ASCT on MS patients with EDSS (Expanded Disability Status Scale) of 5–6.5 using BEAM (BCNU, etoposide, cytosine arabinoside, melphelan) and ATG conditioning regimen. They observed that 18 of 19 patients achieved some degree of improvement or disease stability. Daumer *et al.* (2006) performed a large retrospective study on 2185MS patients among those, they selected 285 patient with RRMS and SPMS (EDSS: 3–5.5) who satisfied study inclusion criteria. They studied factors such as survival rate and time to disease progression and evaluated the effect of AHSCT after high-dose immunosuppressive therapy (HDIT). Sixty-three of the selected patients had received AHSCT accompanied by HDIT. Their results showed treatment-related mortality declined in this group of patients. They also found that the time to disease progression was delayed. However, they did not arrive at any conclusive answers as to whether this treatment is significantly more effective when compared to other treatment options. Therefore, they suggested designing a clinical trial for AHSCT with HDIT and either case matching or utilizing a "virtual placebo or control group" in ongoing cohort studies in hopes of arriving at more definitive conclusions. Because of the heterogeneity of MS pathology, defining a control group for transplanted patients has been very difficult. So, we must find methods where a relatively similar untreated case could be compared to a treated individual, providing an opportunity for more accurate comparisons in order to determine its effectiveness.

Studies performed to date using AHSCT have not been able to make a definite decision about the use of this therapy in MS patients. However, they all concur on the feasibility and safety of this method. All of these investigators point out the need for properly designed studies that should include a control group before reliable results can be obtained (Fassas and Kimiskidis, 2003; Muraro and Martin, 2003; Muraro *et al.*, 2003).

VI. Mesenchymal Stem Cells

A. Their Characterization

MSCs are spindle shaped and resemble fibroblasts in culture (Mohyeddin *et al.*, 2005). Unlike HSCs, they are adherent and readily multiply *in vitro* (Fig. 1). They have no known specific and unique cell surface marker. MSCs are negative for HSC surface markers (CD34, CD45) and HLA class II antigens, but are positive for markers such as Stro1, CD105, CD106, CD166, CD44, CD13, and HLA class I. They create a stromal scaffolding and provide an appropriate microenvironment to support the development and differentiation of blood-forming BM cells (Devine and Hoffman, 2000). MSCs are capable of long-term engraftment as well as differentiation into various mesodermal lineages (cartilage, bone, muscle, tendon, ligament, and adipose) (Barry and Murphy, 2004; Pittenger *et al.*, 1999) and also neurons, astrocytes, and oligodendrocytes (CNS cell lineages)

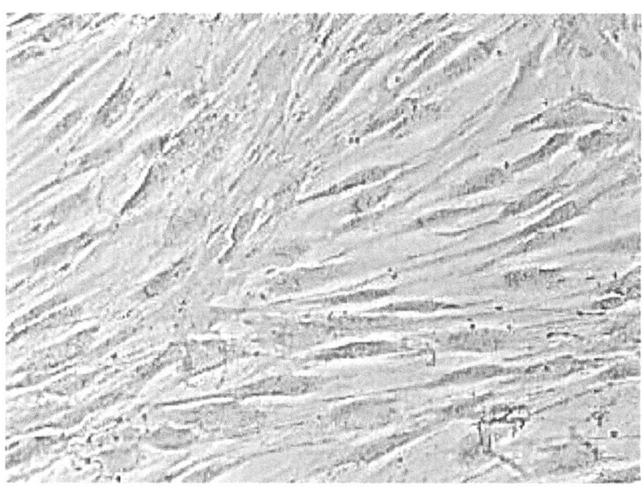

FIG. 1. Expanded mesenchymal stem cells in culture. (See Color Insert.)

(Crigler *et al.*, 2006; Keilhoff *et al.*, 2006; Zemchikhina and Golubeva, 2005). In addition, MSCs comprise a heterogeneous population (Baksh *et al.*, 2004), which can be observed as either small, rapidly self-renewing cells or larger, slow replicating cells (Colter *et al.*, 2001). Moreover, their characteristic features may differ depending on their source (organ) of isolation.

B. MSCs and the Immune System

1. *MSCs and Antigen-Presenting Cells*

MSCs have the potential to downregulate and inhibit immune response directly and indirectly. This effect encompasses all levels of the immune reaction:

- Antigen presentation for recognition by antigen-presenting cells (APCs)
- Recognition through lymphocytes
- Effector actions through B cells, cytotoxic T lymphocyte (CTL), and natural killer (NK) cells. When mononuclear cells (MNC) are isolated from PB and cultured *in vitro* along with a stimulator (antigen or mitogen), these cells (lymphocytes) proliferate and secrete proinflammatory cytokines (INF-γ, INF-α, IL-12, and IL-10). This marks the onset of the immune reaction, replicating the *in vivo* process. The outcome of this process is maintenance of the body's homeostatic balance, while the element which triggered this cascade is being eradicated. In experimental approaches to isolate and characterize MSC, a reduction in lymphocyte proliferation and IFN-γ secretion has been observed when MSCs are added to stimulated PB-MNC culture (Beyth *et al.*, 2005). This inhibitory effect of human MSCs (hMSCs) has a direct relationship to the number of MSCs present in the vicinity, but is not related to the concentration of the primary stimulator (Beyth *et al.*, 2005). PB-MNC used in culture comprise a heterogeneous mixture of T and B lymphocytes and a variety of CD14+ APCs. This report showed when purified CD4+ T cells are cultured with hMSCs, a notable amount of IFN-γ is secreted. However, addition of CD14+ APCs (monocyte and/or DC) to this culture, while hMSC is still there, diminished secretion of IFN-γ, interleukin (IL)-5, and lymphocyte proliferation significantly. This inhibitory effect is observed regardless of the source of APCs, meaning whether they are autologous or allogeneic. If we substitute CD8+ cells for CD4+ cells in the culture media containing hMSCs, the addition of APCs inhibits CD8+ proliferative and cytotoxic activity as well. This reduction of cellular cytotoxicity more likely results from inhibition of CTL proliferation and not the direct inhibition of cytolytic activity.

We know that necrotic cells stimulate the immune system and initiate an inflammatory reaction. Unlike their necrotic counterparts, apoptotic cells block the maturation and maintain tolerance toward self-antigens produced by the apoptotic bodies. Uptake of apoptotic products reduces secretion of proinflammatory cytokines such as TNF-α, IL-1β, and IL-12, while upregulates secretion of anti-inflammatory cytokines such as IL-10.

MSC contact with immune cells has an effect similar to an apoptotic cell encounter. In the presence of hMSCs, immature or partially immature APCs are produced (Beyth *et al.*, 2005). Immature APCs induce regulatory T characteristics (T_{reg}) in the lymphocyte population.

APCs can be divided into two main classes: monocyte/macrophage and the DC. DCs are a major population of the APCs and play a key role in propagating humoral and cellular immune responses against nonself (foreign) antigen by effectively causing the development and proliferation of effector T-cell populations. Immature DCs, on the other hand, silence T cells or induce them to enter an anergic state either by eliminating T cells or by modulating them toward a regulatory (CD4+CD25+) phenotype (Mills, 2004; Steinman and Nussenzweig, 2002).

2. *MSCs and T cells*

Differentiation of CD4+CD25+ T_{reg} cells caused by immature DC (Jonuleit *et al.*, 2000; Roncarolo *et al.*, 2001) inhibits proliferation of effector lymphocytes (Sakaguchi, 2004; Sakaguchi *et al.*, 2001; Vitiello *et al.*, 1995). Studies demonstrate that a heterogeneous mixture of all T_{reg} subsets impose even a more potent effect on cell-to-cell contact or secretion of cytokines such as IL-10 (Zheng *et al.*, 2004). When an MSC population, from an allogeneic third party, is applied to mixed lymphocyte culture (MLC), the effect is more pronounced. Maccario *et al.* (2005) hypothesize that such increased effectiveness results from stimulation of a higher percentage of regulatory lymphocyte cell subset (CD4+CD25+, CD4+CTLA+, CD4+CD25+CTLA+) when the MSCs are allogeneic as opposed to autologous. CTLA4 is a T lymphocyte surface glycoprotein expressed on stimulation or induction of the cell and known to induce an inhibitory signal in T cells that drives the cell toward apoptosis before it has a chance to become active (Gribben *et al.*, 1995).

3. *MSCs and Soluble Factors*

The soluble factors MSCs secret under varying conditions are another means by which these cells impose their inhibitory effect on the microenvironment they occupy. MSCs secrete TGF-β which inhibits the immunosuppressive effects of T cells (Di Nicola, 2002). MSCs constitutively express and secrete ecosinoid prostaglandin E2 (PGE2) (Arikawa *et al.*, 2004). PGE2 imposes a variety of effector functions within the immune system including inhibition of B-cell function (Roper *et al.*, 1994) and stimulation of regulatory cell production (Akasaki *et al.*, 2004). MSCs also secrete a factor known as indoleamine 2,3-dioxygenase (IDO). This factor is produced and secreted by these cells on their activation by proinflammatory cytokines released by active T cells (Meisel *et al.*, 2004). IDO is an enzyme that catalyzes L-tryptophan (Trp). This catalysis depletes the microenvironment of a necessary amino acid (Mellor and Munn, 2001; Munn *et al.*, 2005). IDO seems to be expressed by MSC when they are influenced by IFN-γ (Meisel *et al.*, 2004). So, MSC can also inhibit allogeneic T-cell response through depletion of Trp. However, another study reports that enrichment of mixed lymphocyte reaction (MLR) with IDO, Trp, or IDO inhibitor does not restore the proliferative ability of PB-MNCs (Tse *et al.*, 2003).

4. *MSCs and the Cell Cycle*

Another mechanism MSCs employ to induce immunosuppression on contact with immune cells is to inhibit lymphocyte proliferation through cell cycle inhibition. In 2005, a study was conducted in mice to unravel the mechanism of MSC inhibition on T lymphocytes. The authors reported that although MSCs significantly decrease IFN-γ+/CD8+ T cells, the natural activity of T cells is not impaired. Experimental observations show almost no cell division within the T-cell population after 72 h of incubation with MSCs, although a small degree of protein synthesis is detectable (Glennie *et al.*, 2005). The decreased number of IFN-γ+/CD8+ T cells results from a manipulation of the proliferative ability of T cells. These findings suggest that while MSCs might have a limited effect on T-cell function, they pose a significant inhibitory effect at the proliferative level. MSC downregulate cyclin D2 while upregulating p27kip1, thus preventing cell cycle progression. Cyclin D2 is an early G1 phase regulatory molecule which promotes protein synthesis; p27kip1 is an inhibitory protein which prevents cell cycle progression from the G1 to the S phase. The progression is achieved through cyclin E expression, a protein which downregulates inhibitory proteins like P27kip1 and allows for the cell's entrance into the S phase of the cycle and thus for DNA synthesis to ensue. Accumulation of P27kip1 can inhibit cell cycle progress through the later phases G1, S, G2, and M.

5. *MSCs and B cells*

Researchers have reported that inhibition of humoral immunity through MSCs occurs by inhibition of B-cell proliferation through CD40L and IL-4 induction. This is not surprising since MSCs target cyclin D2 and can thus control proliferation of B-cell progenitors without impairing their activity (Mohamedali *et al.*, 2003). Corcione *et al.* (2006) have shown that hMSCs inhibit B cells by cell cycle manipulation. To demonstrate this, they cultured activated B cells with and without hMSCs and observed that most B cells in MSC-free culture expressed Ki67, while B cells cocultured with MSCs at a ratio of 1:1 displayed significantly diminished amounts of Ki67. Ki67 is a protein present during the later part of all active phases of the cell cycle, but absent from resting cells (Go phase) (Scholzen and Gerde, 2000). This demonstrates that MSCs drives B cells from an active

G1 phase to a resting Go phase where Ki67 is no longer detected. This effect is produced by soluble factors. Among the immune cells, it seems B cells are the only cell type that does not require cell-to cell contact with MSC to be affected. MSC and B cell cocultures inhibit IgM, IgA, and IgG production only when cultured at a ratio of 1:1 and not less (1:2, 1:5, 1:10), so the effect is obviously dose dependent. It seems the decrease in immunoglobulin (Ig) production is due to a diminished number of Ig-producing cells, not inhibition of Ig production itself. The factors thus far identified by which MSCs impose their inhibitory effect on B cells include TGFβ1, HGF, PGE2, and IDO. However, the initiation of this inhibitory effect depends on a stimulatory factor from the B cell itself. This supports the theory that hMSCs inhibit proliferation and differentiation of B cells into Ab-producing plasma cells but also the chemotactic ability of B cells *in vitro*. Corcione *et al.* (2006) examined the effect of hMSC on B cell chemotaxis via inhibition of CXCR4 and CXCR5 expression. CXCR4 and CXCR5 are chemokine receptors continuously expressed on B cells and induce the homing effect in secondary lymphoid organs (Muller *et al.*, 2002).

C. MSC and GVHD

One of the greatest concerns that overshadow HSCT is the risk of an unsuccessful transplant manifested as GVHD. To avoid GVHD, immunosuppressive therapy is induced. Sometimes this method is not effective, particularly in cases of severe acute GVHD. Recently, it has been reported that coinfusion of *ex vivo* expanded MSC and HSC in transplant patients leads to a lower incidence of severe GVHD. To this end, there have been case reports of MSCs applied as a third party haploidentical treatment strategy to reduce severe acute GVHD in recipients of HSCT (Le Blanc *et al.*, 2004). The result of a multicenter analysis on the outcome of MSC coinfused with HSC was reported in 2002. They treated patients with advanced hematological malignancies and observed a significantly lower overall incidence of acute and chronic GVHD when compared to the control group (Krampera *et al.*, 2003).

At the time, these investigators had no idea what the underlying mechanism might be. However, 3 years later and after a plethora of complied experimental data advanced our understanding, we have a fairly good idea of the mechanism behind this unexpected success rate. All in all, MSCs seem to be a suitable alternative to a chemotherapeutic immunosuppression in cases utilizing stem cell therapy. In this way, profound immune cells loss due to chemo-immunosuppression and exposure of patients to the toxic by-products and bio-products of chemotherapeutic agents are avoided. Indeed, MSCs appear to be one of nature's solutions to silence unwanted reactions by the intact immune system.

MSCs have stimulator potential as well. MSCs produce IL-7 which promotes lymphocyte proliferation. While the inhibitory effects of MSCs are dose dependent, the levels of IL-7 produced by only 10 cells (MSCs) is sufficient to activate T cells (Le Blanc *et al.*, 2003b). Even with this effect, proliferation of lymphocytes is not a constitutively observed phenomenon when MSCs are present.

Numerous reports confirm the presence of low numbers of HLA class II molecules expressed on MSC surface (Le Blanc, 2003; Le Blanc *et al.*, 2003a; Tsai *et al.*, 2004). Some studies suggest that MSCs exhibit an APC profile when exposed to IFN-γ (Stagg *et al.*, 2006). Others claim that in the presence of proinflammatory cytokines, MSCs express HLA II and perform phagocytosis (Conget and Minguell, 1999; Silva *et al.*, 2003). Chan *et al.* (2006) showed that MSCs require low levels of autocrine and endogenous IFN-γ in order to express HLA II and act as APC, and this APC activity only seems to affect CD4+ T cells which were previously challenged with the involved antigen. MSCs appear to act as "gate keepers" for a short period, protecting the BM against infectious and inflammatory responses. However, when IFN levels rise above a certain threshold, their APC property is lost and they act as immunomodulatory cells. Therefore, hMSCs induce inhibitory phenotype in APC, upregulate the T_{reg} subset and downregulate differentiation of alloantigen-induced lymphocytes, and also downregulate proliferation of CTL and inhibit NK-cell activity. Thus, MSCs act as a sort of immunomodulating radar, keeping the body alert to the needs of the microenvironment it occupies. When the immune system shows signs of activation, MSCs impose inhibitory signals to suppress reactions, guiding the regional framework toward a tolerant state. This occurs on encountering adaptive immune cells, and not in the naive immune system. MSC properties in relation to immune cells are summarized in Fig. 2. However, when MSCs come in contact with other tissues (nonimmune cells), especially under degenerating and injured conditions, they display a stimulatory property, recruiting the potentials in the environment and composing them into a regenerative power.

D. Regenerative Potential of MSCs

MSCs exhibit the ability and potential to repair tissue. They can ameliorate tissue damage repair by secreting soluble factors which lead to stimulation of local cell proliferation and differentiating into the respective cell type in the damaged area with the aid of microenvironmental effectors. Evidence shows that tissue-specific progenitor cells that participate in tissue repair are actually replenished via BM-derived MSCs that have migrated to the respective organ and either differentiated (Shake *et al.*, 2002) or built the provisional stromal support network (Alexanian and Kurpad, 2005). Thus, MSCs are the driving force in local tissue repair.

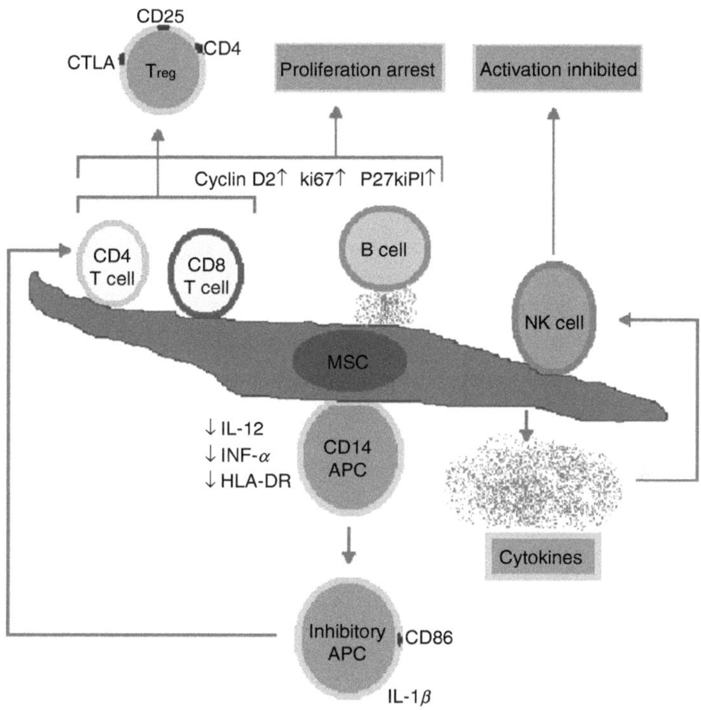

FIG. 2. The role of mesenchymal stem cell in relationship with immune cells. (See Color Insert.)

Animal and human studies show that MSCs injected into the brain migrate to various sites, survive for long time and differentiate into neural cells (Azizi *et al.*, 1998; Horwitz *et al.*, 1999; Jin *et al.*, 2002). Eglitis and Mezey (1997) have shown that when stem cells derived from BM are injected into the mouse tail, they can be traced to the brain while expressing the microglial and astrocyte markers. It has been observed that following tissue damage, such as from cerebral ischemia or traumatic brain injury, they migrate to the damaged region and aid in repair of the respective tissue. Repair is described as resulting from replacement for neuronal loss accompanying growth factor secretion by these transplanted cells (Li *et al.*, 2002; Mahmood *et al.*, 2004). Abouelfetouh *et al.* (2004) cocultured MSCs with hippocampal brain slices and observed directed differentiation of MSCs. They concluded that beside trophic factors and cytokines, cell–cell contact is essential for full differentiation of stem cells into the desired cell type, a point we have stressed earlier in this chapter.

The neuroprotective effect of MSCs stems from their potential ability to replace damaged neurons through cellular differentiation (Black and Woodberg, 2001). *In vitro*, MSCs can differentiate into cells resembling neurons in that they display similar biochemical, anatomical, and electrophysiological characteristics (Sanchez-Ramos, 2002; Woodbury *et al.*, 2000). MSCs support cell survival by producing neurotrophic factors. Thus, they can induce cell proliferation and stimulate regeneration of nerve fibers in the damaged region (Li *et al.*, 2002). Crigler *et al.* (2006) reported that MSCs secrete a variety of neuron-regulating proteins and other factors such as BDNF (brain-derived neurotropic factor) and β-NGF (nerve growth factor-β). These factors promote survival of neural cells, formation of neurites from neuroblastoma cells, and induce the growth of nerves in the lumbar spine. It has been reported that MSCs secrete factors such as hepatocyte growth factor (HGF) (Neuss *et al.*, 2004). HGF induces mitogenic and antiapoptotic activity in various systems, and plays a definite role in repair of damaged tissue (Chunmeng *et al.*, 2004; Ono *et al.*, 2004; Xin *et al.*, 2001). Figure 3 schematically shows the regenerative potential of MSC.

The fact that these cells expand *in vivo* without losing their differentiation potential (Mohyeddin *et al.*, 2006) has highlighted their importance as a therapeutic

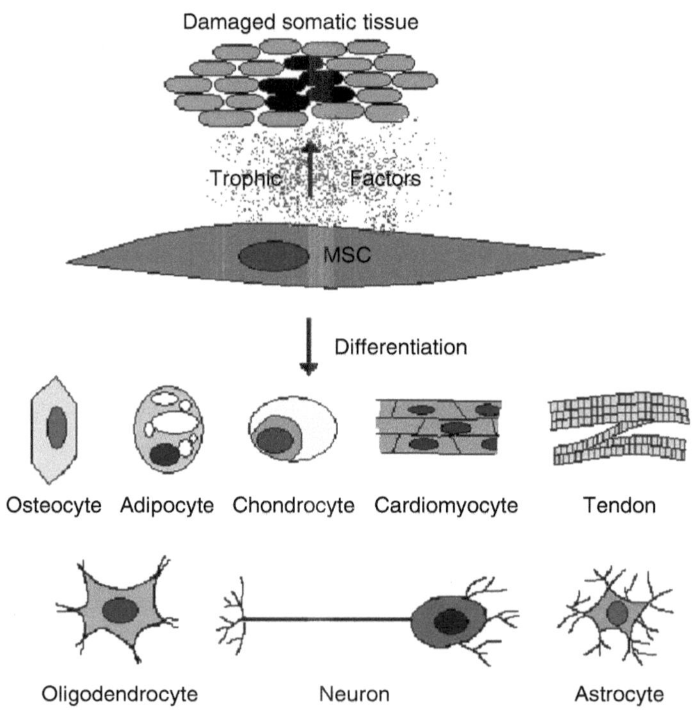

FIG. 3. The role of mesenchymal stem cells in relationship with injured somatic tissue and nonimmune cells. (See Color Insert.)

option. So, it appears plausible to suggest that MSCs can be utilized as a therapeutic option in MS, both autologously and allogenically for the following reasons: MSCs, due to their immunosuppressive/immunomodulatory potential, downregulate activated immune cell reactivity and thereby reduce tissue damage. MSCs induce a local inhibitory milieu through direct contact with certain immune cells such as APC and T cells and indirect contact with other cells like B cells. MSCs are, at least theoretically, safe to use allogenically or from third party donors because we know the primary success in allogeneic transplantation success results from tolerance of donor T cells toward recipient alloantigen (Pawelec *et al.*, 1996). MSCs are unable to generate lymphocyte proliferation in response to alloantigen stimulation (Le Blanc *et al.*, 2003a). They evade allogeneic rejection because they are hypoimmunogenic and their hypoimmunogenicity stems from absence of the surface HLA class II antigen(s) and costimulatory molecule(s). MSCs avoid innate immune activation as well. They accomplish this due to the presence of cell surface HLA class I, thereby evading NK-cell activation and innate immune reactions (Ryan *et al.*, 2005). MSCs play the role of stimulators and "cell factories" in injured and inflamed tissues when in contact with the local microenvironment. They promote tissue repair by differentiating into the injured cell types, thus compensating for their loss as well as secretion of trophic factors. The microenvironment induces this directed differentiation in the damaged tissue. The local region effectively directs the recruited stem cells in the repair of that specific organ and microenvironment.

To our knowledge, we are the first group to embark on MSC therapy in humans. Early in 2004, we were granted permission from the ethics committee of Tehran University of Medical Sciences to begin a pilot study on the effect of hMSCs on 10 MS patients with progressive disease who had failed conventional therapies. We separated MSC from the BM of our MS patients and expanded them *in vitro* for about 2 months (10–13 population doubling). When our desired concentration of MSCs was achieved, we injected them back into the patient intrathecally. Mean follow-up time has been achieved at this time is 17 months. However, we believe that follow-up for patients suffering from chronic illnesses should be much longer.

Our results indicate that injection of these cells is safe and feasible. One of our patients improved remarkably. His EDSS declined from 5 to 2.5. Over all, 30% of our patients have shown improvement in their function and EDSS status, 30% have attained stable (nonprogressive) condition, and 40% have not improved. Brain MR scans of patients who completed their 1 year follow-up did not show significant difference before and after treatment. So, taken together, this treatment may have substantial benefit and at least this treatment modality does not appear to have caused disease progression.

We are at the beginning of a new age in therapeutic science, with biologic solutions based on natural processes expanding out therapeutics. Obviously, there are many components which require modification, including the number of injections required and cellular dose needed, especially now that we know MSCs act in a dose-dependent manner; timing of injection based on the patients condition; the proper use of costimulators to enhance cell proliferation in *in vitro* utilization; determining the best subtype of stem cells to be injected (should we use one of the heterogeneous populations of MSC for better results?). Finally, we also need to devise a method to label and track these cells *in vivo*.

E. Aging and MSC

In order to apply MSCs therapeutically, we must first expand their number *in vitro* because there are very few naturally occurring MSCs in the BM and their inhibitory effect is hypothesized to be dose dependent. Long-term culture of stem cells may have disadvantages. Therefore, before they can be used as therapeutic agents, we should ask: What are the effects of long-term culture on MSC?

Until recently most scientists believed stem cells self-renew indefinitely. Our experience with long-term *in vitro* culture of MSC (Mohyeddin *et al.*, 2006) proved otherwise. Our MSC cultures showed alterations at functional, physiological, and molecular levels. These included the Hayflick (1965) phenomenon of cellular aging (cytoplasm granulation, debris formation in culture), telomeres shortening (Baxter *et al.*, 2004; Zimmermann *et al.*, 2003), decline in differentiation potential (Colter *et al.*, 2000), and decline in proliferative activity (Stenderup *et al.*, 2003).

One disadvantage of long-term *in vitro* culturing which makes cells unsuitable for therapy is the possibility of transformation. Wang *et al.* (2005) showed that subgroups of MSCs exist that appear to differ from typical MSCs. Cytogenetic analysis of such MSC subgroups yielded chromosomal abnormalities and higher telomerase activity compared to typical MSCs. After 120 days, we observed some changes including elongated adherent cells and rounded nonadherent cells, marks of transformation. Serakinci *et al.* (2004) showed that MSC alterations in cell culture can be seen after 70 population doubling but not before that. Rubio *et al.* (2005) demonstrated that MSC-derived adipose tissue can become transformed and immortalized after 4–5 months of culture.

On the basis of the above information, MSCs, like all other cells of an organism undergo the phenomenon called senescence. Therefore, a safe period for culturing needs to be established, particularly when these cells are intended for clinical use. Of course this is not to say these cells necessarily become harmful after such a period (number of passages). Other indicators (gene and protein level variations) need to be examined in order to determine the safety of such cells for *in vivo* administration. We have only taken the first steps in establishing this novel approach of biologic, immune adaptive therapy.

VII. Concluding Remarks

The dynamics of the natural sciences are mesmerizing. The "quest for knowledge" has long attempted to simplify our understanding of nature's mechanisms, applying the lessons learned to "improve human life and alleviate their suffering." As understanding broadened, new problem become apparent. For example, psychological conditions were found to have a potent effect on the immune system, giving rise to the science of psychoneuroimmunology.

On a lesser scale, we begin to see intricate connections between events occurring within the body and the therapeutic processes. We have presented, for example, the link between MC, an inherent natural phenomenon, and stem cell transplantation. In our opinion, significant concepts emerge from advances in the understanding of these two somewhat isolated areas which can be applied to devise innovative therapeutic intervention. Microchimeric cells are naturally engrafted into the host body. Understanding the mechanisms behind this successful event could give us the tools to improve stem cell therapy. MC appears to be a "natural model of transplantation." On the other hand, MSCs, a subpopulation of the BM multipotential cells, have been found to possess innate immunosuppressive qualities. Immunosuppression is a primary need in nonhost engraftment such as pregnancy, a primary means of microchimeric induction which promotes natural immunosuppression. And among the first fetal cells to enter and engraft into the maternal body are the fetal MSCs.

If we are to apply this knowledge to artificial transplantation, we need a better understanding of the immunobiology of MSC. Finally, expanding MSCs *ex vivo* and using them to treat MS (or any immune dysregulation disorders)—after understanding how to protect these cells from potential culture and environmental hazards—could prove as effective therapy combining immunosuppression and regeneration.

References

Abouelfetouh, A., Kondoh, T., Ehara, K., and Kohmura, E. (2004). Morphological differentiation of bone marrow stromal cells into neuron-like cells after co-culture with hippocampal slice. *Brain Res.* **1029**(1), 114–119.

Abramsky, O., Lubetzki-Korn, I., Evron, S., and Brenner, T. (1984). Suppressive effect of pregnancy on MS and EAE. *Prog. Clin. Biol. Res.* **146,** 399–406.

Akasaki, Y., Liu, G., Chung, N. H., Ehtesham, M., Black, K. L., and Yu, J. S. (2004). Induction of a CD4+ T regulatory type 1 response by cyclooxygenase-2-overexpressing glioma. *J. Immunol.* **173**(7), 4352–4359.

Alexanian, A. R., and Kurpad, S. N. (2005). Quiescent neural cells regain multipotent stem cell characteristics influenced by adult neural stem cells in co-culture. *Exp. Neurol.* **191**(1), 193–197.

Arikawa, T., Omura, K., and Morita, I. (2004). Regulation of bone morphogenetic protein-2 expression by endogenous prostaglandin E2 in human mesenchymal stem cells. *J. Cell Physiol.* **200**(3), 400–406.

Artlett, C. M. (2002). Microchimerism in health and disease. *Curr. Mol. Med.* **2**(6), 525–535.

Artlett, C. M. (2005). Pathophysiology of fetal microchimeric cells. *Clin. Chim. Acta* **360**(1–2), 1–8.

Artlett, C. M., Rasheed, M., Russo-Stieglitz, K. E., Sawaya, H. H., and Jimenez, S. A. (2002). Influence of prior pregnancies on disease course and cause of death in systemic sclerosis. *Ann. Rheum. Dis.* **61**(4), 346–350.

Azizi, S. A., Stokes, D., Augelli, B. J., DiGirolamo, C., and Prockop, D. J. (1998). Engraftment and migration of human bone marrow stromal cells implanted in the brains of albino rats—similarities to astrocyte grafts. *Proc. Natl. Acad. Sci. USA* **95**(7), 3908–3913.

Bailo, M., Soncini, M., Vertua, E., Signoroni, P. B., Sanzone, S., Lombardi, G., Arienti, D., Calamani, F., Zatti, D., Paul, P., Albertini, A., Zorzi, F., *et al.* (2004). Engraftment potential of human amnion and chorion cells derived from term placenta. *Transplantation* **78**(10), 1439–1448.

Baksh, D., Song, L., and Tuan, R. S. (2004). Adult mesenchymal stem cells: Characterization, differentiation, and application in cell and gene therapy. *J. Cell. Mol. Med.* **8**(3), 301–316.

Barry, F. P., and Murphy, J. M. (2004). Mesenchymal stem cells: Clinical applications and biological characterization. *Int. J. Biochem. Cell Biol.* **36**(4), 568–584.

Baxter, M. A., Wynn, R. F., Jowitt, S. N., Wraith, J. E., Fairbairn, L. J., and Bellantuono, I. (2004). Study of telomere length reveals rapid aging of human marrow stromal cells following *in vitro* expansion. *Stem Cells* **22**(5), 675–682.

Bettens, F., Tiercy, J. M., Campanile, N., Giostra, E., Majno, P., Rubbia, L., Roosnek, E., Mentha, G., and Villard, J. (2005). Microchimerism after liver transplantation: Absence of rejection without abrogation of anti-donor cytotoxic T-lymphocyte-mediated alloreactivity. *Liver Transpl.* **11**(3), 290–297.

Beyth, S., Borovsky, Z., Mevorach, D., Liebergall, M., Gazit, Z., Aslan, H., Galun, E., and Rachmilewitz, J. (2005). Human mesenchymal stem cells alter antigen-presenting cell maturation and induce T-cell unresponsiveness. *Blood* **105**(5), 2214–2219.

Bianchi, D. W., Zickwolf, G. K., Weil, G. J., Sylvester, S., and DeMaria, MA. (1996). Male fetal progenitor cells persist in maternal blood for as long as 27 years postpartum. *Proc. Natl. Acad. Sci. USA* **93**(2), 705–708.

Black, I. B., and Woodbury, D. (2001). Adult rat and human bone marrow stromal stem cells differentiate into neurons. *Blood Cells Mol. Dis.* **27**(3), 632–636.

Blanco, Y., Saiz, A., Carreras, E., and Graus, F. (2005). Autologous haematopoietic-stem-cell transplantation for multiple sclerosis. *Lancet Neurol.* **4**(1), 54–63.

Burnet, F. M. (1957). A modification of Jerne's theory of antibody production using the concept of clonal selection. *Aust. J. Sci.* **20,** 67–69.

Burt, R. K., Traynor, A. E., Pope, R., Schroeder, J., Cohen, B., Karlin, K. H., Lobeck, L., Goolsby, C., Rowlings, P., Davis, F. A., Stefoski, D., Terry, C., *et al.* (1998). Treatment of autoimmune disease by intense immunosuppressive conditioning and autologous hematopoietic stem cell transplantation. *Blood* **92**(10), 3505–3514.

Campagnoli, C., Roberts, I. A., Kumar, S., Bennett, P. R., Bellantuono, I., and Fisk, N. M. (2001). Identification of mesenchymal stem/progenitor cells in human first-trimester fetal blood, liver, and bone marrow. *Blood* **98**(8), 2396–2402.

Cat, H., and My, D. D. (2003). Obviousness for transdifferentiation of human bone marrow-derived stem cells: Recent progress and controversies. *Pathology* **35**(1), 6–13.

Chan, J. L., Tang, K. C., Patel, A. P., Bonilla, L. M., Pierobon, N., Ponzio, N. M., and Rameshwar, P. (2006). Antigen presenting property of mesenchymal stem cells occurs during a narrow window at low levels of interferon-{gamma}. *Blood* **107**(12), 4817–4824.

Chunmeng, S., Tianmin, C., Yongping, S., Xinze, R., Yue, M., Jifu, Q., Shufen, L., Hui, X., and Chengji, L. (2004). Effects of dermal multipotent cell transplantation on skin wound healing. *J. Surg. Res.* **121**(1), 13–19.

Colter, D. C., Class, R., DiGirolamo, C. M., and Prockop, D. J. (2000). Rapid expansion of recycling stem cells in cultures of plastic-adherent cells from human bone marrow. *Proc. Natl. Acad. Sci. USA* **97**(7), 3213–3218.

Colter, D. C., Sekiya, I., and Prockop, D. J. (2001). Identification of a subpopulation of rapidly self-renewing and multipotential adult stem cells in colonies of human marrow stromal cells. *Proc. Natl. Acad. Sci. USA* **98**(14), 7841–7845.

Compston, A., Ebers, G., Lassmann, H., McDonald, I., Matthews, B., and Wekerle, H. (1999). "McAlpine's Multiple Sclerosis," 3rd ed. Churchil Livingstone, London.

Conget, P. A., and Minguell, J. J. (1999). Phenotypical and functional properties of human bone marrow mesenchymal progenitor cells. *J. Cell Physiol.* **181,** 67–73.

Corcione, A., Benvenuto, F., Ferretti, E., Giunti, D., Cappiello, V., Cazzanti, F., Risso, M., Gualandi, F., Mancardi, G. L., Pistoia, V., and Uccelli, A. (2006). Human mesenchymal stem cells modulate B-cell functions. *Blood* **107**(1), 367–372.

Coutinho, A. (1989). Beyond clonal selection and network. *Immunol. Rev.* **110,** 63–67.

Crigler, L., Robey, R. C., Asawachaicharn, A., Gaupp, D., and Phinney, D. G. (2006). Human mesenchymal stem cell subpopulations express a variety of neuro-regulatory molecules and promote neuronal cell survival and neuritogenesis. *Exp. Neurol.* **198**(1), 54–64.

Daumer, M., Griffith, L. M., Meister, W., Nash, R. A., and Wolinsky, J. S. (2006). Survival, and time to an advanced disease state or progression, of untreated patients with moderately severe multiple sclerosis in a multicenter observational database: Relevance for design of a clinical trial for high dose imunosuppressive therapy with autologous hematopoietic stem cell transplantation. *Mult. Scler.* **12**(2), 174–179.

Devine, S. M., and Hoffman, R. (2000). Role of mesenchymal stem cells in hematopoietic stem cell transplantation. *Curr. Opin. Hematol.* **7**(6), 358–363.

Di Nicola, M., Carlo-Stella, C., Magni, M., Milanesim, Longoni, P. D., Matteucci, P., Grisanti, S., and Gianni, A. M. (2002). Human bone marrow stromal cells suppress T-lymphocyte proliferation induced by cellular or nonspecific mitogenic stimuli. *Blood* **99,** 3838–3843.

Eglitis, M. A., and Mezey, E. (1997). Hematopoietic cells differentiate into both microglia and macroglia in the brains of adult mice. *Proc. Natl. Acad. Sci. USA* **94**(8), 4080–4085.

Evans, P. C., Lambert, N., Maloney, S., Furst, D. E., Moore, J. M., and Nelson, J. L. (1999). Long-term fetal microchimerism in peripheral blood mononuclear cell subsets in healthy women and women with scleroderma. *Blood* **93**(6), 2033–2037.

Fassas, A., and Kimiskidis, V. K. (2003). Stem cell transplantation for multiple sclerosis: What is the evidence? *Blood Rev.* **17**(4), 233–240.

Fassas, A., Anagnostopoulos, A., Kazis, A., Kapinas, K., Sakellari, I., Kimiskidis, V., and Tsompanakou, A. (1997). Peripheral blood stem cell transplantation in the treatment of progressive multiple sclerosis: First results of a pilot study. *Bone Marrow Transplant.* **20**(8), 631–638.

Ferrari, G., Cusella-De Angelis, G., Coletta, M., Paolucci, E., Stornaiuolo, A., Cossu, G., and Mavilio, F. (1998). Muscle regeneration by bone marrow-derived myogenic progenitors. *Science* **279**(5356), 1528–1530.

Friedenstein, A. J., Chailakhjan, R. K., and Lalykina, K. S. (1970). The development of fibroblast colonies in monolayer cultures of guinea-pig bone marrow and spleen cells. *Cell Tissue Kinetics* **3,** 393–403.

Glennie, S., Soeiro, I., Dyson, P. J., Lam, E. W., and Dazzi, F. (2005). Bone marrow mesenchymal stem cells induce division arrest anergy of activated T cells. *Blood* **105**(7), 2821–2827.

Gribben, J. G., Freeman, G. J., Boussiotis, V. A., Rennert, P., Jellis, C. L., Greenfield, E., Barber, M., Restivo, V. A., Jr., Ke, X., Gray, G. S., *et al.* (1995). CTLA4 mediates antigen-specific apoptosis of human T cells. *Proc. Natl. Acad. Sci. USA* **92**(3), 811–815.

Hayflick, L. (1965). The limited *in vitro* lifetime of human diploid cell strains. *Exp. Cell Res.* **37,** 614–636.

Hohlfeld, R., and Wekerle, H. (2004). Autoimmune concepts of multiple sclerosis as a basis for selective immunotherapy: From pipe dreams to (therapeutic) pipelines. *Proc. Natl. Acad. Sci. USA* **1**(Suppl. 2), 14599–14606.

Horwitz, E. M., Prockop, D. J., Fitzpatrick, L. A., Koo, W. W., Gordon, P. L., Neel, M., Sussman, M., Orchard, P., Marx, J. C., Pyeritz, R. E., and Brenner, M. K. (1999). Transplantability and therapeutic effects of bone marrow-derived mesenchymal cells in children with osteogenesis imperfecta. *Nat. Med.* **5**(3), 309–313.

Ichinohe, T., Maruya, E., and Saji, H. (2002). Long-term feto-maternal microchimerism: Nature's hidden clue for alternative donor hematopoietic cell transplantation? *Int. J. Hematol.* **76**(3), 229–237.

Ichinohe, T., Teshima, T., Matsuoka, K., Maruya, E., and Saji, H. (2005). Fetal-maternal microchimerism: Impact on hematopoietic stem cell transplantation. *Curr. Opin. Immunol.* **17**(5), 546–552.

Imaizumi, M., Pritsker, A., Unger, P., and Davies, T. F. (2002). Intrathyroidal fetal microchimerism in pregnancy and postpartum. *Endocrinology* **143**(1), 247–253.

Jin, H. K., Carter, J. E., Huntley, G. W., and Schuchman, E. H. (2002). Intracerebral transplantation of mesenchymal stem cells into acid sphingomyelinase-deficient mice delays the onset of neurological abnormalities and extends their life span. *J. Clin. Invest.* **109**(9), 1183–1191.

Johnson, K. L., Samura, O., Nelson, J. L., McDonnell, M., and Bianchi, D.,W. (2002). Significant fetal cell microchimerism in a nontransfused woman with hepatitis C: Evidence of long-term survival and expansion. *Hepatology* **36,** 1295–1297.

Jonuleit, H., Schmitt, E., Schuler, G., Knop, J., and Enk, A. H. (2000). Induction of interleukin 10-producing, nonproliferating CD4(+) T cells with regulatory properties by repetitive stimulation with allogeneic immature human dendritic cells. *J. Exp. Med.* **192**(9), 1213–1222.

Keilhoff, G., Goihl, A., Langnase, K., Fansa, H., and Wolf, G. (2006). Transdifferentiation of mesenchymal stem cells into Schwann cell-like myelinating cells. *Eur. J. Cell Biol.* **85**(1), 11–24.

Khosrotehrani, K., Stroh, H., Bianchi, D. W., and Johnson, K. L. (2003). Combined FISH and immunolabeling on paraffin-embedded tissue sections for the study of microchimerism. *Biotechniques* **34,** 242–244.

Khosrotehrani, K., Johnson, K. L., Cha, D. H., Salomon, R. N., and Bianchi, D. W. (2004). Transfer of fetal cells with multilineage potential to maternal tissue. *JAMA* **292**(1), 75–80.

Khosrotehrani, K., Johnson, K. L., Guegan, S., Stroh, H., and Bianchi, D. W. (2005). Natural history of fetal cell microchimerism during and following murine pregnancy. *J. Reprod. Immunol.* **66**(1), 1–12.

Krabbe, C., Zimmer, J., and Meyer, M. (2005). Neural transdifferentiation of mesenchymal stem cells—a critical review. *APMIS* **113**(11–12), 831–844.

Krampera, M., Glennie, S., Dyson, J., Scott, D., Laylor, R., Simpson, E., and Dazzi, F. (2003). Bone marrow mesenchymal stem cells inhibit the response of naive and memory antigen-specific T cells to their cognate peptide. *Blood* **101**(9), 3722–3729.

Kraus, K. H., and Kirker-Head, C. (2006). Mesenchymal stem cells and bone regeneration. *Vet. Surg.* **35**(3), 232–242.

Lambert, N. C., Pang, J. M., Yan, Z., Erickson, T. D., Stevens, A. M., Furst, D. E., and Nelson, J. L. (2005). Male microchimerism in women with systemic sclerosis and healthy women who have never given birth to a son. *Ann. Rheum. Dis.* **64**(6), 845–848.

Le Blanc, K. (2003). Immunomodulatory effects of fetal and adult mesenchymal stem cells. *Cytotherapy* **5**(6), 485–489.

Le Blanc, K., Tammik, C., Rosendahl, K., Zetterberg, E., and Ringden, O. (2003a). HLA expression and immunologic properties of differentiated and undifferentiated mesenchymal stem cells. *Exp. Hematol.* **31**(10), 890–896.

Le Blanc, K., Tammik, L., Sundberg, B., Haynesworth, S. E., and Ringden, O. (2003b). Mesenchymal stem cells inhibit and stimulate mixed lymphocyte cultures and mitogenic responses independently of the major histocompatibility complex. *Scand. J. Immunol.* **57**(1), 11–20.

Le Blanc, K., Rasmusson, I., Sundberg, B., Gotherstrom, C., Hassan, M., Uzunel, M., and Ringden, O. (2004). Treatment of severe acute graft-versus-host disease with third party haploidentical mesenchymal stem cells. *Lancet* **363**(9419), 1439–1441.

Li, Y., Chen, J., Chen, X. G., Wang, L., Gautam, S. C., Xu, Y. X., Katakowski, M., Zhang, L. J., Lu, M., Janakiraman, N., and Chopp, M. (2002). Human marrow stromal cell therapy for stroke in rat: Neurotrophins and functional recovery. *Neurology* **59**(4), 514–523.

Loubiere, L. S., Lambert, N. C., Flinn, L. J., Erickson, T. D., Yan, Z., Guthrie, K. A., Vickers, K. T., and Nelson, J. L. (2006). Maternal microchimerism in healthy adults in lymphocytes, monocyte/macrophages and NK cells. *Lab. Invest.* **86**(11), 1185–1192.

Maccario, R., Podesta, M., Moretta, A., Cometa, A., Comoli, P., Montagna, D., Daudt, L., Ibatici, A., Piaggio, G., Pozzi, S., Frassoni, F., and Locatelli, F. (2005). Interaction of human mesenchymal stem cells with cells involved in alloantigen-specific immune response favors the differentiation of CD4+ T-cell subsets expressing a regulatory/suppressive phenotype. *Haematologica* **90**(4), 516–525.

Mahmood, A., Lu, D., and Chopp, M. (2004). Marrow stromal cell transplantation after traumatic brain injury promotes cellular proliferation within the brain. *Neurosurgery* **55**(5), 1185–1193.

Meisel, R., Zibert, A., Laryea, M., Gobel, U., Daubener, W., and Dilloo, D. (2004). Human bone marrow stromal cells inhibit allogeneic T-cell responses by indoleamine 2,3-dioxygenase-mediated tryptophan degradation. *Blood* **103**(12), 4619–4621.

Mellor, A. L., and Munn, D. H. (2001). Tryptophan catabolism prevents maternal T cells from activating lethal anti-fetal immune responses. *J. Reprod. Immunol.* **52**(1–2), 5–13.

Mendez-Ferrer, S., Ellison, G. M., Torella, D., and Nadal-Ginard, B. (2006). Resident progenitors and bone marrow stem cells in myocardial renewal and repair. *Nat. Clin. Pract. Cardiovasc. Med.* **3**(Suppl. 1), S83–S89.

Mezey, E., Chandross, K. J., Harta, G., Maki, R. A., and McKercher, S. R. (2000). Turning blood into brain: Cells bearing neuronal antigens generated *in vivo* from bone marrow. *Science* **290** (5497), 1779–1782.

Mills, K. H. (2004). Regulatory T cells: Friend or foe in immunity to infection? *Nat. Rev. Immunol.* **4**(11), 841–855.

Moalem, G., Leibowitz-Amit, R., Yoles, E., Mor, F., Cohen, I. R., and Schwartz, M. (1999). Autoimmune T cells protect neurons from secondary degeneration after central nervous system axotomy. *Nat. Med.* **5**(1), 49–55.

Mohamedali, A., Soeiro, I., Lea, N. C., Glassford, J., Banerji, L., Mufti, G. J., Lam, E. W., and Thomas, N. S. (2003). Cyclin D2 controls B cell progenitor numbers. *J. Leukoc. Biol.* **74**(6), 1139–1143.

Mohyeddin Bonab, M., Alimoghaddam, K., Talebian, F., Ghaffari, S. H., Ghavamzadeh, A., and Nikbin, B. (2005). In search of mesenchymal stem cells: Bone marrow, cord blood or peripheral blood. *Int. J. Hem. Onc. BMT* **2**(4), 17–22.

Mohyeddin Bonab, M., Alimoghaddam, K., Talebian, F., Ghaffari, S. H., Ghavamzadeh, A., and Nikbin, B. (2006). Aging of mesenchymal stem cell *in vitro*. *BMC Cell Biol.* **10,** 7–14.

Muller, G., Hopken, U. E., Stein, H., and Lipp, M. (2002). Systemic immunoregulatory and pathogenic functions of homeostatic chemokine receptors. *J. Leukoc. Biol.* **72**(1), 1–8.

Munn, D. H., Mellor, A. L., Rossi, M., and Young, J. W. (2005). Dendritic cells have the option to express IDO-mediated suppression or not. *Blood* **105**(6), 2480–2486.

Muraro, P. A., and Martin, R. (2003). Immunological questions on hematopoietic stem cell transplantation for multiple sclerosis. *Bone Marrow Transplant* **32**(Suppl. 1), S41–S44.

Muraro, P. A., Cassiani Ingoni, R., and Martin, R. (2003). Hematopoietic stem cell transplantation for multiple sclerosis: Current status and future challenges. *Curr. Opin. Neurol.* **16**(3), 299–305.

Nelson, J. L. (1998). Pregnancy, persistent microchimerism, and autoimmune disease. *J. Am. Med. Womens Assoc.* **53**(1), 31–32, 47.

Neuss, S., Becher, E., Woltje, M., Tietze, L., and Jahnen-Dechent, W. (2004). Functional expression of HGF and HGF receptor/c-met in adult human mesenchymal stem cells suggests a role in cell mobilization, tissue repair, and wound healing. *Stem Cells* **22**(3), 405–414.

O'Donoghue, K., Choolani, M., Chan, J., de la Fuente, J., Kumar, S., Campagnoli, C., Bennett, P. R., Roberts, I. A., and Fisk, N. M. (2003). Identification of fetal mesenchymal stem cells in maternal blood: Implications for non-invasive prenatal diagnosis. *Mol. Hum. Reprod.* **9**(8), 497–502.

O'Donoghue, K., Chan, J., de la Fuente, J., Kennea, N., Sandison, A., Anderson, J. R., Roberts, I. A., and Fisk, N. M. (2004). Microchimerism in female bone marrow and bone decades after fetal mesenchymal stem-cell trafficking in pregnancy. *Lancet* **364**(9429), 179–182.

Oh, S. K., and Choo, A. B. (2006). Human embryonic stem cells: Technological challenges towards therapy. *Clin. Exp. Pharmacol. Physiol.* **33**(5–6), 489–495.

Ono, I., Yamashita, T., Hida, T., Jin, H. Y., Ito, Y., Hamada, H., Akasaka, Y., Ishii, T., and Jimbow, K. (2004). Local administration of hepatocyte growth factor gene enhances the regeneration of dermis in acute incisional wounds. *J. Surg. Res.* **120**(1), 47–55.

Orlic, D. (2005). BM stem cells and cardiac repair: Where do we stand in 2004? *Cytotherapy* **7**(1), 3–15.

Pawelec, G., Rehbein, A., Schlotz, E., Friccius, H., and Pohla, H. (1996). Cytokine modulation of TH1/TH2 phenotype differentiation in directly alloresponsive CD4+ human T cells. *Transplantation* **62**(8), 1095–1101.

Pittenger, M. F., Mackay, A. M., Beck, S. C., Jaiswal, R. K., Douglas, R., Mosca, J. D., Moorman, M. A., Simonetti, D. W., Craig, S., and Marshak, D. R. (1999). Multilineage potential of adult human mesenchymal stem cells. *Science* **284**(5411), 143–147.

Roncarolo, M. G., Levings, M. K., and Traversari, C. (2001). Differentiation of T regulatory cells by immature dendritic cells. *J. Exp. Med.* **193**(2), F5–F9.

Roper, R. L., Ludlow, J. W., and Phipps, R. P. (1994). Prostaglandin E2 inhibits B lymphocyte activation by a cAMP-dependent mechanism: PGE-inducible regulatory proteins. *Cell Immunol.* **154**(1), 296–308.

Rubio, D., Garcia-Castro, J., Martin, M. C., de la Fuente, R., Cigudosa, J. C., Lloyd, A. C., and Bernad, A. (2005). Spontaneous human adult stem cell transformation. *Cancer Res.* **65**(8), 3035–3039.

Ryan, J. M., Barry, F. P., Murphy, J. M., and Mohan, B. P. (2005). Mesenchymal stem cells avoid allogeneic rejection. *J. Inflamm. Lond.* **2,** 2–8.

Saccardi, R., Mancardi, G. L., Solari, A., Bosi, A., Bruzzi, P., Di Bartolomeo, P., Donelli, A., Filippi, M., Guerrasio, A., Gualandi, F., La Nasa, G., Murialdo, A, *et al.* (2005). Autologous HSCT for severe progressive multiple sclerosis in a multicenter trial: Impact on disease activity and quality of life. *Blood* **105**(6), 2601–2607.

Sakaguchi, S. (2004). Naturally arising CD4+ regulatory T cells for immunologic self-tolerance and negative control of immune responses. *Annu. Rev. Immunol.* **22,** 531–562.

Sakaguchi, S., Sakaguchi, N., Shimizu, J., Yamazaki, S., Sakihama, T., Itoh, M., Kuniyasu, Y., Nomura, T., Toda, M., and Takahashi, T. (2001). Immunologic tolerance maintained by CD25+CD4+ regulatory T cells: Their common role in controlling autoimmunity, tumor immunity, and transplantation tolerance. *Immunol. Rev.* **182,** 18–32.

Sanchez-Ramos, J. R. (2002). Neural cells derived from adult bone marrow and umbilical cord blood. *J. Neurosci. Res.* **69**(6), 880–893.

Scholzen, T., and Grede, J. (2000). The Ki-67 protein: From the known and the unknown. *J. Cell. Physiol.* **182**(3), 311–322.

Semb, H. (2005). Human embryonic stem cells: Origin, properties and applications. *APMIS* **113**(11–12), 743–750.

Serakinci, N., Guldberg, P., Burns, J. S., Abdallah, B., Schrodder, H., Jensen, T., and Kassem, M. (2004). Adult human mesenchymal stem cell as a target for neoplastic transformation. *Oncogene* **23**(29), 5095–5098.

Shake, J. G., Gruber, P. J., Baumgartner, W. A., Senechal, G., Meyers, J., Redmond, J. M., Pittenger, M. F., and Martin, B. J. (2002). Mesenchymal stem cell implantation in a swine myocardial infarct model: Engraftment and functional effects. *Ann. Thorac. Surg.* **73**(6), 1919–1925.

Shizuru, J. A., Negrin, R. S., and Weissman, I. L. (2005). Hematopoietic stem cell and progenitor cells: Clinical and pre-clinical regeneration of the hematolymphoid system. *Annu. Rev. Med.* **56,** 509–538.

Silman, A. J. (1992). Pregnancy and Scleroderma. *Am. J. Reprod. Immunol.* **28,** 238–240.

Silva, W. A., Covas, D. T., Panepucci, R. A., Proto-Siqueira, R., Siufi, J. L., Zanette, D. L., Santos, A. R., and Zago, M. A. (2003). The profile of gene expression of human marrow mesenchymal stem cells. *Stem Cells* **21,** 661–669.

Spector, T. D., and Da Silva, J. A. (1992). Pregnancy and rheumatoid arthritis: An overview. *Am. J. Reprod. Immunol.* **28,** 222.

Stagg, J., Pommey, S., Eliopoulos, N., and Galipeau, J. (2006). Interferon-gamma-stimulated marrow stromal cells: A new type of nonhematopoietic antigen-presenting cell. *Blood* **107**(6), 2570–2577.

Starzl, T. E., and Lakkis, F. G. (2006). The unfinished legacy of liver transplantation: Emphasis on immunology. *Hepatology* **43**(2 Suppl. 1), S151–S163.

Steinman, R. M., and Nussenzweig, M. C. (2002). Avoiding horror autotoxicus: The importance of dendritic cells in peripheral T cell tolerance. *Proc. Natl. Acad. Sci. USA* **99**(1), 351–358.

Stenderup, K., Justesen, J., Clausen, C., and Kassem, M. (2003). Aging is associated with decreased maximal life span and accelerated senescence of bone marrow stromal cells. *Bone* **33**(6), 919–926.

Talebian, F., Amirzargar, A., Ghaffari, S. H., Khosravi, F., Lotfi, J., and Nikbin, B. (2005). Association of microchimerism and multiple sclerosis patients. *J. Neurol. Sci.* **238**(Suppl. 1), S251.

Tan, X. W., Liao, H., Sun, L., Okabe, M., Xiao, Z. C., and Dawe, G. S. (2005). Fetal microchimerism in the maternal mouse brain: A novel population of fetal progenitor or stem cells able to cross the blood-brain barrier? *Stem Cells* **23**(10), 1443–1452.

Tsai, M. S., Lee, J. L., Chang, Y. J., and Hwang, S. M. (2004). Solation of human multipotent mesenchymal stem cells from second-trimester amniotic fluid using a novel two-stage culture protocol. *Hum. Reprod.* **19**(6), 1450–1456.

Tse, W. T., Pendleton, J. D., Beyer, W. M., Egalka, M. C., and Guinan, E. C. (2003). Suppression of allogeneic T-cell proliferation by human marrow stromal cells: Implications in transplantation. *Transplantation* **75**(3), 389–397.

Umeda, K., Adachi, S., Ishihara, H., Higashi, Y., Shiota, M., Watanabe, K. I., Hishizawa, M., Ichinohe, T., Kitoh, T., Maruya, E., Saji, H., Uchiyama, T., *et al.* (2003). Successful T-cell-replete peripheral blood stem cell transplantation from HLA-haploidentical microchimeric mother to daughter with refractory acute lymphoblastic leukemia using reduced-intensity conditioning. *Bone Marrow Transplant.* **31**(11), 1061–1063.

Van Gelder, M., and van Bekkum, D. W. (Gelder 1996). Effective treatment of relapsing experimental autoimmune encephalomyelitis with pseudoautologous bone marrow transplantation. *Bone Marrow Transplant.* **18**(6), 1029–1034.

Van Haaften, T., and Thebaud, B. (2006). Adult bone marrow-derived stem cells for the lung: Implications for pediatric lung diseases. *Pediatr. Res.* **59**(4 Pt. 2), 94R–99R.

Vats, A., Tolley, N. S., Polak, J. M., and Buttery, L. D. (2002). Stem cells: Sources and applications. *Clin. Otolaryngol. Allied Sci.* **27**(4), 227–232.

Vescovi, A., Gritti, A., Cossu, G., and Galli, R. (2002). Neural stem cells: Plasticity and to their transdifferentiation potential. *Cells Tissues Organs* **171**(1), 64–76.

Villaron, E. M., Almeida, J., Lopez-Holgado, N., Alcoceba, M., Sanchez-Abarca, L. I., Sanchez-Guijo, F. M., Alberca, M., Perez-Simon, J. A., San Miguel, J. F., and Del Canizo, M. C. (2004). Mesenchymal stem cells are present in peripheral blood and can engraft after allogeneic hematopoietic stem cell transplantation. *Haematologica* **89**(12), 1421–1427.

Vitiello, A., Ishioka, G., Grey, H. M., Rose, R., Farness, P., LaFond, R., Yuan, L., Chisari, F. V., Furze, J., Bartholomeuz, R., *et al.* (1995). Development of a lipopeptide-based therapeutic vaccine to treat chronic HBV infection. I. Induction of a primary cytotoxic T lymphocyte response in humans. *J. Clin. Invest.* **95**(1), 341–349.

Wang, Y., Iwatani, H., Ito, T., Horimoto, N., Yamato, M., Matsui, I., Imai, E., and Hori, M. (2004). Fetal cells in mother rats contribute to the remodeling of liver and kidney after injury. *Biochem. Biophys. Res. Commun.* **325**(3), 961–967.

Wang, Y., Huso, D. L., Harrington, J., Kellner, J., Jeong, D. K., Turney, J., and McNiece, I. K. (2005). Outgrowth of a transformed cell population derived from normal human BM mesenchymal stem cell culture. *Cytotherapy* **7**(6), 509–519.

Wilder, R. (1998). Hormones, pregnancy, and autoimmune diseases. *Ann. N. Y. Acad. Sci.* **840,** 45–50.

Willer, C. J., Sadovnick, A. D., and Ebers, G. C. (2002). Microchimerism in autoimmunity and transplantation: Potential relevance to multiple sclerosis. *J. Neuroimmunol.* **126**(1–2), 126–133.

Woodbury, D., Schwarz, E. J., Prockop, D. J., and Black, I. B. (2000). Adult rat and human bone marrow stromal cells differentiate into neurons. *J. Neurosci. Res.* **61**(4), 364–370.

Xin, X., Yang, S., Ingle, G., Zlot, C., Rangell, L., Kowalski, J., Schwall, R., Ferrara, N., and Gerritsen, M. E. (2001). Hepatocyte growth factor enhances vascular endothelial growth factor-induced angiogenesis *in vitro* and *in vivo*. *Am. J. Pathol.* **158**(3), 1111–1120.

Yan, Z., Lambert, N. C., Guthrie, K. A., Porter, A. J., Loubiere, L. S., Madeleine, M. M., Stevens, A. M., Hermes, H. M., and Nelson, J. L. (2005). Male microchimerism in women without sons: Quantitative assessment and correlation with pregnancy history. *Am. J. Med.* **118**(8), 899–906.

Zemchikhina, V. N., and Golubeva, O. N. (2005). The capacity of mesenchymal stem cells for neural differentiation *in vitro*. *Tsitologiia* **47**(7), 644–648.

Zheng, S. G., Wang, J. H., Gray, J. D., Soucier, H., and Horwitz, D. A. (2004). Natural and induced CD4+CD25+ cells educate CD4+CD25− cells to develop suppressive activity: The role of IL-2, TGF-beta, and IL-10. *J. Immunol.* **172**(9), 5213–5221.

Zimmermann, S., Voss, M., Kaiser, S., Kapp, U., Waller, C. F., and Martens, U. M. (2003). Lack of telomerase activity in human mesenchymal stem cells. *Leukemia* **17**(6), 1146–1149.

THE INSULIN-LIKE GROWTH FACTOR SYSTEM IN MULTIPLE SCLEROSIS

Daniel Chesik, Nadine Wilczak, and Jacques De Keyser

Department of Neurology, University Medical Center Groningen, Hanzeplein 1, 9713 GZ Groningen, The Netherlands

Multiple sclerosis (MS) is a chronic disorder of the central nervous system characterized by inflammation, demyelination, and axonal degeneration. Present therapeutic strategies for MS reduce inflammation and its destructive consequences, but are not effective in the progressive phase of the disease. There is a need for neuroprotective and restorative therapies in MS. Insulin-like growth factor-1 (IGF-1) is of considerable interest because it is not only a potent neuroprotective trophic factor but also a survival factor for cells of the oligodendrocyte lineage and possesses a potent myelinogenic capacity. However, the IGF system is complex and includes not only IGF-1 and IGF-2 and their receptors but also modulating IGF-binding proteins (IGFBPs), of which six have been identified. This chapter provides an overview of the role of the IGF system in the pathophysiology of MS, relevant findings in preclinical models, and discusses the possible use of IGF-1 as a therapeutic agent for MS.

INTERNATIONAL REVIEW OF NEUROBIOLOGY, VOL. 79
DOI: 10.1016/S0074-7742(07)79009-8

0074-7742/07 $35.00

I. Introduction

Multiple sclerosis (MS) is the most common human demyelinating disease of the central nervous system (CNS), for which there is currently no effective treatment. While contemporary therapeutic approaches target inflammatory responses, the issue of myelin repair and neuroprotection in damaged tissue remains unresolved. Pathological examination of MS lesions suggests that endogenous repair does occur, yet because this restoration is generally limited remyelination remains inefficient (Prineas and Connell, 1979; Prineas *et al.*, 1993; Raine and Wu, 1993; Raine *et al.*, 1981). It is believed that the highly plastic axonal response following demyelination might allow for axonal recovery, yet inefficiency of endogenous myelin sheath repair yields only restricted functional improvement (Kerschensteiner *et al.*, 2004). In addition, neurological impairment and subsequent disability partly result as a consequence of repeated inflammatory insults arising during relapse phases of disease, which are grounds for irreversible focal demyelination and axonal injury (Raine, 1997). Magnetic resonance studies have provided evidence of degeneration of axons not only within the focal lesions but also diffusely throughout the white matter (Fu *et al.*, 1998; Minagar *et al.*, 2005). This axonal damage ultimately progresses and correlates with the progressive clinical disability occurring during the progressive phase of the disease.

Of equal importance to immunomodulatory and immunosuppressive treatments are therapeutic approaches addressing the issue of neuroprotection and induction of myelin repair. There is considerable interest in growth factor-based therapy aimed at guiding remyelination in MS lesions. Effective application of growth factors aimed at repairing tissue damage caused by immune invasion would, however, require stringent control of appropriate stimuli restricting particular cellular processes to occur in a specific succession of events. For example, the task of replenishing oligodendrocyte progenitor cell (OPC) populations as well as triggering the differentiation of these cells into mature myelin producing oligodendrocytes or stimulating persisted mature oligodendrocytes at the rim of lesions to remyelinate axons (Prineas *et al.*, 1989) would demand a spatial and temporal orchestration of growth factor-driven actions.

We presently do not know how spontaneous myelin sheath repair occurs or which factors are involved in promoting or inhibiting this process. If we are to exploit mechanisms involved in a myelination-favoring signaling environment, it is necessary to identify these factors and apply a proper strategy in order to obtain beneficial clinical results. Insulin-like growth factors (IGFs) are potent inducers of oligodendrocyte differentiation and myelin production in the CNS (Carson *et al.*, 1993). Because these factors play essential roles in myelinogenesis and have pleiotropic functions in nervous tissue, they possess potential for therapeutic application in MS.

II. Growth Factors Guide Oligodendrocyte Behavior

To guide effective remyelination in MS, it is essential to understand the roles of factors influencing cell behavior processes of oligodendrocytes. Since these cells are responsible for myelin production, they are the target cells of a myelination-based therapy. An array of intrinsic growth factors has been implicated in inducing biological events in oligodendrocytes (Barres *et al.*, 1994; Gard and Pfeiffer, 1993). It is known that locally derived OPCs proliferate and differentiate into mature myelinating oligodendrocytes in the rodent (Armstrong *et al.*, 1990; Godfraind *et al.*, 1989; Ludwin, 1979). During developmental myelinogenesis, growth factors extensively coordinate the proliferation, migration, and differentiation of OPCs by providing important signals cueing CNS cytoarchitecture development and inducing myelination of axons (Pfeiffer *et al.*, 1993; Woodruff and Franklin, 1997). Individual growth factors generate important signaling events alone, synergically or can counteract the function of other factors. Mitogens prompting perinatal rodent OPCs growth *in vitro* include platelet-derived growth factor (PDGF) (Noble *et al.*, 1988; Raff *et al.*, 1988), basic fibroblast growth factor (bFGF) (Bogler *et al.*, 1990; McKinnon *et al.*, 1990), neurotrophin 3 (NT3) (Barres *et al.*, 1993; Cohen *et al.*, 1996), and glial growth factor-2 (GGF-2) (Canoll *et al.*, 1996; Flores *et al.*, 2000). Although PDGF triggers only a limited number of OPC cell divisions (Noble *et al.*, 1988), further stimulation with FGF has been shown to upregulate the expression of PDGF receptors and primes oligodendrocyte progenitors for continuous proliferative response to PDGF (McKinnon *et al.*, 1990). By contrast, transforming growth factor-β (TGF-β) is thought to interfere with PDGF signaling reducing PDGF-driven proliferation, a mechanism that can be alleviated in the presence of bFGF. This process is likely to be responsible for the promotion of differentiation in these cells (McKinnon *et al.*, 1993).

Although growth factors can regulate multiple aspects of oligodendrocyte behavior, signaling pathways are also responsive to other receptor signaling, as has been shown for cell surface integrin receptors, which can alter, for example, the role of PDGF from inducing proliferation to enhancing survival in oligodendrocytes (Baron *et al.*, 2003).

IGF-1 and IGF-2 are key factors driving oligodendrocyte differentiation and are important for enhancing the production of myelin and increasing cell survival of mature oligodendrocytes (Barres *et al.*, 1993; McMorris and Dubois-Dalcq, 1988; McMorris *et al.*, 1986). Because of its potency to drive oligodendrocyte myelin production, IGF-1 is an attractive candidate for therapeutic application in treating neurodegenerative demyelinating diseases such as MS. However, the complex nature of IGF-1 regulation in the CNS has not yet been completely defined. In order to obtain the desired effects from an IGF-1-based therapy, a rational therapeutic approach must be designed that takes into consideration the complexity of IGF regulation and subsequent actions.

III. Overview of the IGF System

A. The IGFs

IGF-1 and IGF-2, as well as insulin, are genetically related ubiquitous single-chain polypeptides with similar tertiary structure and amino acid sequence homology (Jones and Clemmons, 1995). IGF-1 is one of the major trophic factors in the circulation and acts on many tissues and organs (Dore *et al.*, 1997). Its expression is influenced by hormonal, nutritional, and tissue-specific factors (Bichell *et al.*, 1992; Underwood *et al.*, 1994). Systemic IGF-1 is predominately synthesized in the liver and is regulated by growth hormone (GH) (Juul, 2003). IGF-1 is also synthesized in many other tissues, including in the brain, where GH does not appear to regulate its production (Aberg *et al.*, 2006). For these reasons, IGF-1 is believed, in addition to its classical endocrine functions, to exert autocrine and paracrine effects (Jones and Clemmons, 1995). While insulin primarily serves as an immediate response hormone regulating systemic glucose levels, IGF-1 plays a role in promoting acute anabolic effects on protein and carbohydrate metabolism, as well as longer-term effects on cell replication, survival, and differentiation. Because this pleiotropic protein exerts such diverse effects on various cell types, its actions on peripheral tissues support the maintenance of tissue integrity and homeostasis with especially profound effects observed in musculoskeletal and nervous systems (Minuto *et al.*, 2005).

IGF-1 is essential for many stages of development of the CNS (Anlar *et al.*, 1999). Since IGF-1 is not thought to easily cross the human blood–brain barrier (BBB), local production is considered the main source of this factor in the brain (Russo *et al.*, 2005). Expression of IGF-1 is high in neuronal rich regions of the midbrain, cerebral cortex, hippocampus, and olfactory bulb (Garcia-Segura *et al.*, 1991; Rotwein *et al.*, 1988). IGF-2 is the most abundant IGF in the CNS and is found in high levels in myelin tissue (Bondy *et al.*, 1992). Functions of IGFs in the CNS are discussed in more detail in later sections.

1. *IGF Receptors and Their Functions*

The biological functions of IGF-1 and IGF-2 are mediated by a specific cell surface membrane receptor, designated as the IGF-1 receptor. The IGF-1 receptor gene is transcriptionally regulated by growth factors and tumor suppressors and encodes a 210-kDa $\alpha\beta$-chain, which is proteolytically cleaved and disulfide bonded to produce an $\alpha\beta$-structure (LeRoith, 2000). After posttranslational modifications, the dimer of an α-extracellular and a β-transmembrane subunit form an α-β half-receptor which binds to another half-receptor to form a mature $\alpha_2\beta_2$-heterotetramer (Prager and Melmed, 1993; Rechler and Nissley, 1985). The glycosylated α-subunits contain the ligand-binding domain and β-subunits contain tyrosine kinase domains (Ullrich *et al.*, 1986). On ligand binding, this cell surface receptor transmits a highly integrated intracellular signaling system as discussed in the following section (Kim *et al.*, 2004).

The IGF-2 (type-2 IGF) receptor is identical to the cation-independent mannose-6-phosphate receptor which is involved in the sorting of newly synthesized lysosomal enzymes (Morgan *et al.*, 1987). On the cell surface, this receptor has been proposed to provide a degradative pathway for IGF-2 via receptor-mediated endocytosis (Kornfeld, 1992). Although the IGF-1 receptor is the primary mediator of IGF-1 and IGF-2 actions, the insulin receptor, the insulin/IGF-1 hybrid receptor, and an orphan insulin receptor-related receptor can mediate ligand-binding effects adding further to the complexity of the IGF system (Jones and Clemmons, 1995; Watt *et al.*, 1993).

2. *Signaling*

The most widely studied effect of IGF-1 *in vitro* is the stimulation of DNA synthesis and cell growth. A cellular action of IGF-1 that is complementary to its stimulation of cell proliferation is its capacity in certain cells to inhibit apoptosis (Barres *et al.*, 1993; Kooijman, 2006). Importantly, IGF-1 is a potent inducer of differentiation of cells as shown for neurons and oligodendrocytes (Feldman *et al.*, 1997; Ye and D'Ercole, 2006). These extracellular factors attribute their biological roles to a complex network of intracellular signaling cascades, which ultimately regulate gene transcription and cell cycle promoting factors. Several components of IGF-mediated signal transduction pathways have been identified and functions are being elucidated (Jones and Clemmons, 1995). On ligand binding to the IGF-1 receptor, activation of tyrosine kinase domains results in receptor autophosphorylation. A pivotal role of propagation of the cytoplasmic signal is played by a 185-kDa substrate, termed the insulin receptor substrate (IRS-1/3) (Baxter *et al.*, 1989). Once activated, the receptor phosphorylates IRS-1/3 on multiple tyrosines. Phosphorylation of this large docking protein allows binding of proteins with src homology 2 (SH2) domains such as the p85 regulatory subunit of phosphatidyl inositol-3 kinase (PI3 kinase) and growth factor receptor bound protein 2 (Grb 2). Subsequently, IGF-1 is known to initiate at least two signaling cascades; the first involves activation of PI3 kinase, subsequent formation of phosphatidyl inositol-3 phosphate (PIP3), and the activation of Akt pathways which are involved in survival and inhibition of apoptosis. The second pathway involves the extracellular signal-related kinases ERK-1 and ERK-2 (also termed MAP kinases). Once activated, MAPKs are translocated to the nucleus where they phosphorylate a variety of transcription factors including c-fos, c-jun, and c-myc. This signaling cascade is important for cellular growth and mitogenesis (De Meyts, 1994).

B. IGF-Binding Proteins

In the circulation and throughout the extracellular space of all tissues, the IGFs are present almost entirely bound to members of a family of high affinity IGF-binding proteins (IGFBPs). In the human circulation, a further component includes a 150-kDa protein known as the acid-labile-subunit (ALS), which forms a complex with IGF-1 and IGFBP-3 (Baxter *et al.*, 1989). The concentration of this storage form of IGF-1 is under GH control (Blum *et al.*, 1993). The six known IGFBPs possess an 80% sequence homology and share conserved N- and C-terminal regions that are cystein rich. The IGFBPs have been proposed to have four major functions that are essential to coordinate and regulate the biological activities of the IGFs. These are (1) to act as transport carrier proteins of IGFs; (2) to stabilize and prolong the half-lives of IGFs, thereby regulating their metabolic clearance; (3) to provide a means of tissue and cell type-specific localization; and (4) to directly stimulate or inhibit interactions of the IGFs with their receptors (Jones and Clemmons, 1995). Several IGFBP-specific proteases have been identified, which cleave the peptides into fragments and reduce their affinity for the IGFs (Maile and Holly, 1999). Hence, the biological functions of the IGFBPs in regulating IGFs are strongly influenced by proteases which can induce tissue-specific IGF release.

Although the IGFBPs share common properties in regulating IGFs, they are individual proteins with unique properties and functions. IGFBPs possess individual structural characteristics such as phosphorylation, glycosylation, and cleavage sites which enable specific functional properties of these proteins. Because of these distinctions, several of these proteins have been shown to demonstrate IGF-independent actions (Mohan and Baylink, 2002). Through interactions with specific cell surface structures, such as integrins and glycosaminoglycans, IGFBPs have been shown to regulate cell cycle, apoptosis, and cell migration. IGFBP-1, for example, has been reported to bind to the $\alpha5\beta1$ integrin receptor and influences cell motility and adhesion (Jones *et al.*, 1993). Independent actions of IGFBP-2 include mitogenic effects on epithelial cells (Badinga *et al.*, 1999), whereas IGFBP-3 has been demonstrated to enhance as well as inhibit cell growth (Cohen *et al.*, 2000; Valentinis *et al.*, 1995).

IV. Role of IGF-1 in the CNS

IGF-1 is one of the major growth factors in the circulation and exerts a plethora of effects on a multitude of tissues. This growth factor demonstrates a wide variety of actions during CNS development and maintenance of the adult CNS. Effects of IGFs are dependent on cell type, developmental stage, and microenvironmental milieu, exerting its actions in concert with other growth factors. The following sections focus on IGF-1 effects on neurons and glial cells.

A. IGFs as Neurotrophic Factors

The critical period of cell proliferation in the brain is during fetal and neonatal life. During nervous system development, IGF-1 plays a crucial role in processes such as cell proliferation, differentiation, and survival (Andersson *et al.*, 1988; Anlar *et al.*, 1999; Bondy and Cheng, 2004; Russo *et al.*, 2005). In this period, there is high expression of IGF-1 receptors, suggesting an important role for IGFs in developmental processes (Anlar *et al.*, 1999; Underwood and D'Ercole, 1984). In the developing brain, local production of IGF-1 is principally found in neuronal rich regions such as in the spinal cord, midbrain, cerebral cortex, hippocampus, and olfactory bulb (Beck *et al.*, 1988; Garcia-Segura *et al.*, 1991). The necessity of this factor in CNS development is shown in transgenic animal models. Mice carrying null mutations of the genes encoding IGF-1 displayed reduced brain size, hypomyelination, reduced density of oligodendrocytes, loss of neuron populations, as well as reduced glucose uptake (Beck *et al.*, 1995; Cheng *et al.*, 2000; Liu *et al.*, 1993). By comparison, transgenic mice overexpressing IGF-1 had 55% larger brains with increased cell number and size as compared to controls (Carson *et al.*, 1993). Myelin content in these animals was increased by 130%. In accordance with this, transgenic mice overexpressing the IGF inhibitory IGFBP-1, the percentage of myelinated axons, and the thickness of myelin sheaths were reduced, a feature that was accompanied by reduction of myelin protein expression (Ye *et al.*, 1995).

In addition to its developmental roles, IGFs are of equal importance in adult life where they exert neurotrophic and neuroprotective effects (Trejo *et al.*, 2004). They promote the survival and differentiation of neural cells, including sensory, sympathetic, and motor neurons (Oorschot and McLennan, 1998; Sendtner, 1995). In ageing humans, there is evidence for a strong correlation between reduced levels of circulating IGF-1 and deterioration of cognitive functions (van Dam and Aleman, 2004).

In normal CNS, IGF-2 receptors are found in a variety of neuronal regions, including choroid plexus, granular layer of the cerebellar cortex, dentate gyrus and pyramidal layer of the hippocampus, striatum, and cerebral cortex, whereas white matter is devoid of IGF-2 receptors (Wilczak *et al.*, 2000). Regional distribution of this receptor suggests involvement in functions associated with specific neuronal pathways. This pattern of IGF-2 receptor expression is also interesting because IGF-2 protein is mostly associated with white matter (Logan *et al.*, 1994). This may be due to the fact that the IGF-2 receptor is not present in white matter and therefore ligand levels remain high because they are not rapidly internalized and degraded. In this respect, lower levels of IGF-2 protein found in the gray matter may be due to rapid internalization by IGF-2 receptors.

Many reports have demonstrated mitogenic and protective effects of IGF-1 on neurons *in vitro*. IGFs enhance growth in a variety of neuronal cell types, including sympathetic and dorsal root ganglia and primary and cloned neuronal cell lines (Lauterio, 1992; Meghani *et al.*, 1993). The beneficial effects of IGF-1 on motor and sensory neurons include growth cone motility and long-term neurite outgrowth (Feldman *et al.*, 1997; Kim *et al.*, 1997). Data suggest that IGF-1 enhanced growth cone motility is mediated by promotion of reorganization of actin and activation of focal adhesion proteins via the PI3K pathway, whereas long-term neurite outgrowth is mediated by MAPK activation. As neurogenic factors, IGFs have been shown to induce the differentiation of adult stem cells (Brooker *et al.*, 2000). Differentiation of neurons is in part morphologically defined by axonal and dendritic extensions, which are essential for proper synapse formation and neuronal signaling. IGF-1 is known to initiate neurite outgrowth in many neuronal subtypes of the nervous system, including neurons of the cortex, retina, motor neurons, and sensory neurons (Aizenman and de Vellis, 1987; Politi *et al.*, 2001; Zackenfels *et al.*, 1995).

IGFs protect neurons from various forms of damage. Calcium-mediated hypoglycemic damage of hippocampal and septal neurons has been shown to be rescued by IGF-1 treatment (Cheng and Mattson, 1992). Enhanced movement of IGF-1 from the CSF to damaged regions in the cerebrum occurs following hypoxic-ischemic injury and has been shown to be associated with the rescue of neurons (Gluckman *et al.*, 1992; Guan *et al.*, 1993, 1996a). These beneficial effects have been proposed to be IGFBP dependent, in a manner in which IGFBPs may facilitate IGF-1 targeting to the neuron (Guan *et al.*, 1996b). In summary, IGF-1 boasts a diverse array of effects on neuronal cells; principally differentiation, growth, and cell survival are strongly influenced by this factor.

B. IGF Actions on Glial Cells

In addition to the trophic support of neuronal cells, IGF-1 is a strong inducer of proliferation, differentiation, and survival of glial cells. Although predominantly expressed in neurons, glial cells have also been shown to be an important source of IGF-1 (Chernausek, 1993; Shinar and McMorris, 1995; Ye *et al.*, 2004). *In vitro*, the effects of IGF-1 on oligodendrocytes range from enhancing survival by reducing apoptosis to supporting cell differentiation which is accompanied by enhanced process branching, increased production of myelin proteins, and development of myelin membrane sheaths (Goddard *et al.*, 1999; Kuhl *et al.*, 2002; McMorris and Dubois-Dalcq, 1988; Mozell and McMorris, 1991; Shinar and McMorris, 1995). The well-documented effects of IGF-1 on oligodendrocytes are not only restricted to these cells but also apply to Schwann cells, which demonstrate enhanced differentiation, myelin production, and increased survival in response to IGF-1 stimulation (Cheng *et al.*, 1999; Ogata *et al.*, 2004; Syroid *et al.*, 1999). In this respect, IGF-1 also promotes actions which are involved in peripheral nervous system myelination. The importance of IGF signaling in myelin production has been demonstrated in the cuprizone model of demyelination. Treatment of mice with cuprizone leads to demyelination in the corpus callosum and superior cerebellar peduncles. When treatment is terminated, remyelination ensues. Cuprizone-induced demyelination in mice deficient of the IGF-1 receptor has demonstrated inadequate remyelination and lack of OPC accumulation in the site of injury; furthermore, OPCs did not proliferate or survive as well as OPCs in the wild-type control mice (Mason *et al.*, 2003).

Astrocytes are known to express the IGF-1 receptor and treatment of these cells with IGF-1 enhances DNA synthesis and cell proliferation *in vitro*. The rapid activation of PKC, MAP kinases, and Akt pathways have been shown to be involved in the mitogenic and survival effects of IGF-1 on astrocytes (Han *et al.*, 1987; Tranque *et al.*, 1992). Transgenic mice with astrocyte-specific expression of IGF-1 exhibited an increase in the concentration of glial fibrillary acidic protein, whereas transgenic mice with neuron-specific IGF-1 expression demonstrated marked increases in concentrations of oligodendrocyte and myelin-specific proteins (Ye *et al.*, 2004). For these reasons, astrocyte-derived IGF-1 is believed to have strong paracrine as well as autocrine effects *in vivo*. In this context, astrocyte-derived IGFBP-2 proteases have been suggested to degrade astrocyte-derived IGFBP-2 in the vicinity of these cells, releasing free IGF-1 from a complex with IGFBP-2 (Chesik *et al.*, 2004b). This may serve as a means of enhancing IGF signaling in these cells in an autocrine manner. Indeed, studies from our laboratory indicate a mechanism of targeting of IGF-1 to astrocytes via IGFBP-2, since complexes of IGF-1 with IGFBP-2 inhibited IGF-1 actions on oligodendrocytes, but did not inhibit IGF-1 signaling on astrocytes (Chesik *et al.*, 2006). The upregulation of IGFBP-2 in reactive astrocytes, as observed in MS, experimental autoimmune encephalomyelitis (EAE) and penetrating injury of the CNS, may provide a mechanism of enhancing IGF-1 signaling in these cells, providing the necessary cues contributing to the induction or maintenance of astrogliosis, a phenomenon characterized by proliferation, hypertrophy, and cell migration.

Like astrocytes, microglia express the IGF-1 receptor and are a source of IGF-1 and IGF-2 in the CNS (Kaur *et al.*, 2006). These immunocompetent cells of the CNS are activated in response to inflammatory stimuli, accompanied by morphological changes from ramified process-bearing to rounded ameboid macrophage-like morphology. Among the factors produced by T cells, as key players in the adaptive arm of the immune system, is IFN-γ, a proinflammatory cytokine and activator of microglia. When present in moderate amounts, IFN-γ has a beneficial effect on neural tissues (Shaked *et al.*, 2005; Wildbaum *et al.*, 2003). Thus, it is believed that well-controlled microglial activity is supportive of CNS tissue maintenance and repair. Survival of neurons can be supported by microglia that are stimulated by IFN-γ in well-controlled amounts; however, uncontrolled amounts of IFN-γ can interfere with neural survival (Butovsky *et al.*, 2006). IGF-1 has been shown to act as an antagonist of IFN-γ-induced microglial activation of the hippocampus (Maher *et al.*, 2006). Microglial activation is also accompanied by increases in interleukin (IL)-1β concentrations, which is thought to be involved in the deficit in long-term potentiation. This effect has been shown to be reduced by IGF-1 treatment (Maher *et al.*, 2006).

Neurogenesis and oligodendrogenesis of adult neural progenitor cells are blocked by microglial production of tumor necrosis factor (TNF)-α under inflammatory conditions. However, stimulation of microglia with IL-4 induces IGF-1 expression in these cells, a mechanism which has been correlated with the induction of oligodendrogenesis (Butovsky *et al.*, 2006).

V. Clinical Uses of IGF-1 in Models of Neurodegenerative Demyelinating Disease

The IGF-1 receptor is expressed in most tissues granting IGF-1, a broad range of effects throughout human physiology. This wide range of expression indicates the potential therapeutic use of IGF-1 for the treatment of several disorders. IGF-1 has already been successfully used in treating growth deficiency, osteoporosis, kidney disorders, catabolic disorders, and diabetes, as well as topically to accelerate wound healing (Bondy *et al.*, 1994; Savage *et al.*, 2001). Importantly, IGF-1 is thought to cause limited side effects, making its use for long-term clinical treatment safe even when used at high doses (Bondy *et al.*, 1994). Our present understanding of the physiological importance of IGF-1 in sustaining normal CNS and PNS functions supports its potential therapeutic use in treatment of neurological disorders (Lewis *et al.*, 1993; Yuen and Mobley, 1996). Although functional deficits of neurological diseases vary, most are pathologically marked by neurodegeneration, a feature that can potentially be impeded by IGF-1 of both sensory and motor neurons of the PNS. For a profile of therapeutic potential of IGF-1 in neurodegenerative diseases, such as ALS, Alzheimer's disease, and stroke, the reader is referred to other reports (Carro *et al.*, 2002; De Keyser *et al.*, 1994; Dore *et al.*, 1997; Gasparini and Xu, 2003; Lackey *et al.*, 2000). In the following section, we will focus on the role of IGF-1 in animal models of MS.

A. Animal Models of Demyelination

Therapeutic application of IGF-1 in experimental allergic encephalomyelitis (EAE), the most studied animal model used to define mechanisms underlying immune-mediated CNS demyelinating diseases (Lassmann, 1983; Raine, 1984) has demonstrated its potential value in treating inflammatory disorders of the CNS. Treatment with an emulation of guinea pig spinal cord induces EAE in Lewis rats. In this model, it has been demonstrated that reactive astrocytes upregulate the expression of IGF-1 and IGFBP-2, and that lesions in which remyelination is observed promyelinating oligodendrocytes expressed the IGF-1 receptor (Liu *et al.*, 1994). Similarly, in cuprizone-induced experimental demyelination, IGF-1 expression was upregulated in reactive astrocytes (Komoly *et al.*, 1992). In this study, immature oligodendroglia in areas of demyelination exhibited increased expression of the IGF-1 receptor during early recovery. It has been suggested that astrocytic IGF-1 upregulation might function to regulate myelin metabolism during recovery. In contrast to these observations, reactive astrocytes in experimental autoimmune neuritis (EAN) did not respond with increased expression of IGF-1. Investigators suggested that type and severity of neuronal injury are important regulators which might account for discrepancies between astrocytic responses in EAE as compared to EAN (Gehrmann *et al.*, 1995).

Several studies have shown that injection of IGF-1 into Lewis rats with EAE significantly reduced the number and area of the demyelinating lesions in the spinal cord (Liu *et al.*, 1995; Yao *et al.*, 1995). Accompanying this observation, IGF-1 treatment increased the number of axons containing regenerating myelin segments, reduced permeability of the blood–spinal cord barrier, and enhanced expression of myelin genes. These beneficial effects resulted in rapid clinical and pathological recovery in the treated animals. Similarly, IGF-1 treatment of acute nondemyelinating EAE, induced by passive transfer of myelin basic protein (MBP)-reactive T lymphocytes, as well as in chronic relapsing EAE, decreased clinical deficits, lesion numbers, and areas and reduced the number of CD4-positive T cells and ED-1-positive macrophages (Li *et al.*, 1998; Liu *et al.*, 1997).

Interestingly, in chronic-relapsing murine EAE models, administration of complexes of IGF-1 with IGFBP-3 before disease begin delayed the onset of disease; however, treatment after onset of disease resulted in increased severity of relapses (Lovett-Racke *et al.*, 1998). Investigators suggested that delayed disease onset might be due to downregulation of ICAM-1 in the CNS, whereas increased disease severity following treatment after disease onset might be due to expansion of encephalitogenic T cells. In line with these results, a study has revealed no enhancing effect of IGF-1 treatment on remyelination in EAE (Cannella *et al.*, 2000). Results from this study revealed transient clinical amelioration and low-level remyelination after IGF-1 administration during the acute phase of EAE. However, when tissues from acute phase-treated animals were sampled at chronic time points, no beneficial effects were observed. Furthermore, chronic phase-treated animals showed no beneficial effects of IGF-1 administration. This finding raises doubts on the true effectiveness of systemic IGF-1 in chronic forms of demyelinating diseases such as MS.

VI. IGF System in MS

In order to exploit the IGF system to clinical advantage, an in-depth account of the IGF system in the pathology of MS is crucial. This includes a complete characterization of molecules affecting IGF-1 actions in the CNS as well as which cell types are expressing these molecules and on which cell types IGF-1-mediated effects will ultimately occur. The following section provides an overview of our current understanding of IGF system in the pathology of MS.

A. Serum and Cerebrospinal Fluid

Examination of serum levels of circulating IGF-1, IGF-2, IGFBP-1, IGFBP-2, IGFBP-3, and IGFBP-4 in MS patients has demonstrated that none of these components are regulated, whereas insulin levels are reduced in the circulation of MS patients (Torres-Aleman *et al.*, 1998; Wilczak *et al.*, 1998). Similarly, levels of IGF-1, IGF-2, IGFBP-1, IGFBP-2, and IGFBP-3 in CSF were not different between MS patients and controls. However, GH, a regulator of IGF-1 expression, was demonstrated to be reduced in MS CSF (Poljakovic *et al.*, 2006). Only one report has indicated an alteration in the IGF-axis in serum of MS patients showing a significant correlation between circulating IGFBP-3 levels and both the progression index of disability and the MS Severity Score in patients with primary progressive MS (Wilczak *et al.*, 2005). Patients with increased MS Severity Scores were found to have elevated levels of IGFBP-3. It was suggested that IGFBP-3, as an inhibitor of IGF-1 actions, might be a factor influencing neurodegeneration and the rate of progressive disability in primary progressive MS.

B. Central Nervous System

Postmortem investigations of CNS from MS patients have demonstrated that IGF-1 receptor densities and binding characteristics in normal appearing white matter were not different between MS and control brain (Wilczak and De Keyser, 1997). Similarly, chronic plaques contained IGF-1 receptor densities in the same range as those measured in normal appearing white matter. In contrast, IGF-2 receptors were undetectable in plaques of MS patients, suggesting that this receptor is not involved in processes of astrogliosis (Wilczak *et al.*, 2000). In normal CNS, IGF-2 receptors were detected in a variety of neuronal regions, including plexus choroideus, granular layer of the cerebellar cortex, gyrus dentatus, and pyramidal layer of the hippocampus, striatum and cerebral cortex, whereas white matter was demonstrated to be devoid of IGF-2 receptors (Wilczak *et al.*, 2000).

Although there appears to be no regulation in levels of IGFs or IGFBPs in the circulation of MS patients, cellular expression of these factors within the CNS do show convergent patterns of regulation associated with this pathology. In active MS lesions, IGF-1 and IGF-2 were detected in hypertrophic astrocytes and foamy microglia, respectively (Gveric *et al.*, 1999). In control human adult CNS, immunoreactivity for IGFBP-1–6 was demonstrated in astrocytes (Chesik *et al.*, 2006). Assessment of reactive astrocytes in chronic active MS lesions showed increased levels of IGFBP-2 and IGFBP-4 in these cells. By comparison, control microglia were not shown to express any of the IGFBPs; however, in chronic active MS lesions reactive microglia expressed IGFBP-2 only (Chesik *et al.*, 2004a). Similar to astrocytes, the examination of oligodendrocytes has revealed the expression of all six IGFBPs in control CNS. Interestingly, at the rim of chronic active MS plaques, mature oligodendrocytes demonstrated an upregulation of IGFBP-1 and IGFBP-6 (D. C., N. W., and J. D. K., unpublished results). Contrary to these investigations, one study reported the expression of IGFBP-1 in MS lesions, which was associated with reactive astrocytes and oligodendrocytes surrounding active plaques (IGFBP-4–6 were not studied, Gveric *et al.*, 1999). However, the absence of IGFBP-2 and IGFBP-3 in astrocytes in MS lesions, as reported in that study, may be due to proteolytic degradation of IGFBPs. In the study from Gveric *et al.*, the postmortem time was relatively long (36 h), whereas for material for the above-mentioned studies the average death to snap-freezing time was 8.5 h. Proteolytic degradation plays an important role in IGFBP regulation (Maile and Holly, 1999) and proteases degrading IGFBP-2 have been found to be expressed in astrocytes (Chesik *et al.*, 2004b). On the other hand, Gveric and coworkers examined active lesions as well as subacute/chronic lesions and demonstrated a correlation between lesion activity and expression of IGF-1, IGF-2, and IGFBPs. Differences in cellular expression patterns may, therefore, be dependent on environmental composition of particular lesion types.

C. Role of IGFBPs in MS

Interestingly, the above-mentioned investigations of glial cell-associated IGFBPs have demonstrated only increases in IGFBP levels in MS lesions, as compared to basal expression levels in control CNS. Since IGFBPs are known to be upregulated in response to increased levels of IGF-1 or IGF-2 (Conover, 1990; Conover *et al.*, 1995), upregulation as observed in MS may be consequence of increased levels of ligand, which is enhanced in chronic active MS plaques (Gveric *et al.*, 1999; D. C., N. W., and J. D. K., unpublished results). However, other factors, in particular cytokines such as IL-1β, IL-6, TNF-α, IFN-γ, are known to regulate IGFBP levels (Fernandez-Celemin and Thissen, 2001; Katz *et al.*, 1995; Liu *et al.*, 1999; Price *et al.*, 2002). TNF-α and IL-1 contribute to neurodegeneration and may further influence cell survival by inhibiting the protective effects of growth factors, such as IGFs, by increasing IGFBP levels. Regardless of stimulant, the pattern of IGFBP expression in MS lesions suggests roles for individual IGFBPs in coordinating cell-specific actions of IGFs.

Unfortunately, knowledge about the precise functions of IGFBPs in regulating IGFs remains vague and investigations are only beginning to resolve this issue. Studies on glial cells have indicated both inhibitory as well as stimulatory roles for IGFBPs in regulating IGF-1 actions. *In vitro* treatment of astrocytes with complexes of IGF-1 and IGFBP-2 demonstrated no inhibiting effect on IGF-1-induced proliferation, whereas the same complexes did significantly inhibit IGF-1-enhanced survival of oligodendrocytes (Chesik *et al.*, 2006). IGFBP-2 is the major IGFBP in the CNS and CSF and is upregulated in astrocytes in EAE, CNS injury models, and MS. Although previous reports suggest that IGFBP-2 might target IGF-1 to oligodendrocytes and facilitate IGF-1 effects on remyelination (Liu *et al.*, 1994), *in vitro* studies indicate that the upregulation of IGFBP-2 in MS might target IGF-1 to astrocytes, a mechanism that could enhance the process of astrogliosis. Supporting this notion, in an *in vitro* model of astrogliosis highly proliferative astrocytes expressed higher levels of IGFBP-2 as compared to more differentiated cells (Chesik *et al.*, 2004b). Furthermore, high levels of IGFBP-2 in these "reactive" astrocytes were accompanied by the expression of an IGFBP-2 protease. We conjectured that high levels of IGFBP-2 could sequester IGF-1, whereas proteases could release IGF-1 in the vicinity of the cell, a mechanism may serve to target IGF-1 to the astrocyte and enhance reactivity. On the other hand, higher IGFBP levels, as shown in MS lesions, may lower free levels of IGFs in general, restricting protective effects on neurons in the injured brain.

Although IGF-1 stimulates differentiation and myelin production in oligodendrocytes, other factors, for example FGF and PDGF, stimulate proliferation of OPCs. Untimely triggering of differentiation could have detrimental effects in MS lesions since IGF-1 might prevent further proliferation of OPCs, thereby limiting the number of these progenitor cells in the plaque vicinity. This in mind, one of the biological roles of IGFBPs in MS may be to regulate the timing of oligodendrocyte differentiation by inhibiting IGF signaling and allowing OPC to proliferate for a number of divisions, while preventing premature differentiation induced by IGF-1. In this sense, IGFBPs themselves indirectly serve as growth factors by inhibiting IGF-induced differentiation in a phase when OPCs should proliferate.

In MS, increased levels of IGFBPs may also serve to regulate IGF-1-induced release of cytokines. Reactive astrocytes and microglia/macrophages have been shown to release several cytokines, such as TNF-α, TGF-β, IL-10, and IL-1β (Eddleston and Mucke, 1993; Giulian *et al.*, 1994; Streit *et al.*, 1999). TNF-α has been implicated in the mechanism of several demyelinating disorders, including MS. In rat spinal cord culture, TNF-α selectively damages oligodendrocytes and myelin, resulting in swelling of myelin sheets and oligodendrocyte death

(Selmaj and Raine, 1988). Macrophages respond to IGF-1 with an increased production of the inflammatory cytokine TNF-α (Renier *et al.*, 1996). In this respect, production of IGFBPs in MS lesions may serve to protect cells from cytokine release mediated by IGFs.

Also deserving consideration are the well-known IGF-independent effects of IGFBPs (Mohan and Baylink, 2002). Through their individual properties and capacity to bind to extracellular structures, such as proteoglycans, integrins, and putative receptors, IGFBPs are known to exert a variety of effects that are independent of IGF-1 or IGF-2 presence. Such effects include stimulation of migration, enhancing as well as inhibiting cell proliferation, and inducing apoptosis (Jones *et al.*, 1993; Oh *et al.*, 1993; Perks *et al.*, 1999). It is uncertain whether IGF-independent effects of IGFBPs play a role in cellular processes involved in MS.

VII. Is IGF-1 a Good Candidate for MS Treatment?

The first clinical trail with IGF-1 was performed on seven MS patients who received recombinant human IGF-1 for 6 months (Frank *et al.*, 2002). No significant difference between baseline and treatment periods for any MRI or clinical measures of disease activity could be detected. It was reported, however, that IGF-1 use was well-tolerated and had no adverse side reactions. Authors of that study suggested a lack of passage into the brain parenchyma due to BBB restrictions as well as the short treatment period of 24 weeks may explain the negative results.

Growth factor therapy has many potential pitfalls due to the short half-life of these factors, delivery problems, and targeting to designated tissues and cells. Because IGF signaling is such a complex interacting system, it is simplistic to expect that artificial alteration of the levels of IGF-1 or IGF analogues will yield predictable outcomes. Assuming a therapeutic design achieves its objective of proper delivery by evading obstructions, such as those imposed by the BBB, and assuming that the same design achieves targeting to OPCs; it is not assured that the desired intent, namely remyelination of denuded axons, will be attained. As mentioned above, induction of IGF-1 signaling in OPCs might stimulate untimely cell differentiation in a phase in which cell proliferation is necessary to repopulate lesions with myelin producing cells.

One of the most limiting obstacles preventing beneficial application of IGF-1 in man appears to be the poor penetration of the CNS due to restrictions of the BBB that prevents free passage of IGF-1 to the damaged site. Although IGF-1 is thought to enter the CNS by a saturable transport system at the BBB (Pan and Kastin, 2000; Pardridge, 1993; Sonntag *et al.*, 2005), it remains questionable whether systemic treatment with IGF-1 would indeed raise ligand levels in the parenchyma of the CNS.

Another major issue is the targeting of IGF-1 to the oligodendrocyte in order to promote remyelination. This is limited, for example, by high expression of type-1 IGF receptors on neurons and other glial cells (Anlar *et al.*, 1999) as well as levels of IGFBPs which regulate IGF actions and possibly target IGF-1 to other cell types. In this context, it is worthy to note that IGF-1 levels have a regulatory effect on IGFBP expression. It has been shown that an increase in IGF-1 levels can lead to an increase of IGFBP levels, thereby maintaining levels of free, biologically available IGF-1 (Conover, 1990). Treatment with IGF-1 in MS, therefore, is likely to further affect IGFBP expression which would regulate basal ligand levels, a mechanism that would block therapeutic efforts of rising IGF-1 levels in the CNS.

Further doubts on the effectiveness of growth factors as a therapy have arisen due to conflicting hypothesis of the role of astrogliosis in MS. Whether, the induction of astrogliosis in MS is a response that facilitates or hinders the repair process remains a central question. It is, therefore, difficult to state whether therapies with growth factor application may cause further damage by promoting astrogliosis and thereby aggravating the situation in an MS lesion. However, treatment of EAE with IGF-1 demonstrated positive clinical outcome. Administration of IGF-1 in EAE was not targeted to particular tissues or cell types suggesting that any effects of this growth factor on astrocytes did not impede oligodendrocyte production of myelin. In a study in which astrocytes were transplanted into spinal cord after ethidium bromide demyelination, oligodendrocyte recruitment and remyelination were enhanced, suggesting a beneficial role of astrocytes in remyelination processes (Franklin *et al.*, 1991). The presence of astrocytes in MS lesions may at some crucial point provide support for remyelination processes, yet at later stages of lesion development the glial scar impairs repair processes.

Therapeutic interventions with IGF-1 must therefore find a balance in the dual effects of astrogliosis by alleviating glial scarring while not eliminating supportive functions of astrocytes. Furthermore, an attempt to restrict astrocyte proliferation should be careful not to restrict proliferation of other cells involved in the repair processes, for example OPCs. It is questionable whether a single factor like IGF-1 could adequately overcome barriers in order to sufficiently induce beneficial effects on tissue repair, thereby inducing improved clinical outcome.

VIII. Summary

The rationale for inadequate endogenous remyelination in MS is poorly understood. A sensible therapeutic approach would be to exploit intrinsic control systems, which regulate myelination processes in order to restore myelin that is lost in disease. Since IGF-1 is an essential factor involved in proper myelinogenesis and exerts neurotrophic effects, this molecule holds potential for applications aimed at repairing tissue in MS. In addition, its neuroprotective properties may be useful in protecting axons and slowing down the neurodegenerative process that is mainly responsible for the cumulative irreversible disability. Although the characterization of the IGF-axis in MS is beginning to be elucidated, we do not fully understand the many functions of IGFs in the CNS and their roles in the pathology of MS. Cell type-specific expression and distribution of the IGF-axis during lesion development suggest complex regulation of IGF actions. Although potentially promising, strategies for delivery and targeting of IGF signaling to particular cell types or damaged tissue remain to be established.

References

Aberg, N. D., Brywe, K. G., and Isgaard, J. (2006). Aspects of growth hormone and insulin-like growth factor-I related to neuroprotection, regeneration, and functional plasticity in the adult brain. *Sci. World J.* **6,** 53–80.

Aizenman, Y., and d de Vellis, J. (1987). Brain neurons develop in a serum and glial free environment: Effects of transferrin, insulin, insulin-like growth factor-I and thyroid hormone on neuronal survival, growth and differentiation. *Brain Res.* **406,** 32–42.

Andersson, I. K., Edwall, D., Norstedt, G., Rozell, B., Skottner, A., and Hansson, H. A. (1988). Differing expression of insulin-like growth factor I in the developing and in the adult rat cerebellum. *Acta Physiol. Scand.* **132,** 167–173.

Anlar, B., Sullivan, K. A., and Feldman, E. L. (1999). Insulin-like growth factor-I and central nervous system development. *Horm. Metab. Res.* **31,** 120–125.

Armstrong, R., Friedrich, V. L., Jr., Holmes, K. V., and Dubois-Dalcq, M. (1990). *In vitro* analysis of the oligodendrocyte lineage in mice during demyelination and remyelination. *J. Cell Biol.* **111,** 1183–1195.

Badinga, L., Song, S., Simmen, R. C., Clarke, J. B., Clemmons, D. R., and Simmen, F. A. (1999). Complex mediation of uterine endometrial epithelial cell growth by insulin-like growth factor-II (IGF-II) and IGF-binding protein-2. *J. Mol. Endocrinol.* **23,** 277–285.

Baron, W., Decker, L., Colognato, H., and ffrench-Constant, C. (2003). Regulation of integrin growth factor interactions in oligodendrocytes by lipid raft microdomains. *Curr. Biol.* **13,** 151–155.

Barres, B. A., Schmid, R., Sendnter, M., and Raff, M. C. (1993). Multiple extracellular signals are required for long-term oligodendrocyte survival. *Development* **118,** 283–295.

Barres, B. A., Raff, M. C., Gaese, F., Bartke, I., Dechant, G., and Barde, Y. A. (1994). A crucial role for neurotrophin-3 in oligodendrocyte development. *Nature* **367,** 371–375.

Baxter, R. C., Martin, J. L., and Beniac, V. A. (1989). High molecular weight insulin-like growth factor binding protein complex. Purification and properties of the acid-labile subunit from human serum. *J. Biol. Chem.* **264,** 11843–11848.

Beck, F., Samani, N. J., Byrne, S., Morgan, K., Gebhard, R., and Brammar, W. J. (1988). Histochemical localization of IGF-I and IGF-II mRNA in the rat between birth and adulthood. *Development* **104,** 29–39.

Beck, K. D., Powell-Braxton, L., Widmer, H. R., Valverde, J., and Hefti, F. (1995). Igf1 gene disruption results in reduced brain size, CNS hypomyelination, and loss of hippocampal granule and striatal parvalbumin-containing neurons. *Neuron* **14,** 717–730.

Bichell, D. P., Kikuchi, K., and Rotwein, P. (1992). Growth hormone rapidly activates insulin-like growth factor I gene transcription *in vivo*. *Mol. Endocrinol.* **6,** 1899–1908.

Blum, W. F., Albertsson-Wikland, K., Rosberg, S., and Ranke, M. B. (1993). Serum levels of insulin-like growth factor I (IGF-I) and IGF binding protein 3 reflect spontaneous growth hormone secretion. *J. Clin. Endocrinol. Metab.* **76,** 1610–1616.

Bogler, O., Wren, D., Barnett, S. C., Land, H., and Noble, M. (1990). Cooperation between two growth factors promotes extended self-renewal and inhibits differentiation of oligodendrocyte-type-2 astrocyte (O-2A) progenitor cells. *Proc. Natl. Acad. Sci. USA* **87,** 6368–6372.

Bondy, C. A., and Cheng, C. M. (2004). Signaling by insulin-like growth factor 1 in brain. *Eur. J. Pharmacol.* **490,** 25–31.

Bondy, C., Werner, H., Roberts, C. T., Jr., and LeRoith, D. (1992). Cellular pattern of type-I insulin-like growth factor receptor gene expression during maturation of the rat brain: Comparison with insulin-like growth factors I and II. *Neuroscience* **46,** 909–923.

Bondy, C. A., Underwood, L. E., Clemmons, D. R., Guler, H. P., Bach, M. A., and Skarulis, M. (1994). Clinical uses of insulin-like growth factor I. *Ann. Intern. Med.* **120,** 593–601.

Brooker, G. J., Kalloniatis, M., Russo, V. C., Murphy, M., Werther, G. A., and Bartlett, P. F. (2000). Endogenous IGF-1 regulates the neuronal differentiation of adult stem cells. *J. Neurosci. Res.* **59,** 332–341.

Butovsky, O., Landa, G., Kunis, G., Ziv, Y., Avidan, H., Greenberg, N., Schwartz, A., Smirnov, I., Pollack, A., Jung, S., and Schwartz, M. (2006). Induction and blockage of oligodendrogenesis by differently activated microglia in an animal model of multiple sclerosis. *J. Clin. Invest.* **116,** 905–915.

Cannella, B., Pitt, D., Capello, E., and Raine, C. S. (2000). Insulin-like growth factor-1 fails to enhance central nervous system myelin repair during autoimmune demyelination. *Am. J. Pathol.* **157,** 933–943.

Canoll, P. D., Musacchio, J. M., Hardy, R., Reynolds, R., Marchionni, M. A., and Salzer, J. L. (1996). GGF/neuregulin is a neuronal signal that promotes the proliferation and survival and inhibits the differentiation of oligodendrocyte progenitors. *Neuron* **17,** 229–243.

Carro, E., Trejo, J. L., Gomez-Isla, T., LeRoith, D., and Torres-Aleman, I. (2002). Serum insulin-like growth factor I regulates brain amyloid-beta levels. *Nat. Med.* **8,** 1390–1397.

Carson, M. J., Behringer, R. R., Brinster, R. L., and McMorris, F. A. (1993). Insulin-like growth factor I increases brain growth and central nervous system myelination in transgenic mice. *Neuron* **10,** 729–740.

Cheng, B., and Mattson, M. P. (1992). IGF-I and IGF-II protect cultured hippocampal and septal neurons against calcium-mediated hypoglycemic damage. *J. Neurosci.* **12,** 1558–1566.

Cheng, C. M., Reinhardt, R. R., Lee, W. H., Joncas, G., Patel, S. C., and Bondy, C. A. (2000). Insulin-like growth factor 1 regulates developing brain glucose metabolism. *Proc. Natl. Acad. Sci. USA* **97,** 10236–10241.

Cheng, H. L., Russell, J. W., and Feldman, E. L. (1999). IGF-I promotes peripheral nervous system myelination. *Ann. N. Y. Acad. Sci.* **883,** 124–130.

Chernausek, S. D. (1993). Insulin-like growth factor-I (IGF-I) production by astroglial cells: Regulation and importance for epidermal growth factor-induced cell replication. *J. Neurosci. Res.* **34,** 189–197.

Chesik, D., De Keyser, J., and Wilczak, N. (2004a). Involvement of insulin-like growth factor binding protein-2 in activated microglia as assessed in post mortem human brain. *Neurosci. Lett.* **362,** 14–16.

Chesik, D., Kuhl, N. M., Wilczak, N., and De Keyser, J. (2004b). Enhanced production and proteolytic degradation of insulin-like growth factor binding protein-2 in proliferating rat astrocytes. *J. Neurosci. Res.* **77,** 354–362.

Chesik, D., De Keyser, J., Glazenburg, L., and Wilczak, N. (2006). Insulin-like growth factor binding proteins: Regulation in chronic active plaques in multiple sclerosis and functional analysis on glial cells. *Eur. J. Neurosci.* **24,** 1645–1652.

Cohen, P., Rajah, R., Rosenbloom, J., and Herrick, D. J. (2000). IGFBP-3 mediates TGF-beta1-induced cell growth in human airway smooth muscle cells. *Am. J. Physiol. Lung Cell. Mol. Physiol.* **278,** L545–L551.

Cohen, R. I., Marmur, R., Norton, W. T., Mehler, M. F., and Kessler, J. A. (1996). Nerve growth factor and neurotrophin-3 differentially regulate the proliferation and survival of developing rat brain oligodendrocytes. *J. Neurosci.* **16,** 6433–6442.

Conover, C. A. (1990). Regulation of insulin-like growth factor (IGF)-binding protein synthesis by insulin and IGF-I in cultured bovine fibroblasts. *Endocrinology* **126,** 3139–3145.

Conover, C. A., Clarkson, J. T., and Bale, L. K. (1995). Effect of glucocorticoid on insulin-like growth factor (IGF) regulation of IGF-binding protein expression in fibroblasts. *Endocrinology* **136,** 1403–1410.

De Keyser, J., Wilczak, N., and Goossens, A. (1994). Insulin-like growth factor-I receptor densities in human frontal cortex and white matter during aging, in Alzheimer's disease, and in Huntington's disease. *Neurosci. Lett.* **172,** 93–96.

De Meyts, P. (1994). The structural basis of insulin and insulin-like growth factor-I receptor binding and negative co-operativity, and its relevance to mitogenic versus metabolic signalling. *Diabetologia* **37,** 135–148.

Dore, S., Kar, S., and Quirion, R. (1997). Rediscovering an old friend, IGF-I: Potential use in the treatment of neurodegenerative diseases. *Trends Neurosci.* **20,** 326–331.

Eddleston, M., and Mucke, L. (1993). Molecular profile of reactive astrocytes-implications for their role in neurologic disease. *Neuroscience* **54,** 15–36.

Feldman, E. L., Sullivan, K. A., Kim, B., and Russell, J. W. (1997). Insulin-like growth factors regulate neuronal differentiation and survival. *Neurobiol. Dis.* **4,** 201–214.

Fernandez-Celemin, L., and Thissen, J. P. (2001). Interleukin-6 stimulates hepatic insulin-like growth factor binding protein-4 messenger ribonucleic acid and protein. *Endocrinology* **142,** 241–248.

Flores, A. I., Mallon, B. S., Matsui, T., Ogawa, W., Rosenzweig, A., Okamoto, T., and Macklin, W. B. (2000). Akt-mediated survival of oligodendrocytes induced by neuregulins. *J. Neurosci.* **20,** 7622–7630.

Frank, J. A., Richert, N., Lewis, B., Bash, C., Howard, T., Civil, R., Stone, R., Eaton, J., McFarland, H., and Leist, T. (2002). A pilot study of recombinant insulin-like growth factor-1 in seven multiple sclerosis patients. *Mult. Scler.* **8,** 24–29.

Franklin, R. J., Crang, A. J., and Blakemore, W. F. (1991). Transplanted type-1 astrocytes facilitate repair of demyelinating lesions by host oligodendrocytes in adult rat spinal cord. *J. Neurocytol.* **20,** 420–430.

Fu, L., Matthews, P. M., De Stefano, N., Worsley, K. J., Narayanan, S., Francis, G. S., Antel, J. P., Wolfson, C., and Arnold, D. L. (1998). Imaging axonal damage of normal-appearing white matter in multiple sclerosis. *Brain* **121,** 103–113.

Garcia-Segura, L. M., Perez, J., Pons, S., Rejas, M. T., and Torres-Aleman, I. (1991). Localization of insulin-like growth factor I (IGF-I)-like immunoreactivity in the developing and adult rat brain. *Brain Res.* **560,** 167–174.

Gard, A. L., and Pfeiffer, S. E. (1993). Glial cell mitogens bFGF and PDGF differentially regulate development of O4+ GalC− oligodendrocyte progenitors. *Dev. Biol.* **159,** 618–630.

Gasparini, L., and Xu, H. (2003). Potential roles of insulin and IGF-1 in Alzheimer's disease. *Trends Neurosci.* **26,** 404–406.

Gehrmann, J., Yao, D. L., Bonetti, B., Brenner, M., Bondy, C., Wekerle, H., Kreutzberg, G. W., Webster, H., and de, F. (1995). Astrocytes upregulate glial fibrillary acidic protein (GFAP), but not insulin-like growth factor-I (IGF-I) during experimental autoimmune neuritis (EAN). *Brain Pathol.* **5,** 1–10.

Giulian, D., Li, J., Li, X., George, J., and Rutecki, P. A. (1994). The impact of microglia-derived cytokines upon gliosis in the CNS. *Dev. Neurosci.* **16,** 128–136.

Gluckman, P., Klempt, N., Guan, J., Mallard, C., Sirimanne, E., Dragunow, M., Klempt, M., Singh, K., Williams, C., and Nikolics, K. (1992). A role for IGF-1 in the rescue of CNS neurons following hypoxic-ischemic injury. *Biochem. Biophys. Res. Commun.* **182,** 593–599.

Goddard, D. R., Berry, M., and Butt, A. M. (1999). *In vivo* actions of fibroblast growth factor-2 and insulin-like growth factor-I on oligodendrocyte development and myelination in the central nervous system. *J. Neurosci. Res.* **57,** 74–85.

Godfraind, C., Friedrich, V. L., Holmes, K. V., and Dubois-Dalcq, M. (1989). *In vivo* analysis of glial cell phenotypes during a viral demyelinating disease in mice. *J. Cell. Biol.* **109,** 2405–2416.

Guan, J., Williams, C., Gunning, M., Mallard, C., and Gluckman, P. (1993). The effects of IGF-1 treatment after hypoxic-ischemic brain injury in adult rats. *J. Cereb. Blood Flow Metab.* **13,** 609–616.

Guan, J., Skinner, S. J., Beilharz, E. J., Hua, K. M., Hodgkinson, S., Gluckman, P. D., and Williams, C. E. (1996a). The movement of IGF-I into the brain parenchyma after hypoxic-ischaemic injury. *Neuroreport* **7,** 632–636.

Guan, J., Williams, C. E., Skinner, S. J., Mallard, E. C., and Gluckman, P. D. (1996b). The effects of insulin-like growth factor (IGF)-1, IGF-2, and des-IGF-1 on neuronal loss after hypoxic-ischemic brain injury in adult rats: Evidence for a role for IGF binding proteins. *Endocrinology* **137,** 893–898.

Gveric, D., Cuzner, M. L., and Newcombe, J. (1999). Insulin-like growth factors and binding proteins in multiple sclerosis plaques. *Neuropathol. Appl. Neurobiol.* **25,** 215–225.

Han, V. K., Lauder, J. M., and D'Ercole, A. J. (1987). Characterization of somatomedin/insulin-like growth factor receptors and correlation with biologic action in cultured neonatal rat astroglial cells. *J. Neurosci.* **7,** 501–511.

Jones, J. I., and Clemmons, D. R. (1995). Insulin-like growth factors and their binding proteins: Biological actions. *Endocr. Rev.* **16,** 3–34.

Jones, J. I., Gockerman, A., Busby, W. H., Jr., Wright, G., and Clemmons, D. R. (1993). Insulin-like growth factor binding protein 1 stimulates cell migration and binds to the alpha 5 beta 1 integrin by means of its Arg-Gly-Asp sequence. *Proc. Natl. Acad. Sci. USA* **90,** 10553–10557.

Juul, A. (2003). Serum levels of insulin-like growth factor I and its binding proteins in health and disease. *Growth Horm. IGF Res.* **13,** 113–170.

Katz, J., Weiss, H., Goldman, B., Kanety, H., Stannard, B., LeRoith, D., and Shemer, J. (1995). Cytokines and growth factors modulate cell growth and insulin-like growth factor binding protein secretion by the human salivary cell line (HSG). *J. Cell. Physiol.* **165,** 223–227.

Kaur, C., Sivakumar, V., Dheen, S. T., and Ling, E. A. (2006). Insulin-like growth factor I and II expression and modulation in amoeboid microglial cells by lipopolysaccharide and retinoic acid. *Neuroscience* **138,** 1233–1244.

Kerschensteiner, M., Bareyre, F. M., Buddeberg, B. S., Merkler, D., Stadelmann, C., Bruck, W., Misgeld, T., and Schwab, M. E. (2004). Remodeling of axonal connections contributes to recovery in an animal model of multiple sclerosis. *J. Exp. Med.* **200,** 1027–1038.

Kim, B., Leventhal, P. S., Saltiel, A. R., and Feldman, E. L. (1997). Insulin-like growth factor-I-mediated neurite outgrowth *in vitro* requires mitogen-activated protein kinase activation. *J. Biol. Chem.* **272,** 21268–21273.

Kim, B., van Golen, C. M., and Feldman, E. L. (2004). Insulin-like growth factor-I signaling in human neuroblastoma cells. *Oncogene* **23,** 130–141.

Komoly, S., Hudson, L. D., Webster, H. D., and Bondy, C. A. (1992). Insulin-like growth factor I gene expression is induced in astrocytes during experimental demyelination. *Proc. Natl. Acad. Sci. USA* **89,** 1894–1898.

Kooijman, R. (2006). Regulation of apoptosis by insulin-like growth factor (IGF)-I. *Cytokine Growth Factor Rev.* **17,** 305–323.

Kornfeld, S. (1992). Structure and function of the mannose 6-phosphate/insulin-like growth factor II receptors. *Annu. Rev. Biochem.* **61,** 307–330.

Kuhl, N. M., De Keyser, J., De Vries, H., and Hoekstra, D. (2002). Insulin-like growth factor binding proteins-1 and -2 differentially inhibit rat oligodendrocyte precursor cell survival and differentiation *in vitro*. *J. Neurosci. Res.* **69,** 207–216.

Lackey, B. R., Gray, S. L., and Henricks, D. M. (2000). Actions and interactions of the IGF system in Alzheimer's disease: Review and hypotheses. *Growth Horm. IGF Res.* **10,** 1–13.

Lassmann, H. (1983). Chronic relapsing experimental allergic encephalomyelitis: Its value as an experimental model for multiple sclerosis. *J. Neurol.* **229,** 207–220.

Lauterio, T. J. (1992). The effects of IGF-I and IGF-II on cell growth and differentiation in the central nervous system. *Adv. Exp. Med. Biol.* **321,** 31–36.

LeRoith, D. (2000). Insulin-like growth factor I receptor signaling—overlapping or redundant pathways? *Endocrinology* **141,** 1287–1288.

Lewis, M. E., Vaught, J. L., Neff, N. T., Grebow, P. E., Callison, K. V., Yu, E., Contreras, P. C., and Baldino, F., Jr. (1993). The potential of insulin-like growth factor-I as a therapeutic for the treatment of neuromuscular disorders. *Ann. N. Y. Acad. Sci.* **692,** 201–208.

Li, W., Quigley, L., Yao, D. L., Hudson, L. D., Brenner, M., Zhang, B. J., Brocke, S., McFarland, H. F., and Webster, H. D. (1998). Chronic relapsing experimental autoimmune encephalomyelitis: Effects of insulin-like growth factor-I treatment on clinical deficits, lesion severity, glial responses, and blood brain barrier defects. *J. Neuropathol. Exp. Neurol.* **57,** 426–438.

Liu, J. P., Baker, J., Perkins, A. S., Robertson, E. J., and Efstratiadis, A. (1993). Mice carrying null mutations of the genes encoding insulin-like growth factor I (Igf-1) and type 1 IGF receptor (Igf1r). *Cell* **75,** 59–72.

Liu, X., Yao, D. L., Bondy, C. A., Brenner, M., Hudson, L. D., Zhou, J., and Webster, H. D. (1994). Astrocytes express insulin-like growth factor-I (IGF-I) and its binding protein, IGFBP-2, during demyelination induced by experimental autoimmune encephalomyelitis. *Mol. Cell. Neurosci.* **5,** 418–430.

Liu, X., Yao, D. L., and Webster, H. (1995). Insulin-like growth factor I treatment reduces clinical deficits and lesion severity in acute demyelinating experimental autoimmune encephalomyelitis. *Mult. Scler.* **1,** 2–9.

Liu, X., Linnington, C., Webster, H. D., Lassmann, S., Yao, D. L., Hudson, L. D., Wekerle, H., and Kreutzberg, G. W. (1997). Insulin-like growth factor-I treatment reduces immune cell responses in acute non-demyelinative experimental autoimmune encephalomyelitis. *J. Neurosci. Res.* **47,** 531–538.

Liu, Y., Tsushima, T., Miyakawa, M., Isozaki, O., Yamada, H., Xu, Z. R., and Iwamoto, Y. (1999). Effect of cytokines on production of insulin-like growth factor binding proteins (IGFBPs) from human fibroblasts in culture. *Endocr. J.* **46,** S63–S66.

Logan, A., Gonzalez, A. M., Hill, D. J., Berry, M., Gregson, N. A., and Baird, A. (1994). Coordinated pattern of expression and localization of insulin-like growth factor-II (IGF-II) and IGF-binding protein-2 in the adult rat brain. *Endocrinology* **135,** 2255–2264.

Lovett-Racke, A. E., Bittner, P., Cross, A. H., Carlino, J. A., and Racke, M. K. (1998). Regulation of experimental autoimmune encephalomyelitis with insulin-like growth factor (IGF-1) and IGF-1/IGF-binding protein-3 complex (IGF-1/IGFBP3). *J. Clin. Invest.* **101,** 1797–1804.

Ludwin, S. K. (1979). An autoradiographic study of cellular proliferation in remyelination of the central nervous system. *Am. J. Pathol.* **95,** 683–696.

Maher, F. O., Clarke, R. M., Kelly, A., Nally, R. E., and Lynch, M. A. (2006). Interaction between interferon gamma and insulin-like growth factor-1 in hippocampus impacts on the ability of rats to sustain long-term potentiation. *J. Neurochem.* **96,** 1560–1571.

Maile, L. A., and Holly, J. M. (1999). Insulin-like growth factor binding protein (IGFBP) proteolysis: Occurrence, identification, role and regulation. *Growth Horm. IGF Res.* **9,** 85–95.

Mason, J. L., Xuan, S., Dragatsis, I., Efstratiadis, A., and Goldman, J. E. (2003). Insulin-like growth factor (IGF) signaling through type 1 IGF receptor plays an important role in remyelination. *J. Neurosci.* **23,** 7710–7718.

McKinnon, R. D., Matsui, T., Dubois-Dalcq, M., and Aaronson, S. A. (1990). FGF modulates the PDGF-driven pathway of oligodendrocyte development. *Neuron* **5,** 603–614.

McKinnon, R. D., Piras, G., Ida, J. A., Jr., and Dubois-Dalcq, M. A. (1993). Role for TGF-beta in oligodendrocyte differentiation. *J. Cell. Biol.* **121,** 1397–1407.

McMorris, F. A., and Dubois-Dalcq, M. (1988). Insulin-like growth factor I promotes cell proliferation and oligodendroglial commitment in rat glial progenitor cells developing *in vitro. J. Neurosci. Res.* **21,** 199–209.

McMorris, F. A., Smith, T. M., DeSalvo, S., and Furlanetto, R. W. (1986). Insulin-like growth factor I/somatomedin C: A potent inducer of oligodendrocyte development. *Proc. Natl. Acad. Sci. USA* **83,** 822–826.

Meghani, M. A., Martin, D. M., Singleton, J. R., and Feldman, E. L. (1993). Effects of serum and insulin-like growth factors on human neuroblastoma cell growth. *Regul. Pept.* **48,** 217–224.

Minagar, A., Gonzalez-Toledo, E., Pinkston, J., and Jaffe, S. L. (2005). Neuroimaging in multiple sclerosis. *Int. Rev. Neurobiol.* **67,** 165–201.

Minuto, F., Palermo, C., Arvigo, M., and Barreca, A. M. (2005). The IGF system and bone. *J. Endocrinol. Invest.* **28,** 8–10.

Mohan, S., and Baylink, D. J. (2002). IGF-binding proteins are multifunctional and act via IGF-dependent and -independent mechanisms. *J. Endocrinol.* **175,** 19–31.

Morgan, D. O., Edman, J. C., Standring, D. N., Fried, V. A., Smith, M. C., Roth, R. A., and Rutter, W. J. (1987). Insulin-like growth factor II receptor as a multifunctional binding protein. *Nature* **329,** 301–307.

Mozell, R. L., and McMorris, F. A. (1991). Insulin-like growth factor I stimulates oligodendrocyte development and myelination in rat brain aggregate cultures. *J. Neurosci. Res.* **30,** 382–390.

Noble, M., Murray, K., Stroobant, P., Waterfield, M. D., and Riddle, P. (1988). Platelet-derived growth factor promotes division and motility and inhibits premature differentiation of the oligodendrocyte/type-2 astrocyte progenitor cell. *Nature* **333,** 560–562.

Ogata, T., Iijima, S., Hoshikawa, S., Miura, T., Yamamoto, S., Oda, H., Nakamura, K., and Tanaka, S. (2004). Opposing extracellular signal-regulated kinase and Akt pathways control Schwann cell myelination. *J. Neurosci.* **24,** 6724–6732.

Oh, Y., Muller, H. L., Pham, H., and Rosenfeld, R. G. (1993). Demonstration of receptors for insulin-like growth factor binding protein-3 on Hs578T human breast cancer cells. *J. Biol. Chem.* **268,** 26045–26048.

Oorschot, D. E., and McLennan, I. S. (1998). The trophic requirements of mature motoneurons. *Brain Res.* **789,** 315–321.

Pan, W., and Kastin, A. J. (2000). Interactions of IGF-1 with the blood-brain barrier *in vivo* and *in situ. Neuroendocrinology* **72,** 171–178.

Pardridge, W. M. (1993). Transport of insulin-related peptides and glucose across the blood-brain barrier. *Ann. N. Y. Acad. Sci.* **692,** 126–137.

Perks, C. M., Newcomb, P. V., Norman, M. R., and Holly, J. M. (1999). Effect of insulin-like growth factor binding protein-1 on integrin signalling and the induction of apoptosis in human breast cancer cells. *J. Mol. Endocrinol.* **22,** 141–150.

Pfeiffer, S. E., Warrington, A. E., and Bansal, R. (1993). The oligodendrocyte and its many cellular processes. *Trends Cell Biol.* **3,** 191–197.

Politi, L. E., Rotstein, N. P., Salvador, G., Giusto, N. M., and Insua, M. F. (2001). Insulin-like growth factor-I is a potential trophic factor for amacrine cells. *J. Neurochem.* **76,** 1199–1211.

Poljakovic, Z., Zurak, N., Brinar, V., Korsic, M., Basic, S., and Hajnsek, S. (2006). Growth hormone and insulin growth factor-I levels in plasma and cerebrospinal fluid of patients with multiple sclerosis. *Clin. Neurol. Neurosurg.* **108,** 255–258.

Prager, D., and Melmed, S. (1993). Insulin and insulin-like growth factor I receptors: Are there functional distinctions? *Endocrinology* **132,** 1419–1420.

Price, W. A., Moats-Staats, B. M., and Stiles, A. D. (2002). Pro- and anti-inflammatory cytokines regulate insulin-like growth factor binding protein production by fetal rat lung fibroblasts. *Am. J. Respir. Cell. Mol. Biol.* **26,** 283–289.

Prineas, J. W., and Connell, F. (1979). Remyelination in multiple sclerosis. *Ann. Neurol.* **5,** 22–31.

Prineas, J. W., Kwon, E. E., Goldenberg, P. Z., Ilyas, A. A., Quarles, R. H., Benjamins, J. A., and Sprinkle, T. J. (1989). Multiple sclerosis. Oligodendrocyte proliferation and differentiation in fresh lesions. *Lab. Invest.* **61,** 489–503.

Prineas, J. W., Barnard, R. O., Kwon, E. E., Sharer, L. R., and Cho, E. S. (1993). Multiple sclerosis: Remyelination of nascent lesions. *Ann. Neurol.* **33,** 137–151.

Raff, M. C., Lillien, L. E., Richardson, W. D., Burne, J. F., and Noble, M. D. (1988). Platelet-derived growth factor from astrocytes drives the clock that times oligodendrocyte development in culture. *Nature* **333,** 562–565.

Raine, C. S. (1984). Biology of disease. Analysis of autoimmune demyelination: Its impact upon multiple sclerosis. *Lab. Invest.* **50,** 608–635.

Raine, C. S. (1997). The Norton Lecture: A review of the oligodendrocyte in the multiple sclerosis lesion. *J. Neuroimmunol.* **77,** 135–152.

Raine, C. S., and Wu, E. (1993). Multiple sclerosis: Remyelination in acute lesions. *J. Neuropathol. Exp. Neurol.* **52,** 199–204.

Raine, C. S., Scheinberg, L., and Waltz, J. M. (1981). Multiple sclerosis. Oligodendrocyte survival and proliferation in an active established lesion. *Lab. Invest.* **45,** 534–546.

Rechler, M. M., and Nissley, S. P. (1985). The nature and regulation of the receptors for insulin-like growth factors. *Annu. Rev. Physiol.* **47,** 425–442.

Renier, G., Clement, I., Desfaits, A. C., and Lambert, A. (1996). Direct stimulatory effect of insulin-like growth factor-I on monocyte and macrophage tumor necrosis factor-alpha production. *Endocrinology* **137,** 4611–4618.

Rotwein, P., Burgess, S. K., Milbrandt, J. D., and Krause, J. E. (1988). Differential expression of insulin-like growth factor genes in rat central nervous system. *Proc. Natl. Acad. Sci. USA* **85,** 265–269.

Russo, V. C., Gluckman, P. D., Feldman, E. L., and Werther, G. A. (2005). The insulin-like growth factor system and its pleiotropic functions in brain. *Endocr. Rev.* **26,** 916–943.

Savage, M. O., Camacho-Hubner, C., Dunger, D. B., Ranke, M. B., Ross, R. J., and Rosenfeld, R. G. (2001). Is there a medical need to explore the clinical use of insulin-like growth factor I? *Growth Horm. IGF Res.* **11,** S65–S69.

Selmaj, K. W., and Raine, C. S. (1988). Tumor necrosis factor mediates myelin and oligodendrocyte damage *in vitro*. *Ann. Neurol.* **273,** 34543–34550.

Sendtner, M. (1995). Molecular biology of neurotrophic factors. *Baillieres Clin. Neurol.* **4,** 575–579.

Shaked, I., Tchoresh, D., Gersner, R., Meiri, G., Mordechai, S., Xiao, X., Hart, R. P., and Schwartz, M. (2005). Protective autoimmunity: Interferon-gamma enables microglia to remove glutamate without evoking inflammatory mediators. *J. Neurochem.* **92,** 997–1009.

Shinar, Y., and McMorris, F. A. (1995). Developing oligodendroglia express mRNA for insulin-like growth factor-I, a regulator of oligodendrocyte development. *J. Neurosci. Res.* **42,** 516–527.

Sonntag, W. E., Ramsey, M., and Carter, C. S. (2005). Growth hormone and insulin-like growth factor-1 (IGF-1) and their influence on cognitive aging. *Ageing Res. Rev.* **4,** 195–212.

Streit, W. J., Walter, S. A., and Pennell, N. A. (1999). Reactive microgliosis. *Prog. Neurobiol.* **57,** 563–581.

Syroid, D. E., Zorick, T. S., Arbet-Engels, C., Kilpatrick, T. J., Eckhart, W., and Lemke, G. (1999). A role for insulin-like growth factor-I in the regulation of Schwann cell survival. *J. Neurosci.* **19,** 2059–2068.

Torres-Aleman, I., Barrios, V., and Berciano, J. (1998). The peripheral insulin-like growth factor system in amyotrophic lateral sclerosis and in multiple sclerosis. *Neurology* **50,** 772–776.

Tranque, P. A., Calle, R., Naftolin, F., and Robbins, R. (1992). Involvement of protein kinase-C in the mitogenic effect of insulin-like growth factor-I on rat astrocytes. *Endocrinology* **131,** 1948–1954.

Trejo, J. L., Carro, E., Lopez-Lopez, C., and Torres-Aleman, I. (2004). Role of serum insulin-like growth factor I in mammalian brain aging. *Growth Horm. IGF Res.* **14,** 39–43.

Ullrich, A., Gray, A., Tam, A. W., Yang-Feng, T., Tsubokawa, M., Collins, C., Henzel, W., Le Bon, T., Kathuria, S., and Chen, E. (1986). Insulin-like growth factor I receptor primary structure: Comparison with insulin receptor suggests structural determinants that define functional specificity. *EMBO J.* **5,** 2503–2512.

Underwood, L. E., and D'Ercole, A. J. (1984). Insulin and insulin-like growth factors/somatomedins in fetal and neonatal development. *Clin. Endocrinol. Metab.* **13,** 69–89.

Underwood, L. E., Thissen, J. P., Lemozy, S., Ketelslegers, J. M., and Clemmons, D. R. (1994). Hormonal and nutritional regulation of IGF-I and its binding proteins. *Horm. Res.* **42,** 145–151.

Valentinis, B., Bhala, A., DeAngelis, T., Baserga, R., and Cohen, P. (1995). The human insulin-like growth factor (IGF) binding protein-3 inhibits the growth of fibroblasts with a targeted disruption of the IGF-I receptor gene. *Mol. Endocrinol.* **9,** 361–367.

van Dam, P. S., and Aleman, A. (2004). Insulin-like growth factor-I, cognition and brain aging. *Eur. J. Pharmacol.* **490,** 87–95.

Watt, V. M., Shier, P., Chan, J., Petrisor, B. A., and Mathi, S. K. (1993). IRR: A novel member of the insulin receptor family. *Adv. Exp. Med. Biol.* **343,** 125–132.

Wilczak, N., and De Keyser, J. (1997). Insulin-like growth factor-I receptors in normal appearing white matter and chronic plaques in multiple sclerosis. *Brain Res.* **772,** 243–246.

Wilczak, N., Schaaf, M., Bredewold, R., Streefland, C., Teelken, A., and De Keyser, J. (1998). Insulin-like growth factor system in serum and cerebrospinal fluid in patients with multiple sclerosis. *Neurosci. Lett.* **257,** 168–170.

Wilczak, N., De Bleser, P., Luiten, P., Geerts, A., Teelken, A., and De Keyser, J. (2000). Insulin-like growth factor II receptors in human brain and their absence in astrogliotic plaques in multiple sclerosis. *Brain Res.* **863,** 282–288.

Wilczak, N., Ramsaransing, G. S., Mostert, J., Chesik, D., and De Keyser, J. (2005). Serum levels of insulin-like growth factor-1 and insulin-like growth factor binding protein-3 in relapsing and primary progressive multiple sclerosis. *Mult. Scler.* **11,** 13–15.

Wildbaum, G., Nahir, M. A., and Karin, N. (2003). Beneficial autoimmunity to proinflammatory mediators restrains the consequences of selfdestructive immunity. *Immunity* **19,** 679–688.

Woodruff, R. H., and Franklin, R. J. (1997). Growth factors and remyelination in the CNS. *Histol. Histopathol.* **12,** 459–466.

Yao, D. L., Liu, X., Hudson, L. D., and Webster, H. D. (1995). Insulin-like growth factor I treatment reduces demyelination and up-regulates gene expression of myelin-related proteins in experimental autoimmune encephalomyelitis. *Proc. Natl. Acad. Sci. USA* **92,** 6190–6194.

Ye, P., and D'Ercole, A. J. (2006). Insulin-like growth factor actions during development of neural stem cells and progenitors in the central nervous system. *J. Neurosci. Res.* **83**(1), 1–6.

Ye, P., Carson, J., and D'Ercole, A. J. (1995). *In vivo* actions of insulin-like growth factor-I (IGF-I) on brain myelination: Studies of IGF-I and IGF binding protein-1 (IGFBP-1) transgenic mice. *J. Neurosci.* **15,** 7344–7356.

Ye, P., Popken, G. J., Kemper, A., McCarthy, K., Popko, B., and D'Ercole, A. J. (2004). Astrocyte-specific overexpression of insulin-like growth factor-I promotes brain overgrowth and glial fibrillary acidic protein expression. *J. Neurosci. Res.* **78,** 472–484.

Yuen, E. C., and Mobley, W. C. (1996). Therapeutic potential of neurotrophic factors for neurological disorders. *Ann. Neurol.* **40,** 346–354.

Zackenfels, K., Oppenheim, R. W., and Rohrer, H. (1995). Evidence for an important role of IGF-I and IGF-II for the early development of chick sympathetic neurons. *Neuron* **14,** 731–741.

CELL-DERIVED MICROPARTICLES AND EXOSOMES IN NEUROINFLAMMATORY DISORDERS

Lawrence L. Horstman,* Wenche Jy,* Alireza Minagar,† Carlos J. Bidot,* Joaquin J. Jimenez,* J. Steven Alexander,‡ and Yeon S. Ahn*

*Wallace H. Coulter Platelet Laboratory, Department of Medicine University of Miami, Miami, Florida 33136, USA

†Department of Neurology, Louisiana State University Health Sciences Center, Shreveport, Louisiana 71103, USA

‡Department of Cellular and Molecular Physiology, Louisiana State University Health Sciences Center, Shreveport, Louisiana 71103, USA

INTERNATIONAL REVIEW OF
NEUROBIOLOGY, VOL. 79
DOI: 10.1016/S0074-7742(07)79010-4

0074-7742/07 $35.00

All blood cells and the vascular endothelium shed microparticles (MP) from their plasma membranes when suitably stimulated, and assay of MP in patient blood has found increasing application to the monitoring of disease states. In addition, mounting evidence suggests that MP are not mere epiphenomena but play significant roles in the pathophysiology of thromboses, inflammation, and cancers.

This chapter endeavors to summarize the limited number of studies thus far done on MP in neurological disorders such as multiple sclerosis (MS), transient ischemic attacks, and the neurological manifestations of antiphospholipid syndrome (APS). In addition, the chapter offers some plausible hypotheses on possible roles of MP in the pathophsyiology of these disorders, chiefly, the hypothesis that MP are indeed important participants in some neuropathologies, especially those which are ischemic in nature, but probably also inflammatory ones.

The chapter also goes over the history and general principles of MP studies (e.g., assay methods and pitfalls), comparison with alternative methods (e.g., soluble markers of disease states), subclasses of MP (such as exosomes), and other topics aimed at helping readers to consider MP studies in their own clinical fields. Tables include a listing of bioactive agents known to be carried on MP, many of which were heretofore considered strictly soluble, and some of which can be transferred from cell to cell via MP vectors, for example certain cytokine receptors.

I. General Considerations

A. Aims, Scope, and Terminology

This chapter departs from others on cell-derived microparticles (MP) chiefly in its expanded treatment of topics relevant to neurological disorders, as opposed to the better known functions of MP in hemostasis, thrombosis, angiogenesis, and other processes. However, the relationship between inflammation and hemostasis is not overlooked, nor is the likely relevance of MP to autoimmune states, since both are implicated in some neuropathologic syndromes.

Over the last two decades, measurement of circulating MP has matured to an accepted and popular method of monitoring thrombotic, inflammatory, and other conditions. Accordingly, the focus of interest is now shifting to the question of their function as pro- or anti-inflammatory agents (Ahn *et al.*, 2004b; Freyssinet, 2003). This chapter surveys current thinking on the functional significance of MP with emphasis on inflammation/immune modulation as it pertains to neuropathologic syndromes.

In addition, an apparently distinct class of MP known as *exosomes* has recently come to wide attention, differing in their smaller and more uniform size, and egress from the cell interior (endosomes) rather than from the plasma membrane. Therefore, another aim of this chapter is to consider how exosomes fit in with developing concepts of MP from the plasma membrane, as there may be considerable overlap of detection (Section IV).

MP are now implicated not only in hemostasis and thrombosis, as earlier thought, but increasingly in inflammation, immunologic disorders, cancers, neurological impairments, atherosclerosis, wound healing, and other physiological processes. They issue from all types of blood cells as well as from endothelial cells (EC), that is, from platelets (PMP), leukocytes (LMP), red cells (RMP), and EC (EMP). Although they occur at low levels in normal circulation, their release is greatly augmented by cell activation, apoptosis, high shear rates, immunologic events such as complement (C) deposition, or other stresses, in a process called *vesiculation* or sometimes other names, for example, *ectocytosis*.

Although heterogeneous in size and other phenotypic properties, they are usually defined to be $<1\ \mu m$. Some workers employ alternative names for them, such as *ectosomes* (Gasser and Schifferli, 2005), *exocytic vesicles* (Greenwalt, 2006), or *microvesicles* (Heijnen *et al.*, 1999), but MP is most common and convenient for present purposes.

B. Historical Development

Early work. The era of MP analysis may be dated from Wolf's outstanding report in 1967 of platelet microparticles (PMP), which he termed "platelet dust" (Wolf, 1967). This was followed by observations by Zucker-Franklin, Karpatkin, and colleagues in 1975 and 1977 of high levels of PMP and MP from red blood cells (RMP) in immune thrombocytopenic purpura (ITP) (Kahn *et al.*, 1975; Zucker-Franklin and Karpatkin, 1977). Thereafter, our knowledge of MP was much advanced during the 1980s by a series of publications by Sims *et al.*, often in collaborations (Gilbert *et al.*, 1991; Sims *et al.*, 1989; Zwaal *et al.*, 1992). Those studies concerned mainly PMP and restricted their likely function to hemostasis and thrombosis, as detailed in our 1999 review of PMP (Horstman and Ahn, 1999).

Neurological connection. Platelets are remarkably similar to neurons in many respects (Stahl, 1985), and mechanisms of release of small PMP (exosomes) from platelets are very similar to release of vesicles of synaptic transmission (Lemons *et al.*, 1997), as updated (Giraudo *et al.*, 2006; Pobbati *et al.*, 2006). Probably the first connection between PMP and neurological impairment was shown by Jy *et al.* of Ahn's laboratory in 1992, reporting that high levels of PMP in ITP patients appeared to protect against bleeding in a subset of thrombocytopenic patients, suggesting that the PMP functioned as surrogate platelets in the patients with high PMP. Moreover, patients with the highest levels frequently suffered slowly progressive cognitive impairments and transient ischemic attacks (TIA)-like syndromes (Jy *et al.*, 1992). Ahn *et al.* (2002) have further detailed this syndrome documented by MRI of the central nervous system (CNS). During the 1990s, Ahn's laboratory was one of several that vigorously promoted PMP assay for monitoring various clinical conditions, as reviewed (Horstman and Ahn, 1999).

Endothelial microparticles (EMP). EMP had been observed in tissue culture experiments, such as by Hamilton *et al.* (1990), but it was Combes *et al.* who first showed in 1999 that EMP could be identified and counted in patient plasma (Combes *et al.*, 1999). Their seminal work, together with that of Mallat *et al.* (1999, 2000), was a major advance since it offered a new approach to the otherwise difficult task of assessing endothelial dysfunction, with important advantages over alternative methods (Section I.D). The result was a flurry of work by several laboratories, including ours (Ahn's), showing applications of EMP analysis to diverse clinical conditions, with steadily improving methodologies, as reviewed (Horstman *et al.*, 2004). The great sensitivity of EMP assay to endothelial perturbation is illustrated by the fact that a single high-fat meal is sufficient to elicit a pulse of circulating EMP (Ferreira *et al.*, 2004; Peter *et al.*, 2004), as is application of a blood pressure cuff (Bernal-Mizrachi *et al.*, 2002).

C. Assay Methodologies

A serious obstacle to the more widespread use of MP analysis for clinical applications is the variety of methods employed by different laboratories active in the field. This makes it difficult or impossible to directly compare results between some laboratories.

Almost all methods rely on the fact that MP carry antigenic markers characteristic of the parent cell, and this fact is exploited for identifying the cellular origin of the MP, usually by fluorescent-labeled monoclonal antibodies (mAbs) in flow cytometry but sometimes by ELISA (solid phase) or bead-capture methods.

Table IA summarizes example methods of preparing the sample, Table IB lists the detection antigens commonly labeled, and Table IC cites some ELISA-based methods. It is obvious that methods depart widely, and this accounts for frequently divergent reports. Our view is that there is no right or wrong method, the value of any of them lying in results, some providing insights missed by others.

Two considerations are of special importance. First are the wide differences in centrifuge speeds used to eliminate whole cells or to sediment MP. Laboratories employing the higher centrifuge speeds to eliminate whole cells may discard materials which in other laboratories are measured as MP. The possibility that exosomes predominate in some methods but not others is discssed in Section IV.

Antigenic markers. The second consideration is the many different antigenic markers employed for counting and identifying MP cell origin (Table IB), causing

TABLE IA
EXAMPLE ASSAY METHODOLOGIES: PREPARING THE MP SAMPLE

	Methods using isolated MP		
References	Boulanger *et al.*, 2001	Shet *et al.*, 2004	VanWijk *et al.*, 2002
Preparation of PPP	11,000 *g* 2 min	13,000 *g* 10 min	1550 *g* 20 min
Store?	–	−80°C	−80°C
Sediment MP	13,000 *g* 45 min	100,000 *g* 60 min, twice	17,570 *g* 30 min
Store?	−80°C	–	–
Wash step?	Yes	Yes	Yes
Suspend in buffer	5% original volume RPMI		30% original volume PBS/citrate

	Methods using whole plasma	
References	Combes *et al.*, 1999	Minagar *et al.*, 2001; Jimenez *et al.*, 2001
Preparation of PPP	13,000 *g* 1 min	500 *g* 6 or 10 min

TABLE IB
EXAMPLE ASSAY METHODOLOGIES: FLUORESCENT MARKERS

PMP	CD41 (many); CD31+/CD41+ (Jimenez *et al.*, 2001; Minagar *et al.*, 2001); CD31+/CD54− (Ogura *et al.*, 2004); CD61 (Biro *et al.*, 2004; VanWijk *et al.*, 2002); CD63 (activation) (Biro *et al.*, 2004); CD62P (activation) (Biro *et al.*, 2004)
EMP	CD31+/CD41− (Jimenez *et al.*, 2001; Minagar *et al.*, 2001); CD31+/CD54+ (Ogura *et al.*, 2004); CD31+/CD51+ (Combes *et al.*, 1999); CD51+ (Minagar *et al.*, 2001; Sabatier *et al.*, 2002a); CD62E (Jimenez *et al.*, 2003c; VanWijk *et al.*, 2002); CD144 (Biro *et al.*, 2004; VanWijk *et al.*, 2002); CD146+/annexin V+ (Shet *et al.*, 2003); CD54, see text
LMP	CD45 (many, pan-leukocyte) (Amabile *et al.*, 2005; Faure *et al.*, 2005); CD11b (activation) (Amabile *et al.*, 2005); CD3 (T helper) (Amabile *et al.*, 2005); CD4 (T helper) (VanWijk *et al.*, 2002); CD8 (T suppressor) (VanWijk *et al.*, 2002); CD14 (monocyte) (Biro *et al.*, 2004; VanWijk *et al.*, 2002); CD20 (B cells) (VanWijk *et al.*, 2002); CD66b (granulocytes) (Amabile *et al.*, 2006; Biro *et al.*, 2004; VanWijk *et al.*, 2002); CD66e (granulocytes) (Biro *et al.*, 2004; VanWijk *et al.*, 2002); CD31+/CD51/61+ (aVβ3) (Combes *et al.*, 1999)
RMP	CD235a (glycophorin A) (Amabile *et al.*, 2005; Biro *et al.*, 2004)
EMP–platelet complexes	CD41a+/CD105+ (Heloire *et al.*, 2003)
EMP–leukocyte complexes	CD45+/CD62E+ (Jy *et al.*, 2004b); CD45+/CD54+ (Jy *et al.*, 2004b)
Generic	Annexin V (Amabile *et al.*, 2005; Faure *et al.*, 2005); *Ulex europaeus* (crude) (Jy *et al.*, 2004b); dyes, see text

TABLE IC
EXAMPLE ASSAY METHODOLOGIES: ELISA-BASED METHODS

Kokawa *et al.*, 1993; Mallat *et al.*, 2000; Nomura *et al.*, 2003; Osumi *et al.*, 2001

of inconsistent findings. One reason for this is that MP from the same cell type may occur in multiple antigenic phenotypes, some of which are enriched in some antigens relative to the others and may reflect different clinical states (Section I.E). There has also been a mistaken view held by some investigators that all cell-derived MP may be counted by fluorescent annexin V (AnV), on the assumption that all MP express the anionic phospholipid (PL), phosphatidylserine (PS), but this is incorrect; the fraction of MP which binds AnV (in the presence of calcium) varies widely, from a few percent to >80% for MP from apoptotic cells (Horstman *et al.*, 2004).

Numerous seemingly minor details not shown in the table may be important under some circumstances, for example, buffers used, which may or may not include citrate, Mg^{2+}, Ca^{2+}, albumin, prostaglandin, and so on. Some laboratories routinely add azide or a cocktail of antibiotics to the MP (Amabile *et al.*, 2005; Boulanger *et al.*, 2001). Procedures for antibody labeling may also be critical, such as whether fixation with paraformaldehyde is used, either before or after labeling. Nonspecific binding of label can be avoided by a washing step to remove excess, or by including dilute mouse serum, and/or can be reduced by using $F(ab)^2$ fragment if available. Biotinylated mAb are often used and may provide a brighter signal but in our experience the avidin-conjugated fluorophore may bind nonspecifically to some MP species, and similar problems occur with labeled secondary antibodies (antimouse).

Methods for MP analysis are inadequately detailed in many publications and for that reason we have omitted from the table several otherwise worthy papers. For example, in flow cytyometry, events are recorded only if triggered by a specified signal, which may be forward or side light scatter, or fluorescence, this choice being the primary detection criterion. But this crucial datum is often not specified. Similarly, if multiple labels are used (two or more colors), such as AnV and anti-CD41, and the instrument is triggering on fluorescence, it must be specified which is primary since, for example, if events are triggered by AnV fluorescence, all events counted will be AnV positive (by definition), hence CD41+ but AnV negative events will be missed.

Other markers sometimes used or proposed include the lectin, *Ulex europaeus* (Ulex) with flourescein isothiocyanate (FITC) label. Although traditionally assumed to be specific for endothelium (Jackson *et al.*, 1990; Woywodt *et al.*, 2004), we were unable to find original literature documenting that, and found that it is a sensitive marker not only of EMP (Jy *et al.*, 2004b) but of PMP and RMP

(Horstman *et al.*, 2004). Isolectin B4 from *Bandeiraea simplicifolia* (aka *Griffonia* sp.) is said to be specific for brain microvessels (Ghazanfari and Stewart, 2001). Monoclonal PAL-E is of potential interest as an EMP marker since its target is now identified (Xu *et al.*, 2004) and reacts only with micro- not macrovessels (Schlingemann *et al.*, 1985), and is sensitive to the integrity of the blood–brain barrier (BBB) (Leenstra *et al.*, 1993; Schlingemann *et al.*, 1999). The idea of using lipophilic organic dyes such as Nile Red or FM-143 for labeling total MP is appealing because the expected bright signal should reveal even very small particles. Unfortunately, we had less success in the use of dyes due to artifactual appearance of dye micelles as false MP even in buffer alone. However, leukocyte microparticles (LMP of neutrophils) were usefully labeled using a dye linker to aid in observing their phagocytosis (Gasser and Schifferli, 2004), and Furie's group labeled MP with calcein acetoxymethyl ester (AM) by prelabeling the cells used for generating MP (Falati *et al.*, 2003).

Although detailed discussion of methods is beyond the scope of this chapter, the above remarks suffice to show that many details are critically important. We organized a forum that briefly compared and discussed the methodologies used by six laboratories active in the field (Jy *et al.*, 2004a). Although there is not yet a consensus on methods for MP analysis, it is hoped that future meetings will lead to agreement on at least one basic procedure and MP standards to facilitate comparing and reproducing results between laboratories.

D. Assay of MP Versus Soluble Markers in Clinical Studies

The traditional and still the main alternative methodology for monitoring endothelial activation, prothrombotic, and inflammatory states is measurement of soluble markers of these states in plasma or serum by ELISA techniques.

Table II lists representative studies employing soluble markers (prefix s = soluble, as stated by the authors cited), including some neuropathologic syndromes. A pitfall of that method is the dawning recognition that some or many of the agents listed may not be truly soluble, but instead are released and transported on MP, at least in part. This is important because different mechanisms are involved with the release of MP (vesiculation) versus true soluble forms (proteolytic cleavage, alternative splicing), therefore they reflect different pathophysiological mechanisms. Furthermore, MP-bound agents may have different biologic activities and half-lives in circulation compared to their true soluble counterparts, since the latter generally lack transmembrane domains and have altered conformations and activities.

The fractional extent to which these agents may be MP-bound versus free depends on the agent in question, and probably also on the clinical state of the patient. For the example of circulating tissue factor (TF), it is now accepted that

TABLE II
REPRESENTATIVE CLINICAL STUDIES USING SOLUBLE MARKERS

Disease	Soluble markers measured	References
ACS	sCD40L	Cipollone *et al.*, 2003; Heeschen *et al.*, 2003; Varo *et al.*, 2003
APS	TF	Atsumi *et al.*, 1997
APS	TM	Martinuzzo *et al.*, 1996
APS	IL-6; WF; sVCAM-1; sICAM-1; ET-1; CD62P; sTF; VEGF	Williams *et al.*, 2000
Atherosclerosis	ICAM-1, VCAM-1, CD62E	Oishi *et al.*, 2000
Atherosclerosis	sTM	John *et al.*, 1999
Atherosclerosis/MI	sCD62E; vWF	Blann *et al.*, 1997
Atherosclerosis	sTM; vWF (antigen and activation)	Blann *et al.*, 1998
ACS/PAD	sCD62P	Blann and McCollum, 1998
CAD	TM	Salomaa *et al.*, 1999
CAD	ICAM-1; VCAM-1; CD62E	Seeman *et al.*, 2000
CAD/ atherosclerosis	IVAM-1, VCAM-1, CD62E	Hwang *et al.*, 1997
CAD (stable)	ICAM-1; CD62E	Nasuno *et al.*, 2002
CAD (Review)	CD62E	Roldan *et al.*, 2003
Angina (unstable)	CD62P	Ikeda *et al.*, 1995
MI acute	ICAM-1; VCAM-1; CD62E	Li *et al.*, 1997
MS	PECAM-1 (CD31)	Losy *et al.*, 1999
MS	sVCAM-1; sICAM-1; sCD62E	McDonnell *et al.*, 1999
PAD	sCD40L	Blann *et al.*, 2005
PVD	ICAM-1; VCAM-1; TM; vWF; CD62P, CD62E	Blann *et al.*, 2002
Preeclampsia	TNF, IL-1β	Heyl *et al.*, 1999
Preeclampsia	vWF	He *et al.*, 1999
Preeclampsia	ICAM-1, VCAM-1	Haller *et al.*, 1997
RA	ICAM-1, ICAM-3; VCAM-1; CD62P, CD62E, CD62L	Littler *et al.*, 1997
Stroke	tPA; PAI-1	Margaglione *et al.*, 1994
TTP, HUS	CD62P	Katayama *et al.*, 1993
TTP	TM	Takahashi *et al.*, 1991
TTP, DIC	TM, TAT, D-dimer, others	Wada *et al.*, 1998
TTP	tPA; PAI-1; TAT; D-dimer; others	Wada *et al.*, 1993
TTP, DIC	TF	Koyama *et al.*, 1994
TTP, HUS, DIC HIT, ITP	β-TG; CD62P	Chong *et al.*, 1994
Vasculatides	TM; VCAM-1	Boehme *et al.*, 2000
Review	vWF	Mannucci, 1998
Vasculitis	TM	Ohdama and Aoki, 1997
Vasculitis	TM	Haznedaroglu *et al.*, 1996
Vasculitis	CD62E	Triolo *et al.*, 1999
Liver injury	TM	Takatori *et al.*, 1999

essentially all of it is MP-bound (Giesen and Nemerson, 2000; Giessen *et al.*, 1999). Considering CD40L, it was partly removed by filtration through 0.1 μm (Ahn *et al.*, 2004a) and was completely removed by moderate centrifugation (Varo *et al.*, 2006), hence contrary to wide belief, CD40L is not a true soluble agent. Platelet-activating factor (PAF) may also be largely MP-bound (Iwamoto *et al.*, 1996), as is its degrading enzyme, PAF-AH (Mitsios *et al.*, 2006). These facts alert us to the possibility that many other agents in plasma assumed at present to be soluble may be MP-bound, at least in part.

Since intracellular adhesion molecule (ICAM-1) (CD54), vascular cell adhesion molecule (VCAM-1) (CD106), platelet-endothelial cell adhesion molecule (PECAM)-1 (CD31), P-selectin (CD62P), and others are membrane proteins regularly used to identify MP (Table IB), it follows that they are all MP-bound, at least to some degree. On the other hand, it appears that P-selectin (CD62P) occurs mainly in true soluble form (Katayama *et al.*, 1993; Osmanovic *et al.*, 2000; Polgar *et al.*, 2005). Similarly, although von Willebrand factor (vWF) occurs in active form on MP (Jimenez *et al.*, 2003c), the majority of it is soluble in most samples. However, vWF was depleted in some patient specimens as a result of filtration (Favaloro, 1999; Favaloro *et al.*, 1995, 2001), so there may be individual variation with this and other agents. It remains to be seen what fraction of circulating E-selectin (CD62E) and L-selectin (CD62L) occur in true soluble form, as these are also used as markers of MP. Likewise for vascular endothelial growth factor (VEGF), thrombomodulin (TM), and others since these have been detected on MP (Section III).

Some cases are ambiguous. For example, it was known that IL-1 is secreted from platelets (Hawrylowicz *et al.*, 1991; Loppnow *et al.*, 1998), but was then found to be released within PMP, from which it slowly leaks out (MacKenzie *et al.*, 2001). On the other hand, many proteins released from platelets are indeed soluble, as shown by proteomic analysis of the 100,000 × *g* supernatant (Coppinger *et al.*, 2004). Ordinary centrifugation of blood in the clinical laboratory does not remove most MP from plasma or serum, therefore MP will be detected by ELISA methods intended to detect only soluble agents. Conversely, true soluble proteins are not detected in fluorescence-activated flow cytometry since approximately 300 fluorescent molecules must be closely adjacent to trigger a signal.

In summary, it is important to be aware of this distinction and to compare the effect of high-speed centrifugation with ordinary centrifugation as a means of estimating the fraction of the biomarker that could potentially be MP-bound. It is possible that measurement of both species (true soluble and MP-bound) will result in improved correlations with clinical states, and reduction of inconsistencies such as up to 50-fold differences reported for some soluble markers (Andrew and Newman, 1993).

E. Multiple Phenotypes of MP

Among the significant recent developments in MP studies has been recognition that not all MP from a given cell type are alike. This came to light for EMP but similar considerations are now known to apply to LMP and PMP. In the course of exploring alternative EC-specific markers for EMP, it became clear that counts of EMP were not the same with all markers employed and that the proportion counted by a given marker compared to other markers differed sharply depending on how the cells were stimulated to release EMP. In particular, it was shown in tissue culture that EMP released from brain microvascular EC showed clear differences in counts between markers depending on whether the cells were undergoing apoptosis or activation induced by TNF-α (Jimenez *et al.*, 2003a). Jy *et al.* has shown that EMP positive for CD54 or CD62E do not occur on the same particle with CD31, as judged by double labeling with CD31-PE and CD54-FITC or CD62E-FITC (Jy *et al.*, 2006).

These *in vitro* findings were subsequently reflected in patient studies since certain disease conditions such as multiple sclerosis (MS), cited in Section II, revealed good correlation with some EMP markers but not others. Similarly, some markers of EMP were chronically elevated in a given disorder while others reflected only acute phases of disease, for example, exacerbation of MS (Section II) and coronary artery disease (CAD) (Bernal-Mizrachi *et al.*, 2003).

Taken together, these findings demonstrate that EMP are released in multiple antigenic phenotypes depending on agonist (or disease condition). There is reason to believe that PMP also occur in multiple phenotypes. For example, in a study of PMP in patients with ITP, we observed significant differences between patients with and without symptoms (purpura), and with and without CNS dysfunction, using PMP labeled at CD42b (GP Ib) (Jy *et al.*, 1992), but no such differences were seen when PMP were counted by label of CD41 (GP IIb), not published because it seemed of little interest. Mitsios *et al.* (2006) identified the PAF-degrading enzyme (PAF-AH) on PMP from platelets stimulated by thrombin, but not on PMP from platelets stimulated by adenosine diphosphate (ADP), again indicating phenotypes of PMP. Similarly, Lo *et al.* (2006), in studying the binding of PMP to polymorphonuclear leukocytes (PMN), found no inhibition of binding by either of two antibodies (AP1, SZ2) when detecting the bound PMP by CD42a (GP IX), but significant inhibition was seen when detecting by CD62P (P-selectin), again pointing to at least two PMP phenotypes.

Platelets themselves exist in multiple phenotypes (Behnke and Forer, 1993; Dhar and Ganguly, 1988; Frojmovic and Wong, 1991; Holmes *et al.*, 2003; Jobe *et al.*, 2005; Johnston *et al.*, 1987; Jung *et al.*, 1985; Jy *et al.*, 1995a; Kunicki *et al.*, 1993; Michelson, 1988), possibly differing in release of PMP. Our experience of observing EMP release by fluorescent microscopy indicates that they are shed following clustering of proteins (such as CD31) on the cell surface in response to specific agonists of different activation pathways (Jy *et al.*, 2002).

It has been proposed that MP shedding involves lipid raft domains on the cell surface but data is often conflicting. For example, some but not all GPI-anchored proteins are enriched on RMP (Greenwalt, 2006). It remains unsettled whether there is just one kind of lipid raft or multiple kinds having distinctive PL compositions (Bodin *et al.*, 2003; Gassart *et al.*, 2003; Gousset *et al.*, 2004; Salzer and Prohaska, 2001; Simons and Ikonen, 1997).

The finding of distinctive EMP phenotypes arising from endothelial apoptosis versus activation raises the possibility of determining whether apoptosis is occurring in a given disease state. For example, it has been controversial whether endothelial apoptosis is central to the pathology of thrombotic thrombocytopenic purpura (TTP) (Jimenez *et al.*, 2004). However, extending the *in vitro* observation (Jimenez *et al.*, 2003a) to patient plasma is complicated by the fact that some EMP (notably CD54+) strongly associate with neutrophils and are therefore effectively eliminated from the cell-free supernatant (Jy *et al.*, 2001, 2004b).

II. MP Studies Relevant to Neuropathologies

A. Introduction

Although clinical studies based on MP analyses now number in the hundreds, comparatively few have specifically addressed neurological conditions.

There are two purposes of MP studies in neurological disorders—first, as a means for the clinical monitoring of disease condition, progression, or response to therapy, and second, as a research tool for gaining insights into pathophysiology. The first speaks for itself in following paragraphs. The second often assumes the *working hypothesis* that MP are indeed active players in some neurological disorders, as opposed to being purely epiphenomenal markers. The basis of this hypothesis is (1) the frequent association of elevated MP in disease states, reviewed in this section, (2) the fact that many inflammatory mediators and other bioactive agents are carried on MP (Section III), and (3) *in vitro* studies demonstrating functions likely to apply *in vivo*, not a focus of this chapter but several studies are cited, including in Section IV on exosomes. On the basis of this hypothesis, cell-derived MP may act to transport the pathologic agents believed to be responsible.

B. EMP in MS

It was shown by Minagar *et al.* (2001) that EMP levels correlate closely with clinical state in MS. That study was important for several reasons, apart from confirming endothelial activation in MS. It was the first indication in a clinical setting that different phenotypes of EMP (Section I.E) could be detected *in vivo*, and bore distinctive clinical significances. In particular, EMP counted by marker CD51 were chronically elevated in MS patients, regardless of state of exacerbation, while EMP detected by CD31+/CD42− were elevated only in exacerbations (Minagar *et al.*, 2001). A similar EMP phenotype distinction was later seen also in CAD (Bernal-Mizrachi *et al.*, 2003, 2004).

Subsequently, Minagar *et al.* have extended those findings to examine the role of EMP–monocyte complexes in MS and their effects on transendothelial migration (Jy *et al.*, 2004b) and modulation by interferon-β1b (Jimenez *et al.*, 2005) using an *in vitro* model of the endothelial layer of the BBB. The latter paper also showed that plasma from MS patients in exacerbation but not in remission induced release of EMP and augmented transendothelial migration of monocytes.

The first study (Minagar *et al.*, 2001) illustrates MP analysis chiefly for purposes of clinical monitoring of disease state, for example as may be useful for assessing responses to therapy. However, by clearly confirming endothelial activation in MS, it also illustrates application of MP measurement to gain insights into pathophysiology. Studies are ongoing to better define the nature of that involvement by more refined phenotypic analysis of the EMP released, that is, those released in chronic MS compared to those from exacerbations. The other two cited studies (Jimenez *et al.*, 2005; Jy *et al.*, 2004b) approach more closely to questions of mechanism or pathophysiology, pointing as they do to a role of EMP in facilitating passage of monocytes through the endothelial barrier of the BBB (Jimenez *et al.*, 2005).

For monitoring purposes, we believe that EMP assay is a more sensitive and specific marker of endothelial disturbance than alternative methodologies such as soluble markers (Section I.D), although this remains to be rigorously demonstrated.

C. Platelet Activation in Alzheimer's Disease

As referenced below (Section III), it has been shown that PMP bind and actively transport β-amyloid, prion, and other agents implicated in neurodegenerative disorders. We reported in 1998, a significant degree of platelet activation in Alzheimer's disease (AD) ($p = 0.001$) along with increased platelet microaggregates ($p < 0.001$) and platelet–leukocyte complexes ($p = 0.0001$) (Sevush *et al.*, 1998). Although elevated PMP were not detected in that study, this may be partly explained by the use of a newer flow cytometer with weaker laser and lower detection efficiency; for discussion of limits of MP detection by flow cytometry, see review by Horstman and Ahn (1999). Other studies have shown that platelet–leukocyte complexes include PMP–leukocyte complexes and that substantial amounts of PMP can circulate bound to leukocytes (Jy *et al.*, 1995b). It has been shown that platelet activation is tightly coupled to PMP release (Zwaal *et al.*, 1992) so that evidence of platelet activation is *prima facie* evidence of PMP release, although degree varies with agonist.

Other circumstantial evidence could suggest a role for platelets/PMP in AD. First, for example, amyloid proteins are principally carried in and released from platelets (Baskin *et al.*, 2000; DiLuca *et al.*, 1998, 2000). It has been shown that platelets can ferry amyloid proteins across the BBB (Davies *et al.*, 2000). More pointedly, amyloid β precursor protein (APP) is known to be carried on PMP (Nomura *et al.*, 1994). Also consistent is the therapeutic benefit of non-steroidal anti-inflammatory drugs (NSAID) (Helmuth, 2002; Wyss-Coray and Mucke, 2000) since a main action of NSAID is inhibition of platelet activation.

D. PMP/Platelets in Vascular Dementia

We earlier made reference to the findings of Jy *et al.* (1992) and Ahn *et al.* (2002) on variably progressive cognitive impairment in a subset of ITP patients, documented by MRI, associated with exceptionally high levels of PMP, sometimes progressing to vascular dementia (Section I.B). Lee *et al.* (1993) extended the work of Jy *et al.* (1992) to show that elevated PMP was not a feature unique to ITP-associated vascular dementia, as it was also seen in TIA, lacunar infarcts, and multiinfarct dementias. The syndrome identified by Ahn and colleagues was distinguished clinically by comparative absence of bleeding despite thrombocytopenia, small vessel ischemia of brain white matter in MRI, and marked platelet activation with high PMP (Ahn *et al.*, 2002; Jy *et al.*, 1992). It was shown that plasma from patients with this syndrome activates EC and enhances transendothelial migration of leukocytes across them (Jimenez *et al.*, 2003b). A number of platelet secretory products have been implicated in neuronal injury (Joseph, 1995; Joseph *et al.*, 1991) and several of them are transported on PMP (Section III).

E. MP in Cognitive Impairment by Coronary Procedures

It has long been known that persistent cognitive impairment often follows coronary artery bypass surgery (Ahlgren *et al.*, 2003; Shaw *et al.*, 1985, 1986; Smith, 1988), angioplasty (Blumenthal *et al.*, 1991), and even cardiac catheterization (Schroeder, 1980; Thomas and Troost, 1989), but the explanation for this has been obscure.

It has been observed that blood exposed to a foreign surface resulted in platelet activation and release of PMP (Dewanjee *et al.*, 1993; Gemmell *et al.*, 1995). It was subsequently shown by several laboratories that blood cells are activated during bypass—for example, platelets (Muriithi *et al.*, 2000) and leukocytes (Rinder,

1994)—and more specifically, confirming that PMP become markedly elevated (Miyamoto *et al.*, 1998; Nieuwland *et al.*, 1997; Sturk-Maquelin *et al.*, 2001). Thus, in light of the foregoing articles showing association between cell-derived MP levels and some neurological disorders, the hypothesis that MP play a role in the observed impairment following the procedures seems worth considering. However, since not all cases of elevated MP are associated with neurological impairment, this hypothesis requires that only specific phenotypes of MP, or very high levels, lead to cognitive impairments.

F. Concluding Remarks

Some other neurological disorders in which MP may participate are considered in the next section, for example the neurological manifestations of antiphospholipid syndrome (APS, Section 3.E). Meanwhile, the foregoing articles show that MP analysis is a useful tool for monitoring at least some neurological disorders. Not mentioned are the high levels of MP seen in less subtle conditions such as cerebral malaria (Combes *et al.*, 2004) or meningococcal sepsis (Nieuwland *et al.*, 2000).

Second, MP analysis can provide insights into underlying pathophysiology, as illustrated by the endothelial involvement and MP-mediated leukocyte activation documented in the studies of MS. We expect improved resolution as additional more specific MP markers are identified by two-color flow cytometry. Third and more frankly conjectural is the working hypothesis that MP are active causal agents in at least some neurological impairments, based on their elevations in association with the disorders mentioned, plus the additional suggestive evidence that MP interact functionally with leukocytes and are vectors of potentially pathological bioactive agents, reviewed next.

III. Inflammatory MP-Borne Agents Relevant to Neuropathologies

A. Introduction

This section reviews cell-derived MP as vectors of inflammatory and immune mediators known or believed to be important in one or more neuropathologic syndromes. Table III lists a number of agents of interest carried on MP and is the point of departure for this section. The current list is certainly the tip of an iceberg as it only includes those agents definitely established to be MP associated. Some proteomic studies of MP catalog dozens of such agents (Banfi *et al.*, 2005; Jin *et al.*, 2005), but they are not comprehensive, due to methodological limitations. We have noted the absence of several agents known to be present on EMP from the proteomic listing (Banfi *et al.*, 2005). Platelets contain a number of agents acquired

TABLE III
SOME AGENTS OF INTEREST CARRIED ON MP

PMP	
TF	Siddiqui *et al.*, 2002
PAF	Iwamoto *et al.*, 1996
PAF-AH	Mitsios *et al.*, 2006
β2-GPI	Nomura *et al.*, 1993, 1995b; Vallar *et al.*, 2001
CXCR4	Rozmyslowicz *et al.*, 2003b
CD40/CD40L	Amirkhosravi *et al.*, 2002; Anand *et al.*, 2003; Varo *et al.*, 2006
Amyloid β precursor protein	Nomura *et al.*, 1994
Complement (C9)	Sims and Wiedmer, 1986; Sims *et al.*, 1988
Calpain	Pasquet *et al.*, 1996
EMP	
TF	Kagawa *et al.*, 1998; Sabatier *et al.*, 2002b; Satta *et al.*, 1994; Shet *et al.*, 2003
Prion proteins	Simak *et al.*, 2002
EPCR	Perez-Casal *et al.*, 2005
vWF	Jimenez *et al.*, 2003c; Jy *et al.*, 2003
uPA/uPAR	Basire *et al.*, 2006
Fas	Albanese *et al.*, 1998
LMP	
TF	Satta *et al.*, 1994; Scholz *et al.*, 2002; Siddiqui *et al.*, 2002
TM	Satta *et al.*, 1997
IL-1β	MacKenzie *et al.*, 2001
CD11a/CD18	Satta *et al.*, 1994
CD14	Satta *et al.*, 1994
CCR5	Mack *et al.*, 2000
Complement (C9)	Stein and Luzio, 1991
RMP	
Complement (C)	Pascual *et al.*, 1993
AChE	Butikofer *et al.*, 1989
DAF	Butikofer *et al.*, 1989
MP mixed (patient plasma)	
TF	Aras *et al.*, 2002; Bonderman *et al.*, 2002; Chou *et al.*, 2004; Conde *et al.*, 2004; Muller *et al.*, 2003; Sturk-Maquelin *et al.*, 2003; Welterman *et al.*, 2004
TFPI	Steppich *et al.*, 2005
Protein S	Gris *et al.*, 1996
mRNA	Anker and Stroun, 2002; Hasselman *et al.*, 2001; Ng *et al.*, 2002
Glut-S-transferase	Jin *et al.*, 2005
HSP 27, 60	Jin *et al.*, 2005
C3	Jin *et al.*, 2005
Protein disulfide isomerase	Jin *et al.*, 2005
MP universal	
PS/PF3 (variable)	many, e.g., Gasser *et al.*, 2003 (Absent from exosomes)
Caspase-3	Hussein *et al.*, 2005
MP of oligodendrocytes	CNS, complement (Scolding *et al.*, 1989)
Exosomes	
DAF (CD55)	Rabesandratana *et al.*, 1998
MIRL (CD59)	Rabesandratana *et al.*, 1998

from plasma which therefore will not show up in RNA-based analyses (Bugert *et al.*, 2003), for example immunoglobulins, factor V (Camire *et al.*, 1998).

There is further reason to believe that many agents not listed in Table III occur on MP. For example, it has been shown that secreted platelet factor 4 (PF4) has affinity for the activated platelet membrane (Capitanio *et al.*, 1985) and therefore likely binds to PMP, since platelet-secreted TF pathway inhibitor (TFPI) (Novotny *et al.*, 1988), which binds to TF, is now known to occur on MP (Steppich *et al.*, 2005) and on high-density lipoproteins (HDL)/low-density lipoproteins (LDL) (Hansen *et al.*, 1994) where it was first discovered (and described as the lipoprotein-associated coagulation inhibitor, LACI). Relatedly, since Fas (CD95) is known to occur on EMP (Albanese *et al.*, 1998), it is anticipated that its ligand, FasL (CD178, CD95L), also binds EMP. Thus, many secretory products and essentially all membrane-associated antigens of the cell may be expected to occur on MP to some degree.

The table includes mainly cytokines, cytokine receptors, and related agents variously (and somewhat vaguely) termed inflammatory immunomodulatory mediators, and so on. Excluded from the list in Table III are the many adhesins and other surface proteins routinely used for identifying the cellular origins of the MP (Table IB), as well as coagulation factors, immunoglobulins, and so on.

B. Interleukin-1β

We earlier mentioned the finding by MacKenzie *et al.* that interleukin-1β (IL-1β) is secreted from platelets in PMP, from which it is slowly meted out (MacKenzie *et al.*, 2001). IL-1β is now recognized as a pivotal player in the *inflammasome*, the workings of which are still being unraveled (Mariathasan *et al.*, 2006).

C. Chemokines, Their Receptors, and Their Transfer via MP

Chemokines are central players in inflammation (Murdoch and Finn, 2000). Nomura *et al.* were among the first to seek and find chemokines on MP (Nomura, 2001; Nomura *et al.*, 1995a, 1999). The receptor CXCR4 was found to be abundant on PMP (Rozmyslowicz *et al.*, 2003a), and more remarkably, the PMP were capable of transferring this key receptor to other cells (Rozmyslowicz *et al.*, 2003a). Although the focus of their studies was HIV infection, the phenomenon may be quite widespread since an earlier study by Mack *et al.* established a similar transfer of receptor CCR5 from LMP, and from MP derived from Chinese hamster ovary (CHO) cells (Mack *et al.*, 2000). CXCR4 is of interest also in carcinogenesis and other processes (Burger and Kipps, 2006; Strieter *et al.*, 2006). It is likely that other members of this family of receptors are also transported on MP since they have homologous sequences. For example, CX_3CR1, a pertussis toxin-sensitive receptor of fractalkine, is now found on platelets and platelet activation is induced when fractalkine binds to CX_3CR1 (Schafer *et al.*, 2004). The general concept that MP can transfer bioactive agents from cell to cell is further illustrated by the transfer of TF from leukocytes to platelets carried on LMP (Rauch *et al.*, 2000). Another example is the transfer of integrin CD41 from platelets to cancer cell lines on PMP (exosomes) (Janowska-Wieczorek *et al.*, 2005).

These studies support the general conclusion expressed by several other workers in MP studies (Freyssinet, 2003; Nomura, 2001) that MP can serve as vectors for the transfer of many bioactive mediators between cell types. In addition, these observations again raise the question: to what extent are circulating cytokine receptors truly soluble (Fernandez-Botran, 1991; Heaney and Golde, 1996) and which are MP-bound? (Section I.D).

Other cytokines/chemokines likely carried on MP are implied by the correlations shown in the study by Kim *et al.* (2003), though only some of them are thus far established to occur on MP, most notably VEGF (Brill *et al.*, 2005). The VEGF receptor-2 (Flk-1/KDR, CD309) has now been identified on EMP from EC precursor cells (Pirro *et al.*, 2006). EC precursor cells and their MP are an exciting new topic in the repair and recovery of ischemic damaged tissue in cardiology and it is expected that this will extend to recovery from ischemic brain injury.

D. Caspase-3, Fas, FasL, and Apoptosis

Apoptosis (also called programmed cell death) plays a central role in neurodegenerative, infectious, traumatic, ischemic, and demyelinating disorders (Bredesen *et al.*, 2006). The enzyme caspase-3 is a key player in apoptosis (Lakhani *et al.*, 2006). Hussein *et al.* (2005) found that not only EMP, but all MP examined were rich in caspase-3, and suggested that it could be transferred by the MP vector to other cells to induce their apoptosis. Alternatively, it had been earlier proposed that release of caspase-3 in MP (exosomes) functions to eliminate it and thereby rescue the cell from apoptosis (Gassart *et al.*, 2003), in a manner analogous to the release of active C from EC by shedding of EMP laden with bound C (Hamilton *et al.*, 1990). This is also observed for other cell types (Section III.H). Fas, another key player in apoptosis, has also been reported on MP (Albanese *et al.*, 1998), but whether it can be transferred to other cells was not addressed. Fas on MP may function as a decoy receptor, acting to adsorb and neutralize its ligand, FasL (Pitti *et al.*, 1998), among other conceivable roles.

E. β2-GPI, APS, and MS

The neurological manifestations of APS are well-known and have often been confused with MS (Cuadrado *et al.*, 2000; Hughes, 1999; Ijdo *et al.*, 1999; Karussis *et al.*, 1998). Antiphospholipid antibodies (APLA) are frequently detected in MS

(Cordoliani *et al.*, 1998; Fukazawa *et al.*, 1993; Marchiori *et al.*, 1990; Scott *et al.*, 1994; Sugiyama and Yamamoto, 1996; Tanaka *et al.*, 1987; Tintore *et al.*, 1996; Tsukada *et al.*, 1989). Reported frequencies of APLA in active MS vary widely, as they do for other conditions, due to different assay methodologies, but range from approximately 30% to 60% in the studies mentioned above, and in our own studies (cited below), similar to systemic lupus erythematosus (SLE) and ITP.

Several lines of evidence implicate MP in APS. Nomura *et al.* (1992) found that PMP were elevated in APS, and more pointedly, that APLA interact with PMP. The interaction of APLA with PMP and whole platelets was further studied by Galli *et al.* (1993). EMP levels were also elevated in APS and SLE patients with APLA, and were associated with lupus anticoagulant (LAC) activity (Dignat-George *et al.*, 2004). In addition, they showed that plasma obtained from patients with APS and other immune-mediated disorders stimulated release of EMP from EC *in vitro*. However, only APS plasma and not plasma from patients with other immune-mediated disorders induced EMP with procoagulant activity.

MP have been implicated in pathogenesis of APS since β2-GPI, the principal antigen of anticardiolipin (aCL) in APS, was found to be carried on PMP (Nomura *et al.*, 1993; Vallar *et al.*, 2001). This was subsequently studied by Nomura *et al.* (1995b) in relation to aCL in SLE, where it commonly occurs. β2-GPI is also known as apolipoprotein H. Although it has many potential functions, its true physiological role is not established, but is best known as the target antigen of aCL, most closely associated with LAC and with thrombosis in APS (McIntyre *et al.*, 1997).

We have confirmed a high prevalence of APLA in MS and, in addition, made the novel observations that (1) APLA rose with exacerbation and declined or disappeared with approach to remission, (2) all of the APLA types detected in MS were of IgM class, (3) anti-β2-GP-I was positive in 80% of cases in exacerbation but none in remission (Jy *et al.*, 2005). Interestingly, this pattern was similar to what we observed in ITP (Bidot *et al.*, 2004), except that APLA in ITP included IgG, and anti-β2-GP-1 was less prevalent. Other APLA antigens known to occur on MP include TFPI (Forastiero *et al.*, 2003), endothelial protein C receptor (EPCR) (Hurtado *et al.*, 2004), coagulation factor VII (FVII) (Bidot *et al.*, 2003), PAF (Barquinero *et al.*, 1993), prothrombin, and others.

Some mention of antiendothelial antibodies (aEC) is appropriate in this context since some APLA induce release of EMP from EC (Combes *et al.*, 1999), or cross-react with EC (Meroni *et al.*, 2001; Pierangeli *et al.*, 2001; Simantov *et al.*, 1996), and aEC falls within the broad definition of APLA: antibodies against proteins that bind or are bound to PLs/cell membranes. aEC have been detected in MS and have been proposed to contribute to its pathogenesis. In 1987, Tanaka *et al.* found aEC in 75% of active MS but in only 4% of remission (Tanaka *et al.*, 1987) and later attempted to identify their target antigens (Tsukada *et al.*, 1989).

However, a 1992 report found only 13% positive (Souberbielle *et al.*, 1992) and a 1996 report found only 10% reactive to human umbilical vein EC (HUVEC) (Tintore *et al.*, 1996). Yet, another 1996 report, differing in using brain microvessels, confirmed Tanaka's original work, finding that 12/16 active versus 0/15 inactive MS patients reacted to EC (Trojano *et al.*, 1996). Such discrepant findings may be attributable to differences in assay methodologies.

F. CD40/CD40L

Many readers may be aware of growing interest in CD40 and its ligand, CD40L (CD154), in autoimmune disorders such as APS (Vlachoyiannopoulos *et al.*, 2004), in hemostasis (Langer *et al.*, 2005) and inflammation, and MP may support the traffic of these agents in these conditions. CD40L was originally described for its pivotal role in the immune biology of B and T lymphocytes (Grewal and Flavell, 1998) but was subsequently shown to be widely distributed, but derived mainly from activated platelets (Andre *et al.*, 2002; Hermann *et al.*, 2001; Inwald *et al.*, 2003). Blann *et al.* (2005) has argued that other cells may also contribute significant CD40L since they found good correlation between all pairs of platelet activation markers (CD62P, CD63, and PMP), but no correlation of any of them with soluble CD40L (sCD40L) in plasma.

Measurement of sCD40L *in vivo* has been widely misinterpreted. First, sCD40L in serum, which has been the usual specimen, is grossly higher than in plasma, demonstrated in our laboratory (Ahn *et al.*, 2004a), reported again later in 2004 (Thom *et al.*, 2004), and again in 2006 (Varo *et al.*, 2006). The explanation for this is that platelets become activated during clotting and secrete their contents of CD40L; thus, the level in serum reflects almost the total available, not the level circulating in plasma *in vivo*, and is mainly an artifact *ex vivo*. Second, some or possibly all CD40L is MP-associated, not truly soluble. To our knowledge, PMP-associated CD40L was first identified in 2002 (Amirkhosravi *et al.*, 2002). Ahn *et al.* (2004a) demonstrated that sCD40L could be partly removed from plasma by filtration through 0.1 μm (Ahn *et al.*, 2004a) and Varo *et al.* (2006) showed that it could be entirely removed by high-speed centrifugation. These facts indicate that sCD40L is associated mainly with very small MP, probably exosomes (Section IV).

In regard to the mechanism of expression of platelet CD40L, the peroxisome proliferator-activated receptor (PPAR) has now been identified on platelets, and agonists of it completely block CD40L expression as well as ADP-induced platelet aggregation (Akbiyik *et al.*, 2004). Incidentally, the popular dietary polyphenolic antioxidants found in red wine, quercetin, and catechin are inhibitors of CD40L expression (Pignatelli *et al.*, 2005).

G. Inflammation, MP, and the C System

The C system is fundamental to inflammation and overlaps with the coagulation system (Blajchman and Ozge-Anwar, 1986), including in its anticoagulant actions. For example, protein S in circulation associates with the C4b-binding protein (Dahlback and Stenflow, 1981), and both are partly MP-bound (Gris *et al.*, 1996). Fixed C is selectively released bound to MP from EC (Hamilton *et al.*, 1990), platelets (Sims, 1989; Sims and Wiedmer, 1986), PMN (Stein and Luzio, 1991), red blood cells (RBC) (Pascual *et al.*, 1993), neutrophils (Stein and Luzio, 1991), and oligodendrocytes of the CNS (Scolding *et al.*, 1989). The effect of surface charge of synthetic MP (liposomes) on C activation has been investigated (Chonn *et al.*, 1991). Morgan has reviewed the membrane proteins which protect against self-attack by autologous C (Morgan and Meri, 1994) and some of them are known to occur on MP, for example DAF on RMP (Butikofer *et al.*, 1989).

A central role for C in the pathogenesis of APS has been proposed (Arnout *et al.*, 1997) and this hypothesis has been supported by Munakata *et al.* (2000), who showed that essentially all (96.4%) aCL from patients with APS were capable of fixing C. They conclude that C-fixing aCL may specifically occur in association with episodes of thrombosis in APS (Munakata *et al.*, 2000). Fischetti *et al.* (2005) have shown that thrombus formation attributed to anti-β2-GPI in APS is C-dependent (Fischetti *et al.*, 2005). Thus, since β2-GPI occurs on PMP (Nomura *et al.*, 1993; Vallar *et al.*, 2001), there now exist links between APS, C, and cell-derived MP. Although the chain of causation remains unclear, it may be possible to detect C and/or IgG/IgM or immune complex on the MP, as was done *in vitro* (Hamilton *et al.*, 1990; Sims *et al.*, 1988). Both aCL and the LAC have been shown to interact with PMP much as they do with other anionic PL surfaces (Galli *et al.*, 1993).

H. Summary

In Sections III.B–H, we have briefly reviewed some of the agents listed in Table III with respect to their presence on MP, in relation to neurological disorders, and/or general physiological processes known or believed to underlie neuropathologic syndromes—for example autoimmunity, inflammation, and C. On the other hand, it should hardly be necessary to point out that as yet, there is no proven causal connection between MP and neurological dysfunction, a difficult challenge. All of the evidence summoned in this review is, like that for many other theories current in neurology, circumstantial or based on studies *in vitro*. The task of this chapter is to exhibit the evidence in a favorable light, leaving the reader to judge its merits.

IV. Exosomes

A. Background

During the last decade, a seemingly distinct class of circulating cell-derived MP has come to attention, termed *exosomes* (e.g., not to be confused with this term in molecular biology) (Hilleres *et al.*, 2001). Their main distinguishing feature is a more uniform and smaller size, 40–80 nm (Denzer *et al.*, 2000) or 60–90 nm (Wolfers *et al.*, 2001) compared to the more heterogeneous and usually larger MP discussed above, which extend up to roughly 1500 nm (1.5 μm) (Heijnen *et al.*, 1999). Exosomes are released from endosomes of the cell interior (multivesicular bodies of the late endocytic compartment) rather than from the plasma membrane (Manoury *et al.*, 1999; Segura *et al.*, 2005).

The topic of exosomes was well reviewed by Denzer *et al.* (2000), at that time they were cautious about assigning function to them, but subsequent work cited below leaves little doubt of their many key roles. A 2005 symposium on exosomes found little agreement on their defining features (Couzin, 2005).

According to Gasaart *et al.* (2003), exosomes were first recognized in 1983, but electron micrographs by Allen *et al.* in 1980 clearly show uniform 60-nm particles from RBC enriched in acetylcholinesterase activity, which they termed *nanovesicles* (Allen *et al.*, 1980), and which were enriched in sphingomyelin and depleted in phosphatidylethanolamine. These appear to be exosomes. This size range 60–90 nm, for example, is about twice that of HDL/LDL particles (Barzilai *et al.*, 2003), and similar methods are used in the isolation for study of both high-speed differential and density centrifugations.

B. Overlap with MP of Plasma Membrane?

Here we addressed the question of whether or to what extent exosomes and overlap with MP of plasma membrane (MP_{PM}) may be confused in flow cytometric assays. This is important in view of the various methodologies in use, discussed in Section I.C. That is, some laboratories which study MP_{PM} employ rather high-speed preliminary centrifugations and isolate a much higher-speed fraction than others, implying that they could be measuring predominately exosomes, whether by flow cytometry, ELISA, or bead capture. This makes it impossible to compare results between these laboratories since they may be measuring different classes of MP.

In one study of exosomes, the isolated size fraction of >100 nm was excluded from further study on the rather arbitrary grounds that larger particles are MP_{PM}, not exosomes, yet both expressed the same exosomes' signature antigens CD81 target for proliferative antigen-1 (TAPA-1) and lysosomal marker CD63 lysosome integrate membrane protein (LIMP) (Fritzsching *et al.*, 2002). Thus, there is some overlap of antigenic markers since CD63 has often been used to detect PMP (Abrams and Shattil, 1991; Abrams *et al.*, 1990; Israels *et al.*, 2001; Kannon *et al.*, 1995).

Several GPI-anchored plasma membrane proteins of the reticulocytes can appear on exosomes, such as acetylcholinesterase, decay-accelerating factor, and the lipid composition of exosomes (cholesterol to PL ratio) also reflects that of the plasma membrane of reticulocytes (Gassart *et al.*, 2003; Rabesandratana *et al.*, 1998). Thus, it is not possible to distinguish exosomes from MP_{PM} based on their chemical composition. The exosomes from reticulocytes are said to be devoid of AnV-binding sites, but this is not a distinctive feature since we observe that only a fraction of RBC MP_{PM} bind AnV, these variations likely reflecting the conditions under which they were produced.

Nevertheless, the two types of particles, exosomes and MP_{PM}, from platelets appear distinctive in EM (Heijnen *et al.*, 1999). According to Heijnen *et al.*, the exosomes could not be detected in flow cytometry by biotinylated anti-CD63 and streptavidin FITC, even though CD63 was most abundant, because of their very small size. However, although it is not clear from their description of methods, it appears they are triggering on the light-scatter signal rather than fluorescent events; light-scatter is much less sensitive than fluorescence on most clinical flow cytometer machines.

Further, another reason for overlap of detection between MP_{PM} and exosomes is that exosomes often occur in aggregates or clusters, making them as large as MP_{PM}. In electron microscopy (EM) photos of RMP by Lutz *et al.*, the majority are 185 ± 23 nm, but a significant number are aggregated to size of about 500 nm (Lutz *et al.*, 1977). Likewise, Polasek describes such particles from platelets as multivesicular, since many of them appear in scanning EM as tight clusters of about 20 smaller ones, again totaling about 500 nm (Polasek, 1982). Transmission EM by George *et al.* reveals a range of sizes from about 60 to 500 nm and some of the larger ones may be aggregrates of smaller ones (George *et al.*, 1982). Several photos of these have been published by Hughes *et al.* (2000).

In regard to signal brightness, we have recently been using FITC-labeled lectin *Ulex europaeus I* (UEA-1) to label and count "total MP," since our attempts to use fluorescent lipid dyes to visualize small MP gave inconsistent results. Using FITC-UEA-1 we consistently observed up to 20-fold or more MP in a given sample than by any other marker. This leads us to suspect that the bright signal of FITC-UEA-1 brings into view a larger number of MP than usually detected (triggering on fluorescence), almost certainly including exosomes. If so, then UEA-1 may be a useful alternative to tedious density fractionations for many purposes, as it has been pointed out that studies of exosomes are much hindered by the difficulty of isolating them (Couzin, 2005).

In summary, it is likely that there is overlap of detection of MP_{PM} with exosomes but to variable extent, depending on such factors as state of aggregation of exosomes and method or marker of detection. When MP are detected by ELISA or by bead capture, it is likely that both species are detected, but again depending on capture antibody. Accordingly, it is likely that different laboratories are looking at different collections of MP phenotypes. This may explain some discrepant findings.

C. Overlap of Physiological Functions

Persuasive evidence supports a role of exosomes in angiogenesis, signaling, transport, antigen presentation, cancer metastasis, viral infections, and perhaps other physiological and pathophysiological roles (Blanchard *et al.*, 2002; Gould *et al.*, 2003; Raposo *et al.*, 1996; Segura *et al.*, 2005; Wolfers *et al.*, 2001; Zitvogel *et al.*, 1998). However, nearly all of these functions (except antigen presentation) have also been claimed for MP_{PM}. Thus, it is not clear at present if some of these roles are unique to exosomes or may overlap, to an uncertain extent, with roles now attributed to MP_{PM}. For example, it is clear that MP are active in angiogenesis (Brill *et al.*, 2005; Janowska-Wieczorek *et al.*, 2005; Kim *et al.*, 2002, 2004; Mezentsev *et al.*, 2005), but it is not clear if this is the result of exosomes as has been suggested, or MP_{PM}, since most of the experiments are with supernatants of stimulated platelets and thus may include both types. Sorting out these functions by particle type is an important consideration for future research.

Meanwhile, conceding that exosomes function in antigen presentation (Raposo *et al.*, 1996), this would constitute a further parallel between the "immunologic synapse" and the neuronal synapse, which is also mediated by vesicular transport, and involves similar release mechanisms (Giraudo *et al.*, 2006; Pobbati *et al.*, 2006).

V. Summary and Conclusions

We have observed that an enormous body of circumstantial evidence exists, supporting a central role of cell-derived MP and exosomes in many physiological and disease processes relevant to immunology, neurobiology, hemostasis, thrombosis, cell signaling, cancer, and angiogenesis. However, this is a new frontier, not a well-charted area of secure knowledge, and much work remains to be done in order to establish which subtypes of MP are responsible for which activities.

Of particular interest to readers of this chapter is the question of how MP are related to neurological diseases of various kinds. We feel that the many strong associations are more than epiphenomena, but pinning them down is not easy.

At the very least, it is now clear that assay of MP in clinical samples is a valuable tool for the monitoring of disease progression.

If this chapter succeeds in generating interest or arousing readers to further explore this new territory, bringing back more secure understanding, it will have achieved its goal. The ultimate hope, of course, is that such exploration may lead to new insights that can bring relief to those afflicted with one or more of the diseases considered.

Abbreviations

Ab, antibody; aCL, anticardiolipin Ab; ACS, acute coronary syndrome; Ag, antigen; AIDS, acquired immune deficiency syndrome; AIHA, autoimmune hemolytic anemia; AnV, annexin V; APLA, antiphospholipid antibodies; APP, amyloid β precursor protein; APS, antiphospholipid syndrome; ARDS, acute respiratory distress syndrome; AD, Alzheimer's disease; C, complement; CAD, coronary artery disease; CAEC, coronary artery endothelial cells; CEC, circulating endothelial cells; CEPC, circulating endothelial progenitor cells; DAF, decay accelerating factor; DDAVP, desmopressin; DIC, disseminated intravascular coagulation; EC, endothelial cells; ELISA, enzyme-linked immunosorbant assay; EMP, endothelial microparticles; FITC, FL label, green: flourescein isothiocyanate; FL, fluorescence; GPI, glycosyl phosphatidylinositol; HIT, heparin-induced thrombocytopenia; HITT, HIT with thrombosis; HUS, hemolytic uremic syndrome; HUVEC, human umbilical vein EC; ITP, immune thrombocytopenic purpura; LAC, lupus anticoagulant; LMP, leukocyte microparticles; mAb, monoclonal antibody; MAC, membrane attack complex (complement); MI, myocardial infarction; MIRL, membrane inhibitor of reactive lysis; MMP, matrix metalloproteinase; MP, microparticles; MS, multiple sclerosis; PAF, platelet-activating factor; PBMC, peripheral blood mononuclear cells; PCA, procoagulant activity; PECAM-1, platelet-endothelial cell adhesion molecule; PF3, platelet factor 3 (procoagulant PL); PL, phospholipid; PMN, polymorphonuclear leukocytes (neutrophils); PMP, platelet microparticles; PNH, paroxysmal nocturnal hemoglobinuria; PPP, platelet-poor plasma; PRP, platelet-rich plasma; PS, phosphatidylserine, (an anionic PL); PVD, peripheral vascular disease; RA, rheumatoid arthritis; RBC, red blood cells (erythrocytes); RMP, MP from red blood cells; SEM, scanning electron microscopy; SLE, systemic lupus erythematosus; TEM, transmission electron microscopy; TF, tissue factor; TFPI, tissue factor pathway inhibitor; TIA, transient ischemic attacks; TIMP, tissue inhibitors of metalloproteinase; TM, thrombomodulin; TSP, thrombospondin; TTP, thrombotic thrombocytopenic purpura; uPA, urokinase type plasminogen activator; uPAR, uPA receptor; VEGF, vascular endothelial growth factor; VTE, venous thromboembolism; vWD, von Willebrand disease; vWF, von Willebrand factor; WBC, white blood cells (leukocytes).

Glossary

A. Equivalent Names of CD Numbers Used

CD	11	a	LFA-1 (integrin)
CD	11	b	Mac-1 (integrin)
CD	14		LPSR (GPI-linked)
CD	15	s	sialyl Lewis x; see also CD162
CD	16	a, b	Fc-γ-RIII, A, B
CD	18		Integrin β2
CD	27		TNF-R
CD	31		PECAM-1
CD	36		GP IV
CD	41		GP IIb (integrin family)
CD	42	a	GP IX
CD	42	b	GP 1b
CD	42	d	GP V
CD	45		LCA (leukocyte common antigen)
CD	51		Vitronectin receptor, an integrin
CD	54		ICAM-1
CD	61		GP IIIa; integrin β3
CD	62	E	E-selectin
CD	62	L	L-selectin
CD	62	P	P-selectin
CD	63		LIMP
CD	81		TAPA-1
CD	87		uPAR (GPI-anchored)
CD	102		ICAM-2
CD	105		Endoglin, MUC-1
CD	106		VCAM-1
CD	141		TM
CD	142		Tissue factor
CD	144		VE-cadherin or cadherin-5
CD	146		S-endo; MUC-18
CD	154		CD40L
CD	162		PSGL-1
CD	183		CXCR3

CD	184	CXCR4
CD	195	CCR5
CD	197	CCR7
CD	201	EPCR
CDw	210	IL-10R
CD	212	IL-12R
CD	213	IL-13R (two forms)
CDw	217	IL-17R
CD	220	Insulin R
CD	230	Prion proteins
CD	235	Glycophorin (several; & 236)

References

Abrams, C., and Shattil, S. J. (1991). Immunological detection of activated platelets in clinical disorders. *Thromb. Haemost.* **65,** 467–473.

Abrams, C. S., Ellison, N., Budzynski, A. Z., and Shattil, S. J. (1990). Direct detection of activation platelet and platelet-derived microparticles in humans. *Blood* **75,** 128–138.

Ahlgren, E., Lundqvist, A., Nordlund, A., Aren, C., and Rutberg, H. (2003). Neurocognitive impairment and driving performance after coronary artery bypass surgery. *Eur. J. Cadiothorac. Surg.* **23,** 334–340.

Ahn, E. R., Lander, G., Jy, W., Bidot, C., Jimenez, J. J., Horstman, L. L., and Ahn, Y. S. (2004a). Differences of soluble CD40L in sera and plasma: Implications on CD40L assay as a marker of thrombotic risk. *Thromb. Res.* **114,** 143–148.

Ahn, Y. S., Horstman, L. L., Jy, W., Jimenez, J. J., and Bowen, B. (2002). Vascular dementia in patients with immune thrombocytopenic purpura (ITP). *Thromb. Res.* **107,** 337–344.

Ahn, Y. S., Jy, W., Jimenez, J. J., and Horstman, L. L. (2004b). More on: Cellular microparticles: What are they bad or good for? [Commentary]. *J. Thromb. Haemost.* **2,** 1215–1216.

Akbiyik, F., Ray, D. M., Gettings, K. F., Blumberg, N., Francis, C. W., and Phipps, R. W. (2004). Human bone marrow megakaryocytes and platelets express PPARγ, and PPARγ agonists blunt platelet release of CD40 ligand and thromboxanes. *Blood* **104,** 1361–1368.

Allen, D., Thomas, P., and Limbrick, A. R. (1980). The isolation and characterization of 60 nm vesicles ("nanovesicles") produced during ionophore A23187-induced budding of human erythrocytes. *Biochem. J.* **188,** 881–887.

Albanese, J., Meterissian, S., Kontogiannea, M., Dubreuil, C., Hand, A., Sorba, S., and Dainiak, N. (1998). Biologically active Fas antigen and its cognate ligand are expressed on plasma membrane-derived extracellular vesicles. *Blood* **91,** 3862–3874.

Amabile, N., Guerin, A. P., Leroyer, A., Mallat, Z., Nguyen, C., Boddaert, J., London, G. M., Tedgui, A., and Boulanger, C. M. (2005). Circulating endothelial microparticles are associated with vascular dysfunction in Patients with end-stage renal failure. *J. Am. Soc. Nephrol.* **16,** 3381–3388.

Amirkhosravi, A., Meyer, T., Sackel, D., Desai, H., Biddinger, R., Amaya, M., and Francis, J. L. (2002). Platelet microparticles upregulate TF and VEGF in endothelial and melanoma cells in a

CD40 ligand-dependent manner: Possible role in angiogenesis and metastasis. *Blood* **100**(11), 63b (Abstract 3721).

Anand, S. X., Viles-Gonzalez, J. F., Badimon, J. J., Cavusoglu, E., and Marmur, J. D. (2003). Membrane-associated CD40L and sCD40L in atherosclerotic disease. *Thromb. Haemost.* **90,** 377–384.

Andre, P., Prasad, K. S. S., Denis, C. V., He, M., Papalia, J. M., Hynes, R. O., Phillips, D. R., and Wagner, D. D. (2002). CD40L stabilizes arterial thrombi by a β-3 integrin dependent mechanism. *Nat. Med.* **8,** 247–252.

Andrew, A. J. H., and Newman, W. (1993). Circulating adhesion molecules in disease (review). *Immunol. Today* **14,** 506–512.

Anker, P., and Stroun, M. (2002). Progress in the knowledge of circulating nucleic acids: Plasma RNA is particle-associated. Can it become a general detection marker for a cancer blood test? (Editorial). *Clin. Chem.* **48,** 1210–1211.

Aras, O., Shet, A., Hysjulien, J. L., Escolar, G., Slungaard, A., Hebbel, R. P., Bach, R. R., Jilma, B., and Key, N. S. (2002). Induction of microparticle and cell-associated intravascular tissue factor in human endotoxemia. *Blood* **100,** 55a (Abstract 198).

Arnout, J., Wittevrongel, C., and Vermylen, J. (1997). Phospholipid binding proteins and the antiphospholipid syndrome: A role for complement factor H? *Thromb. Haemost. Suppl. 1* (Abstract SC-1).

Atsumi, T., Khamashta, M. A., Amengual, O., and Hughes, G. R. V. (1997). Up-regulated tissue factor expression in antiphospholipid syndrome. *Thromb. Haemost.* **77,** 222–223.

Banfi, C., Brioschi, M., Wait, R., Begum, S., Gianazza, E., Pirillo, A., Mussoni, L., and Tremoli, E. (2005). Proteome of endothelial cell-derived procoagulant microparticles. *Proteomics* **5,** 4443–4455.

Barquinero, J., Ordi-Ros, J., Selva, A., Perez-Peman, P., Vilardell, M., and Kamashta, M. (1993). Antibodies against platelet activating factor in patients with antiphospholipid antibodies. *Lupus* **3,** 55–58.

Barzilai, N., Atzmon, G., Schechter, C., Schaefer, E. J., Cupples, A. J., Lipton, R., Cheng, S., and Shuldiner, A. R. (2003). Unique lipoprotein phenotype and genotype associated with exceptional longevity. *JAMA* **290,** 2030–2040.

Basire, A., Sabatier, F., Ravet, S., Lamy, E., Mialhe, A., Zabouo, G., Paul, P., Gurewich, V., Sampol, J., and Dignat-George, F. (2006). High urokinase expression contributes to the angiogenic properties of endothelial cells derived from circulating progenitors. *Thromb. Haemost.* **95,** 678–688.

Baskin, F., Rosenberg, R. N., Iyer, L., Hynan, L., and Cullum, C. M. (2000). Platelet APP isoform ratios correlate with declining cognition in AD. *Neurology* **54,** 1907–1909.

Behnke, O., and Forer, A. (1993). Blood platelet heterogeneity: Evidence for two classes of platelets in man and rat. *Br. J. Haematol.* **84,** 686–693.

Bernal-Mizrachi, L., Soriano, A., Valle, M., Jy, W., Jimenez, J. J., Horstman, L. L., and Ahn, Y. S. (2002). Repeated ischemia reduces the capacity of endothelium to release endothelial microparticles (EMP): *In vivo* experiment and clinical observations. *Blood* **100,** 71b (Abstract 3756).

Bernal-Mizrachi, L., Jy, W., Jimenez, J. J., Pastor, J., Mauro, L., Horstman, L. L., deMarchena, E., and Ahn, Y. S. (2003). High levels of circulating endothelial microparticles in patients with acute coronary syndromes. *Am. Heart J.* **145,** 962–970.

Bernal-Mizrachi, L., Jy, W., Fiero, C., McDonough, E., Purow, J., Velasquez, H. S., Jimenez, J. J., Horstman, L. L., Ahn, Y. S., and DeMarchena, E. (2004). Endothelial microparticles correlate with high-risk angiographic lesions in acute coronary syndromes. *Int. J. Cardiol.* **97,** 439–446.

Bidot, C. J., Jy, W., Horstman, L. L., Huang, H., Jimenez, J. J., Yaniz, M., and Ahn, Y. S. (2003). Factor VII/VIIa: A new antigen in the antiphospholipid antibody syndrome. *Br. J. Haematol.* **120,** 618–626.

Bidot, C. J., Jy, W., Horstman, L. L., Ahn, E. R., Jimenez, J. J., Yaniz, M., Lander, G., and Ahn, Y. S. (2004). Antiphospholipid antibodies in immune thrombocytopenic purpura tend to emerge in exacerbation and decline in remission. *Br. J. Haematol.* **128,** 366–372.

Biro, E., Nieuwland, R., and Sturk, A. (2004). Measuring circulating cell-derived microparticles. *J. Thromb. Haemost.* **2,** 1843–1844.

Blajchman, M. A., and Ozge-Anwar, A. H. (1986). The role of the complement system in hemostasis. *Prog. Hematol.* **XIV,** 149–182.

Blanchard, N., Lankar, D., Faure, F., Regnault, A., Dumont, C., Raposo, G., and Hivroz, C. (2002). TCR activation of human T cells induces the production of exosomes bearing the TCR/CD3/zeta complex. *J. Immunol.* **168,** 3235–3241.

Blann, A. D., and McCollum, C. N. (1998). Increased soluble P-selectin in peripheral artery disease: A new marker for the progression of atherosclerosis. *Thromb. Haemost.* **80,** 1031–1032.

Blann, A. D., Faragher, E. B., and McCollum, C. N. (1997). Increased soluble P-selectin following myocardial infarction: A new marker for progression of atherosclerosis. *Blood Coagul. Fibrinolysis* **8,** 383–390.

Blann, A. D., deRomeuf, C., Mazurier, C., and McCollum, C. N. (1998). Circulating von Willebrand factor antigen II in atherosclerosis: A comparison with von Willebrand factor and soluble thrombomodulin. *Blood Coagul. Fibrinolysis* **9,** 261–266.

Blann, A. D., Lip, G. Y. H., and McCollum, C. N. (2002). Influence of the risk factors for atherosclerosis on levels of soluble adhesion molecules and endothelial markers in peripheral vascular disease (Letter). *Thromb. Haemost.* **88,** 366–367.

Blann, A. D., Tan, K. T., Tayebjee, M. H., Davagnanam, I., Moss, M., and Lip, G. Y. H. (2005). Soluble CD40L in peripheral artery disease. *Thromb. Haemost.* **93,** 578–583.

Blumenthal, J. A., Madden, D. J., Burker, E. J., Croughwell, N., Schniebolk, S., Smith, R., White, W. D., Hlatky, M., and Reves, J. G. (1991). A preliminary study of the effects of cardiac procedures on cognitive performance. *Int. J. Psychosom.* **38,** 13–16.

Bodin, S., Viala, C., Ragab, A., and Payrastre, B. (2003). A critical role of lipid rafts in the organization of a key Fc-γ-RIIa-mediated signaling pathway in human platelets. *Thromb. Haemost.* **89,** 318–330.

Boehme, M. W. J., Raeth, U., Scherbaum, W. A., Galle, P. R., and Stremmel, W. (2000). Interaction of endothelial cells and neutrophils *in vitro*: Kinetics of thrombomodulin, intercellular adhesion molecule-1 (ICAM-1): Implications for relevance as serological disease activity markers in vasculitides. *Clin. Exp. Immunol.* **119,** 250–254.

Bonderman, D., Teml, A., Jakowitsch, J., Adlbrecht, C., Gyongyosi, M., Sperker, W., Lass, H., Mosgoeller, W., Glogar, D. H., Probst, P., Maurer, G., Nemerson, Y., *et al.* (2002). Coronary no-reflow is caused by shedding of active tissue factor from dissected atherosclerotic plaque. *Blood* **99,** 2794–2800.

Boulanger, C. M., Scoazec, A., Ebrahimian, T., Henry, P., Mathieu, E., Tedgui, A., and Mallat, Z. (2001). Circulating microparticles from patients with myocardial infarctions cause endothelial dysfunction. *Circulation* **104,** 2649–2652.

Bredesen, D. E., Rao, R. V., and Mehlen, P. (2006). Cell death in the nervous system. *Nature* **443,** 796–802.

Brill, A., Dashevsky, O., Rivo, J., Gozal, Y., and Varon, D. (2005). Platelet-derived microparticles induce angiogenesis and stimulate post-ischemic revascularization [with editorial, p1–3, by Boulanger and Tedgui]. *Cardiovasc. Res.* **67,** 30–38.

Bugert, P., Dugrillon, A., Gunaydin, A., Eichler, H., and Kluter, H. (2003). Messenger RNA profiling in human platelets by microarrary hybridization. *Thromb. Haemost.* **90,** 738–748.

Burger, J. A., and Kipps, T. J. (2006). CXCR4: A key receptor in the crosstalk between tumor cells and their microenvironment. *Blood* **107,** 1761–1767.

Butikofer, P., Kuypers, F. A., Xu, C. M., Chiu, D. T. Y., and Lubin, B. (1989). Enrichment of two glycosyl-phosphatidylinositol-achored proteins, acetylcholinesterase and decay accelerating factor, in vesicles released from human red blood cells. *Blood* **74,** 1481–1485.

Camire, R. M., Pollak, E. S., Kaushansky, K., and Tracy, P. B. (1998). Secretable human platelet-derived factor V originates from the plasma pool. *Blood* **92,** 3035–3041.

Capitanio, A. M., Niewiarowski, S., Ruciniski, B., Tuszynski, G. P., Cierniewski, C. C., Hershock, D., and Kornecki, E. (1985). Interaction of platelet factor 4 with human platelets. *Biochim. Biophys. Acta* **839,** 161–173.

Chong, B. H., Murray, B., Berndt, M. C., Dunlop, L. C., Brighton, T., and Chesterman, C. N. (1994). Plasma P-selectin is increased in thrombotic platelet consumptive disorders. *Blood* **83,** 1535–1541.

Chonn, A., Cullis, P. R., and Devine, D. V. (1991). The role of surface charge in the activation of the classical and alternative pathways of complement by liposomes. *J. Immunol.* **146,** 4234–4241.

Chou, J., Mackman, N., Skoloff, G. M., Pedersen, B., Furie, B. C., and Furie, B. (2004). Hematopoietic cell-derived microparticle tissue factor contributes to fibrin formation during thrombus propagation. *Blood* **104,** 3190–3197.

Cipollone, F., Ferri, C., Desideri, G., Paloscia, L., Materazzo, G., Mascellanti, M., Fazia, M., Iezzi, A., Cuccurullo, C., Pini, B., Bucci, M., Santucci, A., *et al.* (2003). Preprocedural level of soluble CD40L is predictive of enhanced inflammatory response and restenosis after coronary angioplasty. *Circulation* **108,** 2276–2282.

Combes, V., Simon, A. C., Grau, G. E., Arnoux, D., Camoin, L., Sabatier, F., Mutin, M., Sanmarco, M., Sampol, J., and Dignat-George, F. (1999). *In vitro* generation of endothelial microparticles and possible prothrombotic activity in patients with lupus anticoagulant. *J. Clin. Invest.* **104,** 93–102.

Combes, V., Taylor, T. E., Juhan-Vague, I., Mege, J. L., Mwenechanya, J., Tembo, M., Grau, G. E., and Molyneux, M. E. (2004). Circulating endothelial microparticles in Malawaian children with sever falciparum malaria complicated with coma. *JAMA* **291,** 2542–2544.

Conde, I. D., Bahrwani, L., Pendurthi, U., Thiagarajan, P., and Lopez, J. (2004). Extremely high levels of microvesicle-associated tissue factor in patient with cancer-related thrombosis. *Blood* **104,** 713a (Abstract 2605).

Coppinger, J. A., Cagney, G., Toomey, S., Kislinger, T., Belton, O., McRedmond, J. P., Cahill, D. J., Emili, A., Fitzgerald, D. J., and Maguire, P. B. (2004). Characterization of the proteins released from activated platelets leads to localization of novel platelet proteins in human atherosclerotic lesions. *Blood* **103,** 2096–2105.

Cordoliani, M. A., Michon-Pasturel, U., Arvieux, R. K., Masy, E., Hachulla, E., and Vermersch, P. (1998). Multiple sclerosis and antiphosphilipid antibodies: Study of 62 consecutive patients [Abstract; article French]. *Rev. Med. Interne* **19,** 635–639.

Couzin, L. (2005). The ins and outs of exosomes: Highlights from symposium, *Exosomes: Biological Significance*, Montreal, May 20–21. *Science* **308,** 1862–1863.

Cuadrado, M. J., Khamashta, M. A., Ballesteros, A., Godfrey, T., Simon, M. J., and Hughes, G. R. (2000). Can neurological manifestations of Hughes (antiphospholipid) syndrome be distinguished from multiple sclerosis? Analysis of 27 patients and review of the literature. *Medicine (Baltimore)* **79,** 57–68.

Dahlback, B., and Stenflow, J. (1981). High molecular weight complex in human plasma between vitamin K-dependent protein S and complement component C4b-binding protein. *Proc. Natl. Acad. Sci. USA* **78,** 2512–2516.

Davies, T. A., Long, H. J., Eisenhauer, P. B., Hastey, R., Cribbs, D. H., Fine, R. E., and Simons, E. R. (2000). β Amyloid fragments derived from activated platelets deposit in cerebrovascular endothelium: Usage of a novel blood brain barrier endothelial cell model system. *Amyloid* **7,** 153–165.

Denzer, K., Kleijmeer, M. J., Heijnen, H. F. G., Stoorvogel, W., and Geuze, H. J. (2000). Exosomes: From internal vesicle of the multivesicular body to intercellular signaling device (Review). *J. Cell Sci.* **113,** 3365–3374.

Dewanjee, M. K., Kapadvanjwala, M., Mao, W. W., Jy, W., Ahn, Y. S., Ruzius, K., Ghafouripoura, A. K., Serafini, A. N., and Sfajianakis, G. N. (1993). A higher blood flow window of reduced thrombogenicity and acceptable fragmentation in a hollow fiber hemodialyzer. *ASAIO J.* **39,** M363–M367.

Dhar, A., and Ganguly, P. (1988). Lectin fractionation of rabbit blood platelets: Isolation of subpopulations of different ages. *In* "Platelet Membrane Receptors" (G. A. Jamieson, Ed.), Vol. 283, pp. 77–81. Alan R. Liss, New York.

Dignat-George, F., Camoin-Jau, L., Sabatier, F., Arnoux, D., Anfosso, F., Bardin, N., Veit, V., Combes, V., Gentile, S., Moal, V., Sanmarco, M., and Sampol, J. (2004). Endothelial microparticles: A potential contribution to the thrombotic complications of the antiphospholipid syndrome [with editorial, pp. 636–638]. *Thromb. Haemost.* **91,** 667–673.

DiLuca, M., Pastorino, L., Bianchetti, A., Perez, J., Vignolo, L. A., Lenzi, G. L., Trabbucchi, M., Cattabeni, M., and Padovani, A. (1998). Differential level of platelet amyloid β precursor protein isoforms. *Arch. Neurol.* **55,** 1195–1200.

DiLuca, M., Colciaghi, F., Pastorino, L., Borroni, B., Padovani, A., and Cattabeni, F. (2000). Platelets as a peripheral district to study pathogenetic mechanisms of Alzheimer disease: The case of amyloid precursor protein. *Eur. J. Pharmacol.* **405,** 277–283.

Falati, S., Lin, Q., Gross, P., Merrill-Skoloff, G., Chou, J., Vandendries, E., Celi, A., Croce, K., Furie, B. C., and Furie, B. (2003). Accumulation of tissue factor into developing thrombi *in vivo* is dependant upon microparticle P-selectin glycoprotein ligand 1 and platelet P-selectin. *J. Exp. Med.* **197,** 1585–1598.

Faure, V., Dou, L., Sabatier, F., Cerini, C., Sampol, J., Berland, Y., Brunet, P., and Dignat-George, F. (2005). Elevation of circulating endothelial microparticles in patients with chronic renal failure. *J. Thromb. Haemost.* **4,** 566–572.

Favaloro, E. J. (1999). Laboratory assessment as a critical component of the appropriate diagnosis and sub-classification of von Willebrand's disease. *Blood Rev.* **13,** 185–204.

Favaloro, E. J., Mohammed, A., Coombs, R., and Mehrabani, P. A. (1995). Filtered plasma as a potential cause of clinical misdiagnosis: Inappropriate testing in a haematology laboratory. *Br. J. Biomed. Sci.* **52,** 243–248.

Favaloro, E. J., Nair, S. C., and Forsyth, C. J. (2001). Collection and transport of samples for laboratory testing in von Willebrand's disease (VWD): Time for a reappraisal? *Thromb. Haemost.* **86,** 1589–1590.

Fernandez-Botran, R. (1991). Soluble cytokine receptors: Their role in immune regulation. *FASEB J.* **5,** 2567–2574.

Ferreira, A. C., Peter, A. A., Mendez, A. J., Jimenez, J. J., Mauro, L. M., Chirinos, J. A., Ghany, R., Virani, S., Garcia, S., Horstman, L. L., Purow, J., Jy, W., *et al.* (2004). Postprandial hypertriglyceridemia increases circulating levels of endothelial cell microparticles. *Circulation* **110,** 3599–3603.

Fischetti, F., Durigutto, P., Pellis, V., Debeus, A., Macor, P., Bulla, R., Bossi, F., Ziller, F., Sblattero, D., Meroni, P., and Tedesco, F. (2005). Thrombus formation induced by antibodies to β-2-glycoprotein-1 is complement dependent and requires a priming factor. *Blood* **106,** 2340–2346.

Forastiero, R. R., Maetinuzzo, M. E., and Broze, G. J., Jr. (2003). High titers of to tissue factor pathway inhibitor are associated with the antiphospholipid syndrome. *J. Thromb. Haemost.* **1,** 718–724.

Freyssinet, J. M. (2003). Cellular microparticles: What are they bad or good for? *J. Thromb. Haemost.* **1,** 1655–1662.

Fritzsching, B., Schwer, B., Kartenbeck, J., Pedal, A., Horejsi, V., and Ott, M. (2002). Release and intercellular transfer of cell surface CD81 via microparticles. *J. Immunol.* **169,** 5531–5537.

Frojmovic, M., and Wong, T. (1991). Dynamic measurements of the platelet membrane glycoprotein IIb-IIIa receptor for fibrinogen by flow cytometry: II. Platelet size-dependent subpopulations. *Biophys. J.* **59,** 828–837.

Fukazawa, T., Moriwaka, F., Mukai, M., Hamada, T., Koike, T., and Tashiro, K. (1993). Anticardiolipin antibodies in Japanese patients with multiple sclerosis. *Acta Neurol. Scand.* **88,** 184–189.

Galli, M., Bevers, E. M., Comfurius, P., Barbui, T., and Zwaal, R. F. A. (1993). Effects of antiphospholipid antibodies on procoagulant activity of activated platelets and platelet-derived microvesicles. *Br. J. Haematol.* **83,** 466–472.

Gassart, A., Geminard, C., Fevrier, B., Raposo, G., and Vidal, M. (2003). Lipid raft-associated protein sorting in exosomes. *Blood* **102,** 4336–4344.

Gasser, O., and Schifferli, J. A. (2004). Activated polymorphonuclear neutrophils disseminate anti-inflammatory microparticles by ectocytosis. *Blood* **104,** 2543–2548.

Gasser, O., and Schifferli, J. A. (2005). Microparticles released by human neutrophils adhere to erythrocytes in the presence of complement. *Exp. Cell Res.* **307,** 381–387.

Gasser, O., Hess, C., Miot, S., Deon, C., Sanchez, J. C., and Schifferli, J. A. (2003). Characterization and properties of ectosomes release by human polymorphonuclear neutrophils. *Exp. Cell Res.* **285,** 243–257.

Gemmell, C. H., Ramirez, S. M., Yeo, E. L., and Sefton, M. V. (1995). Platelet activation in whole blood by artificial surface: Identification of platelet-derived microparticles and activated platelet binding to leukocytes as material-induced activation events. *J. Lab. Clin. Med.* **125,** 276–287.

George, J. N., Thoi, L. L., McManus, L. M., and Reimann, T. A. (1982). Isolation of human platelet membrane microparticles from plasma and serum. *Blood* **60,** 834–840.

Ghazanfari, F. A., and Stewart, R. R. (2001). Characteristics of endothelial cells derived from the blood-brain barrier and of astrocytes in culture. *Brain Res.* **890,** 49–65.

Giesen, P. L. A., and Nemerson, Y. (2000). Tissue factor on the loose. *Semin. Thromb. Haemost.* **26,** 379–383.

Giessen, P. L. A., Rauch, U., Bohrmann, B., Kling, D., Roque, M., Fallone, J. T., Badimon, J. J., Himber, J., Riederer, M. A., and Nemerson, Y. (1999). Blood-borne tissue factor: Another view of thrombosis. *Proc. Natl. Acad. Sci. USA* **96,** 2311–2315.

Gilbert, G. E., Sims, P. J., Wiedmer, T., Furie, B., Furie, B. C., and Shattil, S. J. (1991). Platelet-derived microparticles express high affinity receptors for factor VIII. *J. Biol. Chem.* **266,** 17261–17268.

Giraudo, C. G., Eng, W. S., Melia, T. J., and Rothman, J. E. (2006). A clamping mechanism involved in SNARE-dependent exocytosis. *Science* **313,** 676–679.

Gould, S. J., Booth, A. M., and Hildreth, J. E. K. (2003). The Trojan exosome hypothesis. *Proc. Natl. Acad. Sci. USA* **100,** 10592–10597.

Gousset, K., Tsvetkova, N. M., Crowe, J. H., and Tablin, F. (2004). Important role of raft aggregation in the signaling events of cold-induced platelet activation. *Biochim. Biophys. Acta* **1660,** 7–15.

Greenwalt, T. J. (2006). The how and why of exocytic vesicles. *Transfusion* **46,** 143–152.

Grewal, I. S., and Flavell, R. A. (1998). CD40 and CD154 in cell mediated immunity. *Annu. Rev. Immunol.* **16,** 111–135.

Gris, J. C., Toulon, P., Brun, S., Maugard, C., Sarlat, C., Schved, J. F., and Berlan, J. (1996). The relationship between plasma microparticles, protein S and anticardiolipin antibodies in patients with human immunodeficiency virus infection. *Thromb. Haemost.* **76,** 38–45.

Haller, H., Ziegler, E. M., Homuth, V., Drab, M., Eichorn, J., Nagy, Z., Busjahn, A., Vetter, K., and Luft, F. C. (1997). Endothelial adhesion molecules and leukocyte integrins in preeclamptic patients. *Hypertension* **29,** 291–296.

Hamilton, K. K., Hattori, R., Esmon, C. T., and Sims, P. J. (1990). Complement proteins C5b-9 induce vesiculation of the endothelial plasma membrane and expose catalytic surface for assembly of the prothrombinase enzyme complex. *J. Biol. Chem.* **265,** 3809–3814.

Hansen, J. B., Huseby, N. E., Sandset, P. M., Svensson, B., Lyngmo, V., and Nordoy, A. (1994). Tissue-factor pathway inhibitor and lipoproteins. Evidence for association with and regulation by LDL in human plasma. *Arterioscler. Thromb.* **14,** 223–229.

Hasselman, D. O., Rappl, G., Tilgen, W., and Reinhold, U. (2001). Extracellular tyrosinase mRNA within apoptotic bodies is protected from degradation in human serum. *Clin. Chem.* **47,** 1488–1489.

Hawrylowicz, C. M., Howells, G. L., and Feldmann, M. (1991). Platelet-derived interleukin 1 induces human endothelial adhesion molecule expression and cytokine production. *J. Exp. Med.* **174,** 785–790.

Haznedaroglu, I. C., Ozdemir, O., Ozcebe, O., Dundar, S. V., and Kirazl, S. (1996). Circulating thrombomodulin as a clue of endothelial damage in Behcet's disease (Letter). *Thromb. Haemost.* **75,** 974–975.

He, S., Silveira, A., Hamsten, A., Blomback, M., and Bremme, K. (1999). Haemostatic, endothelial and lipoprotein parameters and blood pressure levels in women with a history of preeclampsia. *Thromb. Haemost.* **81,** 538–542.

Heaney, M. L., and Golde, D. W. (1996). Soluble cytokine receptors (Review). *Blood* **87,** 847–857.

Heeschen, C., Dimmeler, S., Hamm, C. W., van den Brand, M. J., Boersma, E., Zeiher, A. M., Simoons, M. L., and CAPTURE Study Investigators (2003). Soluble CD40 ligand in acute coronary syndromes. *N. Engl. J. Med.* **348,** 1104–1111.

Heijnen, H. F. G., Schiel, A. E., Fijheer, R., Geuze, H., and Sixma, J. J. (1999). Activated platelets release two types of membrane vesicles: Microvesicles by surface shedding and exosome derived from exocytosis of multivesicular bodies and alpha-granules. *Blood* **94,** 3791–3799.

Helmuth, L. (2002). Protecting the brain while killing the pain? *Science* **297,** 1262–1263.

Heloire, F., Weill, B., Weber, S., and Batteux, F. (2003). Aggregates of endothelial microparticles and platelets circulate in peripheral blood. Variations during stable coronary disease and acute myocardial infarction. *Thromb. Res.* **110,** 173–180.

Hermann, A., Rauch, B. H., Braun, M., Schror, K., and Weber, A. A. (2001). Platelet CD40 ligand (CD40L)—subcellular localization, regulation of expression, and inhibition by clopidogrel. *Platelets* **12,** 74–82.

Heyl, W., Handt, S., Reister, F., Gelen, J., Schroder, W., Mittermayer, C., and Rath, W. (1999). Elevated soluble adhesion molecules in women with pre-eclampsia: Do cytokines like tumor necrosis factor-alpha and interleukin-1 β cause endothelial activation. *Eur. J. Obstet. Gynecol. Reprod. Biol.* **86,** 35–41.

Hilleres, P., McCarthy, T., Rosbash, M., Parker, R., and Jensen, T. H. (2001). Quality control of mRNA 3′-end processing is linked to the nuclear exosome. *Nature* **413,** 528–542.

Holmes, C. E., Bouchard, B. A., Barbick, R. R., Safdar, H., and Tracy, P. B. (2003). The existence of platelet subpopulations in patients with myeloproliferative disease: Preliminary correlation with clinically abnormal hemostasis. *Blood* **102**(11), 787a (Abstract 2909).

Horstman, L. L., and Ahn, Y. S. (1999). Platelet microparticles: A wide-angle perspective (Review). *Crit. Rev. Oncol. Haematol.* **30,** 111–142.

Horstman, L. L., Jy, W., Jimenez, J. J., and Ahn, Y. S. (2004). Endothelial microparticles as markers of endothelial dysfunction [Review]. *Front. Biosci.* **9,** 1118–1135.

Hughes, G. R. V. (1999). The antiphospholipid syndrome and 'multiple sclerosis' [with editorial, pp. 109–115]. *Lupus* **8,** 89.

Hughes, M., Hayward, C. P. M., Warkentin, T. E., Horsewood, P., Chimeyko, K. A., and Kelton, J. G. (2000). Morphological analysis of microparticle generation in heparin-induced thrombocytopenia. *Blood* **96,** 188–194.

Hurtado, V., Montes, R., Gris, J. C., Bertolaccini, M. L., Alonso, A., Martinez-Gonzalez, M. A., Khamashta, M. A., Fukudome, F., Lane, D. A., and Hermida, J. (2004). Autoantibodies against EPCR are found in antiphospholipid syndrome and are a risk factor for fetal death. *Blood* **104,** 1369–1374.

Hussein, M. N. A., Nieuwland, R., Hau, C. M., Evers, L. M., Meesters, E. W., and Sturk, A. (2005). Cell-derived microparticles contain caspase-3 *in vitro* and *in vivo* [with commentary by Ahn *et al.*, pp. 884–887]. *J. Thromb. Haemost.* **3**(5), 888–896.

Hwang, S. J., Ballantyne, C. M., Sharrett, R., Smith, L. C., Davis, C. E., Gotto, A. M., and Boerwinkle, E. (1997). Circulating adhesion molecules VCAM-1, ICAM-1, and E-selectin in carotid atherosclerosis and incident coronary heart disease cases. *Circulation* **96,** 4219–4225.

Ijdo, J. W., Conti-Kelly, A. M., Greco, P., Abedi, M., Amos, M., Provenzale, J. M., and Greco, T. P. (1999). Anti-phospholipid antibodies in patients with multiple sclerosis and MS-like illnesses: MS or APS? *Lupus* **8,** 109–115.

Ikeda, H., Takajo, Y., Ichiki, K., and Imaizumi, T. (1995). Increased soluble form of P-selectin in patients with unstable angina. *Circulation* **92,** 1693–1696.

Inwald, D. P., McDowall, A., Peters, M. J., Callard, R. E., and Klein, N. J. (2003). CD40 is constitutively expressed on platelets and provides a novel mechanism for platelet activation. *Circ. Res.* **92,** 1041–1048.

Israels, S. J., McMillan-Ward, E. M., Easton, J., Robertson, C., and McNicol, A. (2001). CD63 associates with the (alpha IIb β 3) integrin-CD9 complex on the surface of activated platelets. *Thromb. Haemost.* **85,** 134–141.

Iwamoto, S., Kawasaki, T., Kambayashi, J., Ariyoshi, H., and Monden, M. (1996). Platelet microparticles: A carrier of platelet-activating factor? *Biochem. Biophys. Res. Com.* **218,** 940–944.

Jackson, C. J., Garbett, P. K., Nissen, B., and Schrieber, L. (1990). Binding of human endothelium to Ulex europaeus I-coated Dynabeads: Application to the isolation of microvascular endothelium. *J. Cell Sci.* **96,** 257–262.

Janowska-Wieczorek, A., Wysoczynski, M., Kijowski, J., Marquez-Curtis, L., Machalinski, B., Ratajczak, J., and Ratajczak, M. Z. (2005). Microvesicles derived from activated platelets induce metastasis and angiogenesis in lung cancer. *Int. J. Cancer* **113,** 752–760.

Jimenez, J. J., Jy, W., Mauro, L., Horstman, L., and Ahn, Y. (2001). Elevated endothelial microparticles in thrombotic thrombocytopenic purpura (TTP): Findings from brain and renal microvascular cell culture and patients with active disease. *Br. J. Haematol.* **112,** 81–90.

Jimenez, J. J., Jy, W., Mauro, L., Soderland, C., Horstman, L. L., and Ahn, Y. S. (2003a). Endothelial cells release phenotypically and quantitatively distinct microparticles in activation and apoptosis. *Thromb. Res.* **109,** 175–180.

Jimenez, J. J., Jy, W., Mauro, L. M., Ahn, E. R., Horstman, L. L., and Ahn, Y. S. (2003b). Plasma from patients with ITP-associated dementia (ITPAD) activates endothelial cells (EC) and enhances transendothelial migration of leukocytes (TEML). *Blood* **102,** 294a (Abstract 1052).

Jimenez, J. J., Jy, W., Mauro, L. M., Horstman, L. L., Soderland, C., and Ahn, Y. S. (2003c). Endothelial microparticles released in thrombotic thrombocytopenic purpura express von Willebrand factor and markers of endothelial activation. *Br. J. Haematol.* **123,** 896–902.

Jimenez, J. J., Jy, W., Mauro, L. M., Horstman, L. L., Soderland, C., and Ahn, Y. S. (2004). Response to letter-to-editor by Laurence [Invited]. *Br. J. Haematol.* **125,** 416.

Jimenez, J. J., Jy, W., Mauro, L., Horstman, L. L., Ahn, E. R., Ahn, Y. S., and Minagar, A. (2005). Elevated endothelial microparticle-monocyte complexes induced by multiple sclerosis plasma and the inhibitory effects of interferon-β 1b on release of endothelial microparticles, formation and transendothelial migration of monocyte-endothelial microparticle complexes. *Mult. Scler.* **11,** 310–315.

Jin, M., Drwal, G., Bourgeois, T., Salts, J., and Wu, H. M. (2005). Distinct proteome features of plasma microparticles. *Proteomics* **5,** 1940–1952.

Jobe, S. M., Leo, L., Eastvold, J. S., Dickneite, G., Ratliff, T. L., Lentz, S. R., and Di Paola, J. (2005). Role of FcR-γ and factor XIIIA in coated platelet formation. *Blood* **106,** 4146–4151.

John, S., Drobnik, W., Lackner, K., and Schmieder, R. E. (1999). Soluble thrombomodulin and endothelial dysfunction in early atherosclerosis. *Lancet* **354,** 1647.

Johnston, G. I., Pickett, E. B., McEver, R. P., and George, J. N. (1987). Heterogeneity of platelet secretion in response to thrombin demonstrated by fluorescence flow cytometry. *Blood* **69,** 1401–1403.

Joseph, M. (1995). The generation of free radicals by blood platelets (Chapter 11). *In* "Immunopharmacology of Platelets" (M. Joseph, Ed.), pp. 209–223. Academic Press, London/New York.

Joseph, R., Tsering, C., Grunfeld, S., and Welch, K. M. A. (1991). Platelet secretory products may contribute to neuronal injury. *Stroke* **22,** 1448–1451.

Jung, S. M., Tanoue, K., and Yamazaki, H. (1985). The electrophoretic mobility heterogeneity of human platelet subpopulations of different bouyant densities. *Thromb. Haemost.* **53,** 188–194.

Jy, W., Horstman, L. L., Arce, M., and Ahn, Y. S. (1992). Clinical significance of platelet microparticles in autoimmune thrombocytopenias. *J. Lab. Clin. Med.* **119,** 334–345.

Jy, W., Horstman, L. L., Homolak, D., and Ahn, Y. S. (1995a). Electrophoretic properties of platelets from normal, thrombotic and ITP patients by doppler electrophoretic light scattering analysis. *Platelets* **6,** 354–358.

Jy, W., Mao, W. W., Horstman, L. L., Tao, J., and Ahn, Y. S. (1995b). Platelet microparticles bind, activate and aggregate neutrophils *in vitro*. *Blood Cell Mol. Dis.* **21,** 217–231.

Jy, W., Jimenez, J. J., Mauro, L., Horstman, L. L., Bernal-Mizrachi, L., Minagar, A., Soderland, C., and Ahn, Y. S. (2001). Interaction of endothelial microparticles (EMP) with leukocytes: Potential roles of EMP in thrombosis and inflammation. *Blood* **98,** 226a.

Jy, W., Jimenez, J. J., Mauro, L. M., Ahn, Y. S., Newton, K. R., Mendez, A. J., Arnold, P. L., and Schultz, D. R. (2002). Agonist-induced capping of adhesion proteins and microparticle shedding in culture of human renal vascular endothelial cells. *Endothelium* **9,** 179–189.

Jy, W., Jimenez, J. J., Mauro, L. M., Horstman, L. L., Bidot, C. J., Tiede, M. P., Ahn, E., and Ahn, Y. S. (2003). Endothelial microparticles (EMP) interact with platelets via a vWF dependent pathway to form platelet aggregates, more resistant to dissociation than those induced by soluble vWF. *Blood* **102**(11), 783a (Abstract 2896).

Jy, W., Jimenez, J. J., Mauro, L. M., Bidot, C. J., Horstman, L. L., Yaniz, M., Gonzalez, M., and Ahn, Y. S. (2006). Multiple species of endothelial microparticles (EMP) released from different membrane protein clustering domains exhibit distinctive phenotypes and activities. *Blood* **108,** 511a.

Jy, W., Horstman, L. L., Jimenez, J. J., and Ahn, Y. S. (2004a). Measuring circulating cell-derived microparticles [Forum article]. *J. Thromb. Haemost.* **2,** 1842–1851.

Jy, W., Minagar, A., Jimenez, J. J., Sheremata, W. A., Mauro, L., Horstman, L. L., Bidot, C. J., and Ahn, Y. S. (2004b). Endothelial microparticles (EMP) bind to monocytes to activate and enhance transmigration: Elevated circulating EMP-monocyte conjugates in multiple sclerosis. *Front. Biosci.* **9,** 3137–3144.

Jy, W., Jimenez, J. J., Bidot, C. J., Horstman, L. L., Ahn, Y. S., and Minagar, A. (2005). Antiphospholipid antibodies in multiple sclerosis are exclusively IgM class, and occur predominately in exacerbations. *Neurology* **64,** A194 (Abstract PQ3.123).

Kagawa, H., Komiyama, Y., Nakamura, S., Miyaki, T., and Miyazaki, Y. (1998). Expression of functional tissue factor on small vesicles of lipopolysaccharide-stimulated human vascular endothelial cells. *Thromb. Res.* **91,** 297–304.

Kahn, I., Zucker-Franklin, D., and Karpatkin, S. (1975). Microthrombocytosis and platelet fragmentation associated with idiopathic/autoimmune thrombocytopenic purpura. *Br. J. Haematol.* **31,** 449–460.

Kannon, K., Divers, S. G., Lurie, A. A., Chervanak, R., Fukuda, M., and Holcombe, R. F. (1995). Cell surface expression of lysosome-associated membrane protein-2 (lamp-2) and CD63 as markers of *in vivo* platelet activation. *Eur. J. Haematol.* **55,** 145–151.

Karussis, D., Leker, R. R., Ashkenazi, A., and Abramsky, O. (1998). A subgroup of multiple sclerosis patients with anticardiolipin antibodies and unusual clinical manifestations: Do they represent a new nosological entity? *Ann. Neurol.* **44,** 629–634.

Katayama, M., Araki, M. H., Ambo, H., Kawai, Y., Watanabe, K., and Ikeda, Y. (1993). Soluble P-selectin is present in normal circulation and its plasma level is elevated in patients with thrombotic thrombocytopenic purpura and hemolytic uremic syndrome. *Br. J. Haematol.* **84,** 702–710.

Kim, C. W., Lee, H. M., Lee, T. H., Kang, C., Kleinman, H. K., and Gho, Y. S. (2002). Extracellular membrane vesicles from tumor cells promote angiogenesis via sphingomyelin. *Cancer Res.* **62,** 6312–6317.

Kim, H. K., Song, K. S., Park, Y. S., Kang, Y. H., Lee, Y. J., Lee, K. R., Kim, H. K., Ryu, K. W., Bae, J. M., and Kim, S. (2003). Elevated levels of circulating microparticles, VEGF, IL-6 and RANTES in patients with gastric cancer: Possible role of a metastasis predictor. *Eur. J. Cancer.* **39,** 184–191.

Kim, H. K., Song, K. S., Chung, J. H., Lee, K. R., and Lee, S. N. (2004). Platelet microparticles induce angiogenesis *in vitro*. *Br. J. Haematol.* **124,** 376–384.

Kokawa, T., Nomura, S., Yanabu, M., and Yasunaga, K. (1993). Detection of platelet antigen for antiplatelet antibodies in idiopathic thrombocytopenic purpura by flow cytometry, antigen-capture ELISA, and immunoblotting: A comparative study. *Eur. J. Haematol.* **50,** 74–80.

Koyama, T., Nishida, K., Ohdama, S., Sawada, M., Murakami, N., Hirosawa, S., Kuriyama, R., Matsuzawa, K., Hasegawa, R., and Aoki, N. (1994). Determination of plasma tissue factor antigen and its clinical significance. *Br. J. Haematol.* **87,** 343–347.

Kunicki, T. J., Orchekowski, R., Annis, D., and Honda, Y. (1993). Variability of integrin alpha-2-β1 activity on human platelets. *Blood* **82,** 2693–2703.

Lakhani, S. A., Masud, A., Kuida, K., Porter, G. A., Jr., Booth, C. J., ZMehal, W., Inayat, I., and Flavell, R. (2006). Caspases 3 and 7: Key mediators of mitochondrial events and apoptosis. *Science* **311,** 847–851.

Langer, F., Ingersoll, S. B., Amirkhosravi, A., Meyer, T., Siddiqui, F. A., Ahmad, S., Walker, J. M., Amaya, M., Desai, H., and Francis, J. L. (2005). The role of CD40 in CD40L- and antibody-mediated platelet activation [with editorial, p. 1011]. *Thromb. Haemost.* **93,** 1137–1146.

Lee, Y. J., Horstman, L. L., Janania, J., Reyes, Y., Kelley, R., and Ahn, Y. S. (1993). Elevated platelet microparticles in transient ischemic attacks, lacunar infarcts, and multiinfarct dementias. *Thromb. Res.* **72,** 295–304.

Leenstra, S., Troost, D., Das, P. K., Claessen, N., Becker, A. E., and Bosch, D. A. (1993). Endothelial cell marker PAL-E reactivity in brain tumor, developing brain, and brain disease. *Cancer* **72,** 3061–3067.

Lemons, P. P., Chen, D., Bernstein, A. M., Bennett, M. K., and Whiteheart, S. W. (1997). Regulated secretion in platelets: Identification of elements of the platelet exocytosis machinery [see also letter, 92:2191]. *Blood* **90,** 1490–1500.

Li, Y. H., Teng, J. K., Tsai, W., Tsai, L. M., Lin, L. J., and Chen, J. H. (1997). Elevation of soluble adhesion molecules is associated with the severity of myocardial damage in acute myocardial infarction. *Am. J. Cardiol.* **80,** 1218–1221.

Littler, A. J., Buckley, C. D., Wordsworth, P., Collins, I., Martinson, J., and Simmons, D. L. (1997). A distinct profile of six soluble adhesion molecules (ICAM-1, ICAM-3, VCAM-1, E-selectin, L-selectin and P-selectin) in rheumatoid arthritis. *Br. J. Rheumatol.* **36,** 164–169.

Lo, S. C., Hung, C. Y., Lin, D. T., Peng, H. C., and Huang, T. F. (2006). Involvement of platelet glycoprotein Ib in platelet microparticle mediated neutrophil activation. *J. Thromb. Haemost.* **13** (6), 787–796.

Loppnow, H., Bil, R., Hirt, S., Schonbeck, U., Herzberg, M., Werdan, K., Rietschel, E. T., Brandt, E., and Flad, H. D. (1998). Platelet-derived interleukin-1 induces cytokine production, but not proliferation of human vascular smooth muscle cells. *Blood* **91,** 134–141.

Losy, J., Niezgoda, A., and Wender, M. (1999). Increased serum levels of soluble PECAM-1 in multiple sclerosis patients with brain gadolinium-enhancing lesions. *J. Neuroimmunol.* **99,** 169–172.

Lutz, H. U., Liu, S.-C, and Palek, J. (1977). Release of spectrin-free vesicles from human erythrocytes during ATP depletion. I. Characterization of spectrin-free vesicles. *J. Cell Biol.* **73,** 548–560.

Mack, M., Kleinschmidt, A., Bruhl, H., Klier, C., Nelson, P. J., Cihak, J., Plachy, J., Stangassinger, M., Erfle, V., and Schlondorff, D. (2000). Transfer of the chemokine receptor CCR5 between cells by membrane-derived microparticles: A mechanism for cellular human immunodeficiency virus 1 infection. *Nat. Med.* **6,** 769–775.

MacKenzie, A., Wilson, H. L., Kiss-Toth, E., Dower, S. K., North, R. A., and Suprenant, A. (2001). Rapid secretion of interleukin-1β by microvesicle shedding. *Immunity* **8,** 825–835.

Mallat, Z., Hugel, B., Ohan, J., Leseche, G., Freyssinet, J. N., and Tedgui, A. (1999). Shed membrane microparticles with procoagulant potential in human atherosclerotic plaques: A role for apoptosis in plaque thrombogenicity. *Circulation* **99,** 348–353.

Mallat, Z., Benamer, H., Hugel, B., Benessiano, J., Steg, P. G., Freyssinet, J. M., and Tedgui, A. (2000). Elevated levels of shed membrane microparticles with procoagulant potential in the peripheral circulating blood of patients with acute coronary symptoms. *Circulation* **101,** 841–843.

Mannucci, P. M. (1998). Von Willebrand factor: A marker of endothelial damage? *Arterioscler. Thromb. Vasc. Biol.* **18,** 1359–1362.

Manoury, B., Hewitt, E. W., Morrice, N., Dando, P. M., Barrett, A. J., and Watts, C. (1999). An asparaginyl endopeptidase processes a microbial antigen for class I MHC presentation. *Nature* **397,** 695–699.

Marchiori, P. E., DosReis, M., Quevedo, M. E., Callegaro, D., Hirata, M. T., Scaff, M., and DeOlivera, R. M. (1990). Cerebrospinal fluid and serum antiphospholipid antibodies in multiple sclerosis, Guillain-Barre syndrome and systemic lupus erythematosus. *Arq. Neuropsiquiatr.* **48,** 465–468.

Margaglione, M., DiMinno, G., Grandone, E., Vecchione, G., Celentano, E., Cappucci, G., Grilli, M., Somone, P., Panico, S., and Mancini, M. (1994). Abnormally high circulation levels of tissue plasminogen activator and plasminogen activator inhibitor-1 in patients with a history of ischemic stroke. *Arterioscler. Thromb.* **14,** 1741–1745.

Mariathasan, S., Weiss, D. S., Newton, K., McBride, J., O'Rourke, K., Roose-Girma, M., Lee, W. P., Weinrauch, Y., Monack, D. M., and Dixit, V. M. (2006). Cryopyrin activates the inflammasome in response to toxins and ATP. *Nature* **440,** 228–232.

Martinuzzo, M. E., Forastiero, R. R., and Carreras, L. O. (1996). Increased plasma thrombomodulin in different subgroups of patients with antiphospholipid and anti β2 glycoprotein 1 antibodies. *Thromb. Haemost.* **75,** 972–973.

McDonnell, G. V., McMillan, S. A., Douglas, J. P., Droogan, A. G., and Hawkins, S. A. (1999). Serum soluble adhesion molecules in multiple sclerosis: Raised sVCAM-1, sICAM-1 and sE-selectin in primary progressive disease. *J. Neurol.* **246,** 87–92.

McIntyre, J. A., Wagenknecht, D. R., and Sugi, T. (1997). Phospholipid binding plasma proteins required for antiphospholipid antibody detection: An overview. *Am. J. Reprod. Immunol.* **37,** 101–110.

Meroni, P. L., Raschi, E., Testoni, C., Tincani, A., and Balestrieri, G. (2001). Antiphospholipid antibodies and the endothelium. *Rheum. Dis. Clin. North Am.* **27,** 587–602.

Mezentsev, A., Merks, R. M., O'Riordan, E., Chen, J., Mendelev, N., Goligorsky, M. S., and Brodsky, S. V. (2005). Endothelial microparticles affect angiogenesis *in vitro*: Role of oxidative stress. *Am. J. Physiol.* **289,** H1106–H1114.

Michelson, A. D. (1988). Flow cytometric detection of distinct subpopulations of platelets of individual patients. *In* "Platelet Membrane Receptors" (G. A. Jamieson, Ed.), Vol. 283, pp. 131–136. Alan R. Liss, New York.

Minagar, A., Jy, W., Jimenez, J. J., Mauro, L. M., Horstman, L. L., Ahn, Y. S., and Sheremata, W. A. (2001). Elevated plasma endothelial microparticles in multiple sclerosis. *Neurology* **56,** 1319–1324.

Mitsios, J. V., Vini, M. P., Stengel, D., Ninio, E., and Tselepis, A. D. (2006). Human platelets secrete the plasma type of platelet activating acetylhydrolase primarily associated with microparticles. *Arterioscler. Thromb. Vasc. Biol.* **26,** 1907–1913.

Miyamoto, S., Marcinkiewicz, C., Edmunds, L. H., Jr., and Niewiarowski, S. (1998). Measurement of platelet microparticles during cardiopulmonary bypass by means of captured ELISA for GP IIb/IIIa. *Thromb. Haemost.* **80,** 225–230.

Morgan, B. P., and Meri, S. (1994). Membrane proteins that protect against complement lysis. *Springer Semin. Immunopathol.* **15,** 369–396.

Muller, I., Klocke, A., Alex, M., Kotzsch, M., Luther, T., Morgenstern, E., Zieseniss, S., Zahler, S., Preissner, K., and Engelmann, B. (2003). Intravascular tissue factor initiates coagulation via circulating microvesicles and platelets. *FASEB J.* **17,** 476–478.

Munakata, Y., Saito, T., Mutsada, K., Seino, J., Shibata, S., and Sasoki, T. (2000). Detection of complement-fixing antiphospholipid antibodies in associated with thrombosis. *Thromb. Haemost.* **83,** 728–731.

Murdoch, C., and Finn, A. (2000). Chemokine receptors and their role in inflammation and infectious diseases. *Blood* **95,** 3032–3043.

Muriithi, E. W., Belcher, P. R., Rao, J. N., Chaudhry, M. A., Nicol, D., and Weatley, D. J. (2000). The effects of heparin and extracorporeal circulation on platelet counts and platelet microaggregation during cardiopulmonary bypass. *J Thorac. Cardiovasc. Surg.* **120,** 538–543.

Nasuno, A., Matsubara, T., Hori, T., Higuchi, K., Imai, S., Nakagawa, I., Tsuchida, K., Ozaki, K., Mezaki, T., Tanaka, T., Fuse, I., and Aizawa, Y. (2002). Levels of soluble E-selectin and ICAM-1 in the coronary circulation of patients with stable coronary artery disease: Association with the severity of coronary atherosclerosis. *Jpn. Heart J.* **43,** 93–101.

Ng, E. K. O., Tsui, N. B. Y., Lam, N. Y. L., Chiu, R. W. K., Yu, S. C. H., Wong, S. C. C., Lo, E. S. F., Rainer, T. H., Johnson, P. J., and Lo, Y. M. D. (2002). Presence of filterable and nonfilterable mRNA in the plasma of cancer patients and healthy individuals. *Clin. Chem.* **48,** 1212–1217.

Nieuwland, R., Berckmans, R. J., Maquelin, K. N., Roozendaal, K. J., Jansen, P. G. M., Have, K., Eijsman, L., Hack, C. E., and Sturk, A. (1997). Cell-derived microparticles generated in patients during cardiopulmonary bypass are highly procoagulant. *Circulation* **96,** 3534–3541.

Nieuwland, R., Berckmans, R. J., McGregor, S., Boing, A. N., Romijn, F., Westendorp, R. G. J., Hack, C. E., and Sturk, A. (2000). Cellular origin and procoagulant properties of microparticles in meningococcal sepsis. *Blood* **95,** 930–935.

Nomura, S. (2001). Function and clinical significance of platelet-derived microparticles (Review). *Int. J. Hematol.* **74,** 397–404.

Nomura, S., Yanabu, M., Fukuroi, T., Kido, H., Kawakatsu, T., Yamaguchi, K., Suzuki, M., Kokawa, T., and Yasunaga, K. (1992). Anti-phospholipid antibodies bind to platelet microparticles in idiopathic (autoimmune) thrombocytopenic purpura. *Ann. Hematol.* **65,** 46–49.

Nomura, S., Komiyama, Y., Matsuura, E., Kokawa, T., Takahashi, H., and Koike, T. (1993). Binding of β2-glycoprotein 1 to platelet-derived microparticles (Letter, with critique). *Br. J. Haematol.* **85,** 639–640.

Nomura, S., Komiyama, Y., Miyake, T., Miyazaki, Y., Kido, H., Suzuki, M., Kagawa, H., Yanabu, M., Takahashi, H., and Fukuhara, S. (1994). Amyloid β-protein precursor-rich platelet microparticles in thrombotic disease. *Thromb. Haemost.* **72,** 519–522.

Nomura, S., Yanabu, M., Kido, H., Gui, L. X., Ichiyoshi, H., Katsura, K., Miyake, T., Miyazaki, Y., Kagawa, H., and Fukuhara, S. (1995a). Significance of cytokines and CD68-positive microparticles in immune thrombocytopenic purpura. *Eur. J. Haemat.* **55,** 49–56.

Nomura, S., Yanabu, M., Miyake, T., Miyazaki, Y., Kido, H., Kagawa, H., Fukuhara, S., Komiyama, Y., Matsuura, E., and Koike, T. (1995b). Relationship of microparticles with $\beta 2$-glycoprotein I and P-selectin positivity to anticardiolipin antibodies in immune thrombocytopenic purpura. *Ann. Hematol.* **70,** 25–30.

Nomura, S., Ohtani, T., Kagawa, H., Fukuhara, S., Fujimura, Y., Ikeda, Y., and Kambayashi, J. (1999). Relationship between cytokines and platelet-derived microparticles. *Thromb. Haemost. Suppl.* **363** (Abstract 1152).

Nomura, S., Uehata, S., Saito, S., Osumi, K., Ozeki, Y., and Kimura, Y. (2003). Enzyme immunoassay detection of platelet-derived microparticles and RANTES in acute coronary syndromes. *Thromb. Haemost.* **89,** 506–512.

Novotny, W. F., Girard, T. J., Miletich, J. P., and Broze, G. J. (1988). Platelets secrete a coagulation inhibitor functionally and antigenically similar to the lipoprotein-associated coagulation inhibitor. *Blood* **72,** 2020–2025.

Ogura, H., Tanaka, H., Koh, T., Fujita, K., Fujimi, S., Nakamori, Y., Hosotsubo, H., Kuwagata, Y., Shimazu, T., and Sugimoto, H. (2004). Enhanced production of endothelial microparticles with increased binding to leukocytes in patients with severe systemic inflammatory response. *J. Trauma* **56,** 823–830.

Ohdama, S., and Aoki, N. (1997). Increase of plasma thrombomodulin in systemic vasculitis [Letter]. *Thromb. Haemost.* **77,** 609.

Oishi, Y., Wakatsuki, T., Nishkado, A., Oki, T., and Ito, S. (2000). Circulating adhesion molecules and severity of coronary atherosclerosis. *Coron. Artery Dis.* **11,** 77–81.

Osmanovic, N., Romijn, F. P. H., Joop, K., Sturk, A., and Nieuland, R. (2000). Soluble selectins in sepsis: Microparticle-associated but only to a minor degree. *Thromb. Haemost.* **84,** 731–732.

Osumi, K., Ozeki, Y., Saito, S., Nagamura, Y., Ito, H., Kimura, Y., Ogura, H., and Nomura, S. (2001). Development and assessment of enzyme immunoassay for platelet-derived microparticles. *Thromb. Haemost.* **85,** 326–330.

Pascual, M., Lutz, H. U., Steiger, G., Stammler, P., and Schifferli, J. A. (1993). Release of vesicles enriched in complement receptor 1 from human erythrocytes. *J. Immunol.* **151,** 397–404.

Pasquet, J.-M., Toti, F., Nurden, A. T., and Dachery-Prigent, J. (1996). Procoagulant activity and active calpain in platelet-derived microparticles. *Thromb. Res.* **82,** 509–522.

Perez-Casal, M., Downey, C., Fukudome, K., Marx, G., and Toh, C. H. (2005). Activated protein C induces the release of microparticle-associated endothelial protein C receptor. *Blood.* **105,** 1515–1522.

Peter, A., Virani, S., Ghani, R., Purow, J., Marchena, E.d., Ferreira, A., Jimenez, J. J., Mauro, L. M., Mendez, A. J., Jy, W., Horstman, L. L., and Ahn, Y. S. (2004). Effect of a single high-fat meal on endothelial microparticles as a marker of endothelial injury. *J. Am. Coll. Cardiol. Suppl.* **518A.**

Pierangeli, S. S., Espinola, R. G., Ferrara, D. E., and Harris, E. N. (2001). Activation of endothelial cell nuclear factor kB by antiphospholipid antibodies. *Blood* **98,** 56b (Abstract 3837).

Pignatelli, P., DiSanto, S., Carnevale, B., and Violi, F. (2005). The polyphenols quercetin and catechin synergize in inhibiting platelet CD40L expression [Letter]. *Thromb. Haemost.* **94,** 889–890.

Pirro, M., Schillaci, G., Paltriccia, R., Bagaglia, F., Menecali, C., Mannarino, M. R., Capanni, M., Velardi, A., and Mannarino, E. (2006). Increased ratio of CD31+/CD42-microparticles to endothelial progenitors as a novel marker of atherosclerosis in hypercholesterolemia. *Arterioscler. Thromb. Vasc. Biol.* **26,** 2530–2535.

Pitti, R. M., Marsters, S. A., Lawrence, D. A., Roy, M., Kischkel, F. C., Dowd, P., Huang, A., and Donnahue, C. L. (1998). Genomic amplification of decoy receptor for Fas ligand in lung and colon cancer. *Nature* **396,** 699–703.

Pobbati, A. V., Stein, A., and Fasshauer, D. (2006). N- to C-terminal SNARE complex assembly promotes rapid membrane fusion. *Science* **313,** 673–676.

Polasek, J. (1982). The appearance of multivesicular structures during platelet activation as observed by scanning electron microscopy. *Thromb. Res.* **28,** 433–437.

Polgar, J., Matuskova, J., and Wagner, D. D. (2005). The P-selectin, tissue factor, coagulation triad. *J. Thromb. Haemost.* **3,** 1590–1596.

Rabesandratana, H., Toutant, J. P., Reggio, H., and Vidal, M. (1998). Decay accelerating factor (CD55) and membrane inhibitor of reactive lysis (CD59) are released within exosomes during *in vitro* maturation of reticulocytes. *Blood* **91,** 2573–2580.

Raposo, G., Nijman, H. W., Stoorvogel, W., Leijendekker, R., Harding, C. V., Melief, C. J. M., and Geuze, H. J. (1996). B lymphocytes secrete antigen-presenting vesicles. *J. Exp. Med.* **183,** 1161–1172.

Rauch, U., Bonderman, D., Bohrmann, B., Badimon, J. J., Himber, J., Riederer, M. A., and Nemerson, Y. (2000). Transfer of tissue factor from leukocytes to platelets is mediated by CD15 and tissue factor. *Blood* **96,** 170–175.

Rinder, C. S. (1994). Platelet-leukocyte activation and modulation of adhesion receptors in pediatric patients with congenital heart disease undergoing cardiopulmonary bypass. *J. Thorac. Cardiovasc. Surg.* **107,** 280–288.

Roldan, V., Marin, F., Lip, G. Y. H., and Blann, A. D. (2003). Soluble E-selectin in cardiovascular disease and its risk factors [Review]. *Thromb. Haemost.* **90,** 1007–1020.

Rozmyslowicz, T., Majka, M., Kijowski, J., Murphy, S. L., Conover, D. O., Poncz, M., Ratajczak, J., Gaulton, G. N., and Ratajczak, M. Z. (2003a). Platelet- and megakaryocyte-derived microparticles transfer CXCR4 receptor to CXCR4-null cells and make them susceptible to infection by X4-HIV. *AIDS* **17,** 33–42.

Rozmyslowicz, T., Murphy, S. L., Conover, D. O., Wroblewski, K., Ratajczak, M. Z., and Gaulton, G. N. (2003b). The role of microparticles in HIV-1 infection. *Blood* **102**(11), 278a (Abstract 987).

Sabatier, F., Darmon, P., Hugel, P., Combes, V., Sanmarco, M., Velut, J. G., Arnoux, D., Charpiot, P., Freyssinet, J. M., Oliver, C., Sampol, J., and Dignat-George, F. (2002a). Type 1 and type 2 diabetic patients display different patterns of cellular microparticles. *Diabetes* **51,** 2840–2845.

Sabatier, F., Roux, V., Anfosso, F., Camoin, L., Sampol, J., and Dignat-George, F. (2002b). Interaction of endothelial microparticles with monocytic cells *in vitro* induces tissue factor-dependent procoagulant activity. *Blood* **99,** 3962–3970.

Salomaa, V., Matei, C., Aleksic, N., Sansores-Garcia, L., Folsom, A. R., Juneja, H., Chambless, L. E., and Wu, K. K. (1999). Soluble thrombomodulin as a predictor of incident coronary heart disease and symptomless carotid artery atherosclerosis in the Atherosclerosis Risk in Communities (ARIC) Study: A case-cohort study. *Lancet* **353,** 1729–1734.

Salzer, U., and Prohaska, R. (2001). Stomatin, flotillin-1, and flotillin-2 are major integral proteins of erythrocyte lipid rafts. *Blood* **97,** 1141–1143.

Satta, N., Freyssinet, J.-M., and Toti, F. (1997). The significance of human monocyte thrombomodulin during membrane vesiculation and after stimulation by lipopolysaccharide. *Br. J. Haematol.* **96,** 534–542.

Satta, N., Toti, F., Feugeas, O., Bohbot, A., Dachary-Prigent, J., Eschwege, V., Hedman, V., and Freyssinet, J.-M. (1994). Monocyte vesiculation as a possible mechanism of dissemination of membrane-associated procoagulant activities and adhesion molecules after stimulation by lipopolysaccharide. *J. Immunol.* **153,** 3245–3255.

Schafer, A., Schulz, C., Eigenthaler, M., Fraccarollo, D., Kobsar, A., Gawaz, M., Ertl, G., Walter, U., and Bauersachs, J. (2004). Novel role of the membrane-bound chemokine fractalkine in platelet activation and adhesion. *Blood* **103,** 407–412.

Schlingemann, R. O., Dingiam, G. M., Emeis, J. J., Blok, J., Warnaar, S. O., and Ruiter, D. J. (1985). Monoclonal antibody PAL-E specific for endothelium. *Lab. Invest.* **52,** 71–76.

Schlingemann, R. O., Hofman, P., Vrensen, G. F., and Blaauwgeers, H. G. (1999). Increased expression of endothelial antigen PAL-E in human diabetic retinopathy correlates with microvascular leakage. *Diabetologia* **42,** 596–602.

Scholz, T., Temmler, U., Krause, S., Heptinstall, S., and Losche, W. (2002). Transfer of tissue factor from platelets to monocytes: Role of platelet-derived microvesicles and CD62P. *Thromb. Haemost.* **88,** 1033–1038.

Schroeder, S. A. (1980). The complications of coronary arteriography: A problem that won't go away [Editorial]. *Am. Heart J.* **99,** 139–141.

Scolding, N. J., Morgan, B. P., Houston, W. A. J., Linington, C., Campbell, A. K., and Compston, D. A. S. (1989). Vesicular removal by oligodendrocytes of membrane attack complexes formed by activated complement. *Nature* **339,** 620–622.

Scott, F., Brillman, I., Romanelle, I., and Hess, D. (1994). Anticardiolipin antibodies in MS patients (Letter). *Acta Neurol. Scand.* **90,** 374–375.

Seeman, H. B., Gurbel, P. A., Anderson, J. L., Muhlestein, J. B., Carlquist, J. F., Horne, B. D., and Serebruany, V. L. (2000). Soluble VCAM-1 and E-selectin, but not ICAM-1, discriminate endothelial injury in patients with documented coronary artery disease. *Cardiology* **93,** 7–10.

Segura, E., Nicco, C., Lombard, B., Veron, P., Raposo, G., Batteux, F., Amigorena, S., and Thery, C. (2005). ICAM-1 in exosomes from mature dendritic cells is critical for efficient naive T-cell priming. *Blood* **106,** 216–223.

Shaw, P. J., Bates, D., Cartlidge, N. E. F., Heaviside, D., Julian, D. G., and Shaw, D. A. (1985). Early neurological complications of coronary artery bypass surgery. *Br. Med. J.* **291,** 1384–1387.

Shaw, P. J., Bates, D., Cartlidge, N. E. F., French, J. M., Heaviside, D., Julian, D. G., and Shaw, D. A. (1986). Early intellectual dysfunction following coronary bypass surgery. *Q. J. Med.* **58,** 59–68.

Shet, A. S., Aras, O., Gupta, K., Hass, M. J., Rausch, D. J., Saba, N., Koopmeiners, L., Key, N. S., and Hebbel, R. P. (2003). Sickle blood contains tissue factor-positive microparticles derived from endothelial cells and monocytes. *Blood* **102,** 2678–2683.

Shet, A. S., Key, N. S., and Hebbel, R. P. (2004). Measuring circulating cell-derived microparticles. *J. Thromb. Haemost.* **2,** 1848–1850.

Siddiqui, F. A., Desai, H., Amirkhosravi, A., Siddiqui, T. F., Meyer, T., Amaya, M., and Francis, J. L. (2002). Platelet microparticles from healthy donors contain tissue factor. *Blood* **100**(11), 497a (Abstract 1936).

Simak, J., Holada, K., D'Agnillo, F., Janota, J., and Vostol, J. G. (2002). Cellular prion protein is expressed on endothelial cells and is released during apoptosis on membrane microparticles found in human plasma. *Transfusion* **42,** 334–342.

Simantov, R., Lo, S. K., Gharavi, A., Sammaritano, L. R., Salmon, J. E., and Silverstein, R. L. (1996). Antiphospholipid antibodies activate vascular endothelial cells. *Lupus* **5,** 440–441.

Simons, K., and Ikonen, E. (1997). Functional rafts in cell membranes (Review). *Nature* **387,** 569–572.

Sims, P. J. (1989). Interaction of human platelets with the complement system (Chapter 18) [see references pp. 373–374]. *In* "Platelet Immunobiology" (T. J. Kunicki and J. N. George, Eds.), pp. 354–383. J. B. Lippincott, Philadelphia, PA.

Sims, P. J., and Wiedmer, T. (1986). Repolarization of the membrane potential of blood platelets after complement damage: Evidence for a Ca^{2+}-dependant exocytotic elimination of C5b-9 pores. *Blood* **68,** 556–561.

Sims, P. J., Faioni, E. M., Wiedmer, T., and Shattil, S. J. (1988). Complement proteins C5b-9 cause release of membrane vesicles from the platelet surface that are enriched in the membrane receptor for coagulation factor Va and express prothrombinase activity. *J. Biol. Chem.* **263,** 18205–18212.

Sims, P. J., Wiedmer, T., Esmon, C. T., Weiss, H. L., and Shattil, S. J. (1989). Assembly of the platelet prothrombinase complex is linked to vesiculation of the platelet plasma membrane. *J. Biol. Chem.* **264,** 17049–17057.

Smith, P. L. (1988). The cerebral complications of coronary artery bypass surgery. *Ann. R. Coll. Surg. Engl.* **70,** 212–216.

Souberbielle, B. E., Swingler, R. J., Davodson, D. L., Cull, R. E., Atkinson, S., Davison, I., Anderson, J., Bell, J. E., and Russell, W. C. (1992). Western blotting analysis in patients with MS using human brain vessels as antigens. *Acta Neurol. Scand.* **86,** 397–402.

Stahl, S. M. (1985). Platelets as pharmacologic models for the receptors and biochemistry of monoaminergic neurons. *In* "The Platelets" (G. L. Longnecker, Ed.), pp. 308–340. Academic Press, New York.

Stein, J. M., and Luzio, J. P. (1991). Ectocytosis caused by sublytic autologous complement attack on human neutrophils. *Biochem. J.* **274,** 381–386.

Steppich, B., Mattisek, C., Sobczyk, D., Kastrati, A., Schomig, A., and Ott, I. (2005). Tissue factor pathway inhibitor on circulating microparticles in acute myocardial infarction. *Thromb. Haemost.* **93,** 35–39.

Strieter, R. M., Burdick, M. D., Mestas, J., Gomperts, B., Keane, M. P., and Belperio, J. A. (2006). Cancer CXC chemokine networks and tumor angiogenesis. *Eur. J. Cancer* **42,** 768–778.

Sturk-Maquelin, K. N., Nieuwland, R., Romijn, F. P., Eijsman, L., Hack, C. E., and Sturk, A. (2003). Pro- and non-coagulant forms of non-cell-bound tissue factor *in vivo*. *J. Thromb. Haemost.* **1,** 1920–1926.

Sugiyama, Y., and Yamamoto, Y. (1996). Characterization of serum anti-phospholipid antibodies in patients with multiple sclerosis. *Tohoku J. Exp. Med.* **178,** 203–215.

Takahashi, H., Hanano, M., Wada, K., Tatewaki, W., Niwana, H., Tsubouchi, J., Nakano, M., Nakamura, T., and Shibata, A. (1991). Circulating thrombomodulin in thrombotic thrombocytopenic purpura. *Am. J. Hematol.* **38,** 174–177.

Takatori, M., Iwabuchi, S., Ro, S., Murayama, M., Maeyama, S., Uchikoski, T., Nakano, M., and Ishii, H. (1999). Increased serum levels and sinusoidal expression of thrombomodulin in acute liver damage. *Thromb. Res.* **93,** 113–120.

Tanaka, T., Tsukada, N., Koh, C. S., and Yanagisawa, N. (1987). Anti-endothelial cell antibodies and circulating immune complexes in the sera of patients with multiple sclerosis. *J. Neuroimmunol.* **17,** 49–59.

Thom, J., Gilmore, G., Yi, Q., Hankey, G. J., and Eikelboom, J. W. (2004). Measurement of soluble P-selectin and soluble CD40 ligand in serum and plasma [Letter]. *J. Thromb. Haemost.* **2,** 2067–2069.

Thomas, T. D., and Troost, B. T. (1989). Permanent cerebral blindness after cardiac catheterization. *Cathet. Cardiovasc. Diagn.* **4,** 228–230.

Tintore, M., Fernandez, A. L., Rovira, A., Martinez, X., Direskeneli, H., Kamashta, M., Schwartz, S., Codina, A., and Montalban, X. (1996). Antibodies against endothelial cells in patients with multiple sclerosis. *Acta Neurol. Scand.* **93,** 416–420.

Triolo, G., Accardo-Palumbo, A., Triolo, G., Carbone, M. C., Ferrante, A., and Giardina, E. (1999). Enhancement of endothelial cell E-selectin expression by sera from patients with active Behcet's disease: Moderate correlation with anti-endothelial cell antibodies and serum myeloperoxidase levels. *Clin. Immunol.* **91,** 330–337.

Trojano, M., Defazio, G., Ricchiuti, F., DeSalvia, R., and Livrea, P. (1996). Serum IgG to brain microvascular endothelial cells in multiple sclerosis. *J. Neurol. Sci.* **143,** 107–113.

Tsukada, N., Tanaka, Y., Miyagi, K., Yanagisawa, N., and Okana, A. (1989). Autoantibodies to each protein fraction extracted from cerebral endothelial cell membrane in the sera of patients with multiple sclerosis. *J. Neuroimmunol.* **24,** 41–46.

Vallar, L., Regnault, V., Latger-Cannard, V., and Lecompte, T. (2001). β 2-glycoprotein 1 binding to platelet microparticle membrane specifically reduces immunoreactivity of glycoproteins IIb/IIIa. *Thromb. Haemost.* **85,** 314–319.

VanWijk, M. J., Nieuwland, R., Boer, K., VanDerPost, J. A. M., VanBavel, E., and Sturk, A. (2002). Microparticle subpopulations are increased in preeclampsia: Possible involvement in vascular dysfunction? *Am. J. Obstet. Gynecol.* **187,** 450–456.

Varo, N., de Lemos, J. A., Libby, P., Morrow, D. A., Murphy, S. A., Nuzzo, R., Gibson, C. M., Cannon, C. P., Braunwald, E., and Schönbeck, U. (2003). Soluble CD40 L: Risk prediction after acute coronary syndromes. *Circulation* **108,** 1049–1052.

Varo, N., Nuzzo, R., Natal, C., Libby, P., and Schonbeck, U. (2006). Influence of pre-analytical and analytical factors on soluble CD40L measurements. *Clin. Sci. (Lond.)* **111,** 341–347.

Vlachoyiannopoulos, P. G., Mavragani, C. P., Bourazopoulou, E., Balitsari, A. V., and Routsias, J. G. (2004). Anti-CD40 antibodies in antiphospholipid syndrome and systemic lupus erythematosus. *Thromb. Haemost.* **92,** 1303–1311.

Wada, H., Kaneko, T., Onwa, M., Tanigawa, M., Hayashi, T., Tamaki, S., Minami, N., Deguchi, K., Suzuki, T., Nakano, T., and Shirakawa, S. (1993). Increased levels of vascular endothelial cell markers in thrombotic thrombocytopenic purpura. *Am. J. Hematol.* **44,** 101–105.

Wada, H., Mori, Y., Shimura, M., Hiyoyama, K., Ioka, M., Nakasaki, T., Nishikawa, M., Nakano, M., Kumeda, K., Kaneka, T., Nakamura, S., and Shiku, H. (1998). Poor outcome in disseminated intravascular coagulation or thrombotic thrombocytopenic purpura patients with severe vascular endothelial injuries. *Am. J. Hematol.* **58,** 189–194.

Welterman, A., Hron, G., Kollars, M., Kornek, G., Quehenberger, P., Eichenger, S., and Kyrle, P. A. (2004). The association between tissue factor-exposing microparticles and the activation of the coagulation system in cancer patients. *Blood* **104,** 709a (Abstract 2589).

Williams, F. M. K., Parmar, K., Hughes, G. R. V., and Hunt, B. J. (2000). Systemic endothelial cell markers in primary antiphospholipid syndrome. *Thromb. Haemost.* **84,** 742–746.

Wolf, P. (1967). The nature and significance of platelet products in human plasma. *Br. J. Haematol.* **13,** 269–288.

Wolfers, J., Lozier, A., Raposo, G., Regnault, A., Thery, C., Masurier, C., Flament, C., Pouzieux, S., Faure, F., Tursz, T., Angevin, E., Amigorena, S., *et al.* (2001). Tumor-derived exosomes are a source of shared tumor rejection antigens for CTL cross priming. *Nat. Med.* **7,** 297–303.

Woywodt, A., Goldberg, C., Scheer, J., Regelsberger, H., Haller, H., and Haubitz, M. (2004). An improved assay for enumeration of circulating endothelial cells. *Ann. Haematol.* **83,** 491–494.

Wyss-Coray, T., and Mucke, L. (2000). Ibuprofen, inflammation and Alzheimer disease. *Nat. Med.* **6,** 973–974.

Xu, B., deWaal, R. M., Mo-Vaknin, N., Hibbard, C., Markovitz, D. M., and Kahn, M. L. (2004). The endothelial cell-specific antibody PAL-E identifies a secreted form of vimentin in the blood vasculature. *Mol. Cell. Biol.* **24,** 9198–9206.

Zitvogel, L., Regnault, A., Lozier, A., Wolfers, J., Flament, C., Tenza, D., Ricciardi-Castagnoli, P., Raposo, G., and Amigorena, S. (1998). Eradication of established murine tumors using a novel cell-free vaccine: Dendritic cell-derived exosomes. *Nat. Med.* **4,** 594–600.

Zucker-Franklin, D., and Karpatkin, S. (1977). Red cell and platelet fragmentation in idiopathic thrombocytopenic purpura. *N. Engl. J. Med.* **297,** 517–523.

Zwaal, R. F. A., Comfurius, P., and Bevers, E. M. (1992). Platelet procoagulant activity and microvesicle formation: Its putative role in homeostasis and thrombosis (Review). *Biochim. Biophys. Acta* **1180,** 1–8.

MULTIPLE SCLEROSIS IN CHILDREN: CLINICAL, DIAGNOSTIC, AND THERAPEUTIC ASPECTS

Kevin Rostásy

Department of Pediatrics, Division of Pediatric Neurology
Medical University of Innsbruck, Innsbruck 6020, Austria

Multiple sclerosis (MS) is the most common inflammatory disease of human central nervous system (CNS), which is characterized by inflammatory demyelination and neuroaxonal injury/loss. The majority of MS patients are diagnosed in early- to mid-adulthood; however, the onset of MS in childhood is being increasingly recognized. Although adults and children share important aspects of the disease, several features including course of the disease and a broader differential diagnosis are unique to children. This chapter summarizes recent insights and emphasizes that children with MS should be started on immunomodulatory therapies early in order to prevent future disability.

I. Introduction

Multiple sclerosis (MS) is the most common demyelinating disease of the CNS, which is characterized by inflammatory demyelination and neuroaxonal injury/loss. The pathological hallmarks of MS are demyelinated plaques within the white matter combined with inflammatory infiltrates that predominantly consist of lymphocytes and activated macrophages/microglia. In addition, there is now mounting evidence that neuroaxonal injury and loss within the white matter and cortical regions is widespread and already occurs in the beginning of the disease, characterized by neurons undergoing apoptosis, loss of dendritic arborization, transected and demyelinated axons (Kuhlmann *et al.*, 2002; Peterson *et al.*, 2001; Trapp *et al.*, 1998).

INTERNATIONAL REVIEW OF
NEUROBIOLOGY, VOL. 79
DOI: 10.1016/S0074-7742(07)79011-6

0074-7742/07 $35.00

Although no detailed histopathological studies in children with MS exist, several clinical observations indicate that neuroaxonal injury can also occur. Elevated levels of Tau, a neuronal cytoskeletal protein, are found in CSF in children with a second attack (Rostasy *et al.*, 2005). MRI and MRS data suggest that neuroaxonal structures are also affected in children, but persistent damage develops over a longer period of time probably due to a higher capability of regeneration (Balassy *et al.*, 2001; Bruhn *et al.*, 1992).

The events leading to MS are still unknown. Environmental factors including infections with neurotropic viruses and a genetic predisposition are thought to contribute to the initiation of the disease, which usually has begun years prior to overt clinical manifestation. With regard to risk of developing MS, exposure to environmental factors appears to be especially critical in a vulnerable phase before the age of 15 years (Dean and Elian, 1997). Apart from other viruses, Epstein-Barr virus (EBV) has been proposed as a possible trigger for development of MS. We showed that children with MS have a near-complete seropositivity for EBV capsid antigen (98%) in contrast to a hospital-based control group (72%), supporting the notion that EBV might play a role in the disease onset (Pohl *et al.*, 2006). This example underlines that the study of children with MS may offer an opportunity to unravel the pathophysiologic mechanisms triggering the onset of the disease.

MS most commonly affects young adults between the ages of 20 and 40 and has been regarded as rare before the age of 16 years. However, several studies have shown that children with MS represent at least about 5% of the whole MS population (Duquette *et al.*, 1987; Ghezzi *et al.*, 1997), with the majority of all children between 13 and 16 years of age. MS in children below 10 only represents a small, but important subgroup (Hanefeld *et al.*, 1991; Ruggieri *et al.*, 2004).

Recently, important clinical aspects of children with MS have been published thereby increasing the awareness of pediatricians and pediatric neurologists for this unique disease entity. This chapter summarizes important aspects of children with MS and places more emphasis on the concept of early treatment of MS with immunomodulatory agents in order to delay onset of disability.

II. Clinical Aspects of Children with Early-Onset MS

Traditionally, early-onset MS (EOMS)—defined as onset of disease prior to the age of 16—has been regarded as being the same disease as adult-onset MS (AOMS). However, since several studies have been reported on the natural disease course and clinical characteristics of EOMS, it has become apparent that although children and adults with MS share important similarities, they also differ in important aspects. For example, most adult patients experience a relapsing-remitting (80%) and less frequently a primary progressive course (20%) (Amato *et al.*, 1999; Hohol *et al.*, 1999). On the other hand, the overwhelming majority of children with MS have a relapsing-remitting course. No reports on a primary progressive course below the age of 10 years exist. According to our and others experience even between 10 and 16 years of age, a primary progressive course of the disease is exceedingly rare (Boiko *et al.*, 2002; Deryck *et al.*, 2006; Gusev *et al.*, 2002; Simone *et al.*, 2002). In contrast to the relapsing-remitting form, the manifestation of a primary progressive course increases with age and points to the fact that other mechanisms are involved in the disease process (Weinshenker *et al.*, 1989).

Another striking difference is that children with MS have a longer disease duration until they enter a phase of secondary progressive as compared to adults (Boiko *et al.*, 2002; Ghezzi *et al.*, 2002; Simone *et al.*, 2002). Simone *et al.* (2002), for example, found that the estimated median time to reach an expanded disability status scale (EDSS) 4, was significantly longer in the EOMS than the AOMS group (20.2 vs 10.7 years), but that the age at EDSS 4 was lower in EOMS patients (median 31.6 years; range 18–58 years) than adults with a relapsing-remitting course (median 41.1 years; range 21–69 years). Boiko and colleagues followed 113 children with MS and also found that children with MS have a longer time interval until they enter a phase of secondary progression. More than 50% of their pediatric MS patients had an EDSS of 3 at a mean age of 28.5 years. More importantly, 38% of all subjects had an EDSS of 6 at the age of 32.3 years, indicating a rapid progression over a short period of time (Boiko *et al.*, 2002). These findings are of utmost significance when dealing with children with MS and their families because these data illustrate that a substantial number of children will suffer from physical and mental difficulties at a much younger age than adults with MS.

Several other clinical aspects are of note. It appears that children more often than adults present initially with more than one functional system involved, which AOMS carries a worse outcome compared to patients who primarily present with only sensory symptoms (Amato *et al.*, 1999; Noseworthy *et al.*, 2000). In a large cohort of 132 patients reported by Pohl *et al.* (2007), 67% of children with MS had a polysymptomatic manifestation, most frequently with cerebellar (44%), sensory (39%), or visual (36%) symptoms. Interestingly, cerebromental complaints including learning difficulties were reported in up to 22% of all children. Cognitive difficulties at presentation or as squeal of the disease in children have not been studied in detail so far and are certainly areas which need to be more carefully researched in the future. Only 33% of children with MS in this study presented with one functional system involved. The leading symptoms in the monosymptomatic presentation were visual (42%), followed by brainstem (21%), sensory (16%), and cerebellar (9%) symptoms (Pohl *et al.*, 2006). Mikaeloff *et al.* (2004) also described a higher percentage of children with MS with a polysymptomatic presentation, which was associated with a worse outcome in their study. Simone *et al.* (2002) also found that cerebellar and brainstem symptoms were the most frequent initial presentation of children with MS, significantly more frequent than in AOMS. Other studies detected a lower incidence of pyramidal symptoms than in AOMS (Boiko *et al.*, 2002; Gusev *et al.*, 2002; Ozakbas *et al.*, 2003).

Equally important is that certain clinical findings are of prognostic value. A high relapse rate in the first 2 years is associated with a high risk of entering a secondary progressive course (Ghezzi *et al.*, 2002) and a short interval between the first attacks is associated with the development of an EDSS of 4 (Simone *et al.*, 2002). On the other hand, a low initial relapse rate is associated with a better prognosis in pediatric MS (Boiko *et al.*, 2002; Simone *et al.*, 2002).

Another interesting finding in EOMS is the observation that the ratio of female to male is higher compared to AOMS. Before the age of 10 years a male preponderance exists. However, during (between 10 and 14 years) and after puberty (between 15 and 16 years), several studies report a rise in the female/male ratio to 2.1:1–2.4:1, depending on the study reported (Ghezzi *et al.*, 1997; Pohl *et al.*, 2006; Simone *et al.*, 2002). These observations suggest that sexual hormones during puberty might play an important role in triggering or modifying the disease onset in EOMS.

III. Differential Diagnostic Aspects

The differential diagnostic considerations, when confronted with a child, who has a first episode of a presumed inflammatory CNS disease, are much wider compared to adults with MS due to the higher frequency of other neurometabolic white matter diseases and acute disseminated encephalomyelitis (ADEM) (Dale and Branson, 2005; Menge *et al.*, 2005). Since the introduction of the McDonald criteria, which offer the possibility to diagnose MS after a first clinical episode, the pressure on the clinician to make a diagnosis of MS in children and at the same time exclude a vast number of other diseases has increased, partly because of the implication to start immunomodulatory treatment early (McDonald *et al.*, 2001; Polman *et al.*, 2005). In most cases the diagnosis can be made after a thorough clinical history, careful physical examination, and focused investigations including imaging of the brain/spine and multimodal evoked potentials (EPs). A clinical presentation, for example, of a female between 12 and 16 years of age with focal neurological signs, the constellation of CSF oligoclonal bands and MRI findings with multiple corpus callosum and white matter lesions is highly suggestive of a first episode of EOMS.

It is important to realize that, in contrast to AOMS, children with MS often do not meet the McDonald MRI criteria for dissemination in space because children often have fewer lesions at the time of their first attack or their MS diagnosis (Hahn *et al.*, 2004). In a recent study, Pohl and colleagues reported that nearly half of all children after the first episode have evidence of dissemination in space as suggested by abnormalities in visual, auditory, and somatosensory EP. They conclude that EP may constitute an important tool for earlier diagnosis, in particular in situations in which MRI abnormalities are few (Pohl *et al.*, 2006).

Another important finding is that 92% of children with MS already after the first attack had CSF oligoclonal immunoglobulin G (IgG), as an expression of a polyspecific intrathecal antibody production against a wide range of antigens representing another sensitive marker for the early diagnosis of EOMS (Pohl *et al.*, 2004).

Children who present with atypical clinical features including fever, seizures, or bilateral visual loss; absence of CSF oligoclonal bands; or a primary progressive course, should alert the clinician to carry out a detailed work-up in order to exclude other diagnosis.

As aforementioned the differential diagnosis is broad and a detailed discussion beyond the scope of this chapter. Nevertheless, for practical reasons the author would like to point out some diseases, which are more frequent than others and easily mistaken for EOMS.

The main differential diagnosis in EOMS remains ADEM, which is primarily regarded as a monophasic illness. If in rare instances a relapse occurs with a further infection or shortly after the withdrawal of steroids then the term of multiphasic disseminated encephalomyelitis should be used (Dale and Branson, 2005; Menge *et al.*, 2005). Important criteria that help to differentiate both disease entities are that subjects with ADEM are usually younger, more often have preceding viral illnesses, headache, vomiting, and seizures, all of which are unusual in EOMS. Furthermore, patients with ADEM frequently have raised inflammatory serum markers and lymphopenia, features that are also rarely seen in EOMS. CSF studies often reveal absent/transient oligoclonal bands and an increased CSF cell count. In contrast, children with MS nearly always have CSF oligoclonal bands and a CSF cell count that is rarely raised above 61 cells/μl (Dale and Branson, 2005; Pohl *et al.*, 2004).

MRI of the brain in ADEM can be very variable, but lesions have often poorly defined margins, tend to affect the deeper white matter with sparing of the periventricular area. The gray matter in the cortex and basal ganglia is more frequently involved in ADEM than EOMS (Dale *et al.*, 2000; Mikaeloff *et al.*, 2004; Singh *et al.*, 2000). In general, children with ADEM recover without neurological squeal combined with follow-up MRI scans that show partial resolution of lesions in the majority of cases. New lesions would be highly suggestive of EOMS. The timing of follow-up scans is controversial. Some authors recommend repeating the scan only after 6 months as long as clinically indicated in order to avoid the case of ADEM which is still in a subacute phase (Dale and Branson, 2005).

Despite some obvious clinical differences, it is important to realize that in many cases only the evolution over time will give the final answer. As long as no new biomarkers have been identified which separates these two most likely heterogeneous groups of diseases, the distinction will remain a diagnostic challenge.

Another emerging important differential diagnosis in children with MS is neuromyelitis optica (NMO), which is characterized by recurrent attacks of optic neuritis and transverse myelitis. Current diagnostic criteria require absence of clinical disease outside the optic nerve and spinal cord. MRI criteria include that spine lesions should extend over three segments and brain parenchyma should be spared. An autoantibody, NMO-IgG, which is highly specific and sensitive for the diagnosis of NMO was discovered (Lennon *et al.*, 2004). In validating the clinical utility, Pittock *et al.* (2006) identified a subset of patients with symptomatic brain lesions, who were classified with MS. The distinction between the two disease entities is important because NMO is in the majority of cases a progressive disease and only responds to immunosuppressive treatment, including steroids and azathioprine.

Another important disease to consider is Leber's hereditary optic neuropathy (LHON), a mitochondrial disorder often mistaken for MS in children. It is characterized by acute or subacute bilateral visual loss, occasionally associated with other neurological manifestations. Recovery is often protracted and not hastened by steroids. More importantly, children with LHON can have white matter abnormalities (Jansen *et al.*, 1996). We therefore recommend a genetic/mutational analysis for the three most common LHON mutations in cases with bilateral optic neuritis and poor response to steroids in view of the therapeutic implications.

Other diseases that should always be considered according to the author's experience are tumors of the CNS (e.g., astrocytoma) and hemophagocytic lymphohistiocytosis (Rostasy *et al.*, 2004). Both diseases, when treated with corticosteroids, simulate a relapsing-remitting course and are difficult to distinguish from a tumefactive MS. In these circumstances, a cerebral biopsy may be the only conclusive diagnostic procedure. MS variants such as Marburg's disease or Balo's concentric sclerosis are not only exceedingly rare in children, but also can be distinguished by MRI from the more common relapsing-remitting form.

IV. Treatment of Children with MS

Although several prognostic factors such as severity of initial manifestation(s) or short interval between attacks exist, it remains often difficult to predict the disease course in each individual patient. Nevertheless, it is important to realize that the majority of children with MS will develop marked disability at a younger age compared to adults. The current progress in unraveling the pathogenesis of MS has led to the appreciation that inflammation is prominent at the beginning of the disease driving events, which lead to demyelination and axonal injury. Later in the disease course, inflammation decreases and neurodegenerative changes become more apparent. In contrast to adults, the initial phase of inflammation appears to be longer in children for reasons that are currently unknown.

As in adults, the acute episode is treated with high-dose methylprednisolone (MP) for 3–5 days (20 mg/kg/day), which usually leads to improvement of symptoms. Children who continue to have significant symptoms should be reevaluated

2 weeks later, and if necessary receive a second course of high-dose MP. Children who present with significant disabling symptoms and do not respond to high-dose MP should be offered a therapeutic plasma exchange. A study in adult patients with MS and fulminant demyelinating attacks unresponsive to steroids has shown that therapeutic plasma exchange lead to a sustained improvement in the majority of patients (Keegan *et al.*, 2005). A German study including three children has demonstrated a clear benefit also in two children (Schilling *et al.*, 2006).

Children with two clinical episodes or those who fulfill the McDonald criteria should be offered disease-modifying therapies (DMT) despite the fact that none of these therapies has been approved for children. Four FDA approved immunomodulating agents for reducing MS relapses in adults are available: three preparations of interferon beta (IFNB) and glatiramer acetate (GA). Both groups of medications have been shown to induce a 30% decrease of the relapse rate in adults with MS (Johnson *et al.*, 2001; PRISM, 1998, 2001). IFNB and GA have been already administered in several centers specialized in treating children with MS. Different reports have been published which indicate that IFNB and GA are well tolerated by pediatric MS patients and more importantly appear to reduce the number of attacks (Ghezzi *et al.*, 2004; Kornek *et al.*, 2003; Mikaeloff *et al.*, 2001; Pohl *et al.*, 2005; Tenembaum and Segura, 2006).

General agreement is that immunomodulatory treatment should be continued at least for 2 years in order to judge treatment response. During that time children should be carefully followed and monitored. Change of DMT should be considered in the presence of severe side effects, poor compliance, or in patients who have frequent episodes of MS and/or have evidence of significant progression on MRI. In our institution, children who receive IFNB, demonstrate significant clinical progression, and have contrast-enhancing MRI lesions are switched to GA, which has proven beneficial in some patients. Other options include the monthly administration of intravenous immunoglobulins or daily azathioprine, although no studies exist in children to support their efficacy or side effect profile.

Children who initially respond to IFNB, but later have further evidence of progression, should be tested for presence of neutralizing antibodies (NAbs). Studies in adults suggest that NAbs develop in a substantial number of patients, who are more likely to experience an increase in new MRI lesions (Namaka *et al.*, 2006).

Similar to adults who still demonstrate high disease activity (more than two annual relapses, multiple contrast-enhancing lesions on MRI), children might benefit from the monoclonal antibody. Natalizumab, which inhibits the migration of activated T cells across the blood–brain barrier into the CNS, leads to a dramatic reduction in the number of lesions on brain MRI scans and the frequency of attacks (Miller *et al.*, 2003; Polman *et al.*, 2006; Rudick *et al.*, 2006). It is again available for routine application, after it was temporarily withdrawn from the market because of the development of progressive multifocal leukoencephalopathy in patients who were treated in combination with other immunomodulatory drugs. This treatment should be reserved for those who fail to respond to the currently available immunomodulatory drugs.

Another treatment option is the administration of regular pulses of cyclophosphamide or mitoxantrone, the only FDA-approved immunosuppressive medication in MS. Adult patients receiving mitoxantrone experienced significant benefits at 24 months on disability progression, number of treated relapses, and number of new T2-weighted MRI lesions (Hartung *et al.*, 2002). Experience in children are limited and treatment should be carried out in centers specialized in MS, in view of the substantial side effects such as bone marrow suppression and cardiac toxicity.

V. Conclusions

Early-onset MS is increasingly more recognized as a disease entity. Children with MS share important aspects of the disease but also exhibit important differences, which may have important pathophysiologic and therapeutic implications. Most importantly, the majority of children have a disease onset shortly after puberty with a relapsing-remitting course. Children with MS suffer from physical and mental difficulties after much longer disease duration compared to adults, but in view of the early-onset experience significant health problems at a much younger age. On one side this implies that children with MS in the initial phase of the disease may have a higher potential to limit the extent of the disease combined with the ability to recover more readily from immune attacks to the white matter for reasons that are currently unknown. On the other hand, it appears mandatory to initiate immunomodulatory treatment early, after a thorough work-up has been conducted in order to prevent future disability.

Abbreviations

ADEM, acute disseminated encephalomyelitis; AOMS, adult-onset multiple sclerosis; CNS, central nervous system; CSF, cerebralspinal fluid; DMT, disease-modifying therapies; EOMS, early-onset multiple sclerosis; EBV, Epstein-Barr virus; EDSS, expanded disability status scale; EP, evoked potential; GA, glatiramer acetate; IgG, mmunoglobulin G; IFNB, nterferon beta; LHON, Leber's hereditary optic neuropathy; MRI, magnetic resonance imaging; MRS, magnetic resonance spectroscopy; MP, methylprednisolone; MS, multiple sclerosis; NMO, neuromyelitis optica.

References

Amato, M. P., Ponziani, G., Bartolozzi, M. L., and Siracusa, G. (1999). A prospective study on the natural history of multiple sclerosis: Clues to the conduct and interpretation of clinical trials. *J. Neurol. Sci.* **168,** 96–106.

Balassy, C., Bernert, G., Wober-Bingol, C., Csapo, B., Kornek, B., Szeles, J., Fleischmann, D., and Prayer, D. (2001). Long-term MRI observation of childhood-onset relapsing-remitting multiple sclerosis. *Neuropediatrics* **32,** 28–37.

Boiko, A., Vorobeychik, G., Paty, D., Devonshire, V., Sadovnick, D., and the UBC MS Clinic Neurologists (2002). Early onset multiple sclerosis. A longitudinal study. *Neurology* **59,** 1006–1010.

Bruhn, H., Frahm, J., Merboldt, K. D., Hänicke, W., Hanefeld, F., Christen, H. J., Kruse, B., and Bauer, H. J. (1992). Multiple sclerosis in children: Cerebral metabolic alterations monitored by localized proton magnetic resonance spectroscopy *in vivo*. *Ann. Neurol.* **32,** 140–150.

Dale, R. C., and Branson, J. A. (2005). Acute disseminated encephalomyelitis or multiple sclerosis: Can the initial presentation help in establishing a correct diagnosis. *Arch. Dis. Child* **90,** 636–639.

Dale, R. C., de Sousa, C., Chong, W. K., Cox, T. C., Harding, B., and Nevile, B. G. (2000). Acute disseminated encephalomyelitis, multiphasic disseminated encephalomyelitis and multiple sclerosis in children. *Brain* **123,** 2407–2422.

Dean, G., and Elian, M. (1997). Age at immigration to England of Asian and Caribbean immigrants and the risk of developing multiple sclerosis. *J. Neurol. Neurosurg. Psychiatry* **63,** 565–568.

Deryck, O., Ketelaer, P., and Dubois, B. (2006). Clinical characteristic and long term prognosis in early onset multiple sclerosis. *J. Neurol.* **253,** 720–723.

Duquette, P., Murray, T. J., Pleines, J., Ebers, G. C., Sadovnick, D., Weldon, P., Warren, S., Paty, D. W., Upton, A., Hader, W., Nelson, R., Auty, A., *et al.* (1987). Multiple sclerosis in childhood: Clinical profile in 125 patients. *J. Pediatr.* **111,** 359–363.

Ghezzi, A., Deplano, V., Faroni, J., Grasso, M. G., Liguori, M., Marrosu, G., Pozzilli, C., Simone, I. L., and Zaffaroni, M. (1997). Multiple sclerosis in childhood: Clinical features of 149 patients. *Mult. Scler.* **3,** 43–46.

Ghezzi, A., Pozzilli, C., Liguori, M., Marrosu, M. G., Milani, N., Milanese, C., Simone, I., and Zaffaroni, M. (2002). Prospective study of multiple sclerosis with early onset. *Mult. Scler.* **8,** 115–118.

Ghezzi, A., Ruggieri, M., Trojano, M., Filippi, M, and the ITEMS Study Group (2004). Italian studies on early-onset multiple sclerosis: The present and the future. *Neurol. Sci.* **24,** S346–S349.

Gusev, E., Boiko, A., Bikova, O., Maslova, O., Guseva, M., Boiko, S., Vorobeichik, G., and Paty, D. (2002). The natural history of early onset multiple sclerosis: Comparison of data from Moscow and Vancouver. *Clin. Neurol. Neurosurg.* **104,** 203–207.

Hahn, C. D., Shroff, M. M., Blaser, S. I., and Banwell, B. L. (2004). MRI criteria for multiple sclerosis. Evaluation in a pediatric cohort. *Neurology* **62,** 806–808.

Hanefeld, F., Bauer, H. J., Christen, H. J., Kruse, B., Bruhn, H., and Frahm, J. (1991). Multiple sclerosis in childhood: Report of 15 cases. *Brain Dev.* **13,** 410–416.

Hartung, H. P., Gonsette, R., Konig, N., Kwiecinski, H., Guseo, A., Morrissey, S. P., Krapf, H., and Zwingers, T. (2002). Mitoxantrone in progressive multiple sclerosis: A placebo-controlled, double-blind, randomised, multicentre trial. *Lancet* **360,** 2018–2025.

Hohol, M. J., Olek, M. J., Orav, E. J., Stazzone, L., Hafler, D. A., Khoury, S. J., Dawson, D. M., and Weiner, H. L. (1999). Treatment of progressive multiple sclerosis with pulse cyclophosphamide/methylprednisolone: Response to therapy is linked to the duration of progressive disease. *Mult. Scler.* **5,** 403–409.

Jansen, P. H., van der Knaap, M. S., and de Coo, I. F. (1996). Leber's hereditary optic neuropathy with the 11778 mtDNA mutation and white matter disease resembling multiple sclerosis: Clinical, MRI and MRS findings. *J. Neurol. Sci.* **135,** 176–180.

Johnson, K. P., Brooks, B. R., Cohen, J. A., Ford, C. C., Goldstein, J., Lisak, R. P., Myers, L. W., Panitch, H. S., Rose, J. W., Schiffer, R. B, and Copolymer 1 Multiple Sclerosis Study Group (2001). Copolymer 1 reduces relapse rate and improves disability in relapsing-remitting multiple sclerosis: Results of a phase III multicenter, double-blind, placebo-controlled trial. *Neurology* **57,** S16–S24.

Keegan, M., Konig, F., McClelland, R., Bruck, W., Morales, Y., Bitch, A., Panitch, H., Lassmann, H., Weinshenker, B., Rodriguez, M., Parisi, J., and Lucchinetti, C. F. (2005). Relation between humoral pathological changes in multiple sclerosis and response to therapeutic plasma exchange. *Lancet* **366,** 579–582.

Kornek, B., Bernert, G., Balassy, C., Geldner, J., Prayer, D., and Feucht, M. (2003). Glatiramer acetate treatment in patients with childhood and juvenile onset multiple sclerosis. *Neuropediatrics* **34,** 120–126.

Kuhlmann, T., Lingfeld, G., Bitsch, A., Schuchardt, J., and Brueck, W. (2002). Acute axonal damage in multiple sclerosis is most extensive in early disease stages and decrease over time. *Brain* **125,** 2202–2212.

Lennon, V. A., Wingerchuk, D. M., Krytzer, T. J., Lennon, V. A., Wingerchuk, D. M., Kryzer, T. J., Pittock, S. J., Lucchinetti, C. F., Fujihara, K., Nakashima, I., and Weinshenker, B. G. (2004). A serum autoantibody marker of neuromyelitis optica. *Lancet* **364,** 2106–2112.

McDonald, W. I., Compston, A., Edan, G., Goodkin, D., Hartung, H.-P., Lublin, F. D., McFarland, H. F., Paty, D. W., Polman, C. H., Reingold, S. C., Sandberg-Wollheim, M., Sibley, M., *et al.* (2001). Recommended diagnostic criteria multiple sclerosis: Guidelines from the international panel on the diagnosis of multiple sclerosis. *Ann. Neurol.* **50,** 121–127.

Menge, T., Hemmer, B., Nessler, S., Wiendl, H., Neuhaus, O., Hartung, H. P., Kieseier, C., and Stüve, O. (2005). Acute disseminated encephalomyelitis: An update. *Arch. Neurol.* **62,** 1673–1680.

Mikaeloff, Y., Moreau, T., Debouverie, M., Pelletier, J., Lebrun, C., Gout, O., Pedespan, J.-M., Van Hulle, C., Vermersch, P., and Ponsot, G. (2001). Interferon β treatment in patients with childhood-onset multiple sclerosis. *J. Pediatr.* **139,** 443–446.

Mikaeloff, Y., Suissa, S., Vallee, L., Lubetzki, C., Ponsot, G., Confavrfux, C., and Tardieu, M. (2004). First episode of acute CNS inflammatory demyelination in childhood: Prognostic factors for multiple sclerosis and disability. *J. Pediat.* **144,** 246–252.

Miller, D. H., Khan, O. A., Sheremata, W. A., Blumhardt, L. D., Rice, G. P., Libonati, M. A., Willmer-Hulme, A. J., Dalton, C. M., Miszkiel, K. A., O'Connor, P. W., and International Natalizumab Multiple Sclerosis Trial Group (2003). A controlled trial of natalizumab for relapsing-remitting multiple sclerosis. *N. Engl. J. Med.* **348,** 15–23.

Namaka, M., Pollitt-Smith, M., Gupta, A., Klowak, M., Vasconcelos, M., Turcotte, D., Gong, Y., and Melanson, M. (2006). The clinical importance of neutralizing antibodies in relapsing-remitting sclerosis. *Curr. Med. Res. Opin.* **22,** 223–239.

Noseworthy, J. H., Lucchinetti, C., Rodriguez, M., and Weinshenker, B. G. (2000). Multiple sclerosis. *N. Engl. J. Med.* **343,** 938–952.

Ozakbas, S., Idiman, E., Baklan, B., and Yulug, B. (2003). Childhood and juvenile onset multiple sclerosis: Clinical and paraclinical features. *Brain Dev.* **25,** 233–236.

Peterson, J. W., Bo, L., Mork, S., Chang, A., and Trapp, B. D. (2001). Transected neurites, apoptotic neurons, and reduced inflammation in cortical multiple sclerosis lesions. *Ann. Neurol.* **50,** 389–400.

Pittock, S. J., Lennon, V. A., Krecke, K., Wingerchuk, D. M., Lucchinetti, C. F., and Weinshenker, B. G. (2006). Brain abnormalities in neuromyelitis optica. *Arch. Neurol.* **63,** 390–396.

Pohl, D., Rostasy, K., Reiber, H., and Hanefeld, F. (2004). CSF characteristics in early-onset multiple sclerosis. *Neurology* **63,** 1966–1967.

Pohl, P., Rostasy, K. M., Gärtner, J., and Hanefeld, F. (2005). Treatment of early onset multiple sclerosis with subcutaneous interferon-beta-1a. *Neurology* **64,** 88–890.

Pohl, P., Krone, B., Rostasy, K., Kahler, E., Brunner, E., Lehnert, M., Wagner, H. J., Gärtner, J., and Hanefeld, F. (2006). Pediatric multiple sclerosis and Epstein-Barr virus: High seroprevalence and antibody titers. *Neurology* **67**(11), 2063–2065.

Pohl, P., Hennemuth, I., von Kries, R., and Hanefeld, F. (2007). Pediatric multiple sclerosis and acute disseminated encephalomyelitis in Germany: Results of a nationwide survey. *Eur. J. Pediatr.* (Epub ahead of print).

Pohl, P., Rostasy, K., Treiber-Held, S., Brockmann, K., Gärtner, J., and Hanefeld, F. (2006). Pediatric multiple sclerosis: Detection of clinically silent lesions by multimodal evoked potentials. *J. Pediatr.* **149,** 125–127.

Polman, C. H., Reingold, S. C., Edan, G., Filippi, M., Hartung, H. P., Kappos, L., Lublin, F. D., Metz, L. M., McFarland, H. F., O'Connor, P. W., Sandberg-Wollheim, M., Thompson, M., *et al.* (2005). Diagnostic criteria for multiple sclerosis: 2005 revisions to the "McDonald criteria." *Ann. Neurol.* **58,** 840–846.

Polman, C., O'Connor, P., Havrdova, E., Hutchinson, M., Kappos, L., Miller, D. H., Phillips, J. T., Lublin, F. D., Giovannoni, G., Wajgt, A., Toal, M., Lynn, F., *et al.* (2006). A randomized, placebo-controlled trial of natalizumab for relapsing multiple sclerosis. *N. Engl. J. Med.* **354,** 899–910.

PRISMS Study Group (1998). Randomized double-blind placebo-controlled study of interferon-1a in relapsing/remitting multiple sclerosis. *Lancet* **352,** 1498–1504.

PRISMS Study Group and the University of British Columbia MS/MRI Analysis Group (2001). PRISMS-4: Long-term efficacy of interferon-1a in relapsing MS. *Neurology* **56,** 1628–1636.

Rostasy, K., Kolb, R., Pohl, D., Mueller, H., Fels, C., Moers, A. V., Bergmann, M., Hanefeld, F., Pekrun, A., and Schulz-Schaeffer, W. (2004). CNS disease as the main manifestation of hemophagocytic lymphohistiocytosis in two children. *Neuropediatrics* **35,** 45–49.

Rostasy, K., Withut, E., Pohl, D., Lange, P., Ciesielcyk, B., Diehm, R., Gärtner, J., and Otto, M. (2005). Tau phospho-tau and S-100B protein in cerebrospinal fluid of children with multiple sclerosis. *J. Child Neurol.* **20,** 822–825.

Rudick, R., Stuart, W., Calabresi, P., Confavreux, C., Galetta, S. L., Radue, E. W., Lublin, F. D., Weinstock-Guttman, B., Wynn, D. R., Lynn, F., Panzara, M. A., Sandrock, A. W., *et al.* (2006). Natalizumab plus interferon beta-1a for relapsing multiple sclerosis. *N. Engl. J. Med.* **354,** 911–923.

Ruggieri, M., Iannetti, P., Polizzi, A., Pavone, L., and Grimaldi, L. M. E. (2004). Multiple sclerosis in children under 10 years of age. *Neurol. Sci.* **25,** S326–S335.

Schilling, S., Linker, R. A., Konig, F. B., Koziolek, M., Bahr, M., Muller, G. A., Paulus, W., Gartner, J., Bruck, W., Chan, A., and Gold, R. (2006). Plasma exchange therapy for steroid-unresponsive multiple sclerosis relapses: Clinical experience with 16 patients. *Nervenarzt* **77,** 430–438.

Simone, I. L., Carrara, D., Tortorella, C., Liguori, M., Lepore, V., Pellegrini, F., Bellacosa, A., Ceccarelli, A., Pavone, I., and Livrea, P. (2002). Course and prognosis in early-onset MS. Comparison with adult-onset forms. *Neurology* **59,** 1922–1928.

Singh, S., Prabhakar, S., Korah, I. P., Warade, S. S., and Alexander, M. (2000). Acute disseminated encephalomyelitis and multiple sclerosis: Magnetic resonance imaging findings. *Australas Radiol.* **44,** 404–411.

Tenembaum, S. N., and Segura, M. J. (2006). Interferon beta-1a treatment in childhood and juvenile-onset multiple sclerosis. *Neurology* **67,** 511–513 (expedited E-Pub).

Trapp, B. D., Jeterson, J., Ransohoff, R. M., Rudick, R., Moerk, S., and Boe, L. (1998). Axonal transection in the lesions of multiple sclerosis. *N. Engl. J. Med.* **338,** 278–285.

Weinshenker, B. G., Bass, B., Rice, G. P., Noseworthy, J., Carriere, W., Baskerville, J., and Ebers, G. C. (1989). The natural history of multiple sclerosis: A geographically based study. 1. Clinical course and disability. *Brain* **112,** 133–146.

MIGRAINE IN MULTIPLE SCLEROSIS

Debra G. Elliott

Department of Neurology, Louisiana State University Health Sciences Center
Shreveport, Louisiana 71103, USA

Headache and migraine are common features in multiple sclerosis (MS) and can influence the diagnosis, radiological evaluation, treatment, and quality of life of these patients. Similarities in symptoms between patients with migraine and MS presenting with headache can lead to misdiagnosis. Likewise, MRI lesions which may be found in migraine patients without other neurological symptoms or signs may cause diagnostic confusion and patient anxiety. Studies addressing the pathophysiology of these comorbid conditions have not found a clear link, but brainstem lesions and inflammatory processes have been proposed. Management of *de novo* or treatment-induced headache in the MS patient is fairly straightforward if the physician is aware of the problem.

INTERNATIONAL REVIEW OF
NEUROBIOLOGY, VOL. 79
DOI: 10.1016/S0074-7742(07)79012-8

I. Epidemiology

Patients with MS frequently have pain (Ehde *et al.*, 2005; Solaro *et al.*, 2003), including headache, dysesthetic limb pain, back pain, painful spasms (Pöllmann *et al.*, 2004), glossopharyngeal neuralgia (Minagar and Sheremata, 2000), and trigeminal neuralgia (Zakrzewska, 2002). Pöllmann *et al.* (2004) found that in 61% of 157 consecutive clinically definite MS patients had severe pain. In 12%, the pain was the worst symptom of MS as perceived by the patients. Sixty-eight percent reported insufficient care by the physicians consulted. The authors commented that there is an urgent need for physicians to keep pain symptoms in mind when treating MS patients.

Österberg *et al.* (2004) reported that 57.5% of 364 relapsing-remitting multiple sclerosis (RRMS) patients had central pain, defined as pain caused by a lesion or dysfunction in the brain or spinal cord. They eliminated back pain, tension-type headache, migraine, and optic neuritis-related pain from the central pain category. Pain was described as burning and aching felt both superficially and in deep tissues. Similar to central poststroke pain, abnormal pain, and temperature sensibility was often present. They noted that MS-related central pain is severe and causes much patient suffering.

Early reports of migraine in MS patients date back to at least the 1960s (Poser *et al.*, 1966; Watkins and Espir, 1969), but reveal variable prevalence. For instance, Warren and Warren (1981) found no significant association of migraine in 100 hospitalized MS patients. This discrepancy may have been due to lack of standardization for the diagnosis of migraine prior to International Headache Society classification was published in 1988 (Headache classification committee International Headache Society, 1988).

In a retrospective review, Gee *et al.* (2005) reported headache in 55.6% of MS patients, of which 61.7% had migraine, 25.3% had tension-type, and 13% had both migraine and tension-type headaches. D'Amico *et al.* (2004) prospectively studied 137 patients with clinically definite MS (RRMS or progressive form). Eighty-eight total (64.2%) reported headache. Of those patients whose headache was not temporally related to interferon (IFN) treatment, 57.7% of the MS patients had headache, 25.0% migraine without aura (none had aura), and 31.9% tension-type headache (TTH). One male patient suffered cluster headaches. Presence of headache was not related to illness duration or disability. Those with RRMS were more likely to have migraine, whereas the progressive form of MS was associated with TTH. Only 1.7% of patients reported headache as a presenting symptom of MS onset in this study. This result is close to the 1% reported by Freedman and Gray (1989). Both are much lower than the 13% of MS patients presenting with headache reported by Rolak and Brown (1990).

Migraine has also been found to be among the risk factors for MS in some studies. Martinez Sobrepera *et al.* (2001) reported migraine among Cuban MS patients as a risk factor with an odds ratio of 2.5 (95% CI). Other risk factors included high educational level, ≥two children, four or more pregnancies, three or more deliveries, previous measles infection, stress, and insomnia. Zorzon *et al.* (2003) also reported migraine as the second highest associated risk factor for developing MS (OR 8.7, 95% CI) among 140 patients in Italy compared to sex- and age-matched controls. Migraine fell behind family history (OR 12.1, 95% CI) as the second highest risk factor. They also reported comorbidity with migraine (OR 13.5, 95% CI), as well as with autoimmune disease (OR 6.8, 95% CI), and measles vaccination (OR 92.2, 95% CI) in patients with MS. They concluded that these results supported the popular hypothesis that the etiology of MS includes both genetic and environmental risk factors.

Other epidemiological studies do not report migraine as a risk factor but rather cite a variety of possible causes. Genetic risk factors for MS are fairly consistent and include DR(2)15 and DR(3)17 alleles (Alviggi *et al.*, 2006) and apolipoprotein (HLA)-DRB1 allele (Kantarci and Wingerchuk, 2006), whereas the genetics of migraine are unclear except for familial hemiplegic migraine (FHM) (see below). Environmental risk factors reported for MS include viral infection (Poser, 2006), specifically Epstein–Barr virus/mononucleosis (Cook, 2004; Thacker *et al.*, 2006), human herpesvirus 6 (Alvarez-Lafuente *et al.*, 2006), and varicella zoster (Perez-Cesari *et al.*, 2005). Of these, the Epstein–Barr virus has also been a reported possible cause of migraine and chronic daily headache (Leavell *et al.*, 1986; Mack, 2004). Inflammatory bowel disease +/− antitumor necrosis factor-α (TNF-α) therapy was cited as a risk factor for MS by Gupta *et al.* (2005), but was only loosely associated with migraine (Crawford *et al.*, 1994) if at all. Smoking may increase the risk for MS (Lauer, 2006; Nortvedt *et al.*, 2005), but not for migraine or tension-type headache (Takeshima *et al.*, 2004). Vitamin D deficiency may predispose patients to developing MS (Kantarci and Wingerchuk, 2006; Lips, 2006; Mark and Carson, 2006), but the association with migraine is lacking except for anecdotal reports of improvement in migraine frequency and duration with vitamin D therapy (Thys-Jacobs, 1994a,b).

II. Problems in Diagnosing MS and Migraine

A. Case Reports

An unfortunate man residing in a nursing home was recently identified in our clinic with end-stage MS whose weakness and ataxia rendered him wheelchair bound unable to perform many activities of daily living. He came with a previous diagnosis of alcoholism, psychogenic polydipsia with recurrent hyponatremia, and "possible movement disorder" due to his ataxia. His MRI lesion load was very heavy, but was read as "metabolic white matter (WM) changes." He had never been diagnosed with MS and therefore had never been offered treatment (personal experience). The tragedy of this scene is obvious but fortunately is an extreme and unusual case.

Young (1997) commented on a different problem which is more common, that of a man with multiple nonspecific WM changes on MRI but with no history or physical symptoms or signs consistent with MS. The man was "reduced to a psychological cripple" until the diagnostic uncertainty was cleared by a second opinion evaluation.

Finally, a man in Denmark who was originally diagnosed with MS obtained reevaluation. Because the man had MRI findings which are unusual for MS (hyperintensities of the external capsule and anterior temporal lobes) and negative cerebral spinal fluid (CSF), the man and his family had further testing. This resulted in the diagnosis of cerebral autosomal dominant arteriopathy with subcortical infarcts and leukoencephalopathy (CADASIL), which changed the treatment plan considerably and also, affected other members of the family (Kimper-Karl *et al.*, 2006).

These three cases illustrate the difficulty patients and physicians encounter in the quest for a definitive diagnosis of MS.

B. Misdiagnosis of MS

Misdiagnosis of MS occurs in 5–10% of affected patients, with an average delay in diagnosis of 3.5 years (Carmosino *et al.*, 2005; Levin *et al.*, 2003; Trojano and Paolicelli, 2001). Levin *et al.* (2003) reported that of 50 consecutive MS patients presenting to their clinic, 58% were initially given 41 wrong diagnoses. They found an interesting gender difference among those who had been previously misdiagnosed: women were considered to have a psychiatric disorder, whereas men were offered orthopedic work-up. Apparently, patients of both genders emphasized that this initial period of diagnostic confusion increased their anxiety.

Clinical diagnosis of MS depends on the presence of lesions disseminated in both time and space (McDonald *et al.*, 2001). Since migraine is a paroxysmal disorder and can be accompanied by focal neurological deficit (aura) in 15% of patients, this is a common diagnostic error. Diseases which may be confused with or mimic MS include inflammatory conditions such as systemic lupus erythematosus, Sjogren's syndrome, progressive multifocal leukoencephalopathy, HTLV-1 infection, herpes zoster; genetic aberrations such as lysosomal disorders, adrenoleukodystrophy, mitochondrial disorders, and CADASIL (some of which include migraine); metabolic problems such as vitamin B12 deficiency, neoplastic conditions such as central nervous system (CNS) lymphoma; and spinal degenerative disease or spinal vascular malformation (Trojano and Paolicelli, 2001).

Other CNS diseases which may be in the differential of both migraine with aura and MS include antiphospholipid syndrome (Chapman *et al.*, 2003; Hughes, 2003; Kitigawa, 2003), neurosarcoidosis (Charlier *et al.*, 2005; Ferreira *et al.*, 2005; La Mantia and Erbetta, 2004; Marangoni *et al.*, 2006), and systemic lupus erythematosis (Jacobi *et al.*, 2006; Ordonez *et al.*, 2007). In these patients, if serological, CSF, and MRI findings do not corroborate a specific diagnosis, biopsy may be required.

Carmosino *et al.* (2005) studied 281 patients referred to a tertiary MS center during a 30-month period. A total of 186 were misdiagnosed. They found that 88 of these patients had an alternative neurological diagnosis, and 63 had symptoms best explained by a psychiatric diagnosis (note: no information concerning gender was given). Migraine was the most common neurological diagnosis comprising over one-fourth of this group. Of those patients with an alternative neurological diagnosis and positive MRI WM abnormalities on T2 or FLAIR images, 37% were attributable to migraine; 8% were noted to have nonspecific headache with no other explanation for the lesions.

C. Clinical Symptoms of Demyelination

Diagnostic confusion can exist when a patient presents with transient focal, or even worse, vague symptoms which are not accompanied by clear neurological findings on examination. Since MS is a demyelinating disorder, clinical clues to this pathological feature are important. Painful visual loss with optic neuropathy, signifying optic neuritis, a brainstem syndrome such as diplopia, vertigo, or dysarthria, or symptoms of myelopathy such as paraplegia, bladder symptoms, and sensory level are suggestive of demyelination. Symptoms are not as likely to be demyelinating if they are all generalized (i.e., fatigue, global weakness), do not last at least 24 h (i.e., transient visual blurring lasting minutes), or reflect disease outside of the CNS (i.e., radiculopathy) (Carmosino *et al.*, 2005).

D. Transient Focal Neurological Deficits in Migraine

MS is clinically diagnosed on the basis of signs or symptoms of transient focal neurological deficit separated by time and space. Migraine patients also can experience transient neurological symptoms in the form of the migraine aura. An aura is present in 15–20% of migraineurs. This is a focal neurological symptom which can have both "positive" (extra sensations) and "negative" (lack of sensation) symptoms (Silberstein *et al.*, 2002). Each aura symptom typically lasts 5–60 min. The symptoms are usually stereotypical from attack to attack in most patients, and do not suggest neurological foci separated in space, in contrast with MS.

Ninety-nine percent of the migraine patients with aura experience visual symptoms in at least some of their auras. Visual aura occurs in 99% of migrane with aura (MWA) patients, the most common of which is the scintillating scotoma, consisting of a field deficit circumscribed by shimmering lights (Kirchmann, 2006). If multiple auras occur, they come in quick succession. Sensory and visual auras occur on one side of the body. Sensory phenomenon (54%) and speech/aphasic symptoms (32%) are the most common nonvisual auras, though motor (Thomsen and Olesen, 2004) and vertiginous symptoms (Eggers, 2006) have been described. The International Headache Society Classification diagnostic criteria (2004) for MWA are: gradual development of the symptom over 4 min, duration less than 60 min, consisting of one or more reversible brain symptom and the headache should follow within 60 min of it.

FHM and sporadic hemiplegic migraine are characterized by episodes of typical migraine aura symptoms including motor weakness (ICHD-II, 2004) followed by migraine headache. They may have multiple and prolonged auras (motor symptoms for up to 12 h, visual, sensory, or aphasic symptoms up to 24 h) and basilar-type symptoms (dysarthria, vertigo, tinnitus, hearing loss, diplopia, bilateral visual symptoms, decreased level of consciousness, bilateral paresthesia, bilateral paresis) (Kirchmann Eriksen *et al.*, 2006). In the sporadic type, aura may not be followed by headache in some cases (acephalgic migraine). Although the main differential diagnosis includes cerebral ischemia (stroke or transient ischemic attack), epilepsy with Todd's paralysis, and psychogenic attacks, some physicians may also consider MS due to the separation of these focal symptoms in time and space.

Acephalgic migraine occurs in a small but clinically relevant number of migraineurs in which the entire process aborts before any headache occurs. If there is generally a stable migraine pattern in the face of a normal neurological exam, no further diagnostic testing is necessary. Late onset migraine accompaniment presents in middle aged or older patients, and is a benign occurrence (Freedom and Jay, 2003). Migraine equivalents can also occur in the pediatric and adolescent population without associated headache. These are periodic, paroxysmal syndromes consisting of benign paroxysmal torticollis, benign paroxysmal vertigo, abdominal migraine/cyclical vomiting, acephalgic migraine, and confusional migraine (Al-Twaijri and Shevell, 2002). Acephalgic migraine features typical migraine aura lasting <10 min, which is most often visual (Shevell, 1996), and may include micropsia or metamorphopsia (the so-called "Alice in Wonderland syndrome") (Evans and Rolak, 2004).

Alternating hemiplegia of childhood may be a migraine variant, although the true etiology is unknown, with both familial and sporadic forms known to exist. Patients become symptomatic in infancy with repeated hemiplegic attacks shifting from one side to the other which last minutes up to several days (Di Rosa *et al.*, 2006). Confusion with MS is less likely in this disorder due to normal MRI scans, normal motor central conduction time on motor evoked potential testing (Rinalduzzi *et al.*, 2006), and the very young age at presentation.

III. MRI Findings in MS and Migraine

Whereas MRI lesions are almost always present and rule *in* a diagnosis of MS, in clinical practice neuroimaging is used to rule *out* other intracranial pathology rather than rule migraine *in*. However, radiologists' interpretations of WM lesions on MRI can sometimes be noncommittal, and often mention the possibility of "demyelinating disorder such as MS, or can be seen in patients with migraine" (personal experience). Thus, the diseases are often considered together in this scenario, and the clinician is encouraged to correlate the findings clinically. In this section, I will discuss the major features of the MRI findings in MS and migraine and their implications as to the clinical and pathophysiological picture of the diseases.

A. Multiple Sclerosis

Magnetic resonance imaging (MRI) is an important tool for the diagnosis of MS. Indeed, MRI lesions are necessary for the diagnosis (McDonald criteria) if the patient has objective clinical evidence of only one lesion (McDonald *et al.*, 2001; Polman *et al.*, 2005). MRI is also used extensively in follow-up of the clinical course of the disease, which has proven to be especially important in therapeutic trials (Minagar *et al.*, 2005).

WM lesions in MS are related to the MS plaque, a neuroimmune-mediated inflammatory breakdown of the oligodendrocyte/myelin complex which is highly correlated with specific focal neurological deficits. Acutely, there is breakdown of the blood–brain barrier (BBB) with edema and the plaque enhances with gadolinium contrast on T1-weighted images. At this stage, the lesion can enhance homogenously or in a ring or partial "open-ring" fashion which is most consistent with demyelination, but which includes infection, neoplasm, and adrenoleukodystrophy in the differential diagnosis. This acute phase is followed by gliosis and atrophy as demyelination proceeds to axonal loss with abnormal macrophages and astrocytosis (Minagar *et al.*, 2005).

T2-weighted images and especially FLAIR sequences are helpful in identifying MS WM lesions regardless of age or activity. Demyelinating plaques occur in the periventricular regions, the corpus callosum, brainstem, and cerebellum. The periventricular lesions are situated adjacent to the ependymal surface and extend out in a perpendicular axis to the lateral ventricle following perivenular demyelination patterns (Dawson's fingers) (Minagar *et al.*, 2005).

Evans and Rolak (2001) pointed out that MS lesions have the following characteristics which can help to delineate them from other CNS diseases: WM lesions that are more periventricular rather peripheral, oval in shape rather than round or punctate, with margins which are "fuzzy" or irregular rather than sharply defined, and are orientated perpendicular to the ventricles as though radiating away from them. Lesion location in the corpus callosum (as identified on sagittal sections) and/or the infratentorium are further clues to a diagnosis of MS (Evans and Rolak, 2001).

B. Migraine

MRI abnormalities are much less common in migraine. Only approximately 16% of patients with migraine have abnormal brain MRI images, the range being from 12% to 46% (Cooney *et al.*, 1996; Jacome and Leborgne, 1990). Controlling for cardiovascular risk factors and use of vasoconstrictor medications, Kruit *et al.* (2004) found an increase in deep WM MRI lesions among 295 women with migraine (especially those with ≥ 1 migraine attack per month) as compared to nonmigraine controls. Cerebellar lesions were significantly more prevalent in both genders (5.4% migraine vs 0.7% control group). Higher attack frequency (≥ 1 per month) and MWA further increased the risk of such lesions, which they considered to be subclinical ischemic infarcts. The same group found that in 435 migraine subjects, 60 lesions were found (almost 14%), and 65% of these occurred in the posterior circulation, mainly in the cerebellum (Kruit *et al.*, 2005). In a subsequent study of 295 migraineurs, infratentorial lesions (seen in 4.4% of patients and 0.7% of controls) were predominantly pontine involving pontocerebellar fibers, pontine nuclei, the nucleus reticularis tegmenti pontis, the corticopontine tract or medial lemniscus, and sparing areas traditionally associated with migraine such as the dorsal rostral pons and periaqueductal gray (PAG) (Kruit *et al.*, 2006).

Brain lesions on MR images which occur in migraine patients are not nearly as well understood or uniform in their presentation as in MS. Cooney *et al.* (1996) found that among patients with migraine and WM MRI changes, other significant risk factors were also present. In those patients over 50 years old these included hypertension and diabetes, demyelinating disease (not otherwise specified in the chapter), and autoimmune disease. Such WM abnormalities in these migraineurs were thought to be due to vasospastic ischemia or perhaps vasculitis (Cooney *et al.*, 1996). More investigators favor "migrainous infarction" as an explanation for WM lesions citing excitotoxic neural cell death due to excessive release of excitatory peptides in the extracellular space resulting in protracted cerebral spreading depression and oligemia (Longoni and Ferrarese, 2006), or small-vessel disease due to arteriosclerosis, repetitive perfusion deficits, or both (Kruit *et al.*, 2006). A possible association with FHM and mutations of the P/Q-type calcium channel α-subunit gene (CACNA1A) was noted in discussion of those with cerebellar pathology (Kruit *et al.*, 2005).

Gray matter (GM) lesions on MRI can also occur in migraine patients. Resnick *et al.* (2006) described a patient with prolonged migraine aura and reversible parietooccipital cortical MRI abnormalities. They argued against an ischemic cause because of the absence of several factors, including a low apparent diffusion coefficient value, exclusive involvement of the cortical ribbon, and the lack of correlation within a vascular distribution. They felt that FLAIR sequence hyperintensity along the parietooccipital cortex may be due to vasogenic edema and possibly correlated with prolonged hyperperfusion leading to cortical impairment and persistent neuronal suppression during the aura which subsequently resolved after 18 h.

Studies by Rocca *et al.* (2006a,b) performed on a 3.0-T MRI machine have focused on interictal GM changes in migraine patients (excluding hypertension, hypercholesterolemia, diabetes mellitus, vascular or heart disease, and other major systemic and neurological conditions and controlling for age) who previously had at least four (up to 76) discrete WM lesions and normal GM on conventional MRI scans. They found that migraine patients had reduced mean diffusivity histograms on diffusion tensor MRI indicating subtle GM injury. Three possible explanations were posited by the authors for the changes they found in the GM: (1) ischemia due to reduction in cerebral blood flow (CBF); (2) retrograde degeneration stemming from axonal loss due to WM lesions; and (3) increased susceptibility to age-related changes, specifically repeated neuronal vascular insufficiency (Rocca *et al.*, 2006a). The second study localized areas of reduced GM density to the frontal and temporal lobes, bilaterally, as well as increased PAG matter density, a very interesting finding given fMRI, PET, and nociceptive research data linking the PAG region of the brainstem to migraine (Knight and Goadsby, 2001; Welch *et al.*, 2001).

In summary of the above, migraine-associated MRI WM and chronic GM changes may be due to oligemia; acute cortical (GM) MRI findings may be due to hyperperfusion. No nonvascular explanation of migraine-associated MRI abnormalities was found in the literature. Studies of CBF changes utilizing multiple techniques [Xenon blood flow, single photon emission computerized tomography (SPECT), positron emission tomography (PET), and functional MRI] show: (1) a reduction in CBF during visual aura which travels from occiput anteriorly and is not great enough to cause ischemic injury, (2) migraine without aura is not associated with significant hemodynamic changes, and (3) interictal CBF is similar between migraineurs and nonmigraineurs (Bartolini *et al.*, 2005). Since WM foci thought to be due to migraine do not enhance (Bekieslińska-Figatowska, 2004), they are probably not associated with a breach of the BBB or focal intraparenchymal inflammation in contradistinction to MS-associated WM lesions.

C. Migraine and MS Occurring Concurrently

In 1993, an MS patient was described with severe headache that was found to have a single lesion in the PAG (Haas *et al.*, 1993). Since then additional studies have been published describing brainstem lesions in MS patients who have migraine. Gee *et al.* (2005) conducted a retrospective analysis of MRI plus clinical data and questionnaire analysis of patients with MS and headache identifying 277 patients. Those with a plaque in the midbrain/PAG area had a fourfold increase in migraine, and a 2.5- to 2.7-fold increase in tension-type headache and mixed headache, respectively. A trend associating migraine headache with number of MS plaques overall was not statistically significant.

Another retrospective study (Tortorella *et al.*, 2006) of 137 patients with known MRI hyperintense lesions involved 58 with migraine, 42 with MS, and 37 with both migraine and MS (excluding subjects with hypertension, hypercholesterolemia, diabetes, vascular or heart disease, and other major system and neurological condition). In keeping with the inclusion criteria, all patients had multiple hyperintense lesions on dual-echo scans. Supratentorially, these were mainly found in the frontal lobes and periventricular WM. Lesions of the infratentorium occurred in the red nucleus (RN), substantia nigra (SN), and PAG. The brainstem lesions were represented in 72.3% of non-MS patients with MWA, 57.5% of non-MS patients with migraine without aura, 73.8% of MS patients without migraine, and 86.5% in MS patients with migraine. Results comparing lesion prevalence among MS patients with migraine and non-MS patients with migraine without aura reached statistical significance. No differences between migraine with and without aura were observed, leading the authors to suggest that these regions may not be important in the genesis of migraine aura. The relationship of these findings to changes in iron homeostasis in the same regions observed in chronic migraine by Welch *et al.* (2001) was felt to be important. The group further concluded that lesions in the RN, SN, and PAG may be causative of migraine, regardless of the pathogenesis of the lesion. Finally, they added that this study also supports the finding that GM structures are frequently involved in MS.

In those patients with WM changes on MRI in which the physician is unsure of the clinical diagnosis, as stated above the presence of enhancement with contrast should steer them away from a diagnosis of migraine (Bekieslińska-Figatowska, 2004). Furthermore, if an MR study of the cervical cord in a migraine patient shows a hyperintense lesion, an alternative diagnosis such as MS or other pathology should be strongly considered (Rovaris *et al.*, 2001).

IV. Headache and IFN-β

The IFNs have been an important component of the treatment of MS for over a decade. One of the more common adverse effects of the IFNs is a flu-like syndrome (Neilley *et al.*, 1996; Pöllmann *et al.*, 2002; Walther and Hohlfeld, 1999) which includes headache. In the PRISM study (Gold *et al.*, 2005), treatment with IFN-β1a was generally well tolerated for up to 4 years. IFN-β1a therapy was associated with headache in 71% (separated from flu-like symptoms which occurred in 69%) which tended to be mild and most frequently reported during the first month of treatment.

Headache does not seem to be a significant contributing factor to discontinuation of IFN treatment (Neilley *et al.*, 1996; Pöllmann *et al.*, 2006) but does factor into quality of life for these patients (Nortvedt *et al.*, 1999). D'Amico *et al.* (2004) reported that 15.3% of 88 MS patients with headache developed this symptom after starting IFN. Characteristics of the IFN-related headache were 50% migraine and 50% tension-type headache.

Another study compared headache in response to treatment with IFN-β1a 22 μg once per week and IFN-β1b 250 μg every other day. Flu-like symptoms, fever, headache injection site reactions, and depressive symptoms were equally reported in both treatment arms, suggesting no benefit of one over the other in terms of such adverse events (Koch-Henriksen *et al.*, 2006). Indeed, headache seems to be a "class effect" of IFN therapy, as IFN-αII therapy for hepatitis C may also be associated with migraine, which may be associated with vasospastic phenomena including diffuse periventricular WM lesions (Detry-Morel *et al.*, 1995). However, in a letter to the editor by Bräu *et al.* (2003), severe migraine headaches associated with IFN plus ribavirin combination therapy may have been caused by the ribavirin. Finally, treatment with recombinant human albumin-IFN-α for hepatitis C often causes headache associated with chills and fever (Bain *et al.*, 2006; Balan *et al.*, 2006).

The safety and tolerability of immunomodulatory therapies in the pediatric population were studied in a multicenter retrospective review by Banwell *et al.* (2006). Flu-like syndrome occurred in 25% of patients $\leq$10 years, and 37.1% $>$10 years, but headache and migraine were not listed as a separate variable. No child discontinued treatment because of this adverse event. Ghezzi (2005) reported on 81 relapsing-remitting MS (RRMS) patients under age 16 treated with IFN-β1a, IFN-β1b, or glatiramer acetate (GA). Almost 66% of the IFN-β group had side effects; 26% with transient headache and 10.8% with persistent headache. No headache or flu-like symptoms were reported in the GA group, although injection site reaction caused discontinuation of the drug in 1 of 11 patients.

Pöllmann *et al.* (2006) noted that in 65% of 82 patients with RRMS treated with GA had preexisting headache. Only 6% of patients complained of an increased headache frequency of $\geq$50%, which is lower than those treated with IFN-α. Although patients did not discontinue treatment due to headache regardless of type, 75% of those that ultimately stopped treatment were in the preexisting headache group. They emphasized the importance of identifying a history of headache in patients early in the course of management and monitoring them closely for headache exacerbation to optimize therapy. Of interest is the fact that they emphasized that GA should not be considered the only treatment for patients with RRMS and preexisting headache and that the existence of headache is "not at all" a contraindication for IFN-β therapy. However, GA can be considered in patients in whom IFN-β-related headache seems to be a problem.

Natalizumab is a non-IFN-α4 integrin antagonist which is also available for the treatment of RRMS. The original trial reported a relatively high incidence of headache in patients receiving natalizumab (38%) and placebo (33%), perhaps reflecting the baseline association of headache in MS as discussed previously (Polman *et al.*, 2006).

The management of IFN-induced headache and flu-like symptoms has been reviewed (Galetta and Markowitz, 2005; Walther and Hohlfeld, 1999) with the following suggestions to minimize such effects: (1) administer IFN-β before bedtime in the hopes that the patient will sleep through the worst of it; (2) gradually titrate the dose starting at 25% of the recommended dose initially, increasing by 25% weekly or biweekly; and (3) administer oral steroids, nonsteroidal anti-inflammatory drugs (NSAIDs), and/or acetaminophen or paracetamol. They cautioned that this advice was "largely based on anecdotal evidence." There are apparently no double-blind trials, but one randomized study (Río *et al.*, 2004) found no differences in flu-like symptoms overall during the first month of treatment with IFN-β1a between treatment with acetaminophen 500 mg, or ibuprofen 400 mg prior to and twice after injection, or prednisone 60 mg daily for 1 week plus a taper. The ibuprofen group did best immediately postinjection. Reess *et al.* (2002) also found equal efficacy between paracetamol and ibuprofen, whereas Leuschen *et al.* (2004) found that the NSAIDs naproxen and ibuprofen were more effective than acetaminophen in reducing headache, fever, chills, and injection site pain after 5 weeks, but none of these helped IFN-β-induced fatigue or muscle/joint pain.

The impact of IFN-induced migraine can have consequences other than the headache itself. Khromov *et al.* (2005) described a patient who suffered MS exacerbations following her frequent migraine coincident with IFN-β administration. The administration of amitriptyline controlled the migraines and subsequent MS symptoms. In many patients, the tendency for treatment-induced headache subsides over time, but this case points out the benefit of migraine preventive treatment in those patients in which migraine persists.

V. Pathophysiology

A. Neuronal Pathways

When a patient with MS experiences a migraine, the acute clinical picture is dictated by migraine pathophysiology. Migraine headache pain employs activation of the trigeminovascular system peripherally by nociceptive afferent fibers from the ophthalmic division of the trigeminal nerve and upper cervical spinal cord segments, innervating the proximal parts of the large cerebral vessels, pial vessels, large venous sinuses, and dura mater. Pain information is secondarily carried from the trigeminal nucleus caudalis via the quintothalamic tract to the thalamus and cortex (Ferrari, 1998). Cortical-spreading depression, the accepted neuronal event underlying migraine aura may also be responsible for triggering the meningeal and brainstem events underlying headache pain (Buzzi and Moskowitz, 2005).

In contrast, nonmigrainous, central pain in MS (and stroke) is considered to be due to direct lesions affecting the spino- and quintothalamic pathways, especially those projecting to the ventroposterior thalamus rendering spinothalamic neurons hyperexcitable, in turn felt to be due to reduced tonic inhibition (Bowsher, 1995, 1996; Österberg *et al.*, 2004) or perhaps ephaptic transmission between demyelinated axons in proximity (Smith and McDonald, 1999).

B. Brainstem Demyelination

The more relevant question is: why do patients with MS have a greater prevalence of migraine than in the general population? Those who have studied this problem have postulated that, as we have seen in imaging studies, demyelination of brainstem structures, particularly the midbrain areas involved in pain modulation, underlies the development of headache (D'Amico *et al.*, 2004; Gee *et al.*, 2005). These structures are the dorsal raphe nucleus (serotonergic), the locus ceruleus (noradrenergic), and the PAG (the modulatory center for nociceptive information between the spinal cord and thalamocortical structures), and may also include the RN, SN, and hypothalamic areas as well (Gee *et al.*, 2005). This seems to be supported by the high incidence of brainstem plaques in MS patients with migraine.

C. Serotonin

Sandyk and Awerbuch (1994) noted that migraine symptoms can be more prominent during exacerbations of MS and may herald the onset of a relapse. This prompted them to suggest that dysregulation of serotonergic pathways may contribute to an initial breakdown of the BBB leading to neuronal damage. Since then, 5-HT4 agonists have been studied in cultured astrocytes from rats and found to reduce MHC class II activity which may translate into a potential immune therapy for MS (Zeinstra *et al.*, 2006).

Studies in humans with cerebellar syndrome (i.e., tremor) due to MS showing improvement with 5-HT3 antagonist therapy have been conflicting (Monaca-Charley *et al.*, 2003; Rice *et al.*, 1997), and the treatment effect may also be nonspecific to MS. Treatment of MS-associated depression with serotonin reuptake inhibitors (SSRIs) is considered to be as equally effective as with tricyclic antidepressants (Rickards, 2005). This effect, as well as the resulting improvement in fatigue is also likely not MS-specific but rather related to changes in mood (Mohr *et al.*, 2003).

The importance of 5-HT1 receptors in the acute migraine process at the meningeal trigemino-vascular junction is well elucidated, and provides a cornerstone for the acute treatment of a migraine attack with triptans/ergotamine. Preventive treatment with the 5-HT2 antagonists, pizotifen and methysergide (neither of which is currently available in the United States) is also effective to a degree (Ferrari, 1998). The SSRIs have at best, a nonspecific effect on headache by reducing depression as occurs in MS-related fatigue, and are not considered to be an effective migraine prophylactic therapy (Moja *et al.*, 2005).

D. Genetic Factors

Genetic susceptibility has been more apparent in migraine than in MS, although it is likely important in both. Migraine is known predominantly as a familial autosomal dominant disease with variable penetration, with a relative risk among first-degree family members from 1.4 to 1.9 (Russell and Olesen, 1995). This risk increases to 2.5 and 2.38 in family members of patients with onset of migraine younger than age 16 and very severe migraine pain, respectively (Stewart *et al.*, 2006). Spurred by the early findings associating specific syndromes such as FHM and the calcium channel gene CACNA1a, the possibility of migraine as a paroxysmal disorder due to a channelopathy (particularly of the neuronal calcium channel) has been entertained (Tepper *et al.*, 2001). FHM patients have chromosomal aberrations in the CACNA1A related to the calcium channel (80% of cases) or ATP1A2 gene for Na/K ATPase (20% of cases) on chromosomes 19 and 1, respectively. However, these loci are not associated with dominantly inherited migraine with aura (Kirchmann *et al.*, 2006), migraine without aura, or sporadic migraine with hemiplegic motor aura (Jen *et al.*, 2004). The true nature of the genetics of the common migraine is therefore still a mystery.

The genetics of MS, on the other hand, is more closely associated with the autoimmune system. An association with MS and the HLA-DR 15 haplotype (DRB1*1501, DRB5*0101, DQA*0102, and DQB1*0602 alleles) has been reported (Kantarci and Wingerchuk, 2006; Prat *et al.*, 2005). The two HLA-DR heterodimers DR2b and DR2a may allow antigen presentation to myelin-specific T lymphocytes resulting in the autoimmune process of MS, and are able to be suppressed by IFN-γ and interleukin-4 (IL-4) (Prat *et al.*, 2005). Interestingly, recent HLA genotyping in migraine patients for HLA-DRB1 alleles revealed a decrease in the DRB1*12 allele, and an increase of the DRB1*16 allele among patients with migraine without aura compared to healthy controls (Rainero *et al.*, 2005, 2006). Further studies in this region may help us elucidate the genetic susceptibility for migraine.

E. Neuroimmunologic Mechanisms

MS has long been considered a neuroimmunologic event. Autoimmune demyelination was addressed by Aktas *et al.* (2006) as a function of death ligands of the TNF family. In MS, these death ligands induce apoptosis causing the cell to participate in its own "cellular suicide," leading to tissue damage resulting in neurological dysfunction (Aktas *et al.*, 2006). Such devastating destruction does not occur in migraine, but D'Amico *et al.* (2004) observed that a reduction in T8 lymphocytes was reported in both migraine and MS in the 1980s, leading him to suggest therefore that a shared immune mechanism may explain the clinical association between the two (D'Amico *et al.*, 2004).

It has been suggested that the pathogenesis of migraine includes an inflammatory component (Longoni and Ferrarese, 2006); neurogenic inflammation of the meninges is an accepted component of the migraine process (Buzzi *et al.*, 1991; Moskowitz and Cutrer, 1993) with release of vasogenic substances such as calcitonin gene-related peptide, substance P, neurokinin A, vasoactive intestinal peptide, and nitric oxide. Thus, researchers have searched for evidence of these and other inflammatory markers in migraine.

Mueller *et al.* (2001) found that an increase of urinary IL-6 was not statistically significant during menstrual migraine, whereas urinary levels of TNF-α, a proinflammatory cytokine, were significantly low. In contrast, a transient increase in TNF-α in the internal jugular blood of patients with migraine without aura during the ictus was demonstrated along with increased levels of intercellular adhesion molecule-1 and IL-6 by Sarchielli *et al.* (2006). Fidan *et al.* (2006) also reported higher serum levels of IL-6, IL-10, RANTES, and nitric oxide (NO) during migraine attacks. Conflicting results among these studies may be explained by differences in testing procedures (urine testing vs brachial vs jugular sites for blood draws) or due to the sensitivity of the assay or stability of the molecules during testing.

TNF-α and IL-6 cause hyperalgesia when administered centrally or peripherally, adding support to the theory that cytokines are mediators of inflammatory pain (Cunha *et al.*, 1992; Oka *et al.*, 1995) and migraine pain due to neurovascular inflammation (Empl *et al.*, 2003). The cytokines that are found to be elevated in migraine may be released from activated trigeminal endings and stimulate the sterile inflammation of the dura during a migraine attack (Sarchielli *et al.*, 2006). Chemokines and NO may alternatively or simultaneously stimulate the activation of trigeminal nerves and the release of vasoactive peptides (possibly by activation of NF-$\kappa\beta$) causing inflammation and migraine (Chrisopherson and Hromas, 2001; Fidan *et al.*, 2006; Pöllmann *et al.*, 2006). These works support the hypothesis that migraine associated with MS and/or IFN therapy may be caused by the acute increase in proinflammatory cytokines and mediators such as IL-6, IFN-γ, and prostaglandins (Khromov *et al.*, 2005; Walther and Hohlfeld, 1999).

VI. Conclusions

Migraine is comorbid in patients with MS. The exact etiology and pathogenesis of these two seemingly disparate disorders is not completely understood. Research as to how and why they co-occur leads to improvement in the understanding of each. This research has contributed to the recognition of migraine symptoms and headache in MS, improving our diagnostic skills and care of these patients. It has heightened our understanding of features of MS such as the precipitation of exacerbation/relapse and GM involvement. In addition, it has strengthened our knowledge of the role of pain-modulating structures of the brainstem, and neuroimmunologic and inflammatory mechanisms which contribute to migraine. Such collaboration may one day culminate in the prevention and cure of two currently incurable and debilitating diseases.

References

Aktas, O., Prozorovski, T., and Zipp, F. (2006). Death ligands and autoimmune dysfunction. *Neuroscientist* **12,** 305–316.

Al-Twaijri, W. A., and Shevell, M. I. (2002). Pediatric migraine equivalents: Occurrence and clinical features in practice. *Pediatr. Neurol.* **26,** 365–368.

Alvarez-Lafuente, R., De Las Heras, V., Bartolome, M., Garcia-Montojo, M., and Arroyo, R. (2006). Human herpesvirus 6 and multiple sclerosis: A one-year follow-up study. *Brain Pathol.* **16,** 20–27.

Alviggi, C., Carrieri, P. B., Pivonello, R., Scarano, V., Pezzella, M., De Placido, G., Colao, A., and Matarese, G. (2006). Association of pelvic endometriosis with alopecia universalis, autoimmune thyroiditis and multiple sclerosis. *J. Endocrinol. Invest.* **29,** 182–189.

Bain, V. G., Kaita, K. D., Yoshida, E. M., Swain, E. J., Heathcote, M. G., Neumann, A. U., Fiscella, M., Yu, R., Osborn, B. L., Cronin, P. W., Freimuth, W. W., and McHutchison, J. G. *et al.* (2006). A phase 2 study to evaluate the antiviral activity, safety, and pharmacokinetics of recombinant human albumin-interferon alfa fusion protein in genotype 1 chronic hepatitis C patients. *J. Hepatol.* **44,** 671–678.

Balan, V., Nelson, D. R., Sulkowski, M. S., Everson, G. T., Lambiase, L. R., Wiesner, R. H., Dickson, R. C., Post, A. B., Redfield, R. R., Davis, A. U., Neumann, B. L., Osborn, G. L., *et al.* (2006). A phase I/II study evaluating escalating doses of recombinant human albumin-interferon-alpha fusion protein in chronic hepatitis C patients who have failed previous interferon-alpha-based therapy. *Antivir. Ther.* **11,** 35–45.

Banwell, B., Reder, A. T., Krupp, L., Tenembaum, S., Eraksoy, M., Alexey, B., Pohl, D., Freedman, M., Schelensky, L., and Antonijevic, I. (2006). Safety and tolerability of interferon beta-1b in pediatric multiple sclerosis. *Neurologyy* **66,** 472–476.

Bartolini, M., Baruffaldi, R., Paolino, I., and Silvestrini, M. (2005). Cerebral blood flow changes in the different phases of migraine. *Funct. Neurol.* **20,** 209–211.

Bekieslińska-Figatowska, M. (2004). T2-hyperintense foci on brain MR imaging. *Med. Sci. Monit.* **10**(Suppl. 3), 80–87.

Bowsher, D. (1995). The management of central post-stroke pain. *Postgrad. Med. J.* **71,** 598–604.

Bowsher, D. (1996). Central pain: Clinical and physiological characteristics. *J. Neurol. Neurosurg. Psychiatry* **61,** 62–69.

Bräu, N., Bini, E. J., Stancic, S., Finch, D. A., and Aytaman, A. (2003). Severe migraine headaches are caused by ribavirin but not by interferon α-2B in combination therapy for chronic hepatitis C. *J. Hepatol.* **38,** 871–872.

Buzzi, M. G., and Moskowitz, M. A. (2005). The pathophysiology of migraine: Year 2005. *J. Headache Pain* **6,** 105–111.

Buzzi, M. G., Moskowitz, M. A., Shimizu, T., and Heath, H. H. (1991). Dihydroergotamine and sumatriptan attenuate levels of CGRP in plasma in rat superior sagittal sinus during electrical stimulation of the trigeminal ganglion. *Neuropharmacology* **30,** 1193–1200.

Carmosino, M. J., Brousseau, K. M., Arciniegas, D. B., and Corboy, J. R. (2005). Initial evaluations for multiple sclerosis in a university multiple sclerosis center. *Arch. Neurol.* **62,** 585–590.

Chapman, J., Rand, J. H., Brey, R. L., Levine, S. R., Blatt, I., Khamashta, M. A., and Shoenfeld, Y. (2003). Non-stroke neurological syndromes associated with antiphospholipid antibodies: Evaluation of clinical and experimental studies. *Lupus* **12,** 514–517.

Charlier, C., Nunes, H., Trinchet, J. C., Roullet, E., Mouthon, L., Beaugrand, M., and Valeyre, D. (2005). Evolution of previous sarcoidosis under type 1 interferons given for severe associated disease. *Eur. Respir. J.* **25,** 570–573.

Chrisopherson, K., and Hromas, R. (2001). Chemokine regulation of normal and pathologic immune responses. *Stem Cells* **19,** 388–396.

Cook, S. D. (2004). Does Epstein-barr virus cause multiple sclerosis? *Rev. Neurol. Dis.* **1,** 115–123.

Cooney, B. S., Grossman, R. I., Farber, R. E., Goin, J. E., and Galetta, S. L. (1996). Frequency of magnetic resonance imaging abnormalities in patients with migraine. *Headache* **36,** 616–621.

Crawford, S. G., Kaplan, B. J., and Kinsbourne, M. (1994). Are families of children with reading difficulties at risk for immune disorders and nonrighthandedness? *Cortex* **30,** 281–292.

Cunha, F. Q., Poole, S., Lorenzetti, B. B., and Ferreira, S. H. (1992). The pivotal role of tumour necrosis factor alpha in the development of inflammatory hyperalgesia. *Br. J. Pharmacol.* **107,** 660–664.

D'Amico, D., La Mantia, L., Rigamonti, A., Usai, S., Mascoli, N., Milanese, C., and Bussone, Gl. (2004). Prevalence of primary headaches in people with multiple sclerosis. *Cephalalgia* **24,** 980–984.

Detry-Morel, M., Boschi, A., Gehenot, M., and Geubel, A. (1995). Bilateral transient visual obscurations with headache during alpha-II interferon therapy: A case report. *Eur. J. Ophthalmol.* **5,** 271–274.

Di Rosa, G., Spanò, M., Pustorino, G., Ferrari, M. D., Stam, A. H., Sgrò, D. L., Mannarino, E., Bonsignore, M., and Tortorella, G. (2006). Alternating hemiplegia of childhood successfully treated with topiramate: 18 months of follow-up. *Neurology* **66,** 146.

Eggers, S. D. (2006). Migraine-related vertigo: Diagnosis and treatment. *Curr. Neurol. Neurosci. Rep.* **6,** 106–115.

Ehde, D. M., Osborne, T. L., and Jensen, M. P. (2005). Chronic pain in persons with multiple sclerosis. *Phys. Med. Rehabil. Clin. N. Am.* **16,** 503–512.

Empl, M., Sostak, P., Riedel, M., Schwarz, M., Muller, N., Forderreuther, S., and Straube, A. (2003). Decreased sTNF-RI in migraine patients? *Cephalalgia* **23,** 55–58.

Evans, R. W., and Rolak, L. A. (2001). Migraine versus multiple sclerosis. *Headache* **41,** 97–98.

Evans, R. W., and Rolak, L. A. (2004). The Alice in Wonderland syndrome. *Headache* **44,** 624–625.

Ferrari, M. D. (1998). Migraine. *Lancet* **351,** 1043–1051.

Ferreira, S., D'Cruz, D. P., and Hughes, G. R. (2005). Multiple sclerosis, neuropsychiatric lupus and antiphospholipid syndrome: Where do we stand? *Rheumatology* **44,** 434–442.

Fidan, I., Yüksel, S., Ýmir, T., Irkeç, C., and Nur Aksakal, F. (2006). The importance of cytokines, chemokines and nitric oxide in pathophysiology of migraine. *J. Neuroimmunol.* **171,** 184–188.

Freedman, M. S., and Gray, T. A. (1989). Vascular headache: A presenting symptom of multiple sclerosis. *Can. J. Neurol. Sci.* **16,** 63–66.

Freedom, T., and Jay, W. M. (2003). Migraine with and without headache. *Semin. Ophthalmol.* **18,** 210–217.

Galetta, S. L., and Markowitz, C. (2005). US FDA-approved disease-modifying treatments for multiple sclerosis. Review of adverse effect profiles. *CNS drugs* **19,** 239–252.

Gee, J. R., Chang, J., Dublin, A. B., and Vijayan, N. (2005). The association of brainstem lesions with migraine-like headache: An imaging study of multiple sclerosis. *Headache* **45,** 670–677.

Ghezzi, A. (2005). Immunomodulatory treatment of early onset multiple sclerosis: Results of an Italian co-operative study. *Neurol. Sci.* **26**(Suppl. 4), S183–S186.

Gold, R., Rieckmann, P., Chang, P., and Abdalla, J and PRISMS Study Group (2005). The long-term safety and tolerability of high-dose interferon beta-1a in relapsing-remitting multiple sclerosis: 4-year data from the PRISMS study. *Eur. J. Neurol.* **12,** 649–656.

Gupta, G., Gelfand, J. M., and Lewis, J. D. (2005). Increased risk for demyelinating diseases in patients with inflammatory bowel disease. *Gastroenterology* **129,** 819–826.

Haas, D. C., Kent, P. F., and Friedman, D. I. (1993). Headache caused by a single lesion of multiple sclerosis in the peri-aqueductal gray area. *Headache* **33,** 452–455.

Headache Classification Committee of the International Headache Society (1988). Classification and diagnostic criteria for headache disorders, cranial neuralgia and facial pain, 1st ed. *Cephalalgia* **8** (Suppl. 7), 9–96, 1st ed.

Headache Classification Subcommittee of the International Headache Society (2004). Classification and WHO ICD-10 codes—IHS ICHD-II. *Cephalagia* **24,** (Suppl. 1), S1–S16.

Headache Classification Committee of the International Headache Society (2004). The international classification of headache disorders, 2nd ed. *Cephalalgia* **24,** 1–159.

Hughes, G. R. V. (2003). Migraine, memory loss, and "multiple sclerosis." Neurologic features of the antiphospholipid (Hughes') syndrome. *Postgrad. Med. J.* **79,** 81–83.

Jacobi, C., Stingele, K., Kretz, R., Hartmann, M., Storch-Hagenlocher, B., Breitbart, A., and Wildemann, B. (2006). Neuromyelitis optica (Devic's syndrome) as first manifestation of systemic lupus erythematosis. *Lupus* **15,** 107–109.

Jacome, D. E., and Leborgne, J. (1990). MRI studies in basilar artery migraine. *Headache* **30,** 88–90.

Jen, J. C., Kim, G. W., Dudding, K. A., and Baloh, R. W. (2004). No mutations in CACNA1A and ATP1A2 in probands with common types of migraine. *Arch. Neurol.* **61,** 926–928.

Kantarci, O., and Wingerchuk, D. (2006). Epidemiology and natural history of multiple sclerosis: New insights. *Curr. Opin. Neurol.* **19,** 248–254.

Kimper-Karl, A. C., Jensen, H. B., and Kristensen, O. (2006). CADASIL versus multiple sclerosis. *Ugeskr. Laeger.* **168,** 702–703.

Kirchmann, M. (2006). Migraine with aura: New understanding from clinical epidemiologic studies. *Curr. Opin. Neurol.* **19,** 286–293.

Kirchmann, M., Thomsen, L. L., and Olesen, J. (2006). The CACNA1A and ATP1A2 genes are not involved in dominantly inherited migraine with aura. *Am. J. Med. Genet. B Neuropsychiatr. Genet.* **141,** 250–256.

Kirchmann Eriksen, M., Thomsen, L. L., and Olesen, J. (2006). Implications of clinical subtypes of migraine with aura. *Headache* **46,** 286–297.

Kitigawa, Y. (2003). Neurological aspects in antiphospholipid syndrome. *Rinsho Shinkeigaku* **43,** 744–747.

Khromov, A., Segal, M., Nissinoff, J., and Fast, A. (2005). Migraines linked to interferon-β treatment of multiple sclerosis. *Am. J. Phys. Med. Rehabil.* **84,** 644–647.

Knight, Y. E., and Goadsby, P. J. (2001). The peri-aqueductal grey matter modulates trigeminovascular input: A role in migraine? *Neuroscience* **106,** 793–800.

Koch-Henriksen, N., Sørensen, P. S., Christensen, T., Frederiksen, J., Ravnborg, M., Jensen, K., Heltberg, A., Kristensen, O., Stenager, E., Petersen, T., and Hansen, T. (2006). A randomized study of two interferon-beta treatments in relapsing-remitting multiple sclerosis. *Neurology* **66,** 1056–1060.

Kruit, M. C., van Buchem, M. A., Hofman, P. A., Bakkers, J. T., Terwindt, G. M., Ferrari, M. D., and Launer, L. J. (2004). Migraine as a risk factor for subclinical brain lesions. *JAMA* **291,** 427–434.

Kruit, M. C., Launer, L. J., Ferrari, M. D., and van Buchem, M. A. (2005). Infarcts in the posterior circulation territory in migraine. The population-based MRI CAMERA study. *Brain* **128,** 2068–2077.

Kruit, M. C., Launer, L. J., Ferrari, M. D., and van Buchem, M. A. (2006). Brain stem and cerebellar hyperintense lesions in migraine. *Stroke* **37,** 1109–1112.

La Mantia, L., and Erbetta, A. (2004). Headache and inflammatory disorders of the central nervous system. *Neurol. Sci.* **25**(Suppl. 3), S148–S153.

Lauer, K. (2006). Divergent risk of multiple sclerosis in two Anabaptist communities in America. *Med. Hypotheses* **67,** 969–974.

Leavell, R., Ray, C. G., Ferry, P. C., and Minnich, L. L. (1986). Unusual acute neurologic presentations with Epstein-Barr virus infection. *Arch. Neurol.* **43,** 186–188.

Leuschen, M. P., Filipi, M., and Healey, K. (2004). A randomized open label study of pain medications (naproxen, acetaminophen and ibuprofen) for controlling side effects during initiation of IFN b_{β}-1a therapy and during its ongoing use for relapsing-remitting multiple sclerosis. *Mult. Scler.* **10,** 636–642.

Levin, N., Mor, M., and Ben-Hur, T. (2003). Patterns of misdiagnosis of multiple sclerosis. *Isr. Med. Assoc. J.* **5,** 489–490.

Lips, P. (2006). Vitamin D physiology. *Prog. Biophys. Mol. Biol.* **92,** 4–8.

Longoni, M., and Ferrarese, C. (2006). Inflammation and excitotoxicity: Role in migraine pathogenesis. *Neurol. Sci.* **27**(Suppl. 2), S107–S110.

Mack, K. J. (2004). What incites new daily persistent headache in children? *Pediatr. Neurol.* **31,** 122–125.

Marangoni, S., Argentiero, V., and Tavolato, B. (2006). Neurosarcoidosis: Clinical description of 7 cases with a proposal for a new diagnostic strategy. *J. Neurol.* **253,** 488–495.

Mark, B. L., and Carson, J. A. (2006). Vitamin D and autoimmune disease—implications for practice from the multiple sclerosis literature. *J. Am. Diet. Assoc.* **106,** 418–424.

Martinez Sobrepera, H. J., Cabrera Gomez, J. A., and Tuero Iglesias, A. (2001). Exogenous factors in the aetiology of multiple sclerosis in Cuba. A study of cases and controls. *Rev. Neurol.* **33,** 931–937.

McDonald, W. I., Compston, A., Edan, G., Goodkin, D., Hartung, H. P., Lublin, F. D., McFarland, H. F., Paty, D. W., Polman, C. H., Reingold, S. C., Sandberg-Wollheim, M., Sibley, M., *et al.* (2001). Recommended diagnostic criteria for multiple sclerosis: Guidelines from the International Panel on the diagnosis of multiple sclerosis. *Ann. Neurol.* **50,** 121–127.

Minagar, A., and Sheremata, W. A. (2000). Glossopharyngeal neuralgia and MS. *Neurology* **54,** 1368–1370.

Minagar, A., Gonzalez-Toledo, E., Pinkston, J., and Jaffe, S. L. (2005). Neuroimaging in multiple sclerosis. *Int. Rev. Neurobiol.* **67,** 165–201.

Mohr, D. C., Hart, S. L., and Goldberg, A. (2003). Effects of treatment for depression on fatigue in multiple sclerosis. *Psychosom. Med.* **65,** 542–547.

Moja, P. L., Cusi, C., Sterzi, R. R., and Canepari, C. (2005). Selective serotonin re-uptake inhibitors (SSRIs) for preventing migraine and tension-type headaches. *Cochrane Database Syst. Rev.* **3,** CD002919.

Monaca-Charley, C., Stojkovic, T., Duhamel, A., De Seze, J., Ferriby, D., and Vermersch, P. (2003). Double-blind crossover study with dolasetron mesilate, a 5-HT3 receptor antagonist in cerebellar syndrome secondary to multiple sclerosis. *J. Neurol.* **250,** 1190–1194.

Moskowitz, M. A., and Cutrer, F. M. (1993). Sumatriptan: A receptor-targeted treatment for migraine. *Ann. Rev. Med.* **44,** 145–154.

Mueller, L., Gupta, A. K., and Stein, T. P. (2001). Deficiency of tumor necrosis factor alpha in a subclass of menstrual migraineurs. *Headache* **41,** 129–137.

Neilley, L. K., Goodin, D. S., Goodkin, D. E., and Hauser, S. L. (1996). Side effect profile of interferon beta-1b in MS: Results of an open label trial. *Neurology* **46,** 552–554.

Nortvedt, M. W., Riise, T., Myhr, K. M., Nyland, H. I., and Hanestad, B. R. (1999). Type I interferons and the quality of life of multiple sclerosis patients. Results from a clinical trial on interferon alfa-2a. *Mult. Scler.* **5,** 317–322.

Nortvedt, M. W., Riise, T., and Maeland, J. G. (2005). Multiple sclerosis and lifestyle factors: The Hordaland health study. *Neurol. Sci.* **26,** 334–339.

Oka, T., Oka, K., Hosoi, M., and Hori, T. (1995). Intracerebroventricular injection of interleukin-6 induces thermal hyperalgesia in rats. *Brain Res.* **692,** 123–128.

Ordonez, L., Skromne, E., Ontaneda, D., and Rivera, V. M. (2007). Multiphasic disseminated encephalomyelitis, systemic lupus erythematosus and antiphospholipid syndrome. *Clin. Neurol. Neurosurg.* **109,** 102–105.

Österberg, A., Boivie, J., and Thuomas, K.-Å. (2004). Central pain in multiple—prevalence and clinical characteristics. *Eur. J. Pain* **9,** 531–542.

Perez-Cesari, C., Saniger, M. M., and Sotelo, J. (2005). Frequent association of multiple sclerosis with varicella and zoster. *Acta Neurol. Scand.* **112,** 417–419.

Pöllmann, W., Erasmus, L.-P., Feneberg, W., Then Bergh, F., and Straube, A. (2002). Interferon beta but not glatiramer acetate therapy aggravates headaches in MS. *Neurology* **59,** 636–639.

Pöllmann, W., Feneberg, W., and Erasmus, L.-P. (2004). Pain in multiple sclerosis—a still underestimated problem. The 1 year prevalence of pain syndromes, significance and quality of care of multiple sclerosis inpatients. *Nervenarzt* **75,** 135–140.

Pöllmann, W., Erasmus, L.-P., Feneberg, W., and Straube, A. (2006). The effect of glatiramer acetate treatment on pre-existing headaches in patients with MS. *Neurology* **66,** 275–277.

Polman, C. H., Reingold, S. C., Edan, G., Filippi, M., Hartung, H.-P., Kappos, L., Lublin, F. D., Metz, L. M., McFarland, H. F., O'Connor, P. W., Sandberg-Wollheim, M., Thompson, A. J., *et al.* (2005). Diagnostic criteria for multiple sclerosis: 2005 revisions to the "McDonald Criteria." *Ann. Neurol.* **58,** 840–846.

Polman, C. H., O'Connor, P. W., Havrdova, E., Hutchinson, M., Kappos, L., Miller, D. H., Phillips, J. T., Lublin, F. D., Giovannoni, G., Wajgt, A., Toal, M., and Lynn, F., AFFIRM Investigators, *et al.* (2006). A randomized, placebo-controlled trial of natalizumab for relapsing multiple sclerosis. *N. Engl. J. Med.* **354,** 899–910.

Poser, C. M. (2006). The multiple sclerosis trait and the development of multiple sclerosis: Genetic vulnerability and environmental effect. *Clin. Neurol. Neurosurg.* **108,** 227–233.

Poser, C. M., Presthus, J., and Horsda, O. (1966). Clinical characteristics of autopsy-proven multiple sclerosis. *Neurology* **16,** 791–798.

Prat, E., Tomaru, U., Sabater, L., Park, D. M., Granger, R., Kruse, N., Ohayon, J. M., Bettinoti, M. P., and Martin, R. (2005). HLA-DRB5*0101 and -DRB1*1501 expression in the multiple sclerosis-associated HLA-DR15 haplotype. *J. Neuroimmunol.* **167,** 108–119.

Rainero, I., Fasano, E., Rubino, E., Rivoiro, C., Valfre, W., Gallone, S., Savi, L., Gentile, S., Lo Giudice, R., De Martino, P., Dall'Omo, P., and Pinessi, L. (2005). Association between migraine and HLA-DRB1 gene polymorphisms. *J. Headache Pain* **6,** 185–187.

Rainero, I., Dall'Omo, A. M., Rubino, E., Valfre, W., Fasano, M. E., Rivoiro, C., Brancatello, F., Savi, L., Gallone, S., and Pinessi, L. (2006). HLA-DRB1 genotyping in Italian migraine patients. *Neurosci. Lett.* **393,** 90–93.

Reess, J., Haas, J., Gabriel, K., Fuhlrott, A., and Fiola, M. (2002). Both paracetamol and ibuprofen are equally effective in managing flu-like symptoms in relapsing-remitting multiple sclerosis patients during interferon beta-1a (Avonex®) therapy. *Mult. Scler.* **8,** 15–18.

Rinalduzzi, M., Valeriani, M., and Vigevano, F. (2006). Brainstem dysfunction in alternating hemiplegia of childhood: A neurophysiological study. *Cephalalgia* **26,** 511–519.

Resnick, S., Reyes-Iglesias, Y., Carreras, R., and Villalobos, E. (2006). Migraine with aura associated with reversible MRI abnormalities. *Neurology* **66,** 946–947.

Rice, G. P., Lesaux, J., Vandervoort, P., Macewan, L., and Ebers, G. C. (1997). Ondansetron, a 5-HT3 antagonist, improves cerebellar tremor. *J. Neurol. Neurosurg. Psychiatry* **62,** 282–284.

Rickards, H. (2005). Depression in neurological disorders: Parkinson's disease, multiple sclerosis, and stroke. *J. Neurol. Neurosurg. Psychiatry* **76**(Suppl. I), i48–i52.

Río, J., Nos, C., Bonaventura, I., Arroyo, R., Genis, D., Sureda, B., Ara, J. R., Brieva, L., Martín, J., Saiz, A., Sánchez López, A., Prieto, J. M., *et al.* (2004). Corticosteroids, ibuprofen, and acetaminophen for IFNβ-1a flu symptoms in MS. *Neurology* **63,** 525–528.

Rocca, M. A., Ceccarelli, A., Falini, A., Tortorella, P., Colombo, B., Pagani, E., Comi, G., Scotti, G., and Filippi, M. (2006a). Diffusion tensor magnetic resonance imaging at 3.0 tesla shows subtle cerebral grey matter abnormalities in patients with migraine. *J. Neurol. Neurosurg. Psychiatry* **77,** 686–689.

Rocca, M. A., Ceccarelli, A., Falini, A., Colombo, B., Tortorella, P., Bernasconi, L., Comi, G., Scotti, G., and Filippi, M. (2006b). Brain gray matter changes in migraine patients with T2-visible lesions. A 3-T MRI study. *Stroke* **37**(7), 1765–1770.

Rolak, L. A., and Brown, S. (1990). Headaches and multiple sclerosis: A clinical study and review of the literature. *J. Neurol.* **237,** 300–302.

Rovaris, M., Bozzali, M., Rocca, M. A., Colombo, B., and Filippi, M. (2001). An MR study of tissue damage in the cervical cord of patients with migraine. *J. Neurol. Sci.* **183,** 43–46.

Russell, M. B., and Olesen, J. (1995). Increased familial risk and evidence of genetic factor in migraine. *BMJ* **311,** 541–544.

Sandyk, R., and Awerbuch, G. I. (1994). The co-occurrence of multiple sclerosis and migraine headache: The serotoninergic link. *Int. J. Neurosci.* **76,** 249–257.

Sarchielli, P., Alberti, A., Baldi, A., Coppola, F., Rossi, C., Pierguidi, L., Floridi, A., and Calabresi, P. (2006). Proinflammatory cytokines, adhesion molecules, and lymphocyte integrin expression in the internal jugular blood of migraine patients without aura assessed ictally. *Headache* **46,** 200–207.

Shevell, M. I. (1996). Acephalgic migraine of childhood. *Pediatr. Neurol.* **14,** 211–215.

Silberstein, S. D., Lipton, R. B., and Goadsby, P. J. (2002). Migraine: Diagnosis and treatment. *In* "Headache in Clinical Practice," 2nd ed., p. 76. Martin Dunitz Ltd., London, United Kingdom.

Smith, K. J., and McDonald, W. I. (1999). The pathophysiology of multiple sclerosis: The mechanisms underlying the production of symptoms and the natural history of the disease. *Phil. Trans. R. Soc. Lond. B* **354,** 1649–1673.

Solaro, C., Lunardi, G. L., and Mancarki, G. L. (2003). Pain and MS. *Int. MS J.* **10,** 14–19.

Stewart, W. F., Bigal, M. E., Kolodner, K., Dowson, A., Liberman, J. N., and Lipton, R. B. (2006). Familial risk of migraine: Variation by proband age at onset and headache severity. *Neurology* **66,** 344–348.

Takeshima, T., Ishizaki, K., Fukuhara, Y., Ijiri, T., Kusumi, M., Wakutani, Y., Mori, M., Kawashima, M., Kowa, H., Adachi, Y., Urakami, K., and Nakashima, K. (2004). Population-based door-to-door survey of migraine in Japan: The Daisen study. *Headache* **44,** 8–19.

Tepper, S. J., Rapoport, A., and Sheftell, F. (2001). The pathophysiology of migraine. *Neurologist* **7,** 279–286.

Thacker, E. L., Mirzaei, F., and Ascherio, A. (2006). Infectious mononucleosis and risk for multiple sclerosis: A meta-analysis. *Ann. Neurol.* **59,** 499–503.

Thomsen, L. L., and Olesen, J. (2004). Sporadic hemiplegic migraine. *Cephalagia* **24,** 1016–1023.

Thys-Jacobs, S. (1994a). Vitamin D and calcium in menstrual migraine. *Headache* **34,** 544–546.

Thys-Jacobs, S. (1994b). Alleviation of migraines with therapeutic vitamin D and calcium. *Headache* **34,** 590–592.

Tortorella, P., Rocca, M. A., Colombo, B., Annovazzi, P., Comi, G., and Filippi, M. (2006). Assessment of MRI abnormalities of the brainstem from patients with migraine and multiple sclerosis. *J. Neurol. Sci.* **244,** 137–141.

Trojano, M., and Paolicelli, D. (2001). The differential diagnosis of multiple sclerosis: Classification and clinical features of relapsing and progressive neurological syndromes. *Neurol. Sci.* **22**(Suppl. 2), S98–S102.

Walther, E. U., and Hohlfeld, R. (1999). Multiple sclerosis. Side effects of interferon beta therapy and their management. *Neurology* **53,** 1622–1627.

Warren, S., and Warren, K. G. (1981). Multiple sclerosis and associated diseases: A relationship to diabetes mellitus. *Can. J. Neurol. Sci.* **8,** 35–39.

Watkins, M. S., and Espir, M. (1969). Migraine and multiple sclerosis. *J. Neurol. Neurosurg. Psychiatry* **32,** 35–37.

Welch, K. M., Nagesh, V., Aurora, S. K., and Gelman, N. (2001). Periaqueductal gray matter dysfunction in migraine: Cause or the burden of illness? *Headache* **41,** 629–637.

Young, A. C. (1997). Misdiagnosis of MS. *Lancet* **350,** 522.

Zakrzewska, J. M. (2002). Diagnosis and differential diagnosis of trigeminal neuralgia. *Clin. J. Pain* **18,** 14–21.

Zeinstra, E. M., Wilczak, N., Wilschut, J. C., Glazenburg, L., Chesik, D., Kroese, F. G., and De Keyser, J. (2006). 5HT(4) agonists inhibit interferon-gamma-induced MHC class II and B7 costimulatory molecules expression on cultured astrocytes. *J. Neuroimmunol.* **179,** 191–195.

Zorzon, M., Zivadinov, R., Nasuelli, D., Dolfini, P., Bosco, A., Bratina, A., Tommasi, M. A., Locatelli, L., and Cazzato, G. (2003). Risk factors of multiple sclerosis: A case-control study. *Neurol. Sci.* **24,** 242–247.

MULTIPLE SCLEROSIS AS A PAINFUL DISEASE

Meghan Kenner, Uma Menon, and Debra G. Elliott

Department of Neurology, Louisiana State University Health Sciences Center
Shreveport, Louisiana 71103, USA

Pain is a common problem of patients with multiple sclerosis (MS) and may be due to central/neuropathic or peripheral/somatic pathology. Rarely MS may present with pain, or pain may herald an MS exacerbation, such as in painful tonic spasms or Lhermitte's sign. In other patients, pain may become chronic as a long-term sequela of damage to nerve root entry zones (trigeminal neuralgia) or structures in central sensory pathways. Migraine headache may develop as a consequence of MS, and headache can also be a side effect of interferon treatment. The pathophysiology of pain in MS may be linked to certain plaque locations which disrupt the spinothalamic and quintothalamic pathways, abnormal impulses through motor axons, development of an acquired channelopathy in affected nerves, or involve glial cell inflammatory immune mechanisms. At this time, the treatment of pain in MS

INTERNATIONAL REVIEW OF
NEUROBIOLOGY, VOL. 79
DOI: 10.1016/S0074-7742(07)79013-X

employs the use of antiepileptic drugs, muscle relaxers/antispasmodic agents, anti-inflammatory drugs, and nonpharmacological measures. Research concerning cannabis-based treatments shows promising results, and substances which block microglial or astrocytic involvement in pain processing are also under investigation.

I. Introduction

Pain is defined as "an unpleasant sensation associated with actual or potential tissue damage and mediated by specific nerve fibers to the brain where its conscious appreciation may be modified by various factors" in Stedman's Medical Dictionary (Dirckx, 2001). It is a common complaint among many patients encountered in a clinical setting including those with multiple sclerosis (MS). Pain in MS may be from different causes and may even be one of the heralding symptoms of the disease. The World Health Organization has classified pain using a symptom/sign-related approach (Kanner, 1999) and has divided pain in MS as:

1. Neuropathic pain: due to an injury in the central nervous system (CNS),
2. Somatic pain: due to an inappropriate physiologic response to a nociceptive stimulus, and
3. Psychogenic pain: an inappropriate and exaggerated reaction to a painful stimulus. Note: this category should be considered only with extreme caution due to hyperalgesia and allodynia: the experience of pain in reaction to a usually nonnoxious stimulus, as discussed below.

Pain in MS patients may be severe, acute, or chronic and is sometimes poorly responsive to treatment. Its impact on the lives of patients can be widespread and may correlate with both functional as well as psychological disability. MS patients are a group of subjects whose pain symptoms are beginning to gain attention with neurologists. It is important to remember the whole picture when treating MS patients. As physicians, we are not only trying to reduce the number of exacerbations in these patients, but we must also address the issue of pain in order help manage the entire disease process.

Pain is not only a symptom that is seen after the diagnosis of MS is made. In an article by Marchettini *et al.*, 2006, several cases were presented with pain as an initial sign of MS. Most of these patients reported pain as their only symptom for some time before further signs of MS began to appear. In these cases, central neuropathic pain was their first symptom of MS.

II. Measuring Pain

Pain is a difficult symptom to quantify. It tends to be subjective; however, there are several scales used by physicians and researchers to better measure severity, chronicity, and character of pain symptoms. Several recognized reliable and valid rating scales include the Visual Analogue Scale (VAS) for pain severity from the McGill Pain Questionnaire, the Medical Outcomes 36-Item Short-Form Health Survey (SF-36), Kurtzke's Expanded Disability Status Scale (EDSS), and the Hospital Anxiety and Depression Scale (HADS.). These are currently accepted scales used to measure pain and disability; however, most have limitations (Hobart *et al.*, 2001). The EDSS is a method to quantify physical disability in MS, while the HADS is used to assess mental health. Since the complaint of pain is gaining more attention in the population of MS patients, studies have used questionnaires to ask these patients about their pain symptoms, apply current rating scales, and provide a statistical measure of their pain symptoms.

III. Epidemiology

There are several studies estimating the occurrence of pain in patients with MS. The estimates range from 29% to 86% (Archibald *et al.*, 1994; Moulin *et al.*, 1988; Stenager *et al.*, 1991a, 1995b; Svendsen *et al.*, 2003; Vermote *et al.*, 1996; Warnell, 1991). A study by Solaro *et al.* (2004) found that in their large epidemiological face to face structured interviews with MS patients, 43% report at least one type of pain. This correlated with disease course, disability, duration, and age. Pollmann *et al.* (2004) found that 61% of 157 MS patients with clinically definite MS had severe pain. Pain was the worst symptom of MS as perceived by 12% of the patients. Insufficient care by the physicians they consulted was reported by 86% of the patients.

Many of the pain sensations that MS patients commonly describe are headache (seen in over 50%), followed by dysesthetic limb pain, back pain, tonic spasms, Lhermitte's sign (LS), visceral pain, trigeminal neuralgia, and pain associated with optic neuritis (Moulin *et al.*, 1988; Pollmann *et al.*, 2004; Solaro *et al.*, 2004). Solaro *et al.* (2004) found a positive correlation between pain in MS and the age of the patient, higher EDSS score and longer disease duration, and no significant difference in sex of the patient. However, Hadjimichael *et al.* (2006) found that increasing pain severity was associated with female sex, multiple sites of pain, and constancy of pain. They also found that pain severity was significantly associated with increasing levels of MS disability and the relapsing-worsening type of MS.

Osterberg *et al.* (2005) reported that out of 364 patients with relapsing-remitting MS, 57.5% had central pain, defined as pain caused by a lesion or dysfunction in the brain or spinal cord. Back pain, tension-type headache, migraine, and optic neuritis-related pain were not included in the central pain category. They comment that MS-related central pain is severe and causes much patient suffering.

Another study, by Svendsen *et al.* (2003), assessed pain prevalence and its clinical characteristics and impact on daily life by reviewing questionnaires from 627 MS patients and 487 controls. There was no difference in pain prevalence in the month preceding the assessment between MS patients and controls (79.4% vs 74.7%, respectively). A verbal rating scale and VAS for pain were both used to assess the subject's pain intensity. On the verbal rating scale, a higher prevalence of MS patients reported their pain as being moderate or severe as compared to controls. On the VAS, MS patients reported higher pain intensity than controls when "pain is at its least" and "pain is at its worst," and reported pain more often in the extremities, joints, muscles, eyes, and face than did controls. MS patients reported that pain interfered with daily life "all of the time" (19.3%) or "most of the time" (22.3%).

In order to determine the severity of chronic pain and its relationship to quality of life in MS patients, Kalia and O'Connor (2005) compared pain severity in MS patients to that in other chronic painful conditions and examined the relationships between chronic pain and health-related quality of life (HRQOL). Using the SF-36 bodily pain score, they found that pain severity was not different between MS patients with pain and rheumatoid arthritis or osteoarthritis patients. Chronic pain in these MS patients was found to have no significant relationship to age, disease duration, or disease course. The researchers divided types of pain into neurogenic (defined in the study as pain having a burning or sharp quality without an underlying structural basis) versus nonneurogenic (all other pain). Interestingly, 56.6% of MS patients had nonneurogenic pain whereas 17.2% had neurogenic pain. Pain prevalence and severity were found to strongly correlate with reduced social functioning and mental health. Pain in MS was associated with reduced HRQOL, especially mental health.

Rae-Grant *et al.* (1999) found that 12% of patients with MS reported pain to be the worst symptom of MS, and at the time of the study 31% of patients were using analgesics actively. However, Brichetto *et al.* (2003) found that the use of symptomatic analgesics was significantly lower than the frequency of pain. Adverse effects from treatment of MS may also contribute to the presence of pain in MS. For example, subcutaneous injections of interferon-β (IFN-β) can cause systemic flu-like side effects, including myalgia (Calabresi, 2004; Neilley *et al.*, 1996; Walther and Hohlfeld, 1999).

It is clear from all of these studies that MS is indeed a painful disease.

IV. Pain and Disability

Two recent publications have addressed the possible association of pain and disability. Solaro *et al.* (2004) found a correlation between pain and EDSS. Participants in the pain group had a higher EDSS score than those with no pain (mean EDSS 3.5 vs 2.5). This correlation was present for both typical neurogenic pain (i.e., dysesthetic pain) as well as nonneurogenic pain (i.e., back pain and tonic spasms).

Hadjimichael *et al.* (2006) found that pain severity was significantly associated with increasing levels of MS disability and relapsing-worsening MS subtype. The researchers used the Medical Outcomes Study Pain Effects Scale (PES) as a measure of interference of pain with six quality of life factors. With higher levels of pain severity, there was a significant association with participants' reduction in recreational activities, mood, and enjoyment of life.

V. Pain Syndromes in MS

Several pain syndromes have been described in patient with MS. Knowing what pain symptoms are most prevalent can help physicians anticipate patient needs. Moulin *et al.* (1988) reported that 55% of their 159 MS patients surveyed reported either an acute or chronic pain syndrome at some time during the course of their disease. Acute pain syndromes included paroxysmal tic-like pain in which several of the patients were diagnosed with trigeminal neuralgia. Chronic pain syndromes included dysesthetic extremity pain (29%), back pain (14%), painful leg spasms (13%), and abdominal pain (2%).

In a multicenter cross-sectional study by Solaro *et al.* (2004), pain was assessed in 1672 patients with MS. They found the most common types of pain reported were dysesthetic pain (18%), back pain (16%), painful tonic spasms (11%), LS (9%), visceral pain (3%), and trigeminal neuralgia (2%).

Sensory symptoms that can be uncomfortable to the point of being painful are also a common complaint in MS patients. A study by Rae-Grant *et al.* (1999) looked at questionnaires completed by 224 patients with MS and 93 controls in order to assess the frequency and quality of sensory symptoms. They found that sensory symptoms were more common in MS patients than controls and differed in severity and quality. Pain was present at some time in both groups in a similar percentage; however, 45% of MS patients described having active pain problems as compared to controls reporting only 22%. These painful sensations were qualified as itching, burning, electric type, "insects crawling," sharp, throbbing, and muscular. The site of pain differed between the two groups. Controls complained of head, back, and neck sites more often, while MS patients reported arms, legs trunk, hands, feet, or face significantly more often.

A. Headache

Headache and migraine are common features in MS patients. Gee *et al.* (2005) performed a retrospective review of MS patients in which they found that 55.6% reported headache. Of these, 61.7% had migraine, 25.3% had tension type, and 13% had both types. In a prospective study by D'Amico *et al.* (2004), 64.2% of the 137 patients with clinically definite MS reported headache. Of those patients whose headache was not temporally related to IFN treatment, 57.7% had headache (25% migraine without aura, 31.9% tension-type headache, and one patient with cluster headache).

The above headache studies looked at frequency and type in MS patients only. Rae-Grant *et al.* (1999) reported that two-thirds of the MS patients and 67% of controls in their survey complained of headache and there was no difference in frequency or severity. On the other hand, Rolak and Brown (1990) found that 52% of the MS patients they surveyed had headaches versus 18% of the general neurology clinic patients. This confirms that headache is a common problem in the general population which extends to MS patients.

It is also important to note that IFN-β treatment in MS patients can also be a factor contributing to headache. In the PRISM study (Gold *et al.*, 2005), IFN-β1a therapy was associated with headache in 71%, which tended to be mild, was separate from flu-like symptoms, and was most frequently reported during the first month of treatment.

B. Dysesthetic Pain

Central pain (CP) is pain initiated or caused by a primary lesion or dysfunction of the CNS (Mersky and Bogduk, 1994). Dysesthetic pain is described as a constant burning discomfort which can be symmetric or asymmetric usually affecting the lower limbs. It is commonly classified as a type of central pain. Several studies have reported that dysesthetic pain is the most common type of pain in MS patients, aside from headache (Rizzo, 2003; Solaro *et al.*, 2004).

C. Painful Tonic Spasms

An estimated 40–75% of patients with MS experience spasticity (Cervera-Deval *et al.*, 1994; MS Society Symptom Management Survey, 1997; Paisley *et al.*, 2002). Spasticity has been defined in an article by O'Brien (2002) as an abnormal increase in muscle contraction often caused by damage to central motor pathways that control voluntary movement. The patterns of spasticity vary depending on the distribution of lesions in the CNS. Patients often describe painful muscle contractures, stiffness, and reduced mobility. This symptom may affect the daily functioning of these patients in areas involving gross motor activities used for ambulation (Brar *et al.*, 1991) and can also cause pain.

D. Lhermitte's Sign

LS is the sudden sensation of electric shock-like sensation that spreads through the body on flexion of the neck lasting about 2 s or less. According to Gutrecht (1989), LS was first described by Pierre Marie and Chatelin in 1917 and reported by Lhermitte in 1920. Lhermitte published the seminal article on LS in 1924.

A clinical survey and review of literature done by Kanchandani and Howe (1982) revealed that LS occurred as the first symptom in 16% of patients and that LS rarely occurred in other conditions except subacute combined degeneration of the cord, neck trauma, radiation myelitis, and prolapsed cervical disk. Gutrecht *et al.* (1993) confirmed the presumption that a lesion in the posterior columns of cervical spinal cord is the cause of LS in MS in their radiological review. Rae-Grant *et al.* (1999) evaluated the demographics, disease course, and presence of various symptoms in MS. He found that LS was present in two-third of patients at some time during their disease course and it was present in 55% of patients at the time of their survey. Although LS is commonly stereotyped in individual patients (Al-Araji and Oger, 2005), it has a variable natural course and correlates significantly with cervical MRI abnormalities.

E. Trigeminal Neuralgia

Painful paroxysmal symptoms like TN occur in MS and there are differences when compared to controls. On the basis of a retrospective review of surgical records during a 4-year period, Berk *et al.* (2003) found a higher incidence of TN in patients with MS than in general population. The age of onset of MS-associated TN in patients has been noted to be about 10 years younger than in TN patients without MS, and when TN was the first symptom of MS (De Simone *et al.*, 2005; Hooge and Redekop, 1995; Jensen *et al.*, 1982). It was also noted that bilateral TN occurred more frequently in MS, but no difference were noted in the duration or quality of pain, triggering factors, or type of MS.

VI. Pain Management in MS Patients

A. Treatment of TN

Various options of treatment are available for treatment of TN in MS and these include pharmacological and surgical options. The mainstay of pharmacological treatment of TN in MS are the antiepileptic drugs (AEDs), mainly carbamazepine

(Joy and Johnston, 2001; Solaro, 2006). Adjunctive agents may be required as the frequency of pain increases over time, and include the following agents: oxcarbazepine, gabapentin, levetiracetam, and lamotrigine. In cases of refractory TN, topiramate may be used and other agents such as baclofen and misoprostol have also been used with varying success (Crayton *et al.*, 2004; Kassirer, 2000).

Surgical treatments are used in medically refractory TN, and include microvascular decompression (MVD), radio surgery, and radiofrequency ablation. Berk *et al.* (2003) reported that following rhizotomy, complete pain relief was obtained without the need for any type of analgesia in 81% of patients with refractory TN in MS and that adding analgesics resulted in pain control in the remaining patients. However, the group reported that pain was found to recur in 50% of patients during a 52-month follow-up.

Cheng *et al.* (2005) found radio surgery to be an effective procedure resulting in fewer retreatments and longer pain-free intervals compared to MVD or radio frequency ablation. Resnick *et al.* (1996) reported that patients with TN in MS may benefit from exploration of the cerebellopontine angle. Partial sectioning of the nerve and MVD alone did not provide adequate and reliable analgesia in this study. Complete pain relief was obtained with a single treatment of any of these options in 70–85% of patients with refractory MS-related TN.

B. Treatment of Other Types of Pain in MS

The mainstay of treatment of acute MS-related pain are the AEDs, especially for stereotypical paroxysmal episodes characterized by intense pain and worsening with disease progression and age. For pain precipitated by spasms, various agents like baclofen, tizanidine, benzodiazepines, and dantrolene may be tried (Crayton *et al.*, 2004; Paisley *et al.*, 2002).

Management of chronic pain also involves AEDs, as well as nonsteroidal anti-inflammatory drugs (NSAIDs), opioids, tricyclic antidepressants, and nerve blocks (Kalman *et al.*, 2002; Solaro, 2006). Local administration of botulinum toxin in patients with MS was shown to have positive effects on spasticity by Jost (2006), although the effect on pain was not mentioned.

Pharmacological therapy for spasticity includes baclofen and tizanidine. Baclofen acts by stimulating γ -aminobutyric acid (GABA) receptors, while tizanidine is a centrally acting $\alpha2$ adrenergic receptor agonist. Dantrolene can also be used to treat spasms as it acts directly on skeletal muscle. A systemic review of oral agents concluded that baclofen, tizanidine, and dantrolene were effective compared with placebo (Chou *et al.*, 2004).

Recent studies have been performed to determine if cannabis-based medicine is efficacious in treating central pain such as spasticity. The use of cannabinoids was shown to have a modest but significant analgesic effect on central pain in patients with MS by Svendsen *et al.* (2004). In the randomized, double-blind, placebo-controlled trial done by Rog *et al.* (2005), 66 patients with central pain states including seven with painful spasms received either a cannabis-based medicine via oromucosal spray or a placebo. Of the patients with painful spasms, they found that their 11-point Numerical Rating Scale (NRS-11) decreased by 5.7 points in the cannabis-based medicine group and by 2.1 in the placebo group. Sativex® is a cannabis-based product which is approved as an adjunctive treatment for neuropathic pain in patients with MS (Perras, 2005).

A pilot study done by Al-Smadi *et al.* (2003) using transcutaneous electric nerve stimulation (TENS) showed that active TENS was more effective than placebo for low back pain in MS, although the results were not statistically significant. Other nonpharmacological measures including rehabilitation and psychological intervention have been demonstrated to be efficacious for various persistent pain conditions and should be considered for patients with pain in MS (Flor *et al.*, 1992).

VII. Pathophysiology of MS-Related Pain

The pathology underlying central MS pain is not completely understood. Researchers have pointed out certain plaque locations as important associated with pain perception or modulation (Osterberg *et al.*, 2005). Other plaques responsible for causing pain may do so by causing abnormal impulses through motor axons (Spissu *et al.*, 1999). Dysfunction of injured nerves may cause pain secondary to an acquired sodium channelopathy (Waxman, 2001). The role of glial cells in the immune response to CNS inflammation associated with MS may also be important in the development of pain states (Watkins and Maier, 2000; Zhang and Huang, 2006).

A. Plaque Location and Pain

Central pain (CP) has been described in patients who have disruption of spinothalamic and quintothalamic pathways that carry pain and temperature sensation (Boivie, 1999). An MRI study relating plaque location to pain complaints by MS patients found a preponderance of lesions in the periventricular gray matter, corpus callosum, pons, brachia pontis, cerebellum, medulla oblongata, and thalamus (Osterberg *et al.*, 2005). Thalamic lesions were seen in one-third of these patients. Those lesions that affected fibers projecting to the ventroposterior thalamus (neospinothalamic projections) resulting in reduced tonic inhibition were considered most crucial to the development of pain. Spinal cord lesions were present in well over half of the patients, many of which had pain confined to both lower extremities, similar to that seen in spinal cord injuries.

Peripheral mechanisms of neuropathic pain include collateral sprouting, sympathetic-sensory nerve coupling, ectopic discharges, and ephaptic conduction. The latter is one of the mechanisms proposed for patients with painful tonic spasms associated with plaques in the contralateral posterior limb of the internal capsule or the cerebral peduncle (Spissu *et al.*, 1999). The authors point out that the close proximity of the motor fibers in these areas may involve large numbers of axons with a single demyelinating lesion and allow radial spread of ephaptic activation resulting in painful signals. The pathogenic mechanism of TN in both MS and non-MS patients may be the same: demyelination at the TN entry root in the pons (De Simone *et al.*, 2005).

Scherder *et al.* (2005) summarized pain pathology related to the medial and lateral pain systems in MS and Parkinson's disease. The medial system serves the motivational-affective and cognitive-evaluative aspects of pain, autonomic responses to pain, and memory of painful events. It consists of the parabrachial nucleus and locus coeruleus in the reticular formation, periaqueductal gray, the intralaminar, medial, ventral caudal parvocellular, and ventral caudal portae thalamic nuclei, the insula, parietal operculum, secondary somatosensory cotex (SII), anterior-cingulated cortex (ACC), amygdala, hippocampus, and the tuberomammillary nucleus and paraventricular nucleus of the hypothalamus. The lateral pain system processes the sensory discriminative aspects of pain and consists of the lateral thalamus, primary somatosensory area (S1), SII, the parietal operculum, and the insula. The authors reported that clinical physical symptoms in MS roughly corresponded to the location of white matter lesions or cortical atrophy.

Descending inhibitory monoaminergic fibers of pain modulation may also be interrupted by MS plaques. White *et al.* (1989) reported this finding in animals with experimental autoimmune encephalomyelitis (EAE) in studies of the ventral horn and fibers to motor neurons. A similar scenario involving loss of inhibitory fibers to the dorsal horn was considered to be a possible mechanism of hyperalgesia in EAE by Aicher *et al.* (2004).

B. Remote Effects of MS on Sensory Function

Besides overt plaques in the periaqueductal gray, neuropathological changes of medial pain system structures were also reviewed by Scherder *et al.* (2005), including enhanced activation of the insula during sensorimotor activity (Rocca *et al.*, 2002) and SII cortex (Rocca *et al.*, 2003a), enhanced glucose metabolism in the anterior cingulate cortex (Roelcke *et al.*, 1997), and increased thalamic activity associated with simple motor activity (Rocca *et al.*, 2003b) correlated with structural changes in normal appearing gray and white matter. These findings point out the far-reaching generalized physiological effects of the disease process, beyond specific plaque formation. Some changes may be an adaptive compensation mechanism geared toward minimizing the clinical symptoms (Scherder *et al.*, 2005). Thus, a patient with sensory loss may show increased activity in sensory-processing structures, which could lead to pain.

C. Channelopathy as a Cause of Pain

Pain arising from damaged neurons in MS may also be due to an acquired channelopathy (Waxman, 2001). Abnormal sodium channel activity can contribute to hyperexcitability of injured peripheral nerves resulting in symptoms of paresthesia and pain. In MS, dysregulated sodium channel gene expression may cause maladaptive healing during remyelination and continued abnormal action potential conduction, resulting in pain. Other examples of painful channelopathies include erythromelalgia (Choi *et al.*, 2006; Waxman and Dib-Hajj, 2005), migraine (Tepper *et al.*, 2001), and pseudomigraine with lymphocytic pleocytosis (Chapman *et al.*, 2003). These conditions may differ from MS in that the channelopathy is linked to genetic mutations.

D. Hyperalgesia in Animal Models of MS

Animal models provide an excellent opportunity to study MS-related pain. EAE induces hyperalgesia in rodents (Aicher *et al.*, 2004). In these animals, an initial hypoalgesia, indicated by increased tail-flick latency to radiant heat stimulus, occurred during the acute disease phase (day 12–17 postinoculation). This was followed by hyperalgesia (decreased latency of tail-flick) on day 17 continuing into the chronic phase of both active and passive EAE. The magnitude and duration of hyperalgesia correlated to the severity of clinical motor symptoms. Hyperalgesia was seen in the tail much more than in the forepaw, similarly to the high percentage of MS patients with lower extremity CP. This corresponds to histological evidence of greater damage in the LS cord compared to the cervical cord, or perhaps implies that longer axons are more severely impacted by central demyelination.

Interestingly, although hyperalgesia was considered to be closely associated with the neuroimmunologic response in a given region, hyperalgesia was not seen following inflammation alone (administration of Freund's complete adjuvant without the myelin peptide). Hyperalgesia also occurred in passive EAE after injection of activated T cells, supporting the idea that this pain response is unrelated to nonspecific inflammation. This finding is in contradiction to other peripheral inflammatory pain states such as arthritis or nerve injury, but may be important in understanding the relationship of pain to CNS inflammatory conditions.

E. The Immune System and Pain

The basic underlying pathology of MS has generally been assumed to be due to autoimmune activity of Th1-type CD4+ lymphocytes due to molecular mimicry, resulting in death of oligodendrocytes, causing demyelination and hypereactivity of microglia. It is now known to be far more complicated (Sospedra and Martin, 2005). Barnett and Prineas (2004) reported the inciting event in newly formed MS plaques, in which the oligodendrocyte underwent seemingly spontaneous apoptosis, followed by vacuolation and edema. Infiltration of the area by T cells, other immune cells, and phagocytes did occur but was delayed. Therefore, the immune activity was considered reactive to the destruction by the apoptotic oligodendroglia, rather than a T-cell-mediated attack against normal appearing myelin. This reactive inflammatory response might play as important a role in the formation of central pain in MS as a primary immunological attack. This is similar to other forms of CNS injury of the spinal cord and thalamic stroke resulting in central pain in which cellular debris serves as an immune stimulus (Moalem *et al.*, 2004, Zhang and Huang, 2006).

Zhang and Huang (2006) provide a review of the role of the immune system in the development of pain. The process starts with a precipitant such as infection (i.e., Herpes Zoster), autoimmune reaction (such as Guillaín-Barré syndrome), or nerve injury in which cellular debris can be a potent immune stimulus. In the case of MS, the enormous amount of debris created by disintegrating myelin may be the attractant. In general, sickness and injury exposes nearby tissue to cell wall fragments, toxins, and irritating chemicals as well. Clinically, there is a resultant "sickness response" associated with fever, altered plasma ion concentrations, neutrophil and lymphocyte proliferation, increased sleep, decreased activity, and social interaction which preserves energy for recovery. The sickness response from the brain is a reaction to proinflammatory cytokines released by activated immune cells (neutrophils, macrophages, T and B cells), including nerve growth factor (NGF), nitric oxide (NO), prostanoids, interleukin-1 (IL-1), IL-6, and tumor necrosis factor (TNF).

For these large proinflammatory proteins to affect the nervous system, they must cross the blood–brain barrier. They may do this via direct action on the nociceptive nerve terminals, which express receptors for TNF, IL-1, IL-6, leukemia inhibitory factor, histamine, bradykinin, and prostinoids. Indirectly, they may initiate a CNS response via a variety of actions such as cAMP activation thus sensitizing nerve terminals, or by inducing algogens such as PGE2 and sensitizing nociceptive terminals via transient receptor potential vanilloid 1 channels. Dorsal root ganglia neurons may react to chemokines causing increased calcium concentrations resulting in release of substance P. Proteins may cross into the CNS at circumventricular organs where the BBB is more permeable, or bind to blood vessels walls and gain access via second messengers such as prostaglandins, or utilizing specialized sensory nerves to carry immune information to the brain such as the vagus nerve (Watkins and Maier, 2000; Zhang and Huang, 2006).

Experimental manipulations eliciting a classic sickness response also elicit hyperalgesia, validating that pain is part of this sickness phenomenon. Neurocircuitry mediating classic sickness responses and hyperalgesia overlap (vagus nerve), and like the sickness response, hyperalgesia is dependent on IL-1 released in the CNS, in addition to proinflammatory cytokines at the site of infection (Watkins and Maier, 2000). Although it is not clear to what extent this sickness response occurs in MS exacerbations, other "constitutional" symptoms such as fatigue suggest that it may play a role.

CNS mechanisms of pain that have been demonstrated include anatomical reorganization of the spinal cord with sensory hyperexcitability. Watkins and Maier (2000) liken these changes to those in the hippocampus which takes place during learning and the formation of memory in the brain. In the presence of inflammation, brain-derived neurotrophic factor (BDNF) released from small sensory afferents may function as a neurotransmitter or neuromodulator in the dorsal horn causing increased excitability. Release of glutamate and substance P in the dorsal horn activates NMDA ion channels resulting in an influx of calcium ions. A cascade of cellular events follows which culminates in the formation of NO and prostaglandins. Upregulation of IL-1 expression and increased immunoreactivity to IL-6 in sciatic nerve injury results in hyperalgesia associated with a dramatic increase in excitability of dorsal horn neurons and incoming pain fibers. An exaggerated pain state is the result of continued hyperalgesia such that allodynia causes the animal to respond to normally innocuous stimuli such as touch, pressure, or warmth with vigorous and emotional responses indicating pain.

Immune system activities regulate spinal modulatory pain responses. Macrophages and other immune cells release proinflammatory cytokines as has been mentioned previously. Of these, TNF has the strongest activity, and IL-1 has more activity than IL-6. TNF and IL-1 are thought to be key mediators of communication between the immune system and the brain in forming the sickness response, as well as chronic pain. Research in animal models and humans with inflammatory pain states has shown promising results using TNF and IL-1 inhibitors (Moreland *et al.*, 1997; Nuki *et al.*, 2002). The TNF inhibitor etanercept is already an established treatment for arthritic conditions.

MULTIPLE SCLEROSIS AND BEHAVIOR

James B. Pinkston,*,† Anita Kablinger,† and Nadejda Alekseeva†

*Department of Neurology, Louisiana State University Medical Center
Shreveport, Louisiana 71103, USA

†Department of Psychiatry, Louisiana State University Medical Center
Shreveport, Louisiana 71103, USA

Multiple sclerosis (MS) is one of the most frequently seen neurological causes of progressive disability in early to middle adulthood. The disease is variable in its presentation and course, affects roughly 100–300 per 100,000 persons within the United States alone, and is slightly more common among females than males. MS places substantial burdens on patients, families, and caregivers. It negatively affects cognitive abilities and psychiatric functioning, and can add a notably deleterious effect on a patient's quality of life. This chapter reviews the recent literature on the behavioral manifestations of MS. Cognitive domains discussed include executive functioning, processing speed, attention, learning and memory, language functioning,

INTERNATIONAL REVIEW OF
NEUROBIOLOGY, VOL. 79
DOI: 10.1016/S0074-7742(07)79014-1

and visual spatial processing. Some attention will also be paid to differential diagnosis and the cognitive effects of treatment. Psychiatric manifestations are also discussed, including symptoms of depression, bipolar disorder, euphoria, pathological laughter and crying, and psychosis, as well as maladaptive personality traits. Finally, the chapter concludes with a discussion of the effects of MS on quality of life including such areas as fatigue, sexual dysfunction, pain, employment, and cognitive functioning.

I. Introduction

Multiple sclerosis (MS) is an unpredictable, progressive, chronic demyelinating disease that affects approximately 100–300 per 100,000 persons within the United States. It is one of the most common neurological causes of disability in middle adulthood, and it places a considerable and a lasting burden on patients and their families. MS most often strikes during the second and third decades of life, affecting slightly more women than men. Approximately 40–70% of individuals with MS demonstrate behavioral changes associated with the diagnosis and its progression. The varied manifestations of MS can affect nearly all facets of an individual's daily functioning. MS can produce notable alterations in functioning that result in significant emotional, lifestyle, and vocational changes. These changes can manifest as cognitive difficulties, psychiatric disorders, and difficulties related to quality of life.

This chapter will cover the various functional effects of MS. Section II will explore the myriad cognitive changes that can occur in individuals with MS, including declinations in executive functioning, processing speed, attention, learning and memory, language functioning, and visual spatial processing. Some attention will also be paid to differential diagnosis and the cognitive effects of treatment. Section III will explore the effect of MS on psychological functioning. Specific areas discussed include depression, bipolar disorder, euphoria, pathological laughter, and crying, symptoms of psychosis, and maladaptive personality traits. Finally, the chapter will conclude with an investigation into the effects of MS on quality of life. Attention is given to areas including fatigue, sexual dysfunction, pain, measures of quality of life, employment, and cognitive functioning.

II. Cognitive Functioning

MS has known and varied effects on cognitive functioning, and fully 50% of individuals with MS can be expected to show some decline in cognitive functioning during the course of the illness (Calabrese, 2006). Research has demonstrated changes in mental processing speed, executive functioning, attention, learning, memory, and visual spatial abilities in individuals with MS. These cognitive changes may occur at any time during the course of the disease, and restrictions in mental processing speed and memory have been seen even during the early stages of MS. In contrast, intellectual reasoning and language abilities typically remain generally preserved (Calabrese, 2006). Although changes in cognitive functioning in MS are not universal, many individuals with MS will experience some change in their cognitive abilities throughout the course of the disease, and these changes may have significant deleterious effects on their vocational functioning, as well as their ability to independently perform their activities of daily living.

A. Executive Functioning

Although some researchers report equivalence between healthy individuals and persons with MS on measures of executive functioning (Denney *et al.*, 2005), this finding is not universal (Benedict *et al.*, 2006), and persons with MS have been shown to demonstrate reductions in executive functioning on formal testing. Deficits may be seen on tasks required planning, problem solving, concept formation, set shifting, and utilization of corrective feedback (Achiron *et al.*, 2005). Individuals with MS have been shown to demonstrate perseveration of unsuccessful response strategies even in the face of corrective feedback, as well as to have difficulty in shifting from one conceptual set to another. Performance on tasks of executive functioning, combined with impairment on measures of memory recall, has been shown to be most predictive of vocational disability in individuals with MS (Benedict *et al.*, 2006).

B. Processing Speed

Individuals within all MS subtypes have repeatedly been described as demonstrating reduced mental processing speed compared to healthy individuals, and restriction in mental processing speed has been described as the most frequently impaired cognitive finding among individuals with MS (Nocentini *et al.*, 2006). Compared to healthy controls, individuals with relapsing-remitting MS are described as 24% slower, whereas those with primary and secondary progressive MS are noted as 50% slower, on tasks of mental processing speed (De Sonneville *et al.*, 2002). These restrictions in processing speed are thought to largely account for the difficulties that individuals with MS demonstrate on tasks of focused, divided, and sustained attention, as well as on some measures of executive functioning. Restrictions in mental processing speed have been found independent of level of physical disability or symptoms of depression (Denney *et al.*, 2005).

Furthermore, reduced mental processing speed has been seen among individuals with MS regardless of whether speed was overtly or covertly assessed (Denney *et al.*, 2004).

C. Attention

One of the most commonly self-reported difficulties among individuals with MS is deficits in short-term memory or attention. A related problem is seen in restrictions in working memory or the ability to hold bits of information in one's attention while manipulating them. Even among individuals with mild MS, subtle cognitive processing deficits can be elicited in the context of a distracting environment, especially auditory distraction (LaPointe *et al.*, 2005). Individuals with MS also demonstrate more difficulties on measures of divided attention than sustained attention (McCarthy *et al.*, 2005). These difficulties with attention and working memory appear to be largely related to reductions in mental processing speed, which in turn are thought to be secondary to the process of demyelination characteristic of MS (Janculjak *et al.*, 2002). Individuals with MS have demonstrated comparable performances on tasks of working memory on some measures with limited processing speed demands, but their performance declines in contrast to healthy individuals when processing speed demands become more salient (Lengenfelder *et al.*, 2006).

D. Learning and Memory

The ability to learn and recall new information has obvious importance in the lives of individuals with MS. Restrictions in memory functioning may cause serious vocational disability and can lead to problems in activities of daily living. Compared to healthy adults, individuals with MS require more repetition or learning trials to master new information. Furthermore, their recall of this information oftentimes falls below that of healthy individuals (Chiaravalloti *et al.*, 2003). As such, simple repetition may not be sufficient to compensate for memory problems in MS, and additional organizational and mnemonic strategies should be utilized. Research has also demonstrated that the subjective memory dysfunction complaints of individuals with MS more accurately represent restrictions in other cognitive domains, including mental processing speed than a primary memory impairment per se (Randolph *et al.*, 2001).

Of interest, individuals with MS appear to show more difficulty on tasks of explicit memory in which they consciously attempt to learn and retain information, as opposed to tasks of implicit memory, where learning occurs more as a product of simple exposure and mostly outside of their conscious awareness (Seinelä *et al.*, 2002). Thus, explicit memory may be more sensitive to decline in MS than implicit memory functioning, and better utilizing implicit memory may be an effective compensatory strategy for reported memory difficulties. Finally, the memory functioning of individuals with MS appears to be significantly related to deficits in executive functioning and metamemory, as well as the often reversible negative effects of mood disturbance and symptoms of depression (Randolph *et al.*, 2004).

E. Language and Visual Spatial Functioning

Although language comprehension is typically resistant to decline in MS, other facets of language may show impairment. Investigation has reported that declines in verbal fluency may be affected early in the course of MS (Calabrese, 2006). A recent meta-analysis on the effects of MS on verbal fluency was conducted utilizing data from 35 studies with over 3000 participants (Henry and Beatty, 2006). Impairment on measures of verbal fluency were substantial for individuals with MS and were more severe than deficits seen on measures of verbal intellectual functioning, word finding ability, and executive functioning. Restrictions in verbal fluency were found to be related to greater duration of illness and increased neurological disability (Henry and Beatty, 2006).

Impairments in visual spatial functioning appear to result from focal deficits, though some individuals with MS demonstrate slowed visual information processing as a result of reduced mental processing speed that further impairs performance on measures of visual spatial perception (Vleugels *et al.*, 2001). The implications toward driving are significant, as individuals with restricted mental processing speed and difficulties with visual spatial perception are at increased risk for accidents and mishaps while operating a motor vehicle.

F. Differential Diagnosis

Researchers have endeavored to use the results of neuropsychological testing to differentiate between the different subtypes of MS. These investigations have met with varying degrees of success, though some differences in cognitive functioning do appear more robust than others. Individuals with progressive subtypes of MS are typically found to evidence more cognitive impairment compared to individuals with the relapsing-remitting subtype of MS (Kraüs *et al.*, 2005). Differences between the two progressive subtypes (primary and secondary) are inconsistently reported.

The manifestations of reduced mental processing speed, difficulties with attention, concentration, and memory, and impaired abstract reasoning are considered evidence of subcortical pathology. Therefore, the cognitive pathology most commonly seen in MS is typically placed under the rubric of subcortical dementia. However, not all patients show this subcortical presentation. A variant of MS referred to as cortical MS has been described. This condition is said to be primarily identified by predominant cortical pathology along with symptoms such as depression, amnesia, or distinct cortical syndromes (Zarei, 2006). This different presentation of MS is important as it may complicate early accurate diagnosis thereby resulting in individuals being untreated or mistreated.

The prevalence of MS has routinely been described as greater in females than in males. However, the effect of MS on cognitive functioning shows an opposite gradient. Specifically, males appear to be more vulnerable to the negative cognitive effects of MS. Specifically, males demonstrate poorer performances on tasks of verbal and visual memory, visual spatial ability, and executive functioning (Beatty and Aupperle, 2002). The increase in cognitive vulnerability among males with MS is reported to be associated with disease duration, level of disability, and low level of education, as well as with the $\varepsilon 4$ allele of the APOE gene. In contrast none of these factors, or any other investigated variable, was found to correlate significantly with cognitive impairment in women (Savettieri *et al.*, 2004).

MS typically strikes during the third decade of life. However, children and adolescents are not immune from the disorder. Pediatric MS also manifests significant negative effects on cognitive functioning. Specifically, children with MS demonstrate deficits on measures of general cognitive ability, language, visual-motor integration, and verbal and visual memory (Banwell and Anderson, 2005). Cognitive impairment in children and adolescents has been found to be related to level of disability, total number of relapses, and disease duration with younger children appearing to be at greater risk for cognitive impairment (MacAllister *et al.*, 2005). In one-third of pediatric MS cases, these cognitive deficits are sufficient to cause academic problems.

G. Treatment

Research continues on the effects that medications used to treat MS have on cognitive functioning. Compared to untreated controls, individuals with MS who were given IFNβ-1b have demonstrated significant improvements on measures of complex attention, concentration, and visual learning and recall after 1 year. It is hoped that immunomodulating drugs can be used to inhibit cognitive decline as well as reduce the rate of relapse and disease progression (Barak and Achiron, 2002). In addition to the effects of these immune modulating therapies, research is progressing on the cognitive effects of anti-inflammatory and immunosuppressive agents including cyclophosphamide combined with methylprednisolone. In a recent study, individuals with primary and secondary progressive MS received cyclophosphamide combined with methylprednisolone via monthly intravenous pulse doses. Following 6 and 12 months of treatment, significant improvements were discovered on measures of global cognitive efficiency, learning, organizational and planning abilities, and inhibition (Zéphir *et al.*, 2005).

Several researchers have also investigated the possible beneficial effects of donepezil on memory functioning in individuals with MS. Current results suggest that acetylcholinesterase inhibitors can be effective in ameliorating some of the memory problems for individuals with MS (Amato *et al.*, 2006). However, studies employing nonpharmacological interventions, such as cognitive rehabilitation, have not consistently demonstrated similar efficacy in the treatment of cognitive disorders in MS.

III. Psychiatric Manifestations

Half of all patients with MS demonstrate cognitive or psychological impairment. The three most common emotional disturbances seen in patients with MS include depression, pathological laughing or crying, and euphoria. These difficulties may result from the stress associated with the illness, or they may be a direct expression of the central nervous system (CNS) pathology of MS itself.

A. Depression

Depression and cognitive problems are often related. Problems in thinking processes can obviously negatively affect an individual's level of activity and participation in domestic, social, and vocational activities. Solari *et al.* (2004) suggest that this MS-related resulting lack of activity and participation can then lead to depression in many individuals. Arnett *et al.* (2002) found that the linkage between depression and cognitive dysfunction is significantly affected by coping style. Patients with the most cognitive dysfunction and depression were found to be those who also showed low levels of active coping or high levels of avoidant (or maladaptive) coping with respect to their illness.

Clinical symptoms of depression, such as negative self-evaluative thoughts (feelings of worthlessness, inferiority, failure, and rejection sensitivity), are also common among MS patients and can predict cognitive difficulties in MS (Arnett, 2005). Arnett (2005), utilizing an extensive set of neuropsychological tests, suggests that the negative self-evaluative thoughts in persons with MS occupy significant amounts of their cognitive capacity directly leading to cognitive dysfunction. Neurovegetative symptoms on the other hand, such as weight loss, sleep disturbance, and fatigue, were less associated with a decline in cognition and may be more correlated with the physical manifestations of MS itself.

Sobel *et al.* (2005) emphasize that depression in MS patients is difficult to diagnose because many of its symptoms overlap with those of other biological conditions. It is not surprising then that stress can exacerbate the physical disease process, leading to exacerbations of MS (Mohr *et al.*, 2004), adding to the difficulties in separating emotional from physical issues within MS. Furthermore, depression may also be the result of various treatments for MS (IFNB Multiple Sclerosis Study Group and the UBC MS/MRI Analysis Group, 1995), although repeated studies have found this to be more related to pretreatment levels of depression than to the specific treatments themselves (Feinstein *et al.*, 2002; Patten *et al.*, 2002).

The most common symptoms in depressed MS patients include irritability, discouragement, frustration, insomnia, poor appetite, memory and concentration problems, as well as fatigue; whereas guilt and poor self-esteem are less likely to be found in patients with MS (Minden *et al.*, 1987). Reported rates of depression in MS patients are as high as 50% lifetime prevalence and 20% annual prevalence (Siegert and Abernethy, 2005). Diagnosing and treating depression in these patients is important because depression and/or stress may adversely affect the course of the disease as well as the individuals' quality of life. Hart *et al.* (2005) demonstrated that treating moderate to severe depression in patients with MS is associated with improvement in multiple quality of life areas. Suicidal ideation is an important item to assess in MS patients (usually associated with depression), as the rate of suicide may account for 15% of all MS deaths (Sadovnik *et al.*, 1991). Identifying depression and aggressively treating this illness is the best preventative intervention to decrease this rate.

Crayton *et al.* (2004) suggest that treatment for depression be individualized for each patient with focus on communication, education, pharmacological intervention, as well as intervention from occupational therapy, physical therapy, and other modalities to help with the physical and emotional well-being and activities. The Goldman Consensus Group (2005) describes the under-recognition and treatment of MS patients suffering from depression, while Ehde and Bombardier (2005) emphasize the dearth of treatment studies for depression in these patients and urge the research community to focus on this important aspect of MS. Finally, the patient's caregiver should be assessed and treated for symptoms of depression as well as their mood can affect the course of treatment and the disease of the patient (Pozzilli *et al.*, 2004).

B. Bipolar Disorder

Bipolar disorder is also seen in MS patients with a rate twice the prevalence of the general population (Schiffer *et al.*, 1986), and does not entirely appear to be a result of the use of steroids (Minden *et al.*, 1988). Impulsivity (often a symptom of bipolar disorder) also appears to be a common manifestation of patients with MS. Hawkes (2005) suggests that this impulsivity may be a premorbid trait of the individual increasing the odds of developing MS through impulsive behavior and exposure to causative factors.

C. Euphoria and Pathological Laughing and Crying

Euphoria refers to feelings of both physical and emotional well-being despite the presence of a debilitating neurological disorder and may be found in as many as 25% of MS patients. This unusual affective expression is associated with advanced stages of the disease (Rabins, 1990). Pathological laughing and crying occurs in approximately 10% of MS patients and is extremely troubling to the affected individual, as well as family members, friends, and caregivers (Feinstein *et al.*, 1997). Patients experience laughter when there is nothing amusing and cry when there is no sadness. This may occur with or without pseudobulbar palsy, and patients frequently exhibit frontal lobe cognitive dysfunction in association with the emotional illness (Feinstein *et al.*, 1999).

D. Psychosis and Maladaptive Personality Traits

Psychosis appears to be a rare phenomenon in MS patients since it is typically associated with gray matter impairment and MS usually manifests as a white matter disease. Finally, maladaptive personality traits are seen as manifestation of MS and have been evaluated using a five-factor model. This model looks at five domains including neuroticism (extent of emotional responsiveness and likelihood of negative affect), extraversion (need to be outgoing or sensation-seeking), openness to experience (desire for new knowledge, ideas, and experience), agreeableness (desire for socialization, honesty, and altruism in relationships), and conscientiousness (tendency to be organized and deliberate). Benedict *et al.* (2001) compared patient and informant reports using the same personality scale. They found that MS patients were more neurotic and less empathic, agreeable, and conscientious as compared to healthy controls. Interestingly, this same pattern has been demonstrated in patients with Alzheimer's disease (Chatterjee *et al.*, 1992).

IV. Quality of Life in MS

MS remains the most common cause of disability in young adults (15–45 years of age). Clinically, MS manifests with fatigue, weakness, spasticity, cognitive decline, bladder and bowel sphincter dysfunction, sexual dysfunction, pain, sensory syndromes, visual loss, ataxia, and mood disorders. In most MS patients, the neurological and psychiatric deficits interfere with daily life activities and adversely affect productivity, vocational competence, and emotional well-being. Indeed, MS reduces quality of life.

A. Fatigue

Fatigue is the most common complaint among patients with MS and can manifest as both physical and mental pain. There is no generally accepted definition for fatigue in MS. The Council for Clinical Practice Guidelines and Paralyzed Veterans of America recommend the following definition: "a subjective lack of physical and/or mental energy which is perceived by the individual or caregiver as interfering with usual and desired activities." This definition outlines the main characteristics of MS fatigue. Fatigue is a subjective symptom and differs from muscle weakness. Some patients describe their fatigue as an overwhelming sense of exhaustion, lassitude, or lethargy. Fatigue may be the first symptom experienced by the patient prior to development of full-blown MS.

Indeed, fatigue more than a plain manifestation of MS is a disabling phenomenon. Fatigue remarkably decreases the individual's energy level, endurance and productivity, and negatively impacts mood and the ability to cope with accompanying limitations. In many MS patients, fatigue may serve as a predictive factor of disability independent from other neurological deficits. In the MS population, there is no currently known potent correlation between fatigue and demographic variables (age, gender) and disease characteristics (disease duration, muscle weakness, MRI lesion).

The exact etiology of fatigue in MS remains unknown, however, a number of hypotheses have been proposed: (1) Anatomic location of CNS lesions: neuroimaging studies using fMRI and PET have indicated that fatigue in MS is associated with hypometabolism in the prefrontal areas, thalamus, and basal ganglia (Roelcke *et al.*, 1997). A recent neuroimaging study (Niepel *et al.*, 2006) assessed the relationship between fatigue severity and detectable changes on MRI, quantified using the mean T1 relaxation time (T1), in deep gray matter structures in patients with relapsing-remitting MS. The investigators reported that abnormalities of subcortical gray matter structures were involved in pathogenesis of MS-related fatigue. (2) An immune-mediated phenomenon: the excessive fatigue which is particularly felt during relapses of MS, the transient increase in fatigue after injections of interferon beta or during an infectious disorders indicates that certain abnormal immune reactions play a role in pathogenesis of MS fatigue; the exact nature of these immune reactions remain unclear. (3) Endocrine abnormalities such as hypothyroidisms and dysfunction of hypothalamic-pituitary-adrenal axis have been implicated in pathogenesis of fatigue in MS. Several factors such sleep disturbances, depression, and the physical limitations which commonly accompany MS can potentially worsen fatigue. Other factors which precipitate fatigue in MS patients include concurrent medications, infections, anemia, concurrent hypothyroidism, exposure to high temperature, and spasticity.

B. Pain

Despite the erroneous common perception, MS is a painful disease. During the last decade, pain and discomfort have been increasingly recognized as clinical manifestations of MS. Up to two-third of MS patients suffer from some form of painful syndrome during their disease course (Clifford and Trotter, 1984; Moulin *et al.*, 1988; Stenager *et al.*, 1991). Pain in MS patients may occur in the context of specific syndromes such as trigeminal neuralgia, glossopharyngeal neuralgia, transverse myelitis, or optic neuritis. In addition, MS patients may experience pain associated with sensory impairment (neuropathic pain), pain and discomfort associated with severe deforming spasticity, radicular pain, pain due to migraine (which is seen more frequently in MS population), and poor body posture. Interestingly, pain can herald relapse of MS. Unfortunately, despite our further understanding of the pathogenesis of MS, chronic pain remains an underdiagnosed manifestation of MS since the severity of pain in MS has been limited. In general, the presence of any of these painful syndromes, in the context of MS, significantly reduces the quality of life of these patients.

C. Sexual Dysfunction

Sexual dysfunction is commonly experienced by both female and male patients. Among patients with MS, sexual dysfunction may originate from disease progression, overwhelming fatigue, profound depression, or it may be a side effect of antidepressants. Two neuroimaging studies (Zivadinov *et al.*, 2003; Zorzon *et al.*, 2003) have assessed the correlation between sexual dysfunction and MRI findings in individuals with MS. The investigators reported an association between sexual dysfunction and destructive lesions in the pons, as well as pontine atrophy. Male MS patients frequently complain of erectile dysfunction (ED), while female patients report decreased sexual desire, anorgasmia, reduced vaginal sensation and lubrication. ED can be treated with oral phosphodiesterase-5 inhibitors such as sildenafil or vardenafil (DasGupta and Fowler, 2003). Should these medications fail to improve ED, the application of injectable prostaglandin known as alprostadil (Caverject), under the supervision of a urologist may be helpful. Treatments for the deleterious effects on sexual functioning among females with MS are also available. For example, loss of vaginal sensation and associated reduced lubrication can be ameliorated with the use of lubricants. Furthermore, an FDA-approved device, called the Eros device, is available and provides the combined effect of vibration and gentle suction to draw blood into the clitoris and improve the sexual response (Billups *et al.*, 2001). Frustration and anger associated with sexual dysfunction in MS patients exert a salient negative effect on the individual's quality of life, and such can have salient untoward effects on relationships.

D. Disease Progression and Quality of Life

MS is the most common neurological cause of disability in young adults and, due to its progressive nature, has a significant impact on the lives of patients and their family members. In many MS patients, the neurological deficits induced by the underlying pathology interfere with and seriously impair daily life activities and adversely affect an individual's job, emotional health, and quality of life (Brunet *et al.*, 1996; Janssens *et al.*, 2003; Kroencke, 2001; Mitchell *et al.*, 2005; Nortvedt *et al.*, 1999; Rothwell *et al.*, 1997; Solari and Radice, 2001).

The majority of MS patients initially present with relapsing-remitting form of MS. Relapsing-remitting MS is clinically characterized by relapses of the disease with the development of new neurological deficits, or worsening of older symptoms, lasting for at least more than 24 h separated by periods of stability (remission) by at least 1 month. With further progression of underlying disease process, and even despite being on treatment, a large number of relapsing-remitting MS patients deteriorate and enter a new phase of disease known as secondary progressive MS (Weinshenker *et al.*, 1989). Secondary progressive MS is a neurodegenerative syndrome which is characterized by less clinical relapses, unfavorable therapeutic response to available treatment modalities, and accumulation of neurological impairment and disability.

In addition, it is important to realize that MS is a "whole brain disease" and affects both white and gray matter areas of the brain. The neurological damage from MS commences early in the course of the disease process and affects every aspect of CNS function with tragic impact on an individual's desire to live an independent life. As an incurable and destructive disease, MS does have profound adverse neuropsychiatric complications which threaten individual's independence, social status, income, dignity, productivity, and plans for a better and brighter life (Boeije *et al.*, 2002). MS clinical course is unpredictable with only partial, or in most cases, marginal response to best available treatments. Such absence of therapeutic response in a large number of MS patients poses several setbacks and disappointments for patients and their family members. Taken together, these negative characteristics indicate that MS potentially endangers the neurological and psychiatric well-being and stability of the individual and reduce his or her quality of life (Confavreux *et al.*, 2003; Mullins *et al.*, 2001).

A large number of clinical studies have addressed and measured health-related quality of life in MS patients (Benito-Leon *et al.*, 2003; Nortvedt and Riise, 2003; Rudick *et al.*, 1992). Hakim *et al.* (2000) performed a population-based survey to evaluate the effects of MS on the patients' ability to fulfill their chosen family and social roles and to assess the impact of MS on their relatives. The results of this study which included 305 MS patients and their relatives demonstrated that 53% of those patients employed at the time of diagnosis gave up their jobs, and the standard of living of 37% of patients and their families declined as a direct consequence of MS. Ford *et al.* (2001) in another study to assess the health status and quality of life in MS patients reported that up to 70% of MS patients were unemployed, half of these cases as a result of their disease. Various studies on the quality of life of MS patients have revealed that MS patients, particularly those with higher degrees of disability, do have lower quality of life scores than healthy controls (McCabe and McKern, 2002; Nortvedt *et al.*, 1999). More recent studies have explored factors associated with lower quality of life in MS (Janardhan and Bakshi, 2002; Janssens *et al.*, 2003). These studies have demonstrated correlations among a number of clinical variables and quality of life indicators. Generally, poor quality of life in MS patients is associated with progressive disease course, physical disability, depression, and cognitive decline.

E. Employment

A large number of MS patients lose their jobs and income as a direct result of the disability and limitations imposed on them by MS. A large number of these patients, despite effort to get another job, remain unemployed, and such failure to obtain a new job, improve one's social status, and reach financial security shatters the individual's self-concept and self-confidence. This in turn leads to more pressure on both patients and their caregivers. In addition, a number of personal features such as older age, more severe physical disability, advanced cognitive impairment, and lower education further complicate the issue of unemployment in the MS population. Loss of job, income, and social status eventually lead to further compromise of quality of life in MS patients.

F. Cognitive Impairment

Another significant factor which greatly adversely impacts quality of life among MS patients is cognitive or neuropsychological impairment. Studies assessing correlations between neuropsychological impairment and findings on brain magnetic resonance imaging reveal that neuropsychological dysfunction is associated with lesion burden, diffuse disease in normal-appearing brain tissue, and brain atrophy (Benedict *et al.*, 2004).

Taken together, various studies which have measured the quality of life in MS patients indicate that MS, through various mechanisms and associations, impairs and reduces the quality of life of these patients. Further efforts are required to develop methods and treatments which can improve the quality of life of individuals with MS.

References

Achiron, A., Polliack, M., Rao, S., Barak, Y., Lavie, M., Appelbaum, N., and Harel, Y. (2005). Cognitive patterns and progression in multiple sclerosis: Construction and validation of percentile curves. *J. Neurol. Neurosurg. Psychiatry* **76,** 744–749.

Amato, M. P., Portaccio, E., and Zipoli, V. (2006). Are there protective treatments for cognitive decline in MS? *J. Neurol. Sci.* **245,** 183–186.

Arnett, P. A., Higginson, C. I., Voss, W. D., Randolph, J. J., and Grandey, A. A. (2002). Relationship between coping, cognitive dysfunction and depression in multiple sclerosis. *Clin. Neuropsychol.* **16,** 341–355.

Arnett, P. A. (2005). Longitudinal consistency of the relationship between depression symptoms and cognitive functioning in multiple sclerosis. *CNS Spect.* **10**(5), 372–382.

Banwell, B. L., and Anderson, P. E. (2005). The cognitive burden of multiple sclerosis in children. *Neurology* **64,** 891–894.

Barak, Y., and Achiron, A. (2002). Effect of interferon-beta-1b on cognitive functions in multiple sclerosis. *Eur. Neurol.* **47,** 11–14.

Beatty, W. W., and Aupperle, R. L. (2002). Sex differences in cognitive impairment in multiple sclerosis. *Clin. Neuropsychol.* **16,** 472–480.

Benedict, R. H., Carone, D. A., and Bakshi, R. (2004). Correlating brain atrophy with cognitive dysfunction, mood disturbances, and personality disorder in multiple sclerosis. *J. Neuroimaging* **14**(3 Suppl.), 36S–45S.

Benedict, R. H., Cookfair, D., Gavett, R., Gunther, M., Munschauer, F., Carg, N., and Weinstock-Guttman, B. (2006). Validity of the minimal assessment of cognitive function in multiple sclerosis (MACFIMS). *J. Int. Neuropsychol. Soc.* **12,** 549–558.

Benedict, R. H., Priore, R. L., Miller, C., Munschauer, F., and Jacobs, L. (2001). Personality disorder in multiple sclerosis correlates with cognitive impairment. *J. Neuropsychiatry Clin. Neurosci.* **13,** 70–76.

Benito-Leon, J., Morales, J. M., Rivera-Navarro, J., and Mitchell, A. (2003). A review about the impact of multiple sclerosis on health-related quality of life. *Disabil. Rehabil.* **25,** 1291–1303.

Billups, K. L., Berman, L., Berman, J., Metz, M. E., Glennon, M. E., and Goldstein, I. (2001). A new non-pharmacological vacuum therapy for female sexual dysfunction. *J. Sex Marital Ther.* **27,** 435–441.

Boeije, H. R., Duijnstee, M. S., Grypdonck, M. H., and Pool, A. (2002). Encountering the downward phase: Biographical work in people with multiple sclerosis living at home. *Soc. Sci. Med.* **55,** 881–893.

Brunet, D. G., Hopman, W. M., Singer, M. A., Edgar, C. M., and MacKenzie, T. A. (1996). Measurement of health-related quality of life in multiple sclerosis patients. *Can. J. Neurol. Sci.* **23,** 99–103.

Calabrese, P. (2006). Neuropsychology of multiple sclerosis: An overview. *J. Neurol.* **253**(Suppl.), 10–15.

Chatterjee, A., Strauss, M., Smyth, K. A., and Whitehouse, P. J. (1992). Personality changes in Alzheimer's Disease. *Arch. Neurol.* **49,** 486–491.

Chiaravalloti, N. D., Demaree, H., Gaudino, E. A., and DeLuca, J. (2003). Can the repetition effect maximize learning in multiple sclerosis? *Clin. Rehabil.* **17,** 58–68.

Clifford, D. B., and Trotter, J. L. (1984). Pain in multiple sclerosis. *Arch. Neurol.* **41,** 1270–1272.

Confavreux, C., Vukusic, S., and Adeleine, P. (2003). Early clinical predictors and progression of irreversible disability in multiple sclerosis: An amnesic process. *Brain* **126,** 770–782.

Crayton, H., Heyman, R. A., Howard, S., and Rossman, H. S. (2004). A multimodal approach to managing the symptoms of multiple sclerosis. *Neurology* **63,** S12–S18.

DasGupta, R., and Fowler, C. J. (2003). Bladder, bowel and sexual dysfunction in multiple sclerosis: Management strategies. *Drugs* **63**(2), 153–166.

De Sonneville, L. M. J., Boringa, J. B., Reuling, I. E. W., Lazeron, R. H. C., Adér, H. J., and Polman, C. H. (2002). Information processing characteristics in subtypes of multiple sclerosis. *Neuropsychologia* **40,** 1751–1765.

Denney, D. R., Lynch, S. G., Parmenter, B. A., and Horne, N. (2004). Cognitive impairment in relapsing and primary progressive multiple sclerosis: Mostly a matter of speed. *J. Int. Neuropsychol. Soc.* **10,** 948–956.

Denney, D. R., Sworowski, L. A., and Lynch, S. G. (2005). Cognitive impairment in three subtypes of multiple sclerosis. *Arch. Clin. Neuropsychol.* **20,** 967–981.

Ehde, D. M., and Bombardier, C. H. (2005). Depression in persons with multiple sclerosis. *Phys. Med. Rehabil. Clin. N. Am.* **16**(2), 437–448.

Feinstein, A., Feinstein, K. J., Gray, T., and O'Connor, P. (1997). The prevalence and neurobehavioral correlates of pathological laughing and crying in multiple sclerosis. *Arch. Neurol.* **54,** 1116–1121.

Feinstein, A., O'Connor, P., Gray, T., and Feinstein, K. J. (1999). The pathogenesis of pathological laughing and crying in multiple sclerosis: A role for the prefrontal cortex? *Mult. Scler.* **5,** 69–73.

Feinstein, A., O'Connor, P., and Feinstein, K. J. (2002). Multiple sclerosis, interferon beta-1b and depression: A prospective investigation. *J. Neurol.* **249,** 815–820.

Ford, H. L., Gerry, E., Johnson, M. H., and Tennant, A. (2001). Health status and quality of life of people with multiple sclerosis. *Disabil. Rehabil.* **23,** 516–521.

Goldman Consensus Group (2005). The Goldman consensus statement on depression in multiple sclerosis. *Mult. Scler.* **11,** 328–337.

Hakim, E. A., Bakheit, A. M., Bryant, T. N., Roberts, M. W., McIntosh-Michaelis, S. A., Spackman, A. J., Martin, J. P., and McLellan, D. L. (2000). The social impact of multiple sclerosis–a study of 305 patients and their relatives. *Disabil. Rehabil.* **22,** 288–293.

Hart, S., Fonareva, I., Merluzzi, N., and Mohr, D. C. (2005). Treatment for depression and its relationship to improvement in quality of life and psychological well-being in multiple sclerosis patients. *Qual. Life Res.* **14**(3), 695–703.

Hawkes, C. H. (2005). Are multiple sclerosis patients risk-takers? *QJM* **98**(12), 895–911.

Henry, J. D., and Beatty, W. W. (2006). Verbal fluency deficits in multiple sclerosis. *Neuropsychologia* **44,** 1166–1174.

Janardhan, V., and Bakshi, R. (2002). Quality of life in patients with multiple sclerosis: The impact of fatigue and depression. *J. Neurol. Sci.* **205,** 51–58.

Janculjak, D., Mubrin, Z., Brinar, V., and Spilich, G. (2002). Changes of attention and memory in a group of patients with multiple sclerosis. *Clin. Neurol. Neurosurg.* **104,** 221–227.

Janssens, A. C., van Doorn, P. A., de Boer, J. B., Kalkers, N. F., van der Meche, F. G., Passchier, J., and Hintzen, R. Q. (2003). Anxiety and depression influence the relation between disability status and quality of life in multiple sclerosis. *Mult. Scler.* **9,** 397–403.

Janssens, A. C., van Doorn, P. A., de Boer, J. B., van der Meche, F. G., Passchier, J., and Hintzen, R. Q. (2003). Impact of recently diagnosed multiple sclerosis on quality of life, anxiety, depression and distress of patients and partners. *Acta Neurol. Scand.* **108,** 389–395.

Kraüs, J. A., Schütze, C., Brokate, B., Kröger, B., Schwendemann, G., and Hildebrant, H. (2005). Discriminant analysis of the cognitive performance profile of MS patients differentiates their clinical course. *J. Neurol.* **252,** 808–813.

Kroencke, D. C., Denney, D. R., and Lynch, S. G. (2001). Depression during exacerbations in multiple sclerosis: The importance of uncertainty. *Mult. Scler.* **7,** 237–242.

LaPointe, L. L., Maitland, C. G., Blanchard, A. A., Kemker, B. E., Stierwalt, J. A. G., and Heald, G. R. (2005). The effects of auditory distraction on visual cognitive performance in multiple sclerosis. *J. Neuroophthalmol.* **25,** 92–94.

Lengenfelder, J., Bryant, D., Diamond, B. J., Kalmar, J. H., Moore, N. B., and DeLuca, J. (2006). Processing speed interacts with working memory efficiency in multiple sclerosis. *Arch. Clin. Neuropsychol.* **21,** 229–238.

MacAllister, W. S., Belman, A. L., Milazzo, M., Weisbrot, D. M., Christodoulou, C., Sherl, W. F., Preston, T. E., Cianciulli, C., and Krupp, L. B. (2005). Cognitive functioning in children and adolescents with multiple sclerosis. *Neurology* **64,** 1422–1425.

McCabe, M. P., and McKern, S. (2002). Quality of life and multiple sclerosis: Comparison between people with multiple sclerosis and people from general population. *J. Clin. Psychol. Med. Settings* **9,** 287–295.

McCarthy, M., Beaumont, J. G., Thompson, R., and Peacock, S. (2005). Modality-specific aspects of sustained and divided attentional performances in multiple sclerosis. *Arch. Clin. Neuropsychol.* **20,** 705–718.

Minden, S. L., Orav, J., and Reich, P. (1987). Depression in multiple sclerosis. *Gen. Hosp. Psychiatry* **9,** 426–434.

Minden, S. L., Orav, J., and Schildkraut, J. J. (1988). Hypomanic reactions to ACTH and prednisone treatment for multiple sclerosis. *Neurology* **38,** 1631–1634.

Mitchell, A. J., Benito-Leon, J., Gonzalez, J. M., and Rivera-Navarro, J. (2005). Quality of life and its assessment in multiple sclerosis: Integrating physical and psychological components of wellbeing. *Lancet Neurol.* **4,** 556–566.

Mohr, D. C., Hart, S. L., Julian, L., Cox, D., and Pelletier, D. (2004). Association between stressful life events and exacerbation in multiple sclerosis: A meta-analysis. *BMJ* **328**(7442), 731–733.

Moulin, D. E., Foley, K. M., and Ebers, G. C. (1988). Pain syndromes in multiple sclerosis. *Neurology* **38,** 1830–1834.

Mullins, L. L., Cote, M. P., Fuemmeler, B. F., Jean, V. M., Beatty, W. W., and Paul, R. H. (2001). Illness intrusiveness, uncertainty, and distress in individuals with multiple sclerosis. *Rehabil. Psychol.* **46,** 139–153.

Niepel, G., Tench, C. R., Morgan, P. S., Evangelou, N., Auer, D. P., and Constantinescu, C. S. (2006). Deep gray matter and fatigue in MS A T1 relaxation time study. *J. Neurol.* **253,** 896–902.

Nocentini, U., Pasqualetti, P., Bonavita, S., Buccafusca, M., De Caro, M. F., Farina, D., Girlanda, P., Le Pira, F., Lugaresi, A., Quattrone, A., Reggio, A., Salemi, G., *et al.* (2006). Cognitive dysfunction in patients with relapsing-remitting multiple sclerosis. *Mult. Scler.* **12,** 77–87.

Nortvedt, M. W., Riise, T., Myhr, K. M., and Nyland, H. I. (1999). Quality of life in multiple sclerosis: Measuring the disease effects more broadly. *Neurology* **53,** 1098–1103.

Nortvedt, M. W., and Riise, T. (2003). The use of quality of life measures in multiple sclerosis research. *Mult. Scler.* **9,** 63–72.

Patten, B., and Metz, L.SPECTRIMS Study Group (2002). Interferon beta-1a and depression in secondary progressive MS. Data from the SPECTRIMS trial. *Neurology* **59,** 744–746.

Pozzilli, C., Palmisano, L., Mainero, C., Tomassini, V., Marinelli, F., Ristori, G., Gasperini, C., Fabiani, M., and Battaglia, M. A. (2004). Relationship between emotional distress in caregivers and health status in persons with multiple sclerosis. *Mult. Scler.* **10,** 442–446.

Rabins, P. V. (1990). Euphoria in multiple sclerosis. *In* "Neurobehavioral Aspects of Multiple Sclerosis" (S. M. Rao, Ed.), pp. 180–185. Oxford University Press, New York.

Randolph, J. J., Arnett, P. A., and Freske, P. (2004). Metamemory in multiple sclerosis: Exploring affective and executive functions. *Arch. Clin. Neuropsychol.* **19,** 259–279.

Randolph, J. J., Arnett, P. A., and Higginson, C. I. (2001). Metamemory and tested cognitive functioning in multiple sclerosis. *Clin. Neuropsychol.* **15,** 357–368.

Roelcke, U., Kappos, L., Lechner-Scott, J., Brunnschweiler, H., Huber, S., Ammann, W., Plohmann, A., Dellas, S., Maguire, R. P., Missimer, J., Radu, E. W., Steck, A., *et al.* (1997). Reduced glucose metabolism in the frontal cortex and basal ganglia of multiple sclerosis patients with fatigue: A 18F-fluorodeoxyglucose positron emission tomography study. *Neurology* **48**(6), 1566–1571.

Rothwell, P. M., McDowell, Z., Wong, C. K., and Dorman, P. J. (1997). Doctors and patients don't agree: Cross sectional study of patients' and doctors' perceptions and assessments of disability in multiple sclerosis. *BMJ* **314,** 1580–1583.

Rudick, R. A., Miller, D., Clough, J. D., Gragg, L. A., and Farmer, R. G. (1992). Quality of life in multiple sclerosis. Comparison with inflammatory bowel disease and rheumatoid arthritis. *Arch. Neurol.* **49,** 1237–1342.

Sadovnik, A. D., Eisen, K., Ebers, G. C., and Paty, D. W. (1991). Cause of death in patients attending multiple sclerosis clinics. *Neurology* **41,** 1193–1196.

Savettieri, G., Messina, D., Andreoli, S., Bonavita, S., Caltagirone, C., Cittadella, R., Farina, D., Fazio, M. C., Girlanda, P., Le Pira, F., Ligouri, M., Lugaresi, M., *et al.* (2004). Gender-related effect of clinical and genetic variables on the cognitive impairment in multiple sclerosis. *J. Neurol.* **251,** 1208–1214.

Schiffer, R. B., Wineman, M., and Weitkamp, L. R. (1986). Association between bipolar affective disorder and multiple sclerosis. *Am. J. Psychiatry* **143,** 94–95.

Seinelä, A., Hämäläinen, P., Koivisto, M., and Ruutiainen, J. (2002). Conscious and unconscious uses of memory in multiple sclerosis. *J. Neurol. Sci.* **198,** 79–85.

Siegert, R. J., and Abernethy, D. A. (2005). Depression in multiple sclerosis: A review. *J. Neurol. Neurosurg. Psychiatry* **76**(4), 469–475.

Sobel, R. M., Lotkowski, S., and Mandel, S. (2005). Update on depression in neurologic illness: Stroke, epilepsy, and multiple sclerosis. *Curr. Psychiatry Rep.* **7,** 396–403.

Solari, A., Motta, A., Mendozzi, L., Pucci, E., Forni, M., Mancardi, G., Pozzilli, C., and CRIMS Trial (2004). Computer-aided retraining of memory and attention in people with multiple sclerosis: A randomized, double-blind controlled trial. *J. Neurol. Sci.* **222,** 99–104.

Solari, A., and Radice, D. (2001). Health status of people with multiple sclerosis: A community mail survey. *Neurol. Sci.* **22,** 307–315.

Stenager, E., Knudsen, L., and Jensen, K. (1991). Acute and chronic pain syndromes in multiple sclerosis. *Acta Neurol. Scand.* **84,** 197–200.

The IFNB Multiple Sclerosis Study Group and the UBC MS/MRI Analysis Group (1995). Interferon beta-1b in the treatment of multiple sclerosis: Final outcome of the randomized controlled trial. *Neurology* **45,** 1277–1285.

Vleugels, L., Lafosse, C., van Nunen, A., Charlier, M., Ketelaer, P., and Vandenbussche, E. (2001). Visuospatial impairment in MS patients: Nature and possible neural origins. *Mult. Scler.* **7,** 389–401.

Weinshenker, B. G., Bass, B., Rice, G. P., Noseworthy, J., Carriere, W., Baskerville, J., and Ebers, G. C. (1989). The natural history of multiple sclerosis: A geographically based study. 2. Predictive value of the early clinical course. *Brain* **112,** 1419–1428.

Zarei, M. (2006). Clinical characteristics of cortical multiple sclerosis. *J. Neurol. Sci.* **245,** 53–58.

Zéphir, H., de Seze, J., Dujardin, K., Dubois, G., Cabaret, M., Bouillaguet, S., Ferriby, D., Stojkovic, T., and Vermersch, P. (2005). One-year cyclophosphamide treatment combined with methylprednisolone improves cognitive dysfunction in progressive forms of multiple sclerosis. *Mult. Scler.* **11,** 360–363.

Zivadinov, R., Zorzon, M., Locatelli, L., Stival, B., Monti, F., Nasuelli, D., Tommasi, M. A., Bratina, A., and Cazzato, G. (2003). Sexual dysfunction in multiple sclerosis: A MRI, neurophysiological and urodynamic study. *J. Neurol. Sci.* **210,** 73–76.

Zorzon, M., Zivadinov, R., Locatelli, L., Stival, B., Nasuelli, D., Bratina, A., Bosco, A., Tommasi, M. A., Pozzi Mucelli, R. S., Ukmar, M., and Cazzato, G. (2003). Correlation of sexual dysfunction and brain magnetic resonance imaging in multiple sclerosis. *Mult. Scler.* **9**(1), 108–110.

CEREBROSPINAL FLUID ANALYSIS IN MULTIPLE SCLEROSIS

Francisco A. Luque* and Stephen L. Jaffe†

*Neurology Service, Overton Brooks VA Medical Center
Shreveport, Louisiana 71101, USA

†Department of Neurology, Louisiana State University School of Medicine
Shreveport, Louisiana 71103, USA

Although the diagnosis of multiple sclerosis (MS) may be clinically suspect and the magnetic resonance imaging findings compatible, cerebrospinal fluid (CSF) analysis remains mandatory in order to support the diagnosis. This is especially important since our understanding of the defining disease pathogenesis remains incomplete. However, there is no specifically diagnostic CSF test. And until recently, laboratory techniques for CSF analysis had not been rigorously standardized. Unconcentrated CSF without fixative should be used for the determinations of cell count and differential, protein and glucose, lactate, myelin basic protein, and the CSF/serum albumin ratio which is an indicator of blood–CSF barrier disruption. Additionally, CSF immunoglobulin-gamma (IgG) determinations are of major importance and are now included in the MS diagnostic criteria. Testing for oligoclonal IgG bands utilizing isoelectric focusing with IgG immunoblotting, the IgG synthesis rate, and the IgG index should be included. CSF analysis for kappa light chains and IGM may be diagnostically helpful. The search for biomarkers including those possibly present in the CSF which could predict and assess the course as well as response to treatment in a particular MS patient has not yet been successful. CSF immunoglobulin and T-cell/B-cell patterns, soluble HLA class I and II antigens, nitrous oxide metabolites, neurofilament and microtubule components and antibodies, tau protein, 14-3-3-protein, neuronal cell and intercellular adhesion molecules, and chemokines are actively being investigated as MS biomarkers.

INTERNATIONAL REVIEW OF
NEUROBIOLOGY, VOL. 79
DOI: 10.1016/S0074-7742(07)79015-3

I. Introduction

Multiple sclerosis (MS) is considered at least in part a T-cell-mediated disorder of the central nervous system (CNS) capable of producing major disability especially in young adults. Approximately one new case per thousand population occurs per year with a 1.5:1 female preponderance, and an increased incidence is found in temperate climates. A complex interaction of multiple susceptibility genes (e.g., certain HLA genotypes) and environmental factors (e.g., viral and bacterial infections, superantigens) appears to trigger an immune cascade directed against the myelin/oligodendrocyte complex leading to demyelination, and now well-recognized, axonal loss. However, the pathogenesis of MS has very recently become more complicated. With the advent of sophisticated magnetic resonance imaging (MRI) techniques, early axonal and neuronal loss as well as myelin changes in normal-appearing white matter (NAWM) with minimal if any accompanying inflammation have been demonstrated. Thus, a degenerative/metabolic or apoptotic process either resulting from the initial autoimmune inflammatory cascade or the initial environmental trigger must also be posited, with one or the other mechanism becoming dominant, and resulting, for example, in either the inflammatory-marked relapsing-remitting MS (RRMS) or the "noninflammatory"-characterized primary progressive MS (PPMS).

MS subtype classification is based on the particular clinical course, neuroimaging, and neuropathological studies. Approximately 85% of MS presentations are classified as RRMS, with clinical episodes of exacerbations and remissions characterized by partial to almost complete recovery, although a variable degree of persistent deficit often remains. Moreover, RRMS can proceed to a secondary progressive phase (SPMS) with increasing clinical disability and little or no evidence of clinical remission. The disease can also present initially with a primary progressive course of disability (PPMS) without clinical relapses. Acute fulminant syndromes can also occur, including demyelinating pseudotumor, Schilder's disease, Marburg's variant, and Balo's concentric sclerosis. Although it appears that these subtype presentations are dependent on the quality and quantity of the immune cascade including brain tissue predilection, no definite biological markers for subtype or course, including response to treatment, have as yet been consistently and dependably defined.

For most of the last century, the diagnosis of MS was based on the natural history and clinical presentation plus the help of cerebrospinal fluid (CSF) analysis for myelin basic protein, immunoglobulin-gamma (IgG) quantification via protein electrophoresis, and oligoclonal bands, the latter two marking the presence of intrathecal inflammation. Evoked potentials, that is, brain stem auditory and somatosensory evoked potentials, and especially visual evoked potentials were utilized for identifying lesions "disseminated in space" within the CNS. With the advent of sophisticated MRI sequences, demyelinating lesions (plaques) in the CNS have become easily identifiable, and gadolinium enhancement marks the acute lesions. In many instances, a few clinical signs and symptoms plus the MRI findings produce a diagnosis of MS, a clinically problematic situation when the pathophysiology of a disease (or in this case, perhaps different diseases equating to the various subtypes) is only partially known. Furthermore, while MRI is now very sensitive, specificity is a problem with the MRI lesions visualized compatible with many pathophysiological mechanisms and thus different diseases, creating "false positives" for the MS diagnosis. Lesions produced by infectious processes as well as other autoimmune diseases can mimic the appearance of a demyelinating plaque, for example, systemic lupus erythematosus, neuroborreliosis; the *West Nile virus*, Nipah virus, JC, and Hendra virus; *Human herpesvirus 6*, SSPE, HIV, HTLV-1, HSV 1 and 2, and so on. Additionally, even certain CSF findings in granulomatous and vasculitic disorders of the brain can be similar to those found in MS.

In order to improve specificity and reduce the number of false positive diagnoses, a new set of clinical and paraclinical criteria were created for the diagnosis of MS. These include MRI, evoked potentials, and CSF analysis, the latter specifically involving the occurrence of CSF oligoclonal bands (not present in serum) or elevation of the CSF IgG index (McDonald *et al.*, 2001).

II. Basic CSF Analysis

The present techniques for the analysis of CSF are not rigorously standardized and are highly variable regarding specificity, sensitivity, accuracy, and reliability. There is no specific test in the CSF analysis battery which will positively diagnose MS. Moreover, there has been significant variability in techniques and reporting among different laboratories.

As a result in 2005, a specially convened study group reviewed and recommended minimum standards for CSF studies in order to ensure the correctness of the MS diagnosis, that is, of course, in relation to our present knowledge. Although basically concerned with CSF IgG investigations, they reiterated that presently for the most valid MS diagnosis all aspects of the CSF panel should be considered, including cell count, protein, glucose, and lactate levels, among others (Freedman *et al.*, 2005). Opening and closing pressures should be normal unless a fulminant variant of MS is present. Thus, if an acute clinical deficit is present, at least a CAT scan without contrast should be done prior to lumbar puncture to avoid possible herniation due to a large focal lesion producing a cerebral shift.

Routinely, CSF without fixative should be used for cell count and differential. With the present techniques, concentration is no longer necessary even for the immunoglobulin study. This cellular analysis should be performed within a short time (2 h) after the sample is obtained in order to minimize cell damage or even cell destruction. More than a few red blood cells per high powered field (hpf), which often can occur with the trauma of lumbar puncture, are inconsistent with an MS diagnosis and suggest other disease states such as herpes encephalitis or hemorrhagic leukoencephalitis. A very high red blood cell count ($>5 \times 10^6$/ml) with diminution between tubes one and three and absence of xanthochromia usually indicates a traumatic tap which will invariably invalidate attempts at other quantitative measurements. WBC counts, mainly lymphocytes, rarely more than 50/mm^3 are detected in most acute MS relapses, while 34% of MS cases in general will demonstrate a pleocytosis, that is, >5 mononuclear cells/mm^3. In neuromyelitis optica (NMO), eosinophils are found in the CSF (Correale and Fiol, 2004). In MS relapses with early age of onset, 66% of cases will have a pleocytosis and 13% blood–CSF barrier dysfunction as measured by the albumin differential (Pohl *et al.*, 2004).

Similarly, the protein and glucose levels in both the CSF and serum must be measured. CSF protein content is often elevated but rarely exceeds 100 mg/ml, as long as serum protein is within normal range. CSF glucose should be not less than 0.4 of the serum measurement done approximately 1 h prior to CSF collection, thus allowing for equilibration with the intravascular space. CSF lactate can also be measured, when such testing is available, and should be normal, mainly being elevated in an infectious process. In contrast to CSF glucose, lactate evaluation does not require a serum comparison. During acute relapses, especially when brain stem and spinal cord long tracts are affected, myelin basic protein (especially the large protein designated component III) levels will be elevated marking the presence of extensive demyelination. The ratio of CSF/serum albumin varies with age, but when elevated indicates disruption of the blood–CSF barrier. This ratio also allows for the quantification of intrathecal immunoglobulin. In childhood MS, mild CSF pleocytosis, <61 cells/μl, is a common finding. A higher range of pleocytosis associated with marked disruption of the blood–CSF barrier indicates the necessity to search for a different diagnosis (Pohl *et al.*, 2004).

III. CSF Analysis for Immunoglobulins

CSF protein electrophoresis has demonstrated the presence of two or more oligoclonal IgG bands (accepted diagnostic range) primary to the CSF (i.e., not serum "leakage") in >95% of "clinically definite" MS utilizing high-resolution agarose gel electrophoresis and concentrated CSF. The newest technique which tests unconcentrated CSF by isoelectric focusing (IEF) with IgG immunoblotting and has a diagnostic threshold of four bands increases diagnostic sensitivity by 50% (to >90%) with minimal change in specificity, 96–94% (Fortini *et al.*, 2003).

The presence of oligoclonal IgG bands (OCBs) is not specific for MS and occurs in other inflammatory disorders, including the infectious and paraneoplastic types. CSF analysis for OCBs is, however, a sensitive ancillary approach to the diagnosis of MS, and is especially helpful for the early diagnosis of MS. The ability to diagnose MS at an early stage has stimulated the development of medications capable of blocking disease progression. Even in childhood, oligoclonal bands have been demonstrated in up to 92% of MS cases, and this may even be higher with follow-up CSF evaluation. In contrast, quantitative CSF IgG fraction analysis was positive in only 69% of these children, again pointing to qualitative demonstration of OCBs as the "gold standard" for MS diagnosis. Moreover, in this age range where ADEM (acute demyelinating encephalomyelitis) is so important in the MS differential, the lack of OCBs is actually supportive of the ADEM diagnosis (Pohl *et al.*, 2004). Moreover, serum protein electrophoresis has demonstrated the β-1 globulin fraction to be significantly higher in ADEM cases, with a 100% negative predictive value; thus levels >6.5% excluded an MS diagnosis (Chopra *et al.*, 2002). Overall, even in cases where clinical suspicion is high, negative tests for OCBs (done in a reliable laboratory) strongly suggest a different neurological disorder rather than a false negative (Zeman *et al.*, 1996).

When serum–CSF sample pairs were run and the CSF was positive for a single band and the serum negative, repeat lumbar punctures demonstrated that one third of these patients converted from a CSF monoclonal to an oligoclonal band pattern within 6 months. All these converters fell into two groups: (1) early disease including a clinically isolated syndrome (CIS) or (2) progressive disease. On the other hand, many of the nonconverters were diagnosed as having alternative inflammatory disorders. On the basis of these findings, patients with a single band pattern cannot be definitively diagnosed with MS and should have repeated CSF analyses unless the diagnosis can be confirmed by other criteria (Davies *et al.*, 2003).

In the case of PPMS, CSF analysis is the main diagnostic tool. Even early studies suggested that in cases of clinical PPMS, 85% were initially positive for oligoclonal bands, and at that time a less sensitive testing base, polyacrylamide gel electrophoresis, was being used (Freedman, 2004). Several studies have been done to assess the utility of CSF analysis in patients with a CIS with or without typical MS lesions on MRI imaging. When using the newer IEF method with immunoblotting, patients with CIS and normal MRI as well as normal CSF analysis have a low probability of developing MS (Soderstrom *et al.*, 1998).

As of this time, there has been no definite association of oligoclonal bands with a consistent CNS antigen in MS patients. The intrathecal synthesis of antibodies to many different neural antigens as well as infectious agents (measles, rubella, varicella, and zoster viruses especially) produces the CSF IgG oligoclonal band pattern detected. These antigens can be identified by utilizing bacteriophage display libraries containing genetically engineered specific amino acid sequences in their surface proteins and antigen-specific immunoblotting, as well as quantitative detection with an antibody index (Rand *et al.*, 1998; Rostasy *et al.*, 2003). However, there has been no association established between the presence of these specific antigens and the pathophysiology of MS. Localized CNS B-cell expansion accounts for raised immunoglobulin levels and oligoclonal bands, and involve both the mechanisms of clonal expansion and somatic hypermutation. Single-cell polymerase chain reaction has found clonal expansion of B cells as well as receptor revision on analysis of light and heavy chain rearrangements (Qin *et al.*, 1998).

In order to maintain a minimal standard for the analysis of CSF as it relates to the diagnosis of MS and especially in regards to OCBs, an international MS study group developed the following summarized recommendations (Freedman *et al.*, 2005): the best CSF analysis for MS diagnosis is qualitative assessment of CSF IgG performed by using agarose gel IEF together with immunoblotting or immunofixation. This analysis should be done on unconcentrated CSF and must be compared to a serum sample run simultaneously in the same assay in an adjacent track. Similar amounts of IgG from paired serum and CSF samples, approximately 20–1200 ng of IgG (about 3–5 μl of CSF), should be utilized. Recognized positive and negative controls should be run on the same gel as the patient specimens, and the entire run rejected if positive controls are poorly developed or negative controls overdeveloped. Oligoclonal banding reports should be made in terms of one of five recognized staining patterns (Fig. 1) by an individual experienced in the technique.

When necessary, light chain immunodetection may be able to resolve equivocal oligoclonal IgG patterns. Although complimentary, quantitative IgG analysis (synthesis rate >3 in >80% of MS patients; index >0.7 in >86%) does not replace qualitative testing diagnostically; and when performed, nonlinear formulas should be used thus taking into account the integrity of the blood–CSF barrier by determining the ratio of CSF to serum albumin (Q_{alb}). The methodologies for IEF and immunoblotting have been standardized and are commercially available. Specialized laboratories with internal and external assessment controls staffed by experienced clinical biochemists with expertise in this area perform the most reliable testing (Reiber *et al.*, 2003). As noted previously and considering the circular reasoning unavoidable at present, the sensitivity for detecting oligoclonal bands by IEF with immunoblotting in "clinically definite" MS has been reported to range between 90% and >95% (McLean *et al.*, 1990; Ohman *et al.*, 1992), and the specificity appears to be >86% (Paolino *et al.*, 1996).

Kappa and lambda light chain staining can help resolve IgG bands by reducing the polyclonal background, and their presence may also be indicative of IgA and IgM in the CSF. Although kappa light chain bands (KLCBs) are usually associated with IgG synthesis and OCBs, they may occur in isolation but usually appear related to MS dissemination. And several investigators have suggested that

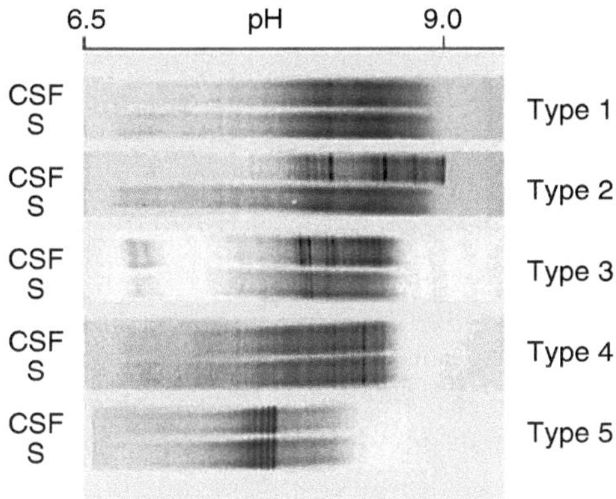

FIG. 1. Isoelectric focusing on agarose gels with immunoblotting. Note that all the oligoclonal bands present are due to IgG. There are five classic patterns: type 1, no bands in cerebrospinal fluid (CSF) and serum (S) sample; type 2, oligoclonal IgG bands in CSF, not in the S sample, indicative of intrathecal IgG synthesis; type 3, oligoclonal bands in CSF (like type 2) and additional identical oligoclonal bands in CSF and the S sample (like type 4), still indicative of intrathecal IgG synthesis; type 4, identical oligoclonal bands in CSF and the S sample, illustrative of a systemic not intrathecal immune reaction, with a leaky or normal or abnormal blood–CSF barrier and oligoclonal bands passively transferred in the CSF; and type 5, monoclonal bands in CSF and the S sample; this is the pattern seen owing to the presence of a paraprotein (monoclonal IgG component). (Courtesy of Freedman *et al.*, 2005.)

KLCBs replace OCBs in McDonald's diagnostic criteria. The isolated finding of lambda light chains in the CSF is presently considered a nonspecific finding. The finding of protein unbound light chains is indicative of intrathecal synthesis since serum unbound light chains are quickly cleared by the kidneys, and thus cannot "spill" into the CSF (Goffette *et al.*, 2004; Krakauer *et al.*, 1998).

The relationship of IgA to MS is unknown, but IgM synthesis occurs at a significantly increased frequency in childhood MS although pathophysiology has not been determined (Pohl *et al.*, 2004). In adults, the finding of increased CSF IgM synthesis has been associated with early conversion to clinically definite MS and more rapid disability accumulation (Villar *et al.*, 2002). Moreover, acutely during MS relapses, increased IgM levels have been reported suggesting the occurrence of an immediately preceding antigenic challenge (Lolli *et al.*, 1991). Correale and Fiol (2004) reported three times the number of IgM-producing cells in the CSF of patients with NMO as compared to RRMS or SPMS (secondary progressive MS). On the other hand, oligoclonal IgG bands are found much less frequently in NMO as compared to the more "typical" MS syndromes; de Seze *et al.* (2003) found OCBs in only 23% of NMO cases versus 88% of MS.

Although the role of OCBs in the pathogenesis of MS remains unresolved, anti-myelin oligodendrocyte glycoprotein (MOG) IgG antibodies appear to be synthesized intrathecally in up to 50% of MS patients, and these individuals have greater pleocytosis and higher IgG indices. Although a role in MS pathogenesis has not been delineated, in the experimental autoimmune encephalomyelitis (EAE) animal model of MS, MOG appears to be the immunodominant target instigating the demyelinating process; and recently MOG-specific IgG has been found in MS plaques. While MBP antibodies occur nonspecifically with many types of CNS damage and are transitory in MS, anti-MOG antibodies have been found to be persistent in MS (Reindl *et al.*, 1999).

OCBs appear to remain stable in MS patients over time even when these patients have been treated with interferon immunomodulation or immunosuppressives or autologous hematopoietic stem cell transplantation. Moreover, these treatments do not appear to affect quantitative IgG production as measured by the IgG index and intrathecal synthesis rate. Moreover, the intrathecal antibody response not only appears to expand as to antigen repertoire over time in MS patients, but also these antibodies then remain persistent. This may be explained by persistence of the intrathecal B-cell clones as a result of constant reactivation by the MS lesion microenvironment as well as increased expression of antiapoptotic proteins found in the CSF. B-cell activity in MS CSF is well established but its clinical significance remains problematic. Several studies have suggested a direct association between the degree of B-cell activation and clinical disability (Petereit and Reske, 2005). In contrast an increased percentage of T cells in the CSF has correlated with decreased MS activity supporting the concept of the protective nature of regulatory T cells (Jensen *et al.*, 2004). Continued analysis of CSF T and B cells in MS utilizing such methodology as flow cytometry, antibodies against surface components, and gene/protein expression via microarray techniques with real-time PCR will hopefully expand diagnostic and prognostic insight as well as the ability to predict and monitor response to therapeutic interventions.

IV. CSF Analysis for Prognostic and Therapeutic Response Biomarkers

Presently biomarkers to assess the course, prognosis, and response to treatment in MS are not available. Although interferon immunomodulators have become part of MS therapy with many experts in the field suggesting their usage at a very early stage of disease in the hopes of blocking progression, these medications are not without side effects and in some cases catastrophic exacerbation of illness results. Therefore, biomarkers that could identify those cases for early therapy, that is those patients who will have a more rapid and severe course, are especially needed. Moreover, biomarkers which will as early as possible (and thus before clinical effectiveness is apparent) signal a positive response to a particular therapeutic mode are also necessary. To further complicate this search, biomarkers valid on a case by case basis rather than statistically valid for an array of patients must be delineated. Sequential serum, saliva as well as CSF samples from MS patients are therefore presently being longitudinally investigated by many institutions including our own, although the required individual case valid biomarker remains elusive.

The ideal marker is one that reflects the unique pathology and/or pathogenesis of MS and thus has very high specificity and sensitivity, is simple to use, has a very reliable and reproducible testing technique which is inexpensive, is minimally invasive, and is consistently valid from patient to patient. Of course, the fact that pathogenesis has not been definitively delineated as yet is a major confounding problem. As discussed in Section III , certain immunoglobulin patterns may have statistical correlations as regards prognosis as well as effects of therapy, but no pattern has been found to approximate universally individual patient validity.

Cell surface HLA class I and especially class II antigens, which are encoded by genes of the major histocompatibility complex (MHC), have been identified as probable mediators of autoimmune pathophysiology. HLA molecules are not only expressed on the cell's surface, but can also be identified in a soluble (shed/secreted) form; and such soluble class I (sHLA-I) and class II (sHLA-II) molecules may play a direct role in immunomodulation or at least act as epiphenomenal markers. Protection from or predisposition toward MS is clearly associated with certain MHC alleles. In addition, CD4+ T lymphocytes, which are present in acute plaques, only recognize antigen(s) in coordination with specific HLA gene products. Moreover, sHLA-I and sHLA-II concentrations correlate quantitatively with specific HLA allotypes. Therefore, soluble HLA molecules may be involved in the pathogenesis of MS (Minagar *et al.*, 2005). sHLA-I and sHLA-G have been found increased in the CSF of MS patients; and CSF sHLA-I appears to increase while serum levels fall during clinical relapses as marked by MRI enhancing lesions. CSF sHLA-G as well as IL-10 increase during remissions raising the possibility of using these CSF molecule fluctuations as markers of immune activation in MS (Fainardi *et al.*, 2002, 2003; Ott *et al.*, 1998). Although CSF sHLA-II has been found present in up to 45% of MS patients, no definite clinical correlation has yet been made (Filaci *et al.*, 1997).

The release of the inflammatory mediator, nitric oxide (NO), by activation of inducible nitrous oxide synthase (iNOS) in microglia and astrocytes, has been implicated in the degeneration of axons as well as block of axonal conduction, disruption of the blood–brain barrier (BBB), and oligodendroglia injury. However, NO also appears to have a salutary immunomodulating effect in MS, and thus iNOS blockade becomes problematic. Electrically active axons degenerate when exposed to low micromolar concentrations of NO, and NO produces reversible inhibition of mitochondrial respiration (Brown *et al.*, 1995; Kapoor *et al.*, 2003; Smith *et al.*, 2001). NO and iNOS are elevated in MS plaques (Bagasra *et al.*, 1995), and high levels of nitric oxide metabolites (nitrates and nitrites) can also be found in the CSF of RRMS and SPMS patients (Johnson *et al.*, 1995; Yuceyar *et al.*, 2001). And CSF cytokines produced by MS patients may induce glial NO production (Xiao *et al.*, 1996). Although correlation with disease activity is weak, CSF NO metabolite concentrations fall after steroid treatment (Yamashita *et al.*, 1997). Recently, CSF NO metabolites were found to be significantly increased during the relapse phase of RRMS as well as in PPMS, and these higher levels correlated with both an elevated IgG index and MRI enhancing lesion volume. Disability progression on follow-up was more severe in those MS patients whose baseline levels of NO metabolites were most elevated (Acar *et al.*, 2003; Rejdak *et al.*, 2004).

Transected, degenerating axons are common features within demyelinated MS plaques. And special MRI techniques such as magnetic resonance spectroscopy and diffusion tensor imaging which can measure axonal loss correlate best with MS clinical disability. Neurofilaments are major axonal cytoskelatal proteins consisting of thee components with different molecular weights: a light chain (NfL) of 68 kDa, an intermediate chain (NfM) of 150 kDa, and a heavy chain (NfH) of 190–210 kDa. The light chain is the core on top of which the other two polymerize to form the neurofilament. The number of neurofilaments as well as their degree of phosphorylation contributes to the diameter of the axon. Nonphosphorylated neurofilaments are seen in active MS lesions (Trapp *et al.*, 1998). Using axonal neurofilament immunostaining, Lovas *et al.* (2000) demonstrated significant reduction in axonal density within plaques and in NAWM. Therefore, neurofilament CSF concentrations may be a potential marker for MS axonal damage and, more generally, neuronal degeneration. Tau protein, a microtubule-associated protein localized in neuronal axons, likewise may be utilizable as a marker of axonal degeneration.

In a 3-year follow-up study of MS patients, baseline CSF NfH levels correlated with EDSS ($r = 0.54$, $p < 0.01$), ambulation index measuring lower limb function ($r = 0.42$, $p < 0.05$), and the nine-hole peg test measuring upper limb function ($r = 0.59$, $p < 0.01$) (Petzold *et al.*, 2005). NfH has been found significantly elevated in the CSF of patients with optic neuritis, although levels were not predictive of MS development; however, levels did correlate with MBP concentrations and MRI enhancing lesion volume in both clinically isolated optic neuritis and clinically definite MS patients supporting previous findings that acute inflammation results in axonal pathology. Moreover, the degree of NfH level change over 3 weeks in MS patients was predictive of clinical outcome at 2 months and 1 year supporting the previously noted correlation between axonal damage and clinical disability (Lim *et al.*, 2004, 2005). NfL CSF levels are increased in all stages of MS but peak almost by a factor of 10 during acute relapses. And glial fibrillary acidic protein (GFAP) CSF levels have been found highest during SPMS probably reflecting profound astrogliosis (Malmestrom *et al.*, 2003; Norgren *et al.*, 2004).

Antibodies against neurofilaments have also been demonstrated in MS CSF. Autoantibodies against NfL correlated with MRI markers of inflammation and especially of atrophy. This correlation was highest in patients with RRMS, again suggesting neurofilament protein disruption and axonal damage early in the course of the disease (Eikelenboom *et al.*, 2003). However, high levels of these antibodies have also been reported in PPMS (Ehling *et al.*, 2004). CSF tau appears significantly elevated in MS, although not specifically since it is found elevated in other inflammatory and noninflammatory neurological diseases; no difference between MS subtypes has been documented. Highest levels were present in acute relapses marked by MRI enhancing lesions and pronounced intrathecal IgG synthesis. A high sensitivity for conversion of a CIS to clinically definite MS was found for a combination of elevated CSF tau and NfH (Brettschneider *et al.*, 2005, 2006).

Microtubules consisting of tubulin subunits are the second major structural components of the axon. Microfilaments, with actin being the major portion, are the main structural components of dendritic spines and presynaptic terminals. Actin and tubulin levels have been found increased in PPMS CSF as compared to RRMS and other controls, and levels appeared to correlate with clinical disability (Semra *et al.*, 2002). However, CSF antibodies to tubulin are not present in MS patients (Silber *et al.*, 2002).

Neuronal cells express 14-3-3 proteins and multiple cellular functions are associated with them including transcriptional and intracellular trafficking control, as well as signal transduction. These proteins are present in the CSF of some patients with MS, seemingly associated with those having greater disability and thus neuronal element destruction (Colucci *et al.*, 2004; Satoh *et al.*, 2003). A study by Martinez-Yelamos *et al.* (2001) suggested that CISs converted more rapidly to clinically definite MS in cases where 14-3-3 protein concentrations were highest.

Neuronal cell adhesion molecule (NCAM) is a member of the immunoglobulin superfamily and an important cell adhesion molecule found in axonal growth cones. Massaro (2002) showed greatly reduced NCAM immunoreactivity in acutely demyelinated plaques, and successfully correlated recovery in CSF NCAM levels in MS patients who improved clinically after treatment with steroids following an acute exacerbation. In contrast, CSF ICAM-1 (intercellular adhesion molecule) as well as TNF-αR (tumor necrosis factor) are elevated in RRMS as compared to controls, without significant concentration changes during clinical relapses and remissions (Alves-Leon *et al.*, 2001).

Finally, the role of cytokines and chemokines (small proinflammatory chemotactic cytokines) in the pathogenesis of MS is under intense investigation with the goals of agonist/antagonist receptor therapeutic intervention, as well as utilization of chemokines as acute relapse markers. The primary inflammatory cytokines appear to activate antigen-specific T cells in the systemic circulation. In support of this, CSF concentrations of TNF-α and interlukins-1, -6, and -8 do not differ from controls. After crossing the BBB as a result of adhesion molecule and matrix metalloproteinase interaction (and MMP-9 levels are elevated in the CSF during acute relapse), these activated T cells in coordination with microglia release several chemokines. Thus, CSF levels of IP-10 (interferon-γ-inducible protein, 10 kDa), Mig (monokine induced by interferon-γ), and RANTES (regulated on activation, normal T cell expressed and secreted) are elevated during an acute MS exacerbation. At the same time as this expression occurs, CSF T cells (CD4+ and CD8+) are found which have expressed receptors for these chemokines, specifically CXCR3 and CCR5 (Th1-type receptors), in significantly greater amount than those T cells in the systemic circulation. This suggests that these chemokine–chemokine receptor interactions may be responsible for bringing proinflammatory cells to specific CNS antigenic sites; and this is supported by the fact that MS acute plaques also contain lymphocytes, systemic macrophages, and microglia expressing these receptors. In contrast, the CSF concentration of chemokine MCP-1 is reduced during acute exacerbations and this chemokine exposure directs T cells toward anti-inflammatory Th2 development, whereas Th1 proinflammatory cell lymphokines appear to trigger the acute MS event. The CCL2 chemokine is consistently reduced in CSF during all stages of MS, while CXCL10 is elevated during the acute relapse; and the receptor for the latter is also the CXCR3 type (Charo and Ransohoff, 2006; Kleine *et al.*, 2003; Mahad *et al.*, 2002; Misu *et al.*, 2001; Sellebjerg and Sorensen, 2003; Sorensen *et al.*, 1999).

References

Acar, G., Idiman, F., Idiman, E., Kirkali, G., Cakmakci, H., and Ozakbas, S. (2003). Nitric oxide as an activity marker in multiple sclerosis. *J. Neurol.* **250,** 588–592.

Alves-Leon, S. V., Batista, E., Papais-Alvarenga, R., and Quirico-Santos, T. (2001). Determination of soluble ICAM-1 and TNFalphaR in the cerebrospinal fluid and serum levels in a population of Brazilian patients with relapsing-remitting multiple sclerosis. *Arq. Neuropsiquiatr.* **59,** 18–22.

Bagasra, O., Michaels, F. H., Zheng, Y. M., Bobroski, L. E., Spitsin, S. V., Fu, Z. F., Tawadros, R., and Koprowski, H. (1995). Activation of the inducible form of nitric oxide synthase in the brains of patients with multiple sclerosis. *Proc. Natl. Acad. Sci. USA* **92,** 12041–12045.

Brettschneider, J., Maier, M., Arda, S., Claus, A., Sussmuth, S. D., Kassubek, J., and Tumani, H. (2005). Tau protein level in cerebrospinal fluid is increased in patients with early multiple sclerosis. *Mult. Scler.* **11,** 261–265.

Brettschneider, J., Petzold, A., Junker, A., and Tumani, H. (2006). Axonal damage markers in the cerebrospinal fluid of patients with clinically isolated syndrome improve predicting conversion to definite multiple sclerosis. *Mult. Scler.* **12,** 143–148.

Brown, G. C., Bolanos, J. P., Heales, S. J., and Clark, J. B. (1995). Nitric oxide produced by activated astrocytes rapidly and reversibly inhibits cellular respiration. *Neurosci. Lett.* **193,** 201–204.

Charo, I. F., and Ransohoff, R. M. (2006). The many roles of chemokines and chemokine receptors in inflammation. *N. Engl. J. Med.* **354,** 610–621.

Chopra, B., Abraham, R., and Abraham, A. (2002). CSF beta-1 globulin—a potential marker in differentiating multiple sclerosis and acute disseminated encephalomyelitis: A preliminary study. *Neurol. India* **50,** 41–44.

Colucci, M., Roccatagliata, L., Capello, E., Narciso, E., Latronico, N., Tabaton, M., and Mancardi, G. L. (2004). The 14-3-3 protein in multiple sclerosis: A marker of disease severity. *Mult. Scler.* **10,** 477–481.

Correale, J., and Fiol, M. (2004). Activation of humoral immunity and eosinophils in neuromyelitis optica. *Neurology* **63,** 2363–2370.

Davies, G., Keir, G., Thompson, E. J., and Giovannoni, G. (2003). The clinical significance of an intrathecal monoclonal immunoglobulin band: A follow up study. *Neurology* **60,** 1163–1166.

de Seze, J., Lebrun, C., Stojkovic, T., Ferriby, D., Chatel, M., and Vermersch, P. (Seze 2003). Is Devic's neuromyelitis optica a separate disease? A comparative study with multiple sclerosis. *Mult. Scler.* **9,** 521–525.

Ehling, R., Lutterotti, A., Wanschitz, J., Khalil, M., Gneiss, C., Deisenhammer, F., Reindl, M., and Berger, T. (2004). Increased frequencies of serum antibodies to neurofilament light in patients with primary chronic progressive multiple sclerosis. *Mult. Scler.* **10,** 601–606.

Eikelenboom, M. J., Petzold, A., Lazeron, R. H., Silber, E., Sharief, M., Thompson, E. J., Barkhof, F., Giovannoni, G., Polman, C. H., and Uitdehaag, B. M. (2003). Multiple sclerosis: Neurofilament light chain antibodies are correlated to cerebral atrophy. *Neurology* **60,** 219–223.

Fainardi, E., Granieri, E., Tola, M. R., Melchiorri, L., Vaghi, L., Rizzo, R., Castellazzi, M., Ceruti, S., Paolino, E., and Baricordi, O. R. (2002). Clinical and MRI disease activity in multiple sclerosis are associated with reciprocal fluctuations in serum and cerebrospinal fluid levels of soluble HLA class I molecules. *J. Neuroimmunol.* **133,** 151–159.

Fainardi, E., Rizzo, R., Melchiorri, L., Vaghi, L., Castellazzi, M., Marzola, A., Govoni, V., Paolino, E., Tola, M. R., Granieri, E., and Baricordi, O. R. (2003). Presence of detectable levels of soluble HLA-G molecules in CSF of relapsing-remitting multiple sclerosis: Relationship with CSF soluble HLA-I and IL-10 concentrations and MRI findings. *J. Neuroimmunol.* **142,** 149–158.

Filaci, G. P., Contini, S., Brenci, S., Gazzola, P., Lanza, L., Scudeletti, M., Indiveri, F., Mancardi, G. L., and Puppo, F. (1997). Soluble HLA class I and class II molecule levels in serum and cerebrospinal fluid of multiple sclerosis patients. *Hum. Immunol.* **54,** 54–62.

Fortini, A. S., Sanders, E. L., Weishenker, M. D., and Katzmann, J. A. (2003). Cerebrospinal fluid oligoclonal bands in the diagnosis of multiple sclerosis. *Am. J. Clin. Pathol.* **120,** 672–675.

Freedman, M. S. (2004). Primary progressive multiple sclerosis: Cerebrospinal fluid considerations. *Mult. Scler.* **10**(Suppl. 1), S31–S34.

Freedman, M. S., Thompson, E. J., Desisenhammer, F., Giovannoni, G., Grimsley, G., Keir, G., Ohman, S., Racke, M. K., Sharief, M., Sindic, C. J., Sellebjerg, F., and Tourtellotte, W. W. (2005). Recommended standard of cerebrospinal fluid analysis in the diagnosis of multiple sclerosis: A consensus statement. *Arch. Neurol.* **62,** 865–870.

Goffette, S., Schluep, M., Henry, H., Duprez, T., and Sindic, C. J. (2004). Detection of oligoclonal free kappa chains in the absence of oligoclonal IgG in the CSF of patients with suspected MS. *J. Neurol. Neurosurg. Psychiatry* **75,** 308–310.

Jensen, J., Langkilde, A. R., Fenst, C., Nicolaisen, M. S., Roed, H. G., Christiansen, M., and Sellebjerg, F. (2004). CD4 T cell activation and disease activity at onset of multiple sclerosis. *J. Neuroimmunol.* **149,** 202–209.

Johnson, A. W., Land, J. M., Thompson, E. J., Bolanos, J. P., Clark, J. B., and Heales, S. J. (1995). Evidence for increased nitric oxide production in multiple sclerosis. *J. Neurol. Neurosurg. Psychiatry* **58,** 107.

Kapoor, R., Davies, M., Blaker, P. A., Hall, S. M., and Smith, K. J. (2003). Blockers of sodium and calcium entry protect axons from nitric oxide-mediated degeneration. *Ann. Neurol.* **53,** 174–180.

Kleine, T. O., Zwerenz, P., Graser, C., and Zofel, P. (2003). Approach to discriminate subgroups in multiple sclerosis with cerebrospinal fluid (CSF) basic inflammation indices and TNF-alpha, IL-1beta, IL-6, IL-8. *Brain Res. Bull.* **61,** 327–346.

Krakauer, M., Nielsen, H. S., Jensen, J., and Sellebjerg, F. (1998). Intrathecal synthesis of free immunoglobulin light chains in MS. *Acta Neurol. Scand.* **96,** 161–165.

Lim, E. T., Grant, D., Pashenkov, M., Keir, G., Thompson, E. J., Soderstrom, M., and Giovannoni, G. (2004). Cerebrospinal fluid levels of brain specific proteins in optic neuritis. *Mult. Scler.* **10,** 261–265.

Lim, E. T., Sellebjerg, F., Jensen, C. V., Altmann, D. R., Grant, D., Keir, G., Thompson, E. J., and Giovannoni, G. (2005). Acute axonal damage predicts clinical outcome in patients with multiple sclerosis. *Mult. Scler.* **11,** 532–536.

Lolli, F., Siracusa, G., Amato, M. P., Fratiglioni, L., Dal Pozzo, G., Galli, E., and Amaducci, L. (1991). Intrathecal synthesis of free immunoglobulin light chains and IgM in initial multiple sclerosis. *Acta Neurol. Scand.* **83,** 239–243.

Lovas, G., Palkovits, M., and Komoly, S. (2000). Increased c-Jun expression in neurons affected by lysolecithin-induced demyelination in rats. *Neurosci. Lett.* **292,** 71–74.

Mahad, D. J., Howell, S. J., and Woodroofe, M. N. (2002). Expression of chemokines in the CSF and correlation with clinical disease activity in patients with multiple sclerosis. *J. Neurol. Neurosurg. Psychiatry* **72,** 498–502.

Malmestrom, C., Haghighi, S., Rosengren, L., Andersen, O., and Lycke, J. (2003). Neurofilament light protein and glial fibrillary acidic protein as biological markers in MS. *Neurology* **61,** 1720–1725.

Martinez-Yelamos, A., Saiz, A., Sanchez-Valle, R., Casado, V., Ramon, J. M., Graus, F., and Arbizu, T. (2001). 14-3-3 protein in the CSF as prognostic marker in early multiple sclerosis. *Neurology* **57,** 722–724.

Massaro, A. R. (2002). The role of NCAM in remyelination. *Neurol. Sci.* **22,** 429–435.

McDonald, W. I., Compston, A., Edan, G., Goodkin, D., Hartung, H. P., Lublin, F. D., McFarland, H. F., Paty, D. W., Polman, C. H., Reingold, S. C., Sandberg-Wolheim, M., Sibley, W., *et al.* (2001). Recommended diagnostic criteria for multiple sclerosis: Guidelines from the international panel on the diagnosis of multiple sclerosis. *Ann. Neurol.* **50,** 121–127.

McLean, B. N., Luxton, R. W., and Thompson, E. J. (1990). A study of immunoglobulin G in the cerebrospinal fluid of 1007 patients with suspected neurological disease using isoelectric focusing and the log IgG-index. *Brain* **113,** 1269–1289.

Minagar, A., Adamashvilli, I., Jaffe, S. L., Glabus, M. F., Gonzalez-Toledo, E., and Kelley, R. E. (2005). Soluble HLA class I and class II molecules in relapsing-remitting multiple sclerosis: Acute response to interferon-beta1a treatment and their use as markers of disease activity. *Ann. N. Y. Acad. Sci.* **1051,** 111–120.

Misu, T., Onodera, H., Fujihara, K., Matsushima, K., Yoshie, O., Okita, N., Takase, S., and Itoyama, Y. (2001). Chemokine receptor expression on T cells in blood and cerebrospinal fluid at relapse and remission of multiple sclerosis: Imbalance of Th1/Th2-associated chemokine signaling. *J. Neuroimmunol.* **114,** 207–212.

Norgren, N., Sundstrom, P., Svenningsson, A., Rosengren, L., Stigbrand, T., and Gunnarsson, M. (2004). Neurofilament and glial fibrillary acidic protein in multiple sclerosis. *Neurology* **63,** 1586–1590.

Ohman, S., Ernerudh, J., Forsberg, P., Henriksson, A., von Schenck, H., and Vrethem, M. (1992). Comparison of seven formulae and isoelectrofocusing for determination of intrathecally produced IgG in neurological disease. *Ann. Clin. Biochem.* **29,** 405–410.

Ott, M., Seidl, C., Westhoff, U., Stecker, K., Seifried, E., Fischer, P. A., and Grosse-Wilde, H. (1998). Soluble HLA class I and class II antigens in patients with multiple sclerosis. *Tissue Antigens* **51,** 301–304.

Paolino, E., Fainardi, E., Ruppi, P., Tola, M. R., Govoni, V., Casetta, I., Monetti, V. C., Granieri, E., and Carreras, M. (1996). A prospective study on the predictive value of CSF oligoclonal bands and MRI in acute isolated neurological syndromes for subsequent progression to multiple sclerosis. *J. Neurol. Neurosurg. Psychiatry* **60,** 572–575.

Petereit, H. F., and Reske, D. (2005). Expansion of antibody reactivity in the cerebrospinal fluid of multiple sclerosis patients—follow-up and clinical implications. *Cerebrospinal Fluid Res.* **2,** 3.

Petzold, A., Eikelenboom, M. J., Keir, G., Grant, D., Lazeron, R. H., Polman, C. H., Uitdehaag, B. M., Thompson, E. J., and Giovannoni, G. (2005). Axonal damage accumulates in the progressive phase of multiple sclerosis: Three year follow up study. *J. Neurol. Neurosurg. Psychiatry* **76,** 206–211.

Pohl, D., Rostasy, K., Reiber, H., and Henafeld, F. (2004). CSF characteristics in early-onset multiple sclerosis. *Neurology* **63,** 1966–1967.

Qin, Y., Duquette, P., Zhang, Y., Talbot, P., Poole, R., and Antel, J. (1998). Clonal expansion and somatic hypermutation of V(H) genes of B cells from cerebrospinal fluid in multiple sclerosis. *J. Clin. Invest.* **102,** 1045–1050.

Rand, K. H., Houck, H., Denslow, N. D., and Heilman, K. M. (1998). Molecular approach to find target(s) for oligoclonal bands in multiple sclerosis. *J. Neurol. Neurosurg. Psychiatry* **65,** 48–55.

Reiber, H., Thompson, E. J., Grimsley, G., Bernardi, G., Adam, P., Monteiro de Almeida, S., Fredman, P., Keir, G., Lammers, M., Liblau, R., Menna-Barreto, M., Sa, M. J., *et al.* (2003). Quality assurance for cerebrospinal fluid analysis: International consensus by an internet-based group discussion. *Clin. Chem. Lab. Med.* **41,** 331–337.

Reindl, M., Linington, C., Brehm, U., Egg, R., Dilitz, E., Deisenhammer, F., Poewe, W., and Berger, T. (1999). Antibodies against the myelin oligodendrocyte glycoprotein and the myelin basic protein in multiple sclerosis and other neurological diseases: A comparative study. *Brain* **122,** 2047–2056.

Rejdak, K., Eikelenboom, M. J., Petzold, A., Thompson, E. J., and Stelmasiak, Z. (2004). CSF nitric oxide metabolites are associated with activity and progression of multiple sclerosis. *Neurology* **63,** 1439–1445.

Rostasy, K., Reiber, H., Pohl, D., Lange, P., Ohlenbusch, A., Eiffert, H., Maass, M., and Hanefeld, F. (2003). *Chlamydia pneumoniae* in children with MS: Frequency and quantity of intrathecal antibodies. *Neurology* **61,** 125–128.

Satoh, J., Yukitake, M., Kurohara, K., Takashima, H., and Kuroda, Y. (2003). Detection of the 14-3-3 protein in the cerebrospinal fluid of Japanese multiple sclerosis patients presenting with severe myelitis. *J. Neurol. Sci.* **212,** 11–20.

Sellebjerg, F., and Sorensen, T. L. (2003). Chemokines and matrix metalloproteinase-9 in leukocyte recruitment to the central nervous system. *Brain Res. Bull.* **61,** 347–355.

Semra, Y. K., Seidi, O. A., and Sharief, M. K. (2002). Heightened intrathecal release of axonal cytoskeletal proteins in multiple sclerosis is associated with progressive disease and clinical disability. *J. Neuroimmunol.* **122,** 132–139.

Silber, E., Semra, Y. K., Gregson, N. A., and Sharief, M. K. (2002). Patients with progressive multiple sclerosis have elevated antibodies to neurofilament subunit. *Neurology* **58,** 1372–1381.

Smith, K. J., Kapoor, R., Hall, S. M., and Davies, M. (2001). Electrically active axons degenerate when exposed to nitric oxide. *Ann. Neurol.* **49,** 470–476.

Soderstrom, M., Ya-Ping, J., Hillert, J., and Link, H. (1998). Optic neuritis: Prognosis for multiple sclerosis from MRI, CSF, and HLA findings. *Neurology* **50,** 708–714.

Sorensen, T. L., Tani, M., Jensen, J., Pierce, V., Lucchinetti, C., Folcik, V. A., Qin, S., Rottman, J., Sellebjerg, F., Strieter, R. M., Frederiksen, J. L., and Ransohoff, R. M. (1998). Expression of specific chemokines and chemokine receptors in the central nervous system of multiple sclerosis patients. *J. Clin. Invest.* **103,** 807–815.

Trapp, B. D., Peterson, J., Ransohoff, R. M., Rudick, R., Mork, S., and Bo, L. (1998). Axonal transection in the lesions of multiple sclerosis. *N. Engl. J. Med.* **338,** 278–285.

Villar, L. M., Masjuan, J., Gonzales-Porque, P., Plaza, J., Sadaba, M. C., Roldan, E., Bootello, A., and Alvarez-Cermeño, J. C. (1998). Intrathecal IgM synthesis predicts the onset of new relapses and a worse disease course in MS. *Neurology* **59,** 555–559.

Xiao, B. G., Zhang, G. X., Ma, C. G., and Link, H. (1998). The cerebrospinal fluid from patients with multiple sclerosis promotes neuronal and oligodendrocyte damage by delayed production of nitric oxide *in vitro*. *J. Neurol. Sci.* **142,** 114–120.

Yamashita, T., Ando, Y., Obayashi, K., Uchino, M., and Ando, M. (1998). Changes in nitrite and nitrate (NO_2/NO_3) levels in cerebrospinal fluid of patients with multiple sclerosis. *J. Neurol. Sci.* **153,** 32–34.

Yuceyar, N., Taskiran, D., and Sagduyu, A. (1998). Serum and cerebrospinal fluid nitrite and nitrate levels in relapsing-remitting and secondary progressive multiple sclerosis patients. *Clin. Neurol. Neurosurg.* **103,** 206–211.

Zeman, A. Z., Kidd, D., McLean, B. N., Kelley, M. A., Francis, D. A., Miller, D. H., Kendall, B. E., Rudge, P., Thompson, E. J., and McDonald, W. I. (1998). A study of oligoclonal band negative multiple sclerosis. *J. Neurol. Neurosurg. Psychiatry* **60,** 27–30.

MULTIPLE SCLEROSIS IN ISFAHAN, IRAN

Mohammad Saadatnia,[*,†,‡,§,¶] Masoud Etemadifar,[*,†,‡,§,¶]
and Amir Hadi Maghzi[§,¶]

[*]Department of Neurology, Isfahan University of Medical Sciences, Isfahan 81744, Iran
[†]Iranian Neurological Science Research Center, Alzahara Hospital, Isfahan 81746, Iran
[‡]Isfahan Medical Education Research Center, Isfahan University of Medical Sciences
Isfahan 81744, Iran
[§]Isfahan Research Committee of Multiple Sclerosis, Isfahan MS Society
Isfahan 8158844799, Iran
[¶]Medical School, Isfahan University of Medical Sciences, Isfahan 81744, Iran

Background: This survey was planned to study the prevalence and incidence of multiple sclerosis (MS) in Isfahan, Iran and to describe the clinical features of MS in general and in specific subgroups of patients (early-onset, late-onset, familial, and conjugal cases) and to compare our results with other reports.

Methods: A cross-sectional study was conducted from April 5, 2003 to July 31, 2006. All patients known to have definite MS according to McDonald's criteria, alive, resident within Isfahan (a large province of Iran) and members of Isfahan

DOI: 10.1016/S0074-7742(07)79016-5

0074-7742/07 $35.00

MS Society (IMSS) were included in the study. Demographic and case-related information were recorded. A total number of 1718 definite MS patients (388 men and 1330 women) were identified from IMSS database.

Results: The overall period prevalence of MS was 43.8/100,000. Among men the prevalence was 19.2 (95% CI: 17.4–21.2)/100,000 and among women 69.6 (95% CI: 66–73.4)/100,000. A female preponderance of 3.4 existed among these patients. In the year 2005, 143 new cases were diagnosed, resulting in an incidence rate of 3.64/100,000. The mean age of onset was 25.36 ± 8.6 years (range 5–63 year), and mean duration of disease was 7.1 (±5.2) years for men and 6.7 (±5) years for women. Sensory and visual disturbances were the most common initial presentations with a prevalence of 51.7% and 47.5%, respectively. Cases identified include: early-onset MS (less than 15 years old at onset) with 87 cases (5%), late-onset MS (over 50 years old at onset) with 20 cases (1.1%), familial MS with 209 cases (12.2%), and conjugal MS with 6 cases (0.5%).

Conclusion: Isfahan is a medium- to high-risk area for MS, with prevalence higher than what has previously been reported, possibly because of an increase in the incidence rate. Clinical and demographic characteristics were similar to other reports; however, some differences existed.

I. Introduction

Multiple sclerosis (MS) is an immune-mediated demyelinating disease of human central nervous system (CNS) of unknown origin (Frohman *et al.*, 2006). Despite intensive epidemiological studies during recent decades the geographical distribution of MS is only marginally known. Several studies of different ethnic groups and over different geographical areas have shown significant differences in the worldwide distribution of MS (Compston, 1998; Kurtzke, 1975, 1980). It is suggested that there is a north-south gradient in the prevalence of MS in the Northern Hemisphere and a south-north gradient in the Southern Hemisphere (Compston, 1998; Kurtzke, 1975, 1980). MS is reported to be rare in tropical areas but relatively frequent in temperate regions; however, some exceptions exist (Marrie, 2004). For MS, the world is divided into high- (30/100,000), intermediate- (5–25/100,000), and low-risk areas with a prevalence of less than 5/100,000 (Kurtzke, 1975, 1980). Asia, Africa, Alaska, Greenland, and Central America are classified as low-risk regions (Compston, 1998; Kurtzke, 1975, 1980; Poser, 1994; Rosati, 2001). As a result, Iran and other Middle East countries are located in the low-risk area (Kurtzke, 1980; Poser, 1994).

Studies have demonstrated that the worldwide prevalence of MS is much higher than what was previously thought (Rosati, 1994) and that exceptions exist to the previously explained north-south gradient (Rosati, 1994, 2001). It appears that the distribution of MS is very uneven, because of great differences not only among areas at the same latitude but also within the countries (Rosati, 1994, 2001).

Whereas abundant epidemiological studies have been performed in Western countries, few studies have been carried out in Iran and Middle East, where there are environmental and ethnical differences (Ale-Yasin *et al.*, 2002; Alshubaili *et al.*, 2005; Daif *et al.*, 1998; El-Salem *et al.*, 2006; Etemadifar *et al.*, 2006; Kahana *et al.*, 1994; Kalanie *et al.*, 2003). According to the north-south gradient of MS distribution, Isfahan (a province of Iran) is located in the low-risk area (prevalence of less than 5/100,000); however, the results of the only population-based study in Isfahan (Etemadifar *et al.*, 2006) demonstrated that the prevalence of MS is 35.5/100,000 and Isfahan could be considered a medium- to high-risk area for MS; considering the geographical latitude of the area, this risk is higher than expected. These data are similar to the prevalence rate of MS in Parsis population, who left Iran and migrated to India between the seventh and tenth centuries AD (Bharucha *et al.*, 1988; Wadia and Bhatia, 1990). Studies in Arab countries also demonstrated that MS prevalence is on the rise and MS risk in these regions is medium (Al-Din *et al.*, 1990, 1995; Al-Rajeh *et al.*, 1993; El-Salem *et al.*, 2006; Yaqub, 1988). However, considering results of studies from Japan and China (Kira *et al.*, 1996; Lau *et al.*, 2002), it can be concluded that not only Asians have a lower prevalence of MS in comparison to Western countries but also differences exist in the pattern of the disease.

Some of the Western Hemisphere epidemiological studies have shed light on some controversial aspects of MS. Familial studies of MS patients suggest that relatives of MS patients are at higher risk of developing MS, providing evidence that the genetic background play a role in triggering the disease cascade (Chataway, 1998, 2001; Ebers, 2000b; Robertson, 1996). Studies focusing on conjugal pairs with MS (both spouses have MS and at least one has developed MS after marriage) suggest that there is no increased risk of developing MS for the spouses of MS patients (Ebers, 2000a); however, controversies exist (Cocco *et al.*, 2001; Hawkes, 2001, 2002; Maghzi *et al.*, 2007). But nearly all studies on offspring of such matings recommend higher risk of developing MS for these cases, compared to offspring with only one affected parent (Ebers, 2000a; Robertson, 1997). Studies about late-onset MS (LOMS) (defined as presentation of first clinical symptoms in patients over 50 years old) indicate poor prognosis and a rapid disability progression for most patients (Delalande *et al.*, 2002; Hoodge and Redekop, 1992; Martinelli *et al.*, 2004). On the other hand, studies on patients with early-onset MS demonstrated that in most cases patients have a relapsing-remitting (RR) course with sensory symptoms as onset symptom, complete recovery, and a slower pace of progression (Duquette *et al.*, 1987). However, long-term follow-up of these patients demonstrated that disease course and prognosis are not different from adult-onset MS (Deryck *et al.*, 2006).

While a great deal is known about the epidemiology of MS in Western countries, relatively few studies have been carried out in developing nations, where genetic factors, availability of medical services, and diagnostic procedures as well as environmental exposures are different (Etemadifar *et al.*, 2006). Further determination of the prevalence of MS in these countries can assist us to learn more about the natural course of MS in these regions of the world and develop more effective treatments. Therefore, this survey was planned to study the prevalence, incidence, and risk of MS in Isfahan, Iran; to describe the clinical features of the disease in general and in specific subgroups of patients (early-onset, late-onset, familial, and conjugal cases); and to compare these data with other reports.

II. Materials and Methods

A. Study Area

This study was performed in Isfahan, a very large province (107,003 km^2) located in central part of Iran. Isfahan is 1590 m above sea level, between latitudes 30° and 34° north of the equator and longitude 49°–55° east, and had a population of almost 3,923,255 in 1999 (2,013,492 males and 1,909,763 females). A high proportion of Isfahan population is young with 51.4% under 20 and only 7.2% older than 60 in 1999, so the population is young. The climate is dry and temperate with a mean daily temperature of 5.3 °C in January and 27.2 °C in August. The mean annual precipitation is 102.5 mm (reference period 1972–2000) (Etemadifar *et al.*, 2006). The study population included all residents of Isfahan Province during the period from April 5, 2003 to July 31, 2006. The population structure and socioeconomic status of Isfahan are similar to the rest of the country. The health services are provided by private physicians and hospitals, district health centers, government, university hospitals, and clinics. Forty neurologists and six neurological departments serve the study area.

B. Data Source

On April 5, 2003, a registry of MS patients was created at the Isfahan MS Society (IMSS), representing the only referral center for MS patients in the province. The target population of IMSS is all residents of Isfahan Province who were diagnosed with MS by neurologists. All neurological services in Isfahan mail discharge letters on all patients who have been admitted with MS to the registry. Information included in the database was also obtained from the local private neurologists. Since 2005, each patient who is diagnosed with definite MS in our province must register with IMSS in order to obtain supports for insurance, laboratory investigations, treatment, and rehabilitation. As a result, nearly all

MS patients in Isfahan Province are now registered. The neurologists of IMSS evaluate each case for diagnostic classification before the case is entered into database. Whenever the IMSS receives new information on a previously registered case, reclassification is considered.

A cross-sectional case register study was conducted from April 5, 2003 to July 31, 2006. All patients with definite MS who were residing within Isfahan were included in the study. The study did not include those MS patients who had moved out of the area, could not be traced, or had passed away. The patients' database included all members of IMSS. Demographic information obtained from patients included name, address, age, gender, marital status, and education. Other pertinent information collected includes clinical features [time of onset, manifesting symptom(s), the results of cerebrospinal fluid (CSF) examination, and magnetic resonance imaging (MRI) findings], occupation, and place of residence. The entire study group was Iranian. We requested participants' permission to review their medical records. All participants consented. The time of onset was defined as the time of the first symptom that could be ascribed to MS. Since MS is a new high-prevalence disease in Iran, there are now few MS associations all over the country; furthermore, other MS societies in other parts of Iran have not yet covered an acceptable percentage of all MS patients in their area; therefore, we decided to use only the database of IMSS.

The Iranian Central Bureau of Statistics (ICBS) regularly conducts population censuses, including in 1999, and also estimates the average annual population for the in-between years based on several registries available in the country. The total population of Isfahan, used for calculating the prevalence rate, was that calculated by the ICBS on the basis of 1999 census data.

C. Diagnostic Criteria

All patients as well as their brain MRI findings were examined and assessed by their treating neurologists. Patients were classified according to the recommended diagnostic criteria for MS by the International panel in MS diagnosis (McDonald *et al.*, 2001). These criteria contain three broad groups: "definite MS," "possible MS," and "not MS." Those patients with sufficient investigations to fulfill the definite MS criteria were included. Primary progressive (PP) cases were diagnosed according to diagnostic criteria for PPMS (Thompson *et al.*, 2000), while disability was assessed by Expanded Disability Status Scale (EDSS) (Kurtzke, 1983). For each patient we also calculated the progression index (PI) by dividing the current EDSS by the disease duration (years), which is a measure to calculate the progression of the disease.

Late-onset and early-onset MS were defined as the first presentation symptom, occurred after the age 50 and before the age 15, respectively. Conjugal MS was defined as a condition in which both spouses have MS and at least one should have developed the disease after marriage. Institutional ethical approval was obtained and completion of the registration questionnaire was considered to imply informed consent.

D. Statistical Analysis

MS prevalence rate/100,000 persons was calculated by gender and age groups. The analyses were done on a personal computer using SPSS for Windows (SPSS Inc., Chicago, IL) and confidence interval analysis software (Gardner and Altman, 1989). All tests for statistical significance were two-tailed, with the level of significance at $p < 0.05$.

III. Results

A. Clinical and Demographical Features

A total number of 1718 patients (1330 females, 77.4%; 388 males, 22.6%) were identified. The mean (±SD) age of onset was 25.36 (±8.6) years (ranging from 5 to 63 years). The mean duration of disease was 7.1 (±5.2) years for men and 6.7 (±5) years for women. Table I summarizes the sociodemographic features and the disease patterns of the patients. Most of the patients had an RR course of the disease (87.8%, $n = 1510$), while 110 (6.4%) had a secondary progressive (SP) course and 98 (5.7%) had a PP course of the disease.

Clinically, the most common initial presentation was sensory impairment found in 51.7% of patients, followed by optic neuritis in 47.5%, weakness in 34.6%, and ataxia and brain stem symptoms (diplopia, vertigo, or dysarthria) in 24.3%. In addition, Seizure, as a presenting symptom, was observed in 0.46% of the patients. About 28% of the patients had multiple system involvement at the onset. The follow-up clinical manifestations of our studied patients were motor disturbances in 45.2%, ataxia and brain stem symptoms in 42.6%, optic neuritis in 42.5%, and sensory symptoms in 23%. The mean current EDSS for all patients was 2.7 (±1.8), so the PI would be 0.4.

B. Prevalence and Incidence

The overall prevalence of MS among men was 19.2/100,000 (95% CI: 17.4–21.2) and 69.6 per 100,000 (95% CI: 66–73.4) among women. Women were 3.4 times more likely to develop MS. Specific prevalence rates for male and female in different age groups are presented in Table II. MS rate was the highest in males (68.3%) and females (72.1%) between 20 and 39 years of age. In the year 2005, 143 new cases were diagnosed, resulting in an incidence of 3.64/100,000.

TABLE I
SOCIODEMOGRAPHIC FEATURES AND DISEASE COURSE FOR STUDIED PATIENTS

Characteristics		Male No. (%)	Female No. (%)
Age at onset (years)	≤15	17 (4.3)	70 (5.2)
	>15	371 (95.6)	1260 (94.7)
	<50	377 (97.1)	1321 (99.3)
	≥50	11 (2.8)	9 (0.6)
Family history of MS	No	325 (85.3)	1150 (88.3)
	Yes	56 (14.7)	153 (11.7)
Pattern of disease	Relapsing-remitting	310 (79.8)	1200 (90.2)
	Primary progressive	28 (7.2)	70 (5.2)
	Secondary progressive	28 (7.2)	82 (6.1)
Education	Primary or below	147 (37.8)	503 (37.8)
	Secondary	131 (33.7)	453 (34)
	Matriculation or above	111 (28.4)	373 (28)
Marital status	Single	122 (32.3)	364 (28.7)
	Married	247 (65.3)	888 (67.9)
	Divorced/widowed	9 (2.4)	56 (4.3)
Occupation	Housewife/unemployed	63 (16.5)	839 (70.2)*
	Student	20 (5.4)	104 (8.7)
	Employed	288 (78)	242 (21.1)
Pregnancy history	Yes	–	815 (72.0)
	No	–	317 (28.0)
Age [Mean (±SD)]	Year	33.7 (±9.6)	31.8 (±9.2)*
Duration [Mean (±SD)]	Year	7.1 (±5.2)	6.7 (±5)
EDSS [Mean (±SD)]		2.7 (±1.7)	2.8 (±2.1)

Note: Total of each variable may vary because of missing value.
*$p < 0.05$.

According to ICBS report in Isfahan there are 725,806 single males (never married) and 548,983 single (never married) females, in addition the total number of ever married males and females is 850,912 and 937,493, respectively. In our study, we found 122 and 364 never married males and females, respectively. Additionally, there were 256 ever married males and 944 females. Therefore, ever married men 1.8 times and ever married women 1.5 times were more likely to develop MS than never married ($p < 0.05$).

C. Laboratory Findings

Among the 26 consecutive MS patients who had undergone lumbar puncture, oligoclonal bands were present only in 57.7% of patients and IgG index was higher than 0.7 in 80.8% of patients.

TABLE II

Age-Specific Prevalence Rates (Case/100,000) for Male, Females, and the Total Number of Patients in Isfahan, Iran in the Year 2003 to July 2006

Age (year)	Male				Female				Total			
	Population	Cases	prevalence	(95% CI)[a]	Population	Cases	Prevalence	(95% CI)	Population	Cases	Prevalence	(95% CI)
10–19	538,461	36	6.68	4.8, 9.25	516,129	92	17.8	14.5, 21.8	1,054,590	128	12.1	10.2, 14.4
20–29	347,458	104	29.9	24.7, 36.3	332,470	496	149.2	136.6, 162.9	679,928	600	88.2	81.4, 95.5
30–39	268,000	161	60	51, 70	251,521	463	184	168, 216	519,521	624	120	111, 130
40–49	172,414	67	38.8	30.6, 49.3	160,334	215	134	117, 153	332,748	282	84.7	75.4, 95.2
≥50	250,000	20	8	5.1, 12	232,323	57	24.5	18.9, 31.7	482,323	77	15.9	12.7, 19.9
Total	2,013,492	388	19.2	17.4, 21.2	1,909,763	1330	69.6	66, 73.4	3,923,255	1718	43.8	41.7, 46

[a]95% Confidence interval.

There were no significant differences in features of MS, gender, onset age, disease duration, disease severity, and onset symptoms between oligoclonal bands positive and negative patients. However, a lower PI was observed in negative oligoclonal bands cases ($p = 0.08$). In addition, IgG index was correlated with PI ($p < 0.05$, $r = 0.42$).

D. Early-Onset MS

Among 1718 patients, 87 (5%) had developed MS under the age 15. The mean age of onset was 13.87 (±3.2) year with a range from 5 to 15 years. The female ($n = 70$) to male ($n = 17$) ratio was 4.11:1. The disease course in 69 patients (79.3%) was RR, in 14 (16.1%) was SP and in 4 (4.5%) was PP. The mean EDSS was 3.1 (±1.1). First presentations in 31 (35.6%) cases were polysymptomatic, whereas in the remaining 56 (64.3%) cases, the initial clinical picture was monosymptomatic. In monosymptomatic group, optic neuritis was observed in 27 patients (31%), ataxia and brain stem-cerebellar disorders in 13 (14.9%), motor disturbances in 11 (12.6%), and sensory symptoms in 8 (9.1%). The most common MRI findings were periventricular white matter lesions in 75 patients (86.2%), followed by brain stem lesions in 14 (16.9%), spinal cord lesions in 12 (13.7%), and finally, cerebellar lesions in 10 (11.5%). The mean PI for the group was 0.45 (±0.35).

E. Late-Onset MS

Among 1718 MS patients, 20 (1.1%) had their first clinical manifestation after the age 50. The mean age of onset was 53.77 (±3.7) with a range of 50–63. The mean age of onset in males and females was not significantly different (54.2 ± 4.4 vs 53.25 ± 2.7, $p > 0.05$). Female ($n = 9$) to male ($n = 11$) ratio was 0.8:1. The disease course was RR in 12 patients (60%), SP in 4 (20%), and PP in 4 (20%). In both sexes, the most prevalent disease course was RRMS (44.4% in female vs 72% in male). In RRMS patients, the mean number of relapses per year was 1.4 (±0.51). The mean EDSS was 2.7 (±1.3). The most common MRI finding was periventricular plaques in 18 patients (90%) followed by brain stem plaques in 6 (30%), cerebellar plaques in 3 (15%), and spinal plaques in 2 (10%). The first clinical presentations were brain stem symptoms in 40%, sensory symptoms in 30%, cerebellar symptoms in 25%, optic neuritis in 20%, motor disturbances in 15%, cognitive impairments in 15%, and spinal symptoms in 5%. Polysymptomatic and monosymptomatic first symptoms were seen in 35% and 65% of LOMS patients, respectively. The mean PI for the group was 0.65 (±0.4).

F. Familial MS

In 209 patients (12.2%), family history of MS existed; of these, 79 (37.8%) had first-degree affected relatives, 34 (16.3%) had second-degree affected relatives, and finally, 96 (45.9%) had third-degree affected relatives.

G. Conjugal MS

Among 1153 (67.1%) married MS patients, 7 conjugal pairs were detected of whom a pair was excluded from the study because the onset of symptoms was before marriage in both spouses. The mean age of participants for wives and husbands was 33.16 (±10.76) and 38 (±5.97) years, respectively. The mean age at onset for wives and husbands was 27.67 (±10.45) and 36.67 (±8.44) years, respectively; only in one case the onset of disease was earlier in the husband; the mean duration of disease was 5.5 (±2.14) years for wives and 1.34 (±1.36) years for husbands; mean EDSS was 3.25 (±2.1) in wives and 1.75 (±1.46) in husbands; and PI for wives and husbands was 0.59 and 1.3, respectively. Disease course was PP in a husband, SP in three wives, and RR in the remaining cases. Initial presentation of disease was visual deficits in two wives and in one husband; vertigo and hearing disturbances in one husband and limb paresis in eight remaining cases. The location of demyelinated plaques shown on MRI scans was periventricular in all cases, followed by brain stem plaques in a wife, and spinal cord plaques in three wives and a husband. The disease course, initial symptom, age at onset, EDSS, PI, and MRI pattern were not similar within couples. The mean age of offspring was 14.53 (±6.95); and none of them (seven sons and six daughters) were diagnosed to have MS. A part of these results have been recently published by Maghzi *et al.* (Maghzi *et al.*, 2007)

IV. Discussion

A. Prevalence and Incidence

Previously, Iran was regarded as a low-risk area for MS (Kurtzke, 1980; Poser, 1994). As it has been shown in a study in Isfahan (Etemadifar *et al.*, 2006), the prevalence of MS was reported to be 35.5/100,000 and now according to this study it is 43.8/100,000, as a result this province is now considered a medium- to high-risk area for MS. Isfahan contains a population of relatively homogeneous racial background. Mostly, all people are white Persian Caucasian known as Middle Eastern.

An underestimation of prevalence rate might have occurred as there was a possibility of missing some cases receiving treatment in the private sector, though the number of such patients was likely to be negligible.

The diagnosis of MS is enhanced by: (1) increased familiarity of physicians and patients with MS, like everywhere else in the world, (2) recent advances in MRI neuroimaging and availability of such technology in Iran, and (3) ordering too many MRIs of the brain and spinal cord by physicians, which may lead to accidental diagnosis of some cases. However, the increasing prevalence rate during recent years in Isfahan is more likely to be because of the increase in the incidence of MS by itself, which is now comparable to the incidence rate in Western countries (Rothwell and Charlton, 1998; Savettieri *et al.*, 1998).

Prevalence rates in various studies from around the world show considerable differences (Table III), depending on variations in their denominator characteristics,

TABLE III
SUMMARY OF MS PREVALENCE AMONG DIFFERENT POPULATION

Site	Prevalence/100,000			References
	Male	Female	Total	
United States	48.0	177.0	113.0	Neuberger *et al.*, 2004
Olmsted County, USA	111.0	239.0	177.0	Mayr *et al.*, 2003
Southampton, UK			99.0	Roberts *et al.*, 1991
Sussex, UK			111.0	Rice-Oxley *et al.*, 1995
Cambridge, UK			130.0	Mumford *et al.*, 1992
North Cambridgeshire, UK			118.0	Robertson *et al.*, 1995
South East Wales, UK			113.0	Swingler *et al.*, 1987
South East Wales, UK			117.0	Swingler and Compston, 1988
Guernsey and Jersey, UK	45.5	139.5	120.2	Sharpe *et al.*, 1995
	37.3	97.5	95.6	
Leeds, UK			84.0	Ford *et al.*, 1998
Rochdale, UK			122.0	Shepherd and Summers, 1996
Devon, UK			118.0	Fox *et al.*, 2004
Vasterbotten, Sweden	105.0	202.0	154.0	Sundstrom *et al.*, 2003
Genoa, Italy	67.0	118.0	94.0	Solaro *et al.*, 2005
Monreale city, Italy	48.5	93.0	71.2	Ragonese *et al.*, 2004
Ferrara, Italy			46.1	Granieri *et al.*, 1985
Belgrade, Yugoslavia	28.2	54.1	41.5	Pekmezovic *et al.*, 2001
Newfoundland and Labrador, Canada			94.4	Sloka *et al.*, 2005
Las Palmas, Spain			73.8	Aladro *et al.*, 2005
Bajo Aragon, Spain			75.0	Modrego and Pina, 2003
Newcastle, Australia			59.1	Barnett *et al.*, 2003
Austria			98.5	Baumhackl *et al.*, 2002
Asahikawa, Japan	7.5	12.6	10.2	Itoh *et al.*, 2003
Hong Kong	–	–	0.77	Lau *et al.*, 2002
Middle East, Jordan	–	–	39	El-Salem *et al.*, 2006
Middle East, Kuwait	–	–	31.15	Alshubaili *et al.*, 2005
Middle East, Israel	–	–	68	Kahana *et al.*, 1994
Middle East, Saudi Arabia	–	–	25	Daif *et al.*, 1998
Middle East, Iran	15.3	56.7	35.5	Etemadifar *et al.*, 2006

case ascertainment accuracy, diagnostic criteria, ethnicity, definitions of prevalence and incidence, and the examination of patients at different stages in the natural history of MS.

Considering previous and recent reports from the Middle East, apart from Israel, a striking increase in the prevalence rate of MS could be observed. For instance, in 1977, a prevalence rate of 7/100,000 was reported in Jordan (Kurdi *et al.*, 1977), while in a study in 1993, it was reported to be 32/100,000 (Al-Din *et al.*, 1995), and a more recent study reported this rate as high as 39/100,000 (El-Salem *et al.*, 2006). There are some reports from Kuwait, which demonstrate an increase in the prevalence rate of MS from 10/100000 in the year 1988 (Al-Din *et al.*, 1990) to 14.77/100,000 in 2000 (Alshubaili *et al.*, 2005). In addition, in Iraq the prevalence rate in 1969 was reported to be 4/100,000 (Hamdi, 1975), but a recent study reports an increase in the frequency of the disease in Iraq (Al-Araji and Mohammed, 2005). In Saudi Arabia, the prevalence rate of MS grew from 8/100,000 in 1977 (Yaqub, 1988) to 25/100,000 in 1998 (Daif *et al.*, 1998). Reports from Israel revealed a prevalence of up to 68/100,000 in some regions (Kahana *et al.*, 1994). Therefore, it can be concluded that along with a general worldwide increase in the incidence of MS, there has been an increase in new cases of MS in Middle East as well.

Clinical and demographical characteristics of MS patients in Iran are similar to those in Western countries, yet some differences exist. Despite such differences, a more prevalent RR course was observed (90.2%). We believe this is biased by the fact that the mean duration of disease is lower than other studies, so the number of RR cases would be more, while the number of SPMS patients would be lower. In addition, the rate of observed PP course among MS patients was lower than reported in other parts of the world (Montalban and Rio, 2001; Thompson *et al.*, 1997). Furthermore, in-line with our results, another study from Iran reported a PP rate of 7% among Iranian patients (Kalanie *et al.*, 2003).

B. Laboratory Findings

The frequency of presence of oligoclonal IgG bands in the CSF was rather low (57.7%) in the present study as compared to that reported in Western countries (Miller *et al.*, 1983). We cannot explain this phenomenon, as all included patients were diagnosed as definite MS according to McDonald's criteria and the method used for analysis was not defective. A possible explanation for the lower frequency of oligoclonal IgG bands among our patients may be ethnic differences. Other studies carried out in Japan and Jordan (El-Salem *et al.*, 2006; Tanaka *et al.*, 2005; Yamasaki *et al.*, 1999) also reported a relatively low percentage of MS patients with oligoclonal IgG bands in their CSF. On the basis of these findings, it may be concluded that the presence of oligoclonal IgG bands is less frequent in the CSF of Asian MS patients than Western MS patients; however, more accurate studies with larger samples are required.

C. Early Onset

Several reports have demonstrated that 0.4–5.6% of all MS patients have their symptom(s) before the age of 16 (Cole and Stuart, 1995; Ghezzi *et al.*, 1997). In our study, a high percentage of early-onset patients (5%) were observed.

Other studies in this field reveals that polysymptomatic disease onset is more frequent in early-onset patients (48.9%) than in adult-onset patients (12%), whereas monosymptomatic onset occurs more frequently in adults (86%) than children (49–62%). (Cole and Stuart, 1995; Ghezzi *et al.*, 1997; Guilhoto *et al.*, 1995; Hanefeld, 1995; Ozakbas *et al.*, 2003; Perniola *et al.*, 1991; Pinhas-Hamiel *et al.*, 1998; Sevon *et al.*, 2001). In this study, polysymptomatic onset was more common in early-onset patients (35.6%) compared to adult-onset patients (12%). However, in this group of patients, monosymptomatic onset (64.3%) was more frequent than polysymptomatic onset (35.6%). In the monosymptomatic group contrary to previous reports in which brain stem syndromes or cerebellar ataxia (Ruggieri *et al.*, 1999) were the most common onset symptom, optic neuritis was the most frequent initial presentation (31%). Similar to other early onset MS studies (Cole and Stuart, 1995; Ghezzi *et al.*, 1997; Guilhoto *et al.*, 1995; Perniola *et al.*, 1991; Ruggieri *et al.*, 1999; Sevon *et al.*, 2001), brain stem syndromes, cerebellar involvement, and optic neuritis were the most common.

In the present study, RRMS was the most common course of the disease, followed by SPMS, and PPMS; our finding was the same as other reports which suggest that RRMS is the most common form (32–84%, mean = 67%), followed by SPMS (7–61%, mean = 24.4%), and finally PPMS (5–15%, mean = 8.6%), (Cole and Stuart, 1995; Ghezzi *et al.*, 1997; Perniola *et al.*, 1991; Ruggieri *et al.*, 1999).

D. Late-Onset MS

In the present study, 1.1% of MS patients had LOMS. Compared to other studies (Delalande *et al.*, 2002; Hoodge and Redekop, 1992; Martinelli *et al.*, 2004), this rate is relatively lower. One study (Polliack *et al.*, 2001) reported that female to male ratio in LOMS patients was the same as in young patients, but our findings suggest that in LOMS group the female to male ratio is nearly equal (0.8:1) and the risk of developing MS in ages over 50 is almost equal for both sexes. This could be due to the effect of menopausal endocrine-immunological changes. Our findings along with those from other studies (Polliack *et al.*, 2001; Weinshenker *et al.*, 1989) suggest that the frequency of PP course in this group of patients was higher than that of adult patients (20% vs 5.7%). The majority of patients had polysymptomatic complaints at onset but the most frequent initial manifestations were brain stem and sensory symptoms and optic neuritis. Similar to another report (Polliack *et al.*, 2001), a higher frequency of brain stem symptoms as onset symptom was observed. Compared to another study (Polliack *et al.*, 2001), optic neuritis was more frequent (10% vs 3%) and PI was lower (0.6 vs 0.8). It appears that patients with LOMS in Isfahan have more benign symptoms.

E. Familial MS

The increased risk of developing MS for twins, first-, second-, and third-degree relatives strongly recommends that genetic background contribute to disease susceptibility (Chataway *et al.*, 1998, 2001; Ebers *et al.*, 2000; Robertson *et al.*, 1996). In some Canadian studies, the familial rate among MS patients was reported to be as high as 20% (Ebers, 2000b); however, it is commonly agreed that this rate is about 15% (Compston, 1997). In our study, it was 12.2% which was not significantly lower than the reported rates in other studies. However, in one study from Middle East, the rate of familial MS was reported to be 9.4%, which is near to our calculated rate (El-Salem *et al.*, 2006). These findings in accordance with other studies on familial MS further support the role of genetic background for MS.

F. Conjugal MS

Conjugal MS is a rare condition in which both spouses have MS and at least one is affected after marriage. Study of these pairs provides information on the possible transmissibility or the effect of exposure to environmental factors in adulthood, and the study concerning the offspring would be helpful in the understanding of the mode of inheritance. We observed no meaningful similarities for disease course, disability, MRI pattern, and mode of presentation within conjugal pairs. Two studies (Finelli, 1991; Fredrikson *et al.*, 1992) reported clinical concordance within conjugal pairs, but our findings support the findings of another study (Robertson *et al.*, 1997) which suggest that there is no clinical concordance between spouses. However, this will not exclude the possibility of transmissibility or the effect of common exposure to the same environmental factors in adulthood because differences in clinical manifestations are more likely to be because of heterogeneity of MS (Sadovnick *et al.*, 1990).

V. Conclusions

Isfahan is a medium- to high-risk region for MS with a prevalence higher than previously reported, possibly because of an increase in the incidence rate. Clinical and demographic characteristics are similar to other reports from other regions. However, a low percentage of late-onset patients and a high percentage of early-onset cases were observed. The familial rate in our study was not significantly different from that from other areas and this could support a genetic basis for the disease. We could not observe any clinical concordance between conjugal pairs, but this will not exclude the possibility of transmission or common exposure to environmental factors in adulthood as a cause of MS. Furthermore, frequency of presence of oligoclonal IgG bands in CSF of our MS patients and some other Asian reports was lower than Western MS patients. These differences may be attributed to geographical or ethnic difference between areas. However, more accurate studies with larger samples are required to further support these conclusions.

References

Al-Araji, A., and Mohammed, A. I. (2005). Multiple sclerosis in Iraq: Does it have the same features encountered in Western countries. *J. Neurol. Sci.* **234**(1–2), 67–71.

Al-Din, A. S., Khogali, M., Poser, C. M., al-Nassar, K. E., Shakir, R., Hussain, J., Behbahani, K., and Chadha, G. (1990). Epidemiology of multiple sclerosis in Arabs in Kuwait: A comparative study between Kuwaitis and Palestinians. *J. Neurol. Sci.* **100,** 137–141.

Al-Din, A. S., el-Khateeb, M., Kurdi, A., Mubaidin, A., Wriekat, A., al-Shehab, A., and Khalil, R. W. (1995). Multiple sclerosis in Arabs in Jordan. *J. Neurol. Sci.* **131,** 144–149.

Al-Rajeh, S., Bademosi, O., Ismail, H., Awada, A., Dawodu, A., al-Freihi, H., Assuhaimi, S., Borollosi, M., and al-Shammasi, S. (1993). A community survey of neurological disorders in Saudi Arabia: The Thughbah study. *Neuroepidemiology* **12,** 164–178.

Aladro, Y., Alemany, M. J., Perez-Vieitez, M. C., Amela, R., Conde, M., Reyes, M. P., Alamo, F., and Angel-Moreno, A. (2005). Prevalence and incidence of multiple sclerosis in Las Palmas Canary Islands Spain. *Neuroepidemiology* **24,** 70–75.

Ale-Yasin, H., Sarai, A., Alaeddini, F., Ansarian, E., Lotfi, J., and Sanati, M. H. (2002). Multiple sclerosis: A study of 318 cases. *Arch. Iranian Med.* **5,** 24–27.

Alshubaili, A. F., Alramzy, K., Ayyad, Y. M., and Gerish, Y. (2005). Epidemiology of multiple sclerosis in, Kuwait: New trends in incidence and prevalence. *Eur. Neurol.* **53,** 125–131.

Barnett, M. H., Williams, D. B., Day, S., Macaskill, P., and McLeod, J. G. (2003). Progressive increase in incidence and prevalence of multiple sclerosis in Newcastle Australia: A 35-year study. *J. Neurol. Sci.* **213,** 1–6.

Baumhackl, U., Eibl, G., Ganzinger, U., Hartung, H. P., Mamoli, B., Pfeiffer, K. P., Fazekas, F., and Vass, K., and Austrian MS Study Committee (2002). Prevalence of multiple sclerosis in Austria. Results of a nationwide survey. *Neuroepidemiology* **21,** 226–234.

Bharucha, N. E., Bharucha, E. P., Wadia, N. H., Singhal, B. S., Bharucha, A. E., Bhise, A. V., Kurtzke, J. F., and Schoenberg, B. S. (1988). Prevalence of multiple sclerosis in the Parsis of Bombay. *Neurology* **38,** 727–729.

Chataway, J., Feakes, R., Coraddu, F., Gray, J., Deans, J., Fraser, M., Robertson, N., Broadley, S., Jones, H., Clayton, D., Goodfellow, P., Sawcer, S., *et al.* (1998). The genetics of multiple sclerosis: Principles, background and updated results of the United Kingdom systematic genome screen. *Brain* **121,** 1869–1887.

Chataway, J., Mander, A., Robertson, N., Sawcer, S., Deans, J., Fraser, M., Broadley, S., Clayton, D., and Compston, A. (2001). Multiple sclerosis in sibling pairs: An analysis of 250 families. *J. Neurol. Neurosurg. Psychiatry* **71,** 757–761.

Cocco, E., Marchi, P., Lai, M., Carboni, F., Spinicci, G., Contu, P., and Marrosu, M. G. (2001). Conjugal multiple sclerosis in Sardinian patients. *J. Neurol.* **248**(Suppl. 2)11/116.

Cole, G. F., and Stuart, C. A. (1995). A long perspective on childhood multiple sclerosis. *J. Dev. Med. Child Neurol.* **37,** 661–666.

Compston, A. (1997). Genetic epidemiology of multiple sclerosis. *J. Neurol. Neurosurg. Psychiatry* **62**(6), 553–561.

Compston, A. (1998). Distribution of multiple sclerosis. *In* "McAlpine's Multiple Sclerosis" (D. A. S. Compston, G. Ebers, H. Lassmann, I. McDonald, B. Matthews, and H. Wekerle, Eds.), pp. 63–100. Churchill Livingston, Edinburgh.

Daif, A. K., Al-Rajeh, S., Awada, A., Al Bunyan, M., Ogunniyi, A., Abdul Jabar, M., and Al Tahan, A. R. (1998). Pattern of presentation of multiple sclerosis in Saudi Arabia: Analysis based on clinical and paraclinical features. *Eur. Neurol.* **39,** 182–186.

Delalande, S., De Seze, J., Ferriby, D., Stojkovic, T., and Vermersch, P. (2002). Late onset MS. *Rev. Neurol. (Paris)* **158,** 1082–1087.

Deryck, O., Ketelaer, P., and Dubois, B. (2006). Clinical characteristics and long term prognosis in early onset multiple sclerosis. *J. Neurol.* **253**(6), 720–723.

Duquette, P., Murray, T. J., Pleines, J., Ebers, G. C., Sadovnick, D., Weldon, P., Warren, S., Paty, D. W., Upton, A., Hader, W., Nelson, R., Auty, A., *et al.* (1987). Multiple sclerosis in childhood: Clinical profile in 125 patients. *J. Pediatr.* **111**(3), 359–363.

Ebers, G. C., Yee, I. M., Sadovnick, A. D., and Duquette, P. (2000a). Conjugal multiple sclerosis: Population-based prevalence and recurrence risks in offspring. Canadian Collaborative Study Group. *Ann. Neurol.* **48**(6), 927–931.

Ebers, G. C., Koopman, W. J., Hader, W., Sadovnick, A. D., Kremenchutzky, M., Mandalfino, P., Wingerchuk, D. M., Baskerville, J., and Rice, G. P. (2000b). The natural history of multiple sclerosis: A geographically based study: 8: Familial multiple sclerosis. *Brain* **123**(Pt. 3), 641–649.

El-Salem, K., Al-Shimmery, E., Horany, K., Al-Refai, A., Al-Hayk, K., and Khader, Y. (2006). Multiple sclerosis in Jordan: A clinical and epidemiological study. *J. Neurol.* **253,** 1210–1216.

Etemadifar, M., Janghorbani, M., Shaygannejad, V., and Ashtari, F. (2006). Prevalence of multiple sclerosis in Isfahan Iran. *Neuroepidemiology* **27,** 39–44.

Finelli, P. F. (1991). Conjugal multiple sclerosis: A clinical and laboratory study. *Neurology* **41**(8), 1320–1321.

Ford, H. L., Gerry, E., Airey, C. M., Vail, A., Johnson, M. H., and Williams, D. R. (1998). The prevalence of multiple sclerosis in the Leeds Health Authority. *J. Neurol. Neurosurg. Psychiatry* **64**(5), 605–610.

Fox, C. M., Bensa, S., Bray, I., and Zajicek, J. P. (2004). The epidemiology of multiple sclerosis in Devon: A comparison of the new and old classification criteria. *J. Neurol. Neurosurg. Psychiatry* **75,** 56–60.

Fredrikson, S., Michelsberg, J., Hillert, J., Wang, Z., Sun, J. B., Olerup, O., Olsson, T., and Link, H. (1992). Conjugal multiple sclerosis: Immunogenetic characterization and analysis of T- and B-cell reactivity to myelin proteins. *Neurology* **42**(3 Pt. 1), 577–582.

Frohman, E. M., Racke, M. K., and Raine, C. S. (2006). Multiple sclerosis—the plaque and its pathogenesis. *N. Engl. J. Med.* **354**(9), 942–955.

Gardner, M. J., and Altman, D. G. (1989). "Statistics with Confidence." British Medical Association, London.

Ghezzi, A., Deplano, V., Faroni, J., Grasso, M. G., Liguori, M., Marrosu, G., Pozzilli, C., Simone, I. L., and Zaffaroni, M. (1997). Multiple sclerosis in children: Clinical features of 149 cases. *Mult. Scler.* **3,** 43–46.

Granieri, E., Tola, R., Paolino, E., Rosati, G., Carreras, M., and Monetti, V. C. (1985). The frequency of multiple sclerosis in Italy: A descriptive study in Ferrara. *Ann. Neurol.* **17,** 80–84.

Guilhoto, L. M., Osorio, C. A., Machado, L. R., de Castro, C. P., Manreza, M. L., Callegaro, D., Kok, F., and Diament, A. (1995). Pediatric multiple sclerosis: Report of 14 cases. *Brain Dev.* **17,** 9–12.

Hamdi, T. I. (1975). Multiple sclerosis in Iraq: A clinical and geomedical survey. *J. Postgrad. Med.* **21,** 1–9.

Hanefeld, F. (1995). Characteristics of childhood multiple sclerosis. *Int. MS J.* **1,** 91–98.

Hawkes, C. (2001). Conjugal multiple sclerosis. *Ann. Neurol.* **50,** 552–553.

Hawkes, C. H. (2002). Is multiple sclerosis a sexually transmitted infection? *J. Neurol. Neurosurg. Psychiatry* **73,** 439–443.

Hoodge, J. P., and Redekop, W. K. (1992). Multiple sclerosis with very late onset. *Neurology* **42,** 1907–1910.

Itoh, T., Aizawa, H., Hashimoto, K., Yoshida, K., Kimura, T., Katayama, T., Koyama, S., Yahara, O., and Kikuchi, K. (2003). Prevalence of multiple sclerosis in Asahikawa a city in northern Japan. *Neurol. Sci.* **214,** 7–9.

Kahana, E., Zilber, N., Abramson, J. H., Biton, V., Leibowitz, Y., and Abramsky, O. (1994). Multiple sclerosis: Genetic versus environmental etiology: Epidemiology in Israel updated. *J. Neurol.* **241,** 341–346.

Kalanie, H., Gharagozli, K., and Kalanie, A. R. (2003). Multiple sclerosis: Report on 200 cases from Iran. *Mult. Scler.* **9,** 36–38.

Kira, J., Kanai, T., Nishimura, Y., Yamasaki, K., Matsushita, S., Kawano, Y., Hasuo, K., Tobimatsu, S., and Kobayashi, T. (1996). Western versus Asian types of multiple sclerosis: Immunogenetically and clinically distinct disorders. *Ann. Neurol.* **40,** 569–574.

Kurdi, A., Ayesh, I., Abdallat, A., and Maayta, U. (1977). Different B lymphocyte alloantigens associated with multiple sclerosis in Arabs and Northern Europeans. *Lancet* **1**(8022), 1123–1125.

Kurtzke, J. F. (1975). A reassessment of the distribution of multiple sclerosis parts 1 and 2. *Acta Neurol. Scand.* **51,** 110–157.

Kurtzke, J. F. (1980). Epidemiologic contributions to multiple sclerosis: An overview. *Neurology* **30,** 61–79.

Kurtzke, J. F. (1983). Rating neurologic impairment in multiple sclerosis: An expanded disability status scale (EDSS). *Neurology* **33,** 1444–1452.

Lau, K. K., Wong, L. K., Li, L. S., Chan, Y. W., Li, H. L., and Wong, V. (2002). Epidemiological study of multiple sclerosis in Hong Kong Chinese: Questionnaire survey. *Hong Kong Med. J.* **8,** 77–80.

Maghzi, A. H., Etemadifar, M., Shaygannejad, V., Saadatnia, M., Salehi, M., and Hassanzadeh, A. (2007). Conjugal multiple sclerosis in Isfahan, Iran: A population-based study. *Mult. Scler.* (in press).

Marrie, R. A. (2004). Environmental risk factors in multiple sclerosis aetiology. *Lancet Neurol.* **3,** 709–718.

Martinelli, V., Rodegher, M., Moiola, L., and Comi, G. (2004). Late onset multiple sclerosis: Clinical characteristics prognostic factors and differential diagnosis. *Neurol. Sci.* **25**(Suppl. 4), S350–S355.

Mayr, W. T., Pittock, S. J., McClelland, R. L., Jorgensen, N. W., Noseworthy, J. H., and Rodriguez, M. (2003). Incidence and prevalence of multiple sclerosis in Olmsted County Minnesota, 1985–2000. *Neurology* **61,** 1373–1377.

McDonald, W. I., Compston, A., Edan, G., Goodkin, D., Hartung, H. P., Lublin, F. D., McFarland, H. F., Paty, D. W., Polman, C. H., Reingold, S. C., Sandberg-Wollheim, M., Sibley, W., *et al.* (2001). Recommended diagnostic criteria for multiple sclerosis: Guidelines from the International Panel on the diagnosis of multiple sclerosis. *Ann. Neurol.* **50,** 121–127.

Miller, J. R., Burke, A. M., and Bever, C. T. (1983). Occurrence of oligoclonal bands in multiple sclerosis and other CNS diseases. *Ann. Neurol.* **13,** 53–58.

Modrego, P. J., and Pina, M. A. (2003). Trends in prevalence and incidence of multiple sclerosis in Bajo Aragon Spain. *J. Neurol. Sci.* **216,** 89–93.

Montalban, X., and Rio, J. (2001). Primary progressive multiple sclerosis. *Neurol. Sci.* **22**(Suppl. 2), S41–S48.

Mumford, C. J., Fraser, M. B., Wood, N. W., and Compston, D. A. (1992). Multiple sclerosis in the Cambridge health district of east Anglia. *J. Neurol. Neurosurg. Psychiatry* **55,** 877–882.

Neuberger, J. S., Lynch, S. G., Sutton, M. L., Hall, S. B., and Schmidt, W. (2004). Prevalence of multiple sclerosis in a residential area bordering an oil refinery. *Neurology* **63,** 1796–1802.

Ozakbas, S., Idiman, E., Baklan, B., and Yulug, B. (2003). Childhood and juvenile onset multiple sclerosis: Clinical and paraclinical features. *Brain Dev.* **25,** 233–236.

Pekmezovic, T., Jarebinski, M., Drulovic, J., Stojsavljevic, N., and Levic, Z. (2001). Prevalence of multiple sclerosis in Belgrade Yugoslavia. *Acta Neurol. Scand.* **104,** 353–357.

Perniola, T., Russo, M. G., Margari, L., Buttiglione, M., and Simone, I. L. (1991). Multiple sclerosis in childhood. Longitudinal study in 14 cases. *Acta Neurol. Napoli* **13**(3), 236–248.

Pinhas-Hamiel, O., Barak, Y., Siev-Ner, I., and Achiron, A. (1998). Juvenile multiple sclerosis: Clinical features and prognostic characteristics. *J. Pediatr.* **132,** 735–737.

Polliack, M. L., Barak, Y., and Achiron, A. (2001). Late-onset multiple sclerosis. *J. Am. Geriatr. Soc.* **49,** 168–171.

Poser, C. M. (1994). The epidemiology of multiple sclerosis: A general overview. *Ann. Neurol.* **36**(Suppl. 2), S180–S193.

Ragonese, P., Salemi, G., D'Amelio, M., Gammino, M., Aridon, P., and Savettieri, G. (2004). Multiple sclerosis in southern Europe: Monreale city Italy. A twenty-year follow-up incidence and prevalence study. *Neuroepidemiology* **23,** 306–309.

Rice-Oxley, M., Williams, E. S., and Rees, J. E. (1995). A prevalence survey of multiple sclerosis in Sussex. *J. Neurol. Neurosurg. Psychiatry* **58,** 27–30.

Roberts, M. H., Martin, J. P., McLellan, D. L., McIntosh-Michaelis, S. A., and Spackman, A. J. (1991). The prevalence of multiple sclerosis in the Southampton and South West Hampshire Health Authority. *J. Neurol. Neurosurg. Psychiatry* **53,** 55–59.

Robertson, N., Deans, J., Fraser, M., and Compston, D. A. (1995). Multiple sclerosis in the north Cambridgeshire districts of East Anglia. *J. Neurol. Neurosurg. Psychiatry* **59,** 71–76.

Robertson, N. P., Fraser, M., Deans, J., Clayton, D., Walker, N., and Compston, D. A. (1996). Age-adjusted recurrence risks for relatives of patients with multiple sclerosis. *Brain* **119**(Pt. 2), 449–455.

Robertson, N. P., O'Riordan, J. I., Chataway, J., Kingsley, D. P., Miller, D. H., Clayton, D., and Compston, D. A. (1997). Offspring recurrence rates and clinical characteristics of conjugal multiple sclerosis. *Lancet* **349**(9065), 1587–1590.

Rosati, G. (1994). Descriptive epidemiology of MS in Europe in the 1980s: A critical overview. *Ann. Neurol.* **36**(Suppl. 2), S164–S174.

Rosati, G. (2001). The prevalence of multiple sclerosis in the world: An update. *Neurol. Sci.* **22,** 117–139.

Rothwell, P. M., and Charlton, D. (1998). Predisposition in south east Scotland: Evidence of a genetic high incidence and prevalence of multiple sclerosis. *J. Neurol. Neurosurg. Psychiatry* **64,** 730–735.

Ruggieri, M., Polizzi, A., Pavone, L., and Grimaldi, L. M. (1999). Multiple Sclerosis in children under 6 years of age. *J. Neurol.* **53,** 478–484.

Sadovnick, A. D., Hashimoto, L. L., and Hashimoto, S. A. (1990). Heterogeneity in multiple sclerosis: Comparison of clinical manifestations in relatives. *Can. J. Neurol. Sci.* **17,** 387–390.

Savettieri, G., Salemi, G., Ragonese, P., Aridon, P., Scola, G., and Randisi, G. (1998). Prevalence and incidence of multiple sclerosis in the city of Monreale Italy. *J. Neurol.* **245,** 40–43.

Sevon, M., Sumelahti, M. L., Tienari, P., Haltia, M., and Iivanainen, M. (2001). Multiple sclerosis in childhood and its prognosis. *Int. MS J.* **8,** 29–33.

Sharpe, G., Price, S. E., Last, A., and Thompson, R. J. (1995). Multiple sclerosis in island populations: Prevalence in the Bailiwicks of Guernsey and Jersey. *J. Neurol. Neurosurg. Psychiatry* **58,** 22–26.

Shepherd, D. I., and Summers, A. (1996). Prevalence of multiple sclerosis in Rochdale. *J. Neurol. Neurosurg. Psychiatry* **61,** 415–417.

Sloka, J. S., Pryse-Phillips, W. E., and Stefanelli, M. (2005). Incidence and prevalence of multiple sclerosis in Newfoundland and Labrador. *Can. J. Neurol. Sci.* **32,** 37–42.

Solaro, C., Allemani, C., Messmer Uccelli, M., Canevari, E., Dagnino, N., Pizio, R., Regesta, G., Tanganelli, P., Battaglia, M. A., and Mancardi, G. L. (2005). The prevalence of multiple sclerosis in the north-west Italian province of Genoa. *J. Neurol.* **252,** 436–440.

Sundstrom, P., Nystrom, L., and Forsgren, L. (2003). Incidence (1988–97) and prevalence (1997) of multiple sclerosis in Vasterbotten County in northern Sweden. *J. Neurol. Neurosurg. Psychiatry* **74,** 29–32.

Swingler, R. J., and Compston, D. A. (1988). The prevalence of multiple sclerosis in south east Wales. *J. Neurol. Neurosurg. Psychiatry* **51,** 1520–1524.

Swingler, R. J., Kirk, P. F., Darke, C., and Compston, D. A. (1987). HLA and multiple sclerosis in south-east Wales. *J. Neurol. Neurosurg. Psychiatry* **50,** 1153–1155.

Tanaka, K., Kujuro, Y., Suzuki, S., Tanahashi, N., Hamada, J., Nogawa, S., and Suzuki, N. (2005). Clinical and laboratory, features of in-patients with multiple sclerosis in a University Hospital in Tokyo from 1988–2002. *Intern. Med.* **44**(6), 560–566.

Thompson, A. J., Polman, C. H., Miller, D. H., McDonald, W. I., Brochet, B., Filippi, M., Montalban, X., and De Sa, J. (1997). Primary progressive multiple sclerosis. *Brain* **120,** 1085–1096.

Thompson, A. J., Montalban, X., Barkhof, F., Brochet, B., Filippi, M., Miller, D. H., Polman, C. H., Stevenson, V. L., and McDonald, W. I. (2000). Diagnostic criteria for primary progressive multiple sclerosis: A position paper. *Ann. Neurol.* **47,** 831–835.

Wadia, N. H., and Bhatia, K. (1990). Multiple sclerosis is prevalent in the Zoroastrians (Parsis) of India. *Ann. Neurol.* **28,** 177–179.

Weinshenker, B. G., Bass, B., Rice, G. P., Noseworthy, J., Carriere, W., Baskerville, J., and Ebers, G. C. (1989). The natural history of multiple sclerosis: A geographically based study. I. Clinical course and disability. *Brain* **112**(Pt. 1), 133–146.

Yamasaki, K., Horiuchi, I., Minohara, M., Kawano, Y., Ohyagi, Y., Yamada, T., Mihara, F., Ito, H., Nishimura, Y., and Kira, J. (1999). HLA-DPB1*0501-associated opticospinal multiple sclerosis: Clinical, neuroimaging and immunogenetic studies. *Brain* **122**(Pt. 9), 1689–1696.

Yaqub, B. A., and Daif, A. K. (1988). Multiple sclerosis in Saudi Arabia. *Neurology* **38,** 621–623.

GENDER ISSUES IN MULTIPLE SCLEROSIS

Robert N. Schwendimann*,† and Nadejda Alekseeva*,†

*Department of Neurology, Louisiana State University
Health Sciences Center, Shreveport
Louisiana 71103, USA

†Department of Psychiatry, Louisiana State University
Health Sciences Center, Shreveport
Louisiana 71103, USA

Multiple sclerosis (MS) varies considerably in the way that it affects females and males. The prevalence of the disease is much greater in women and tends to follow a different clinical course than it follows in the affected male population. It is also well known that MS symptoms often are much less of a problem during pregnancy. This chapter discusses possible explanations for gender differences based on sex hormones as well as the effects of these hormones on cytokines and other factors that may influence the course of MS. Knowledge of these effects may hold some promise in other types of treatment for MS. Since MS is much more prevalent in women of child-bearing age, there are also implications for the use of disease-modifying agents as well as drugs and treatments that may be useful for treatment of MS. MS often causes symptoms of sexual dysfunction, but there may be effective treatment for many of these treatments.

INTERNATIONAL REVIEW OF
NEUROBIOLOGY, VOL. 79
DOI: 10.1016/S0074-7742(07)79017-7

0074-7742/07 $35.00

I. Introduction

Multiple sclerosis (MS) affects female and male patients differently in many ways. There are differences in demographic and etiologic factors as well as clinical features and disease activity between male and female patients. There are marked differences in immunologic make-up of individuals that may explain how MS varies between the two genders. Knowledge about how and why MS affects the sexes in different ways may be beneficial in the management of the disease in females and in males. There are suggestions that therapy may be quite different in female and male patients. There are also many implications for therapy with sex hormones.

There are so many differences between the ways that MS affects females compared to males that it is almost like treating two different diseases. MS affects most commonly young, postpubertal women and most commonly causes relapsing-remitting-type symptomatology. In men, the patients tend to be much older and the disease tends to be much more progressive and debilitating in course, often causing myelopathy. Males tend to have a much worse prognosis than females. There is evidence that males have more cognitive impairment than females, with much poorer performance on standardized neuropsychological testing. In one study cognitive decline in men was associated with disease duration, Expanded Disability Status Scale (EDSS) score, a low level of education, and presence of the $\varepsilon 4$ allele of the apoE gene. Cognitive impairment in women was independent of any investigated variable (Savettieri *et al.*, 2004).

MS is a disease that affects premenopausal women primarily. In the United States, MS affects greater than 350,000 people of both sexes for a prevalence rate of 85/100,000 persons. The prevalence rate among females, however, is estimated at 130/100,000 with an overall prevalence ratio of women to men of 2.6:1. In recent years, there has been a 50% increase in the number of women who have been diagnosed with MS that has increased the prevalence rate among women, while the prevalence rate among men during the same period of time has remained the same (Noonan *et al.*, 2002). Approximately 70–75% of MS patients are female. There is concordance in monozygotic twins in females, but not in males. These prevalence rates are quite likely affected by immune and hormonal, as well as genetic factors. The age of onset in women who are beyond puberty is quite likely to be related to hormonal factors. Less than 1% of MS patients experience onset of symptoms prior to puberty. Only 5% of patients experience onset of symptoms prior to age 16. Typically, onset is in the early third decade, with average age of onset around 28–30 years of age. Ninety percent of patients are between age 15 and 50 years of age. The risk of developing symptoms of relapsing-remitting MS (RRMS) declines with and beyond menopause (Coyle, 2005).

There is an equal sex ratio in the primary progressive form of MS (PPMS). These patients also manifest symptoms of disease about a decade later than those with relapsing-remitting disease. Men with PPMS have a later onset of symptoms and faster rate of disability. Though it is not known while males have a worse prognosis, it may be related to declining testosterone levels which begins at about age 30. Twenty-four percent of men with MS had low testosterone levels compared with age-matched controls in a recent study. Males with MS and sexual dysfunction may benefit from treatment with testosterone (Coyle, 2005).

Certain genes linked to MS may be influenced by gender. A susceptibility gene, the major histocompatibility class II HLA-DR2DQ6 haplotype, is more common in females than in males with MS (Celius *et al.*, 2000). There are reports of a relationship between MS and polymorphisms in the estrogen receptor gene (Nino *et al.*, 2000). In one population, two CD95 polymorphisms may account for 28% of the risk of MS in women but 0% in men. This may be an important factor for MS occurrence in women since CD95 and its ligand mediate activation-induced apoptosis (Kantarci *et al.*, 2004a). In another study, a specific apoE polymorphism is associated with a milder disease course, but only in women (Kantarci *et al.*, 2004b).

There are differences in magnetic resonance imaging (MRI) in males and females that also suggest that sex hormones modulate brain damage in MS. It has been shown that men with MS are more likely to have destructive lesions than the more common inflammatory lesions seen in women. Low testosterone levels in women may correlate with a greater number of gadolinium-enhancing lesions on MRI. In men, there is a positive correlation between estradiol concentrations and brain damage. These findings support the hypothesis that sex hormones play a role in inflammation, damage, and repair mechanisms in MS (Tomassini *et al.*, 2005).

The nervous system, endocrine system, and immune system differ both anatomically and physiologically by gender. Anatomically, men and women show differences in density of neurons, size of regional brain areas, cell receptor expression, neurotransmitter systems, cerebral blood flow patterns, response to stress, hunger and satiety, and cortical activation patterns. MRI studies show that global white matter water content is higher in women. Compared to men, women have larger gray matter volumes but smaller white matter volumes. The brains of males do not age as well as those of females and show greater losses of gray matter volume with time (Rooney and Coyle, 2005). Cervical spinal cords of males have a larger volume than those of women (Rashid *et al.*, 2006).

The most obvious difference between the genders occurs in the endocrine system and the sexual organs. The hypothalamic-pituitary-adrenal (HPA) gland axis is the major link between the endocrine system and the central nervous system. The hypothalamus controls release of gonadotropin-releasing factor (GRF) hormone which is influenced in turn by the menstrual cycle. GRF regulates release of gonadotropins, follicle-stimulating hormone (FSH), and luteinizing hormone (LH) by the anterior pituitary gland. Prolactin is also released by the pituitary gland. FSH and LH promote development of ova and stimulate sex hormone secretions by the ovaries. The ovary also secretes inhibin, a peptide that inhibits FSH release. There are higher levels of cortisone and corticosterone, both of which are important hormones in the HPA axis, in women than in men. Dysfunction of the HPA axis is a fundamental factor in development of autoimmune and immune-related diseases (DaSilva, 1999).

There are gender differences in the immune system. Generally, females of all species show stronger immune responses than males (Kurtz *et al.*, 2000). Compared to males, females have higher base-line immunoglobulin levels and CD4+ T-cell numbers. They have higher production of T-helper cell 1 (Th1) cytokines, and more sustained response to antigenic challenge (Whitacre, 2001). Diseases that are considered to be autoimmune are much more common in women than men. In the United States, 80% of the 8.5 million people with autoimmune disease are female. Many of these diseases are affected by hormonal changes. Like MS, rheumatoid arthritis and psoriasis may improve during pregnancy (Coyle, 2005; Palaszynski *et al.*, 2004a), while systemic lupus erythematosis may worsen (Gilmore *et al.*, 1997). MS and rheumatoid arthritis are both cell-mediated disorders, while SLE is antibody mediated. Fluctuation of symptoms in many of these diseases may also be affected by the menstrual cycle. Treatment with estrogen replacement may be beneficial in some of these diseases as well, that is, perimenopausal rheumatoid arthritis.

II. Sex Hormones

The sex hormones share the same precursor cholesterol and include estrogens (estrone, estradiol, estriol), progestins (progesterone), and androgens [testosterone, dehydroepiandrostenedione (DHEA)] (Palaszynski *et al.*, 2004a). Sex hormones have multiple effects on the nervous system. These include effects on axon growth and myelination, neurotransmission, neural remodeling, and motor coordination. They may enhance cognitive performance and increase transcription of neurotrophic factor genes that are neuroprotective. They regulate brain apoptosis, inhibit nitric oxide production by microglia, and amplify excitotoxic damage to oligodendrocytes. There are estrogen receptors in the basal forebrain, basal ganglia, hippocampus, and locus ceruleus (Coyle, 2005). There are three estrogenreceptors on microglia: α, β, and χ (Drew *et al.*, 2003). Concentrations of receptors may vary according to serum estrogen levels. Estrogen may act as a trophic factor and a neuromodulator. Progesterone can induce expression of myelin proteins in cultured oligodendrocytes (Shomer, 1998).

The sex hormones are potent immunomodulators and may influence Th lymphocytes. The two types of Th cells are Th1 and Th2 cells. Th1 cells are proinflammatory and Th2 cells are anti-inflammatory, depending on the cytokines that they secrete. Estrogen influences cytokine profiles in these cells depending on concentration. In high concentrations (as occurs in pregnancy), levels of tumor necrosis factor-α (TNF-α) are decreased and levels of interleukin (IL)-10 are increased. At lower estrogen concentrations, TNF levels are increased (Gilmore *et al.*, 1997). Estrogen stimulates humoral immunity and Th2 function and inhibits T suppressor and natural killer cell activity. It is a powerful promoter for interferon (IFN)-γ, another Th1 cytokine. Progesterone, in contrast, acts more as an immunosuppressant. Progesterone decreases inflammatory cytokines such as IL-1 and IFN-γ. It stimulates production of the Th2 cytokine IL-4. These hormones act synergistically also and the combined effects of estrogen and progesterone are greater than either one alone. Testosterone also acts as an immunosuppressant. In animal models, castration affects susceptibility to induction of autoimmune disease (Giesser, 2002).

Cytokines can also have an impact on neuroendocrine pathways. Some key cytokines that are involved in the immunopathology of MS can influence secretion and function of sex hormones and gonadotropins. The Th1 cytokine, IL-1, can act directly on the hypothalamus to affect secretion of GRFs and inhibit luteinization (Rivier and Vale, 1990). TNF can inhibit the action of FSH and may inhibit progesterone secretion (Adashi, 1990).

III. Sex Hormones in Animal Models

In Theiler's murine encephalomyelitis, an animal model for MS, testosterone therapy ameliorates disease and shifts lymphocyte response to a Th2 profile. Female animals are more susceptible to disease and show more severe encephalitis. They display a more pronounced response of Th1 proinflammatory cytokines during early infection. Males show a Th2 response, with appropriate higher levels of regulatory cytokines. Treatment with anti-inflammatory cytokines IL-4 and IL-10 is successful in female animals but not in male animals. In the chronic phase of this MS model, male animals develop more severe disease (Alley *et al.*, 2003).

Experimental allergic encephalomyelitis (EAE) is produced in mice, rates, and primates by immunization with myelin, myelin basic protein, myelin oligodendrocyte glycoprotein, proteolipid protein, or by adoptive transfer of CD4+ T cells. Manifestations of disease vary according to animal strain and species. In EAE, females are more susceptible to disease and may manifest more severe disease. Pregnancy in these animals appears to be protective. Administration of exogenous estrogens ameliorates disease and oophorectomy worsens disease. Exogenous estrogen can change treatment nonresponders to responders. Male animals are less susceptible to disease and show a milder form of disease. Exogenous estrogens and testosterone ameliorate disease. Castration increases susceptibility and worsens disease. Castration can also change disease pattern from male to female. Testosterone levels decrease during relapses (Kim *et al.*, 1999; Palaszynski *et al.*, 2004b; Polanczyk *et al.*, 2004).

Other studies in EAE implicate leptin in sex-related EAE vulnerability. Leptin is a pleiotropic cytokine-like hormone. Leptin levels in animals are higher in females than males. Leptin has several actions and affects food intake, body weight, basal metabolism, hematopoiesis, inflammation, and the immune response to pathogens. It is produced by neurons, activated T cells, and local inflammatory cells. Immune effects include increasing CD4+ Th1 cells, costimulating proliferation of naive T cells, and increasing certain proinflammatory cytokines (IFN-γ and IFN-α) and decreasing anti-inflammatory cytokines like IL-4. Animals deficient in leptin are resistant to development of EAE, but become susceptible when given recombinant leptin. In SJL/J mice, a strain often used for EAE, only the females, with high leptin levels compared to males, develop EAE. The male animals become susceptible, however, when injected with recombinant leptin. Hormone levels also correlate with inflammation-related anorexia, weight loss, and EAE susceptibility. Leptin levels are lower in animals that have been starved. These animals show decreased disease severity (Ding *et al.*, 1997; Matarese *et al.*, 2001).

IV. MS and the Menstrual Cycle

In women, there is often variability in symptoms that may fluctuate with the menstrual hormonal cycle. The menstrual cycle is a hormonal state that reflects the actions between the hypothalamus, pituitary gland, ovarian follicle, and endometrium. In the 28-day cycle, bleeding begins on the first day and usually lasts 3–7 days. Total blood loss typically amounts to 80 ml. On about day 14, ovulation occurs in association with an increase in FSH and LH levels. Levels of estrogen peak, at mid-cycle, followed by peak levels of progesterone. As levels of estrogen and progesterone fall, the cycle begins again with menstrual bleeding. Some 80% of female patients with MS report fluctuation of symptoms in association with their menstrual periods. Often, these fluctuations are more pronounced in patients with relapsing-remitting symptoms (Wilson *et al.*, 2004; Zordrager and DeKeyser, 1998). Typical symptoms include fatigue, myalgias, depression, and decreased endurance. On examination, worsening signs include increased spasticity, weakness, incoordination and abnormalities in gait, sensation, vision, and sphincter function.

MRI studies have shown that a relationship exists between the menstrual cycle and MRI changes. One study showed increases in brain water content, consistent with increased blood–brain barrier permeability that was correlated with plasma estradiol levels. Another study showed MRI changes that indicated highest brain lesion activity when estradiol levels were high and progesterone levels were low. This study was performed on 30 patients in early follicular, late follicular, or luteal phase of the cycle. Other studies performed on the follicular and luteal phases of the cycle show no difference in contrast lesion activity, higher estradiol to progesterone level ratios during the luteal phase do seem to correlate with increased number and volume of contrast-enhancing brain lesions. In some studies, women with MS had lower testosterone levels that matched control subjects. Women with less MRI activity also had higher testosterone levels and their DHEA levels correlated with lack of T2 lesion development. This supports the putative protective effect of testosterone in MS (Tomassini *et al.*, 2005).

Though there is limited data on the use of oral contraceptive pills (OCPs) in women with MS, they do not seem to increase either development of MS or worsen the course of MS. There is evidence that OCP use may be associated with fewer symptoms, less disability, and lower risk for developing MS (Wilson *et al.*, 2004). Contraceptive use is less likely to be associated with worsening of signs and symptoms associated with the menstrual cycle. One study suggested that patients with MS who were on OCPs and had one pregnancy before onset of MS showed greater disability. All data available does not suggest any problem associated with use of OCPs in patients who have MS.

V. MS and Pregnancy

Most neurologists are very familiar with the ameliorating effects of pregnancy on MS symptomatology and of significant worsening that may follow in the postpartum period. The use of disease-modifying therapies (DMT) is often affected by pregnancy as well since most of these therapies are contraindicated during pregnancy. Pregnancy in patients with MS is one of the single most important gender issues. There are concerns about the effect of MS on the pregnancy as well as the effects of pregnancy on the patient's disease. MS patients have no difficulty with fertility or the ability to conceive. There appears to be no effect of MS on the pregnant state. MS patients show no increase in spontaneous abortion or fetal malformations. Pregnancy does, however, affect the disease (Tomassini *et al.*, 2001).

It is important to give information regarding genetics prior to pregnancy. Risk for MS in white populations is about 0.2%. With a parent who has been diagnosed with MS, the risk to the child increases to 3–5%. This risk may be even greater when the affected patient is the father (Kantarci *et al.*, 2004a). When both parents have MS, the child's risk increases further to 30.5%.

Onset of MS is much less likely to occur during pregnancy and is associated with a decrease in MS disease activity (Runmaker and Andersen, 1995). Pregnancy is an immunosuppressive state that is produced by hormonal changes such as increases in estrogen, progesterone, and glucocorticoids; increases in human chorionic gonadotropin, norepinephrine and vitamin D3; and changes in pregnancy-related factors and a shift that favors Th2 responses. This results in a decrease in cell-mediated immunity and an increase in humoral immunity, immune complexes, and blocking antibodies. There are also increased levels of cytokines such as IL-10 and IL-4 and decreases in IL-2, IL-12, TNF-α, and IFN-γ. Increases in CD4+, CD25+ regulatory T cells are present (Elenkov *et al.*, 2001; Thompson *et al.*, 1987).

Studies on large numbers of MS patients in pregnancy found that annual relapse rate fell significantly during pregnancy, particularly in the third trimester. In the postpartum period, however, the relapse rate rebounded significantly to 70% above the prepregnancy rate. This rate did not return to baseline for 3-month postpartum. During the immediate postpartum period, 20–40% of women experience a relapse of symptoms (Confavreux *et al.*, 1998).

Even though pregnancy seems to have a protective effect on exacerbations, several MS symptoms may worsen during pregnancy. These may include fatigue, bladder and bowel dysfunction, spasticity, and increased difficulty with mobility/ambulation. Drugs that are often prescribed for these symptoms may be contraindicated during pregnancy. Steroid drugs, baclofen, tizanidine, amantadine, modafinil, tricyclic, and selective serotonin release inhibiting antidepressants, anti-epileptic drugs, oxybutynin, and tolteridine are all risk category C drugs. Many of these are also excreted in breast milk.

DMT are contraindicated in those MS patients who are attempting to conceive and those who actually are pregnant. These drugs are also contraindicated in those who plan to breast-feed their babies. Patients who are on DMT should be advised to take necessary precautions against conception while on therapy. The β-Interferons are considered class C drugs in pregnancy, meaning that there is risk as far as teratogenicity is concerned. They can cause abortion in some laboratory animals, though no negative fetal effects have been documented. They can cause menstrual disturbances. One study of IFN-β in 30 pregnancies reported 45% live births, 27.6% spontaneous abortions, 21% elective abortions, 6.9% prenatal births, and 1 patient lost to follow-up (Coyle *et al.*, 2003). Another study performed in Spain evaluated 36 pregnancies in 31 patients on IFN-β. There were 27 satisfactory deliveries and 9 spontaneous abortions (Sandberg-Wollheim *et al.*, 2005).

A larger database of 245 pregnancies has been reported for patients treated with glatiramer acetate. In this study, there were 66% healthy live births, 20% spontaneous abortions, 11% elective abortions, 2.9% congenital malformations, 0.4% ectopic pregnancies, and 0.4% stillbirths (Coyle *et al.*, 2003).

Mitoxantrone therapy raises both fertility and pregnancy concerns. This chemotherapeutic agent intercalates into DNA and inhibits DNA repair. Use of this drug in women of child-bearing age and in young men raises concerns about banking of ova and sperm for use in future pregnancies. Permanent amenorrhea can occur when treating perimenopausal women with mitoxantrone (Novantrone package insert, 2003).

VI. Effects of Pregnancy on MS Prognosis

There is some controversy as to whether or not pregnancy affects the long-term prognosis of the disease. In the past, many physicians counseled against pregnancy in MS patients; some even recommended abortion. More complete information indicates, however, that pregnancy is not detrimental to relapsing MS. There is an extensive body of literature supporting the view that pregnancy has an overall positive effect on the prognosis in MS, particularly those with relapsing-remitting disease. Patients with progressive forms of MS, however, do not do well during pregnancy. These women who often have a higher level of disability are less likely to pursue pregnancy (Damek and Schuster, 1997).

There appears to be a positive effect on the development of disability in women with MS who have been pregnant. A 5-year prospective study compared the rate of progression of disability between childless women, women who had onset of MS after childbirth, and women who had onset of symptoms before or during their pregnancies. In this small group of patients, the rates of disability increased most rapidly in the group of women who were childless. There was no statistically significant increase in the EDSS over the 5-year period in women who had onset of MS at least 6 months after delivery. This was suggestive of a possible positive effect of pregnancy on the rate of progression of disability. Women who had their onset of MS before or during pregnancy had rates of disability progression that was lower than the childless group, but higher than the women who had onset after pregnancy (Stenager *et al.*, 1994).

How long patients should be off of DMT prior to conception is uncertain, but 1 month is the usual recommendation. Unless the mother plans to breast-feed her child, the drug can be resumed immediately following delivery (Coyle, 2005).

Anesthesia during pregnancy is another issue that must be considered in the MS patient. The Pregnancy in MS Study (PRIMS) analyzed the effect of epidural anesthesia and found no impact on relapse rate or disability with 2 years of the anesthesia. There are cases in the literature which suggest that spinal anesthesia be avoided in patients with MS (Confavreux *et al.*, 1998).

The use of glucocorticoids for treatment of acute exacerbations of disease during pregnancy should be avoided during the first month of pregnancy because of negative effects on organogenesis. (It may be wise to check pregnancy tests prior to treating any patient of child-bearing age prior to high-dose steroid therapy.)

Steroids can be used to treat relapses that occur during the second and third trimesters. Overuse may slow postnatal growth. Steroids have been associated with birth defects in some animal models. Breast-feeding is contraindicated during steroid use since they are excreted in breast milk. Generally, it is best to keep use of any medications to a minimum during pregnancy and postpartum if mother is breast-feeding (Coyle, 2005).

While pregnancy represents some protection for the patient with MS, the postpartum period is associated with a high risk of relapse. Those patients who are on DMT prior to pregnancy should resume their treatments immediately following treatment unless planning to breast-feed. There have been studies also regarding the use of prophylactic treatments such as intravenous immunoglobulin (IVIg). Several studies have shown that use of a standard IVIG regimen of 0.4 g/kg/day for 5 days, following by 0.4 g/kg at later intervals significantly reduces relapse rate in the postpartum period (Achiron *et al.*, 1996, 2004).

Treatment with IV methylprednisolone following delivery has also been evaluated. Patients received monthly treatment for 6 months and were compared to untreated controls. The relapse rate in the first 3 months was 2 ± 0.7 and in treated group was 0.8 ± 0.4. No significant difference between the two groups was seen in months 4–11 (De Seze *et al.*, 2004).

VII. MS and Menopause

There is virtually no information on menopause and MS, though in one small study, 50% of patients reported worsening disability at menopause. This is also a time when many patients with RRMS develop more progressive symptomatology. This is a time when menopausal women develop other problems such as osteoporosis/penia. Both men and women whose mobility is affected or who have received steroids in the past are at even higher risk of developing osteoporosis. Bone scanning may be beneficial in preventing problems in this group of patients. Therapy involves both primary prevention (changes in lifestyle, weight-bearing exercise, calcium and vitamin D supplements) and secondary prevention with use of biphosponate, raloxifene, calcitonin, hormone replacement therapy (HRT), parathyroid hormone therapy, fluoride, ipriflavone, and hydrochlorothiazide. Calcium intake should be 1200 or 1500 mg/day in postmenopausal women. Calcium citrate is best absorbed and is associated with fewer side effects. Absorption may be improved also with vitamin D, magnesium salts, and hydrochlorothiazide. Vitamin D intake should be 800 IU/day and can be increased by exposure to sunlight (Coyle, 2006).

VIII. MS Symptoms in Females and Males

Two major MS symptoms that should be handled differently in male and female patients are bladder dysfunction and sexual dysfunction. Bladder problems typically occur later in the course of the disease, affecting up to 75% of patients. Symptoms tend to fluctuate over time. Patients experience three different types of neurogenic bladder problems. Patients experience symptoms of detrusor hyperreflexia, detrusor sphincter dyssynergia, and detrusor hyporeflexia. Typically, symptoms in women can be managed by changes in voiding schedule, controlled fluid intake, use of bedside commode/urinals, regular emptying techniques, pelvic floor exercises, and pad/special undergarments (DasGupta and Fowler, 2002).

Sexual problems in both genders are also quite common and include decreased libido, sexual aversions, difficulty with sexual arousal, and achieving orgasm. Some of these problems are a direct result of associated clinical symptoms such as diminished sensation in pelvic areas, weakness, and spasticity in the lower extremities as well as fatigue. In males, erectile dysfunction and ejaculatory disorders are common and may be managed with the use of drugs such as sildenifil. Vaginal dryness and pain with intercourse are common in women and may be managed with various types of lubricants and topical estrogen creams (DasGupta and Fowler, 2002).

IX. Sex Hormones in Treatment of MS

There is interest in treating MS with sex hormones because of the observed effects of sex hormones on the nervous and immune systems of animal models of MS and because of the observed effects during pregnancy. In small trials using oral estriol at 8 mg/day (dose duplicates levels of estriol during last trimester of pregnancy), a decrease in number and size of enhancing lesions was observed in patients with RR MS. Lesion size increased when therapy was discontinued. There was an additional positive effect when estriol was combined with progesterone. No effect was noted in patients with primary progressive disease. No significant effect was observed on relapses or EDSS scores either during or following the period of treatment. High-dose estrogen does induce uterine endometrium hyperplasia. All women in this trial suffered menstrual cycle abnormalities and 3/10 had to be explored by endometrial biopsy due to severe menstrual bleeding. Positive effects of estrogen were also noted on cognitive testing in a group of MS patients. DHEA has been used in small trials to treat fatigue in MS patients. The use of progesterone and testosterone has also been investigated in small drug trials as well (Sicotte *et al.*, 2002).

Another possible benefit of sex steroids in MS involves their myelinating and remyelinating effects. It is known that progesterone can be locally synthesized by Schwann cells in the peripheral nervous system (Koenig *et al.*, 1995). These cells express intracellular progesterone receptors (Jung-Testas *et al.*, 1996). Progesterone enhances myelin sheath formation in dorsal root ganglia explants comprising neurons and Schwann cells. Levels of pregnenolone, a precursor of progesterone, have been found to be low in trembler mice, an animal model of Charcot-Marie-Tooth disease, a demyelinating type of neuropathy. This suggests a role for progesterone during myelination. Progesterone can enhance remyelination and thickness of myelin sheaths after cryolesions of male mouse sciatic nerve. This effect can be blocked by the application of RU-486 which is a specific receptor antagonist (Koenig *et al.*, 1995). Progesterone may also be involved in the maintenance of myelin sheaths which is of major importance for neuron preservation.

Another therapeutic approach is the use progestins in the postpartum period to prevent relapses. The POPARTMUS study is a double-blind, placebo-controlled European multicenter trial evaluating oral progestins along with estradiol delivered by patch. This study is still in progress (El-Etr *et al.*, 2005).

X. Preventative Healthcare in MS

Various studies have shown that women with disabilities have difficulty in obtaining basic care, including gynecologic care. One study showed that of 101 adults with cerebral palsy, 90% of the women did not have regular Pap smears, pelvic examinations, or breast examinations. Another revealed that women with paralysis, impaired motor function, or obvious physical deformity were not offered contraceptive counseling. In the same study, less than half of the pregnancies were by choice. Shabas and Weinreb surveyed 220 women with MS by questionnaire. Patients ranged in age from 23–81 years. Forty-seven percent were ambulatory, 38% used assistive devices, and 15% were not ambulatory. Sixty-three percent were premenopausal and 37% were postmenopausal. Of the women surveyed, 50% indicated that they did not have regular medical checkups. Typically, these patients saw their primary care givers when there was an acute problem. Twenty-five percent indicated that they did not have regular Pap smears or pelvic examinations despite many gynecologic complaints such as menstrual irregularities and vaginal discharge. Eleven percent indicated that they had not had a Pap smear in 3–5 years. Forty-eight percent over age 50% and 53% of women over age 40 did not have yearly mammograms as recommended by the current guidelines (Shabas and Weinreb, 2000).

Women with MS may be at higher risk for developing osteoporosis because of decreased activity secondary to fatigue or physical limitations. Their risk may also be increased because of steroid use. Limitation of activity because of heat intolerance may also result in vitamin D deficiency secondary to limited sun exposure. Of women surveyed, 82% had a history of steroid use and 53% had impairment of mobility. Fifty percent of these were not taking calcium supplements and 71% were not taking vitamin D supplements. Only 1% was being treated with bone resorption inhibitors. Eighty-five percent of all women surveyed and 81% of postmenopausal women had never had a bone density screening test. Seventy percent of these were not on HRT (Shabas and Weinreb, 2000).

In this same survey, 14% of patients were advised against having children though 65% were mothers. Sixty-four percent of these reported relapse of symptoms with the first 6-month postpartum. Depression is common in MS with reported prevalence of 50% by the age of 59 years (Sadovnick *et al.*, 1996).

Both men and women with MS have health maintenance issues other than the manifestations of their primary neurological problem. These include getting regular medical checkups as well as having routine screening tests such as stool exams for occult blood and, depending on age, having colonoscopy at recommended intervals. They should be counseled about good nutritional practices, exercise programs, avoiding obesity, and avoided use of tobacco and illicit drugs. Alcohol should be used in moderation. Women should be encouraged to have regular gynecologic examinations, breast examinations with periodic mammography, bone density studies, and be considered for HRT. Men should have an annual prostate examination and regular testicular examinations.

XI. Conclusion

MS is known to affect female and male patients in different ways. Attention to these differences has resulted in new information regarding the pathophysiology of the disease from an immunologic standpoint. A better understanding of the role that sex hormones play as aggravating and ameliorating factors in MS may alter the management of the disease at some future time. From a more practical standpoint, knowledge of gender differences in MS can greatly impact how the neurologist deals with various symptoms of the disease using both DMT and symptomatic therapies.

References

Achiron, A., Rotstein, Z., Noy, S., Mashiach, S., Dulitzky, M., and Achiron, R. (1996). Intravenous immunoglobulin treatment in the prevention of childbirth-associated acute exacerbations in multiple sclerosis: A pilot study. *J. Neurol.* **243,** 25–28.

Achiron, A., Kishner, I., Dolev, M., Stern, Y., Dulitzky, M., Schiff, E., and Achiron, R. (2004). Effect of intravenous immunoglobulin treatment on pregnancy and postpartum related relapses in multiple sclerosis. *J. Neurol.* **251,** 1133–1137.

Adashi, E. Y. (1990). Do cytokines play a role in regulation of ovarian function? *Prog. Neuroendocrinol.* **3,** 11–17.

Alley, J., Khasabov, S., Simone, D., Beitz, A., Rodriguez, M., and Njenga, M. K. (2003). More severe neurologic deficits in SJL/J male than female mice following Theiler's virus-induced CNS demyelination. *Exp. Neurol.* **180,** 14–24.

Celius, E. G., Harbo, H. F., Egeland, T., Vartdal, F., Vandvik, B., and Spurkiand, A. (2000). Sex and age at diagnosis are correlated with the HLS-DR2, DZ6 haplotype in multiple sclerosis. *J. Neurol. Sci.* **178,** 132–135.

Confavreux, C., Hutchinson, M., Hours, M. M., Cortinovis-Tourniaire, P., and Moreau, T. (1998). Rate of pregnancy-related relapses in multiple sclerosis. *N. Engl. J. Med.* **339,** 285–291.

Coyle, P. K. (2005). Gender issues. *Neural Clin.* **23,** 39–60.

Coyle, P. K. (2006). "Menopausal Neurology: Multiple Sclerosis." American Academy of Neurology Annual Courses, San Diego, CA.

Coyle, P. K., Johnson, K., Pardo, L., and Stark, Y. (2003). Pregnancy outcomes in patients with multiple sclerosis treated with glatiramer acetate (Copaxone®). *Neurology* **60**(Suppl. 1), A60.

Damek, D. M., and Schuster, E. (1997). Pregnancy and multiple sclerosis. *Mayo. Clin. Proc.* **72,** 977–989.

DasGupta, R., and Fowler, C. J. (2002). Sexual and urological dysfunction in multiple sclerosis: Better understanding and improved therapies. *Curr. Opin. Neurol.* **15,** 271–278.

DaSilva, J. A. P. (1999). Sex hormones and glucocorticoids: Interactions with the immune system. *Ann. NY Acad. Sci.* **876,** 102–118.

De Seze, J., Chapelotte, M., Delalande, S., Ferriby, D., Stojkovic, T., and Vermersch, P. (2004). Intravenous corticosteroids in the postpartum period for reducing acute exacerbations in multiple sclerosis. *Mult. Scler.* **10,** 596–597.

Ding, M., Wong, J. L., Rogers, N. E., Ignarro, L. J., and Voskuhl, R. R. (1997). Gender differences of inducible oxide production in JSJ/J mice with experimental autoimmune encephalomyelitis. *J. Neuroimmunol.* **77,** 99–106.

Drew, P. D., Chavis, J. A., and Bhatt, R. (2003). Sex steroid regulation of microglial cell activation. Relevance to multiple sclerosis. *Ann. NY Acad. Sci.* **1007,** 329–334.

Elenkov, I. J., Wilder, R. L., Bakalov, V. K., Link, A. A., Dimitrov, M. A., Fisher, S., Crane, M., Kanik, K. S., and Chrousos, G. P. (2001). IL-2, TNF-alpha, and hormonal changes during late pregnancy and early postpartum: Implications for autoimmune disease activity during these times. *J. Clin. Endocrinol. Metab.* **86,** 4933–4938.

El-Etr, M., Vukusic, S., Gignoux, L., Durand-Dubief, F., Achiti, I., Baulieu, E. E., and Confavreux, C. (2005). Steroid hormones in multiple sclerosis. *J. Neurol. Sci.* **233,** 49–54.

Giesser, B. S. (2002). Gender issues in multiple sclerosis. *Neurologist* **8,** 351–356.

Gilmore, W., Weiner, L. P., and Correale, J. (1997). Effect of estradiol on cytokine secretion by proteolipid protein-specific T cell clones isolated from multiple sclerosis patients and normal control subjects. *J. Immunol.* **158,** 446–451.

Jung-Testas, I., Schumacher, M., Robel, P., and Baulieu, E. E. (1996). The neurosteroid progesterone increases the expression of myelin proteins (MBP and CNPase) in rat oligodendrocytes in primary culture. *Cell Mol. Neurobiol.* **16,** 439–443.

Kantarci, O. H., Hebrink, D. D., Achenbach, S. J., Atkinson, E. J., de Andrade, M., McMurray, C. T., and Weinshenker, B. G. (2004a). CD95 polymorphisms are associated with susceptibility to MS women: A population-based study of CD95 and CE95L in MS. *J. Neuroimmunol.* **146,** 62–70.

Kantarci, O. H., Hebrink, D. D., Achenbach, S. J., Pittock, S. J., Altintas, A., Schaefer-Klein, J. L., Atkinson, E. J., De Andrade, M., McMurray, C. T., Rodriguez, M., and Weinshenker, B. G. (2004b). Association of APOE polymorphisms with disease severity in MS is limited to women. *Neurology* **62,** 811–814.

Kim, S., Livam, S. M., Dalal, M. A., Verity, M. A., and Voskuhl, R. R. (1999). Estriol ameliorates autoimmune demyelinating disease. *Neurology* **52,** 1230–1238.

Koenig, H. L., Schumarcher, M., Robel, P., Ferzaz, B., Thi, A. N., Ressouches, A., Guennoun, R., Jung-Testas, I., Robel, P., Akwa, Y., and Baulieu, E. E. (1995). Progesterone synthesis and myelin formation by Schwann cells. *Science* **268,** 1500–1503.

Kurtz, J., Wiesner, A., Gotz, P., and Sauer, K. P. (2000). Gender differences and individual variation in the immune system of the scorpionfly Panorpa vulgaris (Insecta: Mecoptera). *Dev. Comp. Immunol.* **24,** 1–12.

Matarese, G., Sanna, V., Di Giacomo, A., Lord, G. M., Howard, J. K., Bloom, S. R., Lechler, R. I., Fontana, S., and Zappacosta, S. (2001). Leptin potentiates experimental autoimmune encephalomyelitis in SJL female mice and conferes susceptibility to males. *Eur. J. Immunol.* **31,** 1324–1332.

Nino, M., Kikuchi, S., Fukazawa, T., Yabe, I., and Tashiro, K. (2000). Estrogen receptor gene polymorphism in Japanese patients with multiple sclerosis. *J. Neurol. Sci.* **179,** 70–75.

Noonan, C. W., Kathman, S. J., and White, M. C. (2002). Prevalence estimates for MS in the United States and evidence of an increasing trend for women. *Neurology* **58,** 136–138.

Novantrone package insert (2003). Teva Neuroscience, Kansas City (MO).

Palaszynski, K. M., Liu, H., Loo, K. K., and Voskuhl, R. R. (2004a). Estriol treatment ameliorates disease in males with experimental autoimmune encephalomyelitis: Implications for multiple sclerosis. *J. Neuroimmunol.* **149,** 84–89.

Palaszynski, K. M., Loo, K. K., Ashouri, J. F., Liu, H. B., and Voskuhl, R. R. (2004b). Androgens are protective in experimental autoimmune encephalomyelitis: Implications for multiple sclerosis. *J. Neuroimmunol.* **146,** 144–152.

Polanczyk, M. J., Carson, B. D., Subramanian, S., Afentoulis, M., Vandenbark, A. A., Ziegler, S. F., and Offner, H. (2004). Cutting edge: Estrogen drives expansion of the CD4+ CD25+ regulatory T cell compartment. *J. Immunol.* **173,** 2227–2230.

Rashid, W., Davies, G. R., Chard, D. T., Griffin, C. M., Altmann, D. R., Gordon, R., Kapoor, R., Thompson, A. J., and Miller, D. H. (2006). Upper cervical cord area in early relapsing-remitting multiple sclerosis: Cross-sectional study of factors influencing cord size. *J. Magn. Reson. Imaging.* **23**(4), 473–476.

Rivier, C., and Vale, N. (1990). Cytokines act within brain to inhibit LH secretion and ovulation in the rat. *Endocrinology* **127,** 849–859.

Rooney, W. D., and Coyle, P. K. (2005). Recent advances in the neuroimaging of multiple sclerosis. *Curr. Neurol. Neurosci. Rep.* **5,** 217–224.

Runmaker, B., and Andersen, O. (1995). Pregnancy is associated with a lower risk of onset and a better prognosis in multiple sclerosis. *Brain* **118,** 253–261.

Sadovnick, A. D., Remick, R. A., Allen, J., Swartz, E., Yee, I. M., Eisen, K., Farquhar, R., Hashimoto, S. A., Hooge, J., Kastrukoff, L. F., Morrison, W., Nelson, J., *et al.* (1996). Depression and multiple sclerosis. *Neurology* **46,** 628–632.

Sandberg-Wollheim, M., Frank, D., Goodwin, T. M., Giesser, B., Lopez-Bresnaham, M., Stam-Moraga, M., Chang, P., and Francis, G. S. (2005). Pregnancy outcomes during treatment with interferon beta-1a in patients with sclerosis. *Neurology* **65,** 788–799.

Savettieri, G., Messina, D., Andreoli, V., Bonavita, S., Caltagirone, C., Cittadella, R., Farina, D., Fazio, M. C., Girlanda, P., Le Pira, F., Liguori, M., Lugaresi, A., *et al.* (2004). Gender-related effect of clinical and genetic variables on the cognitive impairment in multiple sclerosis. *J. Neurol.* **251,** 1208–1214.

Shabas, D., and Weinreb, H. (2000). Preventative healthcare in women with multiple sclerosis. *J. Womens. Health* **9,** 389–395.

Shomer, D. (1998). Ovarian hormones and the nervous system. *In* "Neurologic Disease in Women" (D. Kaplan, Ed.), pp. 45–53. Demos Press, New York.

Sicotte, N. L., Liva, S. M., Klutch, R., Pfeiffer, P., Bouvier, S., Odesa, S., Wu, T. C., and Voskuhl, R. R. (2002). Treatment of multiple sclerosis with the pregnancy hormone estriol. *Ann. Neurol.* **52,** 421–428.

Stenager, E., Stenager, E. N., and Jensen, K. (1994). Effect of pregnancy on the prognosis for multiple sclerosis, a 5-year follow-up investigation. *Acta Neurol. Scand.* **90,** 305–308.

Thompson, D. S., Nelson, L. M., Burks, A., and Franklin, G. M. (1987). The effects of pregnancy in multiple sclerosis: A retrospective study. *Neurology* **36,** 1097–1099.

Tomassini, V., Giugni, E., Mainere, C., Onesti, E., Giuliani, S., Paolillo, A., Bastianello, S., Salvetti, M., and Pozzilli, C. (2001). Relationship between sex hormones and MRI activity in relapsing-remitting multiple sclerosis. *Neurology* **56**(Suppl. 3), A254–A255.

Tomassini, V., Onesti, E., Mainero, C., Giugni, E., Paolillo, A., Salvetti, M., Nicoletti, F., and Pozzilli, C. (2005). Sex hormones modulate brain damage in multiple sclerosis. *J. Neurol. Neurosurg. Psychiatry* **76,** 272–275.

Whitacre, C. C. (2001). Sex differences in autoimmune disease. *Nat. Immunol.* **2,** 777–780.

Wilson, S., Donnan, P., Swingrem, R., *et al.* (2004). A serial observational study on hormonal influences in MS symptoms [abstract]. *Mult. Scler.* SS188–SS189, Poster 382.

Zordrager, A., and DeKeyser, J. (1998). Premenstrual exacerbations in multiple sclerosis. *J. Neurol. Neurosurg. Psychiatry* **65,** 279–280.

DIFFERENTIAL DIAGNOSIS OF MULTIPLE SCLEROSIS

Halim Fadil, Roger E. Kelley, and Eduardo Gonzalez-Toledo

Department of Neurology, Louisiana State University Health Sciences Center
Shreveport, Louisiana 71103, USA

INTERNATIONAL REVIEW OF
NEUROBIOLOGY, VOL. 79
DOI: 10.1016/S0074-7742(07)79018-9

0074-7742/07 $35.00

There are a number of illnesses that can mimic multiple sclerosis (MS). This pretty much includes any pathological process that can reflect injury to the central nervous system either in a transient or progressive basis. Typically, MS presents itself in individuals in their teens up to their late 30s. On occasion, however, one can see MS present in patients in their 60s. However, in retrospect, many of these patients might have had subtle manifestations of MS in their younger years. Visual obscuration or visual loss can be a manifestation of retinal ischemia, retinal migraine, or optic neuritis which might or might not evolve into a clinical picture compatible with MS. Cranial neuropathy, long tract signs, sensory disturbance, and/or gait ataxia can be related to a number of different processes such as illicit drug use, neurosarcoidosis, neuro-Behcet's disease, neuroborreliosis, HIV-related disease, neurosyphilis, vascular occlusive disease including vasculitis, connective tissue disorders, acute disseminated encephalomyelitis (ADEM), idiopathic transverse myelitis, neuromyelitis optica (NMO), or tropical spastic paraparesis. In addition, a constellation of symptoms, with questionable objective findings, along with normal MRI imaging, normal CSF results, and normal evoked response testing, when indicated, might identify a conversion disorder or possibly malingering. There are now established criteria for the diagnosis of MS, but initial presentations can be less than "textbook" in nature. With the advent of immunomodulating therapy, it has become more important to diagnose MS more effectively earlier on in the course of the illness. Prior to specific therapy for MS, astute clinicians did not necessarily move with alacrity to establish the diagnosis in patients with subtle or transient manifestations. This was in recognition of the fact that little could be offered to alter the course of the illness and a number of patients might never experience further problems if they were lucky enough to have their illness go into permanent remission after one minor exacerbation.

I. Introduction

Multiple sclerosis (MS) is the most common demyelinating disease of the central nervous system (CNS) and it has a genetic susceptibility (Noseworthy *et al.*, 2000). It is also the most common cause of neurological disability in young adults. It affects women more commonly than men with a preference for people of Northern European descent. MS is also more prevalent in populations located further north from the equator in the Northern Hemisphere and further south from the equator in the Southern Hemisphere. It typically manifests itself between the ages of 15 and 45. The disease is characterized pathologically by multifocal areas of demyelination in the white matter. However, there is increasing evidence suggesting that the disease process may include damage of the gray matter and the axons as well (Bitsch *et al.,* 2000). Clinically, it is a very heterogeneous entity, so that in certain cases, the diagnosis can be very challenging.

Clinicians have been striving for many years to establish diagnostic criteria that would allow more ready, and reliable, diagnosis of MS. The fundamental diagnostic principle of the occurrence of signs and symptoms over two distinct periods in time (dissemination in time) as well as the involvement of at least two separate sites in the CNS (dissemination in space) was first laid down by Schumaker *et al.* (1965). New diagnostic criteria were presented by Poser *et al.* (1983) to incorporate the value of a number of laboratory studies including cerebrospinal fluid (CSF) analysis, evoked potentials, and neuroimaging. To reflect the increasing importance of magnetic resonance imaging (MRI) in demonstrating demyelinating lesions, compatible with MS, which are disseminated in space and time, the International Panel on the Diagnosis of MS published modified criteria (McDonald *et al.*, 2001).

These modified criteria have come to be known as the McDonald criteria and they have been found to be useful for classifying the patient's diagnosis as MS, "possible MS," or "not MS." Polman *et al.* (2005) introduced a revision to the McDonald criteria to simplify the diagnosis of primary progressive MS and clarify the role of spinal cord lesions. Even though, the diagnosis has been facilitated by the use of MRI and the establishment of diagnostic criteria, it can still be difficult to be sure about the diagnosis in certain patients. This is particularly important with the recognition of the expense involved in immunomodulating therapy and the potential for not insignificant side effects of these agents which must be administered by injection. To this day, there is no test finding which is pathognomonic for MS. Moreover, many conditions can mimic MS either clinically, immunologically, or radiographically. Therefore, it is important for the clinician to keep a high level of suspicion for these conditions and investigate them, when clinically appropriate. An accurate diagnosis is especially important at a time when disease modifying drugs are available to alter the course of MS and to potentially improve the patients' quality of life.

The differential diagnosis of MS can be divided into several pathophysiological categories: inflammatory/autoimmune diseases, infectious diseases, genetic/hereditary disorders, neoplastic diseases, and other demyelinating diseases (Table I).

II. Differential Diagnosis of Inflammatory/Autoimmune Diseases

A. Systemic Lupus Erythematosus

Systemic lupus erythematosus (SLE) is an autoimmune multisystem connective tissue disease of unknown cause. It affects the nervous system in between 25% and 80% of the cases with a wide spectrum of neurological and psychiatric

TABLE I
DIFFERENTIAL DIAGNOSIS OF MULTIPLE SCLEROSIS

Inflammatory/autoimmune diseases	*Dysmyelinating disorders*
Systemic lupus erythematosus	Metachromatic leukodystrophy
Antiphospholipid antibody syndrome	Krabbe's leukoencephalopathy
Primary Sjogren's syndrome	Multiple sulfatase deficiency
Neurosarcoidosis	Alexander disease
Neuro-Behcet's disease	Adrenoleukodystrophy
Primary angiitis of the CNS	Pelizaeus-Merzbacher disease
Polyarterits nodosa	*Other demyelinating disorders*
Wegener's granulomatosis	Neuromyelitis optica (Devic disease)
Susac syndrome	Acute disseminated encephalomyelitis
Infectious etiologies	Concentric sclerosis
Neuroborreliosis	Schilder's disease
Progressive multifocal leukoencephalopathy	Optic neuritis
	CNS neoplasms
Tropical spastic paraparesis	CNS lymphoma
HIV-related disorders	
Genetic/hereditary disorders	
Migraine	
CADASIL	
Leber's hereditary optic atrophy	

manifestations (Sanna *et al.*, 2003). The most common are headache syndromes, cerebrovascular disease, mood disorders, cognitive dysfunction, seizures, psychosis, anxiety disorder, acute confusional state, polyneuropathy, mononeuritis, cranial neuropathy, and movement disorders. Other less common syndromes are aseptic meningitis, Guillain–Barre syndrome, transverse myelitis, and optic neuritis.

In rare instances, the neuropsychiatric form of SLE (NPSLE) can have a relapsing course and be multifocal in its symptomatology resembling the relapsing-remitting form of MS. MRI of the brain may reveal atrophy, cortical and deep gray matter lesions related to ischemia (Figs. 1 and 2), as well as white matter lesions that are more often in the subcortical area. The location of these lesions usually helps to differentiate NPSLE from MS. The laboratory tests may not be as useful since the CSF in SLE patients may reveal elevated protein, lymphocytic pleocytosis, oligoclonal bands, and increased IgG index just like in MS. On the other hand, the serum may falsely test positive for antinuclear antibodies (ANA) in as many as 81% of MS patients (Dore-Duffy *et al.*, 1982). The clinician should also keep in mind that cases of coexistent MS and SLE have been reported (Hietaharju *et al.*, 2001).

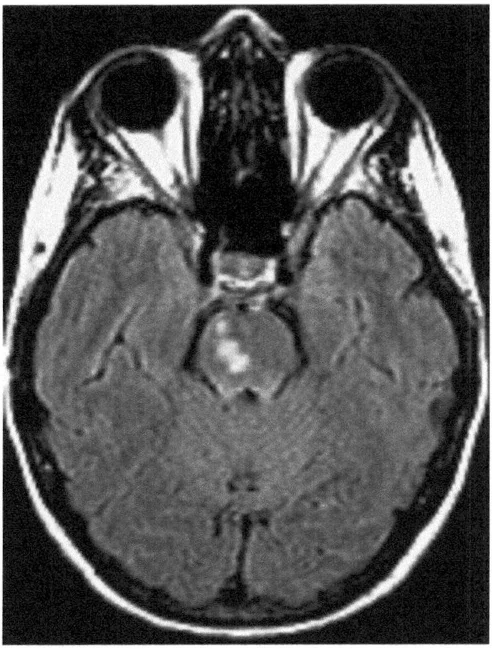

FIG. 1. Axial FLAIR sequence of brain magnetic resonance imaging (MRI) showing a hyperintense lesion in the pons in a patient with SLE.

B. Antiphospholipid (Hughes) Syndrome

The antiphospholipid syndrome (APS) is characterized by arterial and/or venous thrombosis in the presence of antiphospholipid antibodies (anticardiolipin antibodies and/or lupus anticoagulant) (Hughes, 1983). In female patients, it may be associated with significant obstetric complications such as recurrent spontaneous abortions. APS can occur either as a primary disorder or in association with a connective tissue disease, most frequently SLE. As highlighted by Hughes (1983), in the first description of the syndrome, the neurological manifestations can be prominent and varied. The most common are cerebrovascular events such as transient ischemic attacks, ischemic strokes, acute ischemic encepahalopathy, and cerebral venous thrombosis. Other syndromes reported are epilepsy, headache, chorea, transverse myelitis, Guillain–Barre syndrome, cognitive dysfunction, dementia, psychiatric disorders, optic neuropathy, and other ocular syndromes (Levine *et al.*, 1990). Recurrent clinical syndromes with symptoms and signs mimicking MS, mainly in its relapsing-remitting form, can occur (Cuadrado *et al.*, 2004, Scott *et al.*, 1994). In many of these cases MRI studies show white matter T2 hyperintense lesions (Fig. 3). Cortical or deep gray

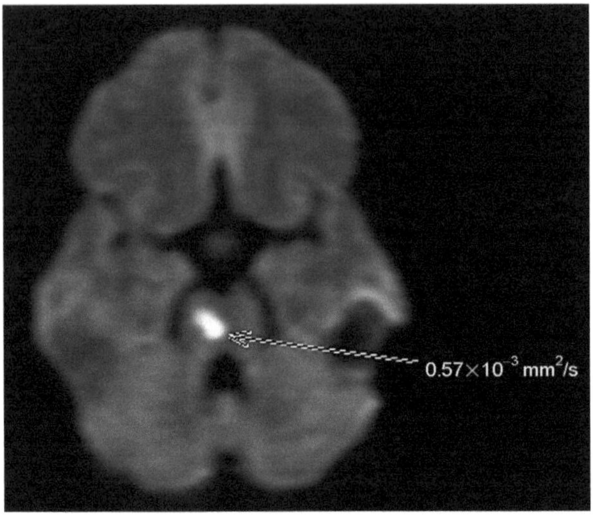

FIG. 2. Diffusion weighted imaging (DWI) sequence of brain MRI showing a hyperintense lesion in DWI with a low ADC value (0.57×10^{-3} mm^2/s) consistent with an acute infarct in the same patient imaged in Fig. 1.

matter lesions, if present, are in favor of an APS diagnosis (Fig. 4). Some authors noted that static lesions seen in a repeat MRI of the brain would suggest APS. In terms of CSF exam, the presence of oligoclonal bands does not exclude the diagnosis of APS. However, their absence is usually considered as evidence against a diagnosis of MS.

In reference to serum antiphospholipid antibodies, they are not specific to APS since they can be found in association with other diseases including MS and ischemic stroke.

The distinction between APS and MS can be so difficult that some authors recommended that every MS patient should routinely be tested for antiphospholipid antibodies and even undergo a trial of anticoagulant therapy, especially if it is an atypical form of MS with any history suggestive of APS (Ferreira *et al.*, 2005).

C. Primary Sjogren's Syndrome

Primary Sjogren's syndrome (pSS) is a chronic inflammatory autoimmune disorder. It is characterized by lymphocytic infiltration of the salivary and lacrimal glands resulting in xerostomia and xerophtalmia. However, various organs may be involved as well. CNS manifestations have been reported in 20–25% of SS patients (Delalande *et al.*, 2004) and can be the initial manifestation of the disease in a good portion of patients. Typical symptoms include motor and sensory deficits, cognitive impairment, aphasia, dysarthria, dysphagia, dizziness, seizures,

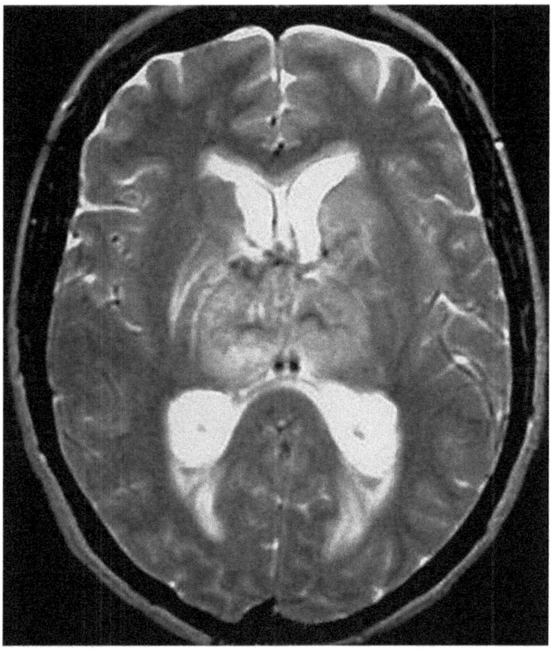

FIG. 3. Axial FLAIR sequence of brain MRI showing bilateral hyperintense demyelinating lesions in postcommissural fornix in a patient presenting with antiphospholipid syndrome.

and cerebellar or brain stem syndromes. Other manifestations include encephalopathy, aseptic meningitis, encephalitis, transverse myelitis, as well as involvement of the peripheral nervous system (PNS) with polyneuropathy, mononeuritis multiplex, and entrapment neuropathy. Initial manifestations with optic neuritis have been reported (Kadota *et al.*, 2002). The MRI findings may simulate MS, even though they tend to be less pronounced and more difficult to interpret in pSS (Morgen *et al.*, 2004). CSF examination may reveal lymphomonocytic pleocytosis with normal or elevated protein and normal or reduced glucose.

The IgG index is increased in up to 50% and oligoclonal bands are present in up to 90% of the cases. However, the latter seem to be less stable than in MS since they can disappear after corticosteroid treatment (Reske *et al.*, 2005). In addition, MS patients most commonly have multiple bands, whereas SS patients usually have one or two bands. In the serum, most SS patients have positive antibodies (ANA, anti-Ro, and anti-La), however their absence does not exclude the diagnosis of SS. pSS can mimic MS either in its relapsing-remitting form or primary progressive one (Alexander *et al.*, 1986). Its typical clinical features (Sicca syndrome) and serological antibodies should be sought before making a definite diagnosis of MS (Pericot *et al.*, 2003). A salivary gland biopsy may be necessary in very suspicious cases.

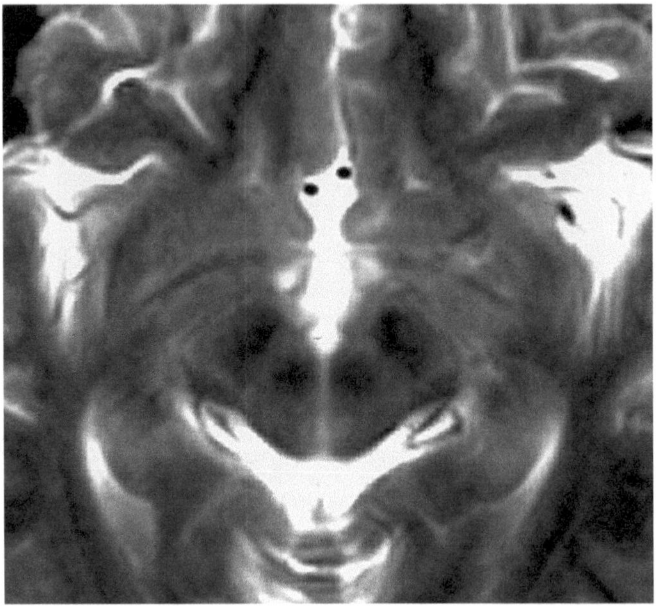

FIG. 4. Axial T2 sequence of brain MRI showing deep venous thrombosis with bilateral thalamic infarction and demyelination in a 30-year-old patient, post-C section, diagnosed with antiphospholipidic syndrome.

D. Neurosarcoidosis

Sarcoidosis is an inflammatory multisystem disease of unknown cause. It is characterized pathologically by the presence of noncaseating epitheloid granulomata. It affects most commonly the lungs, although no organ is immune. The prevalence of the nervous system involvement is estimated to range from 5% to 10% (Stern *et al.*, 1985) and can be the initial manifestation of the disease in about half of these patients (Sharma, 1997). Any part of the nervous system can be affected with a preference for the cranial nerves, the hypothalamus, and the pituitary gland. Optic neuropathy, brainstem, and spinal cord involvement can cause a clinical picture similar to MS. Of particular note, multifocal lesions occur in about one-third of patients with neurosarcoidosis.

The disease can have either a relapsing-remitting course or a progressive one. As for the neuroimaging presentation, it may take the form of an isolated mass lesion, diffuse intraparenchymal lesions in the brain and spinal cord, leptomeningeal enhancement, hydrocephalus, intracranial hemorrhage, or MS-like periventricular white matter lesions (Figs. 5 and 6; Smith *et al.*, 1989). The CSF analysis may not be very useful in distinguishing MS from sarcoidosis since it may reveal similar abnormalities in both conditions (McLean *et al.*, 1995). Nevertheless, evidence of

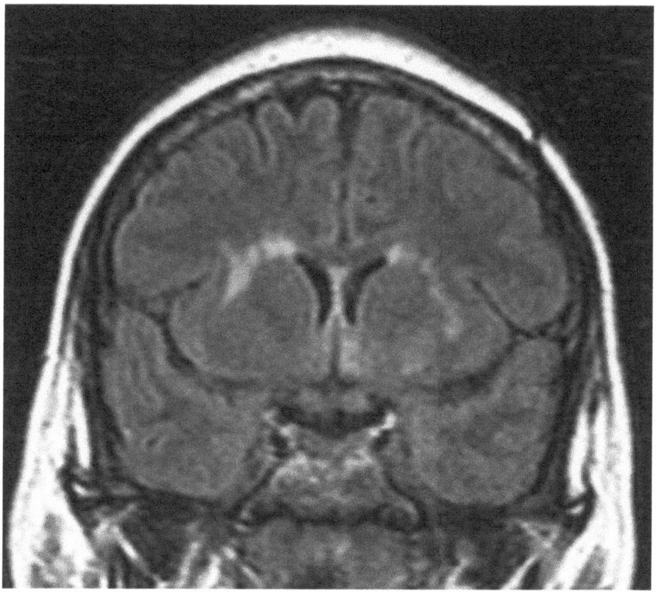

FIG. 5. Coronal FLAIR sequence of brain MRI showing demyelination involving the external capsule bilaterally in a patient with sarcoid.

involvement outside the nervous system, elevated serum and CSF angiotensin coverting enzyme, leptomeningeal enhancement on scans, and tissue biopsy can all help to support or confirm the diagnosis of sarcoidosis. It is also helpful to recognize that this disorder tends to be much more common in African-Americans.

E. Neuro-Behcet's Disease

Neuro-Behcet's disease is a multisystem inflammatory disorder of unknown cause. It is characterized clinically by a classic triad of oral ulcers, genital ulcers, and recurrent anterior uveitis. It can also include vasculopathy and aseptic meningitis. Other organs typically affected are the gastrointestinal tract, the lungs, the skin, and the joints. Although neurological involvement is not uncommon, it is rarely the initial manifestation of the disease (Kidd *et al.,* 1999). The most common manifestations are brainstem syndromes, including intranuclear ophthalmoplegia, psychiatric symptoms, headache, pyramidal signs, cognitive disturbance, as well as sensory and/or motor deficits. Isolated cerebellar syndromes are less frequent and optic neuritis seems to be uncommon.

The course of neuro-Behcet's disease can be relapsing-remitting, although it can also be progressive (Akman-Demir *et al.,* 1999). From a laboratory standpoint,

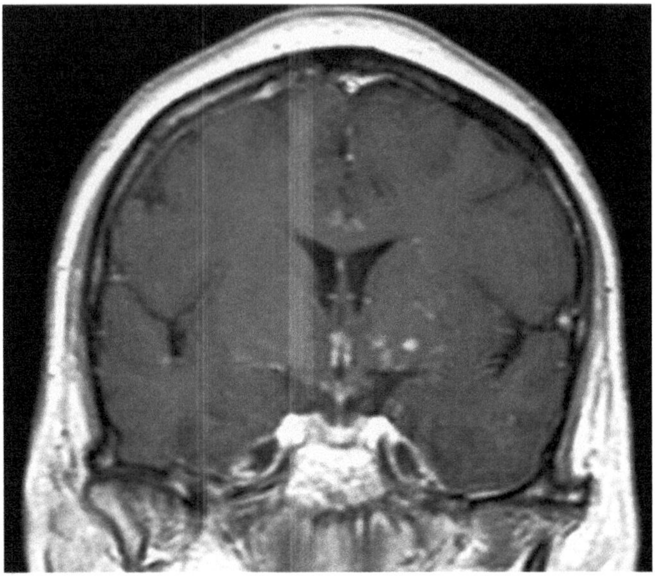

FIG. 6. Coronal T1 sequence of brain MRI after contrast (Gd-DTPA) showing multiple small hyperintensities in the left putamen and ventral striatum with leptomeningeal enhancement in a patient with sarcoid.

the most common abnormalities in the CSF are elevated protein and pleocytosis. The latter is frequently granulocytic unlike that typically seen in MS (Sharief *et al.*, 1997). Oligoclonal bands can be detected. However, they are less stable than in MS and can be suppressed by corticosteroid treatment. On brain MRI, the lesions are typically located in the diencephalon and brainstem (Fig. 7), but sometimes can involve the supratentorial white matter (Tali *et al.,* 1997). In general, neuro-Behcet's disease can easily be diagnosed when other organs are affected and its diagnostic criteria are fulfilled. The confusion arises when the initial manifestation is a neurological attack. In this case, a prominent involvement of the brainstem and basal ganglia, with a granulocytic pleocytosis in the CSF, can help support the diagnosis of neuro-Behcet's disease.

F. Primary Angiitis of the CNS

Primary angiitis of the CNS (PACNS), also known as isolated vasculitis of the CNS or granulomatous angiitis of the CNS, is an inflammatory disorder that affects small- and medium-sized arteries of the meninges and parenchyma in the cerebrum and the spinal cord. Its cause is unknown. Clinically, it may present with nonspecific symptoms or signs such as headache, cognitive deficits, personality

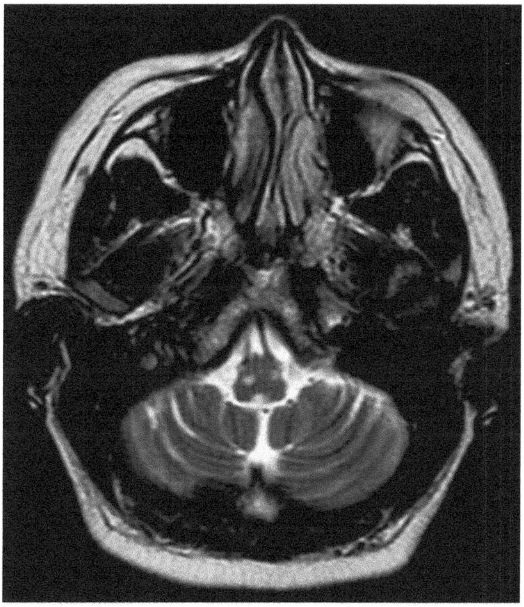

FIG. 7. Axial T2 sequence of brain MRI showing hyperintense lesions in the medulla in a patient with neuro-Behcet's disease.

changes, encephalopathy, paresis, seizures, cranial neuropathy, or myelopathy (Calabrese *et al.,* 1997). Strokelike episodes, as well as intracerebral and subarachnoid hemorrhages, have been reported. The CSF is abnormal in 80–90% of PACNS cases. The CSF most commonly demonstrates an elevated protein and lymphocytic pleocytosis. The IgG synthesis can be increased and oligoclonal bands may be present. Brain MRI may reveal patchy foci or diffuse increased signal in periventricular, deep, and subcortical white matter (Berger *et al.,* 1998, Finelli *et al.,* 1997). Involvement of gray matter structures, or of a specific vascular territory, as well as meningeal or vascular enhancement, is very suggestive of cerebral vasculitis (Shoemaker *et al.,* 1994). The diagnosis usually requires demonstration of vasculitic changes either by cerebral angiography or by meningeal biopsy.

G. Polyarteritis Nodosa

Polyarteristis nodosa (PAN) is a systemic necrotizing vasculitis that affects medium- and small-sized vessels. The nervous system is frequently involved. Neuropathy occurs in 50–75% of the cases, and the CNS is involved in up to 40% of the patients (Siva, 2001). The neurological manifestations vary from encephalopathy with cognitive deficits and seizures to cranial nerve palsies and strokelike episodes (Reichart *et al.*, 2000). MS-like presentations have been reported (Waisberg *et al.*, 1974). MRI of the brain may reveal multiple white matter lesions, and the CSF may show pleocytosis with elevated protein level as well as presence of oliglonal bands (Paula De Carvalho Panzerri *et al.*, 2004). Nevertheless, CNS involvement tends to occur late in the course of the disease, and evidence of other organs involvement will help delineating PAN from MS.

H. Wegener's Granulomatosis

Wegener's granulomatosis is a systemic vasculitis of the small- and medium-sized arteries and veins (DeRemee, 2006). It is a relatively rare disease of unknown etiology. It is usually associated with the presence of circulating antineutrophil cytoplasm antibodies (c-ANCA) which are considered a useful serological marker for the disease. However, ANCA-negative cases have been reported in the medical literature. Histologically, it is characterized by granulomatous inflammation, necrosis, and vasculitis. It primarily affects the head and neck area, the lungs, and the kidneys; although it may involve virtually any organ. Neurological involvement has been reported in a range of 22–54% with mononeuritis multiplex and distal neuropathy being the most common manifestations. CNS involvement occurs less frequently in about 8–18% of the cases. Three main forms have been described: chronic hypertrophic pachymeningitis, cerebral vasculitis, and pituitary gland disease. The symptoms and syndromes that can occur are headache, cranial nerves palsies, limb paresis, seizures, optic neuritis, arterial or venous thrombosis, transient ischemic attacks, strokes, intracranial or subarachnoid hemorrhage, and myelopathy (Seror *et al.*, 2006). The CSF can be normal or show a lymphocytic pleocytosis with elevated protein (Weinberger *et al.*, 1993). The clinical picture can sometimes resemble MS; however, there is usually evidence of other organs involvement. In addition, MR imaging can provide useful diagnostic information by showing an enlarged pituitary gland, focal or diffuse dural thickening and enhancement contiguous with orbital, nasal, or paranasal disease (Murphy *et al.*, 1999; Provenzale *et al.*, 1996). However, it may sometimes show only nonspecific vasculitic disease (Fig. 8). Tissue biopsy may be necessary for diagnosis.

I. Susac Syndrome

Susac syndrome is a rare disorder of unknown etiology. It typically affects females between ages of 20 and 40 years with partial or complete recovery within 2 years. It is characterized by a microangiopathy that affects the arterioles of the brain, retina, and cochlea resulting in a clinical triad of visual loss (related to branch retinal artery occlusions), sensorineural hearing loss, and a subacute encephalopathy. The CSF may reveal a modest pleocytosis with marked protein elevation. In cases lacking all three findings at presentation, the presence of hyperintense lesions white

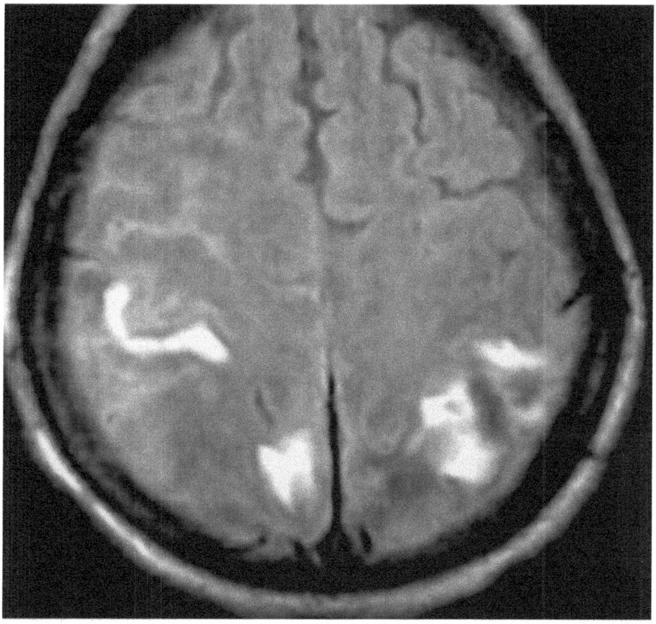

FIG. 8. Axial FLAIR sequence of brain MRI showing gyral hyperintensities in an immunosupressed patient with vasculitis, diagnosed with Wegener's granulomatosis.

matter lesions, with a predilection for the corpus callosum, on T2-weighted MR images, as well as possible deep gray matter and leptomeningeal involvement may be a clue to the diagnosis (Susac *et al.*, 2003).

III. Differential Diagnosis of Infectious Etiologies

A. Lyme Disease (Neuroborreliosis)

This illness is typically preceded by a tick bite, resulting in infection with the spirochete, *Borrelia burgdorferi*. It is usually associated with a characteristic rash (*erythema chronica migrans*). There can be cranial nerve palsies or an encephalitic picture. There is a predilection for the seventh cranial nerve and one generally considers Neuroborreliosis in an endemic area. There is usually little likelihood of confusion with MS as the meningitic involvement, on MRI, is usually easily distinguished from the expected white matter plaques of MS. In addition, the CSF pleocytosis with elevated Lyme titers should allow ready distinction. However, this is not always the case (Drozdowski, 2006).

B. Neurosyphilis

One can see either meningovascular in late secondary early tertiary phase of neurosyphilis or parenchymal involvement, with gummatous lesions, in more advanced tertiary stages. There appears to be an enhanced susceptibility in HIV-infected patients.

An atypical stroke-like presentation tends to be most characteristic of meningovascular syphilis while contrast-enhancing gummas in later stages may potentially look like MS.

Both illnesses can result in CSF pleocytosis with elevated CSF gammaglobulin and the presence of oligoclonal bands. Despite the fact that neurosyphilis can still be the "great imitator" (Fadil *et al.*, 2006), the presence or absence of a serum fluorescent treponemal antibody study (FTA-ABS) should provide essentially 100% sensitivity for infection with syphilis. Furthermore, a positive CSF venereal disease research laboratory (VDRL) titer is quite specific for CNS infection with syphilis, but the sensitivity is not 100%.

C. Progressive Multifocal Leucoencephalopathy

Progressive multifocal leucoencephalopathy (PML) is generally seen in immunocompromised patients and it is a rather rare entity. It is related to activation of JC virus, which is in the papovaviruses family, in susceptible individuals. There has been recent notoriety related to its association with the agent natalizumab, a drug released for the relapsing-remitting form of MS which was not responsive to presently available immunomodulating therapy (Berger and Koralnik, 2005). The risk of PML with this agent is of the order of 1 in 1000. The drug was recently re-released with specific indications and guidelines in term of monitoring. Generally, PML is associated with more extensive white matter lesions which do not contrast enhance on MRI brain scans (Figs. 9 and 10). The lesions tend to be multifocal and become confluent over time. The CSF is usually normal. Unlike MS, this tends to be a rapidly progressive disorder usually resulting in death within several months.

D. Tropical Spastic Paraparesis

Tropical spastic paraparesis is caused by human T-lymphotrophic virus, type 1 (HTLV-1). It is also known as HTLV-1-associated myelopathy or (HTLV-1)-associated myelopathy/tropical spastic paraparesis (HAM/TSP) and can be confused with MS (Howard *et al.*, 2003). This tends to be a sexually transmitted disorder that is

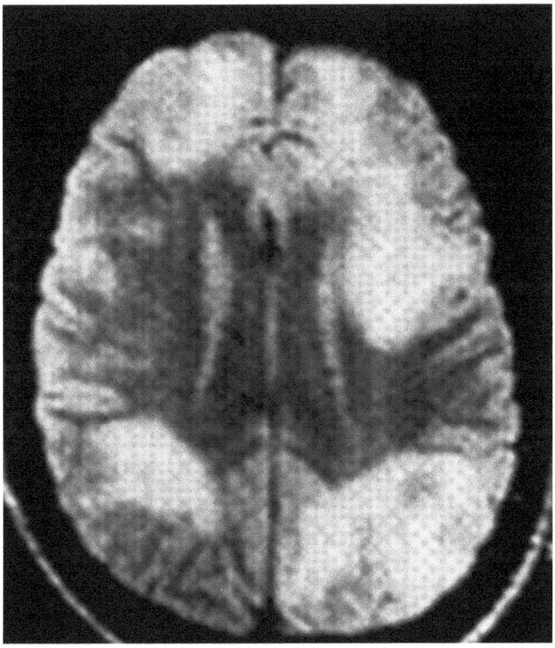

FIG. 9. Axial FLAIR sequence of brain MRI showing multiple large hyperintensities involving mainly the white matter in an HIV patient diagnosed with progressive multifocal leukoencephalopathy (PML).

most prevalent in the tropics. It can mimic MS in terms of spinal cord involvement and characteristic imaging on MRI scan of the spinal cord (Bagnato *et al.*, 2005). However, it tends to be a subacute, progressive myelopathy and can be identified by a positive HTLV-1 titer in the CSF. As in MS, there is often a mild lymphocytic pleocytosis in the CSF along with mildly elevated total protein and the presence of oligoclonal bands. There can be special difficulty in the differentiation when the clinical course is relapsing (Umehara *et al.*, 2006).

E. HIV-Related Disorders of the CNS

There are a number of CNS manifestations of HIV infection. In Stage C of disease, one can see HIV encephalopathy, CNS lymphoma, cerebral toxoplasmosis, PML, cryptococcal meningitis, mycobacterial meningitis, and there appears to be an association with tertiary syphilis both in terms of sexually transmitted disease and enhanced CNS susceptibility because of immunosuppression. In general, unexplained CNS disease in a person at risk for HIV infection should lead to investigation of this possibility.

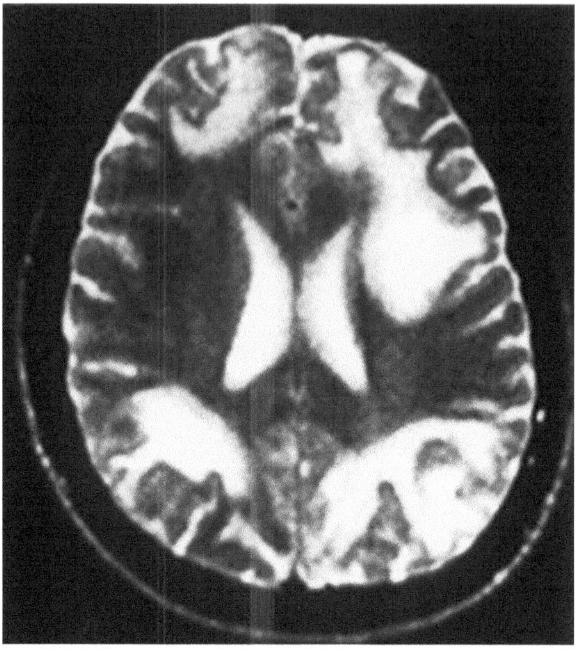

FIG. 10. Axial T2-weighted image of a brain MRI showing multiple large white matter hyperintensities in an HIV patient with PML.

IV. Differential Diagnosis of Genetic/Heriditary Disorders

A. Migraine

Migraine is often familial and affects up to 10% of the population, preferably women. It can be associated with an array of neurological manifestations including sensory loss, visual loss or distortion, loss of speech, ophthalmoparesis, confusion, true vertigo, and it can progress to migrainous (cerebral) infarction. It is a common explanation for transient neurological symptoms and is usually associated with a characteristic throbbing head pain along with light and sound sensitivity and not uncommonly nausea. Generally, the temporal relationship with the head pain identifies migraine as the mechanism for the neurological accompaniments. The latter can occur before the headache in the case of migraine with aura or during the headache "complicated migraine." There are occasionally neurological manifestations without the headache "amigrainous migraine."

The confusion with MS can occur when there are prominent transient neurological symptoms along with white matter areas of increased signal intensity on the

T2-weighted MRI (Igarashi *et al.*, 1991). These areas are typically small and do not enhance, but they tend to be located in the deep frontal white matter or centrum semiovale (Figs. 11 and 12). The MRI report often includes MS in the differential diagnosis, especially when the lesions are prominent. Of pathophysiological interest, there are gray matter changes seen in association with the T2-weighted lesions which appear to correlate with disease duration (Rocca *et al.*, 2006). To add to the confusion, clinically, the age group for developing symptoms of migraine and symptoms of MS tends to overlap and both disorders tend to be more common in women. Furthermore, it is not at all uncommon to have coexistent migraine and MS.

B. Cerebral Autosomal Dominant Arteriopathy with Subcortical Infarcts and Leukoencephalopathy

Cerebral autosomal dominant arteriopathy with subcortical infarcts and leukoencephalopathy (CADASIL) generally is associated with a strokelike picture (Dichgans *et al.*, 1998). However, migraine is also a common accompaniment. This disorder is caused by a mutation in the *Notch 3* gene. Patients often present

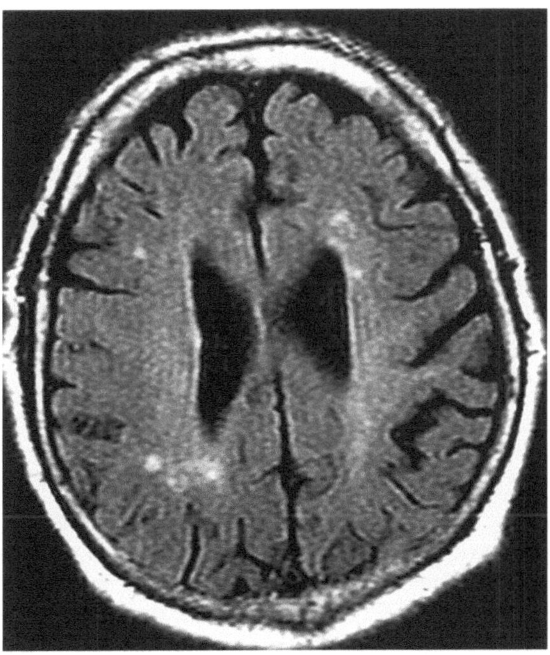

Fig. 11. Axial FLAIR sequence of brain MRI showing multiple nonspecific hyperintensities (UBOs = unidentified bright objects) in a patient with history of migraine headache.

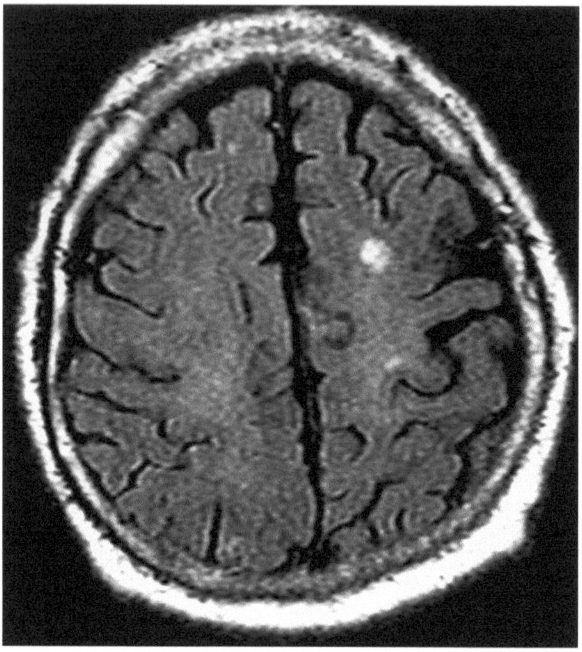

FIG. 12. Axial FLAIR sequence of brain MRI showing multiple nonspecific hyperintensities in another patient with history of migraine headache.

with symptoms of transient ischemic attack or ischemic stroke before the age of 40. As the disease progresses, patients typically manifest long tract signs and dementia. There are prominent areas of white matter ischemic demyelination which progress over time (Fig. 13). This disorder can now be detected by commercially available DNA analysis. Alternatively, characteristic osmophilic granules, within vessels, can be seen by electron microscopy through either skin, muscle, or peripheral nerve biopsy.

C. Leber's Hereditary Optic Neuropathy

This disorder is maternally transmitted and typically affects young males between adolescence and age 30, although women and older men can also be affected. It causes optic atrophy with acute or subacute visual loss (Yen *et al.*, 2006). Other manifestations such as ataxia, dysarthria, spastic weakness, sensory loss, and dystonia have been reported in both male patients and female carriers of the disease. Leber's hereditary optic neuropathy (LHON) is usually associated with one of three primary mitochondrial mutations and can be diagnosed through genetic testing.

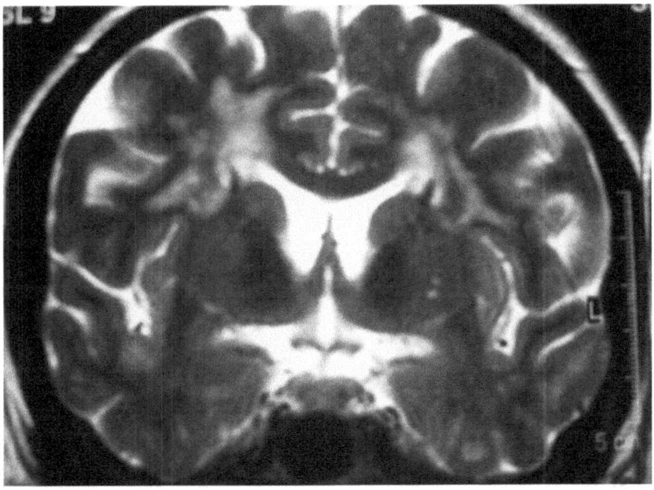

FIG. 13. Coronal T2-weighted MRI brain scan revealing prominent periventricular white matter changes. This is a 30-year-old woman who has history of recurrent transient ischemic attacks with positive genetic testing and a positive family history of cerebral autosomal dominant arteriopathy with subcortical infarcts and leukoencephalopathy (CADASIL).

D. Mitochondrial Encephalomyopathy with Lactic Acidosis and Strokelike Episodes

Mitochondrial encephalomyopathy with lactic acidosis and strokelike episodes (MELAS) may develop in childhood, adolescence, or early adulthood. It usually manifests with recurrent migraine-like headaches, encephalopathy, and acute, partially regressive neurological syndromes, resembling ischemic strokes. Brain MRI scans usually reveal bilateral infarct-like lesions that do not correlate with vascular territories, particularly in the cortex and subcortical white matter of the parietoccipital lobes as well as the basal ganglia. The frontal lobes and brainstem are typically spared (Hausegger *et al.*, 1991). Mitochondrial dysfunction can be evident as lactic acidosis and ragged-red fibers in muscle.

V. Differential Diagnosis of CNS Neoplasms

A primary brain tumor or metastatic lesion can generally be easily distinguished from MS by the clinical presentation and the imaging characteristics typical of either primary or secondary CNS tumor. However, there can be instances such as low-grade glioma when there is the potential for confusion,

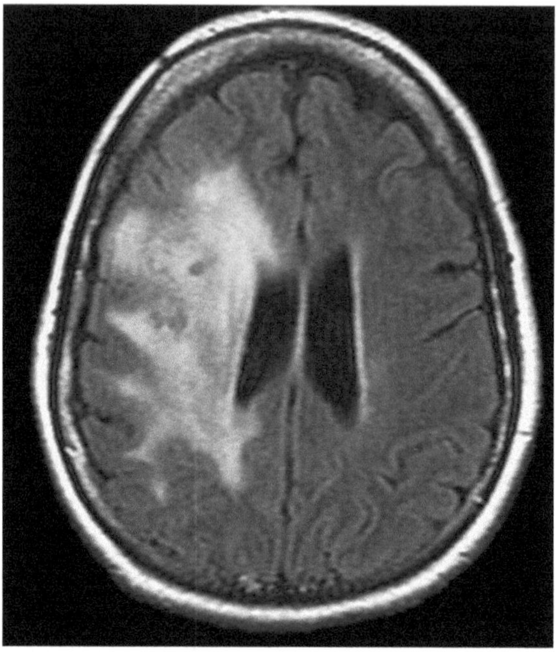

Fig. 14. Axial FLAIR sequence of brain MRI showing a hyperintense lesion involving both white matter and cortex in a patient who presented with seizures, and subsequently diagnosed with low-grade glioma. Deformity of the anterior aspect of right lateral ventricle and subtle mass effect can be noted.

especially with the tumefactive form of MS (Fig. 14). Moreover, CNS lymphoma, for example, can have a fluctuating course and show improvement with steroid therapy adding more to the confusion (Feuerhake *et al.*, 2006). In such circumstances, newer methodologies such as proton magnetic resonance spectroscopy (Raizer *et al.*, 2005) may be useful.

VI. Differential Diagnosis of Dysmyelinating Disorders

A. Overview

Genetic disorders affecting myelin metabolism are often referred to as leukodystrophies. These include: metachromatic leukodystrophy (MLD), Krabbe's (globoid cell) leukodystrophy, multiple sulfatase deficiency, Alexander disease, adrenoleukodystrophy (ALD), and Pelizaeus-Merzbacher disease. All but the latter two have an autosomal recessive inheritance pattern. They all demonstrate demyelinative (white matter) lesions on MRI brain scan which can cause confusion with MS.

B. Metachromatic Leukodystrophy

MLD has either a late infantile, juvenile, or adult form described. It is associated with deficiency of the sulfatide-cleaving enzyme arylsulfatase A, but there is at least one form without arylsulfatase A deficiency, but rather deficiency of arylsulfatase A activator. This latter type can look either like late infantile or juvenile form. The late infantile form of MLD is characterized by loss of motor and coordination function at between 1 and 2.5 years of age in association with progressive cognitive impairment, optic atrophy, and peripheral neuropathy with slow nerve conduction velocities. The form with onset at age 3–10 years of age, termed juvenile MLD, is similar, but more slowly progressive. The adult form presents with a slowly progressive dementia in association with ataxia and long tract signs.

C. Krabbe's Leukodystrophy

Krabbe's leukodystrophy is associated with deficiency of the enzyme galactocerebrosidase and there is either an infantile, juvenile, or adult form described. The infantile form typically presents at 3–6 months of age and there is progressive motor impairment, seizures, blindness secondary to optic atrophy, cognitive impairment evolving into a vegetative state along with prolonged nerve conduction velocities. The juvenile and adult forms are characterized by long tract signs, cognitive impairment, and optic atrophy with slower progression as a function of age of onset.

D. Multiple Sulfatase Deficiency

Multiple sulfatase deficiency, also known as sulfatidosis, has clinical features of both MLD and mucopolysaccharidosis. There is developmental delay with progressive neurological impairment by 1–2 years of age.

E. Alexander Disease

Alexander disease is characterized by demyelinative lesions on brain scan which is especially prominent in the centrum semiovale with a characteristic frontal-to-occipital gradient of white matter involvement (Fig. 15). There is a rapidly progressive infantile variant associated with megalencephaly, spastic quadriparesis, and seizures as well as a more protracted course in adolescents, usually without epilepsy. There can be a form seen in adults with bulbar palsy and ataxia sometimes in association with dementia and spastic weakness.

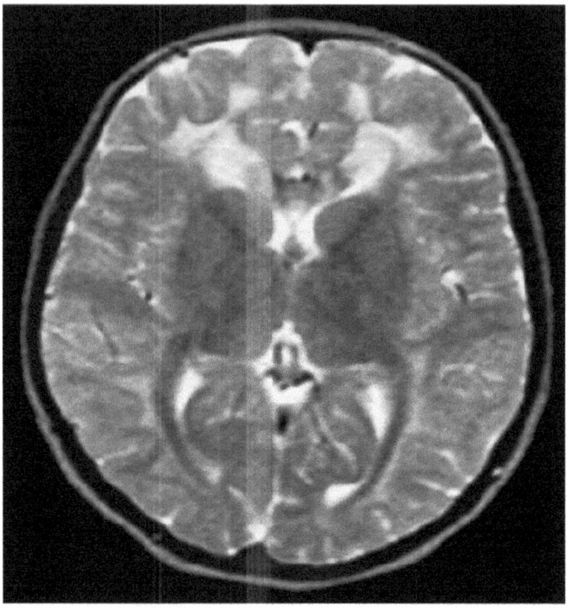

FIG. 15. Axial T2 sequence of brain MRI showing increased signal in the frontal centrum semiovale and decreased signal in a thin periventricular area. This is a 10-year-old female who presented with macrocephaly, delayed psychomotor development, and seizures. She was found to have Alexander disease.

F. Adrenoleukodystropy

ALD is X-linked, appearing only in males, but with variable age of onset ranging from childhood to middle age. There are a number of clinical phenotypes described. However, the most common presentation is spastic paraparesis, demyelinating peripheral neuropathy, and adrenal insufficiency beginning in adolescence.

The brain MRI typically shows diffuse demyelination with sparing of U-fibers in the centrum semiovale (Fig. 16). The brain MRI findings along with adrenal insufficiency and the typical neurological manifestations help in arriving at the diagnosis with confirmation by the finding of very-long-chain fatty acids in the plasma or in cultured skin fibroblasts.

G. Pelizaeus-Merzbacher Disease

There are two forms of Pelizaeus-Merzbacher disease, one form is present at birth and the most common form is noted in infancy with a longer clinical course. The primary suggestive feature is prominent nystagmus with head tremor noted

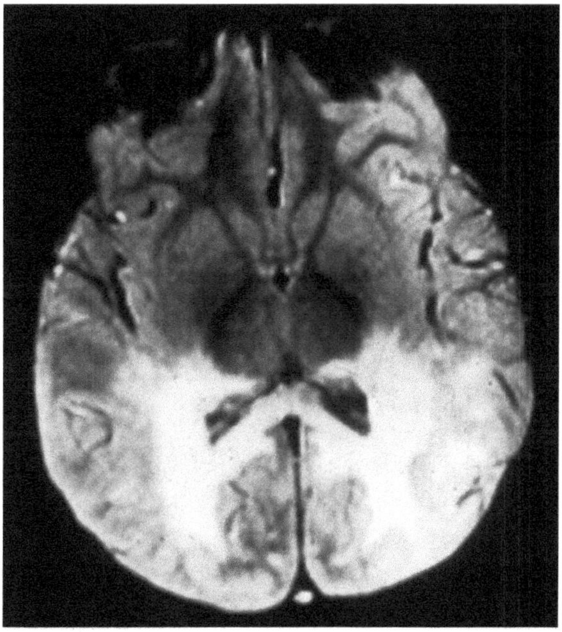

FIG. 16. Axial proton density sequence of brain MRI showing hyperintense signal involving the splenium of the corpus callosum as well as the temporoocccipital white matter in a patient with adreunoleukodystrophy.

at birth or in infancy. The affected males have pronounced failure of neurological development evolving into optic atrophy, severe spastic weakness, and ataxia. Typically the patient succumbs in a completely dependent state within the first 2 years of life. However, milder forms have been described resulting in a mute wheelchair-confined condition by adolescence.

VII. Differential Diagnosis Other Demyelinating Disorders

A. Neuromyelitis Optica (Devic's Disease)

This disorder is characterized by a combination of transverse myelitis and optic neuritis occurring within a finite period of time from each other, typically within weeks. Neuromyelitis optica (NMO) is a noninfectious inflammatory disorder which appears to be pathologically distinguishable from MS and does not tend to follow the relapsing-remitting course characteristic of MS.

The identification of the serum autoantibody marker (NMO-IgG) (Pittock and Lucchinetti, 2006) for this disorder should be helpful in distinguishing it from MS which can certainly present in a similar fashion. A clinical feature which can also be useful is the acute monophasic presentation that is observed in children and adolescents.

B. Acute Disseminated Encephalomyelitis

Acute disseminated encephalomyelitis (ADEM) can look like MS both clinically and radiographically. Some view it as a variant of MS. However, it tends to be a postinfectious process often following a respiratory infection. It can also be postvaccinal. It tends to follow a monophasic course and is not uncommonly associated with headache, encephalopathy, and long tract signs occurring within a month of a systemic illness or a vaccination (Schwarz *et al.*, 2001). CSF findings can be similar to MS, but there tends to be a higher white cell count and protein

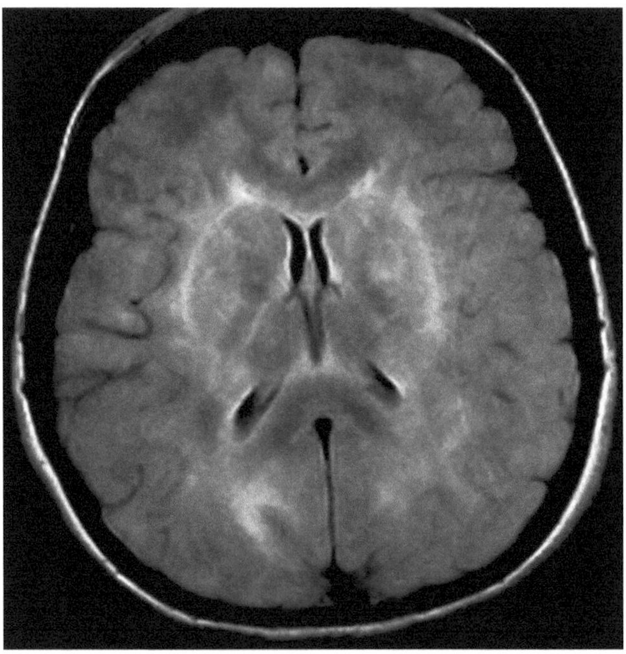

Fig. 17. Axial FLAIR sequence of brain MRI through the basal ganglia showing hyperintense signal in the external capsule, internal capsule, and TO centrum semiovale in a patient with ADEM.

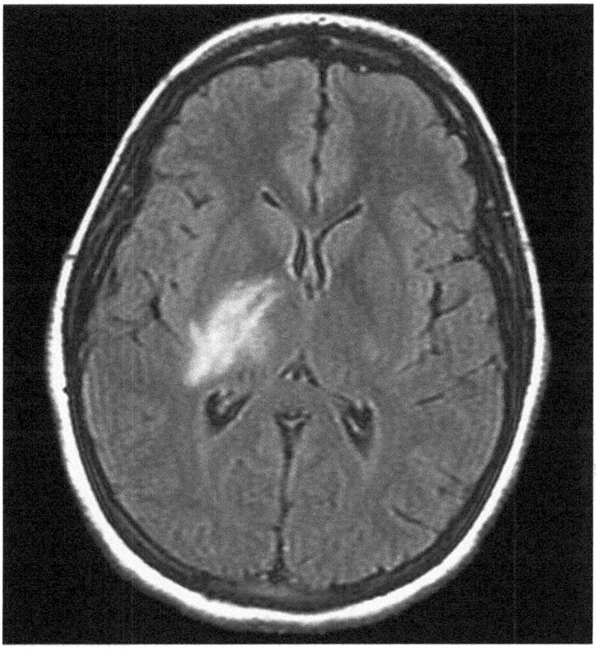

FIG. 18. Axial FLAIR of brain MRI through the basal ganglia showing a hyperintense area involving the posterior limb of the right internal capsule in a patient diagnosed with ADEM.

level with lesser likelihood of the presence of oligoclonal bands. MRI brain (Figs. 17 and 18) scan tends to demonstrate larger demyelinative lesions with a predilection for the cerebral cortex and basal ganglia and less involvement of the periventricular white matter. MR spectroscopy usually shows a low *N*-acetyl aspartate peak with a high glutamate peak (Fig. 19).

There tends to be more homogenous contrast enhancement in ADEM compared to MS, and one does not tend to see the characteristic ovoid lesions perpendicular to the ventricular system, and T1 "holes" that one not uncommonly seen in MS (Hynson *et al.,* 2001).

C. Concentric Sclerosis (Balo's Disease)

This disorder can affect all age groups, but appears to be more common in children. It tends to have a slow, progressive course. There are concentric zones of demyelination seen on brain imaging as well as pathologically (Fig. 20). There is often an insidious onset with focal deficit which evolves to generalized spastic weakness and dementia.

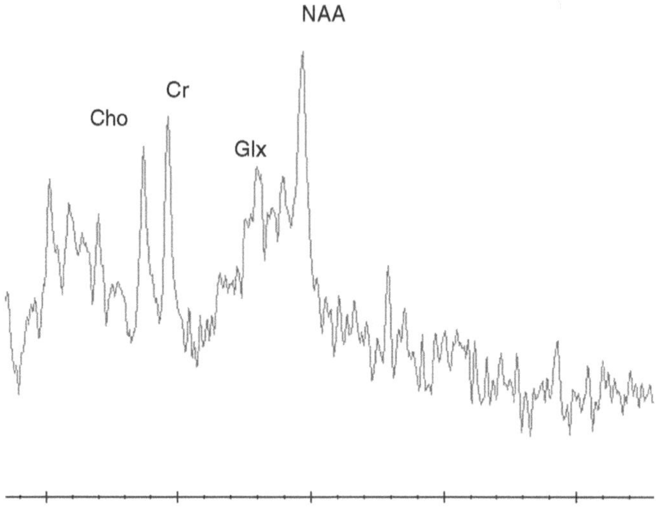

FIG. 19. MR spectroscopy showing low *N*-acetyl aspartate and high glutamate peaks over the region of interest (ROI) in a patient with ADEM. The ROI is the lesion seen in Fig. 18.

D. Schilder's Disease

This is an acute, fulminant demyelinating disease typically seen in children and viewed as a fulminant form of MS in children. However, there tends to be confluent lesions involving both cerebral hemispheres with particular involvement of the centrum semiovale. However, the pathological findings are those associated with MS. This might well be akin to the so-called Marburg variant of MS reported in adults (Jeffery *et al.*, 2004) in which there is a rapidly progressive course with severe disability or death within a year of presentation.

E. Optic Neuritis

Optic neuritis can be a harbinger of MS or can be an isolated self-limited process. It is an idiopathic demyelinating process often associated with a good prognosis for recovery of vision within several weeks. 95% of patients return to at least 20/40 vision within 12 months and the recovery is hastened with intravenous steroid therapy. In the Optic Neuritis Study Group (2003) trial, the risk of MS

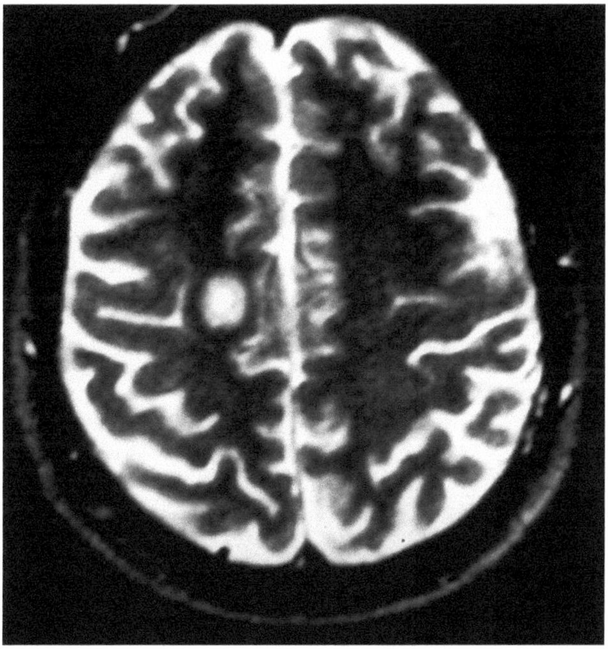

FIG. 20. Axial T2-weighted image showing a round hyperintense rim around the MS plaque in a patient with Balo's concentric sclerosis.

in acute isolated optic neuritis was 56% within 10 years if there was at least one T2-weighted lesion on brain MRI. However, the 10-year risk for MS was only 22% with no lesion on baseline MRI.

Additional features which were associated with no evolution to MS included lack of pain, severe optic disc edema with peripapillary hemorrhages, and the presence of a macular star.

VIII. Conclusions

Despite the wide array of disorders that can potentially mimic MS, most patients present with a characteristic enough clinical picture to avoid unnecessary and expensive testing for alternative diagnoses. This has been augmented considerably by the wide availability of MRI, especially T2-weighted and fluid attenuation inversion recovery (FLAIR) imaging, where one expects a sensitivity of the order of 90–95% for the detection of lesions compatible with MS. This, in combination with typical CSF findings, seen in up to 90% or more of patients with MS, helps considerably in establishing the diagnosis.

References

Akman-Demir, G., Serdaroglu, P., and Tasci, B. (1999). Neuro-Behcet Study Group. Clinical patterns of neurological involvement in Behcet's disease: Evaluation of 200 patients. *Brain* **122,** 2171–2181.

Alexander, E. L., Malinow, K., Lejewski, J. E., Jerdan, M., Provost, T. T. S., and Alexander, G. E. (1986). Primary Sjogren's syndrome with central nervous disease mimicking multiple sclerosis. *Ann. Intern. Med.* **104,** 323–330.

Bagnato, F., Butman, J. A., Mora, C. A., Gupta, S., Yamano, Y., Tasciyan, T. A., Solomon, J. M., Santos, W. J., Stone, R. D., McFarland, H. F., and Jacobson, S. (2005). Conventional magnetic resonance imaging features in patients with tropical spastic paraparesis. *J. Neurovirol.* **11,** 525–534.

Berger, J. R., and Koralnik, I. J. (2005). Progressive mutlifocal encephalopathy and Natalizumab-unforeseen consequences. *New Engl. J. Med.* **353,** 414–416.

Berger, J. R., Wei, T., and Wilson, D. (1998). Idiopathic granulomatosis of the CNS manifesting as diffuse white matter disease. *Neurology* **51,** 1774–1775.

Bitsch, A., Schuchardt, J., Bunkowski, S., Kuhlmann, T., and Bruck, W. (2000). Acute axonal injury in multiple sclerosis: Correlation with demyelination and inflammation. *Brain* **123,** 1174–1183.

Calabrese, L. H., Duna, G. F., and Lie, J. T. (1997). Vasculitis in the central nervous system. *Arthritis Rheumatol.* **40,** 1189–1201.

Cuadrado, M. J., Wakefield, K., Sanna, G., Khamashta, M. A., and Hughes, G. R. (2004). Multiple sclerosis and the antiphospholipid (Hughes) syndrome: A common differential diagnosis? *Clin. Exp. Rheumatol.* **22,** 652–653.

Delalande, S., De Seze, J., Fauchais, A. L., Hachulla, E., Stojkovic, T., Ferriby, D., Dubucquoi, S., Pruvo, J. P., Vermersch, P., and Hatron, P. Y. (2004). Neurologic manifestations in primary Sjogren syndrome: A study of 82 patients. *Medicine (Baltimore)* **83,** 280–291.

DeRemee, R. A. (2006). Wegnener's granulomatosis. *Lancet* **367,** 1362–1366.

Dichgans, M., Mayer, M., Uttner, I., Bruning, R., Muller-Hocker, J., Rungger, G., Ebke, M., Klockgether, T., and Gasser, T. (1998). The phenotypic spectrum of CADASIL: Clinical findings in 102 cases. *Ann. Neurol.* **44,** 731–739.

Dore-Duffy, P., Donaldson, J. O., Rothman, B. L., and Zurier, R. B. (1982). Antinuclear antibodies in multiple sclerosis. *Arch. Neurol.* **39,** 504–506.

Drozdowski, W. (2006). Multifocal central nervous system lesions—multiple sclerosis or neuroborreliosis? *Przegl. Epidemiol.* **60**(Suppl. 1), 39–45.

Fadil, H., Gonzalez-Toledo, E., Kelley, B. J., and Kelley, R. E. (2006). Neuroimaging findings in neurosyphilis. *J. Neuroimaging* **16,** 286–289.

Ferreira, S., D'Cruz, D. P., and Huges, G. R. V. (2005). Multiple sclerosis, lupus and antiphospholipid antibody syndrome: Where do we stand? *Rheumatology* **44,** 434–442.

Feuerhake, F., Baumer, C., Cyron, D., Illerhaus, G., Olschewski, M., Tilgner, J., Ostertag, C. B., and Volk, B. (2006). Primary CNS lymphoma in immunocompetent patients from 1989 to 2001: A retrospective analysis of 164 cases uniformly diagnsosed by stereotactic biopsy. *Act. Neurochir.* **148,** 831–838.

Finelli, P. F., Onykie, H. C., and Uphaff, D. F. (1997). Idiopathic granulomatous angiitis of the CNS manifesting as diffuse white matter disease. *Neurology* **49,** 1696–1699.

Hausegger, K. A., Millner, M. M., Ebner, F., Flückiger, F., and Justich, E. (1991). Mitochondrial encephalomyopathy-two years follow-up by MRI. *Pediatr. Radiol.* **21,** 231–233.

Hietaharju, A., Peltola, J., Luukkainen, R., and Dastidar, P. (2001). The coexistence of systemic lupus erythematosus and multiple sclerosis in a mother and daughter. *Scand. J. Rheumatol.* **30,** 120–122.

Howard, A. K., Li, D. K., and Oger, J. (2003). MRI contributes to the differentiation between MS and HTLV-I associated myelopathy in British Columbian coastal natives. *Can. J. Neurol. Sci.* **30,** 41–48.

Hughes, G. R. (1983). Thrombosis, abortion, cerebral disease, and the lupus anticoagulant. *Br. Med. J.* **287,** 1088–1089.

Hynson, J. L., Kornberg, A. J., Coleman, J. T., Shield, L., Harvey, A. S., and Kean, M. J. (2001). Clinical and neuroradiologic features of acute disseminated encephalomyelitis in children. *Neurology* **56,** 1308–1321.

Igarashi, H., Sakai, F., Kan, S., Okada, J., and Tazaki, Y. (1991). Magnetic resonance imaging of the brain in patients with migraine. *Cephalagia* **11,** 69–74.

Jeffery, D. R., Lefkowitz, D. S., and Crittenden, J. P. (2004). Treatment of Marburg variant of multiple sclerosis with mitoxantrone. *J. Neuroimaging* **14,** 58–62.

Kadota, Y., Tokumaru, A. M., Kamakura, K., Kohyama, S., Okizuka, H., Kaji, T., and Kusano, S. (2002). Primary Sjogren's syndrome initially manifested by optic neuritis: MRI findings. *Neuroradiology* **44,** 338–341.

Kidd, A., Steuer, A., Demman, A. M., and Rudge, P. (1999). Neurological complications in Behcet's syndrome. *Brain* **122,** 2183–2194.

Levine, S. L., Deegan, M. J., Futrell, N., and Welch, K. M. (1990). Cerebrovascular and neurologic disease associated with antiphospholipid antibodies. *Neurology* **40,** 1181–1189.

McDonald, W. I., Compston, A., Edan, G., Goodkin, D., Hartung, H.-P., Lublin, F. D., McFarland, H. F., Paty, D. W., Polman, C. H., Reingold, S. C., Sandberg-Wollheim, M., Sibley, W., *et al.* (2001). Recommended diagnostic criteria for multiple sclerosis: Guidelines from the international panel on the diagnosis of multiple sclerosis. *Ann. Neurol.* **50,** 121–127.

McLean, B. N., Miller, D. H., and Thompson, E. J. (1995). Oligoclonal banding of IgG in CSF, blood brain barrier function, and MRI findings in patients with sarcoidosis, systemic lupus erythematosus, and Behcet's disease involving the nervous system. *J. Neurol. Neurosurg. Psychiatr.* **58,** 548–554.

Morgen, K., McFarland, H. F., and Pillemer, S. R. (2004). Central nervous system disease in primary Sjogren's syndrome: The role of magnetic resonance imaging. *Semin. Arthritis. Rheum.* **34,** 623–630.

Murphy, J. M., Gomez-Anson, B., Gillard, J. H., Antoun, N. M., Cross, J., Elliott, J. D., and Lockwood, M. (1999). Wegener's granulomatosis: MR imaging findings in the brain and meninges. *Radiology* **213,** 794–799.

Noseworthy, J. H., Lucchinetti Rodriguez, M., and Weinshenker, B. G. (2000). Multiple sclerosis. *New Engl. J. Med.* **343,** 938–952.

Optic Neuritis Study Group (2003). High risk and low risk profiles for the development of multiple sclerosis within 10 years after optic neuritis: Experience of the Optic Neuritis Treatment Trial. *Arch. Ophthalmol.* **121,** 944–949.

Paula De Carvalho Panzerri, Y. H., Paes Leme Ferriani, C. A., Tanuri Caldas, V., Celia Cervi, C., Pileqqi, M., Carvalho, C., Carlos Dos Santos, A., and Chanq, D. (2004). Polyarteritis nodosa with central nervous system involvement mimicking meningoencephalitis. *Pediatr. Crit. Care Med.* **5**(3), 286–288.

Pericot, I., Brieva, L., Tintore, M., Rio, J., Sastre-Garriga, J., Nos, C., and Montalban, X. (2003). Myelopathy in seronegative Sjogren syndrome and/or primary progressive multiple sclerosis. *Mult. Scler.* **9,** 256–259.

Pittock, S. J., and Lucchinetti, C. F. (2006). Inflammatory transverse myelitis: Evolving concepts. *Curr. Opin. Neurol.* **19,** 362–368.

Polman, C. H., Reingold, S. C., Edan, G., Filippi, M., Hartung, H. P., Kappos, L., Lublin, F. D., Metz, L. M., McFarland, H. F., O'connor, P. W., Sandberg-Wollheim, M., Thompson, A. J., *et al.* (2005). Diagnostic criteria for multiple sclerosis: 2005 revisions to the "McDonald Criteria." *Ann. Neurol.* **58,** 840–846.

Poser, C. M., Paty, D. W., Scheinberg, L. C., McDonald, W. I., Davis, F. A., Ebers, G. C., Johnson, K. P., Sibley, W. A., Silberberg, D. H., and Tourtellotte, W. W. (1983). New diagnostic criteria for multiple sclerosis: Guidelines for research protocols. *Ann. Neurol.* **13,** 227–231.

Provenzale, J. M., and Allen, N. B. (1996). Wegener's granulomatosis: CT and MR findings . *AJNR* **17,** 785–792.

Raizer, J. J., Koutcher, J. A., Abrey, L. E., Panageas, K. S., DeAngelis, L. M., Lis, E., Xu, S., and Zakian, K. L. (2005). Proton magnetic resonance spectroscopy in immunocompetent patients with primary central nervous system lymphoma. *J. Neurooncol.* **71,** 173–180.

Reichart, M. D., Bogousslavsky, J., and Janzer, R. C. (2000). Early lacunar strokes complicating polyarteritis nodosa thrombotic microangiopathy. *Neurology* **54**(4), 883–889.

Reske, D., Petereit, H. F., and Heiss, W.-D. (2005). Difficulties in the differentiation of chronic inflammatory diseases of the central nervous system-value of cerebrospinal fluid analysis and immunological abnormalities in the diagnosis. *Acta Neurol. Scand.* **112,** 207–213.

Rocca, M. A., Ceccarelli, A., Falini, A., Colombo, B., Tortorella, P., Bernasconi, L., Comi, G., Scotti, G., and Filippi, M. (2006). Brain gray matter changes in migraine patients with T2-visible lesions: A 3-T MRI study. *Stroke* **37,** 1765–1770.

Sanna, G., Bertolaccini, M. L., Cuadrado, M. J., Laing, H., Khamashta, M. A., Mathieu, A., and Hughes, G. R. V. (2003). Neuropsychiatric manifestations in systemic lupus erythematosus: Prevalence and association with antiphospholipid antibodies. *J. Rheumatol.* **30,** 985–992.

Scott, T. F., Hess, D., and Brillman, J. (1994). Antiphospholipid antibody syndrome mimicking multiple sclerosis clinically and by magnetic resonance imaging. *Arch. Intern. Med.* **154,** 917–920.

Schumaker, G. A., Beebe, G., Kibler, R. F., Loewenson, R. B., Silberberg, D. H., and Tourtellotte, W. W. (1965). Problem of experimental trials of therapy in multiple sclerosis: Report of the panel on evaluation of experimental trials of therapy in mutiple sclerosis. *Ann. N.Y. Acad. Sci.* **122,** 552–568.

Schwarz, S., Mohr, A., Knauth, M., Wildemann, B., and Storch-Hagenlocher, B. (2001). Acute disseminated encephalomyelitis: A follow-up study of 40 adult patients. *Neurology* **56,** 1313–1318.

Schumaker, G. A., Beebe, G., Kibler, R. F., Loewenson, R. B., Silberberg, D. H., and Tourtellotte, W. W. (1965). Problems of experimental trials of therapy in multiple sclerosis: Report of the panel on evaluation of experimental trials of therapy in multiple sclerosis. *Ann. N.Y. Acad. Sci.* **122,** 552–568.

Seror, R., Mahr, A., Ramanoelina, J., Pagnoux, C., Cohen, P., and Guillevin, L. (2006). Central nervous system involvement in Wegener Granulomatosis. *Medicine (Baltimore)* **85,** 54–65.

Sharma, O. P. (1997). Neurosarcoidosis: A personal perspective based on the study of 37 patients. *Chest* **112,** 220–228.

Sharief, M. K., Hentges, R., and Thomas, E. (1997). Siginficance of CSF immunoglobulins in monitoring neurologic disease activity in Behcet's disease. *Neurology* **41,** 1398–1401.

Shoemaker, E. I., Lin, Z., Rae-Grant, A. D., and Little, B. (1994). Primary angiitis of the nervous system: Unusual MR appearance. *Ann. J. Neuroradiol.* **15,** 331–334.

Siva, A. (2001). Vasculitis of the nervous system. *J. Neurol.* **248**(6), 451–468.

Smith, A. S., Meisler, D. M., Weinstein, M. A., Tomsak, R. L., Hanson, M. R., Rudick, R. A., Farris, B. K., and Ransohoff, R. M. (1989). High-signal periventricular lesions in patients with sarcoidosis: Neurosarcoidosis or multiple sclerosis? *AJR Am. J. Roentgenol.* **153,** 147–152.

Stern, B. J., Krumholz, A., Johns, C., Scott, P., and Nissim, J. (1985). Sarcoidosis and its neurological manifestations. *Arch. Neurol.* **42,** 909–917.

Susac, J. O., Murtagh, F. R., Egan, R. A., Berger, J. R., Bakshi, R., Lincoff, N., Gean, A. D., Galetta, S. L., Fox, R. J., Costello, F. E., Lee, A. G., Clark, J., *et al.* (2003). MRI findings in Susac's syndrome. *Neurology* **61,** 1783–1787.

Tali, E. T., Attilla, S., Keskin, T., Simonson, T., Isik, S., and Yuh, W. T. (1997). MRI in Neuro-Behcet's disease. *Neuroradiology* **39,** 2–6.

Umehara, F., Tokunaga, N., Hokezu, Y., Hokonohara, E., Yoshishige, K., Shiraishi, T., Okubo, R., and Osame, M. (2006). Relapsing cervical cord lesions on MRI in patients with HTLV-I-associated myelopathy. *Neurology* **66,** 289.

Waisberg, H., Meloff, K. L., and Buncic, R. (1974). Polyarteritis nodosa complicated by a multiple sclerosis like syndrome. *Can. J. Neurol. Sci.* **1**(4), 250–252.

Weinberger, L. M., Cohen, M. L., Rember, B. F., Naheedy, M. H., and Leigh, R. J. (1993). Intracranial Wegener's Granulomatosis. *Neurology* **43,** 1831–1834.

Yen, M. Y., Wang, A. G., and Wei, Y. H. (2006). Leber's hereditary optic neuropathy: A multifactorial disease. *Prog. Retin. Eye Res.* **25,** 381–396.

PROGNOSTIC FACTORS IN MULTIPLE SCLEROSIS

Roberto Bergamaschi

Department of Clinical Neurology, Multiple Sclerosis Center, Neurological Institute "C. Mondino," Via Mondino 2, Pavia 27100, Italy

Prognostic factors to determine the patient's likelihood of developing MS are important for several reasons. Prognostic factors are important to the patient who wants to be informed about his/her prospects, to the clinician who needs to individuate the patients who deserve immune treatments, and to the researcher who needs to improve the design and the analysis of the therapeutic trials. In addition, with the development of new immune therapies, whose early use is strongly encouraged, it is crucial to dispose of reliable clinical predictors to identify the patients who are candidates for early or aggressive therapies.

Several studies have indicated that a poor prognosis is related to male gender; a late age at onset; motor, cerebellar, and sphincter involvement at onset; a progressive course at onset; a short inter-attack interval; a high number of early attacks; and a relevant early residual disability. Paraclinical support for MS prognosis is given by imaging techniques, cerebrospinal fluid analysis, and evoked potential examinations. The most sensitive paraclinical test to predict conversion from suspected demyelinating disease to definite MS is MRI.

INTERNATIONAL REVIEW OF
NEUROBIOLOGY, VOL. 79
DOI: 10.1016/S0074-7742(07)79019-0

0074-7742/07 $35.00

I. Introduction

Prognostic factors to determine the patient's likelihood of developing MS are important for several reasons. Of course, prognostic factors are important to patients who want to be informed about their diagnosis and their prospects. The uncertainty about future disease progression is a predominant factor for all MS patients, even those with low disability (Boeije and Janssens, 2004). Prognostic factors are also important to the clinician who needs to individuate at an early stage of disease those patients who deserve early/aggressive treatments, and to the researcher who necessitates to improve the design and the analysis of clinical therapeutic trials and observational studies. In addition, the trend to start new immunomodulating therapies early in the course of MS (Comi *et al.*, 2001; Goodin *et al.*, 2002; Jacobs *et al.*, 2000; Johnson *et al.*, 1995; The IFNB MS Study Group and the University of British Columbia MS/MRI Analysis Group, 1995) makes it crucial to dispose of reliable predictors of disease's evolution, both unfavorable and favorable. As a matter of fact, the existence of a "benign" form of MS must be taken into consideration for a single patient treatment decision; studies proved that 15–25% of MS patients show a mild or absent disability despite a long duration of the disease (Pittock *et al.*, 2004; Ramsaransing *et al.*, 2001); in addition, if an MS patient has a "benign" course after 10 years with the disease, he/she has a high probability of remaining benign after another 10 years (Pittock *et al.*, 2004).

In the present chapter, we will assess clinical and instrumental variables collected at the onset or during the course of the disease, whose occurrence could predict the conversion from clinically isolated syndrome (CIS) to clinically definite MS (CDMS), the development of relevant disability, and the conversion from relapsing-remitting (RR) to secondary progressive (SP) form of MS.

II. Prognostic Factors at the Disease's Onset

A. Demographic Factors

1. *Gender*

The female sex is an important issue in the pathogenesis and evolution of MS, as well as of other autoimmune diseases. The high prevalence of MS in females (2:1), with a higher susceptibility during puberty (Simone *et al.*, 2002), suggests a possible role of the female hormones in the inflammatory/neurodegenerative/neuroreparative network that characterizes MS (Cardona-Gomez *et al.*, 2002; Levic *et al.*, 1999). In particular, hyperestrogenism is related to improvement in the inflammatory phase (Confavreux *et al.*, 1998; Zang *et al.*, 2002), while hyperandrogenism seems to be related to disease progression (Falaschi *et al.*, 2001). Several authors identified the female gender as a favorable prognostic factor (Bergamaschi *et al.*, 2001; Confavreux *et al.*, 1980, 2003; Detels *et al.*, 1982; Lauer and Firnhaber, 1992; Runmarker and Andersen, 1993; Trojano *et al.*, 1995; Wolfson and Confavreux 1987; Weinshenker *et al.*, 1991). Accordingly, Kantarci *et al.* (1998) reported that the male gender was related to a poor prognosis. However, other studies did not find any prognostic relevance of gender factor (Amato *et al.*, 1999; Broman *et al.*, 1981; Riise *et al.*, 1992; Visscher *et al.*, 1984). Tremlett *et al.* (2006) investigated risk factors for disability progression in a large geographically based MS population (2837 patients; 22,723 patient years). Men progressed 38% more quickly than women from onset, yet both required canes at similar ages (58.8 years for men vs 60.1 for women), demonstrating that male sex was not associated with worse disease outcome.

2. *Age at Onset*

The age at clinical onset, whose distribution shows peaks between 20 and 30 years (Simone *et al.*, 2002), is another demographic factor with prognostic importance in MS. A younger age at the onset of the disease is related to a better prognosis (Confavreux *et al.*, 1980, 2003; Noseworthy *et al.*, 1983; Phadke, 1990; Poser *et al.*, 1982; Runmarker and Andersen, 1993; Simone *et al.*, 2002; Thompson *et al.*, 1986; Trojano *et al.*, 2002; Visscher *et al.*, 1984; Weinshenker *et al.*, 1991), while an older age is an unfavorable prognostic factor (Bergamaschi *et al.*, 2005; Citterio *et al.*, 1989; Confavreux *et al.*, 1980; Detels *et al.*, 1982; Phadke, 1990; Riise *et al.*, 1992; Runmarker and Andersen, 1993; Trojano *et al.*, 2002; Visscher *et al.*, 1984; Weinshenker *et al.*, 1989, 1991, 1996). The disease typically begins as RR in patients <30 years of age (Lublin and Reingold, 1996), while the disease often begins as progressive in patients >50 years of age (Weinshenker *et al.*, 1989). The median time a patient takes to reach a disability of 6.0 quantified according to the Kurtzke's Expanded Disability Status Scale (EDSS) of Kurtzke (1983) is 31 years for patients who were under 19 years old at onset, while it is only 7 years for patients who were over 50 years old (Confavreux *et al.*, 2003).

A Bayesian analysis with Markov chain Monte Carlo simulation technology, used to model the natural history of MS patients with initial RR course (Bergamaschi *et al.*, 2000, 2001), confirms that the risk of unfavorable evolution (to become SP) is significantly related to age at onset. In the last years, particular attention has been pointed to the juvenile forms of MS. The clinical characteristics of MS with onset during childhood or adolescence (under 16 years) are different to those of the adult form (Ghezzi, 2005), with higher female to male ratio, higher relapse rate, and more frequent occurrence of hyperacute forms and of encephalopatic symptoms. Severe disability is reached after a longer interval than in the adult form but, in spite of this, at a given age disability is higher.

The KIDMUS Study Group (Mikaeloff *et al.*, 2004) followed for 3 years a cohort of 296 children having a first episode of acute inflammatory demyelination of the central nervous system (CNS). At the end of the follow-up, 57% of children had CDMS, 29% had a monophasic acute disseminated encephalomyelitis (ADEM), and 14% remained with a single focal episode. The risk of a second attack was higher in patients with age at onset $\geq$10 years. Interestingly, 29% of patients with a second attack, that led to diagnosis of MS, had an initial diagnosis of ADEM.

The counterpart of juvenile MS, that is the late onset MS (defined as the first presentation of clinical symptoms in patients over 50 years), is not a rare phenomenon as thought, since the prevalence ranges between 4% and 9.6%. Late onset MS shows particular clinical characteristics too: the course of the disease is often primary progressive (PP) and pyramidal or cerebellar involvement is observed in 60–70% of the patients at presentation (Martinelli *et al.*, 2004). Late onset MS is usually associated with a faster progression to disability compared to young adult MS.

The previous considerations support the hypothesis that distinct and heterogeneous pathogenetic phenomena might operate at different times in individual patients, since the disability is more attributable to inflammation at a younger age and to degeneration at an older age (Smith *et al.*, 1999), and could reflect age-related changes in the behavior of the immune system (Huseby *et al.*, 2001; Phadke, 1990).

Anyway, researches reject the prognostic role of age at onset (Deryck *et al.*, 2006; Tremlett *et al.*, 2006). Derick's study confirms that in MS patients with early onset of the disease the clinical features, disease course, and long-term prognosis are quite different from the adult onset patients, with a preponderance in the subgroup of early onset MS patients of female sex, of visual and brain stem-cerebellar presentations, and of RR disease course. However, after a long period of follow-up (28.7 years), the overall disease course and prognosis do not seem to differ from that in adult onset MS. Tremlett *et al.* (2006) found that a younger onset age predicted a slower progression, but those patients older at onset were consistently older when reaching severe disability. By consequence, the authors sustain that a young age at onset in MS cannot be considered a factor that predict a long-term favorable evolution.

Finally, even if younger and female patients seem to have lower risk of short-term unfavorable evolution (in terms of early appearance of disability and progressive phase), the same patients have an higher relapse rate, as observed in a large cohort from the placebo arms of the Silvia Lawry Centre for Multiple Sclerosis Research (Held *et al.*, 2005).

B. Clinical Features

1. *Symptoms at Onset*

A poor prognosis was related to a polisymptomatic onset compared with a monosymptomatic onset in adults (Amato *et al.*, 1999; Bergamaschi *et al.*, 2001; Kraft *et al.*, 1981; Phadke, 1990; Runmarker and Andersen, 1993; Wolfson and Confavreux, 1987). The same was also observed in children with MS (Mikaeloff *et al.*, 2004).

a. Clinically Isolated Syndromes. In most instances, the first symptoms and signs of MS indicate a lesion in the optic nerve, brain stem, or spinal cord. These neurological patterns are defined CIS. However, not all patients with CIS progress to CDMS.

The probability for a CIS to convert to CDMS proportionally increases over years: 26–37% of CIS develop CDMS within 1 year (Brex *et al.*, 2002; Tintore *et al.*, 2003), 38% within 2 years (Beck, 1995), 43% within 5 years (Molmissey *et al.*, 1998), 83% within 10 years (O'Riordan *et al.*, 1998a), and 88% within 14 years (Brex *et al.*, 2002). However, even if it is highly probable that a subject with the first neurological symptom suggestive of MS will have a future relapse, it is difficult to predict when the relapse will happen and, above all, the severity of disease evolution. In general, the onset of the disease with sensory (Amato *et al.*, 1999; Phadke, 1990; Riise *et al.*, 1992) or visual (Phadke, 1990; Riise *et al.*, 1992) symptoms is more favorable than the onset with symptoms regarding motor (Amato *et al.*, 1999; Kantarci *et al.*, 1998; Pittock *et al.*, 2004; Riise *et al.*, 1992; Simone *et al.*, 2002; Trojano *et al.*, 1995; Weinshenker *et al.*, 1991), cerebellar (Amato *et al.*, 1999; Lauer and Firnhaber, 1992; Phadke, 1990; Poser *et al.*, 1982; Riise *et al.*, 1992; Weinshenker *et al.*, 1996), or sphincter (Amato *et al.*, 1999; Citterio *et al.*, 1989; Kantarci *et al.*, 1998; Phadke, 1990; Simone *et al.*, 2002) functions. Citterio *et al.* (1989) found that sphincter involvement was associated not only to higher disability but also to the risk of mortality. These findings partially suggest that symptoms from efferent pathways are worse than symptoms from afferent pathways. The Bayesian model (Bergamaschi *et al.*, 2001) also found that the risk of poor prognosis (development of SP course) was associated with early clinical predictors such as a polysimptomatic onset, a motor and a sphincter onset, as well as a motor-sensory involvement.

b. Optic Neuritis. The isolated optic neuritis (ON) is a common first symptom of MS but it can be followed by a relapse after a long time or also remain an isolated episode; therefore, it is important to dispose of long-term follow-up in predicting the outcome of ON.

In a prospective study with shorter mean follow-up time (3.8 years) (Jin *et al.*, 2003), the individual risk of CDMS after monosymptomatic ON was predicted by age and season of clinical onset.

Nilsson *et al.* (2005) studied a cohort of patients with acute monosymptomatic unilateral ON; 38% progressed to CDMS after up to 18 years, and only 1% after up to 31 years. The estimated 15-year risk of MS was 40%; most cases (60%) became CDMS within 3 years. Contrariwise to the previous study, younger patients had greater risk to convert to CDMS.

The recurrent presentation of ON is uncommon, but it has a relevant prognostic role. This kind of presentation can characterize not only the onset of MS but also of neuromyelitis optica (NMO). Among a large series of 1247 ON, 5.7% had recurred ON without intervening symptoms of further disseminated demyelination (Pirko *et al.*, 2004). The 5-year conversion rate of recurrent ON to MS was 14.4%, and to NMO was 12.5%. Women tended to convert to NMO more frequently than men, as did those with a higher annual frequency of ON.

c. Recovery After the First Attack. In adult, a short duration of initial symptoms seems to be a favorable indicator (Phadke, 1990), as is a complete recovery after the first episode of disease (Confavreux *et al.*, 2003; Runmarker and Andersen, 1993), while an incomplete remission is related to a poor prognosis (Amato *et al.*, 1999; Mikaeloff *et al.*, 2004; Phadke, 1990; Runmarker and Andersen, 1993; Trojano *et al.*, 1995; Weinshenker *et al.*, 1991). In Amato's study (1999), the incapacity to recover after relapse, and thus the residual disability in the early phase of the disease, was the most important factor that increased the risk of transition to the SP phase.

In children, occurrence of severe disability was associated with sequel after the first attack (Mikaeloff *et al.*, 2004).

C. Neuroradiological Findings

1. *Conventional Magnetic Resonance Imaging*

Magnetic resonance imaging (MRI) is the most sensitive paraclinical test in MS diagnosis, showing abnormalities in about 95% of patients with CDMS (Paty *et al.*, 1988). Therefore, MRI is currently considered indispensable to the diagnostic process and for monitoring the disease course of MS (Frohman *et al.*, 2003).

Nowadays, the diagnosis of MS is established when the two fundamental criteria of dissociation in space and time are fulfilled, according to the new diagnostic criteria for MS, known as the "McDonald Criteria" (McDonald *et al.*, 2001), revised in 2005 (Polman *et al.*, 2005). In 2001, the International Panel on the Diagnosis of Multiple Sclerosis presented these criteria with the intent to achieve better and more reliable MS diagnosis, balancing early diagnosis with the need to avoid false positive diagnosis. The criteria formally incorporated MRI into the well-established diagnostic workup that focuses on detailed neurological history and examination and a variety of paraclinical laboratory examinations.

MRI can also be a tool for prognostic purposes.

At baseline, 73% of patients had lesions on T2-weighted fast spin-echo (FSE) brain images and 42% had asymptomatic spinal cord lesions (Brex *et al.*, 1999). After a mean follow-up duration of 28 months, 47% of the patients with abnormal MRI progressed to CDMS (Barkhof *et al.*, 1997). Completely normal brain MRI scans were acquired in a low percentage of cases (18%), with a very high negative predictive value as in previous studies (Filippi *et al.*, 1994). Anyway, a negative brain MRI does not rule out definitively the diagnosis of MS. Thorpe *et al.* (1996) described a series of suspected MS patients with a negative brain scan, in whom MRI of the spinal cord revealed lesions in all of them, suggesting that spinal imaging further increases sensitivity and negative predictive value. More specifically, the variables that are obtainable with a conventional MRI and that are related to the disease's evolution are the number, sites and volume of lesions, and the detection of contrast enhancement.

a. Number of MRI Lesions. Morrissey *et al.* (1993) reported conversion to CDMS after 5 years of follow-up in 54% of patients with one to three focal white matter lesions on T2-weighted MRI and in 85% of the patients with four or more lesions.

Among the patients with a high number of T2 lesions (>10) at baseline MRI, the probability of overcoming an EDSS >5.5 within 10 years was 35% versus 20% in patients with less than 10 lesions (O'Riordan *et al.*, 1998a).

The probability to convert from CIS to CDMS within 14 years with the disease was 19% if there was no MRI lesion at onset, but it was 89% if there was at least one MRI lesion (Brex *et al.*, 2002). After 14 years, the median EDSS score was 2.0 when there were 1–3 lesions initially, 4.0 when there were 4–10 lesions initially, and 6.0 when there were more than 10 lesions initially (Brex *et al.*, 2002).

b. Sites of MRI Lesions. As the lesions' location concerns, infratentorial lesions and spinal appear to be more unfavorable than others. Infratentorial lesions best predicted long-term disability and were related to a major risk of conversion to CDMS in patients with initial findings suggestive of MS (Minneboo *et al.*, 2004). To have an asymptomatic spinal cord lesion (T2-weighted and FLAIR images), displayed in 27% patients with an isolated syndrome of the optic nerve, brain stem, or spinal cord, is an MRI factor related the risk of developing CDMS (O'Riordan *et al.*, 1998b). Brex *et al.* (1999) observed that 18% of CIS patients with T2 brain lesions had a second clinical attack within 12 months, but the percentage rose to 48% if additional spinal lesions were detected. These findings suggest that MRI of the spinal cord could increase the sensitivity and negative predictive value of MRI criteria in predicting conversion to CDMS. For these reason, spinal MRI abnormalities have been valorized among the reviewed McDonald's Criteria (Polman *et al.*, 2005).

c. Volume of MRI Lesions. The MRI burden at the time of initial presentation of the disease is predictive of the disease over the next 5–14 years. In particular, 90% of CIS patients shifted to CDMS within 5 years when the initial lesion volume was more than 1.23 cm^3, compared to 55% when the initial lesion volume was less than 1.23 cm^3 (Filippi *et al.*, 1994). After 14 years with the disease, the median EDSS score was 2.0 when initial median lesion load was 0.6 cm^3, 4.0 when initial median lesion load was 0.9 cm^3, and 6.0 when initial median lesion load was 5.6 cm^3 (Brex *et al.*, 2002).

d. Contrast-Enhancing MRI Lesions. Contrast enhancement detects, with high sensitivity disturbance of the blood–brain barrier, an early event in the development of inflammatory lesions in MS. Gadolinium enhancement should always be obtained at presentation in CIS because its ability to show the time dissemination of lesions, and its high predictive value for subsequent evolution to CDMS (Rovaris and Filippi, 2000). The risk for a CIS patient of becoming CDMS is higher if there are T1 gadolinium-enhancing lesions at onset (Barkhof *et al.*, 1997). T1 gadolinium-enhancing lesions at baseline or the appearance of enhancing lesions or new FSE lesions (detected in 43% of patients) 3 months after onset is highly sensitive (92%) and specific (81%) to an early conversion to CDMS within 1 year (Brex *et al.*, 1999).

e. Prognostic Value of MRI Barkhof Criteria. Barkhof *et al.* (1997) investigated the relative contribution of several MRI criteria to better predict conversion to CDMS of patients presenting with a first episode of symptoms suggestive of MS. They studied a cohort of patients with CIS, 45% of which developed CDMS after a median period of 9 months. The authors concluded that a four-parameter dichotomized MRI model including gadolinium enhancement, juxtacortical, infratentorial, and periventricular lesions best predicts conversion to CDMS.

Interestingly, the final model did not include the total number of lesions. This result coincides with the common sense that it is not merely the number of lesions that counts, but rather what sort of lesions are found and where they are found. The neuroradiological finding is also in agreement with the results obtained from the Bayesian modeling of the natural history of the disease, that demonstrated that some clinical events (motor and sphincter involvement, early increase in disability) were important short-term predictors of a long-term outcome, while other events were not, indicating that the type of clinical events has a more relevant prognostic value than their number (Bergamaschi *et al.*, 2001).

A collaborative study of seven centers (Korteweg *et al.*, 2006) studied the baseline MRI and clinical follow-up data for 532 patients with CIS, considering the development of a second clinical event as the main outcome. Overall conversion rate was 32.5% with a median survival time of 85.3 months. After a survival time of 2 years, the CIS patients that fulfilled at least three out of four of the Barkhof criteria (Barkhof *et al.*, 1997) had a risk to convert to MS four times compared with those not fulfilling the criteria. Forty-five percent with asymptomatic lesions versus about 10% in those with no asymptomatic lesions at baseline had a major risk of conversion. These increased risks were not affected by other factors such as sex, presenting symptoms, or source center.

Another study aimed to identify initial MRI factors that were predictive of a second attack and disability after a first episode of acute CNS inflammatory demyelination in childhood (Mikaeloff *et al.*, 2004a). In a cohort of 116 children who had a first episode of acute CNS inflammatory demyelination and a follow-up of about 5 years, the rate of second attack was three times higher in patients with corpus callosum long axis perpendicular lesions (reported in 30% of patients) on the initial MRI and/or with the sole presence of well-defined lesions, reported in 40% (hazard ratio 1.71). Both criteria were more specific predictors (100%) of relapse demonstrating conversion to CDMS, than the three Barkhof criteria (63%), but were less sensitive (21% compared with 52%). None of the MRI criteria was predictive of severe disability. Therefore, the two factors seem to be very specific to predict relapse and conversion to MS after a first episode of acute CNS inflammatory demyelination in childhood, but their lower sensitivity respect to Barkhof criteria, shows that this prediction remains difficult.

f. Brain Atrophy. In spite of the previous belief, axonal damage occurs early (Kornek and Lassmann, 2003). As a consequence, brain atrophy can be detected by MRI at the very early stage, or since the beginning of the disease.

The increase of ventricular volume (expression of progressive brain tissue loss) was significantly larger in CIS converted to CDMS within 1 year, than in patients that remained CIS (Dalton *et al.*, 2002).

Filippi *et al.* (2004) found that brain parenchymal volume loss is present since the onset of the diseases in CIS patients. The progressive axonal loss is predictive of development to CDMS and can be slowed by early therapy with interferon beta-1a (IFN-β1a). Interestingly, parenchymal volume correlated weakly with the number of new T2 lesions. These findings suggest that inflammatory and neurodegenerative processes are, at least partly, dissociated from the earliest clinical stage of MS onward.

g. MRI Findings in ON. The prognostic role of MRI is particularly interesting in evaluation of the evolution of clinically isolated ON. The presence of asymptomatic MRI abnormalities in ON ranged from 51% to 77% (Swanton *et al.*, 2006; Tintore *et al.*, 2005). Studies have shown that 31–82% of patients with ON and some MRI abnormalities will progress to CDMS within 2 years, compared with only 13–16% with normal MRI (CHAMPS Study Group, 2002; Optic Neuritis Study Group, 1997).

After monosymptomatic ON, the presence of MS-like brain MRI lesions was a strong prognostic marker of CDMS within a mean follow-up time of 3.8 years (hazard ratio of 4.68) (Jin *et al.*, 2003).

The Optic Neuritis Treatment Trial (Beck *et al.*, 2003) has shown that the overall risk for CDMS following ON over 10 years is 38%, but that this increases to 56% in those with one or more typical lesions on the baseline MRI.

A study from Barcellona (Tintore *et al.*, 2005) reported that MRI obtained close to the disease onset had an higher frequency of MRI abnormalities in subjects with brain stem syndromes and spinal cord syndromes and a higher conversion rate to CDMS than did those with ON. ON behaved differently from the other CISs for lower conversion and smaller proportion of patients, fulfilling MRI dissemination in space, time, or both. Thus, ON presentation could be thought to have a better prognosis. However, when only those patients with basal abnormalities on MRI were selected, no significant differences for clinical or MRI conversion were found. These data indicate that, even if ON has a smaller risk for conversion to CDMS, baseline MRI abnormalities and not CIS topography seem to be the crucial prognostic issue at MS presentation.

Concerning recurrent presentation of ONs, among patients with two or more MRI lesions consistent with MS, 40% converted to MS and none to NMO, while among the recurrent ON patients without such MRI lesions, none converted to MS and 18% to NMO (Pirko *et al.*, 2004).

Finally, MRI lesions suggestive of demyelinating disease were detected in 66% of ON without clinical manifestations of MS after 19–31 years (Nilsson *et al.*, 2005). These findings indicate that clinically silent MRI lesions suggestive of MS are long-term detectable in a majority of the patients with isolate ON. This finding should be taken into account when discussing prognosis and early intervention in patients with clinically isolated ON.

2. *Nonconventional MRI*

New MRI techniques can provide useful prognostic indications in addition to the findings obtained by conventional MRI.

Iannucci *et al.* (2000) reported that magnetization transfer (MT) imaging was able to detect focal abnormalities in the so-called normal-appearing brain tissue (NABT) before the appearance of lesions on conventional MRI. To create MT histograms of the NABT, macroscopic lesions were segmented from dual-echo images, were superimposed automatically, and were nulled out from the coregistered and scalp-stripped MT ratio (MTR) maps. CIS patients had lower average NABT-MTR and peak position than did control volunteers, but patient brain size was similar to that of volunteers. At 33 months follow-up, 41% patients developed CDMS. Patients who developed CDMS during the follow-up period had higher T2 lesion volume and lower average NABT-MTR and peak position than did those who did not develop CDMS. T2 lesion volume and average NABT-MTR were independent predictors of the subsequent development of CDMS. These findings indicated that subtle changes occur outside lesions visible on conventional MR images among patients with CIS suggestive of MS at presentation; the greater the extent of such abnormalities is, the higher is the risk of subsequent development of CDMS.

D. Biological Markers

1. *Cerebrospinal Fluid Oligoclonal Bands*

A low or absent number of IgG oligoclonal bands (OBs) in the cerebrospinal fluid (CSF) at the time of diagnosis predicts a better prognosis (Avasarala *et al.*, 2001). The absence of OB is also a favorable prognostic factor influencing the good clinical response to IFN-β treatment in RR patients (Annunziata *et al.*, 2006).

Among factors present at onset of acute monosymptomatic unilateral ON, the prognostic importance of the detection of IgG OBs was suggested by Frederiksen (1998) who found OBs in all the ON patients that at 1 year follow-up developed CDMS.

The presence of IgG OB was confirmed as a strong prognostic markers of conversion to CDMS. After monosymptomatic ON, CSF with mononuclear pleocytosis and/or IgG OBs increased the risk for subsequent MS significantly (49%) compared with 23% for those with normal CSF (Nilsson *et al.*, 2005). Within a mean follow-up time of 3.8 years after ON, the hazard ratio to CDMS was 5.39 when IgG OB were initially detected (Jin *et al.*, 2003).

In addition to the "conventional" IgG OBs, also IgM OBs seem to be a prognostic marker in MS. In the study of Villar *et al.* (2005), a considerable increase in CD5(+) B lymphocytes was found in patients with IgM OBs. These cells mostly secrete IgM antibodies directed against myelin lipids (mainly phosphatidylcholine). Disease course of patients with IgM OBs against myelin lipids suffered a second relapse earlier, had more relapses, and showed increased disability compared with those without anti-lipid IgM. The presence of intrathecal anti-myelin lipid IgM antibodies could therefore be a predictor of severe evolution in MS.

Berger *et al.* (2003) sustained that the presence of serum antibodies against myelin oligodendrocyte glycoprotein (MOG) and myelin basic protein (MBP) is predictive of early conversion from CIS to CDMS. They studied patients with CIS, abnormal MRI, and OBs in the cerebrospinal. Patients with anti-MOG and anti-MBP antibodies had relapses more often (95% vs 23%) and earlier (within a mean of 7.5 vs 45.1 months) than patients without these antibodies.

The hazard ratio for the development of CDMS was 76.5 among the patients who were seropositive for both antibodies and 31.6 among the patients who were seropositive only for anti-MOG antibodies, as compared with the seronegative patients.

E. Neurophysiological Examinations

Neurophysiological examinations, particularly evoked potentials (EPs), are useful in MS diagnosis and monitoring. Visual EPs (VEPs) demonstrated clinically silent lesions in 22% of patients, more frequently than somatosensory

EPs (SSEPs) (12%) and brain stem auditory EPs (BAEPs) (5%) (Hume and Waxman, 1988).

There are also a few evidences that EPs are relevant for prognosis.

An abnormality demonstrating a clinically silent lesion in any EP's modality (VEPs, BAEPs, SSEPs) at the onset of the disease was associated with a 48% chance of becoming CDMS within the 2-year follow-up, while normal EPs were associated with a 4% chance of CDMS. VEPs and motor EPs (MEPs) abnormalities correlate with the disability level after 2 years (Fuhr *et al.*, 2001).

Recent interest has drawn by multifocal VEPs (mVEPs), a new technique that let the detection of multiple individual VEP signals across the visual field up to eccentricity of 20–30°, while the conventional VEP is predominated by the central few degrees of vision and may not fully assess peripheral parts of the optic nerve. By providing much more information than conventional VEP, the mVEP was able to give prognostic information about ON evolution: 36% of ON with abnormal mVEPs progressed to MS over 1 year, whereas none with normal mVEPs progressed (Fraser *et al.*, 2006).

III. Prognostic Factors During the Early Phase/Course of the Disease

A. Clinical Features

1. *Time from Onset to the Second Attack*

Studies found that a long inter-attack interval from clinical onset to the second neurological episode was related to a favorable course (Amato *et al.*, 1999; Confavreux *et al.*, 2003; Phadke, 1990; Thompson *et al.*, 1986). Consequently, a short first inter-attack interval has an unfavorable influence on the disease's evolution (Simone *et al.*, 2002; Trojano *et al.*, 1995; Weinshenker *et al.*, 1991).

Even in children, occurrence of severe disability was associated with short interval between first and second attack (Ghezzi, 2005).

2. *Number and Type of Relapses During the First Years*

Studies also found that a high number of relapses in the first 5 years of the disease (Confavreux *et al.*, 2003), in particular the first and second years (Ebers, 2005; Weinshenker *et al.*, 1989), was a predictor of an unfavorable prognosis.

In children, occurrence of severe disability was associated with high relapse rate (Ghezzi, 2005; Mikaeloff *et al.*, 2004b).

The already mentioned Bayesian model (Bergamaschi *et al.*, 2001) confirmed that a high number of relapses in the first years had an unfavorable prognostic value, but the Bayesian approach added a qualitative feature to these findings: the importance of the type of relapses. Specifically, the motor/sphincter relapses had a strong impact on the prognosis, whereas certain types of relapses appeared to have no impact at all. Currently, some clinical therapeutic trials in MS use endpoints such as the "total count of relapses." In the light of Bayesian results, such an endpoint appears to be inappropriate, since it undervalues the fact that certain types of relapse are more important than others.

3. *Time of Development of Moderate/Severe Disability*

Several authors (Confavreux *et al.*, 2003; Runmarker and Andersen, 1993; Trojano *et al.*, 1995) indicated that having a moderate disability (that is EDSS score of 4.0) outside relapse increases the risk of progression. These findings supported Weinshenker's report (Weinshenker *et al.*, 1991), which stated that the time patients spend at different EDSS levels is short when EDSS is between 3.5 and 5.0. The arrival at such a level of disability is related to the higher risk of further increasing the progression. With respect to the initial manifestations of the disease, the median intervals are significantly longer for cases with an isolated ON at onset (28 years) than for cases with an isolated dysfunction of long tracts (16 years) (Confavreux *et al.*, 2003).

Interestingly, Confavreux *et al.* (2003) observed that after the assignment of an irreversible disability with an EDSS score of 4.0, the further assignment of a higher score (6.0 or 7.0) is not influenced by other prognostic factors such as gender, age at onset, initial symptoms, initial course, degree of recovery, number of relapses. This suggests that MS is a two-stage disease, with an initial phase of variable duration that the above-mentioned clinical prognostic factors influence, and a second phase that varies little in comparison with the early characteristics.

The Bayesian model (Bergamaschi *et al.*, 2001) also found that the risk of poor prognosis was associated with a moderate residual disability EDSS score of 4.0 outside relapse) within 1 year's disease. Finally, in children, occurrence of severe sequel was associated with high disability after the first year (Ghezzi, 2005).

4. *Disease Course*

The natural course of MS is characterized by several acute episodes of neurological impairment, followed by a partial or complete remission with clinical stability between attacks. The latter is defined as RR phase, often followed by the SP phase, which is characterized by progressive clinical impairment, with or without superimposed relapses and remissions, that eventually leads to irreversible disability (Confavreux *et al.*, 1980; Lublin and Reingold, 1996). Only 10–15% of MS patients accumulate a progressive disability from the beginning, referred to PP course (Confavreux *et al.*, 2000).

A RR course is followed by SP in some 80% of cases within 2 decades. The initial RR course (Confavreux *et al.*, 1980, 2003; Phadke 1990; Weinshenker *et al.*, 1991) is favorable compared to an PP course (Amato *et al.*, 1999;

Bergamaschi *et al.*, 2005; Citterio *et al.*, 1989; Lauer and Firnhaber, 1992; Mikaeloff *et al.*, 2004; Noseworthy *et al.*, 1983; Phadke, 1990; Runmarker and Andersen, 1993; Weinshenker *et al.*, 1991). In fact, a PP course predicted a more rapid progression (Tremlett *et al.*, 2006). The median time a patient takes to reach EDSS 6.0 is 23 years when the onset is RR, but only 7 years when it is PP (Confavreux *et al.*, 2003). In children, occurrence of severe disability is associated with progressive course from onset MS (Mikaeloff *et al.*, 2004). Even if the clinical features and the prognosis of relapsing and progressive forms of MS are well defined, the relationships among these phases and their long-term outcomes remain uncertain. Kremenchutzky *et al.* (2006) analyzed clinical features within a natural history cohort with mean follow-up of 25 years. Median times to reach disability levels of 6, 8, and 10 according to EDSS were 12.7, 20.6, and 43.9 years, respectively. Among RR patients, the great majority entered a progressive phase with a mean time to progression of 10.4 years. Patients who had a progressive phase were divided in three subgroups: (1) PP disease, (2) attack-onset disease where only a single attack has occurred before onset of progression (SAP), and (3) SP disease where several relapses have occurred before onset of progression. Among the three cohorts, there was no difference in time to EDSS 6, 8, and 10. The site of the original attack was not usually where progression began. These findings demonstrate that the progressive course is independent of relapses preceding the onset of progression. The relatively stereotyped nature (prevalent distal corticospinal tract dysfunction) of the progressive phase seen in all patients suggests a degenerative process with a progressive dying back axonal loss, at least in part independent of plaque location or burden and may entail mechanisms common to all forms of progressive MS (PP, SAP, and SP).

B. Neuroradiological Findings

1. *Conventional MRI*

a. Number of MRI Lesions. T2 lesions in RR correlate strongly with clinical disease severity and with brain tissue loss 13 years later the baseline observation (Rudick *et al.*, 2006).

b. Contrast-Enhancing MRI Lesions. A meta-analysis from the members of European Magnetic Resonance Network in Multiple Sclerosis (MAGNIMS) and additional centers in the USA (Kappos *et al.*, 1999) which collected data from five natural course studies and four placebo groups of clinical trial (307 patients) aimed to investigate the prognostic value of gadolinium-enhanced MRI of the brain. The relapse rate in the first year was predicted by the mean number of gadolinium-enhancing lesions in monthly scans during the first 6 months. The predictive value of the number of gadolinium-enhancing lesions in one baseline scan was less strong. The best predictor for relapse rate was the variation of lesion counts in the first 6 monthly scans which allowed an estimate of relapse in the first and in the second year. Neither the initial scan nor monthly scans over 6 months were predictive of change in the EDSS in the subsequent 12 or 24 months. The mean of gadolinium-enhancing lesion counts in the first 6 monthly scans was weakly predictive of EDSS change after 1 and 2 years. The authors concluded that although detection of gadolinium enhancement in MRI is a predictor of the occurrence of relapses, it is not a strong predictor of the development of cumulative impairment or disability. This discrepancy supports the idea that variant pathogenetic mechanisms are operative in the occurrence of relapses and in the development of long-term disability in MS.

c. MRI in PPMS. PP course characterizes about 10% of MS cases (Confavreux *et al.*, 2000). MRI brain volume, changes in T2 lesion load and cord area, and number of new lesions were predictive for long-term unfavorable outcome in PP (Sastre-Garriga *et al.*, 2005).

2. *Conventional MRI and Clinical Factors*

Scott *et al.* (2000) tried to join clinical and MRI variables at onset, in order to predict the short-term evolution of MS. Data were collected prospectively on patients newly diagnosed with MS. Recorded prognostic factors were: age at onset, symptoms at onset, MRI status at onset and at diagnosis of CDMS, interval between the first and second attack, attack frequency in the first 2 years, and completeness of recovery from initial attacks. Progression was determined by final EDSS and changes in EDSS between initial presentation and final follow-up. Short-term prognosis was influenced by the presence or absence of a high number of the risk factors. In particular, patients predicted to have low risk of progression of MS based on the prognostic factors represented 17% of patients (zero to one risk factor for progression). The patients with high risk of progression (four to six risk factors) represented 24% of patients. Patients with a high number of risk factors did significantly worse than those with a small or medium number of risk factors in terms of final EDSS and progression to higher EDSS. At the time of diagnosis of CDMS, MRI findings suggestive of MS were seen in 84% of patients (suspicious in 13%, negative in 3%).

3. *Unconventional MRI*

Santos *et al.* (2002) demonstrated that changes in MT imaging, which can detect focal abnormalities in normal-appearing white matter (NAWM) before the appearance of lesions on conventional MRI, were able to predict the clinical evolution of MS better than conventional MRI. Data were obtained from MS patients who were followed for 5 years. Patients were divided into stable and worsening groups according to their change in EDSS scores over 5 years. The mean MT ratio (MTr) in gray matter and in NAWM, as well as in a thick central brain showed significant differences between clinically stable and worsening MS patients. On the contrary, the relationship between disability changes in EDSS and T2-weighted MRI lesions was weaker. NAWM MTr very well predicted the clinical evolution, yielding a positive predictive value of 78%, a negative predictive value of 89%, and an odds ratio of 28. These data indicate that the quantification of MTr in the NAWM can predict the clinical evolution of MS. Lower MTr values predict poorer long-term clinical outcome. Abnormalities of MTr values in the NAWM are more relevant to the development of future patient disability than those in the T2-weighted MRI lesions.

As several clinical and instrumental factors seem to be able to predict if a patient will have a poor prognosis or a favorable prognosis, an open question could be: does benign MS exist?

A study with MT (De Stefano *et al.*, 2006) proposes to answer the previous question. Lesional-MTr, perilesional-MTr, NAWM-MTr, and cortical-MTr in "benign" patients (EDSS <3, disease's duration >15 years) were significantly higher than white matter of RR patients at their earliest disease stages (disease's duration <3 years). These findings confirmed the truly existence of benign forms of MS, which are probably characterized by an extraordinary beneficial natural response to demyelination.

C. Biological Markers

1. *Cerebrospinal Fluid Oligoclonal Bands*

Intrathecal IgM bands (ITMB) predicted a worse evolution in MS patients followed up from 5 to 16 years (Villar *et al.*, 2003). Time to conversion to SP MS, time elapsed to reach a disability of 6 in the EDSS, percentage of patients with a benign MS, and changes in EDSS score were evaluated. During the follow-up, 70.8% of patients with ITMB converted to SPMS. None of the patients without ITMB did. At the end of the study, 63.6% of patients with ITMB had reached EDSS 6, whereas none of the patients lacking ITMB reached values above EDSS 3. When patients with benign MS were analyzed, 82% lacked ITMB. All patients with a nonbenign MS had ITMB. At the end of the study, the mean EDSS score was 4.64 in patients with ITMB and 1.31 in those without. The presence of oligoclonal IgM bands in CSF is an unfavorable prognostic marker in MS.

2. *Other Biological Markers*

Recent interest has been voted to possible markers of axonal loss; one of this, the Tau protein, could be helpful to predict short-term outcome in patients with early RRMS. Martinez-Yelamos *et al.* (2004) analyzed the relationship between Tau protein concentration in the CSF (CSF-TAU) of RR patients and disability, time to next relapse, and time to experience a one-point increase on the EDSS. CSF-TAU was correlated with the progression index at the end of follow-up. Patients with higher CSF-TAU experienced a more rapid increase in the EDSS. CSF-TAU was the only independent variable to predict the time to next relapse.

The predictive value of markers of T-cell activation for long-term disease progression was investigated by Killestein *et al.* (2001). Flow cytometry measurements were correlated to changes in EDSS and MR T2 lesion load over 36 months in patients with SP MS. A significant correlation was found between the percentage of tumor necrosis factor-α-producing CD4(+) T cells at baseline and the change in T2 lesion load during 3-year follow-up.

IV. Conclusions

MS is classically labeled as an unpredictable disease. The outcome of MS patients is highly variable, since they can be quite asymptomatic or severely disabled after 10–15 years with the disease. However, the present chapter proves that we dispose of several clinical (Table I) and instrumental (Table II) factors, collected at the onset or during the course of the disease, which can be able to predict a poor MS prognosis.

TABLE I
Clinical Factors Predicting Unfavorable MS Evolution, Collected at Onset or During the Course of the Disease

	Prognostic factors	Unfavorable events
Variables at onset		
Demographic variables	Male gender	Early disability, faster progression
	Late age at onset	Early disability, faster progression
Clinical features	Polisymptomatic onset	Early disability, faster progression
	Motor involvement	Early disability, faster progression
	Motor-sensory involvement	Faster progression
	Sphincter involvement	Early disability, faster progression, mortality
	Cerebellar involvement	Early disability, faster progression
	Incomplete recovery after onset	Early disability, faster progression
Postonset variables		
	Short inter-attack interval	Early disability, faster progression
	Number of relapses	Early disability, faster progression
	Sphincter plus motor relapses	Faster progression
	Motor relapses	Faster progression
	Motor sequel	Faster progression
	EDSS $\geq$ 4 outside relapse	Faster progression
	Progressive course	Early disability

TABLE II
INSTRUMENTAL FACTORS PREDICTING UNFAVORABLE MS EVOLUTION, COLLECTED AT ONSET OR DURING THE COURSE OF THE DISEASE

	Prognostic factors	Unfavorable events
Variables at onset		
Neuroradiological findings		
Conventional MRI	Number of lesions	CIS to CDMS conversion
	Site of lesions (infratentorial, spinal)	CIS to CDMS conversion
	Volume of lesions	CIS to CDMS conversion
	Contrast-enhancing lesions	CIS to CDMS conversion
	Brain atrophy	CIS to CDMS conversion faster progression
Unconventional MRI	Magnetization transfer	CIS to CDMS conversion
Biological markers	CSF IgG and IgM OBs	Early disability, faster progression
	Serum anti-MOG and anti-MBP antibodies	CIS to CDMS conversion
Neurophysiological examinations	cVEPs, mVEPs, MEPs, BAEPs, SSEPs	CIS to CDMS conversion
Postonset variables		
Neuroradiological findings	Conventional MRI	
	Number of lesions	Early disability
	Enhancement of lesions	Higher relapse rate
Unconventional MRI	Magnetization transfer	Early disability
Biological markers	CSF IgG OBs	Severe disability
	CSF IgM OBs	Faster progression, severe disability, conversion RR to SP
	CSF Tau protein	Severe disability, faster progression
	Tumor necrosis factor-α-producing CD4(+) T cells	Larger T2 lesion load
Neurophysiological exams	VEPs, mVEPs, BAEPs, SSEPs	CIS to CDMS conversion
	MEPs	Severe disability

The main clinical factors indicative of long-term bad prognosis are male gender, late age at onset, polisymptomatic onset with initial pyramidal, cerebellar and sphincter signs, progressive onset, incomplete recovery after onset, short inter-attack interval with high relapse rate during the first years, and motor–sphincter involvement. MRI is the best instrumental predictor of conversion to definite MS after a first demyelinating event; the presence of OBs in CSF has a relevant prognostic value too.

The majority of the studies that identified the prognostic factors focused on demographic variables or on clinical and instrumental findings collected at the onset of the disease. The few studies that used classical statistical approaches and that enlarged their investigations to the postonset phases of the disease, often limited their analyses to the count of events instead of the type of events. In addition, the obtained data were exploitable at population level, while a prognostic forecast at individual level, valuable for a treatment decision, was not usually available. Therefore, although the general course of MS is fairly predictable from natural history studies, the determination of a detailed prognosis in a given individual seems to be almost an insurmountable task.

Thus, an essential problem remains to solve: how to combine all the clinical and instrumental negative factors in a single estimate of risk, useful for prediction of the unfavorable evolution of disease at a single patient's level?

In the past only a few studies tempted to face this problem. Scott *et al.* (2000) tried to put together clinical and MRI early at onset, in order to predict the short-term evolution of MS, but the approach they used, limited to onset variables, seems to be an oversimplification of the problem. More sophisticated statistical approaches should be used to better exploit the large amount of variables collectable during the course of MS. In recent years the Bayesian methodology has been promoted as an alternative to the frequentist statistical approach when dealing with medical problems (Berzuini and Larizza, 1996; Spiegelhalter *et al.*, 1999), the MS prognosis included (Weinshenker, 2001). A Bayesian methodology has been already employed to predict the natural history of MS, but only clinical variables were analyzed (Bergamaschi *et al.*, 2001). In our opinion, advanced statistical approaches could represent a different way of analyzing MS prognosis, improving the management of several factors by incorporating complex background knowledge.

Future studies should be able to utilize the follow-up data fully and to produce forecasts of risk at a single patient level, constructing individual risk scores. These analyses should incorporate not only further clinical aspects of the disease (e.g., cognitive impairment and fatigue), but also genetic, immunologic, radiological, and neurophysiologic findings.

References

Amato, M. P., Ponziani, G., Bartolozzi, M. L., and Siracusa, G. (1999). A prospective study on the natural history of multiple sclerosis: Clues to the conduct and interpretation of clinical trials. *J. Neurol. Sci.* **168,** 96–106.

Annunziata, P., Giorgio, A., De Santi, L., Zipoli, V., Portaccio, E., Amato, M. P., Clerici, R., Scarpini, E., Moscato, G., Iudice, A., Vacca, G., Orefice, G., *et al.* (2006). Absence of cerebrospinal fluid oligoclonal bands is associated with delayed disability progression in relapsing-remitting MS patients treated with interferon-beta. *J. Neurol. Sci.* **244,** 97–102.

Avasarala, J. R., Cross, A. H., and Trotter, J. L. (2001). Oligoclonal band number as a marker for prognosis in multiple sclerosis. *Arch. Neurol.* **58**(12), 2044–2045.

Barkhof, F., Filippi, M., Miller, D. H., Scheltens, P., Campi, A., Polman, C. H., Comi, G., Ader, H. J., Losseff, N., and Valk, J. (1997). Comparison of MRI criteria at first presentation to predict conversion to clinically definite multiple sclerosis. *Brain* **120,** 2059–2069.

Beck, R. W. (1995). The optic neuritis treatment trial: Three-year follow-up results. *Arch. Ophthalmol.* **113**(2), 136–137.

Beck, R. W., Trobe, J. D., Moke, P. S., Gal, R. L., Xing, D., and Optic, Neuritis Study Group (2003). High and low risk profiles for the development of multiple sclerosis within 10 years after optic neuritis: Experience of the Optic Neuritis Treatment Trial. *Arch. Ophthalmol.* **121,** 944–949.

Bergamaschi, R., Romani, A., Tonietti, S., Citterio, A., Berzuini, C., and Cosi, V. (2000). Usefulness of Bayesian graphical models for early prediction of disease progression in multiple sclerosis. *Neurol. Sci.* **21**(4 Suppl. 2), S819–S823.

Bergamaschi, R., Berzuini, C., Romani, A., and Cosi, V. (2001). Predicting secondary progression in relapsing-remitting multiple sclerosis: A Bayesian analysis. *J. Neurol. Sci.* **189,** 13–21.

Bergamaschi, R., Montomoli, C., Candeloro, E., Fratti, C., Citterio, A., and Cosi, V. (2005). Disability and mortality in a cohort of multiple sclerosis patients: A reappraisal. *Neuroepidemiology* **25,** 15–18.

Berger, T., Rubner, P., Schautzer, F., Egg, R., Ulmer, H., Mayringer, I., Dilitz, E., Deisenhammer, F., and Reindl, M. (2003). Antimyelin antibodies as a predictor of clinically definite multiple sclerosis after a first demyelinating event. *N. Engl. J. Med.* **349,** 139–145.

Berzuini, C., and Larizza, C. (1996). A unified approach for modelling longitudinal and failure time data, with application in medical monitoring. *IEEE Trans Pattern Anal. Mach. Intell.* **2,** 109–123.

Boeije, H. R., and Janssens, A. C. (2004). 'It might happen or it might not': How patients with multiple sclerosis explain their perception of prognostic risk. *Soc. Sci. Med.* **59,** 861–868.

Brex, P. A., O'Riordan, J. I., Miszkiel, K. A., Moseley, I. F., Thompson, A. J., Plant, G. T., and Miller, D. H. (1999). Multisequence MRI in clinically isolated syndromes and the early development of MS. *Neurology* **53**(6), 1184–1190.

Brex, P. A., Ciccarelli, O., O'Riordan, J. I., Sailer, M., Thompson, A. J., and Miller, D. H. (2002). A longitudinal study of abnormalities on MRI and disability from multiple sclerosis. *N. Engl. J. Med.* **346**(3), 158–164.

Broman, T., Andersen, O., and Bergmann, L. (1981). Clinical studies on multiple sclerosis. I. Presentation of an incidence material from Gothenburg. *Acta Neurol. Scand.* **63,** 6–33.

Cardona-Gomez, G. P., Mendez, P., DonCarlos, L. L., Azcoitia, I., and Garcia-Segura, L. M. (2002). Interactions of estrogens and insulin-like growth factor-I in the brain: Implications for neuroprotection. *Brain Res. Brain Res. Rev.* **37,** 320–334.

CHAMPS Study Group (2002). MRI predictors of early conversion to clinically definite multiple sclerosis in CHAMPS placebo group. *Neurology* **59,** 998–1005.

Citterio, A., Azan, G., Bergamaschi, R., Erbetta, A., and Cosi, V. (1989). Multiple sclerosis: Disability and mortality in a cohort of clinically diagnosed patients. *Neuroepidemiology* **8,** 249–253.

Comi, G., Filippi, M., Barkhof, F., Comi, G., Filippi, M., Barkhof, F., Durelli, L., Edan, G., Fernandez, O., Hartung, H.-P., Seeldrayers, P., Sørensen, P. S., *et al.* (2001). Effect of early interferon treatment on conversion to definite multiple sclerosis: A randomised study. *Lancet* **357** (9268), 1576–1582.

Confavreux, C., Aimard, G., and Devic, M. (1980). Course and prognosis of multiple sclerosis assessed by the computerised data processing of 349 patients. *Brain* **103,** 281–300.

Confavreux, C., Hutchinson, M., Hours, M. M., Cortinovis-Tourniaire, P., and Moreau, T. (1998). Rate of pregnancy-related relapse in multiple sclerosis. Pregnancy in Multiple Sclerosis Group. *N. Engl. J. Med.* **339,** 285–291.

Confavreux, C., Vukusic, S., Moreau, T., and Adeleine, P. (2000). Relapses and progression of disability in multiple sclerosis. *N. Engl. J. Med.* **343**(20), 1430–1438.

Confavreux, C., Vukusic, S., and Adeleine, P. (2003). Early clinical predictors and progression of irreversible disability in multiple sclerosis: An amnesic process. *Brain* **126,** 770–782.

Dalton, C. M., Brex, P. A., Jenkins, R., Fox, N. C., Miszkiel, K. A., Crum, W. R., O'Riordan, J. I., Plant, G. T., Thompson, A. J., and Miller, D. H. (2002). Progressive Ventricular enlargement in patients with clinically isolated syndromes is associated with the early development of multiple sclerosis. *J. Neurol. Neurosurg. Psychiatry* **73,** 141–147.

De Stefano, N., Battaglini, M., Stromillo, M. L., Zipoli, V., Bartolozzi, M. L., Guidi, L., Siracusa, G., Portaccio, E., Giorgio, A., Sorbi, S., Federico, A., and Amato, M. P. (Stefano 2006). Brain damage as detected by magnetization transfer imaging is less pronounced in benign than in early relapsing multiple sclerosis. *Brain* **129,** 2008–2016.

Deryck, O., Ketelaer, P., and Dubois, B. (2006). Clinical characteristics and long term prognosis in early onset multiple sclerosis. Epub 2006 March 6*J. Neurol.* **253,** 720–723.

Detels, R., Clark, V. A., Valdiviezo, N. L., Visscher, B. R., Malmgren, R. M., and Dudley, J. P. (1982). Factors associated with a rapid course of multiple sclerosis. *Arch. Neurol.* **39,** 337–341.

Ebers, G. C. (2005). Prognostic factors for multiple sclerosis: The importance of natural history studies. *J. Neurol.* **252**(Suppl. 3), 15–20.

Falaschi, P., Martocchia, A., Proietti, A., D'Urso, R., and Antonini, G. (2001). High incidence of hyperandrogenism-related clinical signs in patients with multiple sclerosis. *Neuro. Endocrinol. Lett.* **22,** 248–250.

Filippi, M., Horsfield, M. A., Morrissey, S. P., MacManus, D. G., Rudge, P., McDonald, W. I., and Miller, D. H. (1994). Quantitative brain MRI lesion load predicts the course of clinically isolated syndromes suggestive of multiple sclerosis. *Neurology* **44,** 635–641.

Filippi, M., Rovaris, M., Inglese, M., Barkhof, F., De Stefano, N., Smith, S., and Comi, G. (2004). Interferon beta-1a for brain tissue loss in patients at presentation with syndromes suggestive of multiple sclerosis: A randomised, double-blind, placebo-controlled trial. *Lancet* **364,** 1463–1464.

Fraser, C., Klistorner, A., Graham, S., Garrick, R., Billson, F., and Grigg, J. (2006). Multifocal visual evoked potential latency analysis. Predicting progression to multiple sclerosis. *Arch. Neurol.* **63,** 847–850.

Frederiksen, J. L. (1998). Can CSF predict the course of optic neuritis? *Mult. Scler.* **4,** 132–135.

Frohman, E. M., Goodin, D. S., Calabresi, P. A., Corboy, J. R., Coyle, P. K., Filippi, M., Frank, J. A., Galetta, S. L., Grossman, R. I., Hawker, K., Kachuck, N. J., Levin, M. C., *et al.* (2003). The utility of MRI in suspected MS: Report of the Therapeutic and Technology Assessment Subcommittee of the American Academy of Neurology. *Neurology* **61,** 602–611.

Fuhr, P., Borggrefe-Chappuis, A., Schindler, C., and Kappos, L. (2001). Visual and motor evoked potentials in the course of multiple sclerosis. *Brain* **124,** 2162–2168.

Ghezzi, A. (2005). Childhood-juvenile multiple sclerosis: Clinical characteristics and treatment. *Expert Rev. Neurother.* **5,** 403–411.

Goodin, D. S., Frohman, E. M., Garmany, G. P., Jr., Halper, J., Likosky, W. H., Lublin, F. D., Silberberg, D. H., Stuart, W. H., and van den Noort, S. (2002). Disease modifying therapies in multiple sclerosis: Report of the Therapeutics and Technology Assessment Subcommittee of the American Academy of Neurology and the MS Council for Clinical Practice Guidelines. *Neurology* **58,** 169–178.

Held, U., Heigenhauser, L., Shang, C., Kappos, L., and Polman, C., and Lawry Centre for MS Research (2005). Predictors of relapse rate in MS clinical trials. *Neurology* **65,** 1769–1773.

Hume, A. L., and Waxman, S. G. (1988). Evoked potentials in suspected multiple sclerosis: Diagnostic value and prediction of clinical course. *J. Neurol. Sci.* **83,** 191–210.

Huseby, E. S., Sather, B., Huseby, P. G., and Goverman, J. (2001). Age-dependent T cell tolerance and autoimmunity to myelin basic protein. *Immunity* **14**(4), 471–481.

Iannucci, G., Tortorella, C., Rovaris, M., Sormani, M. P., Comi, G., and Filippi, M. (2000). Prognostic value of MR and magnetization transfer imaging findings in patients with clinically isolated syndromes suggestive of multiple sclerosis at presentation. *Am. J. Neuroradiol.* **21,** 1034–1038.

Jacobs, L. D., Beck, R. W., Simon, J. H., Kinkel, R. P., Brownscheidle, C. M., Murray, T. J., Simonian, N. A., Slasor, P. J., and Sandrock, A. W. (2000). Intramuscular interferon beta-1a therapy initiated during a first demyelinating event in multiple sclerosis. CHAMPS study group. *N. Engl. J. Med.* **343,** 898–904.

Jin, Y. P., de Pedro-Cuesta, J., Huang, Y. H., and Soderstrom, M. (2003). Predicting multiple sclerosis at optic neuritis onset. *Mult. Scler.* **9,** 135–141.

Johnson, K. P., Brooks, B. R., Cohen, J. A., Ford, C. C., Goldstein, J., Lisak, R. P., Myers, L. W., Panitch, H. S., Rose, J. W., Schiffer, R. B., Vollmer, T., Weiner, L. P., *et al.* (1995). Copolymer 1 reduces relapse rate and improves disability in relapsing-remitting multiple sclerosis: Results of a phase III multicenter, double-blind, placebo-controlled trial. *Neurology* **45,** 1268–1276.

Kantarci, O., Siva, A., Eraksoy, M., Karabudak, R., Sutlas, N., Agaoglu, J., Turan, F., Ozmenoglu, M., Togrul, E., and Demirkiran, M. (1998). Survival and predictors of disability in Turkish MS patients. *Neurology* **51,** 765–772.

Kappos, L., Moeri, D., Radue, E. W., Schoetzau, A., Schweikert, K., Barkhof, F., Miller, D., Guttmann, C. R., Weiner, H. L., Gasperini, C., and Filippi, M. (1999). Predictive value of gadolinium-enhanced magnetic resonance imaging for relapse rate and changes in disability or impairment in multiple sclerosis: A meta-analysis. Gadolinium MRI Meta-analysis Group. *Lancet* **353,** 964–969.

Killestein, J., Kalkers, N. F., Meilof, J. F., Barkhof, F., van Lier, R. A., and Polman, C. H. (2001). TNFalpha production by CD**4**(+) T cells predicts long-term increase in lesion load on MRI in MS. *Neurology* **57,** 1129–1131.

Kornek, B., and Lassmann, H. (2003). Neuropathology of multiple sclerosis-new concepts. *Brain Res. Bull.* **61,** 321–326.

Korteweg, T., Tintore, M., Uitdehaag, B., Rovira, A., Frederiksen, J., Miller, D., Fernando, K., Filippi, M., Agosta, F., Rocca, M., Fazekas, F., Enzinger, C., *et al.* (2006). MRI criteria for dissemination in space in patients with clinically isolated syndromes: A multicentre follow-up study. *Lancet Neurol.* **5,** 221–227.

Kraft, G. H., Freal, J. E., Coryell, J. K., Hanan, C. L., and Chitnis, N. (1981). Multiple sclerosis: Early prognostic guidelines. *Arch. Phys. Med. Rehabil.* **62**(2), 54–58.

Kremenchutzky, M., Rice, G. P., Baskerville, J., Wingerchuk, D. M., and Ebers, G. C. (2006). The natural history of multiple sclerosis: A geographically based study 9: Observations on the progressive phase of the disease. *Brain* **129,** 584–594.

Kurtzke, J. F. (1983). Rating neurologic impairment in multiple sclerosis: An expanded disability status scale (EDSS). *Neurology* **33,** 1444–1452.

Lauer, K., and Firnhaber, W. (1992). Prognostic criteria in an epidemiological group of patients with multiple sclerosis: An exploratory study. *J. Neurol.* **239,** 93–97.

Levic, Z. M., Dujmovic, I., Pekmezovic, T., Jarebinski, M., Marinkovic, J., Stojsavljevic, N., and Drulovic, J. (1999). Prognostic factors for survival in multiple sclerosis. *Mult. Scler.* **5,** 171–178.

Lublin, F. D., and Reingold, S. C. (1996). Defining the clinical course of multiple sclerosis: Results of an international survey. *Neurology* **46,** 907–911.

Martinelli, V., Rodegher, M., Moiola, L., and Comi, G. (2004). Late onset multiple sclerosis: Clinical characteristics, prognostic factors and differential diagnosis. *Neurol. Sci.* **25**(Suppl. 4), S350–S355.

Martinez-Yelamos, A., Saiz, A., Bas, J., Hernandez, J. J., Graus, F., and Arbizu, T. (2004). Tau protein in cerebrospinal fluid: A possible marker of poor outcome in patients with early relapsing-remitting multiple sclerosis. *Neurosci. Lett.* **363,** 14–17.

McDonald, W. I., Compston, A., Edan, G., Goodkin, D., Hartung, H.-P., Lublin, F. D., McFarland, H. F., Paty, D. W., Polman, C. H., Reingold, S. C., Sandberg-Wollheim, M., Sibley, W., *et al.* (2001). Recommended diagnostic criteria for multiple sclerosis: Guidelines from the International Panel on the diagnosis of multiple sclerosis. *Ann. Neurol.* **50,** 121–127.

Mikaeloff, Y., Adamsbaum, C., Husson, B., Vallee, L., Ponsot, G., Confavreux, C., Tardieu, M., and Suissa, S.KIDMUS Study Group on Radiology (2004a). MRI prognostic factors for relapse after acute CNS inflammatory demyelination in childhood. *Brain* **127,** 1942–1947.

Mikaeloff, Y., Suissa, S., Vallee, L., Lubetzki, C., Ponsot, G., Confavreux, C., and Tardieu, M. KIDMUS Study Group (2004b). First episode of acute CNS inflammatory demyelination in childhood: Prognostic factors for multiple sclerosis and disability. *J. Pediatr.* **144,** 246–252.

Minneboo, A., Barkhof, F., Polman, C. H., Uitdehaag, B. M., Knol, D. L., and Castelijns, J. A. (2004). Infratentorial lesions predict long-term disability in patients with initial findings suggestive of multiple sclerosis. *Arch. Neurol.* **61,** 217–221.

Morrissey, S. P., Miller, D. H., Kendall, B. E., Kingsley, D. P., Kelly, M. A., Francis, D. A., MacManus, D. G., and McDonald, W. I. (1993). The significance of brain magnetic resonance imaging abnormalities at presentation with clinically isolated syndromes suggestive of multiple sclerosis. A 5-year follow-up study. *Brain* **116,** 135–146.

Nilsson, P., Larsson, E. M., Maly-Sundgren, P., Perfekt, R., and Sandberg-Wollheim, M. (2005). Predicting the outcome of optic neuritis: Evaluation of risk factors after 30 years of follow-up. *J. Neurol.* **252,** 396–402.

Noseworthy, J., Paty, D., Wonnacott, T., Feasby, T., and Ebers, G. (1983). Multiple sclerosis after age 50. *Neurology* **33**(12), 1537–1544.

O'Riordan, J. I., Thompson, A. J., Kingsley, D. P., Macmanus, D. G., Kendall, B. E., Rudge, P., Mcdonald, W. I., and Miller, D. H. (1998a). The prognostic value of brain MRI in clinically isolated syndromes of the CNS. A 10-year follow-up. *Brain* **121,** 495–503.

O'Riordan, J. I., Losseff, N. A., Phatouros, C., Thompson, A. J., Moseley, I. F., MacManus, D. G., McDonald, W. I., and Miller, D. H. (1998b). Asymptomatic spinal cord lesions in clinically isolated optic nerve, brain stem, and spinal cord syndromes suggestive of demyelination. *J. Neurol. Neurosurg. Psychiatry* **64,** 353–357.

Optic Neuritis Study Group D. H. (1997). The 5-year risk of MS after optic neuritis: Experience of the Optic Neuritis Treatment Trial. *Neurology* **49,** 1404–1413.

Paty, D. W., Oger, J. J. F., Kastrukoff, L. F., Hashimoto, S. A., Hooge, J. P., Eisen, A. A., Eisen, K. A., Purves, S. J., Low, M. D., Brandejs, V., Robertson, W. D., and Li, D. K. B. (1988). MRI in the diagnosis of MS: A prospective study with comparison of clinical evaluation, evoked potentials, oligoclonal banding, and CT. *Neurology* **38,** 180–185.

Phadke, J. C. (1990). Clinical aspect of multiple sclerosis in North East Scotland with particular reference to its course and prognosis. *Brain* **113,** 1597–1628.

Pirko, I., Blauwet, L. K., Lesnick, T. G., and Weinshenker, B. G. (2004). The natural history of recurrent optic neuritis. *Arch. Neurol.* **61,** 1401–1405.

Pittock, S. J., McClelland, R. L., Mayr, W. T., Jorgensen, N. W., Weinshenker, B. G., Noseworthy, J., and Rodriguez, M. (2004). Clinical implications of benign multiple sclerosis: A 20-year population-based follow-up study. *Ann. Neurol.* **56**(2), 303–306.

Polman, C., Reingold, S., Edan, G., Filippi, M., Hartung, H. P., Kappos, L., Lublin, F. D., Metz, L. M., McFarland, H. F., O'connor, P. W., Sandberg-Wollheim, M., Thompson, A. J., *et al.* (2005). Diagnostic Criteria for Multiple Sclerosis: 2005 Revisions to the "McDonald Criteria." *Ann. Neurol.* **58,** 840–846.

Poser, S., Raun, N. E., and Poser, W. (1982). Age at onset, initial symptomatology and the course of multiple sclerosis. *Acta Neurol. Scand.* **66,** 355–362.

Ramsaransing, G., Maurits, N., Zwanikken, C., and De Keyser, J. (2001). Early prediction of a benign course of multiple sclerosis on clinical grounds: A systematic review. *Mult. Scler.* **7,** 345–347.

Riise, T., Gronnig, M., Fernandez, O., Lauer, K., Midgard, R., Minderhoud, J. M., Nyland, H., Palffy, G., Poser, S., and Aarli, J. A. (1992). Early prognostic factors for disability in multiple sclerosis: A European multicenter study. *Acta Neurol. Scand.* **85,** 212–218.

Rovaris, M., and Filippi, M. (2000). Contrast enhancement and the acute lesion in multiple slcerosis. *Neuroimaging Clin. N. Am.* **10,** 705–716.

Rudick, R. A., Lee, J. C., Simon, J., and Fisher, E. (2006). Significance of T2 lesions in multiple sclerosis: A 13-year longitudinal study. *Ann. Neuro.* [Epub June 2006 (ahead to print)].

Runmarker, B., and Andersen, O. (1993). Prognostic factors in a multiple sclerosis incidence cohort with twenty five years of follow-up. *Brain* **116,** 117–134.

Santos, A. C., Narayanan, S., de Stefano, N., Tartaglia, M. C., Francis, S. J., Arnaoutelis, R., Caramanos, Z., Antel, J. P., Pike, G. B., and Arnold, D. L. (2002). Magnetization transfer can predict clinical evolution in patients with multiple sclerosis. *J. Neurol.* **249,** 662–668.

Sastre-Garriga, J., Ingle, G. T., Rovaris, M., Tellez, N., Jasperse, B., Altmann, D. R., Benedetti, B., Stevenson, V. L., Cercignani, M., Leary, S. M., Barkhof, F., Brochet, B., *et al.* (2005). Long-term clinical outcome of primary progressive MS: Predictive value of clinical and MRI data. *Neurology* **65,** 633–635.

Scott, T. F., Schramke, C. J., Novero, J., and Chieffe, C. (2000). Short-term prognosis in early relapsing-remitting multiple sclerosis. *Neurology* **55,** 689–693.

Simone, I. L., Carrara, D., Tortorella, C., Liguori, M., Lepore, V., Pellegrini, F., Bellacosa, A., Ceccarelli, A., Pavone, I., and Livrea, P. (2002). Course and prognosis in early-onset MS. Comparison with adult-onset forms. *Neurology* **59,** 1922–1928.

Smith, M. E., Eller, N. L., McFarland, H. F., Racke, M. K., and Raine, C. S. (1999). Age dependence of clinical and pathological manifestations of autoimmune demyelination. Implications for multiple sclerosis. *Am. J. Pathol.* **155**(4), 1147–1161.

Spiegelhalter, D. J., Myles, J. P., Jones, D. R., and Abrams, K. R. (1999). Methods in health service research. An introduction to Bayesian methods in health technology assessment. *BMJ* **319,** 508–512.

Swanton, J. K., Fernando, K., Dalton, C. M., Miskiel, K. A., Thompson, A. J., Plant, G. T., and Miller, D. H. (2006). Is the frequency of MRI abnormalities in isolated optic neuritis related to the prevalence of multiple sclerosis? A global comparison. J. Neurol. Neurosurg. Psychiatry [Epub June 2006 (ahead of print)].

The IFNB MS Study Group, and the University of British Columbia, MS/MRI Analysis Group (1995). Interferon beta-1b in the treatment of multiple sclerosis: Final outcome of the randomised controlled trial. *Neurology* **45,** 1277–1285.

Thompson, A. J., Hutchinson, M., Brazil, J., Feighery, C., and Martin, E. A. (1986). A clinical and laboratory study of benign multiple sclerosis. *Q. J. Med New Ser.* **58,** 69–80.

Thorpe, J. W., Kidd, D., Moseley, I. F., Thompson, A. J., MacManus, D. G., Compston, D. A., McDonald, W. I., and Miller, D. H. (1996). Spinal MRI in patients with suspected multiple sclerosis and negative brain MRI. *Brain* **119,** 709–714.

Tintore, M., Rovira, A., Rio, J., Nos, C., Grivé, E., Sastre-Garriga, J., Pericot, I., Sánchez, E., Comabella, M., and Montalban, X. (2003). New diagnostic criteria for multiple sclerosis: Application in first demyelinating episode. *Neurology* **60**(1), 27–30.

Tintore, M., Rovira, A., Rio, J., Nos, C., Grive, E., Tellez, N., Pelayo, R., Comabella, M., and Montalban, X. (2005). Is optic neuritis more benign than the other first attacks in multiple sclerosis? *Ann. Neurol.* **57,** 210–215.

Tremlett, H., Paty, D., and Devonshire, V. (2006). Disability progression in multiple sclerosis is slower than previously reported. *Neurology* **66,** 172–177.

Trojano, M., Avolio, C., Manzari, C., Calo, A., De Robertis, F., Serio, G., and Livrea, P. (1995). Multivariate analysis of predictive factors of multiple sclerosis course with a validated method to assess clinical events. *J. Neurol. Neurosurg. Psychiatry* **58,** 300–306.

Trojano, M., Liguori, M., Bosco Zimatore, G., Bugarini, R., Avolio, C., Paolicelli, D., Giuliani, F., De Robertis, F., Marrosu, M. G., and Livrea, P. (2002). Age-related disability in multiple sclerosis. *Ann. Neurol.* **51,** 475–480.

Villar, L. M., Masjuan, J., Gonzalez-Porque, P., Plaza, J., Sadaba, M. C., Roldan, E., Bootello, A., and Alvarez-Cermeno, J. C. (2003). Intrathecal IgM synthesis is a prognostic factor in multiple sclerosis. *Ann. Neurol.* **53,** 222–226.

Villar, L. M., Sadaba, M. C., Roldan, E., Masjuan, J., Gonzalez-Porque, P., Villarrubia, N., Espino, M., Garcia-Trujillo, J. A., Bootello, A., and Alvarez-Cermeno, J. C. (2005). Intrathecal synthesis of oligoclonal IgM against myelin lipids predicts an aggressive disease course in MS. *J. Clin. Invest.* **115,** 187–194.

Visscher, B. R., Liu, K. S., Clark, V. A., Detels, R., Malmgren, R. M., and Dudley, J. P. (1984). Onset symptoms as predictors of mortality and disability in multiple sclerosis. *Acta Neurol. Scand.* **70,** 321–328.

Weinshenker, B. G., Bass, B., Rice, G. P. A., Noseworthy, J., Carriere, W., Baskerville, J., and Ebers, G. C. (1989). The natural history of multiple sclerosis: A geographically based study. 2. Predictive value of the early clinical course. *Brain* **112,** 133–146.

Weinshenker, B. G., Rice, G. P. A., Noseworthy, J. H., Carriere, W., Basekerville, J., and Ebers, G. C. (1991). The natural history of multiple sclerosis: A geographically based study. 4. Applications to planning and interpretation of clinical therapeutic trials. *Brain* **114,** 1057–1067.

Weinshenker, B. G., Issa, M., and Baskerville, J. (1996). Long-term and short-term outcome of multiple sclerosis: A 3-year follow-up study. *Arch. Neurol.* **53,** 353–358.

Weinshenker, B. G. (2001). Bayesian analysis: What does it add to studies of the natural history of MS? *J. Neurol. Sci.* **189,** 1–2.

Wolfson, C., and Confavreux, C. (1987). A improvements to a simple Markov model of the natural history of multiple sclerosis, I. Short term prognosis. *Neuroepidemiology* **6,** 101–115.

Zang, Y. C., Halder, J. B., Hong, J., Rivera, V. M., and Zhang, J. Z. (2002). Regulatory effects of estriol on T cell migration and cytokine profile: Inhibition of transcription factor NF-kappa B. *J. Neuroimmunol.* **124,** 106–114.

NEUROIMAGING IN MULTIPLE SCLEROSIS

Robert Zivadinov and Jennifer L. Cox

Buffalo Neuroimaging Analysis Center, The Jacobs Neurological Institute, State University of New York at Buffalo, School of Medicine and Biomedical Sciences, Buffalo, New York 14203, USA

Conventional magnetic resonance imaging (MRI) has routinely been used to improve the accuracy of multiple sclerosis (MS) diagnosis and prognosis. Metrics derived from conventional MRI are now routinely used to detect therapeutic effects and extend clinical observations. However, conventional MRI measures, such as the use of lesion volume and count of gadolinium-enhancing and T2 lesions, have insufficient sensitivity and specificity to reveal the true degree of pathological changes occurring in MS. They cannot distinguish between inflammation, edema, demyelination, Wallerian degeneration, and axonal loss. In addition, they do not show a reliable correlation with clinical measures of disability and do not provide a complete assessment of therapeutic outcomes. Recent neuropathologic studies of typical chronic MS brains reveal macroscopic demyelination in cortical and deep gray matter (GM) that cannot be detected by currently available MRI techniques. Therefore, there is a pressing need for the development of newer MRI techniques to detect these lesions. Newer metrics of MRI analysis, including T1-weighted hypointense lesions, central nervous system atrophy measures, magnetization transfer imaging, magnetic resonance spectroscopy, and diffusion tensor imaging,

INTERNATIONAL REVIEW OF
NEUROBIOLOGY, VOL. 79
DOI: 10.1016/S0074-7742(07)79020-7

are able to capture a more global picture of the range of tissue alterations caused by inflammation and neurodegeneration. At this time, they provide the only proof—albeit indirect—that important occult pathology is occurring in the GM. However, evidence is increasing that these nonconventional MRI measures correlate better with both existing and developing neurological impairment and disability when compared to conventional metrics.

I. Introduction

Multiple sclerosis (MS) is thought to be a chronic, inflammatory demyelinating disease of the central nervous system (CNS) with suspected autoimmune origin. It is characterized by myelin destruction and repair and axonal loss, which cause intermittent and accumulating neurological dysfunction. The exact immunopathogenic mechanisms underlying disease initiation and progression are unknown.

In the last two decades, magnetic resonance imaging (MRI) has become an important diagnostic and monitoring tool in MS (McFarland *et al.*, 1992). Gadolinium (Gd)-enhanced MRI is 5–10 times more sensitive than clinical data in the assessment of disease activity (Thompson *et al.*, 1992). Serial T2-weighted images (T2-WI) often detect the number and volume of lesions that are clinically silent. Hyperintensity on T2-WI of MS lesions is related primarily to increased water content and thus cannot distinguish between inflammation, edema, demyelination, Wallerian degeneration, and axonal loss (Markovic-Plese and McFarland, 2001). Lesions that enhance with Gd on T1-weighted images (T1-WI) indicate blood–brain barrier (BBB) permeability associated with active inflammatory activity. Neither Gd enhancing nor T2 lesions provide sufficient information about the extent and severity of the inflammatory and neurodegenerative phases of the disease (Zivadinov and Bakshi, 2004c). MRI has the potential to detect demyelination and axonal loss far earlier than is possible with strictly clinical determinants. Therefore, conventional MRI metrics have become primary endpoints in phase II clinical trials, particularly when an immunomodulatory agent is being tested (IFNB Multiple Sclerosis Study Group and University of British Columbia MS/MRI Analysis Group, 1995; Li and Paty, 1999; Simon *et al.*, 1998). However, conventional MRI is insensitive to tissue damage that occurs in normal-appearing brain tissue (NABT), including normal-appearing gray matter (NAGM) and normal-appearing white matter (NAWM), damage which probably contributes to short- and long-term clinical impairment (Bakshi *et al.*, 2005; Filippi and Rocca, 2004; Miller *et al.*, 2002; Narayana, 2005; Rovaris *et al.*, 2005b; van Waesberghe *et al.*, 1999; Zivadinov and Bakshi, 2004d). Therefore, although conventional MRI is preferable to clinical measures, it has a number of disadvantages that limit its reliability as a surrogate marker for monitoring clinical disease progression (Zivadinov and Leist, 2005).

There is increasing evidence for gray matter (GM) pathology in patients with MS. Systematic neuropathologic analyses of typical chronic MS brains reveal plaques in the periventricular and deep white matter (WM) of the hemispheres, extensive demyelination in the cerebral cortex, and severe atrophy of both WM and GM (Benedict *et al.*, 2006; Bo *et al.*, 2006; Chen *et al.*, 2004; Geurts *et al.*, 2005; Kutzelnigg and Lassmann, 2005, 2006; Kutzelnigg *et al.*, 2005; Vercellino *et al.*, 2005).

In the last decade, a number of nonconventional MRI techniques and analysis methods have been introduced for the assessment of MS (Filippi and Rocca, 2006; Zivadinov and Bakshi, 2005). When compared to conventional metrics, nonconventional MRI techniques appear to be better surrogate markers for monitoring the destructive pathological processes related to disease activity and clinical progression. They are able to reveal the underlying substrate of intrinsic pathology within lesions and NABT that include edema, inflammation, demyelination, axonal loss, and neurodegeneration (Bakshi *et al.*, 2005; Filippi and Rocca, 2006; Zivadinov and Bakshi, 2004c). Hypointense lesions on T1-WI (black holes), magnetization transfer imaging (MTI), diffusion-weighted and tensor imaging (DWI and DTI), proton MRI spectroscopy (MRS), functional MRI (fMRI), and high field MRI are emerging as promising tools for improving our understanding of the pathophysiology of MS (Zivadinov and Bakshi, 2004c). Due to their ability to detect the neurodegenerative aspects of MS, including recent evidence for cortical demyelination, these techniques are receiving increased attention as clinically relevant markers of disease progression. This chapter outlines the major roles of conventional and nonconventional MRI techniques in detecting inflammation and neurodegeneration in MS lesions and NABT.

II. Role of Conventional MRI in Detecting Inflammation and Neurodegeneration

A. Gd Enhancement

Gd enhancement in MS lesions has been correlated with histopathologic findings of BBB breakdown and active inflammation (Fig. 1) (Filippi, 2000). Gd-enhancing lesions on T1-WI usually correspond to areas of high signal intensity on T2-WI and low signal intensity on unenhanced T1-WI, probably due to edema and demyelination (Zivadinov and Bakshi, 2004c). Gd enhancement is a transient phenomenon in MS that usually disappears after an average of 3–6 weeks (Cotton *et al.*, 2003). Gd-enhancement usually precedes or accompanies the appearance of a majority of new lesions on T2-WI in MS patients. Most of

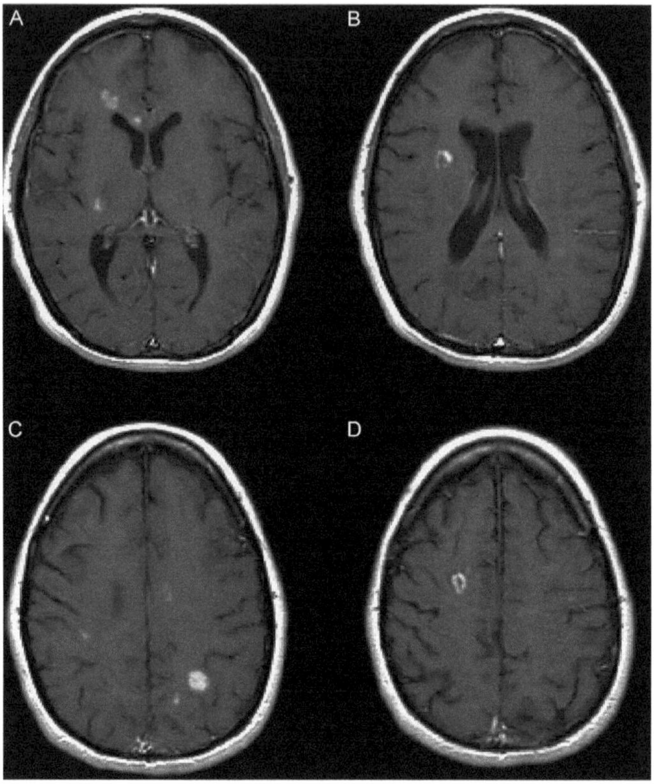

FIG. 1. Single dose (0.1 mmol/kg) gadolinium postcontrast axial T1-weighted image from a 31-year-old female with relapsing-remitting MS showing homogeneously enhancing lesions (A and C), and ring-enhancing lesions (B and D).

the enhancing plaques are not associated with the presentation of clinical symptoms and do not correlate with clinical status in cross-sectional, and especially longitudinal, studies in the midterm and long-term (Kappos *et al.*, 1999; Zivadinov and Leist, 2005). This discrepancy supports the idea that a variety of factors are operative in the occurrence of relapses and in the development of long-term sustained disability in MS. Nevertheless, the presence of continuing enhancement indicates a higher risk of relapses over the short-to-intermediate term and may contribute to long-term clinical dysfunction (Filippi, 2000; Molyneux *et al.*, 1998; Zivadinov and Bakshi, 2004c).

Several strategies were proposed to increase the sensitivity of Gd-enhanced MRI for the detection of active MS lesions. Determination of an enhancement pattern may be indicative of differences in the histopathology of MS plaques. Concentric ring-enhancing lesions with central contrast pallor arise in previously damaged areas or in areas of accelerated local inflammation (Rovira *et al.*, 1999; Roychowdhury *et al.*, 2000; Zivadinov and Leist, 2005). These ring-enhancing lesions are larger, have a shorter duration of enhancement (Minneboo *et al.*, 2005) and a lower apparent diffusion coefficient (ADC) (Leist *et al.*, 2001) and magnetization transfer ratio (MTR) (Morgen *et al.*, 2001) when compared with homogeneously enhancing plaques (Fig. 1B and D). It has also been shown that ring-enhancing lesions strongly predict the development of persisting hypointense lesions on T1-W1 and brain atrophy (Bagnato *et al.*, 2003; Minneboo *et al.*, 2005; van Waesberghe *et al.*, 1998; Zivadinov *et al.*, 2004). Therefore, the appearance of ring-enhancing plaques on Gd-enhanced MRI may be characteristic of a more aggressive form of MS and predictive of long-term deterioration. Other strategies that maximize the information that can be obtained by Gd enhancement include frequent serial monthly scanning (Weiner *et al.*, 2000), scanning spinal cord as well as the brain (Thorpe *et al.*, 1996), a delay of 5 min or more between Gd injection and scanning (Cotton *et al.*, 2003), using doses of higher contrast (e.g., a double or triple dose instead of a standard 0.1-mmol/kg dose) (Filippi *et al.*, 1998b), acquiring thinner tomographic slices (Barkhof *et al.*, 1997), coregistration (Filippi *et al.*, 1998a), reducing the background signal by the application of MTI pulses to T1-WI (Silver *et al.*, 1999), and finally, use of high field strength scanners (Erskine *et al.*, 2005; Sicotte *et al.*, 2003).

B. New Cell-Specific Contrast Agents

Gd enhancement is now considered as contrast for detection of acute inflammatory lesions, although some chronic lesions may also show Gd enhancement (Barnes *et al.*, 1991). In inflammatory and some neurodegenerative diseases of the CNS, increasing evidence supports the role of macrophage infiltration into the brain parenchyma (Raivich and Banati, 2004). Besides BBB permeability, phagocytic activity of activated microglia and infiltrated hematogenous macrophages plays a major role in MS lesion development. Inflammatory lesions in MS are characterized by microglia activation and infiltration of T cells, B cells, and macrophages across the BBB. Some observations in animal models have demonstrated imaging investigations with a new contrast agent that accumulates in phagocytic cells, ultrasmall-particle iron oxide (USPIO), which revealed *in vivo* the presence of macrophage brain infiltration (Deloire *et al.*, 2004). The ability of MR imaging to monitor *in vivo* the macrophagic activity and heterogeneity of inflammatory MS lesions has been investigated (Dousset *et al.*, 2006). Ten patients with MS in acute relapses were included in a prospective MR-imaging study using USPIO and Gd contrast. Twenty-four hours after USPIO injection, 9 of the 10 patients presented 33 USPIO-enhanced acute lesions. Gd enhancement was detected in 55 MS lesions among 7 of 10 patients. The three other patients showed no Gd enhancement. Thirty-one MS lesions were enhanced by both contrast agents. Twenty-four MS lesions were enhanced by Gd only. Two USPIO-enhanced brain lesions were observed in two patients who showed no Gd enhancement. In one patient, no lesions were enhanced either by Gd or USPIO. The discrepancy of both contrast agents' uptake may indicate that cellular inflow and BBB permeability are spatially or temporally independent (Dousset *et al.*, 2006). Gd contrast is considered a predictor of moderate ability for the occurrence of relapses, but it is not predictive for cumulative impairment or disability (Kappos *et al.*, 1999; Zivadinov and Leist, 2005). It could be hypothesized that some inflammatory Gd-enhanced lesions may not experience tissue destruction (concept of neuroprotective immunomodulation). Therefore, cell-specific macrophage MR imaging may be a stronger predictor for MS disease development than the imaging of increased BBB permeability. Further pathological MRI studies are needed to determine the MS destructive phenomenon in Gd- and/or USPIO-enhanced lesions.

C. T2-Weighted MRI Scans

T2-WI is highly sensitive in detection of hyperintense lesions in the WM and, less commonly, the GM. The most typical sites are in the WM: the periventricular region (Fig. 2A), corpus callosum (Fig. 2D), posterior fossa (Fig. 2C), and cortical regions (Fig. 2B). There are a number of techniques for identifying T2 hyperintense lesions; from among them, the most often recommended are conventional spin-echo, fast spin-echo, and fluid-attenuated inversion recovery (FLAIR) (Figs. 2 and 3A and B) (Zivadinov and Bakshi, 2004c). FLAIR is particularly helpful in the evaluation of periventricular and cortical–juxtacortical lesions (Figs. 2B and 3A and B), where cerebrospinal fluid (CSF) may mask the visualization of these plaques on T2-WI (Bakshi *et al.*, 2005; Zivadinov and Bakshi, 2004c). In the last decade, continuous technical improvements in MRI hardware and software have led to the development of new pulse sequences that are more efficient and sensitive. Among them, turbo or fast spin-echo (TSE or FSE), proton density (PD), and fast-FLAIR have already demonstrated their usefulness in a wide variety of neurological diseases, including MS (Filippi *et al.*, 2006; Simon *et al.*, 2006; Zivadinov and Bakshi, 2004c). Better lesion-to-CSF contrast is achieved with PD-WI because of the relatively lower signal intensity of CSF on this sequence and improved lesion-to-tissue contrast. FSE showed greater sensitivity than CSE in detecting areas of T2 prolongation in MS. Fast FLAIR sequences are especially helpful in evaluating periventricular and cortical–juxtacortical lesions where CSF signal may mask these plaques on T2-WI (Fig. 2B) (Bakshi *et al.*, 2001a; Zivadinov and Bakshi, 2004c). Due to fat suppression, areas of T2 prolongation can also be detected using short tau inversion recovery (STIR) sequences. This sequence may be superior to T2-WI (Fig. 4) in detecting spinal cord lesions in MS when using certain scanning platforms (Campi *et al.*, 2000). Using STIR also provides an added advantage when imaging the optic nerves

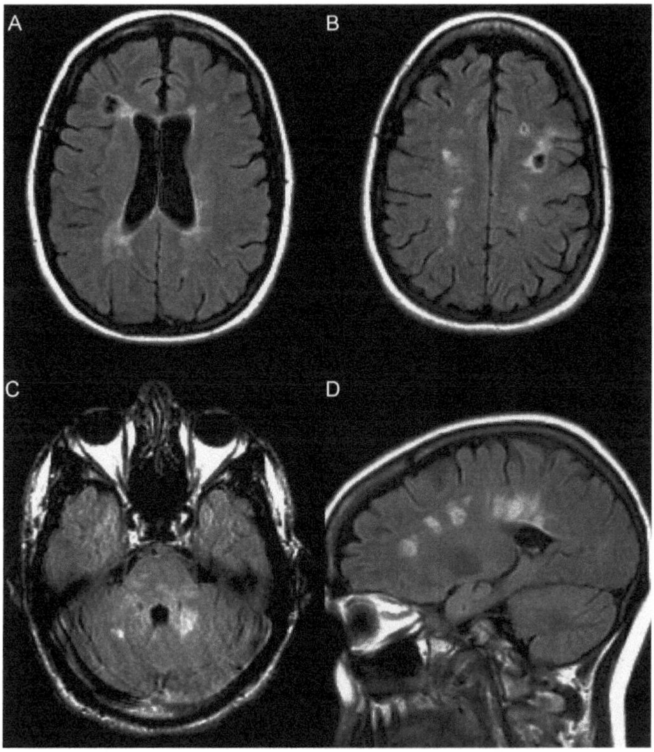

FIG. 2. Axial T2-weighted FLAIR image from a 28-year-old female with relapsing-remitting MS showing periventricular (A) and cortical (B) hyperintense white matter lesions. (C) Axial T2-weighted FLAIR image from a 25-year-old male with secondary progressive MS showing hyperintense white matter lesions in the cerebellum and pons. (D) Sagittal T2-weighted FLAIR image from 28-year-old female with RRMS showing pericallosal hyperintense lesions (Dawson's fingers).

because the contrast is increased between lesions and the surrounding retrobulbar fat (Moseley *et al.*, 1998). MRI consensus guidelines for imaging of brain and spinal cord in patients with MS were proposed by the Consortium of Multiple Sclerosis Centers (CMSC) (Simon *et al.*, 2006). For imaging of the brain, FSE, PD, FLAIR, 3D-T1, and pre- and post-Gd-enhanced T1 sequences were proposed. For imaging of the spinal cord, FSE, PD, and pre- and post-Gd-enhanced T1 sequences were recommended. Similar guidelines have also been provided in Europe by the European Federation of Neurological Science Task Force (Filippi *et al.*, 2006).

Despite the sensitivity of T2-WI to reveal disease activity and lesions over time (Paty and Li, 1993), the correlation between MRI findings and clinical evolution is modest, except in subjects with very early disease (Molyneux *et al.*, 1998; O'Riordan *et al.*, 1998; Rudick *et al.*, 2006; Sailer *et al.*, 1999; Zivadinov

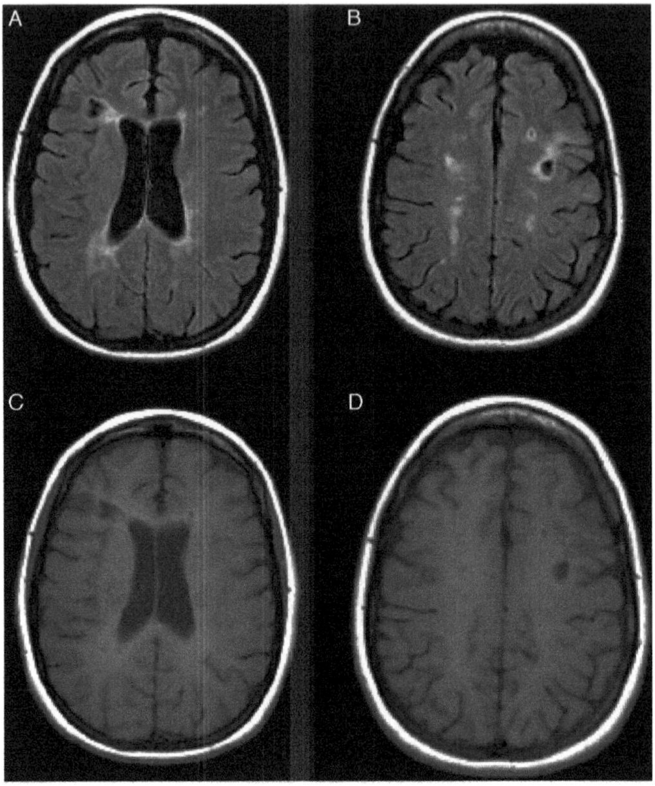

FIG. 3. Comparison of axial T2-weighted FLAIR hyperintense lesions (A and B) and noncontrast axial T1-weighted hypointense (black hole) lesions (C and D) from a 25-year-old female with relapsing-remitting MS (EDSS = 4.0).

et al., 2001b). Several long-term studies have examined the correlation of disability progression and the accumulation of T2-lesion burden. One of the longest MRI studies was from the Queens Square group in London, which followed patients with clinically isolated syndrome (CIS) for up to 14 years (Brex *et al.*, 2002). After 5 years of follow-up, data showed that T2-lesion volume (LV) accumulation predicted 25% of the correlation variance in disability. At 10 years it was down to 16%, and at 14 years, it explained only about 10–12% of the variance. The discrepancy between clinical and MRI evolution in MS patients is often referred to as the "clinical–MRI paradox" (Bakshi *et al.*, 2005; Barkhof, 2002; Zivadinov and Leist, 2005). Several explanations have been proposed in the last decade to explain the "clinical–MRI paradox" (Zivadinov and Leist, 2005). As previously mentioned, T2 scans are pathologically nonspecific; they cannot distinguish among the potential substrates of hyperintensity such as inflammation, edema, demyelination, Wallerian degeneration, and axonal loss (Zivadinov and Bakshi, 2004c).

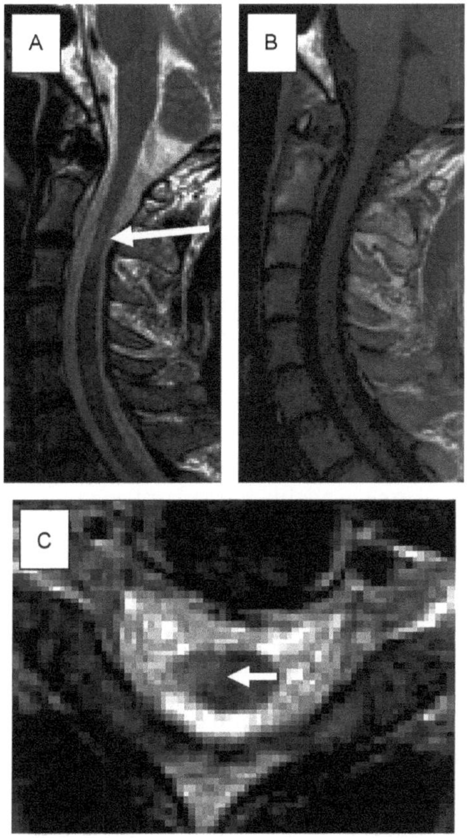

FIG. 4. Spinal T2-weighted images from a 54-year-old male with primary progressive MS. Sagittal (A) and axial (C) T2-weighted images show a hyperintense lesion in the cervical spinal cord. The same region is isointense in the sagittal T1-weighted image.

There are limitations associated with EDSS scores, including its nonlinearity, an artificial reduction of variance caused by design features of major cross-sectional studies, and the generally short duration of longitudinal trials (Sailer *et al.*, 1999). Conventional MRI cannot quantify the severity of tissue damage in NABT (Bakshi *et al.*, 2005; Filippi and Rocca, 2006; Zivadinov and Bakshi, 2004c). The presence of spinal cord disease may directly influence the EDSS score (Zivadinov and Leist, 2005). In fact, spinal cord lesions are typically more symptomatic than brain lesions and correlate better with the degree of physical disability (Zivadinov and Bakshi, 2004c). Regional T2-LV may show a better correlation with EDSS score than the whole-brain disease burden, if located in the strategic areas of the brain responsible for motor function impairment (Zivadinov and Leist, 2005). This phenomenon may be related to the heterogeneity of lesion volume threshold in nonstrategic areas of the brain. Plasticity mechanisms and individual patient capacity to compensate for structural brain damage may also play an important role in these findings. There has been mounting evidence in recent years showing that lesion repair and regression may occur (Barkhof *et al.*, 2003; Chen *et al.*, 2005). Remyelination is not uncommon and may be associated with the presence of Gd-enhancing and T2 abnormalities (concept of neuroprotective immunomodulation). Finally, evidence is increasing that diffuse, and particularly central, brain atrophy as characteristic of late-stage MS may influence this relationship. It is possible that T2-LV may be "artificially" lowered by the loss of lesions along with normal-appearing tissue. A decrease in the relationship between T2-LV and disability in advanced disease stages cautions against the assumptions that T2-LV progression is a function of disease duration alone and that stabilizing T2-LV indicates a reduction in disease activity (Dwyer *et al.*, 2005; Li *et al.*, 2006).

Despite the previously mentioned limits, several strategies for increasing the sensitivity of T2-WI became available in the last few years. Consensus guidelines recommend a $\leq$3-mm slice thickness on two-dimensional (2D) and $\geq$1.5 mm on 3D acquisition sequences for the evaluation of disease burden in MS patients scanned in clinical routine practice (Simon *et al.*, 2006). Thinner slices provide increased lesion detection and higher measurement consistency. Recent consensus guidelines also recommend that any scanner used in clinical routine practice should operate at a field strength higher than 1.0 T. With the introduction of 3-T MRI systems into clinical practice, several questions arise, including the comparability of 3 T versus 1.5 T. It has been demonstrated that scanner field strength has a substantial impact on the measured T2-LV, being about 25–40% higher with standard 3-T magnets than for lower field scanners (Erskine *et al.*, 2005; Keiper *et al.*, 1998; Sicotte *et al.*, 2003). Higher field MRI increases specificity in the correlation between detected lesions and clinical disability. Many reports have shown that more frequent scanning may increase the sensitivity for detecting T2 lesions. According to the CMSC recommendations (Simon *et al.*, 2006), clinical indications for follow-up imaging are as follows: unexpected clinical worsening, reassessment of disease burden for initiating treatment, and suspicion of a secondary diagnosis. Nevertheless, most of the physicians in North America scan patients on a yearly basis and reevaluation of these relatively conservative recommendations will be warranted in the near future.

D. Cortical Demyelination

MS is considered to predominantly affect the WM of the human CNS. However, GM MS lesions were well described in neuropathologic literature more than four decades ago (Brownell and Hughes, 1962). Similar to plaques in the

WM, GM lesions are defined as sharply demarcated areas of demyelination located within the cerebral cortex (Bo *et al.*, 2003, 2006; Geurts *et al.*, 2005; Kutzelnigg and Lassmann, 2005, 2006; Kutzelnigg *et al.*, 2005; Peterson *et al.*, 2001), the basal ganglia, and the gray matter of brain stem and spinal cord (Bakshi *et al.*, 2001b; Cifelli *et al.*, 2002; Wylezinska *et al.*, 2003).

There are three types of cortical lesions present in MS patients. Type I, or cortical–juxtacortical lesions, are located at the edge between the GM and WM, and type II are the vascular cortical lesions. Better detection of these lesions is now available using improved MRI techniques (Bo *et al.*, 2003, 2006; Geurts *et al.*, 2005). Type III, or cortical subpial lesions, are GM lesions discovered using new pathological staining techniques, based mainly on myelin basic protein (Kutzelnigg and Lassmann, 2005, 2006; Kutzelnigg *et al.*, 2005; Peterson *et al.*, 2001). Type III lesions were not previously observed mainly because unspecific myelin staining of WM and GM in the brains of MS patients was not able to detect them. Cortical lesions are less likely to be detected by conventional MRI sequences than WM lesions. This is mainly due to poor contrast resolution between NAGM and lesions in the cortex and to partial volume effects with subarachnoid spaces and surrounding CSF outside the cortex. Moreover, type III lesions are hypocellular compared to WM lesions, and it is believed that they contain a very small amount of T-cells (Bo *et al.*, 2003). Demyelinating lesions in the cerebral cortex of MS patients are common and may contribute to neurological dysfunction in MS patients via axonal and dendritic transection and apoptotic loss of neurons. These cortical subpial lesions cannot be detected on currently available MRI sequences. Pathological studies lead to two important questions: How much do these cortical subpial lesions contribute to the progression of GM and WM atrophy? And what is their relation to neurological and neuropsychological disability in patients with MS?

In 2005, Kutzelnigg and Lassman published a landmark pathological examination of 52 patients with primary progressive (PP), secondary progressive (SP) disease, and acute and relapsing-remitting (RR) disease (Kutzelnigg *et al.*, 2005). Patients with PPMS showed a significantly higher amount of cortical subpial lesions when compared to WM lesions. Patients with SPMS showed, at a minimum, similar amounts of these types of lesions in the brain when compared to WM lesions. This study confirms that cortical demyelination is an important part of the disease process in patients with MS, especially in more advanced disease stages. Although the small number of patients in the very early phase of disease made it difficult to determine the role of cortical demyelination in that phase of the disease, this study represents a breakthrough in the current understanding of the pathogenesis of neurodegeneration in patients with MS. Moreover, it is becoming clear that in patients in advanced disease stages of MS, at least 50% of the lesion burden is not visible using currently available conventional MRI techniques.

III. Role of Nonconventional MRI in Detecting Chronic Inflammation and Neurodegeneration

Several nonconventional MRI measures may reflect chronic inflammation and neurodegeneration in MS patients. These include the creation of T1 hypointense black holes, the development of brain atrophy, and changes on MTI, MRS, and DWI and DTI.

A. Hypointense Lesions on T1-WI

On noncontrast T1-WI, most lesions are isointense in WM. Some are hypointense, and these are known as black holes (Fig. 3C and D) (Paolillo *et al.*, 2000; van Walderveen *et al.*, 1999). T1-weighted black holes are an important MRI metric of neurodegeneration in patients with MS (Bagnato *et al.*, 2003). Most newly formed black holes (40–80%) will revert to isointensity within a few months because of remyelination and the resolution of edema (Brex *et al.*, 2001; Losseff *et al.*, 2001). Some remain hypointense on T1-WI and do not return to an isointense state (about 20–60%), thus representing a more advanced pathological substrate of the lesions—mainly axonal loss, Wallerian degeneration, and gliotic changes (Bakshi *et al.*, 2005; Stevenson *et al.*, 2004). They can predict development of disability better than other lesion-based inflammatory MRI measures (Zivadinov and Leist, 2005). Various longitudinal studies have shown that that the presence of a ring-enhancing lesion pattern is one of the most important contributors to the development of chronic black holes (Bagnato *et al.*, 2003; Minneboo *et al.*, 2005; van Waesberghe *et al.*, 1998; Zivadinov *et al.*, 2004). However, the predictive power is limited because of the variability in black hole evolution and appears to be a patient-specific phenomenon, although the outcome seems to vary according to disease phase and severity (Losseff *et al.*, 2001). Obviously, the goal of any disease-modifying therapy is to inhibit the creation of chronic black holes, that is, the conversion of Gd-enhancing lesions into permanent black holes. Another issue related to the significance of black holes is that the contrast hypointensity ratio of those lesions is not the same across all T1 hypointense lesions. Postmortem studies have shown that the degree of black hole hypointensity correlates strongly with axonal density (van Waesberghe *et al.*, 1998). On T1-weighted precontrast scans, strongly hypointense lesions are those which have lost most of the axons, while the slightly hypointense lesions are those which have preserved most of the axons. It appears that patients who have a similar number of black holes may have different levels of disability depending on the histopathologic changes across these lesions. Absolute intensity information has been of limited clinical use, however, due to scanner/sequence-imposed and natural regional intensity variation. Contrast hypointensity ratio may address these issues, thus providing a better surrogate marker of disability progression than black hole T1-LV alone.

Hypointense T1 lesions have a dynamic and variable natural history that is quite consistent in a given patient and is more variable between patients, as determined by MRI characteristics. Baseline Gd-enhancing lesions only partially predict the development of persistent black holes and additional measures, including MRS, DTI, and MTI, may improve understanding of the pathological processes in black hole evolution. T1-LV shows a stronger relationship with clinical dysfunction than does the T2-LV.

B. CNS Atrophy

Another very important measure of neurodegeneration in MS is brain atrophy (Fig. 5). Mounting evidence supports the concept that brain atrophy is an important biomarker of the disease process in MS (Bermel and Bakshi, 2006; Miller *et al.*, 2002; Zivadinov and Bakshi, 2004a,b,d). A bulk of studies underscore the usefulness of MRI in assessing CNS atrophy and its relationship to long-term neurodegeneration and disability progression (De Stefano *et al.*, 2003; Fisher *et al.*, 2002; Paolillo *et al.*, 2002; Zivadinov *et al.*, 2001a;). It has been established that CNS atrophy is a moderate but significant predictor of neurological impairment (Zivadinov and Bakshi, 2004a,b,d). The association between atrophy and disability is independent of the effect of conventional MRI lesions and the relationship between spinal cord atrophy and disability is strong. Studies suggest that CNS atrophy begins in patients with CIS even before the first clinical symptoms, especially in those at high risk for MS. The estimated percentage change of brain atrophy varies across studies in CIS but seems to be around 0.8% per year (Zivadinov and Bakshi, 2004a). Natural history and therapeutic studies of patients treated with placebo suggest that

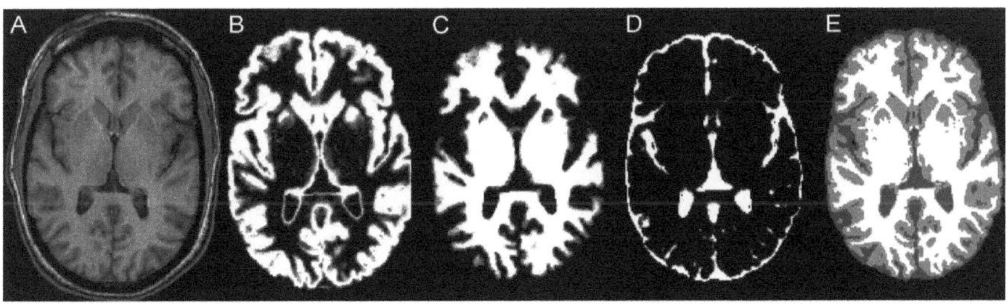

Fig. 5. Axial view of 3D-SPGR image (A) from a 44-year-old female with relapsing-remitting MS. An automated, cross-sectional method (Structural Image Evaluation, including Normalization, of Atropy, SIENAX) is applied to the image to generate separate images of gray matter (B), white matter (C), cerebral spinal fluid (D), and segmented image (E).

CNS atrophy is common in patients with RRMS, even in the earliest stages of the disease. The estimated annual rate of whole-brain atrophy varies across studies but is slightly higher in patients with early RRMS (range, −0.7% to 1.33%) than in those with advanced RRMS (range, −0.61% to 1.2%) (Zivadinov and Bakshi, 2004b). Patients with PPMS have a slightly higher annual rate of ventricular enlargement (range, −2.4% to +7.7%) than patients with SPMS; however, this rate is lower than that in patients with RRMS (range, +2.1% to +29.8%). Spinal cord atrophy develops at a faster rate than brain atrophy in patients with PPMS (Zivadinov and Bakshi, 2004a,b).

Although initial reports indicated that brain atrophy in MS is primarily due to decreases in WM (Fig. 5C and E) (Ge *et al.*, 2000), several more reports have noted diffuse GM atrophy in the brains of patients with MS (Fig. 5B and E) (Benedict *et al.*, 2006; Bermel *et al.*, 2003; Chard *et al.*, 2002; Chen *et al.*, 2004; Dalton *et al.*, 2004; De Stefano *et al.*, 2003; Fabiano *et al.*, 2003; Minagar, 2003; Sanfilipo *et al.*, 2005; Tiberio *et al.*, 2005; Valsasina *et al.*, 2005; Zivadinov *et al.*, 2007). These findings suggest that the disease process in MS is not limited to WM and that inclusion of GM atrophy in the assessment of patients with MS may further improve the usefulness of MRI measurements. Because of the presence of cortical subpial lesions and the increasingly important notion of neurodegeneration, measurement of GM atrophy is becoming an important endpoint in clinical studies. Preliminary data from several short- and long-term studies suggests that GM atrophy develops at a much faster rate than WM or whole-brain atrophy (Dalton *et al.*, 2004; Dolezal *et al.*, 2006; Tiberio *et al.*, 2005; Valsasina *et al.*, 2005; Zivadinov *et al.*, 2006). However, it is not clear whether GM atrophy is a result of the disease process in MS or is secondary to WM atrophy. Our research team used an approach based on regional segmentation called semiautomatic brain region extraction, or SABRE, to detect predilection for brain atrophy development on a region-by-region basis (Carone *et al.*, 2006b). The study compared 66 MS patients and 40 normal controls and found that the most affected regions in the brain were deep GM structures including the posterior basal ganglia and the thalamic regions, as well as the cortical regions in the orbital frontal, superior parietal, superior frontal, and medial superior frontal regions. This suggests that there is a regional specificity for brain atrophy development, prevalent mostly in GM structures, in areas of the brain that do not usually show WM lesions. Wallerian degeneration and independent neuronal degeneration are proposed mechanisms for GM atrophy in MS. We investigated partial correlations between T2 and T1 regional lesion volumes and regional/total GM atrophy in 110 MS patients (Carone *et al.*, 2006a). Partial correlations between regional lesion volume and regional GM atrophy—controlling for total GM atrophy—were not significant for any of the 26 investigated regions. Results suggest that a distinct generalized disease process better accounts for GM atrophy than regionally distinct Wallerian degeneration.

Furthermore, it is not clear whether GM atrophy contributes to neurological impairment in MS because several reports have failed to find an association between GM loss and neurological impairment (Chard *et al.*, 2002; Dalton *et al.*, 2004; Ge *et al.*, 2000; Quarantelli *et al.*, 2003; Sastre-Garriga *et al.*, 2004, 2005). On the other hand, a significant association between GM loss and impairment, in both RRMS and PPMS, has been found in other studies (Chen *et al.*, 2004; Dolezal *et al.*, 2006; Sanfilipo *et al.*, 2005; Zivadinov *et al.*, 2006).

C. Magnetization Transfer Imaging

MTI is an advanced MRI technique that has been widely used to evaluate characteristics and evolution of MS lesions and NABT. It is based on the interactions and exchange between protons that are unbound in a free water pool with those where motion is restricted due to binding with macromolecules (Fig. 6) (Filippi and Rocca, 2004; Horsfield *et al.*, 2003). MT contrast is achieved by applying radiofrequency (RF) power only to the proton magnetization of the macromolecules. Tissue damage in MS is usually reflected by a reduction in this exchange of protons and thus a decrease in MTR. Decreases in MTR indicate a reduced capacity of free water to exchange magnetization with the brain tissue matrix with which the water comes into intimate contact and is not specific to MS pathological substrates. Although MTR decreases are not specific to any of the various MS pathological substrates, a relationship has been shown between MTR and the percentage of residual axons and the degree of demyelination (van Buchem *et al.*, 1996, 1997). The most compelling evidence to support this hypothesis comes

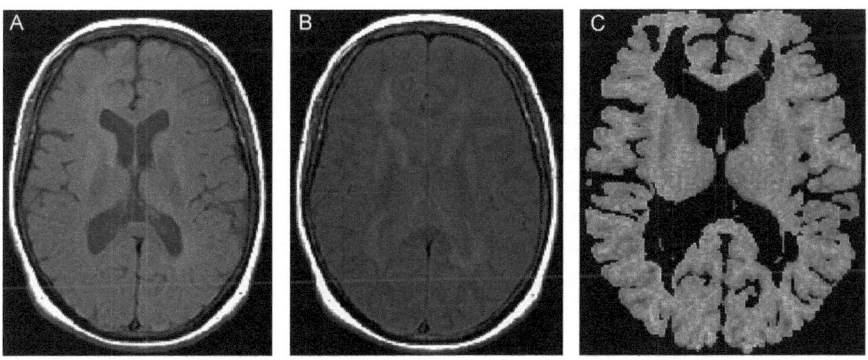

Fig. 6. Axial magnetization transfer (MT) images from a 60-year-old female with secondary progressive MS. PD images without (A) and with (B) on-resonance MT saturation radiofrequency pulse. Images (A) and (B) are coregistered and MT ratio maps are generated after subtracting CSF and lesioned tissue (C).

from a postmortem study showing a strong correlation of MTR values from MS lesions and NABT with the percentage of residual axons and the degree of demyelination (Schmierer *et al.*, 2004).

MTI can be used to assess tissue injury in lesions, in the whole brain, and in specific brain structures (Filippi and Rocca, 2004; Sharma *et al.*, 2006; Zivadinov *et al.*, 2001a). MTI studies have demonstrated two possible evolution paths for new MS lesions: (1) in some lesions, a moderate decrease in MTR with subsequent complete recovery of MTR within a few weeks may reflect edema, demyelination, and subsequent remyelination (Filippi and Rocca, 2004; Inglese *et al.*, 2005); (2) in other lesions, a marked reduction of MTR with only partial recovery at follow-up (Dousset *et al.*, 1998; Inglese *et al.*, 2005; Oreja-Guevara *et al.*, 2006). Different MTI studies have revealed clinically relevant pathological changes in areas of WM and GM that appear normal on conventional images. Such changes in NAWM and NAGM occur early in the disease process and provide prognostic information pertaining to the risk of MS progression (Chen *et al.*, 2005; Filippi *et al.*, 2000; Laule *et al.*, 2003; Rovaris *et al.*, 2003; Santos *et al.*, 2002).

MTI metrics have been correlated with the degree of disability, as was investigated over the midterm (Rovaris *et al.*, 2003; Santos *et al.*, 2002). In general, modest-to-strong correlation was found between baseline MTR and subsequent change in the EDSS disability score. These data support the notion that early MTR abnormalities in NABT can predict the clinical evolution of MS.

D. Magnetic Resonance Spectroscopy

In addition to providing information on tissue structure, MRS offers the potential to investigate tissue metabolism and function. MRS offers a mass of data on the biochemistry of a selected brain tissue volume, which represents potential surrogate markers for the pathology underlying MS (Narayana, 2005). It also provides a quantitative assessment of disease involvement related primarily to two major pathological aspects of MS, active inflammatory demyelination and axonal/neuronal injury (Inglese *et al.*, 2005; Tartaglia and Arnold, 2006). The *N*-acetylaspartate (NAA) peak in an MR spectrum is a putative marker of neuronal and axonal integrity, and the choline peak appears to reflect cell membrane metabolism. While inflammation and demyelination are represented by increases in choline, lactate and lipids, axonal and neuronal injury can be quantified through decreases in NAA (Narayana, 2005). On this basis, a diminished NAA peak is interpreted to represent neuronal/axonal dysfunction. On the other hand, an elevated choline peak represents heightened cell membrane turnover, as seen in demyelination, remyelination, inflammation, or gliosis. Therefore, MRS may provide unique information regarding metabolic, and not just structural, changes in the CNS. It evaluates the severity of MS, may follow disease evolution, and provides insight into its pathogenesis. Other metabolic peaks are of increasing interest in the study of MS. Among these, the glutamate/glutamine peak represents a mixture of amino acids and bioamines used throughout the CNS as excitatory and inhibitory neurotransmitters (Srinivasan *et al.*, 2005; Vrenken *et al.*, 2005), and the myoinositol peak represents a sugarlike molecule thought to be a marker of glial proliferation and now recognized for its importance in osmotic regulation of brain tissue volume (Geurts *et al.*, 2006; Narayanan *et al.*, 2006).

MRS studies have demonstrated that neurodegenerative changes may be detected in cortical lesions and deep GM tissue (Adalsteinsson *et al.*, 2003; Geurts *et al.*, 2006; Inglese *et al.*, 2004). Correlations between disability and decreased NAA–Cr ratio were found in several studies, suggesting that MRS measures of brain metabolites are better predictors of clinical disability than is conventional MRI.

E. Diffusion Imaging

DWI and DTI are unique MRI techniques that allow the measurement of tissue water diffusional motion and, as a consequence, provide information about orientation, size, and geometry of the tissue (Filippi *et al.*, 2001; Rovaris *et al.*, 2005b). The mobility of water molecules is reduced in highly organized tissue like WM and GM because of the interactions with cellular and tissue structures. In the CNS, diffusion is influenced by the microstructural components of tissue, including cell membranes and organelles. As a result, the ADC is lower in those tissues than in pure water. The ADC values depend on the orientation of the tissue relative to the measurement. Conventionally, the average ADC is calculated from three (DWI) or more (DTI) orthogonal directions that provide information on the overall diffusivity in the tissue. Pathological processes that modify tissue organization can cause abnormal water motion, thereby altering ADC values. In MS, the two main pathological processes affecting the brain are demyelination and neurodegeneration. They can alter the geometry of brain tissue orientation, resulting in an increase of water diffusivity measurable with different DWI and DTI indices. These measures include mean diffusivity (MD), fractional anisotropy (FA), entropy, and tractography.

Several DWI–DTI studies have demonstrated abnormal MD or FA values in MS lesions (Castriota-Scanderbeg *et al.*, 2003; Werring *et al.*, 2000) that differ between Gd-enhancing versus nonenhancing lesions. Some studies reported lower MD in Gd-enhancing lesions when compared with nonenhancing lesions (Roychowdhury *et al.*, 2000), whereas others did not observe significant differences (Cercignani *et al.*, 2001). On the contrary, the FA studies showed a consistent decrease of FA in Gd-enhancing lesions, compared to nonenhancing lesions (Werring *et al.*, 1999). Most of the studies have shown higher MD and lower FA values in lesions than in areas of NAWM, the most abnormal being in the hypointense T1 lesions (Filippi *et al.*, 2001; Rovaris *et al.*, 2005b; Werring *et al.*, 1999). Numerous diffusion MRI studies have consistently shown increased diffusivity and reduced anisotropy in the NAWM and NAGM from patients with MS when compared to normal controls (Rovaris *et al.*, 2005b).

Significant cross-sectional correlations between DWI and DTI and clinical findings have been established (Cercignani *et al.*, 2001), indicating that NAWM and NAGM are not spared by the disease process (Fabiano *et al.*, 2003; Oreja-Guevara *et al.*, 2005; Rovaris *et al.*, 2005a). Diffusion studies have confirmed that the severity of damage within T2-visible lesions and in the NAGM, as well as in clinically eloquent WM regions, might have a significant impact on MS patients' neurological disability.

IV. Conclusions

Conventional MRI metrics have a limited ability to accurately determine the full extent of inflammatory and neurodegenerative pathological changes in MS. Neither Gd enhancement nor T2-WI provides complete information about the severity of inflammation and nature of the underlying pathology. Demyelinating lesions in the cerebral cortex and deep GM of MS patients are common and may contribute to neurological dysfunction via axonal and dendritic transection and apoptotic loss of neurons. These lesions cannot be detected on conventional MRI sequences. Newer nonconventional techniques have become more promising tools for monitoring the destructive pathophysiology of MS. In particular, measures of T1-black hole lesions, brain and spinal cord atrophy, spectroscopy, magnetization transfer, and diffusion imaging have been shown to be powerful predictors of clinical impairment, disease progression, and the accumulation of long-term physical and neuropsychological disability.

Acknowledgments

We thank Eve Salczynski for technical support in preparation of this chapter.

References

Adalsteinsson, E., Langer-Gould, A., Homer, R. J., Rao, A., Sullivan, E. V., Lima, C. A., Pfefferbaum, A., and Atlas, S. W. (2003). Gray matter N-acetyl aspartate deficits in secondary progressive but not relapsing-remitting multiple sclerosis. *AJNR Am. J. Neuroradiol.* **24,** 1941–1945.

Bagnato, F., Jeffries, N., Richert, N. D., Stone, R. D., Ohayon, J. M., McFarland, H. F., and Frank, J. A. (2003). Evolution of T1 black holes in patients with multiple sclerosis imaged monthly for 4 years. *Brain* **126,** 1782–1789.

Bakshi, R., Ariyaratana, S., Benedict, R. H., and Jacobs, L. (2001a). Fluid-attenuated inversion recovery magnetic resonance imaging detects cortical and juxtacortical multiple sclerosis lesions. *Arch. Neurol.* **58,** 742–748.

Bakshi, R., Dmochowski, J., Shaikh, Z. A., and Jacobs, L. (2001b). Gray matter T2 hypointensity is related to plaques and atrophy in the brains of multiple sclerosis patients. *J. Neurol. Sci.* **185,** 19–26.

Bakshi, R., Minagar, A., Jaisani, Z., and Wolinsky, J. S. (2005). Imaging of multiple sclerosis: Role in neurotherapeutics. *NeuroRx* **2,** 277–303.

Barkhof, F. (2002). The clinico-radiological paradox in multiple sclerosis revisited. *Curr. Opin. Neurol.* **15,** 239–245.

Barkhof, F., Filippi, M., van Waesberghe, J. H., Molyneux, P., Rovaris, M., Lycklama a Nijeholt, G., Tubridy, N., Miller, D. H., Yousry, T. A., Radue, E. W., and Ader, H. J. (1997). Improving interobserver variation in reporting gadolinium-enhanced MRI lesions in multiple sclerosis. *Neurology* **49,** 1682–1688.

Barkhof, F., Bruck, W., De Groot, C. J., Bergers, E., Hulshof, S., Geurts, J., Polman, C. H., and van der Valk, P. (2003). Remyelinated lesions in multiple sclerosis: Magnetic resonance image appearance. *Arch. Neurol.* **60,** 1073–1081.

Barnes, D., Munro, P. M., Youl, B. D., Prineas, J. W., and McDonald, W. I. (1991). The longstanding MS lesion. A quantitative MRI and electron microscopic study. *Brain* **114**(Pt. 3), 1271–1280.

Benedict, R., Bruce, J., Dwyer, M., Abderlahman, N., and Zivadinov, R. (2006). Neocortical atrophy, third ventricular width, and cognitive dysfunction in multiple sclerosis. *Arch. Neurol.* **63,** 1301–1306.

Bermel, R. A., and Bakshi, R. (2006). The measurement and clinical relevance of brain atrophy in multiple sclerosis. *Lancet Neurol.* **5,** 158–170.

Bermel, R. A., Sharma, J., Tjoa, C. W., Puli, S. R., and Bakshi, R. (2003). A semiautomated measure of whole-brain atrophy in multiple sclerosis. *J. Neurol. Sci.* **208,** 57–65.

Bo, L., Vedeler, C. A., Nyland, H., Trapp, B. D., and Mork, S. J. (2003). Intracortical multiple sclerosis lesions are not associated with increased lymphocyte infiltration. *Mult. Scler.* **9,** 323–331.

Bo, L., Geurts, J. J., Mork, S. J., and van der Valk, P. (2006). Grey matter pathology in multiple sclerosis. *Acta Neurol. Scand. Suppl.* **183,** 48–50.

Brex, P. A., Molyneux, P. D., Smiddy, P., Barkhof, F., Filippi, M., Yousry, T. A., Hahn, D., Rolland, Y., Salonen, O., Pozzilli, C., Polman, C. H., Thompson, A. J., *et al.* (2001). The effect of IFNbeta-1b on the evolution of enhancing lesions in secondary progressive MS. *Neurology* **57,** 2185–2190.

Brex, P. A., Ciccarelli, O., O'Riordan, J. I., Sailer, M., Thompson, A. J., and Miller, D. H. (2002). A longitudinal study of abnormalities on MRI and disability from multiple sclerosis. *N. Engl. J. Med.* **346,** 158–164.

Brownell, B., and Hughes, J. T. (1962). The distribution of plaques in the cerebrum in multiple sclerosis. *J. Neurol. Neurosurg. Psychiatry* **25,** 315–320.

Campi, A., Pontesilli, S., Gerevini, S., and Scotti, G. (2000). Comparison of MRI pulse sequences for investigation of lesions of the cervical spinal cord. *Neuroradiology* **42,** 669–675.

Carone, D., Bruce, J., Benedict, R. H., Yella, V., Dwyer, M. G., Tjoa, C., and Zivadinov, R. (2006a). Regional lesion volume predicts total and then regional gray matter atrophy in MS. *Neurology* **66,** P02.101, A99.

Carone, D. A., Benedict, R. H., Dwyer, M. G., Cookfair, D. L., Srinivasaraghavan, B., Tjoa, C. W., and Zivadinov, R. (2006b). Semi-automatic brain region extraction (SABRE) reveals superior cortical and deep gray matter atrophy in MS. *Neuroimage* **29,** 505–514.

Castriota-Scanderbeg, A., Fasano, F., Hagberg, G., Nocentini, U., Filippi, M., and Caltagirone, C. (2003). Coefficient D(av) is more sensitive than fractional anisotropy in monitoring progression of irreversible tissue damage in focal nonactive multiple sclerosis lesions. *AJNR Am. J. Neuroradiol.* **24,** 663–670.

Cercignani, M., Inglese, M., Pagani, E., Comi, G., and Filippi, M. (2001). Mean diffusivity and fractional anisotropy histograms of patients with multiple sclerosis. *AJNR Am. J. Neuroradiol.* **22,** 952–958.

Chard, D. T., Griffin, C. M., Parker, G. J., Kapoor, R., Thompson, A. J., and Miller, D. H. (2002). Brain atrophy in clinically early relapsing-remitting multiple sclerosis. *Brain* **125,** 327–337.

Chen, J. T., Narayanan, S., Collins, D. L., Smith, S. M., Matthews, P. M., and Arnold, D. L. (2004). Relating neocortical pathology to disability progression in multiple sclerosis using MRI. *Neuroimage* **23,** 1168–1175.

Chen, J. T., Collins, D. L., Freedman, M. S., Atkins, H. L., and Arnold, D. L. (2005). Local magnetization transfer ratio signal inhomogeneity is related to subsequent change in MTR in lesions and normal-appearing white-matter of multiple sclerosis patients. *Neuroimage* **25,** 1272–1278.

Cifelli, A., Arridge, M., Jezzard, P., Esiri, M. M., Palace, J., and Matthews, P. M. (2002). Thalamic neurodegeneration in multiple sclerosis. *Ann. Neurol.* **52,** 650–653.

Cotton, F., Weiner, H. L., Jolesz, F. A., and Guttmann, C. R. (2003). MRI contrast uptake in new lesions in relapsing-remitting MS followed at weekly intervals. *Neurology* **60,** 640–646.

Dalton, C. M., Chard, D. T., Davies, G. R., Miszkiel, K. A., Altmann, D. R., Fernando, K., Plant, G. T., Thompson, A. J., and Miller, D. H. (2004). Early development of multiple sclerosis is associated with progressive grey matter atrophy in patients presenting with clinically isolated syndromes. *Brain* **127,** 1101–1107.

De Stefano, N., Matthews, P. M., Filippi, M., Agosta, F., De Luca, M., Bartolozzi, M. L., Guidi, L., Ghezzi, A., Montanari, E., Cifelli, A., Federico, A., and Smith, S. M. (2003). Evidence of early cortical atrophy in MS: Relevance to white matter changes and disability. *Neurology* **60,** 1157–1162.

Deloire, M. S., Touil, T., Brochet, B., Dousset, V., Caille, J. M., and Petry, K. G. (2004). Macrophage brain infiltration in experimental autoimmune encephalomyelitis is not completely compromised by suppressed T-cell invasion: *In vivo* magnetic resonance imaging illustration in effective anti-VLA-4 antibody treatment. *Mult. Scler.* **10,** 540–548.

Dolezal, O., Tekwe, C. I., Horakova, D., Seidl, Z., Havrdova, E., Balachandran, S., Bergsland, N., Dwyer, M. G., and Zivadinov, R. (2006). Evolution of grey matter and central atrophy in early relapsing-remitting multiple sclerosis is most responsible for development of disability. *J. Neurol.* **253,** P227, 57–58.

Dousset, V., Gayou, A., Brochet, B., and Caille, J. M. (1998). Early structural changes in acute MS lesions assessed by serial magnetization transfer studies. *Neurology* **51,** 1150–1155.

Dousset, V., Brochet, B., Deloire, M. S., Lagoarde, L., Barroso, B., Caille, J. M., and Petry, K. G. (2006). MR imaging of relapsing multiple sclerosis patients using ultra-small-particle iron oxide and compared with gadolinium. *AJNR Am. J. Neuroradiol.* **27,** 1000–1005.

Dwyer, M. G., Abdelrahman, N., Weinstock-Guttman, B., Srinivasaraghavan, B., Prakash, R., Hussein, S., Bergsland, N., Garg, N., Meyer, M. G., Munschauer, F., and Zivadinov, R. (2005). Quantitative analysis of lesion load progression in advanced multiple sclerosis. *Mult. Scler.* **11,** 68, S14.

Erskine, M. K., Cook, L. L., Riddle, K. E., Mitchell, J. R., and Karlik, S. J. (2005). Resolution-dependent estimates of multiple sclerosis lesion loads. *Can. J. Neurol. Sci.* **32,** 205–212.

Fabiano, A. J., Sharma, J., Weinstock-Guttman, B., Munschauer, F. E., III, Benedict, R. H., Zivadinov, R., and Bakshi, R. (2003). Thalamic involvement in multiple sclerosis: A diffusion-weighted magnetic resonance imaging study. *J. Neuroimaging* **13,** 307–314.

Filippi, M. (2000). Enhanced magnetic resonance imaging in multiple sclerosis. *Mult. Scler.* **6,** 320–326.

Filippi, M., and Rocca, M. A. (2004). Magnetization transfer magnetic resonance imaging in the assessment of neurological diseases. *J. Neuroimaging* **14,** 303–313.

Filippi, M., and Rocca, M. A. (2006). The use of modern magnetic resonance techniques to monitor disease evolution in multiple sclerosis. *Adv. Neurol.* **98,** 167–183.

Filippi, M., Horsfield, M. A., Hajnal, J. V., Narayana, P. A., Udupa, J. K., Yousry, T. A., and Zijdenbos, A. (1998a). Quantitative assessment of magnetic resonance imaging lesion load in multiple sclerosis. *J. Neurol. Neurosurg. Psychiatry* **64**(Suppl. 1), S88–S93.

Filippi, M., Rovaris, M., Gasperini, C., Capra, R., Bastianello, S., Kuhne, I., and Yousry, T. A. (1998b). A preliminary study comparing the sensitivity of serial monthly enhanced MRI after standard and triple dose gadolinium-DTPA for monitoring disease activity in primary progressive multiple sclerosis. *J. Neuroimaging* **8,** 88–93.

Filippi, M., Inglese, M., Rovaris, M., Sormani, M. P., Horsfield, P., Iannucci, P. G., Colombo, B., and Comi, G. (2000). Magnetization transfer imaging to monitor the evolution of MS: A 1-year follow-up study. *Neurology* **55,** 940–946.

Filippi, M., Cercignani, M., Inglese, M., Horsfield, M. A., and Comi, G. (2001). Diffusion tensor magnetic resonance imaging in multiple sclerosis. *Neurology* **56,** 304–311.

Filippi, M., Rocca, M. A., Arnold, D. L., Bakshi, R., Barkhof, F., De Stefano, N., Fazekas, F., Frohman, E., and Wolinsky, J. S. (2006). EFNS guidelines on the use of neuroimaging in the management of multiple sclerosis. *Eur. J. Neurol.* **13,** 313–325.

Fisher, E., Rudick, R. A., Simon, J. H., Cutter, G., Baier, M., Lee, J. C., Miller, D., Weinstock-Guttman, B., Mass, M. K., Dougherty, D. S., and Simonian, N. A. (2002). Eight-year follow-up study of brain atrophy in patients with MS. *Neurology* **59,** 1412–1420.

Ge, Y., Grossman, R. I., Udupa, J. K., Wei, L., Mannon, L. J., Polansky, M., and Kolson, D. L. (2000). Brain atrophy in relapsing-remitting multiple sclerosis and secondary progressive multiple sclerosis: Longitudinal quantitative analysis. *Radiology* **214,** 665–670.

Geurts, J. J., Pouwels, P. J., Uitdehaag, B. M., Polman, C. H., Barkhof, F., and Castelijns, J. A. (2005). Intracortical lesions in multiple sclerosis: Improved detection with 3D double inversion-recovery MR imaging. *Radiology* **236,** 254–260.

Geurts, J. J., Reuling, I. E., Vrenken, H., Uitdehaag, B. M., Polman, C. H., Castelijns, J. A., Barkhof, F., and Pouwels, P. J. (2006). MR spectroscopic evidence for thalamic and hippocampal, but not cortical, damage in multiple sclerosis. *Magn. Reson. Med.* **55,** 478–483.

Horsfield, M. A., Barker, G. J., Barkhof, F., Miller, D. H., Thompson, A. J., and Filippi, M. (2003). Guidelines for using quantitative magnetization transfer magnetic resonance imaging for monitoring treatment of multiple sclerosis. *J. Magn. Reson. Imaging* **17,** 389–397.

IFNB Multiple Sclerosis Study Group and University of British Columbia, MS/MRI Analysis Group (1995). Interferon beta-1b in the treatment of multiple sclerosis: Final outcome of the randomized controlled trial. *Neurology* **45,** 1277–1285.

Inglese, M., Ge, Y., Filippi, M., Falini, A., Grossman, R. I., and Gonen, O. (2004). Indirect evidence for early widespread gray matter involvement in relapsing-remitting multiple sclerosis. *Neuroimage* **21,** 1825–1829.

Inglese, M., Grossman, R. I., and Filippi, M. (2005). Magnetic resonance imaging monitoring of multiple sclerosis lesion evolution. *J. Neuroimaging.* **15,** 22S–29S.

Kappos, L., Moeri, D., Radue, E. W., Schoetzau, A., Schweikert, K., Barkhof, F., Miller, D., Guttmann, C. R., Weiner, H. L., Gasperini, C., and Filippi, M. (1999). Predictive value of gadolinium-enhanced magnetic resonance imaging for relapse rate and changes in disability or impairment in multiple sclerosis: A meta-analysis. Gadolinium MRI Meta-analysis Group. *Lancet* **353,** 964–969.

Keiper, M. D., Grossman, R. I., Hirsch, J. A., Bolinger, L., Ott, I. L., Mannon, L. J., Langlotz, C. P., and Kolson, D. L. (1998). MR identification of white matter abnormalities in multiple sclerosis: A comparison between 1.5 T and 4 T. *AJNR Am. J. Neuroradiol.* **19,** 1489–1493.

Kutzelnigg, A., and Lassmann, H. (2005). Cortical lesions and brain atrophy in MS. *J. Neurol. Sci.* **233,** 55–59.

Kutzelnigg, A., and Lassmann, H. (2006). Cortical demyelination in multiple sclerosis: A substrate for cognitive deficits? *J. Neurol. Sci.* **245,** 123–126.

Kutzelnigg, A., Lucchinetti, C. F., Stadelmann, C., Bruck, W., Rauschka, H., Bergmann, M., Schmidbauer, M., Parisi, J. E., and Lassmann, H. (2005). Cortical demyelination and diffuse white matter injury in multiple sclerosis. *Brain* **128,** 2705–2712.

Laule, C., Vavasour, I. M., Whittall, K. P., Oger, J., Paty, D. W., Li, D. K., MacKay, A. L., and Arnold, D. L. (2003). Evolution of focal and diffuse magnetisation transfer abnormalities in multiple sclerosis. *J. Neurol.* **250,** 924–931.

Leist, T. P., Gobbini, M. I., Frank, J. A., and McFarland, H. F. (2001). Enhancing magnetic resonance imaging lesions and cerebral atrophy in patients with relapsing multiple sclerosis. *Arch. Neurol.* **58,** 57–60.

Li, D. K., and Paty, D. W. (1999). Magnetic resonance imaging results of the PRISMS trial: A randomized, double-blind, placebo-controlled study of interferon-beta1a in relapsing-remitting multiple sclerosis. Prevention of Relapses and Disability by Interferon-beta1a Subcutaneously in Multiple Sclerosis. *Ann. Neurol.* **46,** 197–206.

Li, D. K., Held, U., Petkau, J., Daumer, M., Barkhof, F., Fazekas, F., Frank, J. A., Kappos, L., Miller, D. H., Simon, J. H., Wolinsky, J. S., and Filippi, M. (2006). MRI T2 lesion burden in multiple sclerosis: A plateauing relationship with clinical disability. *Neurology* **66,** 1384–1389.

Losseff, N. A., Miller, D. H., Kidd, D., and Thompson, A. J. (2001). The predictive value of gadolinium enhancement for long term disability in relapsing-remitting multiple sclerosis—preliminary results. *Mult. Scler.* **7,** 23–25.

Markovic-Plese, S., and McFarland, H. F. (2001). Immunopathogenesis of the multiple sclerosis lesion. *Curr. Neurol. Neurosci. Rep.* **1,** 257–262.

McFarland, H. F., Frank, J. A., Albert, P. S., Smith, M. E., Martin, R., Harris, J. O., Patronas, N., Maloni, H., and McFarlin, D. E. (1992). Using gadolinium-enhanced magnetic resonance imaging lesions to monitor disease activity in multiple sclerosis. *Ann. Neurol.* **32,** 758–766.

Miller, D. H., Barkhof, F., Frank, J. A., Parker, G. J., and Thompson, A. J. (2002). Measurement of atrophy in multiple sclerosis: Pathological basis, methodological aspects and clinical relevance. *Brain* **125,** 1676–1695.

Minagar, A. (2003). Gray matter involvement in multiple sclerosis: A new window into pathogenesis. *J. Neuroimaging* **13,** 291–292.

Minneboo, A., Uitdehaag, B. M., Ader, H. J., Barkhof, F., Polman, C. H., and Castelijns, J. A. (2005). Patterns of enhancing lesion evolution in multiple sclerosis are uniform within patients. *Neurology* **65,** 56–61.

Molyneux, P. D., Filippi, M., Barkhof, F., Gasperini, C., Yousry, T. A., Truyen, L., Lai, H. M., Rocca, M. A., Moseley, I. F., and Miller, D. H. (1998). Correlations between monthly enhanced MRI lesion rate and changes in T2 lesion volume in multiple sclerosis. *Ann. Neurol.* **43,** 332–339.

Morgen, K., Jeffries, N. O., Stone, R., Martin, R., Richert, N. D., Frank, J. A., and McFarland, H. F. (2001). Ring-enchancement in multiple sclerosis: Marker of disease severity. *Mult. Scler.* **7,** 167–171.

Moseley, I. F., Miller, D. H., and Gass, A. (1998). The contribution of magnetic resonance imaging to the assessment of optic nerve and spinal cord involvement in multiple sclerosis. *J. Neurol. Neurosurg. Psychiatry* **64**(Suppl. 1), S15–S20.

Narayana, P. A. (2005). Magnetic resonance spectroscopy in the monitoring of multiple sclerosis. *J. Neuroimaging* **15,** 46S–57S.

Narayanan, S., Francis, S. J., Sled, J. G., Santos, A. C., Antel, S., Levesque, I., Brass, S., Lapierre, Y., Sappey-Marinier, D., Pike, G. B., and Arnold, D. L. (2006). Axonal injury in the cerebral normal-appearing white matter of patients with multiple sclerosis is related to concurrent demyelination in lesions but not to concurrent demyelination in normal-appearing white matter. *Neuroimage* **29,** 637–642.

Oreja-Guevara, C., Rovaris, M., Iannucci, G., Valsasina, P., Caputo, D., Cavarretta, R., Sormani, M. P., Ferrante, P., Comi, G., and Filippi, M. (2005). Progressive gray matter damage in patients with relapsing-remitting multiple sclerosis: A longitudinal diffusion tensor magnetic resonance imaging study. *Arch. Neurol.* **62,** 578–584.

Oreja-Guevara, C., Charil, A., Caputo, D., Cavarretta, R., Sormani, M. P., and Filippi, M. (2006). Magnetization transfer magnetic resonance imaging and clinical changes in patients with relapsing-remitting multiple sclerosis. *Arch. Neurol.* **63,** 736–740.

O'Riordan, J. I., Thompson, A. J., Kingsley, D. P., MacManus, D. G., Kendall, B. E., Rudge, P., McDonald, W. I., and Miller, D. H. (1998). The prognostic value of brain MRI in clinically isolated syndromes of the CNS. A 10-year follow-up. *Brain* **121**(Pt. 3), 495–503.

Paolillo, A., Pozzilli, C., Gasperini, C., Giugni, E., Mainero, C., Giuliani, S., Tomassini, V., Millefiorini, E., and Bastianello, S. (2000). Brain atrophy in relapsing-remitting multiple sclerosis: Relationship with 'black holes', disease duration and clinical disability. *J. Neurol. Sci.* **174,** 85–91.

Paolillo, A., Pozzilli, C., Giugni, E., Tomassini, V., Gasperini, C., Fiorelli, M., Mainero, C., Horsfield, M., Galgani, S., Bastianello, S., and Buttinelli, C. (2002). A 6-year clinical and MRI follow-up study of patients with relapsing-remitting multiple sclerosis treated with Interferon-beta. *Eur. J. Neurol.* **9,** 645–655.

Paty, D. W., and Li, D. K. (1993). Interferon beta-1b is effective in relapsing-remitting multiple sclerosis. II. MRI analysis results of a multicenter, randomized, double-blind, placebo-controlled trial. UBC MS/MRI Study Group and the IFNB Multiple Sclerosis Study Group. *Neurology* **43,** 662–667.

Peterson, J. W., Bo, L., Mork, S., Chang, A., and Trapp, B. D. (2001). Transected neurites, apoptotic neurons, and reduced inflammation in cortical multiple sclerosis lesions. *Ann. Neurol.* **50,** 389–400.

Quarantelli, M., Ciarmiello, A., Morra, V. B., Orefice, G., Larobina, M., Lanzillo, R., Schiavone, V., Salvatore, E., Alfano, B., and Brunetti, A. (2003). Brain tissue volume changes in relapsing-remitting multiple sclerosis: Correlation with lesion load. *Neuroimage* **18,** 360–366.

Raivich, G., and Banati, R. (2004). Brain microglia and blood-derived macrophages: Molecular profiles and functional roles in multiple sclerosis and animal models of autoimmune demyelinating disease. *Brain Res. Brain Res. Rev.* **46,** 261–281.

Rovaris, M., Agosta, F., Sormani, M. P., Inglese, M., Martinelli, V., Comi, G., and Filippi, M. (2003). Conventional and magnetization transfer MRI predictors of clinical multiple sclerosis evolution: A medium-term follow-up study. *Brain* **126,** 2323–2332.

Rovaris, M., Gallo, A., Valsasina, P., Benedetti, B., Caputo, D., Ghezzi, A., Montanari, E., Sormani, M. P., Bertolotto, A., Mancardi, G., Bergamaschi, R., Martinelli, V., Comi, G., and Filippi, M. (2005a). Short-term accrual of gray matter pathology in patients with progressive multiple sclerosis: An *in vivo* study using diffusion tensor MRI. *Neuroimage* **24,** 1139–1146.

Rovaris, M., Gass, A., Bammer, R., Hickman, S. J., Ciccarelli, O., Miller, D. H., and Filippi, M. (2005b). Diffusion MRI in multiple sclerosis. *Neurology* **65,** 1526–1532.

Rovira, A., Alonso, J., Cucurella, G., Nos, C., Tintore, M., Pedraza, S., Rio, J., and Montalban, X. (1999). Evolution of multiple sclerosis lesions on serial contrast-enhanced T1weighted and magnetization-transfer MR images. *AJNR Am. J. Neuroradiol.* **20,** 1939–1945.

Roychowdhury, S., Maldjian, J. A., and Grossman, R. I. (2000). Multiple sclerosis: Comparison of trace apparent diffusion coefficients with MR enhancement pattern of lesions. *AJNR Am. J. Neuroradiol.* **21,** 869–874.

Rudick, R. A., Lee, J. C., Simon, J., and Fisher, E. (2006). Significance of T2 lesions in multiple sclerosis: A 13-year longitudinal study. *Ann. Neurol.* **60,** 236–242.

Sailer, M., O'Riordan, J. I., Thompson, A. J., Kingsley, D. P., MacManus, D. G., McDonald, W. I., and Miller, D. H. (1999). Quantitative MRI in patients with clinically isolated syndromes suggestive of demyelination. *Neurology* **52,** 599–606.

Sanfilipo, M. P., Benedict, R. H., Sharma, J., Weinstock-Guttman, B., and Bakshi, R. (2005). The relationship between whole brain volume and disability in multiple sclerosis: A comparison of normalized gray vs. white matter with misclassification correction. *Neuroimage* **26,** 1068–1077.

Santos, A. C., Narayanan, S., de Stefano, N., Tartaglia, M. C., Francis, S. J., Arnaoutelis, R., Caramanos, Z., Antel, J. P., Pike, G. B., and Arnold, D. L. (2002). Magnetization transfer can predict clinical evolution in patients with multiple sclerosis. *J. Neurol.* **249,** 662–668.

Sastre-Garriga, J., Ingle, G. T., Chard, D. T., Ramio-Torrenta, L., Miller, D. H., and Thompson, A. J. (2004). Grey and white matter atrophy in early clinical stages of primary progressive multiple sclerosis. *Neuroimage* **22,** 353–359.

Sastre-Garriga, J., Ingle, G. T., Chard, D. T., Cercignani, M., Ramio-Torrenta, L., Miller, D. H., and Thompson, A. J. (2005). Grey and white matter volume changes in early primary progressive multiple sclerosis: A longitudinal study. *Brain* **128**(Pt. 6), 1454–1560.

Schmierer, K., Scaravilli, F., Altmann, D. R., Barker, G. J., and Miller, D. H. (2004). Magnetization transfer ratio and myelin in postmortem multiple sclerosis brain. *Ann. Neurol.* **56,** 407–415.

Sharma, J., Zivadinov, R., Jaisani, Z., Fabiano, A. J., Singh, B., Horsfield, M., and Baski, R. (2006). A magnetization transfer MRI study of deep gray matter involvement in multiple sclerosis. *J. Neuroimaging* **16,** 302–310.

Sicotte, N. L., Voskuhl, R. R., Bouvier, S., Klutch, R., Cohen, M. S., and Mazziotta, J. C. (2003). Comparison of multiple sclerosis lesions at 1.5 and 3.0 Tesla. *Invest. Radiol.* **38,** 423–427.

Silver, N., Lai, M., Symms, M., Barker, G., McDonald, I., and Miller, D. (1999). Serial gadolinium-enhanced and magnetization transfer imaging to investigate the relationship between the duration of blood-brain barrier disruption and extent of demyelination in new multiple sclerosis lesions. *J. Neurol.* **246,** 728–730.

Simon, J. H., Jacobs, L. D., Campion, M., Wende, K., Simonian, N., Cookfair, D. L., Rudick, R. A., Herndon, R. M., Richert, J. R., Salazar, A. M., Alam, J. J., Fischer, J. S., *et al.* (1998). Magnetic resonance studies of intramuscular interferon beta-1a for relapsing multiple sclerosis. The Multiple Sclerosis Collaborative Research Group. *Ann. Neurol.* **43,** 79–87.

Simon, J. H., Li, D., Traboulsee, A., Coyle, P. K., Arnold, D. L., Barkhof, F., Frank, J. A., Grossman, R., Paty, D. W., Radue, E. W., and Wolinsky, J. S. (2006). Standardized MR imaging protocol for multiple sclerosis: Consortium of MS Centers consensus guidelines. *AJNR Am. J. Neuroradiol.* **27,** 455–461.

Srinivasan, R., Sailasuta, N., Hurd, R., Nelson, S., and Pelletier, D. (2005). Evidence of elevated glutamate in multiple sclerosis using magnetic resonance spectroscopy at 3 *T. Brain* **128,** 1016–1025.

Stevenson, V. L., Ingle, G. T., Miller, D. H., and Thompson, A. J. (2004). Magnetic resonance imaging predictors of disability in primary progressive multiple sclerosis: A 5-year study. *Mult. Scler.* **10,** 398–401.

Tartaglia, M. C., and Arnold, D. L. (2006). The role of MRS and fMRI in multiple sclerosis. *Adv. Neurol.* **98,** 185–202.

Thompson, A. J., Miller, D., Youl, B., MacManus, D., Moore, S., Kingsley, D., Kendall, B., Feinstein, A., and McDonald, W. I. (1992). Serial gadolinium-enhanced MRI in relapsing/remitting multiple sclerosis of varying disease duration. *Neurology* **42,** 60–63.

Thorpe, J. W., Kidd, D., Moseley, I. F., Kenndall, B. E., Thompson, A. J., MacManus, D. G., McDonald, W. I., and Miller, D. H. (1996). Serial gadolinium-enhanced MRI of the brain and spinal cord in early relapsing-remitting multiple sclerosis. *Neurology* **46,** 373–378.

Tiberio, M., Chard, D. T., Altmann, D. R., Davies, G., Griffin, C. M., Rashid, W., Sastre-Garriga, J., Thompson, A. J., and Miller, D. H. (2005). Gray and white matter volume changes in early RRMS: A 2-year longitudinal study. *Neurology* **64,** 1001–1007.

Valsasina, P., Benedetti, B., Rovaris, M., Sormani, M. P., Comi, G., and Filippi, M. (2005). Evidence for progressive gray matter loss in patients with relapsing-remitting MS. *Neurology* **65,** 1126–1128.

van Buchem, M. A., McGowan, J. C., Kolson, D. L., Polansky, M., and Grossman, R. I. (1996). Quantitative volumetric magnetization transfer analysis in multiple sclerosis: Estimation of macroscopic and microscopic disease burden. *Magn. Reson. Med.* **36,** 632–636.

van Buchem, M. A., Udupa, J. K., McGowan, J. C., Miki, Y., Heyning, F. H., Boncoeur-Martel, M. P., Kolson, D. L., Polansky, M., and Grossman, R. I. (1997). Global volumetric estimation of disease burden in multiple sclerosis based on magnetization transfer imaging. *AJNR Am. J. Neuroradiol.* **18,** 1287–1290.

van Waesberghe, J. H., Kamphorst, W., De Groot, C. J., van Walderveen, M. A., Castelijns, J. A., Ravid, R., Lycklama a Nijeholt, G. J., van der Valk, P., Polman, C. H., Thompson, A. J., and Barkhof, F. (1999). Axonal loss in multiple sclerosis lesions: Magnetic resonance imaging insights into substrates of disability. *Ann. Neurol.* **46,** 747–754.

van Waesberghe, J. H., van Walderveen, M. A., Castelijns, J. A., Scheltens, P., Lycklama a Nijeholt, G. J., Polman, C. H., and Barkhof, F. (1998). Patterns of lesion development in multiple sclerosis: Longitudinal observations with T1-weighted spin-echo and magnetization transfer MR. *AJNR Am. J. Neuroradiol.* **19,** 675–683.

van Walderveen, M. A., Truyen, L., van Oosten, B. W., Castelijns, J. A., Lycklama a Nijeholt, G. J., van Waesberghe, J. H., Polman, C., and Barkhof, F. (1999). Development of hypointense lesions on T1-weighted spin-echo magnetic resonance images in multiple sclerosis: Relation to inflammatory activity. *Arch. Neurol.* **56,** 345–351.

Vercellino, M., Plano, F., Votta, B., Mutani, R., Giordana, M. T., and Cavalla, P. (2005). Grey matter pathology in multiple sclerosis. *J. Neuropathol. Exp. Neurol.* **64,** 1101–1107.

Vrenken, H., Barkhof, F., Uitdehaag, B. M., Castelijns, J. A., Polman, C. H., and Pouwels, P. J. (2005). MR spectroscopic evidence for glial increase but not for neuro-axonal damage in MS normal-appearing white matter. *Magn. Reson. Med.* **53,** 256–266.

Weiner, H. L., Guttmann, C. R., Khoury, S. J., Orav, E. J., Hohol, M. J., Kikinis, R., and Jolesz, F. A. (2000). Serial magnetic resonance imaging in multiple sclerosis: Correlation with attacks, disability, and disease stage. *J. Neuroimmunol.* **104,** 164–173.

Werring, D. J., Clark, C. A., Barker, G. J., Thompson, A. J., and Miller, D. H. (1999). Diffusion tensor imaging of lesions and normal-appearing white matter in multiple sclerosis. *Neurology* **52,** 1626–1632.

Werring, D. J., Brassat, D., Droogan, A. G., Clark, C. A., Symms, M. R., Barker, G. J., MacManus, D. G., Thompson, A. J., and Miller, D. H. (2000). The pathogenesis of lesions and normal-appearing white matter changes in multiple sclerosis: A serial diffusion MRI study. *Brain* **123**(Pt. 8), 1667–1676.

Wylezinska, M., Cifelli, A., Jezzard, P., Palace, J., Alecci, M., and Matthews, P. M. (2003). Thalamic neurodegeneration in relapsing-remitting multiple sclerosis. *Neurology* **60,** 1949–1954.

Zivadinov, R., and Bakshi, R. (2004a). Central nervous system atrophy and clinical status in multiple sclerosis. *J. Neuroimaging* **14,** S27–S35.

Zivadinov, R., and Bakshi, R. (2004b). Neurologic disability and central nervous system atrophy in multiple sclerosis. *In* "Brain and Spinal Cord Atrophy in Multiple Sclerosis" (R. Zivadinov and R. Bakshi, Eds.), pp. 113–136. Nova Biomedical Books, Hauppauge, NY.

Zivadinov, R., and Bakshi, R. (2004c). Role of MRI in multiple sclerosis I: Inflammation and lesions. *Front Biosci.* **9,** 665–683.

Zivadinov, R., and Bakshi, R. (2004d). Role of MRI in multiple sclerosis II: Brain and spinal cord atrophy. *Front Biosci.* **9,** 647–664.

Zivadinov, R., and Bakshi, R. (2005). Role of magnetic resonance imaging in the diagnosis and prognosis of multiple sclerosis. *In* "Multiple Sclerosis, Etiology, Diagnosis, and New Treatment Strategies" (M. Olek, Ed.), pp. 55–90. Humana Press, Totowa, NJ.

Zivadinov, R., and Leist, T. P. (2005). Clinical-magnetic resonance imaging correlations in multiple sclerosis. *J. Neuroimaging* **15,** 10S–21S.

Zivadinov, R., Rudick, R. A., De Masi, R., Nasuelli, D., Ukmar, M., Pozzi-Mucelli, R. S., Grop, A., Cazzato, G., and Zorzon, M. (2001a). Effects of IV methylprednisolone on brain atrophy in relapsing-remitting MS. *Neurology* **57,** 1239–1247.

Zivadinov, R., Sepcic, J., Nasuelli, D., De Masi, R., Bragadin, L. M., Tommasi, M. A., Zambito-Marsala, S., Moretti, R., Bratina, A., Ukmar, M., Pozzi-Mucelli, R. S., Grop, A., *et al.* (2001b). A longitudinal study of brain atrophy and cognitive disturbances in the early phase of relapsing-remitting multiple sclerosis. *J. Neurol. Neurosurg. Psychiatry* **70,** 773–780.

Zivadinov, R., Bagnato, F., Nasuelli, D., Bastianello, S., Bratina, A., Locatelli, L., Watts, K., Finamore, L., Grop, A., Dwyer, M., Catalan, M., Clemenzi, A., *et al.* (2004). Short-term brain atrophy changes in relapsing-remitting multiple sclerosis. *J. Neurol. Sci.* **223,** 185–193.

Zivadinov, R., Cookfair, D., Locatelli, L., Srinivasaraghavan, B., Ukmar, M., Bratina, A., Maggiore, C., Bosco, A., Grop, A., Pozzi-Mucelli, R. S., Catalan, M., and Zorzon, M. (2007). Interferon-beta-1a slows down progression of brain atrophy in relapsing-remitting multiple sclerosis by reducing predominantly grey matter atrophy progression. *Mult. Scler.* (in press).

Zivadinov, R., Dolezal, O., Tekwe, C. I., Horakova, D., Dwyer, M. G., Balachandran, S., Bergsland, N., and Havrdova, E. (2006). A longitudinal study of cortical atrophy in patients with relapsing-remitting multiple sclerosis over a 2-year period. *J. Neurol.* **253,** P222, 56–57.

DETECTION OF CORTICAL LESIONS IS DEPENDENT ON CHOICE OF SLICE THICKNESS IN PATIENTS WITH MULTIPLE SCLEROSIS

Ondrej Dolezal,*[,†] Michael G. Dwyer,* Dana Horakova,[†] Eva Havrdova,[†]
Alireza Minagar,[‡] Srivats Balachandran,* Niels Bergsland,* Zdenek Seidl,[§]
Manuela Vaneckova,[§] David Fritz,* Jan Krasensky,[§] and Robert Zivadinov*

*Department of Neurology, State University of New York at Buffalo
Buffalo Neuroimaging Analysis Center, The Jacobs Neurological Institute
Buffalo, New York 14203, USA

[†]Department of Neurology, 1st School of Medicine
Charles University, Prague, Czech Republic

[‡]Department of Neurology, Louisiana State University
Health Sciences Center, Shreveport
Louisiana 71103, USA

[§]Department of Radiology, 1st School of Medicine
Charles University, Prague, Czech Republic

Understanding the importance of cortical lesions in MS pathogenesis has changed. Histopathologic studies using new immunohistochemical methods show that cortical lesions can be detected more frequently than previously reported. Newer MRI sequences also detect cortical lesions more accurately. The aim of this study was to evaluate whether the effect of slice thickness (th) is an important factor for detection of cortical lesions in patients with multiple sclerosis (MS). We aimed also to investigate the relationship of cortical lesions with clinical status or other MRI variables. Forty-one patients with relapsing-remitting (RR) MS (11 males,

INTERNATIONAL REVIEW OF
NEUROBIOLOGY, VOL. 79
DOI: 10.1016/S0074-7742(07)79021-9

30 females with mean EDSS 2.3) underwent scans of Two-dimensional (2D)-fluid-attenuated inversion recovery (FLAIR) and 3D-T1-WI at 1.5-, 3-, and 5-mm slice thicknesses on 1.5-T MRI. Cortical and juxtacortical lesions were volumetrically assessed using a semiautomated method. 2D-FLAIR and 3D-T1-WI were coregistered and the matrix of the neocortical volume (NCV) segmentation mask (SIENAX-generated) was used to classify the location of the cortical–subcortical lesions. Cortical lesions fell into three classes: (1) class 1 were defined as lesions located in the NCV, (2) class 2 were juxtacortical lesions in contact with the NCV mask, and (3) class 3 were cortical–juxtacortical lesions situated in both regions. We measured NCV and normalized gray matter (GM) volume as well. We used partial correlation and multiple regressions to investigate the relationship between cortical lesions and other clinical and MRI variables. Of the total T2-lesion volume (T2-LV) measured on 1.5-mm th scans (mean 16108 mm^3), cortical lesions represented 2.4% (276 mm^3), juxtacortical lesions 6.1% (760 mm^3), and cortical–juxtacortical 3.7% (491 mm^3). Compared to 1.5-mm th scan, cortical LV was reduced by -28.3%, $p < 0.001$ on 3-mm th and by -40.78%, $p < 0.001$ on 5-mm th scans. Results for juxtacortical LV were for 3-mm th scans (-17.9%, $p < 0.01$) and for 5-mm th scans (-30.3%, $p < 0.01$). The figures for cortical–juxtacortical LV were also for 3-mm th scans (-16.2%, $p < 0.01$) and for 5-mm th scans (-26.7%, $p < 0.01$). We observed a significant correlation between T2-LV and GM atrophy in all slice thickness ($r = -0.4$ to -0.48, $p = 0.001$–0.003) and a modest relationship between cortical and cortical–juxtacortical LVs and disability, especially at 1.5-mm slice thickness ($r = 0.35$, $p = 0.025$). Use of thinner slices (1.5mm) on 2D-FLAIR images can significantly increase the sensitivity and precision of detecting cortical and juxtracotical lesions in patients with MS. Cortical and juxtacortical lesions contribute more to disability development than total T2-LV alone.

I. Introduction

Multiple sclerosis (MS) is an autoimmune demyelinating disease of the central nervous system (CNS), leading to disability in young adults. Historically, MS has been considered primarily a disease affecting the white matter (WM) of the CNS. Evidence is mounting that MS is also a gray matter (GM) disease (Bo *et al.*, 2003; Brownell and Hughes, 1962; Kutzelnigg and Lassmann, 2005; Peterson *et al.*, 2001; Wegner *et al.*, 2006). GM lesions are similar to those in the WM; they are described as clearly defined areas of demyelination within the cerebral cortex, basal ganglia, and GM of the spinal cord and brain stem that reflect an ongoing neurodegenerative process and predict future clinical disability (Zivadinov, 2006). Research on magnetic resonance imaging (MRI) emphasizes the role of cortical lesions combined with cortical GM volume reduction in early pathogenesis of MS (Dalton *et al.*, 2004; Zivadinov *et al.*, 2006). Regional cortical thinning and cortical volume reduction is frequently detected in patients with MS (Chen *et al.*, 2004; Sailer *et al.*, 2003).

Several histopathologic studies have described different patterns of cortical damage (Bo *et al.*, 2003; Brownell and Hughes, 1962; Kidd *et al.*, 1999; Kutzelnigg and Lassmann, 2005; Peterson *et al.*, 2001) and loss of synaptic density in neocortical lesions (Wegner *et al.*, 2006). Cortical lesions are heterogeneous and may be responsible for development of cortical GM atrophy. Three main subtypes of cortical lesions are currently differentiated in patients with MS: (1) leucocortical (Type 1), which are present in both GM and adjacent WM; (2) intracortical (Type 2), whose existence is limited to the GM region; and (3) cortical subpial "ribbon-like" lesions (Type 3) expanding from and lining the pial surface and representing more than 45% (Wegner *et al.*, 2006) to 60% (Peterson *et al.*, 2001) of the total amount of cortical lesions. The majority of these lesions are associated with transected dendrites, loss of axon neurons, and glial cells (Peterson *et al.*, 2001; Wegner *et al.*, 2006). The pathogenesis of these processes is not yet fully elucidated. The immunologic and pathological background of inflammatory processes in cortical areas is different from WM regions. These lesions are hypocellular and contain fewer T lymphocytes when compared to WM myelinated areas (Bo *et al.*, 2003; Peterson *et al.*, 2001; Wegner *et al.*, 2006).

Conventional MRI has low sensitivity for detecting cortical lesions because of poor contrast resolution between normal-appearing GM and the plaques in question, and because of partial volume effects of the subarachnoid spaces and cerebrospinal fluid (CSF) surrounding the cortex. Therefore, Type 1 and 2 cortical lesions are not adequately visible on routinely used MRI sequences and their "real" volume is generally underestimated. Cortical subpial lesions are undetectable by currently available imaging techniques.

In an early series of studies on the prevalence and distribution of cortical plaques, it was reported (Kidd *et al.*, 1999) that small cortical lesions are common in MS and are underreported by conventional MRI. The majority (about 75%) of the lesions arose within the region of the principal cortical veins, whereas the remaining plaques were found in the area of the small branch or superficial veins. Improved cortical lesion detection became possible using a number of recent advanced imaging modalities. Two-dimensional (2D)-fluid-attenuated inversion recovery (FLAIR) shows T2 tissue prolongation with CSF suppression (Bakshi *et al.*, 2001). Consequently, it is more capable of detecting superficial cortical lesions than T2-, proton density (PD)-, or T1-weighted imaging (WI). 3D-fast FLAIR imaging is increasing sensitivity in detecting cortical and periventricular lesions when compared to 2D-FLAIR (Bink *et al.*, 2006; Ciccarelli *et al.*, 2002;

Mills *et al.*, 2006; Tan *et al.*, 2002a,b; Tubridy *et al.*, 1998a,b). Another study revealed that 3D double-inversion recovery (DIR) imaging and FLAIR identified more intracortical lesions ($n = 1735$ and 1714, respectively) than did T2-weighted spin-echo ($n = 1455$) (Geurts *et al.*, 2005). Moreover, 3D DIR imaging was superior to both FLAIR and T2 spin-echo in depicting intracortical lesions (mean 8.0 vs 3.1 and 1.0, respectively). Notably, 3D DIR imaging produces better definition of juxtacortical lesions because of greater contrast between the lesions and adjacent areas (Pouwels *et al.*, 2006). Coregistration of T1-weighted 3D spoiled gradient (SPGR) volumes and 2D-FLAIR sequences [which generated a high signal-to-noise ratio (SNR) mean image] were used successfully to retrospectively identify cortical lesions in 22 MS patients (Bagnato *et al.*, in press).

Currently available standard scanners using 1.5-T and 3-T field strength sequences are not sufficiently sensitive to detect cortical subpial and deep GM lesions (Bagnato *et al.*, in press; Geurts *et al.*, 2005; Pouwels *et al.*, 2006). Therefore, the use of ultrahigh field strength imaging provides only a modest gain in the ability to identify cortical lesions, though the newer WM suppression inversion recovery techniques improve the detectability of Type 1 and 2 lesions (Zivadinov, 2006). Although these newer MRI techniques have shown the value of high-SNR images in detecting stable cortical brain abnormalities and are also able to detect these lesions more accurately than conventional techniques, most of them are not yet available for clinical practice or prospective clinical trials.

The choice of slice thickness on 2D-FLAIR probably influences the ability to identify pure cortical–juxtacortical lesions. The aim of this study was to evaluate whether the effect of slice thickness—comparing 1.5, 3, and 5 mm—is an important factor for the detection of cortical lesions in patients with MS. Another goal of the study was to define the relationship of cortical lesions with disability and/or other MRI measures.

II. Methods

The current analysis is based on MRI data collected from patients at the Department of Neurology, 1st School of Medicine, Charles University, Prague, Czech Republic. Recruited into the study were 41 subjects (11 males, 30 females) with clinically definite early relapsing-remitting (RR) MS (Poser *et al.*, 1983) confirmed by MRI and the presence of at least two oligoclonal bands in CSF, aged 18–55 years, with Expanded Disability Status Scale (EDSS) (Kurtzke, 1983) $\leq$3.5 on the day of enrollment and active disease defined by 2 relapses in the previous 12 months or 3 relapses in the previous 24 months. All patients had no history of previous IFN-β and/or glatiramer acetate therapy in the previous 12 months or immunosuppressive treatment with either pulse cyclophosphamide

TABLE I
DEMOGRAPHIC, CLINICAL, AND MRI VARIABLES AT BASELINE

Sex (M/F)	11/30		
Age, mean (SD) range	33.6	9	19–56
Relapse rate in year previous to study entry, mean (SD) range	1.8	0.67	1–3
Disease duration (years), mean (SD) range	6.5	6.1	0.9–21.2
EDSS, mean (SD) range	2.3	0.9	0–4.0

or mitoxantrone in the previous 6 months. Mean age of the patients was 33.6 years (age range 18.9–55 years, SD 8.9). Mean disease duration was 6.5 years (range 0.9–21.2 years, SD 6.1) and mean EDSS was 2.3 (range 0–4, SD 0.9) (Table I).

A. MRI Acquisition and Analysis

MRI scans were acquired on a 1.5-T Phillips Gyroscan scanner (Best, The Netherlands). Axial images of the brain were obtained using 2D-FLAIR (TR/TE 1100/14 ms, matrix size 256 × 181, flip angle 30°) without gap and were acquired at 1.5-, 3-, and 5-mm thicknesses, respectively. Axial T1-weighted 3D SPGR images (TR/TE 25/5 ms, matrix size 256 × 256) were obtained with 1-mm slice thickness. All images were nongapped. Patients were positioned in the magnet according to European Community guidelines (Miller *et al.*, 1991). The MRI scan was delayed for a minimum of 30 days after the last steroid treatment to avoid any transient effects of steroids on MRI results.

Image analysis was performed at the Buffalo Neuroimaging Analysis Center, The Jacobs Neurological Institute, Department of Neurology, State University at Buffalo, New York. Buffalo investigators were blinded to the patients' clinical characteristics and status.

Semiautomated edge detection method with manual corrections for region of interest (ROI) definition was used for measurement of total and regional T2-lesion volumes (T2-LV) on 2D-FLAIR images (Zivadinov *et al.*, 2001a,b). Particular attention was devoted to detection of cortical and juxtacortical lesions. All semiautomated measurements were performed by a trained operator with intrarater reproducibility $<1\%$.

We used the brain extraction tool (BET) to remove all non-brain, non-CSF tissue from the image and to identify the outer surface of the skull on 3D-T1 SPGR images (Smith, 2002). Brain and skull images were then used to perform a scaling-constrained registration to a standard brain and skull image set in order to determine a subject-specific normalization factor (Smith *et al.*, 2001). The deskulled image was then processed with an automated image segmentation tool (Zhang *et al.*, 2001). Compartment-specific absolute volumes were then quantified (taking partial volume estimates into account) and multiplied by the subject-specific normalization factor to obtain normalized tissue volume measures. Normalized volumes of the whole brain (NBV), GM (NGMV), WM (NWMV), and CSF were obtained via this process (Benedict *et al.*, 2006; Fritz *et al.*, 2006). To measure normalized neocortical volume (NCV), a standard space mask was used to separate neocortical GM from non-NCV (Benedict *et al.*, 2006) (Fig. 1). The NCV was calculated by multiplying the absolute volume of the neocortex by the standard space normalization factor.

Afterward, 3D-T1-weighted SPGR images were coregistered to 2D-FLAIR images (Jenkinson *et al.*, 2002) and both sets of images were aligned in the same spatial orientation. NCV segmentation mask was then applied to this matrix (Fig. 2). This coregistered mask was then resliced to obtain masks of the same thickness for 1.5, 3, and 5 mm at 2D-FLAIR images. The ROIs were superimposed and coregistered to these NCV masks for all three different slice thicknesses.

With this procedure, we were able to differentiate among three types of cortical lesions: (1) an intracortical lesion was defined as a lesion with its whole volume in the cortical area, and considered analogous to Type 2 of histopathologic classification (Peterson *et al.*, 2001); (2) a juxtacortical lesion was defined as

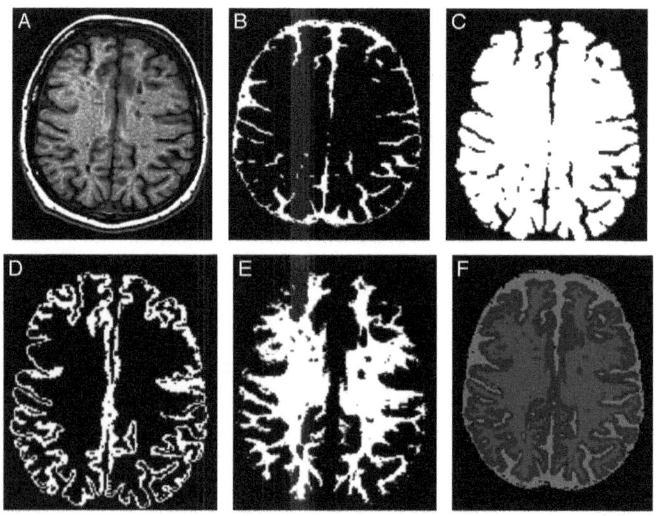

FIG. 1. Representative images of SIENAX segmentation in patient with relapsing-remitting multiple sclerosis. From (A) to (C) in the upper row are shown original 3D-T1-SPGR image, cerebrospinal fluid image, and skull-stripped image, and in the lower row (from D to F), gray matter, white matter, and three tissue class masks showing cerebrospinal fluid (slightly gray), neocortical volume (dark gray), and white matter (mid gray).

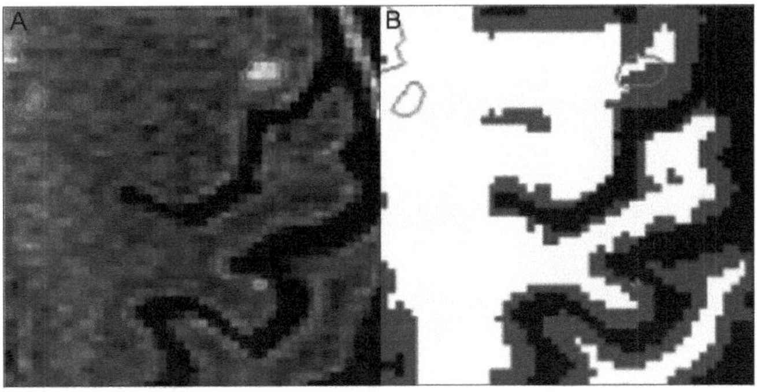

FIG. 2. Original 2D-FLAIR image with superimposed lesion ROIs (A) and these ROIs coregistered on neocortical volume mask (B) are shown. White matter lesions (left side of A and B), cortical lesions (right bottom of A and B), and cortical–juxtacortical lesions (right upper part of A and B) are presented.

a WM lesion in contact with the cortical mask and regarded as equivalent to Type 1 of histopathologic classification; and (3) a cortical–juxtacortical lesion was defined as a lesion situated in both GM and WM areas and deemed equivalent to Type 1 of histopathologic classification (Fig. 2).

B. Statistical Analysis

Statistical analysis was performed using SPSS 13 (SPSS, Chicago, Illinois). Data were first examined statistically and visually to ensure that they met the assumption of normality. Descriptive statistics were performed using parametric and non-parametric statistics, as appropriate. Significant differences in LV changes, according to different slice thicknesses, were investigated by the Student's *t*-test. In order to test correlation between cortical, juxtacortical, and cortical–juxtacortical LVs with different MRI and clinical variables, we used partial correlation (controlling for age) and regression analyses. Two-sided *p* values were used to perform all significance tests.

III. Results

A. Cortical, Juxtacortical, and Cortical–Juxtacortical Lesion Volumes According to the Different Slice Thicknesses

Comparing 1.5-, 3-, and 5-mm slice thicknesses, we found an important difference between the three groups. Percentage distributions for cortical, juxtacortical, and cortical–juxtacortical LVs are shown in Fig. 3 according to slice

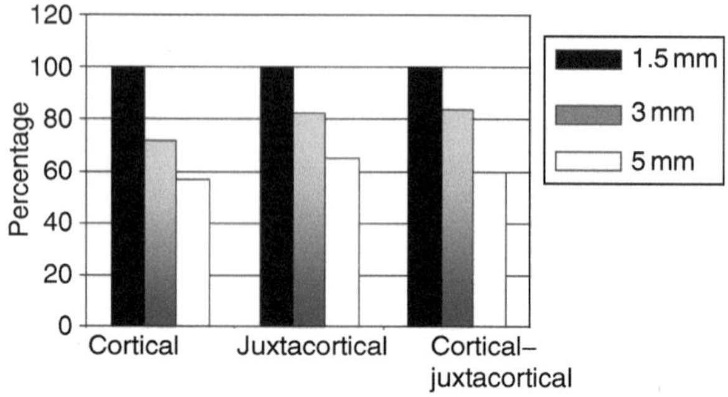

FIG. 3. Percentage of cortical, juxtacortical, and cortical–juxtacortical lesion volumes according to different slice thicknesses (1.5-mm slice thickness was chosen as a gold standard).

thickness. Compared to 1.5-mm scan thickness, cortical LV was lower by −28.3%, $p < 0.001$ at 3 mm and −40.78%, $p < 0.001$, at 5-mm slice thickness scans. Results for juxtacortical LVs were as follows: −17.9%, $p < 0.01$ for the 3-mm scans and −30.3%, $p < 0.01$ for the 5-mm scans. For the cortical–juxtacortical lesions, the results were: −16.22%, $p < 0.01$ for the 3-mm scans and −26.7%, $p < 0.01$ for the 5-mm scans (Fig. 3). Percentage distributions for cortical, juxtacortical, and cortical–juxtacortical LVs—with respect to the total T2-LV—increased with larger slice thickness and were as follows: for 1.5 mm (2.4%, 6.1%, and 3.7%), for 3 mm (2.1%, 5.7%, and 3.5%), and for 5 mm (1.4%, 5.1%, and 2.9%) (Fig. 4).

Comparison of absolute LV between 1.5- and 3-mm slice thicknesses was significant only for cortical LV ($p = 0.045$), but not for juxtacortical LV ($p = 0.25$), cortical–juxtacortical LV ($p = 0.24$), and total T2-LV ($p = 0.45$). On the other hand, a statistically significant difference between 1.5- and 5-mm slice thicknesses was found for all types of absolute LVs (cortical $p = 0.013$, juxtacortical $p = 0.03$, and cortical–juxtacortical $p = 0.027$, respectively), except for the total T2-LV ($p = 0.33$). Absolute LVs are shown in Table II.

B. Gray Matter Atrophy and Lesion Volumes

We observed a significant correlation between total T2-LV and GM atrophy (for NGMV $r = 0.4$, $p = 0.003$ and for NCV $r = 0.48$, $p = 0.001$), but slice thickness did not influence this relationship. This observation was also supported by regression analysis, where higher total T2-LV at 1.5-mm slice thickness predicted lower NCV volume ($R2 = 0.17$ $p = 0.007$). A trend for correlation of higher cortical LV at 1.5-mm thickness and lower NGMV volume was found ($r = 0.31$,

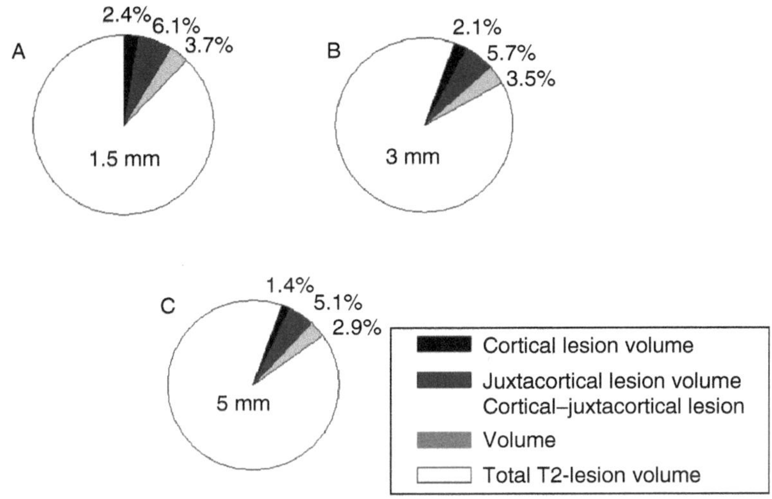

FIG. 4. Percentage of cortical, juxtacortical, and cortical–juxtacortical lesion volumes with respect to total T2-lesion volume (white) presented separately for each slice thickness (A–C).

$p = 0.056$). Higher juxtacortical and cortical–juxtacortical LVs at different slice thicknesses were not related to GM, WM, or global atrophy measures.

C. Relationship of Cortical, Juxtacortical, and Cortical–Juxtacortical Lesion Volumes to Disability

We found a significant relationship between cortical and cortical–juxtacortical LVs and clinical disability (as measured by EDSS) for 1.5- and 3-mm slice thicknesses (cortical–juxtacortical LV at 1.5 mm $r = 0.32$, $p = 0.044$, cortical LV at 3 mm $r = 0.35$, $p = 0.028$, and cortical–juxtacortical LV at 3 mm $r = 0.33$, $p = 0.041$). The cortical–juxtacortical LV at 5-mm slice thickness was not related to disability. There was no correlation between total T2-LV and disability at any slice thickness.

IV. Discussion

Although newer nonconventional MRI techniques can more accurately detect cortical lesions than conventional methods, most of these techniques are not yet available for clinical practice or research studies. Therefore, it is important to determine whether the choice of slice thickness on conventional 2D-FLAIR

TABLE II

Absolute Cortical, Juxtacortical, Cortical–Juxtacortical, and Total T2-Lesion Volumes Assessed on Different Slice Thicknesses

MRI LV measure	Mean	Standard deviation	Range
Cortical LV on 1.5 mm	276.1	277.6	487.5–6753
Cortical LV on 3 mm	176.1	146.1	0–621
Cortical LV on 5 mm	145.9	173.9	0–103.5
Juxtacortical LV on 1.5 mm	760.8	702.4	198–5768
Juxtacortical LV on 3 mm	610.6	594.6	29–3002
Juxtacortical LV on 5 mm	473.6	445.9	23–1988.5
Cortical–juxtacortical LV on 1.5 mm	491.1	546.7	210–5449
Cortical–juxtacortical LV on 3 mm	365.5	409.3	5.5–1875
Cortical–juxtacortical LV on 5 mm	270.7	304	0–1240
T2-LV 1.5 mm	16108.5	15324.8	487.5–67536
T2-LV 3 mm	13713.8	13181.9	198–57681
T2-LV 5 mm	13097.3	12465.5	210–54495

All results are reported in mm^3. LV, lesion volume.

images influences the ability to identify pure cortical–juxtacortical lesions and to define their relationship to disability and/or other MRI measures. Our findings suggest that use of thinner slices (1.5 mm) can significantly increase sensitivity and precision in detecting cortical and juxtracotical lesions in patients with MS.

At this time, the generally accepted consensus in clinical practice regarding use of slice thickness for detection of WM and cortical–juxtacortical lesions on FLAIR sequences favors 5-mm slice thickness. Use of 5-mm slice thickness is not always appropriate for detection of typically small lesions in cortical regions and can lead to loss of important lesion information (Filippi *et al.*, 1998a; Rovaris *et al.*, 1998). Recently published consensus guidelines from the Consortium of Multiple Sclerosis Centers recommend 3-mm slice thickness without gap as an acquisition standard for axial 2D-FLAIR, PD-, T2-, and T1-weighted precontrast sequences (Simon *et al.*, 2006a). European Federation Neurological Society guidelines for use of MRI in MS do not specify the preferred slice thickness for detection of cortical lesions (Filippi *et al.*, 2006).

In our study, the percent distribution of cortical, juxtacortical, and cortical–juxtacortical LVs was significantly higher on 1.5-mm slice thickness scans, when compared to both 3- and 5-mm thicknesses (Fig. 3). The comparison of absolute LVs between 1.5- and 3-mm scans confirmed this finding only for cortical LVs and there was a significant difference for all cortical-related LVs between 1.5- and 5-mm scans. Contrary to other studies, we did not find a difference between the three slice thicknesses in detecting total T2-LV (Filippi *et al.*, 1995a,b, 1998a,b; Molyneux *et al.*, 1998). This could be explained by considerable differences in the percentage of LV from patient to patient, and it could be possible that there was a larger variation in slice thickness for patients with smaller lesions and higher LV in the WM compartment. However, this was not the case for lesions located in the cortical–juxtacortical regions. In a previous study, Molyneux *et al.* (1998) used a 3D-fast FLAIR sequence to obtain brain images at slice thicknesses of 5, 3, and 1 mm. A progressive reduction in slice thickness resulted in a significant (8%) increase in T2-LV on MR images, as the thickness was reduced from 5 to 3 mm. Nevertheless, despite a further increase in lesion detection at a slice thickness of 1 mm, this did not result in an increase in total T2-LV, indicating that the relationship between section thickness and LV on MR images is not linear. Our study shows that, compared to 1.5-mm slice thickness scans, cortical LV was lower by −28.3% on 3 mm and −40.78% on 5 mm, although the ratio of the lesions in cortical–juxtacortical regions to total T2-LV increased only slightly when all three slice thicknesses scans were compared. In fact, lesions in the cortical and juxtacortical regions generally comprised a minor portion (less than 10%) of the total T2-LV and this supports our recommendation that the smallest slice thickness should be used for routine assessment. With application of newer nonconventional MRI techniques (3-T scanners, DIR, 3D-FLAIR), more accurate detection of Type 1 and 2 cortical lesions will become possible. In the meantime, our recommendation for use of thinner slices related to 1.5-mm scans for detection of cortical lesions might be useful.

Evidence is mounting that neurodegenerative mechanisms play an important role in MS pathogenesis and contribute to long-term disability progression. Neuronal apoptosis and neuronal, glial, and synaptic loss predominantly occur in GM regions (both in cortical and in deep GM regions) (Kutzelnigg and Lassmann, 2005; Peterson *et al.*, 2001; Pomeroy *et al.*, 2005; Wegner *et al.*, 2006). Cumulative neuronal damage due to retrograde Wallerian degeneration, with probable temporal delay between focal lesion formation and GM atrophy development, probably also contributes to GM damage (Simon *et al.*, 2006b). This may suggest that Wallerian degeneration from WM lesions may be an important contributor to development of cortical atrophy in the long term. Correlation and regression results from this study suggest that only 15–20% of GM atrophy development can be explained by inflammatory changes in the WM. Our data support the temporal delay between pathological changes occurring in WM and GM. The delay between focal lesion genesis and GM atrophy development in this study may be explained by the lower rate of secondary Wallerian degeneration in the early phases of the disease or by the fact that these processes are truly independent. While we found moderate cross-sectional correlation between T2-LV and GM atrophy markers, the correlation of cortical and juxtacortical lesions with GM atrophy was even lower, explaining at best 10% of the variance in GM atrophy. These findings may indicate that separate neurodegeneration and inflammatory processes are ongoing in GM regions and are only partially dependent on WM pathology (Kutzelnigg and Lassmann, 2005; Wegner *et al.*, 2006).

We found a significant LV relationship in cortical and cortical–juxtacortical LVs and higher disability progression. This relationship was dependent on the choice of slice thickness, favoring use of thinner slices (1.5 and 3 mm). The cortical–juxtacortical LV at 5-mm slice thickness was not related to disability. On the other hand, there was no correlation between total T2-LV and disability at any slice thickness. This indicates that Type 1 and 2 cortical lesions in patients with MS may contribute more to development of disability than total T2-LV.

In summary, our study shows that use of thinner slices (1.5 mm) can significantly increase the sensitivity and precision in detecting cortical and juxtracotical lesions in patients with MS. Cortical and juxtacortical lesions contribute more to disability development than total T2-LV alone. Development of scanning techniques and analysis methods with increased sensitivity for detecting cortical lesions—along with subsequent availability for routine clinical practice—is mandatory. More precise detection of Type 1 and 2 cortical lesions is needed to understand their full contribution to disability development. Development of noninvasive scanning techniques is needed for detection of Type 3 cortical subpial and deep GM lesions.

Acknowledgments

O.D. was supported by the Dr. Larry D. Jacobs Jog-for-the-Jake Fellowship.

References

Bagnato, F., Butman, J. A., Gupta, S., Calabrese, M., Pezawas, L., Ohayon, J. M., Tovar-Moll, F., Riva, M., Cao, M. M., Talagala, L., and McFarland, H. F. (2006). *In vivo* detection of cortical plaques by magnetic resonance imaging in patients with multiple sclerosis. *AJNR Am. J. Neuroradiol.* **27**(10), 2161–2167.

Bakshi, R., Ariyaratana, S., Benedict, R. H., and Jacobs, L. (2001). Fluid-attenuated inversion recovery magnetic resonance imaging detects cortical and juxtacortical multiple sclerosis lesions. *Arch. Neurol.* **58,** 742–748.

Benedict, R. H., Bruce, J. M., Dwyer, M. G., Abdelrahman, N., Hussein, S., Weinstock-Guttman, B., Garg, N., Munschauer, F., and Zivadinov, R. (2006). Neocortical atrophy, third ventricular width, and cognitive dysfunction in multiple sclerosis. *Arch. Neurol.* **63,** 1301–1306.

Bink, A., Schmitt, M., Gaa, J., Mugler, J. P., III, Lanfermann, H., and Zanella, F. E. (2006). Detection of lesions in multiple sclerosis by 2D FLAIR and single-slab 3D FLAIR sequences at 3.0 T: Initial results. *Eur. Radiol.* **16,** 1104–1110.

Bo, L., Vedeler, C. A., Nyland, H., Trapp, B. D., and Mork, S. J. (2003). Intracortical multiple sclerosis lesions are not associated with increased lymphocyte infiltration. *Mult. Scler.* **9,** 323–331.

Brownell, B., and Hughes, J. T. (1962). The distribution of plaques in the cerebrum in multiple sclerosis. *J. Neurol. Neurosurg. Psychiatry* **25,** 315–320.

Chen, J. T., Narayanan, S., Collins, D. L., Smith, S. M., Matthews, P. M., and Arnold, D. L. (2004). Relating neocortical pathology to disability progression in multiple sclerosis using MRI. *Neuroimage* **23,** 1168–1175.

Ciccarelli, O., Brex, P. A., Thompson, A. J., and Miller, D. H. (2002). Disability and lesion load in MS: A reassessment with MS functional composite score and 3D fast FLAIR. *J. Neurol.* **249,** 18–24.

Dalton, C. M., Chard, D. T., Davies, G. R., Miszkiel, K. A., Altmann, D. R., Fernando, K., Plant, G. T., Thompson, A. J., and Miller, D. H. (2004). Early development of multiple sclerosis is associated with progressive grey matter atrophy in patients presenting with clinically isolated syndromes. *Brain* **127,** 1101–1107.

Filippi, M., Horsfield, M. A., Campi, A., Mammi, S., Pereira, C., and Comi, G. (1995a). Resolution-dependent estimates of lesion volumes in magnetic resonance imaging studies of the brain in multiple sclerosis. *Ann. Neurol.* **38,** 749–754.

Filippi, M., Horsfield, M. A., Tofts, P. S., Barkhof, F., Thompson, A. J., and Miller, D. H. (1995b). Quantitative assessment of MRI lesion load in monitoring the evolution of multiple sclerosis. *Brain* **118**(Pt. 6), 1601–1612.

Filippi, M., Horsfield, M. A., Ader, H. J., Barkhof, F., Bruzzi, P., Evans, A., Frank, J. A., Grossman, R. I., McFarland, H. F., Molyneux, P., Paty, D. W., Simon, J., *et al.* (1998a). Guidelines for using quantitative measures of brain magnetic resonance imaging abnormalities in monitoring the treatment of multiple sclerosis. *Ann. Neurol.* **43,** 499–506.

Filippi, M., Horsfield, M. A., Hajnal, J. V., Narayana, P. A., Udupa, J. K., Yousry, T. A., and Zijdenbos, A. (1998b). Quantitative assessment of magnetic resonance imaging lesion load in multiple sclerosis. *J. Neurol. Neurosurg. Psychiatry* **64**(Suppl. 1), S88–S93.

Filippi, M., Rocca, M. A., Arnold, D. L., Bakshi, R., Barkhof, F., De Stefano, N., Fazekas, F., Frohman, E., and Wolinsky, J. S. (2006). EFNS guidelines on the use of neuroimaging in the management of multiple sclerosis. *Eur. J. Neurol.* **13,** 313–325.

Fritz, D. A., Dwyer, M. G., Bagnato, F., Watts, K. L., Bratina, A., Zorzon, M., Durastanti, V., Locatelli, L., Millefiorini, E., and Zivadinov, R. (2006). Effect of MRI coregistration on serial short-term brain volume changes in multiple sclerosis. *Neurol. Res.* **28,** 275–279.

Geurts, J. J., Pouwels, P. J., Uitdehaag, B. M., Polman, C. H., Barkhof, F., and Castelijns, J. A. (2005). Intracortical lesions in multiple sclerosis: Improved detection with 3D double inversion-recovery MR imaging. *Radiology* **236,** 254–260.

Jenkinson, M., Bannister, P., Brady, M., and Smith, S. (2002). Improved optimization for the robust and accurate linear registration and motion correction of brain images. *Neuroimage* **17,** 825–841.

Kidd, D., Barkhof, F., McConnell, R., Algra, P. R., Allen, I. V., and Revesz, T. (1999). Cortical lesions in multiple sclerosis. *Brain* **122**(Pt. 1), 17–26.

Kurtzke, J. F. (1983). Rating neurologic impairment in multiple sclerosis: An expanded disability status scale (EDSS). *Neurology* **33,** 1444–1452.

Kutzelnigg, A., and Lassmann, H. (2005). Cortical lesions and brain atrophy in MS. *J. Neurol. Sci.* **233,** 55–59.

Miller, D. H., Barkhof, F., Berry, I., Kappos, L., Scotti, G., and Thompson, A. J. (1991). Magnetic resonance imaging in monitoring the treatment of multiple sclerosis: Concerted action guidelines. *J. Neurol. Neurosurg. Psychiatry* **54,** 683–688.

Mills, R. J., Young, C. A., and Smith, E. T. (2006). 3D MRI in multiple sclerosis: A study of three sequences at 3 T. *Br. J. Radiol.* DOI:10.1259/bjr/52670770.

Molyneux, P. D., Tubridy, N., Parker, G. J., Barker, G. J., MacManus, D. G., Tofts, P. S., Moseley, I. F., and Miller, D. H. (1998). The effect of section thickness on MR lesion detection and quantification in multiple sclerosis. *AJNR Am. J. Neuroradiol.* **19,** 1715–1720.

Peterson, J. W., Bo, L., Mork, S., Chang, A., and Trapp, B. D. (2001). Transected neurites, apoptotic neurons, and reduced inflammation in cortical multiple sclerosis lesions. *Ann. Neurol.* **50,** 389–400.

Pomeroy, I. M., Matthews, P. M., Frank, J. A., Jordan, E. K., and Esiri, M. M. (2005). Demyelinated neocortical lesions in marmoset autoimmune encephalomyelitis mimic those in multiple sclerosis. *Brain* **128,** 2713–2721.

Poser, C. M., Paty, D. W., Scheinberg, L., McDonald, W. I., Davis, F. A., Ebers, G. C., Johnson, K. P., Sibley, W. A., Silberberg, D. H., and Tourtellotte, W. W. (1983). New diagnostic criteria for multiple sclerosis: Guidelines for research protocols. *Ann. Neurol.* **13,** 227–231.

Pouwels, P. J., Kuijer, J. P., Mugler, J. P., III, Guttmann, C. R., and Barkhof, F. (2006). Human gray matter: Feasibility of single-slab 3D double inversion-recovery high-spatial-resolution MR imaging. *Radiology* **241**(3), 873–879.

Rovaris, M., Rocca, M. A., Capra, R., Prandini, F., Martinelli, V., Comi, G., and Filippi, M. (1998). A comparison between the sensitivities of 3-mm and 5-mm thick serial brain MRI for detecting lesion volume changes in patients with multiple sclerosis. *J. Neuroimaging* **8,** 144–147.

Sailer, M., Fischl, B., Salat, D., Tempelmann, C., Schonfeld, M. A., Busa, E., Bodammer, N., Heinze, H. J., and Dale, A. (2003). Focal thinning of the cerebral cortex in multiple sclerosis. *Brain* **126,** 1734–1744.

Simon, J. H., Li, D., Traboulsee, A., Coyle, P. K., Arnold, D. L., Barkhof, F., Frank, J. A., Grossman, R., Paty, D. W., Radue, E. W., and Wolinsky, J. S. (2006a). Standardized MR imaging protocol for multiple sclerosis: Consortium of MS Centers consensus guidelines. *AJNR Am. J. Neuroradiol.* **27,** 455–461.

Simon, J. H., Zhang, S., Laidlaw, D. H., Miller, D. E., Brown, M., Corboy, J., and Bennett, J. (2006b). Identification of fibers at risk for degeneration by diffusion tractography in patients at high risk for MS after a clinically isolated syndrome. *J. Magn. Reson. Imaging* **24,** 983–988.

Smith, S. M. (2002). Fast robust automated brain extraction. *Hum. Brain Mapp.* **17,** 143–155.

Smith, S. M., De Stefano, N., Jenkinson, M., and Matthews, P. M. (2001). Normalized accurate measurement of longitudinal brain change. *J. Comput. Assist. Tomogr.* **25,** 466–475.

Tan, I. L., Pouwels, P. J., van Schijndel, R. A., Ader, H. J., Manoliu, R. A., and Barkhof, F. (2002a). Isotropic 3D fast FLAIR imaging of the brain in multiple sclerosis patients: Initial experience. *Eur. Radiol.* **12,** 559–567.

Tan, I. L., van Schijndel, R. A., Pouwels, P. J., Ader, H. J., and Barkhof, F. (2002b). Serial isotropic three-dimensional fast FLAIR imaging: Using image registration and subtraction to reveal active multiple sclerosis lesions. *Am. J. Roentgenol.* **179,** 777–782.

Tubridy, N., Barker, G. J., MacManus, D. G., Moseley, I. F., and Miller, D. H. (1998a). Optimisation of unenhanced MRI for detection of lesions in multiple sclerosis: A comparison of five pulse sequences with variable slice thickness. *Neuroradiology* **40,** 293–297.

Tubridy, N., Barker, G. J., Macmanus, D. G., Moseley, I. F., and Miller, D. H. (1998b). Three-dimensional fast fluid attenuated inversion recovery (3D fast FLAIR): A new MRI sequence which increases the detectable cerebral lesion load in multiple sclerosis. *Br. J. Radiol.* **71,** 840–845.

Wegner, C., Esiri, M. M., Chance, S. A., Palace, J., and Matthews, P. M. (2006). Neocortical neuronal, synaptic, and glial loss in multiple sclerosis. *Neurology* **67,** 960–967.

Zhang, Y., Brady, M., and Smith, S. M. (2001). Segmentation of brain MR images through a hidden Markov random field model and the expectation maximization algorithm. *IEEE Trans. Biomed. Imaging* **20,** 45–57.

Zivadinov, R. (2006). Evidence for neuroprotection in multiple sclerosis: Can imaging techniques measure neuroprotection and remyelination? *Neurology* (in press).

Zivadinov, R., Rudick, R. A., De Masi, R., Nasuelli, D., Ukmar, M., Pozzi-Mucelli, R. S., Grop, A., Cazzato, G., and Zorzon, M. (2001a). Effects of IV methylprednisolone on brain atrophy in relapsing-remitting MS. *Neurology* **57,** 1239–1247.

Zivadinov, R., Sepcic, J., Nasuelli, D., De Masi, R., Bragadin, L. M., Tommasi, M. A., Zambito-Marsala, S., Moretti, R., Bratina, A., Ukmar, M., Pozzi-Mucelli, R. S., Grop, A., *et al.* (2001b). A longitudinal study of brain atrophy and cognitive disturbances in the early phase of relapsing-remitting multiple sclerosis. *J. Neurol. Neurosurg. Psychiatry* **70,** 773–780.

Zivadinov, R., Dolezal, O., Tekwe, C. I., Horakova, D., Dwyer, M. G., Balachandran, S., Bergsland, N., and Havrdova, E. (2006). A longitudinal study of cortical atrophy in patients with relapsing-remitting multiple sclerosis over a 2-year period. *J. Neurol.* **253,** P222, 56–57.

THE ROLE OF QUANTITATIVE NEUROIMAGING INDICES IN THE DIFFERENTIATION OF ISCHEMIA FROM DEMYELINATION: AN ANALYTICAL STUDY WITH CASE PRESENTATION

Romy Hoque,* Christina Ledbetter,† Eduardo Gonzalez-Toledo,*,‡ Vivek Misra,* Uma Menon,* Meghan Kenner,* Alejandro A. Rabinstein,§ Roger E. Kelley,* Robert Zivadinov,¶ and Alireza Minagar*

*Department of Neurology, Louisiana State University Health Sciences Center Shreveport, Louisiana 71103, USA
†Department of Psychiatry, Louisiana State University Health Sciences Center Shreveport, Louisiana 71103, USA
‡Department of Radiology, Louisiana State University Health Sciences Center Shreveport, Louisiana 71103, USA
§Department of Neurology, Mayo Clinic, Rochester, Minnesota 55905, USA
¶Buffalo Neuroimaging Analysis Center, The Jacobs Neurological Institute State University of New York at Buffalo, Buffalo, New York 14203, USA

Background and purpose: Differentiation of acute and subacute ischemic stroke lesions from acute demyelinating lesions of multiple sclerosis (MS) may not be possible on conventional magnetic resonance imaging (MRI). Both lesion types enhance on T1 with gadolinium (Gd) contrast and both are hyperintense on diffusion-weighted imaging (DWI). This study is an analysis of two quantitative

DOI: 10.1016/S0074-7742(07)79022-0

0074-7742/07 $35.00

MR indices: (1) calculated apparent diffusion coefficients (ADCs) and (2) T2 relaxation times (T2R) as means toward differentiating acute ischemic lesions from acute demyelinating lesions. Chronic ischemic and demyelinating lesions were evaluated for comparison as well.

Methods: The MRI of nine patients with both acute and chronic ischemic lesions and six patients with both acute and chronic demyelinating lesions were analyzed for ADC and T2Rs. The indices were measured by manually placing regions of interest (ROIs) at the anatomic center of the acute lesion. Acute ischemic lesions were chosen by their hyperintensity on DWI and hypointensity on ADC mapping. Acute demyelinating lesions were selected by peripheral contrast enhancement after the administration of Gd. Computation of the ADC involved the diffusion coefficient on a region by region basis as follows: $D = -(b_0/b_{1000})\ln(S_{b1000}/S_{b0})$, where S_{b1000} is the signal intensity on DWI and S_{b0} is the signal intensity on T2 with diffusion sensitivities of b_0 and b_{1000}, respectively. Computation of the T2R was made as follows: $\mathrm{T2R} = (\mathrm{TE}_{\mathrm{T2}} - \mathrm{TE}_{\mathrm{PD}})/(\ln \mathrm{SI}_{\mathrm{PD}} - \ln \mathrm{SI}_{\mathrm{T2}})$, where TE is the echo time of the different pulse sequences, SI is signal intensity on the different echo sequences, and PD represents proton density sequence.

Results: Twenty-nine acute ischemia, 27 acute demyelination, 28 chronic ischemia, and 43 chronic demyelination image sets were analyzed. The differences between $\mathrm{ADC}_{\mathrm{acute\ infarct}}$ (0.760) versus $\mathrm{ADC}_{\mathrm{acute\ plaque}}$ (1.106) were significant ($p < 0.02$). The differences between $\mathrm{T2R}_{\mathrm{acute\ infarct}}$ (235.5) versus $\mathrm{T2R}_{\mathrm{acute\ plaque}}$ (170.5) were also significant ($p < 0.02$).

Conclusions: ADC in combination with T2R is a useful tool to differentiate acute ischemic from acute demyelinating lesions. The use of these neuroimaging indices along with magnetic resonance spectroscopy metabolite ratios is then demonstrated in elucidating the pathophysiological mechanism for a case of delayed posttraumatic bilateral internuclear ophthalmoplegia.

I. Introduction

Calculated neuroimaging indices have advanced our understanding of the underlying pathophysiological mechanisms of many neurological diseases. In the present study, we describe in detail the role of apparent diffusion coefficients (ADCs), T2 relaxation times (T2Rs), and magnetic resonance spectroscopy (MRS) metabolic ratios in the differentiation of acute ischemia from acute demyelinating lesions. Later, we apply these neuroimaging indices along with MRS to further elucidate the pathophysiology of a case of delayed posttraumatic bilateral internuclear ophthalmoplegia (INO).

A. Apparent Diffusion Coefficients

Diffusion-weighted imaging (DWI) measures the random diffusion of water molecules in brain tissue and can be quantified by calculating its ADC. DWI by itself does neither measure ADC or quantifies water diffusion. The reason is that in a single DWI, the contribution to signal intensity of T2 versus diffusion cannot be differentiated. The simplest way to differentiate a lesion with low ADC from one that is merely T2-weighted hyperintense is to obtain DWI with at least two different values of *b*, where *b* represents the intensity and duration of a gradient. Of note, all diffusion gradients are always accompanied by the acquisition of reference images without active diffusion gradients ($b = 0\ \mathrm{s/mm^2}$) which corresponds to a T2-weighted image.

The isotropic signal intensity (SI) of a voxel of tissue is calculated by the Stejskal–Tanner formula as follows:

$$\mathrm{SI} = \mathrm{SI}_0 \exp(-b \times D) \tag{1}$$

where SI_0 is the signal intensity on the T2-weighted (or $b = 0\ \mathrm{s/mm^2}$) image and D is the diffusion coefficient (Schaefer *et al.*, 2000). According to Fick's law, true diffusion is the net movement of molecules due to a concentration gradient. However, with MRI we cannot correct for the increases in distance traveled by tortuous pathways. Therefore, when measuring molecular motion with DW imaging, only the ADC can be calculated. Therefore, the signal intensity of DW images is best expressed as:

$$\mathrm{SI} = \mathrm{SI}_0 \exp(-b \times \mathrm{ADC}) \tag{2}$$

Obtaining images with $b = 0\ \mathrm{s/mm^2}$ and with $b = 1000\ \mathrm{s/mm^2}$ and solving for ADC allows for the calculation of ADC shown in Eq. (3) (Schaefer *et al.*, 2000).

$$D = -\frac{b_0}{b_0} \ln\left(\frac{S_{b1000}}{S_{b0}}\right) \tag{3}$$

S_{b1000} is the signal intensity on DWI and S_{b0} is the signal intensity on T2 with diffusion sensitivities of b_0 and b_{1000}, respectively. Calculated ADC values using Eq. (3) can be used to characterize lesions of different etiologies on voxel by voxel basis.

B. T2 Relaxometry

T2-weighted images are commonly used because of their high sensitive to a wide range of pathologic syndromes. The signal in a T2 relaxation image contains not only T2-weighting but also information about the amount of water (proton density) in each pixel. Dual spin-echo MRI permits calculation of T2 time, which is influenced by the amount of free water protons in a specific region. Voxel-based T2R indices, first described by Mirowitz *et al.* (1990) in the differentiation of benign cavernous hemangiomas of the liver from malignant neoplasms, quantify the tissue absolute T2R, which translates into the water content of brain tissue in a specific region of interest.

T2 relaxometry provides a sensitive measure of the increased interstitial cerebrospinal fluid (CSF) accumulation and intramyelin edema. Values of T2R in an individual voxel can be calculated on any MRI slice. In a single slice, two separate spin-echo images of this slide are taken at different echo times (TE). The signal intensity of a single voxel of tissue using a specific spin echo sequence (SI_{SE}) can be approximated using a modified form of the Bloch equation as follows:

$$SI_{SE} = (1 - e^{-TR/T1})(e^{-TE/T2}) \tag{4}$$

where TR is repetition time, TE is echo time, T1 is the longitudinal relaxation time, and T2 is the transverse relaxation time. When TR and T1 are maintained, the ratio of signal intensity on short TE sequences (proton density images) to signal intensity on long TE sequences (T2-weighted images) can be expressed as follows:

$$\frac{SI_{shortTE}}{SI_{longTE}} = e^{-TE/2} \tag{5}$$

Solving for T2 in the Eq. (5) gives the T2R equation shown in Eq. (6) (Cittadini *et al.*, 2004; Mirowitz *et al.*, 1990).

$$T2R = \frac{TE_{T2} - TE_{PD}}{\ln SI_{PD} - \ln SI_{T2}} \tag{6}$$

SI_{PD} equals the lesion intensity on proton density sequences and SI_{T2} equals the lesion intensity on T2-weighted images. Using Eq. (6), the rate of signal intensity decay with increasing TE can be represented indirectly by T2R (Jackson and Connelly, 1999).

C. Edema and Neuroimaging Indices

Cerebral edema is classified, based on pathophysiological mechanisms, into cytotoxic and vasogenic edemas (Rabinstein, 2006). In cytotoxic edema, the increased water content is localized intracellularly and is due to the failure of ionic pumps that normally maintain cellular homeostasis. Ischemia and profound metabolic derangement are the most common causes. On the other hand in vasogenic edema, there is a disruption of the blood–brain barrier (BBB), leading to increased permeability and escape of fluid from the intravascular to the extravascular, extracellular space. This is associated with tumors, inflammatory lesions, and traumatic tissue damage.

The increased volume of extracellular water which is present in vasogenic edema should increase the tissue values of both ADC and T2R. In contrast, the intracellular water entrapment observed in cytotoxic edema should produce a decreased ADC value, while the T2R value should remain elevated due to the excess tissue water (Figueroa *et al.*, 2005).

D. Magnetic Resonance Spectroscopy

Water suppressed, proton magnetic resonance spectra of the healthy human brain at long echo times reveal four major resonances: tetramethylamines, mainly choline-containing phospholipids (Cho); creatine and phosphocreatine (Cr); *N*-acetyl groups, mainly *N*-acetylaspartate (NAA); and the methyl resonance of lactate. NAA is indicative of axonal integrity, whereas Cho and lactate are thought to correlate with acute inflammatory and demyelinating changes. Proton MRS with shorter echo times can detect additional metabolites, such as lipids and myoinositol (mI), which are thought to indicate ongoing myelin damage and gliosis.

II. ADC and T2R in Acute Ischemia and Acute Demyelination

A. Introduction

Previous neuroimaging studies of patients with ischemic stroke and multiple sclerosis (MS) have elucidated the diagnostic significance of DWI, ADC, and T2R in exploring the natural history of these common neurological disorders (Christiansen *et al.*, 1993; Horsfield *et al.*, 1996; Larsson *et al.*, 1992; Lovblad *et al.*, 1998; Rosso *et al.*, 2006; Tievsky *et al.*, 1999). However, these diagnostic tools have not been employed previously to differentiate acute ischemic from acute demyelinating lesions given the fact that both lesions may present as hyperintense signals on DWI, have reduced signal on ADC maps, and show enhancement on postcontrast T1-weighted images (Rosso *et al.*, 2006).

We aimed to use the different water content of acute ischemic infarcts and acute demyelinating lesions in MS as a means of quantitatively differentiating these two lesion types on MRI. ADC and T2R values of acute ischemic and acute demyelinating lesions were measured and compared, along with the ADC and T2R values, of chronic ischemic and demyelinating lesions.

B. Methods

1. *Image Acquisition*

Using conventional MRI and DWI, 12 subjects with ischemic stroke and 6 subjects with MS were studied. Axial T1-, T2-, and PD-weighted brain MR scans were obtained on a 1.5-T scanner (GE, Milwaukee, Wisconsin). The following pulse sequences were applied: T1-weighting [600 ms/10.6 ms/1 (repetition time/echo time/excitations)]; T2-weighting (2000 ms/121 ms), PD (2000 ms/15 ms); fluid-attenuated inversion recovery (FLAIR) (8000 ms/143 ms/1); and DWI (2000/121). Slice thickness was 5 mm with an interslice gap of 0.5 mm, a matrix size of 256×224 pixels, and a field of view of $24 \times 24\ cm^2$.

2. *Image Analysis*

The ADC and T2R were measured by manually placing regions of interest (ROIs) within the enhancing area demonstrated on postcontrast T1-weighted images. In accordance with the concept that ischemic lesions tend to grow outward from a central core, the ROI was placed as close as possible to the anatomic center of the lesion. Uniform oval ROIs were drawn in areas large enough to accommodate an 8.0-mm^2 ROI without apparent volume averaging of surrounding tissue. These 8.0-mm^2 ROIs were also small enough to counteract nonlesion related decreases in T2R secondary to atrophy and increased CSF seen in some patients. ROIs of frontal subcortical normal-appearing white matter (NAWM) were used for reference values.

ADC on a region by region basis was calculated using Eq. (3). Measurements of T2R were also obtained by using an operator-defined ROI at the anatomic center of the lesion. The T2R values from dual echo fast spin images were defined by Eq. (6).

Ischemic lesions were determined to be acute if they were hyperintense on DWI and corresponded to the onset of acute clinical symptoms in the 24–48 h previous to the MR scan. MS lesions were classified as acute if they showed enhancement and surrounding edema on T1-weighted postcontrast images after infusion of gadolinium chelate (Gd), or chronic if they showed hyperintensity on T2-weighted images in the absence of acute Gd enhancement on T1-weighted postcontrast images.

All lesions studied were either cortical or subcortical in the frontal or parietal lobes. Hemorrhagic infarcts were excluded given recent MR studies showing increased T2R values with increased iron concentrations in an ROI (House *et al.*, 2006). This was also the rationale for excluding lesions in the putamen—which was shown to have higher iron concentrations in postmortem studies (Hallgren and Sourander, 1958).

The measured ADC and T2R values in the lesions were assessed for statistical significance using a one-tailed Student *t*-test given the narrow range of values. Differences were considered significant when $p < 0.05$.

C. Results

ADC values were calculated for 37 lesions in 6 subjects with MS and for 26 ischemic lesions in 9 stroke subjects. T2R was calculated for 29 plaques in 6 MS patients, and 25 ischemic lesions in 7 patients. Acute demyelinating plaques were identified in five of the six MS patients. These plaques were characterized by high signal intensity on diffusion-weighted images. Figure 1 shows different neuroimaging sequences of an acute plaque. Acute ischemic lesions were identified in 10 of the 13 stroke patients. Figure 2 shows MR images of an acute infarct.

Global results for ischemic and demyelinating lesions are presented in Table I. The acute ischemic lesions showed the lowest mean ADC [$(0.760 \pm 0.595) \times 10^{-3}$ mm^2/s]. The mean ADC value for acute demyelinating lesions was higher

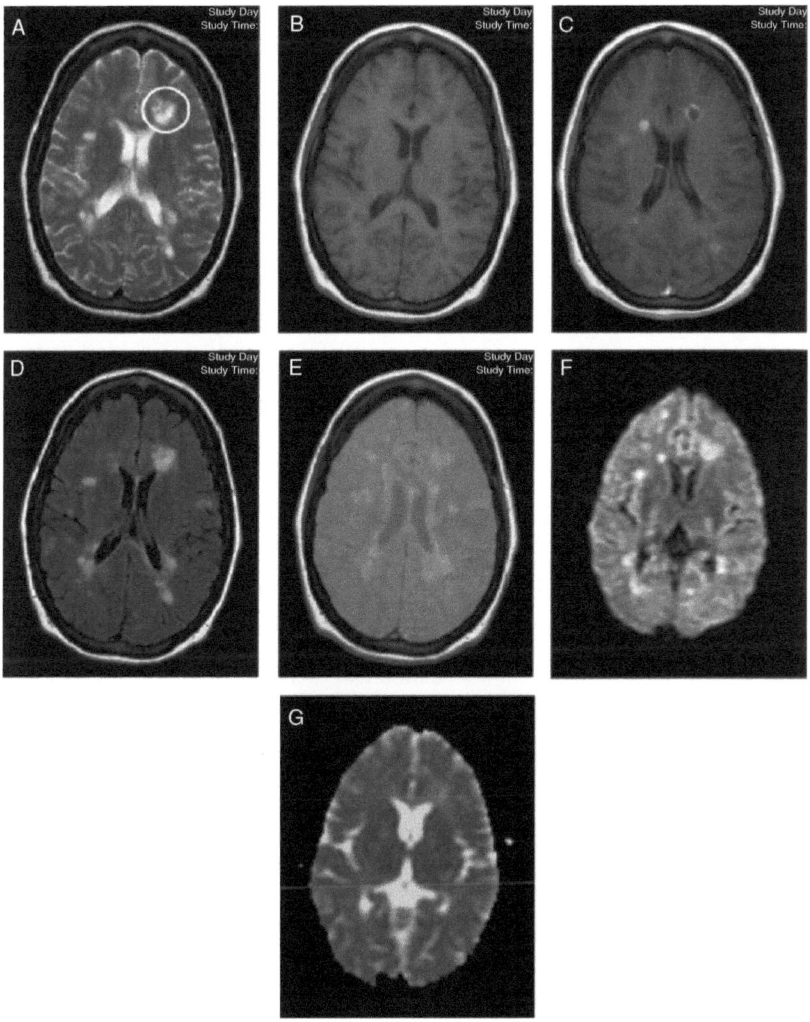

FIG. 1. MRI of an acute demyelinating lesion. (A) T2-weighted image with a circle identifying the lesion. (B) T1-weighted image without contrast. (C) T1-weighted image with contrast. (D) FLAIR. (E) Proton density. (F) Diffusion-weighted image. (G) Apparent diffusion coefficient image.

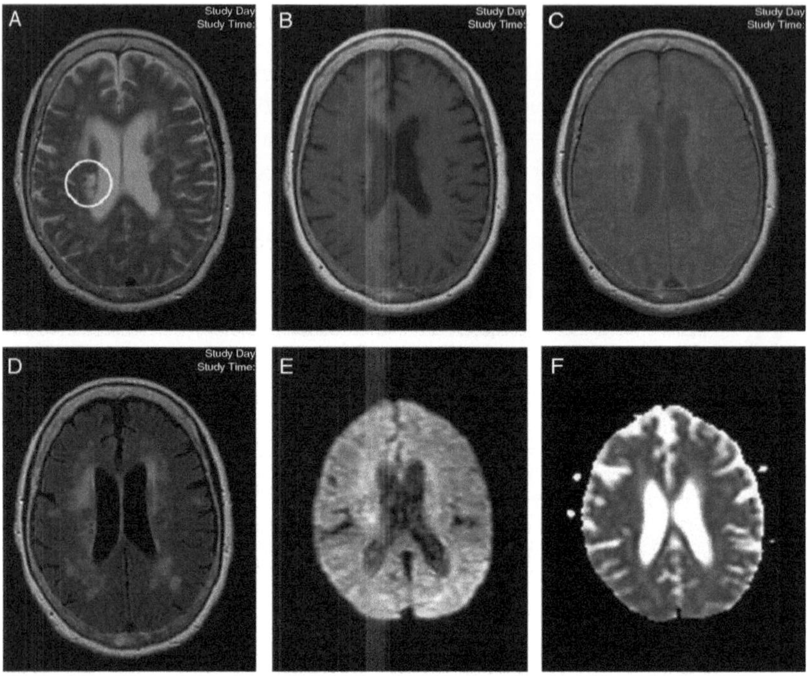

FIG. 2. MRI of an acute ischemic lesion. (A) T2-weighted image with a circle identifying the lesion. (B) T1-weighted image without contrast. (C) T1-weighted image with contrast. (D) Fluid attenuated inversion recovery image. (E) Diffusion-weighted image. (F) Apparent diffusion coefficient image.

TABLE I

MEAN ADC ($\times 10^{-3}$ mm^2/s), AND T2R IN MS FOR ACUTE AND CHRONIC ISCHEMIC LESIONS AND ACUTE AND CHRONIC DEMYELINATING LESIONS

Calculated indices	NAWM	Acute ischemia	Chronic ischemia	Acute plaque	Chronic plaque
ADC					
N	19	17	9	12	25
Mean	0.901	0.760	1.319	1.106	1.233
SD	0.10	0.595	0.986	0.211	0.366
T2R					
N	19	12	13	16	18
Mean	123.96	235.5	344.9	174.50	160.7
SD	59.1	100.5	96.7	34.39	55.5

at $(1.106 \pm 0.211) \times 10^{-3}$ mm^2/s. This difference was significant for $ADC_{acute\ demyelinating\ plaque}$ versus $ADC_{acute\ ischemic\ lesion}$ ($p < 0.02$). The mean T2R values for acute ischemic lesions was 235.5 ± 100.5 ms. The mean T2R value for the acute demyelinating lesions was 170.50 ± 34.39 ms. This difference was significant for $T2R_{acute\ ischemic\ lesion}$ versus $T2R_{acute\ demyelinating\ lesion}$ ($p < 0.02$).

For the chronic demyelinating and chronic ischemic lesions, the mean ADC values were $(1.319 \pm 0.986) \times 10^{-3}$ mm^2/s and $(1.233 \pm 0.366) \times 10^{-3}$ mm^2/s, respectively. This difference was not statistically significant. Chronic ischemic lesions had the highest mean T2R value at 344.9 ± 96.7 ms, and the chronic demyelinating plaques had a mean T2R of 160.7 ± 55.5 ms.

When acute and chronic ischemic lesions were compared, the ADC values in acute lesions were lower ($p < 0.05$). When T2R values for acute and chronic ischemic lesions were compared, there was not a significant difference. ADC values for both acute ischemic and acute demyelinating lesions were lower than chronic lesions: $ADC_{acute\ stroke\ and\ demyelinating\ plaque} < ADC_{chronic\ stroke\ and\ demyelinating\ plaque}$ ($p < 0.01$). Calculated T2R values did not reveal significant differences between $T2R_{acute\ stroke\ and\ demyelinating\ plaque}$ versus $T2R_{chronic\ stroke\ and\ demyelinating\ plaque}$. For both ischemic and demyelinating lesions, T2R values did not show change in relation to time as the differences between acute versus chronic stroke; and the difference between acute versus chronic plaques were not significant. However, independent of time, $T2R_{all\ demyelinating\ plaques} < T2R_{all\ ischemic\ lesions}$ was significant ($p < 0.02$).

D. Discussion

The intracellular water content was different between acute ischemic and demyelinating lesions. As infarcted tissue is cleared by the body's macrophages, CSF is left behind leading to a lower ADC. This low ADC is not indicative of increased intracellular water content, but rather—a lack of neuronal content in the area of interest.

The current data in the literature concerning ADC values of enhancing and nonenhancing lesions are contradictory. Some authors have reported higher ADC in nonenhancing compared to enhancing lesions, whereas studies of larger numbers of patients and lesions have reported no significant difference between the two types of lesions (Droogan *et al.*, 1999; Filippi *et al.*, 2000; Horsfield *et al.*, 1996; Larsson *et al.*, 1992; Rosso *et al.*, 2006; Roychowdhury *et al.*, 2000; Tievsky *et al.*, 1999). Our study showed no significant difference in ADC between enhancing and nonenhancing lesions. Discrepancies among prior studies confirms the high variability of tissue damage in newly enhancing MS lesions and may also reflect different lesion ages (Dousset *et al.*, 1998; Filippi *et al.*, 1998a,b, 2003). The lack of significant changes in ADC values between acute and chronic demyelinating lesions, as opposed to the significant increase in ADC values from acute to chronic ischemia, may reflect the different mechanisms through which these lesions evolve.

ADC values for our acute ischemic lesions did not correspond to ADC values for hyperacute ischemia reported in prior studies. ADC values less than 0.5×10^{-3} mm/s were observed within the first 2 h of symptom onset (Lansberg *et al.*, 2001). However, our slightly higher mean up to 48 h from symptom onset.

Compared to normal appearing white matter, the mean T2R value was elevated in both ischemic and demyelinating lesions. However, the T2R was, in general, significantly lower for demyelination versus ischemic lesions indicative of lower intrinsic water content of demyelinating plaques versus stroke lesions. Furthermore, demyelination leads to decreased intramyelin space associated with a lower T2R. Another possible explanation for a lower T2R in MS plaques may be a decrease in lesion heme-related iron concentration. Further biochemical and MRI studies are needed to assess how iron concentrations differ in ischemic lesions and demyelinating lesions; and how iron concentrations of lesions may evolve over time.

T2R in combination with ADC is a useful tool in the differentiation of acute MS and stroke lesions. The lower ADC of acute stroke versus acute demyelinating lesions may be indicative of a greater cytotoxic edema in the acute infarct. The higher T2R of acute stroke versus acute demyelinating lesions may also reflect higher interstitial water in the ischemic area. Despite the theoretical mechanisms proposed, calculated neuroimaging indices of ADC and T2R may prove to be useful in differentiating ischemic from demyelinating disease (Freedman *et al.*, 2005).

III. ADC, T2R, and MRS in a Case of Posttraumatic Bilateral Intranuclear Ophthalmoplegia

A. Introduction

INO is characterized by ipsilateral medial rectus weakness in combination with dissociated nystagmus of the abducting eye on lateral gaze due to dysfunction of the medial longitudinal fasciculus (MLF). Bilateral INO is most frequent in MS (Gass and Hennerici, 1997). However, bilateral INO can be associated with minor head trauma (Muthukumar *et al.*, 2001). Two different mechanisms for localized injury to the MLF in head injury have been proposed: (1) shearing forces within the brain stem may compress or tear the perforating arteries causing localized ischemic damage to the MLF; or (2) shear forces may damage the white matter within the MLF directly without associated ischemic injury (Muthukumar *et al.*, 2001). The differentiation of infarction from shear force axonal injury can be difficult. Modern neuroimaging techniques such as MRS, calculated ADC, T2R, magnetic resonance angiography (MRA), and interval volumetric measurements have not been applied previously to the evaluation of posttraumatic bilateral INO mechanism. We describe a patient who developed sudden bilateral INO within 2 weeks of minor closed head trauma. Assessment included serial MRS, ADC, T2R, interval volumetrics, and an MRA to explore the nature of posttraumatic bilateral INO in this patient.

B. Case Report

1. *History and Neurological Examination*

A 19-year-old right-handed Caucasian male who sustained a closed head injury to the vertex and occipital regions while swimming. The patient lost consciousness for less than 1 min and was immediately rescued from the water. He was alert and verbally responsive. His neurological examination in the ER was normal. Computerized tomography (CT) of the head and sinuses without contrast did not reveal any pathology. X-rays of the head and cervical spine did not reveal any fractures. Within 2 weeks of the head trauma, the patient noted horizontal diplopia on awakening without any visual loss.

On reevaluation in the ER, he was alert with no cognitive impairment. Subsequent neurological evaluation revealed equally reactive pupils with full visual fields. There was no ptosis. He had bilateral adduction weakness on horizontal gaze with dissociated greater amplitude nystagmus in the abducting eye. The remainder of his neurological examination was unremarkable.

On admission to the hospital he underwent serial neuroimaging, lumbar puncture, and serologic tests. Over 2 weeks of serial examination including that following discharge, there was progressive improvement in bilateral INO (Fig. 3). Repeat MRIs at 20, 40, and 80 days posttrauma were performed in order to assess for any interval change (Figs. 4–6). The patient received treatment with 1-g IV solu-medrol for 3 days starting at day 45 and was continued on a tapering schedule of oral prednisone.

2. *Laboratory Results*

Complete blood counts (CBC), serum electrolytes, liver function tests, erythrocyte sedimentation rate, and C-reactive protein were all normal. Serum testing for Human immunodeficiency virus (HIV), Lyme titres, and venereal disease research laboratory (VDRL) were all negative. The CSF protein was 43 mg/100 ml with a glucose of 60 mg/100 ml. There were 8 red blood cells, and 0 white cells and the CSF myelin basic protein (MBP) was normal. The CSF analysis failed to show oligoclonal bands or increased IgG on electrophoresis. The CSF also was negative for Herpes simplex virus (HSV)-1 and -2, West Nile encephalitis, and St. Louis encephalitis.

Repeat CBC and serum electrolytes were normal. The repeat CSF protein was 42 mg/100 ml, glucose was 70 mg/100 ml, and there were 20 red blood cells. There were 0 white blood cells, no oligoclonal bands, and increase in MBP. CSF cytology for malignant cells, for both specimens, was negative.

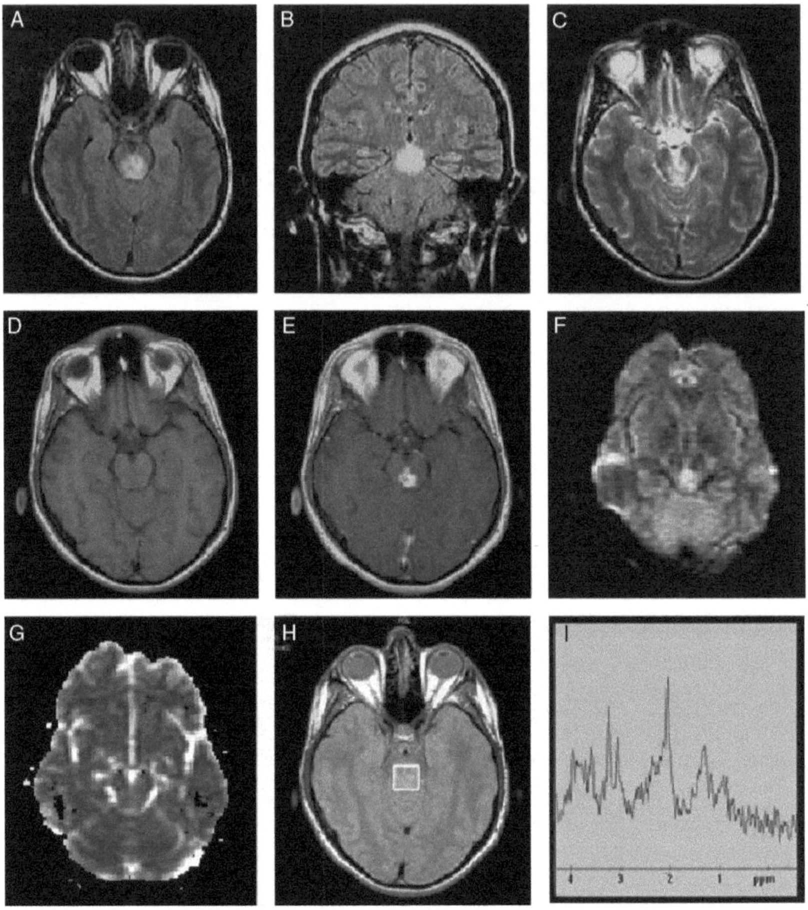

FIG. 3. MRI 14 days posttrauma. (A) Axial fluid attenuated inversion recovery image. (B) Coronal fluid attenuated inversion recovery image. (C) Axial T2-weighted image. (D) Axial T1-weighted image without contrast. (E) Axial T1-weighted image with contrast. (F) Diffusion-weighted image. (G) Apparent diffusion coefficient image. (H) Voxel placement for magnetic resonance spectroscopy. (I) Magnetic resonance spectroscopy.

C. Method

1. *MRI Protocol*

Using a 1.5-T scanner (GE, Milwaukee), brain and cervical spinal cord MR images were obtained. The following axial pulse sequences were applied: T1-weighted [600 ms/10.6 ms/1 (repetition time/echo time/excitations)],

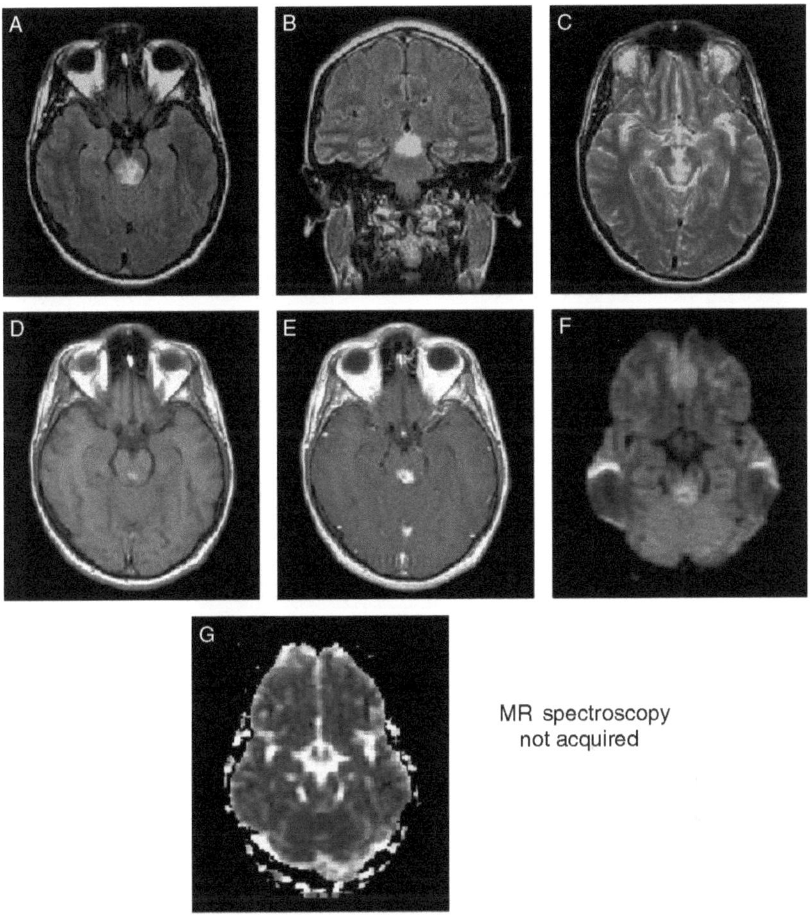

FIG. 4. MRI 20 days posttrauma. (A) Axial fluid attenuated inversion recovery image. (B) Coronal fluid attenuated inversion recovery image. (C) Axial T2-weighted image. (D) Axial T1-weighted image without contrast. (E) Axial T1-weighted image with contrast. (F) Diffusion-weighted image. (G) Apparent diffusion coefficient image.

T2-weighted (2000 ms/121 ms), PD (2000 ms/15 ms), FLAIR (8000 ms/ 143 ms/1), DWI (2000 ms/121 ms), and ADC mapping. Slice thickness was 5 mm, with an interslice gap of 0.5 mm, a matrix size of 256×224 pixels, and a field of view of 24×24 cm^2. MRI protocols with and without Gd contrast were identical between baseline and the three follow-ups. A time-of-flight extra- and intracranial MRA of the head and neck was also performed with the first MRI. The timing of the four MRIs is shown in Table II.

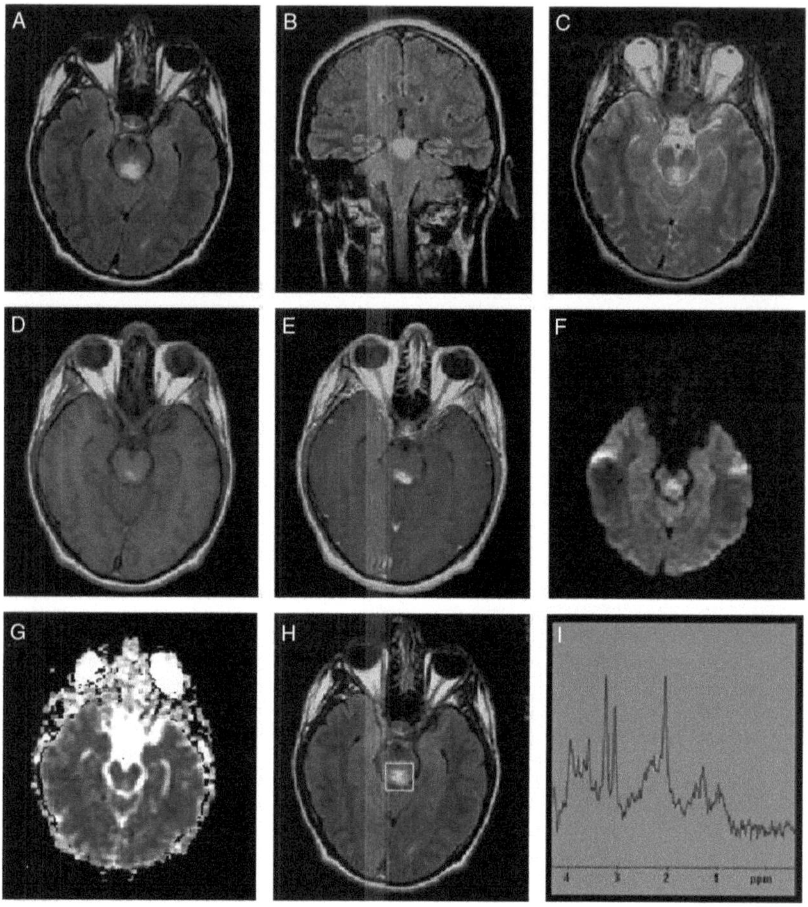

FIG. 5. MRI 40 days posttrauma. (A) Axial fluid attenuated inversion recovery image. (B) Coronal fluid attenuated inversion recovery image. (C) Axial T2-weighted image. (D) Axial T1-weighted image without contrast. (E) Axial T1-weighted image with contrast. (F) Diffusion-weighted image. (G) Apparent diffusion coefficient image. (H) Voxel placement for magnetic resonance spectroscopy. (I) Magnetic resonance spectroscopy.

2. *Lesion Volume Analysis*

We calculated the volume of the lesion by summing the area of each slice in which the lesion was present multiplied by the interslice distance. The change in volume was calculated by subtracting the follow-up volume from the baseline volume (Fig. 7).

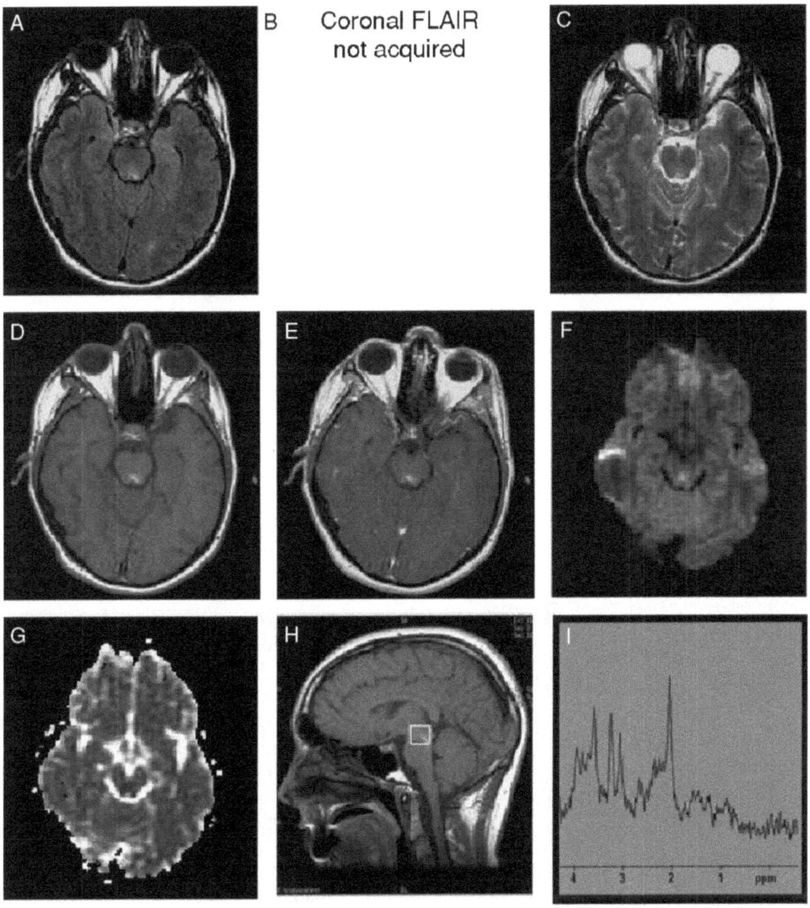

FIG. 6. MRI 80 days posttrauma. (A) Axial fluid attenuated inversion recovery image. (B) Coronal fluid attenuated inversion recovery image not acquired. (C) Axial T2-weighted image. (D) Axial T1-weighted image without contrast. (E) Axial T1-weighted image with contrast. (F) Diffusion-weighted image. (G) Apparent diffusion coefficient image. (H) Voxel placement for magnetic resonance spectroscopy. (I) Magnetic resonance spectroscopy.

3. *Single Voxel Point Resolved Spectroscopy*

Single voxel point resolved spectroscopy (PRESS) sequences with 1500 ms/ 35 ms (TR/TE) with a single voxel size of $2 \times 2 \times 2$ cm^3 were performed on three out of the four MRI studies. Metabolite peaks used in the evaluation of the lesion were as follows: NAA at 2.02 parts per million (ppm), mI at 3.60 ppm, Cho at 3.22 ppm, Cr at 3.01 ppm, glutamate- and glutamine-containing compounds (Glx) from 2.20 to 2.36 ppm, and lactate from 1.24 to 1.59 ppm. Metabolite

TABLE II
Summary of Results

Data	MRI No. 1	MRI No. 2	MRI No. 3	MRI No. 4	Control No. 1, age: 23	Control No. 2, age: 32
Number of days posttrauma	14	20	40	80	N/A	N/A
Number of days postadmission to LSU	0	6	26	66	N/A	N/A
Figure legend	3	4	5	6	8	
Volumetrics	0	0.17	0.64	0.23	N/A	N/A
• Volumetrics on T1 sin contrast in cm^3	N/A	N/A	0.47 (+278%)	0.06 (+35%)		
• Change of volume from baseline in cm^3 (%)						
• Volumetrics on T1 with contrast	1.18	1.09	0.66	0.40	N/A	N/A
in cm^3	N/A	0.09 (−7.6%)	0.52 (−44%)	0.78 (−66%)		
• Change of volume from baseline in cm^3 (%)						
• Volumetrics on FLAIR in	3.46	3.95	1.92	0.19	N/A	N/A
cm^3	N/A	0.49 (+14%)	1.54 (−44%)	3.27 (−95%)		
• Change of volume from baseline in cm^3 (%)						
• Volumetrics on T2 in cm^3	1.24	3.41	1.46	0	N/A	N/A
• Change of volume from baseline in cm^3 (%)	N/A	2.17 (+175%)	0.22 (+17%)	1.24 (−100%)		
ADC	1.21	1.05	1.13	1.13	0.98	0.97
• ADC-enhancing lesion ($\times 10^{-3}$ mm^2/s), ADC left MLF ($\times 10^{-3}$ mm^2/s) for controls						

• $ADC_{nonenhancing\ area}$ ($\times 10^{-3}$ mm^2/s), ADC right MLF ($\times 10^{-3}$ mm^2/s) for controls	1.17	0.93	0.87	0.92	0.96	0.96
• Relative ADC (rADC): $ADC_{enhancing\ lesion}$/ $ADC_{nonenhancing\ area}$, relative ADC (rADC): $ADC_{left\ MLF}$/ $ADC_{right\ MLF}$ for controls	1.03	1.12	1.29	1.22	1.02	1.01
T2R	177	162	159	90	87	93
• T2R-enhancing lesion (ms), T2R left MLF (ms) for controls						
• $T2R_{nonenhancing\ area}$ (ms), T2R right MLF (ms) for controls	98	87	92	86	92	90
• Relative T2R (rT2R): $T2R_{enhancing\ lesion}$/ $T2R_{nonenhancing\ area}$, Relative T2R (rT2R): $T2R_{left\ MLF}$/ $T2R_{right\ MLF}$ for controls	1.80	1.86	1.72	1.04	0.94	1.03
MRS	1.55	N/A	1.20	1.63	1.68	1.90
• NAA/Cr						
• Cho/Cr	1.30	N/A	1.25	1.18	1.49	1.69
• NAA/Cho	1.18	N/A	0.95	1.38	1.13	1.13
• Lactate/Cr	2.12	N/A	1.41	1.10	1.17	1.22
• mI/Cr	0.71	N/A	0.63	1.12	0.69	0.75
• Glx/Cr	1.13	N/A	0.95	1.03	1.18	1.01
Clinical exam	• Bilateral INO • Gait instability	• No change	• Gait instability resolved	• Bilateral INO resolved	N/A	N/A

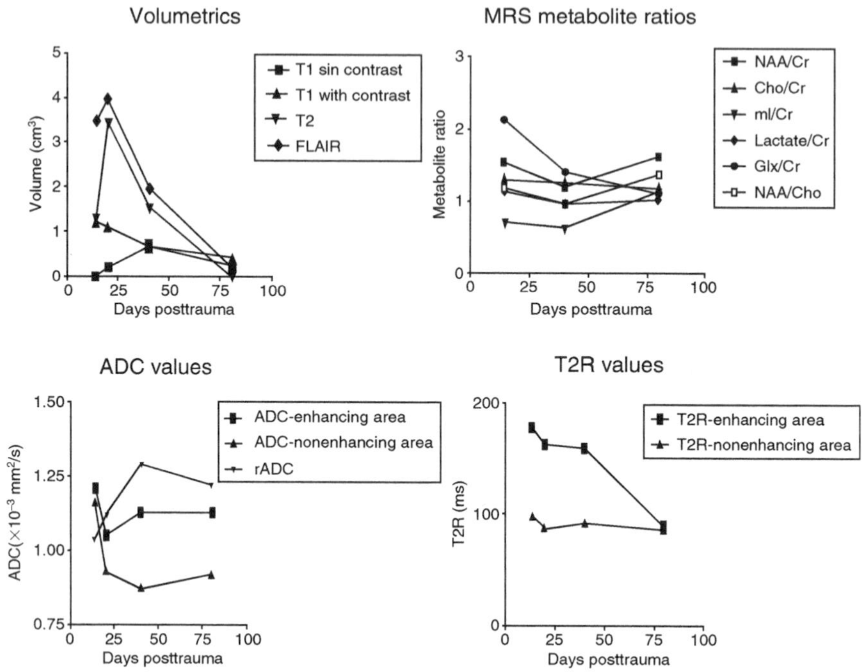

FIG. 7. Graphs of lesion volume analysis, magnetic resonance spectroscopy ratios, apparent diffusion coefficients, and T2 relaxation times.

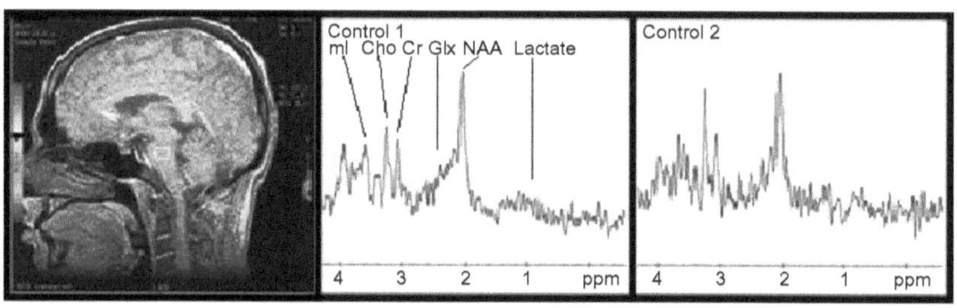

FIG. 8. Magnetic resonance spectroscopy voxel placement and spectra of two control subjects. Point-resolved spectroscopy (PRESS) with a short echo time (TE) of 36 ms allows for better visualization of myoinositol (mI) and glutamate (Glx). Other metabolites labeled include choline (Cho), creatine (Cr), *N*-acetylaspartate (NAA), and lactate.

values were calculated automatically from the area under each metabolite peak using the GE Spectra Analysis program (SAGE). Peak integral values were normalized to the internal Cr peak. Metabolite ratios of NAA/Cr, Cho/Cr, NAA/Cho, Lactate/Cr, Glx/Cr, and mI/Cr were calculated.

Pontomesencephalic junction MRS was performed on two healthy control subjects aged 23 and 32 for comparison (Fig. 8). The MRS protocol, voxel size, and voxel location were identical for our patient and the two controls.

4. *Calculated ADC Values*

The ADC was measured by manually placing ROIs at the anatomic center of the enhancing area on contrast-enhanced T1-weighted MR images. A second ROI was chosen for the nonenhancing area within the pontomesencephalic junction in order to calculate relative ADC values comparing cases of enhancement and nonenhancement within the lesion. Uniform oval ROIs were placed in areas large enough to accommodate a 8.0-mm^2 ROI without apparent volume averaging of surrounding midbrain. Computation of the ADC involved the diffusion coefficient on a region by region basis using Eq (3). ADC values were also calculated in two control subjects.

5. *Calculated T2R Values*

Measurements of T2 signal intensity were obtained by using an operator-defined ROI at the anatomic center of the lesion. The T2R values estimate from dual echo fast spin images is defined by Eq. (6). T2R values were also calculated in the two control subjects.

D. Results

1. *T1-Weighted Images With and Without Gd, T2, PD, DWI, and FLAIR*

T1-weighted images without contrast were unremarkable on the first MRI. On the second MRI, and thereafter, there was T1 hyperintensity. T2, PD, and DWI revealed a focal area of increased signal intensity ventral to the periaqueductal grey matter involving the MLF region bilaterally near the pontomesencephalic junction. This area of hyperintensity enhanced with Gd on T1-weighted images. FLAIR image revealed a larger lesion area than the area outlined on T1-weighted image with contrast. DWI signal was hyperintense in the area of interest on all four MRIs, while ADC mapping was not hypointense at the pontomesencephalic junction. Therefore, the DWI hyperintensity represents "T2 shine through" given the lack of ADC darkness. Cervical spine MRI was normal.

2. *Lesion Volume Analysis*

Volumetric analysis along with other calculated neuroimaging indices are shown in Table II. Figure 1 depicts the volume change over time for the different imaging sequences. The lesion is largest in FLAIR, the smallest on T1-weighted image without contrast, and the most clearly circumscribed on T1 with contrast. The lesion was not present on T1-weighted image without contrast at the time of the first MRI. The lesion almost completely resolved on FLAIR by the fourth MRI and did completely resolve on T2-weighted images. On axial postcontrast T1-weighted images, the lesion volume decreased by 66% over the course of 66% days.

3. *Single Voxel Point Resolved Spectroscopy*

Recent studies have shown that choline level was increased in the midbrain and pons as compared to the cortex (Jacobs *et al.*, 2001). As a result, MRS results need to be compared to region-specific metabolite ratios in normal tissue. Normal metabolic ratios for the midbrain and pons reported in the literature are shown in Table III. Ratios vary depending on data acquisition methods (chemical shift imaging versus single voxel spectroscopy), TR/TE differences, and voxel location (Cwik *et al.*, 1998; Galanaud *et al.*, 2001; Haley *et al.*, 2000; Mascalchi *et al.*, 1998, 2002a,b; Moreno-Torres *et al.*, 2005; Tedeschi *et al.*, 1997; Vion-Dury and Michotey 2005). As a result, no meta-analysis of the data in Table III can be performed to provide standard brainstem normal metabolite ratios. Given the disparity of the acquisition methods, and TR/TE used in the literature, we performed pontomesencephalic spectroscopies on two control subjects with our MR scanner, protocol settings, and voxel locations.

Comparison of our patients values with both the literature and our own controls show that the NAA/Cr and NAA/Cho ratios are decreased in comparison to defined normal values, while the Cho/Cr ratios are very close to normal values. On the second MRS, the NAA/Cr, Cho/Cr, and NAA/Cho continue to decline. The final MRS showed an increase in the NAA/Cr and a decrease in the Cho/Cr. Lactate was present on all three spectroscopy exams and declined with time. mI/Cr for our patient was in the range of the two control subjects on the first two spectroscopies. The mI/CR on the third spectroscopy showed an increase over control values. Glx/Cr decreased from the first to the third MRS.

4. *Calculated ADC Values*

The ADC value on the first MRI was increased over our normal control values and increased over the next two MRIs. Given the extent of lesion enhancement, there was not an ADC value typical of an acute demyelinating plaque.

Table IV displays ADC values of subacute ischemic lesions described in the literature (Copen *et al.*, 2001; Fiebach *et al.*, 2002; Lansberg *et al.*, 2001; Schlaug *et al.*, 1997). Comparing our patients, ADC and rADC to the values listed in Table IV show our patient's values to be higher. ADC values of ischemic lesions have been shown to return to baseline after 1 week—a process known as pseudo-normalization. After this time, ADC and rADC will continue to rise as necrotic debris is phagocytosed leaving CSF in its place. Even though our

TABLE III

NORMAL MRS METABOLITE RATIOS IN THE MIDBRAIN AND PONS FROM THE LITERATURE

Authors, year	Number of patients studied	Location	Data acquisition	MR Magnet	TR (ms)	TE (ms)	NAA/Cr [SD]	Cho/Cr [SD]	NAA/Cho [SD]
Tedeschi *et al.*, 1997	11	Pontomesencephalic junction	Chemical shift imaging	GE 1.5 T	2200	272	2.68 [0.24]	1.43 [0.25]	1.87 [0.30]
Cwik *et al.*, 1998	Variable	Pons	Single voxel, PRESS	Magnex 3.0 T	2000	120	1.95 [0.14], $n = 17$	1.58 [0.28], $n = 10$	1.25 [0.17], $n = 10$
Mascalchi *et al.*, 1998	10	Pons	Single voxel, STEAM	GE 1.5 T	2020	30	1.76 [0.08]	1.61 [0.06]	N/A
Haley *et al.*, 2000	17	Pontomesencephalic junction	Single voxel, SE	Philips 1.5 T	1800	272	6.65 [0.68]	3.96 [0.58]	N/A
Galanaud *et al.*, 2001	37	Midbrain	Chemical shift imaging	Siemens 1.5 T	1500	136	2.11 [0.34]	1.29 [0.19]	1.65 [0.34]
Galanaud *et al.*, 2001	37	Pons	Chemical shift imaging	Siemens 1.5 T	1500	136	2.33 [0.50]	1.47 [0.29]	1.62 [0.41]
Mascalchi *et al.*, 2002a	27	Midbrain	Single voxel, PRESS	Philips 1.5 T	2000	136, 272	2.2 [0.4], 2.8 [0.4]	1.3 [0.3], 1.6 [0.3]	1.6 [0.3], 1.7 [0.4]
Mascalchi *et al.*, 2002a	27	Pons	Single voxel, PRESS	Philips 1.5 T	2000	136, 272	3.0 [0.8], 4.1 [0.6]	1.8 [0.2], 2.6 [0.6]	1.6 [0.6], 1.6 [0.3]
Mascalchi *et al.*, 2002b	20	Pons	Single voxel, STEAM	GE 1.5 T	2000	30	1.75 [0.09]	1.60 [0.07]	N/A
Moreno-Torres *et al.*, 2005	57	Midbrain	Single voxel, PRESS	GE 1.5 T	1600	135	2.03 [0.39]	1.46 [0.26]	N/A
Moreno-Torres *et al.*, 2005	57	Pons	Single voxel, PRESS	GE 1.5 T	160	135	2.36 [0.35]	1.89 [0.28]	N/A
Vion-Dury and Michotey, 2005	20	Brainstem (midbrain, pons, and medulla)	Single voxel, PRESS	Siemens 1.5 T	1500	135	1.97 [0.55]	1.84 [0.92]	2.52 [1.54]

TABLE IV
ADC VALUES FROM THE LITERATURE FOR ISCHEMIC AND DEMYELINATING LESIONS

Authors, year	Lesion type, lesion age	ADC ($\times 10^{-3}$ mm^2/s) [SD]	rADC [SD]
Schlaug *et al.*, 1997	Ischemia, subacute (10 days)	0.984 [0.148]	1.23 [0.07]
Schlaug *et al.*, 1997	Ischemia, chronic (>30 days)	1.472 [0.154]	1.98 [0.22]
Copen *et al.*, 2001	Ischemia, subacute (2 days)	N/A	0.609 [0.221]
Lansberg *et al.*, 2001	Ischemia, subacute—chronic ($\leq$45 days)	0.785 [0.282]	N/A
Lansberg *et al.*, 2001	Ischemia, subacute (5 days)	N/A	0.83 [0.18]
Lansberg *et al.*, 2001	Ischemia, subacute (8 days)	N/A	1.01 [0.17]
Lansberg *et al.*, 2001	Ischemia, chronic ($\geq$14 days)	N/A	1.37 [0.31]
Fiebach *et al.*, 2002	Ischemia, subacute (10 days)	N/A	1.1 [N/A]
Christiansen *et al.*, 1993	Demyelination, acute	1.99 [N/A]	N/A
Christiansen *et al.*, 1993	Demyelination, chronic	1.35 [N/A]	N/A
Larsson *et al.*, 1992	Demyelination, acute	N/A	1.33 [N/A]
Larsson *et al.*, 1992	Demyelination, chronic	N/A	1.14 [N/A]
Horsfield *et al.*, 1996	Demyelination, acute	1.50 [N/A]	N/A
Droogan *et al.*, 1999	Demyelination, acute	1.15 [N/A]	N/A
Droogan *et al.*, 1999	Demyelination, chronic	1.11 [N/A]	N/A
Tievsky *et al.*, 1999	Demyelination, acute	1.59 [0.62]	N/A
Tievsky *et al.*, 1999	Demyelination, subacute	0.99 [0.50]	N/A
Tievsky *et al.*, 1999	Demyelination, chronic	1.08 [0.22]	N/A
Rosso *et al.*, 2006	Demyelination, acute	0.52 [N/A]	N/A

patient's rADC did rise, it did not rise to values normally associated with chronic strokes—that is, in the range of 1.5–3.

Six studies with ADC values reported for acute MS lesions are also listed in Table IV (Christiansen *et al.*, 1993; Droogan *et al.*, 1999; Horsfield *et al.*, 1996; Larsson *et al.*, 1992; Rosso *et al.*, 2006; Tievsky *et al.*, 1999). Acuity for MS plaques is primarily defined by Gd contrast enhancement on T2 imaging. Comparing our patient's ADC values to the literature favors subacute and chronic demyelination rather than subacute ischemia.

In addition, the final rADC calculated for our patient at day 80 more closely resemble rADC reported for demyelinating lesions. However, given the large variation in ADCs reported for acute demyelinating lesions in the literature and the wide range of reported mean rADC for ischemic lesions, ADC alone is not sufficient in differentiating ischemia from demyelination.

5. *Calculated T2R Values*

The T2R value for the lesion was elevated in the first three MRIs compared to controls, but, on the fourth MRI, T2R declined toward normal values approximating those of our two control subjects. T2R values are also compared to T2R values

TABLE V
NORMAL T2R (IN MS) VALUES IN THE LITERATURE

Location	Fletcher *et al.*, 1993	Alfano *et al.*, 1997	Bernarding *et al.*, 2000	Sullivan and Pfefferbaum, 2001
Number of patients	20	50	27	74
T2R white matter: range or mean [SD]	61–100	54–88	92 [8]	N/A
T2R gray matter: range or mean [SD]	61–109	58–119	96 [8]	N/A
T2 pons: range	N/A	N/A	N/A	60–70

from normal white matter and gray matter from the literature (Alfano *et al.*, 1997; Bernarding *et al.*, 2000; Fletcher *et al.*, 1993; Sullivan and Pfefferbaum, 2001). Four studies of T2R in the normal brain using a 1.5-T MR scanner are presented in Table V. The T2R values for the control subjects and in the nonenhancing area of our patient's pontomesencephalic junction correspond to the defined values for normal appearing cortical white matter and gray matter reported in the literature. However, they are higher than the T2R values reported by Sullivan *et al.* for the pons. Perhaps this can be explained by the FLAIR image which shows a larger area of involvement—almost encompassing the entire axial cross-section of the pontomesencephalic junction. As a result, T2R may be diffusely increased—even in areas outside the area of greatest T2R increase, that is, the area demarcated on T1 with contrast.

6. *MRA of the Neck and Head*

The MRA revealed no evidence of vessel occlusion in the arteries imaged. However, this does not exclude potential involvement of smaller perforating arteries that cannot be effectively imaged with MRS.

E. DISCUSSION

Our patient's symptoms developed 2 weeks after head trauma. This delay excludes shear force injury to pontine tegmental vessels resulting in an ischemic infarct as the primary mechanism of MLF injury. Instead, the delayed onset supports potential build up of interstitial fluid perhaps with a delayed component of vasogenic edema. It is quite possible that microvascular compromise was also a contributing factor. A review of the literature reveals one reported case of delayed posttraumatic bilateral INO. The patient reported by Devereaux *et al.* (1979) sustained large subdural hematomas that led to transtentorial herniation and midbrain injury.

Both acute demyelinating and acute ischemic lesions may enhance with the administration of Gd contrast on T1-weighted images. As a result, the presence or absence of enhancement cannot be used to reliably distinguish between possible mechanisms of our patient's delayed posttraumatic INO. In fact, the enhancement simply indicates the disruption of the BBB in the region of the MLF.

Interpretation of T2 hyperintensity as a result of either hemorrhage or ischemia is difficult in the posterior fossa as both injuries may appear hyperintense on DWI. In hemorrhagic lesions on MRI, T2 hyperintensity should correspond with T1 hyperintensity—as both imaging modalities are hyperintense with the presence of metahemogloblin. If an ischemic stroke produces the T2 hyperintense signal, T1-weighted image signal intensity is usually hypointense. In our patient, the MLF on the T1-weighted image without contrast was isointense on the first MRI. However, follow up T1-weighted images without contrast revealed a hyperintense lesion which could reflect hemorrhagic transformation.

The rapid and large reduction in lesion volume over a 66-day interval is less common in ischemia as a possible pathophysiologic mechanism. The extensive area of involvement, by FLAIR hyperintensity, indicates a diffuse pontomesencephalic lesion with the posterior MLF lesion. There was prominent volume reduction prior to corticosteroid treatment. However, corticosteroid therapy was associated with near zero volume of the lesion on all imaging modalities at the time of the fourth brain MRI when there was clinical resolution of bilateral INO. Possible mechanistic etiologies that would be consistent with a lesion that demonstrates such a degree of resolution include demyelination and/or edema. Ischemia would not be expected to resolve on imaging in such a manner.

1. *ADC Values*

Vasogenic edema and demyelination both have ADC values higher than normal brain parenchyma, whereas cytotoxic edema and acute ischemia present with lower ADC values than normal brain parenchyma. The ADC values calculated for the lesion were greater than 1.0 on all four MRIs. In addition, the rADC value more than 1.0 and continued to increase. This supports a mechanism other than acute ischemia or cytotoxic edema as the cause of the MLF injury.

2. *T2R Values*

The increased interstitial water content of the lesion is evident by the increased T2R value. And as the lesion resolves, this interstitial water is removed leading to a lower T2R with each subsequent MRI. Anderson *et al.* showed an inverse relationship between T2Rs and regional cerebral blood volume (CBV) in the putamen and cerebellar vermis of healthy adult patients (Anderson *et al.*, 2005). Even though Anderson's study was not performed in the brain stem, it does allow us to hypothesize about the nature of the T2 increase we initially observed and why it gradually decreased. Perhaps with the edema on the first MRI, the cerebral blood volume in the area of interest was decreased leading to an increase in the T2R. With resolution of the edema, blood flow resumed through the perforating arteries into the tegmental area leading to a gradual decrease of the T2 value.

3. *MRS Metabolite Ratios*

The MRS findings indicate the presence of direction neuronal injury. Measurement of NAA, as a marker for neuronal viability, is useful in this case since decreases in NAA indicate neuronal loss (Tallan *et al.*, 1956). Prior studies have demonstrated that partial reversibility of NAA decreases may be due to transient mitochondrial impairment followed by improvement (Bates *et al.*, 1996; Brooks *et al.* 2001; Davie *et al.*, 1994; De Stefano *et al.*, 1995a,b; Ross and Michaelis, 1994). In our patient, the NAA/Cr increased over the course of the three MR spectra.

Lactate, as the end product of glycolysis, is present at low levels in normal tissue. The concentration of lactate rises when anaerobic metabolism predominates due to a lack of oxygen. With the restoration of blood circulation at the injured area, aerobic metabolism predominates once again leading to a reduction in lactate. Lactate levels at the site of injury in patients with ischemic strokes are elevated for weeks to months (Gideon *et al.*, 1992; Saunders *et al.*, 1995). Our patient's rapid lactate decline is inconsistent with ischemic injury.

The MR visible choline compounds were never elevated over control values, which provides evidence against demyelination as the pathogenic mechanism. In addition, the glial cell marker mI was increased indicating an increase in gliosis at the site of injury. This finding corresponds to previous reports demonstrating posttraumatic midbrain increases in mI (Brooks *et al.*, 2001). In our patient, axonal shearing and demyelination as the primary mechanism of the MLF injury is unlikely to be correct. The MRS data indicated neuronal damage that relatively spared the myelin and did not lead to a release of MRS visible choline compounds. In addition, the lack of a high Cho/Cr excludes the presence of an acute MS lesion.

4. *Interrelation of MRS and T2R Results*

The gradual normalization of T2R values paralleled resolution of edema as indicated by lesion volume on MRI and MRS metabolite ratios, in particular—NAA/Cr and Lactate/Cr. Our patient's MRS metabolite ratios showed improvement that corresponded with improvement of the vasogenic edema aided by the administration of steroid therapy. However, Kamada *et al.* (1994) showed that correction of NAA/Cr values by their corresponding T2 values negates the significance of the decline in NAA/Cr in edematous lesions. Therefore, changes in NAA and its subsequent recovery may reflect either T2 changes in NAA and or its index metabolite Cr. Since the majority of acute brain injury cause edema, one should be cautious in the interpretation of spectroscopic data.

IV. Conclusions

This case analysis demonstrates the use of neuroimaging indices as a means toward understanding pathophysiological mechanism of neurological disorders in order to tailor more effective treatment. Serial analysis of our patient with quantified serial neuroimaging indices provides pathophysiologic evidence that minor head trauma caused vasogenic edema at the midbrain and pons. With expansion of edema, vascular supply was compromised to the bilateral MLF leading to a bilateral INO. Treatment with steroids was guided by the information we obtained from our neuroimaging investigation. Corticosteroids in the setting of vasogenic edema, as opposed to cytotoxic edema, have some proven efficacy (Rabinstein, 2006). This held true for out patient as his neurological function returned to normal after the initiation of steroid therapy. In conclusion, application of advanced neuroimaging techniques can provide the neurologist invaluable information about pathophysiology and potential treatments of various neurological disorders.

References

Alfano, B., Brunetti, A., Covelli, E. M., Quarantelli, M., Panico, M. R., Ciarmiello, A., and Salvatore, M. (1997). Unsupervised, automated segmentation of the normal brain using a multispectral relaxometric magnetic resonance approach. *Magn. Reson. Med.* **37**(1), 84–93.

Anderson, C. M., Kaufman, M. J., Lowen, S. B., Rohan, M., Renshaw, P. F., and Teicher, M. H. (2005). Brain T2 relaxation times correlate with regional cerebral blood volume. *Magma* **18**(1), 3–6.

Bates, T. E., Strangward, M., Keelan, J., Davey, G. P., Munro, P. M., and Clark, J. B. (1996). Inhibition of N-acetylaspartate production: Implications for 1H MRS studies *in vivo*. *Neuroreport* **7** (8), 1397–1400.

Bernarding, J., Braun, J., Hohmann, J., Mansmann, U., Hoehn-Berlage, M., Stapf, C., Wolf, K. J., and Tolxdorff, T. (2000). Histogram-based characterization of healthy and ischemic brain tissues using multiparametric MR imaging including apparent diffusion coefficient maps and relaxometry. *Magn. Reson. Med.* **43**(1), 52–61.

Brooks, W. M., Friedman, S. D., and Gasparovic, C. (2001). Magnetic resonance spectroscopy in traumatic brain injury. *J. Head Trauma Rehabil.* **16**(2), 149–164.

Christiansen, P., Gideon, P., Thomsen, C., Stubgaard, M., Henriksen, O., and Larsson, H. B. (1993). Increased water self-diffusion in chronic plaques and in apparently normal white matter in patients with multiple sclerosis. *Acta Neurol. Scand.* **87**(3), 195–199.

Cittadini, G., Santacroce, E., Giasotto, V., and Rescinito, G. (2004). Focal liver lesions: Characterization with quantitative analysis of T2 relaxation time in TSE sequence with double echo time. *Radiol. Med. (Torino)* **107**(3), 166–173.

Copen, W. A., Schwamm, L. H., Gonzalez, R. G., Wu, O., Harmath, C. B., Schaefer, P. W., Koroshetz, W. J., and Sorensen, A. G. (2001). Ischemic stroke: Effects of etiology and patient age on the time course of the core apparent diffusion coefficient. *Radiology* **221**(1), 27–34.

Cwik, V. A., Hanstock, C. C., Allen, P. S., and Martin, W. R. (1998). Estimation of brainstem neuronal loss in amyotrophic lateral sclerosis with *in vivo* proton magnetic resonance spectroscopy. *Neurology* **50**(1), 72–77.

Davie, C. A., Hawkins, C. P., Barker, G. J., Brennan, A., Tofts, P. S., Miller, D. H., and McDonald, W. I. (1994). Serial proton magnetic resonance spectroscopy in acute multiple sclerosis lesions. *Brain* **117**(Pt. 1), 49–58.

De Stefano, N., Matthews, P. M., and Arnold, D. L. (1995a). Reversible decreases in N-acetylaspartate after acute brain injury. *Magn. Reson. Med.* **34**(5), 721–727.

De Stefano, N., Matthews, P. M., Ford, B., Genge, A., Karpati, G., and Arnold, D. L. (1995b). Short-term dichloroacetate treatment improves indices of cerebral metabolism in patients with mitochondrial disorders. *Neurology* **45**(6), 1193–1198.

Devereaux, M. W., Brust, J. C., and Keane, J. R. (1979). Internuclear ophthalmoplegia caused by subdural hematoma. *Neurology* **29,** 251–255.

Dousset, V., Gayou, A., Brochet, B., and Caille, J. M. (1998). Early structural changes in acute MS lesions assessed by serial magnetization transfer studies. *Neurology* **51**(4), 1150–1155.

Droogan, A. G., Clark, C. A., Werring, D. J., Barker, G. J., McDonald, W. I., and Miller, D. H. (1999). Comparison of multiple sclerosis clinical subgroups using navigated spin echo diffusion-weighted imaging. *Magn. Reson. Imaging* **17**(5), 653–661.

Fiebach, J. B., Jansen, O., Schellinger, P. D., Heiland, S., Hacke, W., and Sartor, K. (2002). Serial analysis of the apparent diffusion coefficient time course in human stroke. *Neuroradiology* **44**(4), 294–298.

Figueroa, R. E., Hoffman, W. H., Momin, Z., Pancholy, A., Passmore, G. G., and Allison, J. (2005). Study of subclinical cerebral edema in diabetic ketoacidosis by magnetic resonance imaging T2 relaxometry and apparent diffusion coefficient maps. *Endocr. Res.* **31**(4), 345–355.

Filippi, M., Rocca, M. A., and Comi, G. (1998a). Magnetization transfer ratios of multiple sclerosis lesions with variable durations of enhancement. *J. Neurol. Sci.* **159**(2), 162–165.

Filippi, M., Rocca, M. A., Rizzo, G., Horsfield, M. A., Rovaris, M., Minicucci, L., Colombo, B., and Comi, G. (1998b). Magnetization transfer ratios in multiple sclerosis lesions enhancing after different doses of gadolinium. *Neurology* **50**(5), 1289–1293.

Filippi, M., Iannucci, G., Cercignani, M., Assunta Rocca, M., Pratesi, A., and Comi, G. (2000). A quantitative study of water diffusion in multiple sclerosis lesions and normal-appearing white matter using echo-planar imaging. *Arch. Neurol.* **57**(7), 1017–1021.

Filippi, M., Rocca, M. A., and Comi, G. (2003). The use of quantitative magnetic-resonance-based techniques to monitor the evolution of multiple sclerosis. *Lancet Neurol.* **2**(6), 337–346.

Fletcher, L. M., Barsotti, J. B., and Hornak, J. P. (1993). A multispectral analysis of brain tissues. *Magn. Reson. Med.* **29**(5), 623–630.

Freedman, M. S., Thompson, E. J., Deisenhammer, F., Giovannoni, G., Grimsley, G., Keir, G., Ohman, S., Racke, M. K., Sharief, M., Sindic, C. J., Sellebjerg, F., and Tourtellotte, W. W. (2005). Recommended standard of cerebrospinal fluid analysis in the diagnosis of multiple sclerosis: A consensus statement. *Arch. Neurol.* **62**(6), 865–870.

Galanaud, D., Le Fur, Y., Nicoli, F., Denis, B., Confort-Gouny, S., Ranjeva, J. P., Viout, P., Pelletier, J., and Cozzone, P. J. (2001). Regional metabolite levels of the normal posterior fossa studied by proton chemical shift imaging. *Magma* **13**(2), 127–133.

Gass, A., and Hennerici, M. G. (1997). Neurological picture. Bilateral internuclear ophthalmoplegia in multiple sclerosis. *J. Neurol. Neurosurg. Psychiatr.* **63**(5), 564.

Gideon, P., Henriksen, O., Sperling, B., Christiansen, P., Olsen, T. S., Jorgensen, H. S., and Arlien-Soborg, P. (1992). Early time course of N-acetylaspartate, creatine and phosphocreatine, and compounds containing choline in the brain after acute stroke. A proton magnetic resonance spectroscopy study. *Stroke* **23**(11), 1566–1572.

Haley, R. W., Marshall, W. W., McDonald, G. G., Daugherty, M. A., Petty, F., and Fleckenstein, J. L. (2000). Brain abnormalities in Gulf War syndrome: Evaluation with 1H MR spectroscopy. *Radiology* **215**(3), 807–817.

Hallgren, B., and Sourander, P. (1958). The effect of age on the non-haemin iron in the human brain. *J. Neurochem.* **3**(1), 41–51.

Horsfield, M. A., Lai, M., Webb, S. L., Barker, G. J., Tofts, P. S., Turner, R., Rudge, P., and Miller, D. H. (1996). Apparent diffusion coefficients in benign and secondary progressive multiple sclerosis by nuclear magnetic resonance. *Magn. Reson. Med.* **36**(3), 393–400.

House, M. J., St Pierre, T. G., Foster, J. K., Martins, R. N., and Clarnette, R. (2006). Quantitative MR imaging R2 relaxometry in elderly participants reporting memory loss. *AJNR Am. J. Neuroradiol.* **27**(2), 430–439.

Jackson, G. D., and Connelly, A. (1999). New NMR measurements in epilepsy. T2 relaxometry and magnetic resonance spectroscopy. *Adv. Neurol.* **79,** 931–937.

Jacobs, M. A., Horska, A., van Zijl, P. C., and Barker, P. B. (2001). Quantitative proton MR spectroscopic imaging of normal human cerebellum and brain stem. *Magn. Reson. Med.* **46**(4), 699–705.

Kamada, K., Houkin, K., Hida, K., Matsuzawa, H., Iwasaki, Y., Abe, H., and Nakada, T. (1994). Localized proton spectroscopy of focal brain pathology in humans: Significant effects of edema on spin-spin relaxation time. *Magn. Reson. Med.* **31**(5), 537–540.

Lansberg, M. G., Thijs, V. N., O'Brien, M. W., Ali, J. O., de Crespigny, A. J., Tong, D. C., Moseley, M. E., and Albers, G. W. (2001). Evolution of apparent diffusion coefficient, diffusion-weighted, and T2-weighted signal intensity of acute stroke. *AJNR Am. J. Neuroradiol.* **22**(4), 637–644.

Larsson, H. B., Thomsen, C., Frederiksen, J., Stubgaard, M., and Henriksen, O. (1992). *In vivo* magnetic resonance diffusion measurement in the brain of patients with multiple sclerosis. *Magn. Reson. Imaging* **10**(1), 7–12.

Lovblad, K. O., Laubach, H. J., Baird, A. E., Curtin, F., Schlaug, G., Edelman, R. R., and Warach, S. (1998). Clinical experience with diffusion-weighted MR in patients with acute stroke. *AJNR Am. J. Neuroradiol.* **19**(6), 1061–1066.

Mascalchi, M., Tosetti, M., Plasmati, R., Bianchi, M. C., Tessa, C., Salvi, F., Frontali, M., Valzania, F., Bartolozzi, C., and Tassinari, C. A. (1998). Proton magnetic resonance spectroscopy in an Italian family with spinocerebellar ataxia type 1. *Ann. Neurol.* **43**(2), 244–252.

Mascalchi, M., Brugnoli, R., Guerrini, L., Belli, G., Nistri, M., Politi, L. S., Gavazzi, C., Lolli, F., Argenti, G., and Villari, N. (2002a). Single-voxel long TE 1H-MR spectroscopy of the normal brainstem and cerebellum. *J. Magn. Reson. Imaging* **16**(5), 532–537.

Mascalchi, M., Cosottini, M., Lolli, F., Salvi, F., Tessa, C., Macucci, M., Tosetti, M., Plasmati, R., Ferlini, A., Tassinari, A., and Villari, N. (2002b). Proton MR spectroscopy of the cerebellum and pons in patients with degenerative ataxia. *Radiology* **223**(2), 371–378.

Mirowitz, S. A., Lee, J. K., and Heiken, J. P. (1990). Cavernous hemangioma of the liver: Assessment of MR tissue specificity with a simplified T2 index. *J. Comput. Assist. Tomogr.* **14**(2), 223–228.

Moreno-Torres, A., Pujol, J., Soriano-Mas, C., Deus, J., Iranzo, A., and Santamaria, J. (2005). Age-related metabolic changes in the upper brainstem tegmentum by MR spectroscopy. *Neurobiol. Aging* **26**(7), 1051–1059.

Muthukumar, N., Veerarajkumar, N., and Madeswaran, K. (2001). Bilateral internuclear ophthalmoplegia following mild head injury. *Childs Nerv. Syst.* **17**(6), 366–369.

Rabinstein, A. A. (2006). Treatment of cerebral edema. *Neurologist* **12**(2), 59–73.

Ross, B., and Michaelis, T. (1994). Clinical applications of magnetic resonance spectroscopy. *Magn. Reson. Q.* **10**(4), 191–247.

Rosso, C., Remy, P., Creange, A., Brugieres, P., Cesaro, P., and Hosseini, H. (2006). Diffusion-weighted MR imaging characteristics of an acute strokelike form of multiple sclerosis. *AJNR Am. J. Neuroradiol.* **27**(5), 1006–1008.

Roychowdhury, S., Maldjian, J. A., and Grossman, R. I. (2000). Multiple sclerosis: Comparison of trace apparent diffusion coefficients with MR enhancement pattern of lesions. *AJNR Am. J. Neuroradiol.* **21**(5), 869–874.

Saunders, D. E., Howe, F. A., van den Boogaart, A., McLean, M. A., Griffiths, J. R., and Brown, M. M. (1995). Continuing ischemic damage after acute middle cerebral artery infarction in humans demonstrated by short-echo proton spectroscopy. *Stroke* **26**(6), 1007–1013.

Schaefer, P. W., Grant, P. E., and Gonzalez, R. G. (2000). Diffusion-weighted MR imaging of the brain. *Radiology* **217**(2), 331–345.

Schlaug, G., Siewert, B., Benfield, A., Edelman, R. R., and Warach, S. (1997). Time course of the apparent diffusion coefficient (ADC) abnormality in human stroke. *Neurology* **49**(1), 113–119.

Sullivan, E. V., and Pfefferbaum, A. (2001). Magnetic resonance relaxometry reveals central pontine abnormalities in clinically asymptomatic alcoholic men. *Alcohol. Clin. Exp. Res.* **25**(8), 1206–1212.

Tallan, H. H., Moore, S., and Stein, W. H. (1956). N-Acetyl-L-aspartic acid in brain. *J. Biol. Chem.* **219**(1), 257–264.

Tedeschi, G., Litvan, I., Bonavita, S., Bertolino, A., Lundbom, N., Patronas, N. J., and Hallett, M. (1997). Proton magnetic resonance spectroscopic imaging in progressive supranuclear palsy, Parkinson's disease and corticobasal degeneration. *Brain* **120**(Pt. 9), 1541–1552.

Tievsky, A. L., Ptak, T., and Farkas, J. (1999). Investigation of apparent diffusion coefficient and diffusion tensor anisotrophy in acute and chronic multiple sclerosis lesions. *AJNR Am. J. Neuroradiol.* **20**(8), 1491–1499.

Vion-Dury, J., and Michotey, P. (2005). Single-voxel proton spectroscopy using a press sequence with a 135 ms echotime: Normal values. *J. Neuroradiol.* **32**(4), 239–246.

HLA-DRB1*1501, -DQB1*0301, -DQB1*0302, -DQB1*0602, AND -DQB1*0603 ALLELES ARE ASSOCIATED WITH MORE SEVERE DISEASE OUTCOME ON MRI IN PATIENTS WITH MULTIPLE SCLEROSIS

Robert Zivadinov,* Laura Uxa,† Alessio Bratina,‡ Antonio Bosco,‡ Bhooma Srinivasaraghavan,* Alireza Minagar,§ Maja Ukmar,¶ Su yen Benedetto,† and Marino Zorzon‡

*Department of Neurology, State University of New York at Buffalo
Buffalo Neuroimaging Analysis Center, The Jacobs Neurological Institute
State University at Buffalo, New York 14203, USA
†Transfusion Center, Trieste, Italy
‡Department of Clinical Medicine and Neurology, University of Trieste
Trieste, Italy
§Department of Neurology, Louisiana State University Health Sciences Center
Shreveport, Louisiana 71103, USA
¶Department of Clinical, Morphological and Technological Sciences
University of Trieste, Trieste, Italy

The most important confirmed genetic factor of susceptibility to multiple sclerosis (MS) has been identified in the HLA class II region. The hypothesis that several genes, including HLA class II, may influence the prognosis of patients with MS has been proposed. In a recent study, using low intermediate resolution typing, we found that some HLA alleles may predict disease severity as assessed by magnetic resonance imaging (MRI) measures. The aim of this study was to examine the relationship between high-resolution typing of HLA alleles and disease severity as measured by brain MRI quantitative markers of demyelinating and destructive pathology in patients with MS. In 41 MS patients (27 relapsing-remitting, 7 secondary progressive,

INTERNATIONAL REVIEW OF
NEUROBIOLOGY, VOL. 79
DOI: 10.1016/S0074-7742(07)79023-2

0074-7742/07 $35.00

and 7 primary progressive), we performed high-resolution typing of alleles HLA-DRB1*04, -DQB1*03, -DRB1*15, -DQB1*06, and of haplotypes -DRB1*04-DQB1*03 and -DRB1*15-DQB1*06. These alleles and haplotypes were associated with higher susceptibility to MS in a recently published case-control study conducted in the Friuli–Venezia–Giulia region, Italy. Of 41 included patients, 13 were men and 28 were women. Mean age was 43.3 (SD 11.4) years, mean disease duration 10.3 (SD 7.8) years, and mean EDSS 2.3. DNA extraction and genomic typing were obtained with the sequence-specific primers method using primer pairs that amplified the HLA alleles. All patients underwent a 1.5-T MRI examination of the brain. Disease severity was assessed by clinical measures [Expanded Disability Status Scale (EDSS)] and MRI measures. T2- and T1-lesion volumes (LVs) and brain atrophy measures [fractions of brain parenchyma (BPF), gray matter (GMF), and white matter (WMF)] were calculated. We used general linear model analysis (GML), controlled for age, disease duration, and treatment status, to compare the MRI measures according to allele and haplotype status. The following significant results were found: HLA-DRB1*1501 positive patients had significantly lower GMF (0.493 vs 0.526, $p < 0.001$), lower BPF (0.784 vs 0.815, $p = 0.018$), and higher T1-LV (2.8 vs 0.7 ml, $p = 0.036$); -DQB1*0301 positive patients had significantly higher T2-LV (34.1 vs 0.7 ml, $p = 0.041$), and showed a trend for lower BPF (0.790 vs 0.846, $p = 0.064$); -DQB1*0302 positive patients had significantly lower T1-LV (2.4 vs 0.9 ml, $p = 0.016$); and -DQB1*0602 positive patients had significantly lower GMF (0.492 vs 0.521, $p = 0.007$) and lower BPF (0.781 vs 0.811, $p = 0.023$). No differences were found in the indices of MRI disease severity according to HLA haplotype associations. Both in correlation and in regression analyses, we observed significant associations between HLA-DRB1*1501 and lower GMF and BPF and higher T1-LV, between -DQB1*0301 and higher T2-LV and disease duration, between -DQB1*0302 and lower GMF and higher T1- and T2-LV, between -DQB1*0602 and lower GMF and BPF, and between -DQB1*0603 and higher T1-LV and EDSS. High-resolution HLA genotyping analysis revealed a robust relationship between alleles HLA-DRB1*1501, -DQB1*0301, -DQB1*0302, -DQB1*0602, and -DQB1*0603, and more severe damage on inflammatory and neurodegenerative MRI measures.

I. Introduction

Susceptibility to multiple sclerosis (MS) is determined by several different genes. The most important confirmed genetic factor of susceptibility to MS has been identified in the human leukocyte antigen (HLA) class II region on the short arm of chromosome 6. Evidence is mounting that the combination of DRB1*1501-DQA1*0102-DQB1*0602 (DR15) in northern Europeans (Haines *et al.*, 1996, 1998) and DRB1*0301-DQA1*0501-DQB1*0301 (DR3) in southern Europeans (Marrosu *et al.*, 2001; Wekerle and Hohlfeld, 2003) increases the risk of MS several times over. The hypothesis that several other genes, including HLA class II, may influence the prognosis of patients with MS has also been proposed (Dyment *et al.*, 2004; Herrera and Ebers, 2003; Kantarci *et al.*, 2002; Wekerle and Hohlfeld, 2003). The influence of these genes on clinical outcome in MS patients is not completely clear and the proposed associations need further investigation. In a published review study on the role of disease-modifying genes in MS, the authors focused on four genes (HLA class II, apoE, IL-1ra, and IL-1β) of probable importance (Kantarci *et al.*, 2002). Specifically, interleukin-1 cluster genes (Crusius *et al.*, 1995), apolipoproteinE (Chapman *et al.*, 2001; Enzinger *et al.*, 2004; Fazekas *et al.*, 2000, 2001), and the MHC and T-cell receptor genes (Schreiber *et al.*, 2002; Sellebjerg *et al.*, 2000) may influence the course of the disease.

The possible relationship between HLA status and magnetic resonance imaging (MRI) findings at disease onset has been extensively examined (Frederiksen *et al.*, 1997; Hauser *et al.*, 2000; Kelly *et al.*, 1993; Nilsson *et al.*, 2005; Soderstrom *et al.*, 1994, 1998). The results from these studies are consistent and suggest that HLA status at the first clinical event is independent of the presence and/or number of brain lesions on MRI. Moreover, HLA-DR2 and -DQ alleles are involved in susceptibility to future demyelinating lesion formation and the subsequent development of clinically definite MS in MRI positive patients.

The association of HLA status with brain quantitative MRI findings has not been extensively investigated. Some studies explored the relationship between HLA status and the presence/number of inflammatory lesions on MRI (Fainardi *et al.*, 2002, 2006, 2003; Luomala *et al.*, 2001; Yamasaki *et al.*, 1999), whereas others examined the correlation with quantitative measures of lesion volumes (LVs) and brain atrophy (Zivadinov *et al.*, 2003). The patients recruited in these studies had variable clinical course, sample size, and disease duration and it is not surprising that the results differ among them.

In a study using low intermediate resolution typing, our group found that some HLA alleles may predict disease severity as assessed by MRI measures (Zivadinov *et al.*, 2003). We studied 100 patients with MS and 122 age, sex-, ethnic-, and residence-matched controls. T2-, T1-LVs, brain atrophy, and disability status were not significantly different according to the carrier status of these HLA alleles, although a trend toward a more severe disease progression was observed in DRB1*15 and DQB1*02 positive patients. A higher T2-LV was associated with the presence of DRB1*04 and B7 alleles. T1-LV was associated with B7 and DRB1*12 alleles, whereas lower brain volume was predicted only by the presence of the DRB1*12 allele. We concluded that some HLA alleles might predict the destructive pathological processes visible on MRI.

In this study, we aimed to examine the relationship between high-resolution typing of HLA alleles and disease severity as measured by brain MRI quantitative markers of demyelinating and destructive pathology in patients with MS.

II. Methods

Forty-one patients with clinically definite MS (McDonald *et al.*, 2001), who have been followed at the Department of Clinical Medicine and Neurology of the University of Trieste, Trieste, Friuli–Venezia–Giulia region, Italy, were included in the study. Exclusion criteria consisted of concurrent exacerbation and steroid treatment in the 3 months preceding study entry, immunomodulating and immunosuppressive treatment in the 6 months preceding study entry, and preexisting concomitant disorders and conditions which can modify radiological outcomes in MS patients. Among 85 consecutive patients screened for the study, 41 patients satisfied the entry criteria [27 with relapsing-remitting (RR), 7 with secondary progressive (SP), and 7 with primary progressive (PP) MS]. Written informed consent was obtained from all the patients and the study was approved by the local Ethical Committee.

A. MRI Outcome Measures

Brain MRI was performed on a Philips Gyroscan S15 ACS II 1.5 T unit, Best, The Netherlands. MRI sequences included axial images of the brain with 3-mm slice thickness using PD/T2-weighted SE sequence (TR 2709/TE 20–80) and unenhanced T1-weighted SE sequence (TR 600/TE 27). A matrix of 179 × 256 pixels was used for a total of 48 sections. Field of view was 220 mm. Patients were positioned in the magnet according to European Community guidelines (Miller *et al.*, 1991).

Image analysis was performed in the Postimaging Analysis Laboratory at the Department of Clinical Medicine and Neurology, University of Trieste, Italy (T2- and T1-LVs analyses) and at the Buffalo Neuroimaging Analysis Center, The Jacobs Neurological Institute, Department of Neurology, State University at Buffalo, New York (brain atrophy calculations). MRI investigators were blinded to patients' clinical characteristics and status.

T2- and T1-LVs were calculated using a highly reproducible semiautomated local thresholding technique for lesion segmentation. Extensive details of this method have been described previously (Zivadinov *et al.*, 2001a,b).

We used statistical parametric mapping (SPM2; Wellcome Department of Cognitive Neurology, Institute of Neurology, Queen Square, London, United Kingdom) (Ashburner and Friston, 2000) to segment voxels into gray matter (GM), white matter (WM), and cerebrospinal fluid (CSF) compartments. The MRI scans were aligned by placing the anterior commissure (AC) at the SPM2 origin and centering the *x*-axis of origin to pass through the posterior commissure (PC) in the mid-sagittal plane. This procedure, known as AC–PC alignment, ensured that brain slices in all subjects were perpendicular to a uniformly and anatomically defined axis of the brain. Brain scans were aligned by matching the interhemispheric fissure to the origin's *x*- and *y*-axes in the anterior–posterior and superior–inferior planes. The scans were realigned, stripped of extracranial tissue (e.g., soft tissue, skull, orbit), and segmented into brain parenchyma and CSF compartments using a maximum level of inhomogeneity correction (Carone *et al.*, 2006). SPM was used to create a measure of normalized brain volume (brain parenchymal fraction, BPF) by dividing the total brain parenchyma (GM + WM) by total intracranial volume (GM + WM + CSF). Similarly, GM fraction (GMF) and WM fraction (WMF) were obtained by dividing each of these variables by total intracranial volume. In a previous study, using the same type of scanning protocol, the scan-rescan reliability (COV) for SPM-based fractions was 0.4% for BPF, 0.3% for GMF, and 0.9% for WMF. Because SPM has been known to misclassify T1 hypointensities in WM as CSF, all BPF and total WMF values were corrected by adding the total volume of T1 hypointensities into the numerator of the BPF and WMF calculation (Sanfilipo *et al.*, 2004). This correction was not performed for total GMF since few T1 hypointensities appear in these regions. T1 hypointensities were traced by expert operators on T1-WI using a semiautomated thresholding-contouring method (Java Image Manipulation 3.0), as previously described (Sanfilipo *et al.*, 2004).

B. HLA Typing

HLA class II typing was performed by using a patient's DNA, extracted from peripheral blood mononuclear cells (Genomix, Talent, Trieste, Italy), amplified by protein chain reaction, and then hybridized with primer-specific oligonucleotide probes by using a commercially available reverse line blot assay (Biopsa, Genovision, Friuli-Venezia-Gulia, Italy). The resolution of typing was at a high level and was performed for alleles HLA-DRB1*04, -DQB1*03, -DRB1*15, and -DQB1*06, and for haplotypes -DRB1*04-DQB1*03 and -DRB1*15-DQB1*06. These alleles and haplotypes were associated with higher susceptibility to MS in a published case-control study conducted in the Friuli–Venezia–Giulia region, Italy (Zivadinov *et al.*, 2003).

C. Statistical Analysis

Statistical analysis was performed using the Statistical Package for Social Sciences (SPSS, Version 14.0, SPSS, Inc., Chicago, Illinois). For comparisons between the groups, parametric and nonparametric statistics were used, as appropriate. Correlation and multiple regression analyses were performed to look for a relationship between the severity of MRI findings and the HLA high-resolution genotype. The effect of possible confounding variables (age at onset, disease duration, years of lifetime disease-modifying treatment, and lifetime total days of steroid use) was taken into account. The minimum significance level for entry and for staying in the equation was 0.05. All p values were based on two-tailed tests.

III. Results

A. Demographic, MRI, and HLA Characteristics

Mean age was 43.1 years (SD 11.4) and mean disease duration 10.3 years (SD 7.8). Mean EDSS was 2.3 (SD 2.9). Mean T2-LV was 15.1 ml (SD 12.1), mean T1-LV was 1.9 ml (SD 3), mean BPF was 0.800 (SD 0.03), mean GMF was 0.510 (0.03), and mean WMF was 0.290 (0.03). Demographic and MRI characteristics according to disease type are shown in Table I.

The frequency of alleles HLA-DRB1*04, -DQB1*03, -DRB1*15, and -DQB1*06 and of haplotypes -DRB1*04-DQB1*03 and -DRB1*15-DQB1*06 is shown in Table II. No differences were found in the frequency of high-resolution HLA alleles and haplotypes when patients were stratified according to disease type.

B. Comparison of MRI Measures According to HLA Status

We used general linear model analysis (GML) to compare the MRI measures according to allele and haplotype status. In these analyses, age, disease duration, and lifetime years of disease-modifying treatment were used as covariates.

TABLE I
Demographic, Clinical, and MRI Characteristics in Patients with Multiple Sclerosis

Variable	MS (41)	RR (27)	SP (7)	PP (7)
Female gender, number (%)	28 (63)	23 (85.2)	3 (42.9)	2 (28.6)
Age (years), mean (SD)	43.3 (11.4)	39.4 (10.2)	49.4 (10.8)	52.4 (9.8)
Disease duration (years), mean (SD)	10.3 (7.8)	10.2 (7.5)	17.4 (9.7)	5.6 (3.7)
EDSS, mean (SD)	2.9 (2.3)	1.9 (1.8)	6.1 (1.7)	3.5 (0.9)
T2-LV (ml), mean (SD)	15.1 (12.1)	14.6 (15.2)	19.1 (14.2)	12.7 (13.7)
T1-LV (ml), mean (SD)	1.9 (3)	2 (3.2)	3.6 (2.7)	2.5 (1)
BPF, mean (SD)	0.800 (0.03)	0.801 (0.04)	0.789 (0.04)	0.795 (0.03)
GMF, mean (SD)	0.510 (0.03)	0.514 (0.03)	0.497 (0.03)	0.499 (0.02)
WMF, mean (SD)	0.290 (0.02)	0.287 (0.03)	0.293 (0.03)	0.305 (0.03)

EDSS, Expanded Disability Status Scale; BPF, brain parenchymal fraction; GMF, gray matter fraction; WMF, white matter fraction; SD, standard deviation.

TABLE II
HIGH-RESOLUTION HLA STATUS IN PATIENTS WITH MULTIPLE SCLEROSIS

Frequency of alleles and haplotypes	Number of patients (%)
DR04	
−01	3 (7.3)
−02	4 (9.8)
−03	1 (2.4)
−04	6 (14.6)
−05	3 (7.3)
−07	1 (2.4)
DR15	
−01	23 (56.1)
−02	2 (4.9)
DQ03	
−01	4 (9.8)
−02	14 (34.1)
DQ06	
−01	3 (7.3)
−02	17 (41.5)
−03	3 (7.3)
−04	2 (4.9)
DRB1*04-DQB1*03	
−0401*0301	2 (4.9)
−0401*0302	1 (2.4)
−0402*0302	3 (7.3)
−0403*0302	1 (2.4)
−0404*0302	4 (9.8)
−0407*0301	2 (4.9)
DRB1*15-DQB1*06	
−1501*0601	1 (2.4)
−1501*0602	13 (31.7)
−1501*0603	2 (4.9)

In order to account for multiple comparisons, we used Bonferroni correction directly in the GML SPSS model. HLA-DRB1*1501 positive patients had significantly lower GMF (0.493 vs 0.526, $p < 0.001$), lower BPF (0.784 vs 0.815, $p = 0.018$), and higher T1-LV (2.8 vs 0.7 ml, $p = 0.036$), when compared to HLA-DRB1*1501 negative patients. HLA-DQB1*0301 positive patients had significantly higher T2-LV when compared to HLA-DQB1*0301 negative patients (34.1 vs 0.7 ml, $p = 0.041$), and showed a trend for lower BPF (0.790 vs 0.846, $p = 0.064$). HLA-DQB1*0302 positive patients had significantly lower T1-LV when compared to HLA-DQB1*0302 negative patients (2.4 vs 0.9 ml, $p = 0.016$), and there was a trend for higher T2-LV in HLA-DQB1*0302 positive patients

(19.2 vs 7.1 ml, $p = 0.055$). HLA-DQB1*0602 positive patients had significantly lower GMF (0.492 vs 0.521, $p = 0.007$) and lower BPF (0.781 vs 0.811, $p = 0.023$) when compared to HLA-DQB1*0602 negative patients. No other differences were found in the indices of MRI disease severity according to HLA carrier status, including different analyses performed with associations of haplotypes HLA-DRB1*04-DQB1*03 and -DRB1*15-DQB1*06.

C. Relationship Between HLA Status and MRI Measures

Spearman correlation analysis showed significant associations between HLA-DRB1*1501 and lower GMF ($r = -0.61$, $p < 0.0001$) and BPF ($r = -0.45$, $p = 0.019$), and higher T1-LV ($r = 0.40$, $p = 0.033$) and disease duration ($r = 0.35$, $p = 0.03$) (Table III). HLA-DQB1*0301 was associated with higher T2-LV ($r = 0.38$, $p = 0.039$), -DQB1*0302 with lower GMF ($r = -0.58$, $p < 0.0001$) and higher T1-LV ($r = 0.45$, $p = 0.014$) and T2-LV ($r = 0.36$, $p = 0.051$), -DQB1*0602 with lower GMF ($r = -0.54$, $p = 0.004$) and BPF ($r = -0,44$, $p = 0.023$), and -DQB1*0603 with higher T1-LV ($r = 0.37$, $p = 0.051$) and EDSS ($r = 0.35$, $p = 0.034$).

Logistic regression analysis was performed in order to investigate the relationship between HLA high-resolution alleles and MRI measures. Age, disease duration, and lifetime years of disease-modifying treatment were entered in the first block, and inflammatory and neurodegenerative MRI measures in the second one. This model was created to investigate which of the MRI measures would independently predict HLA allele status. HLA-DRB1*1501 was predicted independently

TABLE III
Correlation Analysis Between High-Resolution HLA Allele Status and Clinical and MRI Characteristics in Patients with Multiple Sclerosis

	DRB1*1501		DQB1*0301		DQB1*0302	
HLA status	*r*	*p*	*r*	*p*	*r*	*p*
T2-LV	0.15	0.424	**0.38**	**0.039**	**0.36**	**0.051**
T1-LV	**0.40**	**0.033**	0.33	0.085	**0.45**	**0.014**
BPF	**−0.45**	**0.019**	−0.28	0.117	−0.32	0.109
GMF	**−0.61**	**<0.0001**	−0.18	0.652	**−0.58**	**<0.0001**
WMF	−0.15	0.424	−0.09	0.727	−0.22	0.278
EDSS	0.07	0.706	0.07	0.672	0.12	0.483
Disease duration	**0.35**	**0.03**	0.08	0.616	0.11	0.500

LV, lesion volume; BPF, brain parenchymal fraction; GMF, gray matter fraction; WMF, white matter fraction; EDSS, Expanded Disability Status Scale.

only by GMF (adjusted R2 = 0.35, $p = 0.003$). Allele HLA-DQB1*0301 was predicted independently by T2-LV (adjusted R2 = 0.31, $p = 0.008$) and T1-LV (adjusted R2 = 0.16, $p = 0.002$). HLA-DQB1*0302 was associated with T1-LV (adjusted R2 = 0.32, $p = 0.007$), whereas -DQB1*0602 and -DQB1*0603 were not independently related to MRI measures in the regression analysis.

IV. Discussion

This study showed a robust relationship between high-resolution HLA genotyping of alleles -DRB1*1501, -DQB1*0301, -DQB1*0302, -DQB1*0602, and -DQB1*0603, and severity of disease damage on inflammatory and neurodegenerative MRI measures in patients with MS.

Several attempts have been made to establish whether HLA genotypes might determine the prognosis of MS patients who present more severe clinical and MRI outcomes (Dyment *et al.*, 2004; Hauser *et al.*, 2000; Kantarci *et al.*, 2002). HLA-DR2 and, in particular, -DR15 alleles have been associated with a worse prognosis of MS (Hensiek *et al.*, 2002; Soderstrom *et al.*, 1998). Short-term follow-up studies have suggested that genotype HLA-DR2 could predict a higher susceptibility for conversion to clinically definite MS in patients presenting with high signal abnormalities on baseline MRI over a 5-year follow-up (Hauser *et al.*, 2000). This might indicate that HLA genotype status can influence the heterogeneity and severity of MS pathological processes that lead to conversion to clinically definite MS. However, a study in which a cohort of MS patients was followed for more than 30 years failed to find any evidence that DR2 alleles influenced the risk for MS (Nilsson *et al.*, 2005). Dw2 haplotypes also did not influence MS prognosis over the long term (Runmarker and Andersen, 1993). This issue was readdressed in the large datasets of Swedish (Masterman *et al.*, 2000) and UK (Hensiek *et al.*, 2002) MS patients that failed to support the role of HLA-DRB alleles in predicting a worse prognosis of MS.

Nevertheless, genes influencing susceptibility may differ from genes influencing disease severity. More than one gene may interact epistatically in determining disease severity and number of genes, and the degrees of epistasis are not yet fully elucidated (Coraddu *et al.*, 2003; Goedde *et al.*, 2002; Hensiek *et al.*, 2003; Liguori *et al.*, 2003; Sawcer *et al.*, 2002). The role of disease-modifying genes in MS other than HLA class II genes (apoE, IL-1ra, and IL-1β, and so on) has not been determined (Kantarci *et al.*, 2002). In a previous study of 100 MS patients and 122 age, sex-, ethnic-, and residence-matched controls, we found that some HLA genotypes may predict disease severity as assessed by MRI measures (Zivadinov *et al.*, 2003). Low intermediate resolution typing of HLA alleles and haplotypes was used in that study, and it was found that DRB1*15 and DQB1*02 positive patients had more severe MRI and disease outcomes. In particular, alleles HLA-DRB1*04 and -B7 were associated with higher T2-LV, and alleles B7 and DRB1*12 with higher T1-LV, whereas allele -DRB1*12 predicted more severe brain atrophy. Although this study did not establish a robust relationship with MRI markers of disease severity, it did suggest that some HLA genotypes might better predict the destructive pathological processes visible on MRI (Zivadinov *et al.*, 2003).

In the present study, we aimed to expand on our recent findings using high-resolution HLA typing of alleles and haplotypes in a heterogeneous group of MS patients. We focused our work on high-resolution alleles HLA-DRB1*04, -DQB1*03, -DRB1*15, and -DQB1*06 and on haplotypes -DRB1*04-DQB1*03 and -DRB1*15-DQB1*06 that were associated with higher susceptibility and disease severity of MS in a published case-control study conducted in the Friuli–Venezia–Giulia region, Italy (Zivadinov *et al.*, 2003). We confirmed a higher frequency of alleles HLA-DRB1*1501 (56.1%), -DQB1*0602 (41.5%), -DQB1*0302 (34.1%), and -DRB1*0404 (14.6%), and haplotype DRB1*1501 -DQB10602 (31.7%) in this study population.

Newer MRI tools appear to be more sensitive biomarkers for measuring the pathogenetic processes associated with disease activity and progression, including the presence of occult pathology in the GM. Techniques such as measurement of T1-hypointense lesions (black holes), magnetization transfer imaging, diffusion-weighted and tensor imaging, magnetic resonance spectroscopy, and molecular imaging are able to identify the underlying pathological processes within lesions and in normal-appearing brain tissue. These processes range from edema and inflammation to demyelination, axonal loss, and neurodegeneration (Zivadinov, 2006). In the present study, all patients were studied with one or more of these advanced MRI techniques, including measurement of the damage in different tissue atrophy compartments. We used various statistical approaches to explore whether specific alleles or haplotypes were related to more severe MRI and clinical outcomes. Allele HLA-DRB1*1501 showed a particularly strong relation with lower GMF and BPF values and higher T1-LV. This was confirmed by GML and correlation analyses. In addition, multivariate regression analysis showed that the presence of this allele was independently predicted by higher GM atrophy progression. To the best of our knowledge, this is the first study to establish a robust association between a specific HLA status and a destructive MRI measure that reflects neurodegeneration in a patient with MS. It could be hypothesized that microglial activation, defined genetically by a particular HLA genotype, may act as a marker for oligodendrocyte/myelin and axonal pathology in MS and may reflect major differences in the underlying pathogenetic mechanisms over the long term (Lucchinetti *et al.*, 2000, 2001). It has also been reported that a substantial loss of oligodendrocytes accompanied active myelin loss at the plaque margin, with oligodendrocyte recruitment and variable remyelination present at the plaque center (Lucchinetti *et al.*, 1999). The mechanisms responsible for GM atrophy in MS are uncertain, but the most accepted opinion is that the causative mechanism is a combination of demyelination and axonal loss (Zivadinov and Bakshi, 2004). The T1-hypointense lesions represent areas of more severe axonal damage. Therefore, the pathological substrate of black holes seems to represent predominantly axonal damage that leads irreversibly to neuronal dropout. This study shows that allele HLA-DRB1*1501 may increase the susceptibility for early axonal loss and subsequently cause the accumulation of black holes and the progression of brain atrophy in the long term.

This study's results also indicate that alleles HLA-DRB1*0602 and -DQB1*0302 predicted lower GMF and BPF, and higher T1- and T2-LVs. The GML and correlation analyses showed that allele HLA-DRB1*0602 in particular was related to more advanced GM atrophy, and this finding was notably robust. In addition, regression analysis showed that allele HLA-DQB1*0301 predicted 48% of the variance in T2- and T1-LVs. All in all, these findings suggest that other HLA genotypes (HLA-DQB1*0301, -DQB1*0302, and -DRB1*0602) may be related to a more severe MRI outcome.

The relationship with candidate genes and disease severity, as assessed by clinical disability, has been investigated in several studies (Kantarci *et al.*, 2002). In a recent study performed in a Danish population of MS patients, carriers of allele DRB1*15 did not show more severe disease progression by clinical outcome measures. The limited number of patients in our study did not allow us to investigate the relationship with clinical variables. Nevertheless, patients with positive allele HLA-DRB1*1501 showed a modest correlation with disease duration. There was also a modest correlation between HLA-DRB1*0603 and higher disability, as measured by EDSS. These findings do not permit firm conclusions, but suggest that a larger sample size using high-resolution HLA typing is needed to confirm our observations.

Differences in clinical, radiological, and pathological characteristics have led to the suggestion that HLA-status may distinguish different clinical entities in patients with MS. PPMS is characterized by older age at onset, spinal phenotype, more severe disability, less inflammation but prominent axonal pathology, and fewer brain lesions (Thompson *et al.*, 2000). Whereas an increased frequency of alleles HLA-DR15 and -DR3 appeared in RRMS, an association with an increased frequency of -DR4 and a decreased frequency of -DR3 was observed in PPMS (Weinshenker *et al.*, 1998). We performed a separate analysis for RR/SP- and PPMS disease types but did not find significant differences in frequency or relation to MRI and clinical outcomes (data not shown). However, the number of patients was too small to conclude whether a specific HLA allele might be of particular importance in determining more severe MRI and clinical outcomes in PPMS patients.

In conclusion, the study findings suggest that high-resolution HLA genotyping reveals a robust relationship between alleles HLA-DRB1*1501, -DQB1*0301,

-DQB1*0302, -DQB1*0602, and -DQB1*0603, and more severe damage on inflammatory and, particularly, on newer neurodegenerative MRI measures of tissue-compartment-specific destruction (GM atrophy). The size of the sample considered in this study is relatively small and does not permit firm conclusions about genotype correlation in patients with MS, but this research does warrant a larger study to confirm our findings.

Acknowledgments

This study was supported by a grant from the University of Trieste, Italy (DO37S2RI99CG).

References

Ashburner, J., and Friston, K. J. (2000). Voxel-based morphometry—the methods. *Neuroimage* **11,** 805–821.

Carone, D. A., Benedict, R. H., Dwyer, M. G., Cookfair, D. L., Srinivasaraghavan, B., Tjoa, C. W., and Zivadinov, R. (2006). Semi-automatic brain region extraction (SABRE) reveals superior cortical and deep gray matter atrophy in MS. *Neuroimage* **29,** 505–514.

Chapman, J., Vinokurov, S., Achiron, A., Karussis, D. M., Mitosek-Szewczyk, K., Birnbaum, M., Michaelson, D. M., and Korczyn, A. D. (2001). APOE genotype is a major predictor of long-term progression of disability in MS. *Neurology* **56,** 312–316.

Coraddu, F., Lai, M., Mancosu, C., Cocco, E., Sawcer, S., Setakis, E., Compston, A., and Marrosu, M. G. (2003). A genome-wide screen for linkage disequilibrium in Sardinian multiple sclerosis. *J. Neuroimmunol.* **143,** 120–123.

Crusius, J. B., Pena, A. S., Van Oosten, B. W., Bioque, G., Garcia, A., Dijkstra, C. D., and Polman, C. H. (1995). Interleukin-1 receptor antagonist gene polymorphism and multiple sclerosis. *Lancet* **346,** 979–980.

Dyment, D. A., Ebers, G. C., and Sadovnick, A. D. (2004). Genetics of multiple sclerosis. *Lancet Neurol.* **3,** 104–110.

Enzinger, C., Ropele, S., Smith, S., Strasser-Fuchs, S., Poltrum, B., Schmidt, H., Matthews, P. M., and Fazekas, F. (2004). Accelerated evolution of brain atrophy and "black holes" in MS patients with APOE-epsilon 4. *Ann. Neurol.* **55,** 563–569.

Fainardi, E., Granieri, E., Tola, M. R., Melchiorri, L., Vaghi, L., Rizzo, R., Castellazzi, M., Ceruti, S., Paolino, E., and Baricordi, O. R. (2002). Clinical and MRI disease activity in multiple sclerosis are associated with reciprocal fluctuations in serum and cerebrospinal fluid levels of soluble HLA class I molecules. *J. Neuroimmunol.* **133,** 151–159.

Fainardi, E., Rizzo, R., Melchiorri, L., Vaghi, L., Castellazzi, M., Marzola, A., Govoni, V., Paolino, E., Tola, M. R., Granieri, E., and Baricordi, O. R. (2003). Presence of detectable levels of soluble HLA-G molecules in CSF of relapsing-remitting multiple sclerosis: Relationship with CSF soluble HLA-I and IL-10 concentrations and MRI findings. *J. Neuroimmunol.* **142,** 149–158.

Fainardi, E., Rizzo, R., Melchiorri, L., Castellazzi, M., Paolino, E., Tola, M. R., Granieri, E., and Baricordi, O. R. (2006). Intrathecal synthesis of soluble HLA-G and HLA-I molecules are reciprocally associated to clinical and MRI activity in patients with multiple sclerosis. *Mult. Scler.* **12,** 2–12.

Fazekas, F., Strasser-Fuchs, S., Schmidt, H., Enzinger, C., Ropele, S., Lechner, A., Flooh, E., Schmidt, R., and Hartung, H. P. (2000). Apolipoprotein E genotype related differences in brain lesions of multiple sclerosis. *J. Neurol. Neurosurg. Psychiatry* **69,** 25–28.

Fazekas, F., Strasser-Fuchs, S., Kollegger, H., Berger, T., Kristoferitsch, W., Schmidt, H., Enzinger, C., Schiefermeier, M., Schwarz, C., Kornek, B., Reindl, M., Huber, K., *et al.* (2001). Apolipoprotein E epsilon 4 is associated with rapid progression of multiple sclerosis. *Neurology* **57,** 853–857.

Frederiksen, J. L., Madsen, H. O., Ryder, L. P., Larsson, H. B., Morling, N., and Svejgaard, A. (1997). HLA typing in acute optic neuritis. Relation to multiple sclerosis and magnetic resonance imaging findings. *Arch. Neurol.* **54,** 76–80.

Goedde, R., Sawcer, S., Boehringer, S., Miterski, B., Sindern, E., Haupts, M., Schimrigk, S., Compston, A., and Epplen, J. T. (2002). A genome screen for linkage disequilibrium in HLA-DRB1*15-positive Germans with multiple sclerosis based on 4666 microsatellite markers. *Hum. Genet.* **111,** 270–277.

Haines, J. L., Ter-Minassian, M., Bazyk, A., Gusella, J. F., Kim, D. J., Terwedow, H., Pericak-Vance, M. A., Rimmler, J. B., Haynes, C. S., Roses, A. D., Lee, A., Shaner, B., *et al.* (1996). A complete genomic screen for multiple sclerosis underscores a role for the major histocompatability complex. The Multiple Sclerosis Genetics Group. *Nat. Genet.* **13,** 469–471.

Haines, J. L., Terwedow, H. A., Burgess, K., Pericak-Vance, M. A., Rimmler, J. B., Martin, E. R., Oksenberg, J. R., Lincoln, R., Zhang, D. Y., Banatao, D. R., Gatto, N., Goodkin, D. E., *et al.* (1998). Linkage of the MHC to familial multiple sclerosis suggests genetic heterogeneity. The Multiple Sclerosis Genetics Group. *Hum. Mol. Genet.* **7,** 1229–1234.

Hauser, S. L., Oksenberg, J. R., Lincoln, R., Garovoy, J., Beck, R. W., Cole, S. R., Moke, P. S., Kip, K. E., Gal, R. L., and Long, D. T. (2000). Interaction between HLA-DR2 and abnormal brain MRI in optic neuritis and early MS. Optic Neuritis Study Group. *Neurology* **54,** 1859–1861.

Hensiek, A. E., Sawcer, S. J., Feakes, R., Deans, J., Mander, A., Akesson, E., Roxburgh, R., Coraddu, F., Smith, S., and Compston, D. A. (2002). HLA-DR 15 is associated with female sex and younger age at diagnosis in multiple sclerosis. *J. Neurol. Neurosurg. Psychiatry* **72,** 184–187.

Hensiek, A. E., Roxburgh, R., Smilie, B., Coraddu, F., Akesson, E., Holmans, P., Sawcer, S. J., and Compston, D. A. (2003). Updated results of the United Kingdom linkage-based genome screen in multiple sclerosis. *J. Neuroimmunol.* **143,** 25–30.

Herrera, B. M., and Ebers, G. C. (2003). Progress in deciphering the genetics of multiple sclerosis. *Curr. Opin. Neurol.* **16,** 253–258.

Kantarci, O. H., de Andrade, M., and Weinshenker, B. G. (2002). Identifying disease modifying genes in multiple sclerosis. *J. Neuroimmunol.* **123,** 144–159.

Kelly, M. A., Cavan, D. A., Penny, M. A., Mijovic, C. H., Jenkins, D., Morrissey, S., Miller, D. H., Barnett, A. H., and Francis, D. A. (1993). The influence of HLA-DR and -DQ alleles on progression to multiple sclerosis following a clinically isolated syndrome. *Hum. Immunol.* **37,** 185–191.

Liguori, M., Sawcer, S., Setakis, E., Compston, A., Giordano, M., D'Alfonso, S., Mellai, M., Malferrari, G., Trojano, M., Livrea, P., De Robertis, F., Massacesi, L., *et al.* (2003). A whole genome screen for linkage disequilibrium in multiple sclerosis performed in a continental Italian population. *J. Neuroimmunol.* **143,** 97–100.

Lucchinetti, C., Bruck, W., Parisi, J., Scheithauer, B., Rodriguez, M., and Lassmann, H. (1999). A quantitative analysis of oligodendrocytes in multiple sclerosis lesions. A study of 113 cases. *Brain* **122**(Pt. 12), 2279–2295.

Lucchinetti, C., Bruck, W., Parisi, J., Scheithauer, B., Rodriguez, M., and Lassmann, H. (2000). Heterogeneity of multiple sclerosis lesions: Implications for the pathogenesis of demyelination. *Ann. Neurol.* **47,** 707–717.

Lucchinetti, C., Bruck, W., and Noseworthy, J. (2001). Multiple sclerosis: Recent developments in neuropathology, pathogenesis, magnetic resonance imaging studies and treatment. *Curr. Opin. Neurol.* **14,** 259–269.

Luomala, M., Elovaara, I., Ukkonen, M., Koivula, T., and Lehtimaki, T. (2001). The combination of HLA-DR1 and HLA-DR53 protects against MS. *Neurology* **56,** 383–385.

Marrosu, M. G., Murru, R., Murru, M. R., Costa, G., Zavattari, P., Whalen, M., Cocco, E., Mancosu, C., Schirru, L., Solla, E., Fadda, E., Melis, C., *et al.* (2001). Dissection of the HLA association with multiple sclerosis in the founder isolated population of Sardinia. *Hum. Mol. Genet.* **10,** 2907–2916.

Masterman, T., Ligers, A., Olsson, T., Andersson, M., Olerup, O., and Hillert, J. (2000). HLA-DR15 is associated with lower age at onset in multiple sclerosis. *Ann. Neurol.* **48,** 211–219.

McDonald, W. I., Compston, A., Edan, G., Goodkin, D., Hartung, H. P., Lublin, F. D., McFarland, H. F., Paty, D. W., Polman, C. H., Reingold, S. C., Sandberg-Wollheim, M., Sibley, W., *et al.* (2001). Recommended diagnostic criteria for multiple sclerosis: Guidelines from the International Panel on the diagnosis of multiple sclerosis. *Ann. Neurol.* **50,** 121–127.

Miller, D. H., Barkhof, F., Berry, I., Kappos, L., Scotti, G., and Thompson, A. J. (1991). Magnetic resonance imaging in monitoring the treatment of multiple sclerosis: Concerted action guidelines. *J. Neurol. Neurosurg. Psychiatry* **54,** 683–688.

Nilsson, P., Larsson, E. M., Maly-Sundgren, P., Perfekt, R., and Sandberg-Wollheim, M. (2005). Predicting the outcome of optic neuritis evaluation of risk factors after 30 years of follow-up. *J. Neurol.* **252,** 396–402.

Runmarker, B., and Andersen, O. (1993). Prognostic factors in a multiple sclerosis incidence cohort with twenty-five years of follow-up. *Brain* **116**(Pt. 1), 117–134.

Sanfilipo, M. P., Benedict, R. H., Zivadinov, R., and Bakshi, R. (2004). Correction for intracranial volume in analysis of whole brain atrophy in multiple sclerosis: The proportion vs. residual method. *Neuroimage* **22,** 1732–1743.

Sawcer, S., Maranian, M., Setakis, E., Curwen, V., Akesson, E., Hensiek, A., Coraddu, F., Roxburgh, R., Sawcer, D., Gray, J., Deans, J., Goodfellow, P. N., *et al.* (2002). A whole genome screen for linkage disequilibrium in multiple sclerosis confirms disease associations with regions previously linked to susceptibility. *Brain* **125,** 1337–1347.

Schreiber, K., Otura, A. B., Ryder, L. P., Madsen, H. O., Jorgensen, O. S., Svejgaard, A., and Sorensen, P. S. (2002). Disease severity in Danish multiple sclerosis patients evaluated by MRI and three genetic markers (HLA-DRB1*1501, CCR5 deletion mutation, apolipoprotein E). *Mult. Scler.* **8,** 295–298.

Sellebjerg, F., Madsen, H. O., Jensen, C. V., Jensen, J., and Garred, P. (2000). CCR5 delta32, matrix metalloproteinase-9 and disease activity in multiple sclerosis. *J. Neuroimmunol.* **102,** 98–106.

Soderstrom, M., Lindqvist, M., Hillert, J., Kall, T. B., and Link, H. (1994). Optic neuritis: Findings on MRI, CSF examination and HLA class II typing in 60 patients and results of a short-term follow-up. *J. Neurol.* **241,** 391–397.

Soderstrom, M., Ya-Ping, J., Hillert, J., and Link, H. (1998). Optic neuritis: Prognosis for multiple sclerosis from MRI, CSF, and HLA findings. *Neurology* **50,** 708–714.

Thompson, A. J., Montalban, X., Barkhof, F., Brochet, B., Filippi, M., Miller, D. H., Polman, C. H., Stevenson, V. L., and McDonald, W. I. (2000). Diagnostic criteria for primary progressive multiple sclerosis: A position paper. *Ann. Neurol.* **47,** 831–835.

Weinshenker, B. G., Santrach, P., Bissonet, A. S., McDonnell, S. K., Schaid, D., Moore, S. B., and Rodriguez, M. (1998). Major histocompatibility complex class II alleles and the course and outcome of MS: A population-based study. *Neurology* **51,** 742–747.

Wekerle, H., and Hohlfeld, R. (2003). Molecular mimicry in multiple sclerosis. *N. Engl. J. Med.* **349,** 185–186.

Yamasaki, K., Horiuchi, I., Minohara, M., Kawano, Y., Ohyagi, Y., Yamada, T., Mihara, F., Ito, H., Nishimura, Y., and Kira, J. (1999). HLA-DPB1*0501-associated opticospinal multiple sclerosis: Clinical, neuroimaging and immunogenetic studies. *Brain* **122**(Pt. 9), 1689–1696.

Zivadinov, R. (2006). Evidence for neuroprotection in multiple sclerosis: Can imaging techniques measure neuroprotection and remyelination? *Neurology* (in press).

Zivadinov, R., and Bakshi, R. (2004). Role of MRI in multiple sclerosis II: Brain and spinal cord atrophy. *Front. Biosci.* **9,** 647–664.

Zivadinov, R., Rudick, R. A., De Masi, R., Nasuelli, D., Ukmar, M., Pozzi-Mucelli, R. S., Grop, A., Cazzato, G., and Zorzon, M. (2001a). Effects of IV methylprednisolone on brain atrophy in relapsing-remitting MS. *Neurology* **57,** 1239–1247.

Zivadinov, R., Sepcic, J., Nasuelli, D., De Masi, R., Bragadin, L. M., Tommasi, M. A., Zambito-Marsala, S., Moretti, R., Bratina, A., Ukmar, M., Pozzi-Mucelli, R. S., Grop, A., *et al.* (2001b). A longitudinal study of brain atrophy and cognitive disturbances in the early phase of relapsing-remitting multiple sclerosis. *J. Neurol. Neurosurg. Psychiatry* **70,** 773–780.

Zivadinov, R., Uxa, L., Zacchi, T., Nasuelli, D., Ukmar, M., Furlan, C., Pozzi-Mucelli, R., Tommasi, M. A., Locatelli, L., Ulivi, S., Bratina, A., Bosco, A., *et al.* (2003). HLA genotypes and disease severity assessed by magnetic resonance imaging findings in patients with multiple sclerosis. *J. Neurol.* **250,** 1099–1106.

GLATIRAMER ACETATE: MECHANISMS OF ACTION IN MULTIPLE SCLEROSIS

Tjalf Ziemssen and Wiebke Schrempf

Multiple Sclerosis Center Dresden, Neurological University Clinic
Dresden University of Technology, Dresden 01307, Germany

Glatiramer acetate (GA), formerly known as copolymer 1, is a mixture of synthetic polypeptides composed of four amino acids resembling the myelin basic protein (MSP). GA has been shown to be highly effective in preventing and suppressing experimental autoimmune encephalomyelitis (EAE), the animal model of multiple sclerosis (MS). Therefore, it was tested in several clinical studies and so approved for the immunomodulatory treatment of relapsing-type MS.

In contrast to other immunomodulatory MS therapies, GA has a distinct mechanism of action: GA demonstrates an initial strong promiscuous binding to major histocompatibility complex molecules and consequent competition with various (myelin) antigens for their presentation to T cells. In addition, antigen-based therapy generating a GA-specific immune response seems to be the prerequisite for GA

INTERNATIONAL REVIEW OF
NEUROBIOLOGY, VOL. 79
DOI: 10.1016/S0074-7742(07)79024-4

0074-7742/07 $35.00

therapy. GA treatment induces an *in vivo* change of the frequency, cytokine secretion pattern and the effector function of GA-specific CD4+ and CD8+ T cells, probably by affecting the properties of antigen-presenting cells such as monocytes and dendritic cells. As demonstrated extensively in animal experiments, GA-specific, mostly, T helper 2 cells migrate to the brain and lead to *in situ* bystander suppression of the inflammatory process in the brain. Furthermore, GA-specific cells in the brain express neurotrophic factors like the brain-derived neurotrophic factor (BDNF) in addition to anti-inflammatory T helper 2-like cytokines. This might help tip the balance in favor of more beneficial influences because there is a complex interplay between detrimental and beneficial factors and mediators in the inflammatory milieu of MS lesions.

I. Introduction

Multiple sclerosis (MS) is the most common inflammatory demyelinating disease of the central nervous system (CNS). It is believed to be a multifocal immune-mediated disorder in which the myelin sheath or the oligodendrocyte is targeted by the immune system in genetically susceptible people. There is a considerable heterogeneity in terms of clinical, radiological, and pathological changes, mirrored by a high intrapatient and interpatient variability in the clinical course and its manifestations. The disease affects approximately 0.1% of the population in temperate climates. It is a disease of young people with a lifelong and often disabling course. MS is manifested in physical symptoms (relapses and disability progression), CNS inflammation, brain atrophy and cognitive dysfunction.

About 85% of the patients begin with a relapsing-remitting disease, where neurological symptoms and signs develop over several days, plateau, and then usually improve over days to weeks. Approximately two-thirds of patients with relapsing-remitting MS (RRMS) undergo a conversion to a secondary progressive disease course (SPMS), where relapse frequency lessens over time and progressive neurological dysfunction emerges. The remaining 15% of patients begin the disease course with a gradually progressive neurological dysfunction, typically a slowly worsening myelopathy [primary progressive MS (PPMS)].

II. Pathology and Immunology

The pathological hallmark of MS is the demyelinating plaque which consists of infiltrating lymphocytes and macrophages, damage to the blood–brain barrier, and loss of myelin.

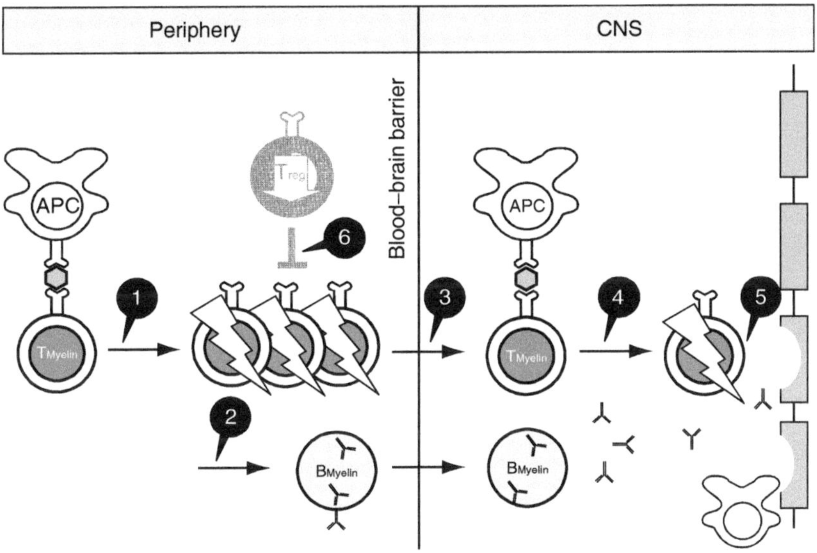

FIG. 1. Hypothetical pathophysiological cascade in MS. Details (inclusive numbers) in the text.

Oligodendrocytes synthesize and maintain the axonal myelin sheath of up to 40 neighboring nerve axons in the CNS. Compact myelin consists of a condensed membrane, spiraled around axons to form the insulating segmented sheath needed for saltatory axonal conduction; voltage-gated sodium channels cluster at the unmyelinated nodes of Ranvier between myelin segments, from where the action potential is propagated and spreads down the myelinated nerve segment to trigger another action potential at the next node.

In MS, the composition of the inflammatory infiltrate varies depending on the stage of demyelinating activity. Early symptoms of MS are widely believed to result from this inflammatory demyelination which leads to slowing or blockade of axonal conduction. The regression of symptoms has been attributed to the resolution of inflammatory edema and to partial remyelination.

Current concepts assume that MS occurs as a consequence of immune tolerance breakdown in genetically susceptible individuals (Hafler, 2004). The major contributing factors thus include genetics, environment, and immune dysregulation. T helper (Th) cells are considered to play a pivotal role in the whole self-reactive immune response of the CNS, primarily characterized by inflammatory demyelination. Putting experimental data from human studies and animal experiments together, the following pathophysiological cascade seems quite attractive which is shown and numbered in Fig. 1:

1. Autoreactive T cells, known to exist in any individual, are activated in the periphery, probably by molecular mimicry (i.e., recognition of epitopes that are common to autoantigens and microbial structures as exogenous triggers) or by self-antigens.
2. T cells may trigger B-cell activation and antibody formation, the latter potentially exerting detrimental effector functions at the myelin sheath (Ziemssen and Ziemssen, 2005).
3. T-cell activation enables transmigration through the blood–brain barrier to the sites of inflammation, a cascade of events influenced by adhesion molecules, chemotactic factors and migration promoters.
4. Recognizing their antigens presented by microglia, the local antigen-presenting cells (APCs), the autoreactive T cells are reactivated.
5. T cells of both CD4+ and CD8+ cytotoxic phenotypes release proinflammatory cytokines. An inflammatory cascade is initiated followed by the further recruitment of inflammatory cells like monocytes.
6. Underlying immunoregulatory defects, such as decrease of regulatory T cells in the circulation of patients with MS, allow the further pathological activation of autoreactive T cells.

III. MS as Neurodegenerative Disease

Although MS seems to be primarily an inflammatory autoimmune disease, it has become evident that axonal loss plays an important role in the pathogenesis of disability in patients with MS. While axonal pathology was elegantly and precisely described in classic MS neuropathologic studies more than a century ago (Charcot, 1877), it has reemerged as a major focus of research (Trapp *et al.*, 1998). The important question to be addressed is not whether there is axonal loss in MS but when and to what extent does the axonal loss occur. The timing and degree of axonal loss is of importance not only in its relationship to the etiology of the disease but may be central to the appearance of clinical symptoms and the progressive deterioration associated with the disease. The fact that axonal loss is irreversible has important implications for when and which therapeutic intervention should be used.

It is likely that various mechanisms contribute to axonal damage during different stages of disease. In active lesions, the extent of axonal transection correlates with inflammatory activity while even there seems to be an inflammation-independent axonal loss (Perry and Anthony, 1999). Hence, axonal loss may be caused by inflammatory products of activated immune and glial cells, including proteolytic enzymes, cytokines, oxidative products, and free radicals, although the precise molecular mechanisms of axonal damage are poorly understood. In addition, the magnitude of axonal loss in chronic MS lesions without pronounced inflammatory infiltrates suggests that mechanisms other than inflammatory demyelination contribute to the degeneration of axons. Several conditions interfere with attempts of axonal regrowth after lesions develop. These include the lack of neurotrophic factors that support growth, the presence of a glial scar (depending on the site of lesion) or the presence of inhibitory molecules that impede axonal growth. Evidence shows that axon degeneration following injury has similarities with the cellular mechanisms underlying programmed cell death (Perry and Anthony, 1999).

The biochemical events underlying the inflammatory and neurodegenerative phases of the disease are not known in detail and may be quite different (Steinman, 2001). Inflammation involves activation of T and B cells in the periphery, crossing the blood–brain barrier and homing to the lesion site. In the lesion, T cells are reactivated by myelin antigens, release cytokines that attract macrophages and activate microglia which start to destroy the myelin sheath. Anti-myelin antibodies bind complement, attract macrophages, and stimulate opsonization of myelin. Demyelination leads to reversible and to some extent irreversible impairment of functionality of the axon whose conduction properties are deteriorated, thus accounting for the clinical symptoms associated with relapses.

Neurodegeneration is likely to be a complex process (Chitnis *et al.*, 2005; Grigoriadis *et al.*, 2004), especially when it takes place in inflammatory disorders like MS. To a significant extent, axonal loss seems to be a major consequence of demyelination and inflammation, for example, by binding of CD8+ T cells to exposed axons and secretion of toxic factors. However, other mechanisms not directly related to demyelination and inflammation are also likely to be important. An example is excitotoxicity: glutamic acid can bind to excitatory amino acid receptors on the cell bodies, dendrites, or axon terminals of neurons and initiate a process of neuronal cell death (Steinman, 2001).

Theoretically, there are several possible relationships between inflammation and neurodegeneration in MS. Three exclusive hypotheses can be envisaged: (a) that neurodegeneration is entirely secondary to inflammation, (b) that inflammation is entirely secondary to neurodegeneration, or (c) that inflammation and neurodegeneration are entirely independent. On the other hand, nonexclusive hypotheses stating that neurodegeneration is partially dependent and partially independent of inflammation or vice versa can also be put forward, and these seem intuitively more likely (Ziemssen, 2005).

There are a number of clinical arguments in favor of some independence between the inflammatory and neurodegenerative processes. For example, in a clinical trial of alemtuzumab (Campath-1H), a monoclonal antibody directed against CD52 which leads to T-cell depletion, in secondary progressive MS (SPMS), a gradual extinction of exacerbations and lesion activity visible on magnetic resonance imaging (MRI) was demonstrated (Coles *et al.*, 1999). However, disability continued to progress in about half the patients in whom progressive brain atrophy and axonal degeneration could be observed using MRI and magnetic resonance spectroscopy (MRS). The investigators concluded firstly that inflammation and demyelination were responsible for relapses of MS and could be prevented by alemtuzumab treatment and secondly that ongoing axonal degeneration accounted for the progressive phase of disability. Even though axonal injury may have been conditioned by prior inflammation, this process can continue despite complete suppression of inflammatory activity.

There is also neuropathologic evidence for a dissociation between inflammatory demyelination and axonal injury from a series of 42 biopsy samples obtained from patients with MS (Bitsch *et al.*, 2000). Acute axonal injury was visualized by amyloid precursor protein (APP) staining. There was no relationship between the expression of APP or axonal density and the extent of demyelination, with axonal injury being observed even in lesions that were successfully remyelinating. Similarly, there was no association between the extent of axonal injury and markers of acute inflammation such as tumor necrosis factor (TNF)-α or inducible nitric oxide synthase. However, axonal injury was correlated to some extent with the extent of infiltration by CD8+ T lymphocytes and macrophages. A dissociation between neurodegeneration and inflammatory demyelination is also observed in lesions within the cortex. These lesions are characterized by a significant degree of axonal transection and apoptosis of neuronal cell bodies. However, the extent of infiltration by T lymphocytes and macrophages and the expression of inflammatory markers are low (Bo *et al.*, 2003; Peterson *et al.*, 2001).

MRI studies in very early disease also suggest that inflammation and neuronal injury are not strictly related. A study of 31 subjects presenting with a clinically isolated syndrome evaluated inflammatory lesion activity with classical T2- and T1-weighted images after gadolinium enhancement and measured a surrogate marker of axonal injury, the size of the *N*-acetylaspartate peak (NAA) determined in the whole brain (Filippi *et al.*, 2003). In these patients, the mean size of the NAA peak was some 20% lower than that observed in matched controls. No correlation was observed between the size of the NAA peak and lesion volume on either T1 or T2 images. The investigators concluded that significant axonal injury occurs early in the disease and that this is only indirectly linked to inflammatory activity.

Studies such as these suggest that treatment strategies for MS need to address both the inflammatory and neurodegenerative components of the disease, and that anti-inflammatory therapies may only be able to control the inflammation-related neurodegenerative process adequately (Chitnis *et al.*, 2005).

IV. The Janus Face of CNS-Directed Autoimmune Inflammation

Inflammation is considered to be a key feature in MS pathogenesis. The neurotoxic effects of inflammation are well established and thought to be at least partially responsible for the observed axonal damage. Recently, an increasing body of experimental evidence supports the view of a dual role of the immune system in CNS-directed autoimmune inflammation. A number of studies have proposed that autoimmune inflammation may have neuroprotective effects in the CNS (Ziemssen, 2005).

On the one hand, MS and its animal models, representing neuroimmunological diseases which arise when immune cells attack the nervous system, provide the paradigm for the deleterious interaction between cells of the immune and nervous systems. EAE, the MS animal model, can be induced by active immunization with CNS autoantigens MBP or by the transfer of autoantigen-specific T cells into naive syngenic recipients (Gold *et al.*, 2006). On the other hand, it was demonstrated that MBP-specific, encephalitogenic T cells may have seemingly neuroprotective (side) effects. The neuroprotective and regenerative potential of immune cells was first coined by Schwartz *et al.* (1999).

They demonstrated that autoimmune T cells could protect neurons in an animal model of secondary degeneration after a partial crush injury of the optic nerve (Moalem *et al.*, 1999). In several experiments, T cells with different specificities [specific for MBP, control antigen ovalbumin (OVA), or a heat-shock protein (hsp) peptide] were activated by restimulation with their respective antigens *in vitro*, and then injected into rats immediately after a unilateral optic nerve injury. Seven days after injury, the optic nerves were analyzed by immunohistochemistry for the presence of T cells. Small numbers of T cells could be found in the intact (uninjured) optic nerves of rats injected with anti-MBP T cells, which is consistent with previous observations that activated MBP-specific T cells home to intact CNS white matter.

A much more pronounced accumulation of T cells, however, was observed in the *crushed optic nerves* of the rats injected with T cells specific for MBP, hsp peptide, or OVA. The degree of primary and secondary damage to the optic nerve axons and their attached retinal ganglion cells was measured by injecting a neurotracer distal to the site of the optic nerve lesion immediately after the injury, and again after 2 weeks. The number of labeled retinal ganglion cells as marker for viable axons was significantly greater in the retinae of the rats injected with anti-MBP T cells than with anti-OVA or anti-hsp peptide T cells. Thus, although all three T-cell lines (TCLs) accumulated at the site of injury, only the MBP-specific, autoimmune T cells had a substantial effect in limiting the extent of secondary degeneration. This neuroprotective effect was confirmed by electrophysiological studies.

The results demonstrate that T-cell autoimmunity can mediate significant neuroprotection after CNS injury. The authors speculate that after injury, "cryptic" epitopes might become available and might be recognized by endogenous non-encephalitogenic (benign) T cells. After local stimulation, these protective autoreactive T cells could exert their neuroprotective effect. The findings further substantiate the idea that "natural autoimmunity" can be benign and may even function as a protective mechanism (Cohen, 1992).

Macrophages seem to represent another type of immune cell which is capable of mediating neuroprotection and/or stimulating recovery of CNS lesions. The injection of activated macrophages into transected rat spinal cord stimulated tissue repair and partial recovery of motor function (Rapalino *et al.*, 1998). The neuroprotective activity of immune cells is not restricted to the CNS. After experimental axotomy of the facial nerve of immunodeficient SCID mice, the survival of facial motor neurons was severely impaired compared to immunocompetent wild-type mice. Reconstitution of SCID mice with wild-type splenocytes containing T and B cells restored the survival of facial motor neurons in these mice to the level of the wild-type controls (Serpe *et al.*, 1999).

It is equally evident that a large number of neurotoxic and proinflammatory mediators are produced and released by immune cells (Kerschensteiner *et al.*, 2003). The neutralization of toxic inflammatory mediators may improve the outcome of MS and experimental models of CNS damage. On the other hand, we know that immune cells can supply a number of neuroprotective mediators, including neurotrophic factors, anti-inflammatory cytokines, and prostaglandins which may reduce tissue damage (Hohlfeld *et al.*, 2006).

The idea that inflammatory reactions may not always be harmful under certain conditions even confers neuroprotection and repair, has important consequences for the design of immunomodulatory therapies for MS (Hohlfeld *et al.*, 2006). Undebatably, there is convincing rationale for immunosuppressive treatment when the noxious effects of the inflammatory reaction prevail. Because nonselective immunosuppressive treatments will suppress both destructive and beneficial components of inflammation, therapy is likely to fail when the beneficial effects of CNS inflammation outweigh its negative consequences. It seems, for example, possible that the lack of beneficial (side) effects of inflammation during the late phase of MS with little inflammation but ongoing axonal loss contributes to the pathogenesis of neurodegeneration. In MS, it is unfortunately unclear whether there is a stage of the disease when the inflammatory reaction is more beneficial than harmful.

The concept of the neuroprotective role of inflammation can be extended to neurodegenerative, ischemic, and traumatic lesions of the CNS, considering that inflammation is a universal tissue reaction crucial for defense and repair (Ziemssen and Ziemssen, 2005).

V. Neurotrophic Factors Are Released by Different Immune Cells

The precise mechanisms involved in immune-mediated neuroprotection remain to be clarified. A number of recent studies have shown that several types of immune cells and hematogenic progenitor cells express one or more neurotrophic factors. For example, nerve growth factor (NGF) is produced by B cells, which also express the trkA receptor and p75 NGF receptor (Torcia *et al.*, 1996). Because neutralization of endogenous NGF caused apoptosis of memory B cells, it was concluded that NGF is an autocrine growth factor for memory B cells.

More recently, another neurotrophin, brain-derived neurotrophic factor (BDNF), was found to be expressed in immune cells. BDNF was originally cloned in 1989 as the second member of the neurotrophin family which includes NGF and neurotrophins (NT)-3, -4/5, -6, and -7 (Lewin and Barde, 1996). Since then, the important role of BDNF in regulating the survival and differentiation of various neuronal populations, including sensory neurons, cerebellar neurons, and spinal motor neurons, has been firmly established. Neurons are the major source of BDNF in the nervous system. BDNF binds to different types of receptors: the tyrosine kinase receptor B (trkB) which exists in two isoforms—the full-length receptor (gp145trkB) and the truncated receptor (gp95trkB) lacking the tyrosine kinase domain—and the p75 neurotrophin receptor (Klein *et al.*, 1991). It is thought that BDNF and NT-4/5 exert their biological function via the full-length form of trkB receptor which expression seems to be restricted to neuronal cell populations.

Immune cells can be a potent source of the neuroprotective factor BDNF in neuroinflammatory disease (Hohlfeld *et al.*, 2006). Activated human T cells, B cells, and monocytes are able to secrete bioactive BDNF after *in vitro* activation (Kerschensteiner *et al.*, 1999). The BDNF secreted by immune cells is bioactive as it supports neuronal survival *in vitro*. In histology, BDNF immunoreactivity was found in T cells and macrophages in active and inactive MS lesions.

Similar observations were recently reported by several other groups of investigators. After experimental injury of the striatum, activated macrophages and microglia cells transcribe mRNA for glial cell line-derived neurotrophic factor (GDNF) and BDNF (Batchelor *et al.*, 1999). This could help to explain the sprouting of dopaminergic neurons observed after experimental injury. Transcripts for BDNF and NT-3 and their receptors trkB and trkC were found in subpopulations of human peripheral blood cells (Besser and Wank, 1999). BDNF protein was secreted by cultured T-cell clones.

Besides BDNF, there is a robust expression of the full-length BDNF receptor gp145trkB in neurons in the vicinity of MS plaques (Stadelmann *et al.*, 2002). Single neurons with clearly pronounced trkB immunoreactivity close to MS lesions can be observed, suggesting an upregulation of trkB in a proportion of damaged neurons. Additionally, full-length trkB immunoreactivity is present in reactive astrocytes within the lesions. The restriction of trkB expression to neural cell types in MS underscores the possibility that there may be BDNF signaling from infiltrating cells to neurons in neuroinflammatory lesions, as would be necessary for immune cells to support neuronal survival or provide axonal protection.

Within the MS lesion, immune cells seem to be the major source of BDNF (Stadelmann *et al.*, 2002). They are likely to release this substance in the immediate vicinity of nerve cell processes, which—according to the observed trkB expression—are likely to be responsive to the neuroprotective effects of BDNF. This neurotrophin-mediated neuroimmune signaling network could be a major factor that helps to preserve axons in a microenvironment that is clearly capable of exerting significant neurotoxicity. Thus, it should be considered as a beneficial aspect of neuroinflammation that could be worth preserving therapeutically, or even reinforcing using tailored immunomodulatory treatment strategies.

VI. Glatiramer Acetate: Historical Remarks

Glatiramer acetate (GA, Copaxone®), formerly known as copolymer 1, is the acetate salt of a standardized mixture of synthetic polypeptides containing the four amino acids L-alanine, L-glutamic acid, L-lysine, and L-tyrosine with a defined molar ratio of 0.14:0.34:0.43:0.09 and an average molecular mass of 4.7–11.0 kDa, that is, an average length of 45–100 amino acids (Arnon *et al.*, 1996; Teitelbaum *et al.*, 1997a,b).

In the 1960s, Drs. Sela, Arnon, and their colleagues at the Weizmann Institute, Israel were involved in studies on the immunologic properties of a series of polymers and copolymers which were developed to resemble MBP, a myelin protein. MBP in Freund's complete adjuvant induces EAE, the best animal model of MS. They were interested in evaluating whether these polypeptides could simulate the ability of MBP and fragments and regions of the MBP molecule to induce EAE (Teitelbaum *et al.*, 1971, 1973, 1974a,b). None of these series was capable of inducing EAE, but several polypeptides were able to suppress EAE in guinea pigs. Copolymer 1, later known as GA, was shown to be the most effective polymer in preventing or decreasing the severity of EAE. The suppressive effect is a general phenomenon and not restricted to a particular species, disease type, or encephalitogen used for EAE induction.

Abramsky *et al.* (1977) were the first who treated a group of severe RRMS patients with intramuscular GA 2–3 mg every 2–3 days for 3 weeks, then weekly for 2–5 months. No conclusions could be drawn regarding drug efficacy but there were no significant undesirable side effects. Three clinical trials in the 1980s were performed showing some evidence of efficacy that was adequate for support of the Food and Drug Administration (FDA) approval and a good safety profile (Bornstein *et al.*, 1982, 1984, 1987). However, the results of these studies must be interpreted with caution because before 1991 the production of the drug was not standardized (Johnson, 1996). Different batches had variable suppressive effects on EAE which could also imply variable effects in MS patients.

In 1991, a phase III multicenter trial with a daily 20-mg dose of an s.c. administered, highly standardized GA preparation was started in the United States. This double-blind, placebo-controlled study demonstrated that GA significantly reduced the relapse rate without significant side effects (Johnson *et al.*, 1995).

In 1996, GA was approved by the US FDA as a treatment of ambulatory patients with active RRMS. Since then GA has been licensed for approval in many other countries (Ziemssen *et al.*, 2002b).

VII. GA: Overview of Clinical Studies

Four early exploratory open studies were performed in the late 1970s and early 1980s in order to obtain indications of dosing and safety (Abramsky *et al.*, 1977; Bornstein *et al.*, 1982, 1984, 1987). In total, 41 patients with relapsing-remitting or SPMS were enrolled in these studies. The treatment schedules, doses of GA, and treatment duration were quite variable from study to study. The maximum dose of 20 mg/day was well tolerated and no severe adverse effects were detected in these studies.

To assess more comprehensively the efficacy of GA in patients with RRMS, a definitive phase III trial was conducted at 11 US medical centers with a total of 251 RRMS patients who received GA at a dosage of 20 mg or placebo by daily s.c. injection for 2 years (Johnson *et al.*, 1995). As the primary end point, the mean annualized relapse rates were 0.59 for the GA-treated group and 0.84 for the placebo group, a 29% reduction was statistically significant ($p = 0.007$). Trends in the proportion of relapse-free patients and median time to first relapse favored GA treatment. Patients in both groups with higher disability at entry, measured by the Expanded Disability Status Scale (EDSS), had a higher relapse rate, while the largest reduction in relapse rate between groups occurred in patients with a baseline EDSS of 0–2 (33% vs a reduction of 22% in patients with an EDSS score at entry >2). When the proportion of patients who improved, were unchanged, or worsened by ≥ 1 EDSS step from baseline to end of study (2 years) was evaluated, significantly more patients on GA improved and more on placebo worsened ($p = 0.037$). The effect of treatment was constant throughout the entire study duration. Patient withdrawals were 19 (15.2%) from the copolymer 1 group and 17 (13.5%) from the placebo group at approximately the same intervals. The treatment was well tolerated. The most common adverse experience was the local injection-site reaction. Rarely, the transient self-limiting immediate post-injection reaction followed the injection in 15.2% of those treated with GA and 3.2% of those treated with placebo.

In an extension of this study up to 35 months with unchanged blinding and study conditions, the clinical benefit of GA for both the relapse rate and for neurological disability was sustained (Johnson *et al.*, 1998). Thus, the reduction of annual relapse rate was 32% in favor of GA ($p = 0.002$). The results of this extension study confirmed the excellent tolerability and safety profile of GA. This study was further on extended as open-label study with all patients receiving active drug. The published data from approximately 10 years of ongoing trial showed a persistent reduced relapse rate of about 1 relapse per 5 years (Ford *et al.*, 2006). After more than 10 years of observation, 108 out of 232 patients (47%) were still participating. Sixty-two percent of the group who stayed on GA without interruption were neurologically unchanged or improved from baseline by at least one step of the EDSS scale in contrast to 28% of the withdrawn patients. This is the first time that the benefits of an immunomodulatory treatment over this clinically highly relevant long-term time period are demonstrated in a planned neurological follow-up study.

Studies using an oral formulation of GA (CORAL) (Filippi *et al.*, 2006) and applying GA to PPMS patients (PROMISE) (Wolinsky and PROMiSe Trial Study Group, 2004) failed to demonstrate significant treatment effects. There are promising new data on using a higher GA dose (40 mg) per injection which are studied in a multicenter, controlled trial (FORTE).

VIII. GA: Imaging Studies

Several early MRI studies indicated a trend towards benefit with GA, but were limited by the small number of evaluated patients. Wolinsky *et al.* were able to demonstrate a definite, but modest effect of GA on MRI enhancements of MS patients in the US open-label GA extension MRI trial for relapsing MS (Ge *et al.*, 2000; Mancardi *et al.*, 1998; Wolinsky *et al.*, 2001).

A large randomized, double-blind, placebo-controlled MRI trial with 239 RRMS patients was conducted to determine the effect, onset, and durability of GA on disease activity monitored with MRI monthly over a 9-month period (Comi *et al.*, 2001). Treatment of GA showed as primary outcome measure a significant reduction of 30% in the total number of enhancing lesions on T1-weighted images ($p = 0.003$) and—as secondary outcome measure—a significant reduction of the number of new enhancing lesions ($p < 0.003$), the monthly change in the volume of enhancing lesions ($p = 0.01$) and the change in the volume ($p = 0.006$) and the number ($p < 0.003$) of new lesions seen on T2-weighted images compared to placebo. The relapse rate was also significantly reduced by 33% for GA-treated patients ($p = 0.012$).

These data were further substantiated in the open-label phase of the trial that extended the study up to 18 months. Especially, the MRI lesion burden at the study's end was significantly lower in patients initially randomized to GA compared to placebo ($p = 0.018$).

In addition to its effect on lesion load and volume, GA treatment inhibited lesion evolution into permanent black holes (Filippi *et al.*, 2001). At each time point, the proportion of new lesions evolving into black holes was lower in the GA-treated group compared to placebo, reaching significance at month 7 (18.9% vs 26.3%; $p = 0.04$). By month 8, the proportion of lesions evolving into black holes was roughly 50% less in the GA treatment group. Using the new SIENA method, Sormani *et al.* (2004) were able to demonstrate a significant effect of GA on brain atrophy in this patient cohort. In a small pilot study, Khan *et al.* (2005) used brain proton MRS which allows *in vivo* examination of axonal integrity by quantifying the resonance intensity of the neuronal marker NAA. In the small group of 18 GA-treated patients, GA leads to axonal metabolic recovery and protection from sublethal axonal injury which supports an *in situ* effect of GA therapy inside the CNS.

IX. GA: Animal Models

Our current knowledge about the mechanism of action of GA is based on the pioneering studies done by Drs. Sela, Arnon, and colleagues on GA in EAE of several species of animals. Their work identified GA-specific Th2-type T cells as key mediators of the beneficial effects of the drug in EAE. These specific Th2 cells isolated from GA-treated animals can transfer protection from EAE to nontreated animals (Arnon and Aharoni, 2004).

Results of *in vitro* studies suggest that GA in some way competes with MBP (Webb *et al.*, 1973). Specifically, GA competes with MBP at the APC level for binding to the major histocompatibility complex (MHC), and GA/MHC competes with MBP/MHC for binding to the T-cell receptor (TCR). GA binds to many different alleles of MHC class II molecules (promiscuous binding) (Fridkis-Hareli *et al.*, 1994). Interestingly, the stereoisomer of GA, D-GA, which is composed of D-amino acids, binds as effectively to MHC class II, but fails to suppress EAE (Webb *et al.*, 1976). Further studies showed that in addition GA might act like an altered peptide ligand by modifying the T-cell response to several myelin antigens (Aharoni *et al.*, 1999).

Results of *in vivo* studies indicate that GA induces regulatory T cells of the Th2 type in the peripheral immune system outside the CNS. When spleen cells from GA-treated mice were adoptively transferred into syngenic animals, these cells protected from EAE induced by different CNS antigens (Aharoni *et al.*, 1993,

1997, 1998, 1999). Therefore GA has been mainly regarded as a T-cell-directed immunotherapeutic agent. Further support for the proposed protective role of GA-reactive regulatory T cells comes from the demonstration that GA-specific Th2 cells are present in the CNS of GA-treated mice (Aharoni *et al.*, 2000). Some of these GA-specific Th2 cells are reactivated in the CNS because they cross-react with locally presented myelin autoantigens (Aharoni *et al.*, 1997). The specific T cells secrete anti-inflammatory cytokines such as interleukin 4, which in turn dampen the activity of any nearby autoaggressive T cells (Aharoni *et al.*, 1997, 2000) even if they recognize unrelated antigens and do not directly cross-react with GA. This bystander suppression is regarded as a central element of the mechanism of action of GA (Aharoni *et al.*, 1998). There is some discussion whether the GA-specific Th2 cells are the only cells accounting for the beneficial effects in EAE (Jee *et al.*, 2006). Future research will demonstrate the variety of involved mechanisms.

In addition to the anti-inflammatory bystander suppression, Aharoni *et al.* (2005) could demonstrate that GA treatment induced a sustained augmentation in the expression of neurotrophic factors like BDNF in various brain regions as demonstrated by histological analysis of immunostained brain sections. GA treatment, even when started 45 days after disease induction, restored the impaired level of neurotrophins to that of healthy mice. These results indicate that the immunomodulator GA exerts not only an anti-inflammatory effect, but also enhances neuroprotection and regeneration of neural elements in the diseased brain.

These results are in line with the paper of Gilgun-Sherki *et al.* (2003), demonstrating neuroprotective effects in myelin oligodendrocyte glycoprotein (MOG)-induced EAE by GA. This form of EAE is characterized by axonal loss and neuronal damage in addition to the EAE-typical inflammatory demyelination. As expected from former EAE studies, the treatment of MOG-EAE with GA lead to a significant suppression of clinical disease and strongly reduced inflammatory changes. In this study, the investigators were able to demonstrate a strong significant reduction of axonal loss and neuronal damage by GA treatment.

As stated above, it was demonstrated that autoimmune T cells directed against CNS-associated myelin antigens protect neurons from secondary degeneration (Moalem *et al.*, 1999). Because it would be desirable to obtain "autoimmune neuroprotection" free of any possible autoimmune disease, the nonencephalitogenic copolymer GA was analyzed in different models of secondary neuronal degeneration. Kipnis *et al.* (2000) demonstrated that the posttraumatic spread of degeneration in the damaged optic nerve can be attenuated by active immunization with GA on the day of injury and by adoptive transfer of GA-specific T cells. In the development of therapies based on autoimmune neuroprotection, the use of GA seems to involve "safe" antigenic epitopes which will not cause the induction of an autoimmune disease.

GA-reactive T cells were also found to be neuroprotective in other models of CNS injury, where myelin-associated antigens are not active, such as the insult caused by direct exposure of retinal ganglion cells to glutamate toxicity or the death of retinal ganglion cells resulting from increased intraocular pressure in a model of high-tension glaucoma (Schori *et al.*, 2001). Following injury, peripheral lymphocytes, regardless of their antigenic specificity, enter the CNS. T cells reactive to myelin antigens are activated at the site of injury or in the cervical lymph nodes which is a prerequisite for neuroprotection. The neuroprotective effects of GA in the neurodegenerative animal models suggest that on passive transfer or active immunization with GA there is an accumulation and local reactivation of GA-specific T cells because of the cross-reactivity with myelin proteins *in vivo*. Activated GA-reactive T cells can then produce neurotrophic factors besides anti-inflammatory factors within the region of injured and damaged neurons and axons. It was shown by Angelov *et al.* (2003) that therapeutic vaccination with GA protects motor neurons against acute and chronic degenerative conditions. In acute degeneration after facial nerve axotomy, the number of surviving motor neurons was almost two times higher in GA-vaccinated mice than in nonvaccinated mice. In mice that express the mutant human gene Cu/Zn superoxide dismutase G93A (SOD1), and therefore simulate the chronic human motor neuron disease amyotrophic lateral sclerosis, GA vaccination prolonged life span compared to untreated matched controls. Butovsky *et al.* (2006) demonstrated that using Alzheimer's disease double-transgenic (APP/PS1) mice, therapy with GA resulted in decreased plaque formation, induction of neurogenesis, and reduction of cognitive decline. The vaccination apparently exerted its effects by causing a phenotype switch of brain microglia to dendritic-like cells producing insulin-like growth factor 1.

A promising approach to improve MS treatment is to combine drugs that impact on different aspects of the disease process. Several candidates have already been tested in combination with GA in EAE experiments. The combination of minocycline and GA resulted in a significant reduction of disease severity and disease burden with attenuation of inflammation, axonal loss, and demyelination (Giuliani *et al.*, 2005). In addition, orally administered cholesterol-lowering HMG-CoA reductase inhibitors were evaluated in combination with GA in the EAE model as atorvastatin can enhance protective Th2 responses and is effective in EAE treatment by itself (Stuve *et al.*, 2006). Combination therapy using suboptimal doses of atorvastatin and GA prevented or reversed clinical and histological EAE. Secretion of proinflammatory Th1 cytokines was reduced. Conversely, Th2 cytokine secretion was increased in these mice, but not in mice treated with each drug alone at the same dose. Monocytes treated with the combination of suboptimal doses of atorvastatin and GA secreted an anti-inflammatory type II cytokine pattern and, when used as APCs, promoted Th2 differentiation of naive myelin-specific T cells. These results in EAE demonstrate that agents with different mechanisms of immune modulation can be combined in a synergistic manner for the treatment of CNS autoimmunity (Giuliani *et al.*, 2005; Stuve *et al.*, 2006). On the other side, there are combinations without such benefits in EAE experiments like the combination of interferon (IFN)-α and GA (Brod *et al.*, 2000).

X. GA in MS: Mechanisms of Action

The EAE animal models have obvious limitations, for example, in EAE, GA is typically given mostly through a single injection before, together with immunostimulants like Freund's adjuvant, or at the time of disease induction. However, in patients with MS, the disease is established before treatment is initiated, which consists of daily injections. Not surprisingly therefore, some unexpected observations have been made on the immune response to GA in human beings. Several new papers, however, have shed light on the mechanisms of GA in MS and suggest several major effects of GA on different components of the immune system.

In contrast to other immunomodulatory MS therapies which exert its effects in an antigen-nonspecific way, GA appears to preferentially affect immune cells in an antigen-specific way.

The peptide mixture is given to patients with a putative autoimmune disease over many years by daily injection. From a broader perspective, GA is one of the few practical examples of therapeutic vaccination, as distinct from prophylactic vaccination against infectious diseases. Comparing different immunomodulatory treatments available today, three main principles can be described beyond immunosuppressive therapy (Fig. 2):

- *Cytokine-based therapies* using IFN-β can modulate all cells carrying an interferon receptor on their cell surface. This fact is responsible for the pleiotropy of different immunologic effects. On the other side, the necessary interaction between the interferon molecule and its receptor implies the risk that the effect can be neutralized by antibodies occurring during therapy.
- *Therapies with monoclonal antibodies* like natalizumab have the advantage that they can target specific molecules and cellular functions which may be of high relevance for disease pathogenesis. The problem is that specificity only means immunologic specificity. It is not the case that only the pathological process is blocked by the monoclonal antibody, but therapy interacts with physiological processes as well which may cause significant adverse effects.
- GA is one of the few positive examples of *an antigen-based, therapeutic vaccination* although there is still a significant discrepancy between the cured EAE animals and only modest effects in MS patients. Two important questions which can only be answered by further scientific work are whether the human protocol of GA vaccination can be improved to increase efficacy and whether all MS patients respond to GA therapy in the same way. A summary of most of the immunologic effects induced by GA treatment is presented in Fig. 3.

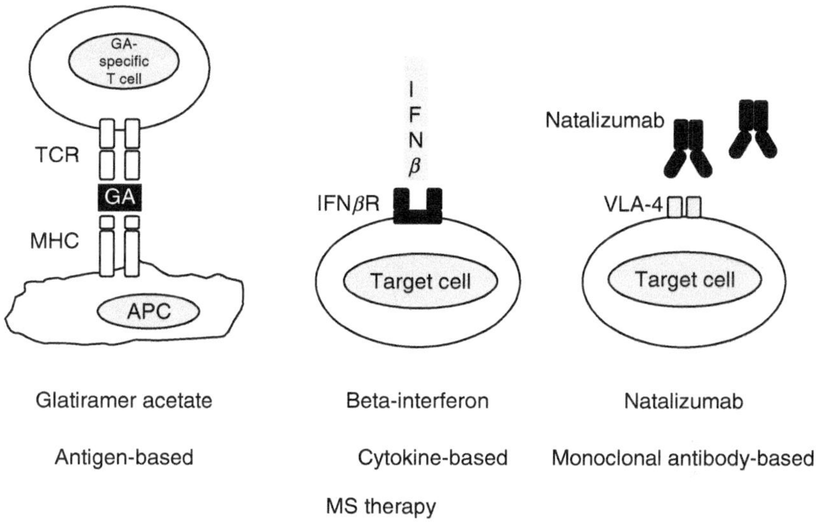

FIG. 2. Three actual principles in immunomodulatory MS therapy. Details in the text.

A. Effects of GA by Binding to MHC Class II Molecules

GA exhibits a very rapid, high and efficient binding to various MHC class II molecules on murine and human APCs without antigen processing and without any obvious preference for particular alleles, and even displaced peptides from the MHC-binding site (Fridkis-Hareli and Strominger, 1998). The competition for binding to the MHC can consequently lead to inhibition of various pathological effector functions. Although distinct binding motifs of GA to MS-associated HLA-DR molecules could be defined, virtually all of the GA polypeptides seem to have a binding capacity to MHC class II, owing perhaps to their random composition. In addition to MHC class II molecules, GA also seems to be able to interact with MHC class I. Clearly, this type of competition could occur with any antigen and is therefore antigen nonspecific (Neuhaus *et al.*, 2001).

B. Effects of GA on the APC Level

As past investigations have focused on GA-induced changes of the adaptive (antigen-specific) immune system, especially T and B cells, the effects of the drug on components of the innate (broadly reactive) immune system, notably monocytes and dendritic cells, have been studied more recently. Even in animal models, there is still a remarkable difference between the beneficial effect of GA-specific

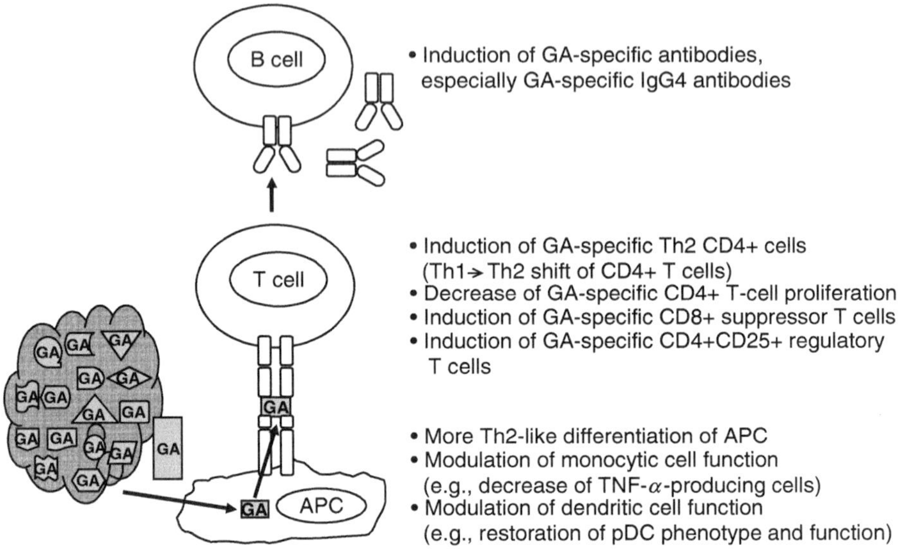

FIG. 3. Overview of the most important immunological changes during GA treatment in MS patients. Details in the text.

T cells and immunization with GA: Adoptive transfer of GA-specific T cells can block induction of EAE by different encephalitogens, but cannot cure ongoing EAE, which shows that T cells alone only have a limited role when disease is established (Aharoni *et al.*, 1993). By contrast, GA injections do not only prevent but also suppress established EAE (Teitelbaum *et al.*, 1971, 1973, 1974a).

The interplay between such APCs and T cells is fundamental in any adaptive immune response. In addition, there is reciprocal feedback between APC and T cells:

- APC influence the differentiation of T cells. So alterations on the APC level may be of high relevance for GA-induced changes of the adaptive immune response: GA treatment leads to generalized, antigen-nonspecific modulation of APC function in such a way that it stimulates Th2-like responses. This may be the critical mechanism to induce the GA-specific Th1–Th2 shift (Kim *et al.*, 2004b; Weber *et al.*, 2004).
- T cells modify the phenotype of APC. Human monocytes exposed to supernatants from Th1-like T cells tend to express a range of adhesion molecules, costimulatory molecules, and cytokines favoring differentiation along the Th1 pathway. By contrast, monocytes cultured in Th2 supernatants push T cells toward Th2 differentiation (Kim *et al.*, 2004b). In case of GA, it could be demonstrated that GA-specific Th2 cells inhibited the production of nitric oxide, interleukin 2 (IL-2), and TNF-α by lipopolysaccharide (LPS)-activated peritoneal macrophages (Kayhan *et al.*, 2003). This T-cell-mediated deviation of APC properties

has also been observed with human microglia cells, which suggests that human CNS APC can be modified *in vivo* into type-2 APC and subsequently favors Th2 differentiation (Kim *et al.*, 2004b). In experiments using an Alzheimer's disease animal model, microglia demonstrated a phenotype switch to dendritic-like (CD11c+) microglial cells by GA treatment (Butovsky *et al.*, 2006).

First *in vitro* experiments demonstrated that GA blocked the activation of a monocytic cell line in the absence of T cells. GA inhibited the induction of HLA-DR and HLA-DQ proteins, the release of TNF-α, and the activity of cathepsin B (Li *et al.*, 1998). Not only monocytic cell lines, but also freshly isolated human monocytes could be modulated *in vitro* by GA. GA blocked LPS-mediated induction of several activation markers (Kim *et al.*, 2004b; Weber *et al.*, 2004). By contrast, GA-treated monocytes produced increased amounts of the Th2-like cytokine IL-10.

On the level of dendritic cells, GA inhibited the production of IL-12 and TNF-α by human *in vitro*-generated dendritic cells (Hussien *et al.*, 2001; Sanna *et al.*, 2006; Vieira *et al.*, 2003). However, GA-treated dendritic cells were still able to stimulate proliferation of naive T cells (Sanna *et al.*, 2006; Vieira *et al.*, 2003). Interestingly, GA-treated dendritic cells shifted the phenotype of naive T cells towards Th2. Using an autologous dendritic cell–lymphocyte coculture system, Sanna *et al.* (2006) demonstrated that in MS patients, pretreatment with GA significantly decreased the *in vitro* proliferative effect of dendritic cells on lymphocytes as compared to healthy controls and to unpulsed or MBP-pulsed dendritic cells from MS patients. In addition, GA-treated dendritic cells from both MS patients and healthy controls significantly increased the lymphocyte production of IL-5 and IL-13 as compared to MBP-treated ones.

The effects of the drug on APC seem to depend on the cell type. For example, after exposure to GA, LPS-stimulated rat peritoneal macrophages produced less TNF-α and more IL-10. By contrast, GA-exposed, bone marrow-derived dendritic cells produced more TNF-α and IL-10 (Jung *et al.*, 2004).

Plasmacytoid dendritic cells (pDCs) represent a dendritic cell subtype that exerts divergent functions in innate and adoptive immunity, including the immediate reaction to microbial factors and the induction of immunoregulatory responses. As Stasiolek *et al.* (2006) have demonstrated, pDCs from MS patients exhibited lower expression levels of costimulatory markers and an inefficient maturation after stimulation compared to controls. In MS, but not in controls, pDCs failed to upregulate proliferative responses and IFN-γ secretion of autologous PBMCs in a coculture system. Interestingly, GA treatment partially restored phenotype and function of pDCs in MS patients. So GA seems to modulate functional abnormalities of this special cell type in MS patients, which might be of importance in immune dysregulation of this disease.

There is, however, an inherent problem in all *in vitro* studies—the difficulty of predicting the concentrations of GA that are reached in the relevant tissues *in vivo* during treatment. The fact that effects on APC in treated patients could be seen in cell preparations directly analyzed *ex vivo* suggests that GA modifies the properties of APC *in vivo* during treatment (Kim *et al.*, 2004b; Stasiolek *et al.*, 2006; Weber *et al.*, 2004). These observations provide evidence that daily subcutaneous administration of GA induces generalized changes in the APC system.

Another unresolved issue relates to the selectivity (antigen specificity) of the effects of GA. One might argue that global inhibition or deviation of APC should favor a Th2 shift not only of GA-reactive T cells, but also of T cells with other specificities. However, several studies have shown that the Th2 shift involves only GA-specific T cells, whereas circulating, preexisting memory T cells with other specificities (e.g., those specific for tetanus toxoid or tuberculin) seem to retain their cytokine profile (Farina *et al.*, 2001; Ragheb *et al.*, 2001).

In conclusion, APC inhibition and deviation seem to be a possible mechanism of action of GA. This possibility would explain the observed Th1–Th2 shift of T cells, which would still be a central element of the mechanism of action, but probably secondary to the effect on APC.

XI. Effects of GA on the B-Cell Level

As an example of antigen-based, therapeutic vaccination, the development of GA-reactive antibodies seems to be a condition *sine qua non*. So an early report, which included data from three different therapeutic trials, demonstrated that all GA-treated patients developed GA-reactive antibodies which declined after 6 months and remained low, but remained higher than baseline values (Brenner *et al.*, 2001).

By contrast, anti-MBP antibody titers were low and remained unchanged during treatment. Antibodies to GA were not detected in patients treated with placebo. All antibodies were of the IgG class with IgG1 isotype levels two- to threefold higher than those of IgG2. A cross-sectional study by Farina *et al.* (2002a) confirmed the presence of antibodies to GA in treated patients. They also found that some untreated individuals have naturally occurring anti-GA antibodies. Owing to the multiepitope nature of GA, the presence of reactive immunoglobulins in some unprimed individuals is not surprising. These naturally occurring antibodies might belong to the low-affinity, polyreactive pool. Some unexposed individuals have GA-reactive IgM, IgG1, and IgG2 antibodies. Treated patients also produce IgG1 and IgG2 antibodies; but by contrast with untreated individuals, they specifically produce IgG4 antibodies to GA. Because the isotype switch to IgG4 is regulated by the Th2 cytokine IL-4, IgG4 reflects the interaction of B cells with T lymphocytes, in this case the GA-induced Th1–Th2 shift. A comparable observation was described by Basile *et al.* (2006) demonstrating that in GA-treated, PPMS patients initially GA-reactive antibodies of the IgG1 subclass predominated, peaking at month 9 of therapy. After 9 months IgG1 decreased while anti-GA antibodies of the IgG4 subclass increased and remained high for the 3 years of follow-up.

As antigen-based therapy, there could be no neutralizing antibodies per definition because GA is presented by APC using the MHC complex, as stated above. So Teitelbaum *et al.* (2003) did not observe neutralizing activity in serum of GA-treated patients *in vitro*. By contrast, when precipitated serum proteins were used, Salama *et al.* (2003) demonstrated some neutralization of GA-specific T-cell reactivity. However, there is no clear evidence that antibodies to GA inhibit clinical efficacy *in vivo*. In fact, in one study, relapse-free patients seemed to develop higher antibody titers, which would be consistent with a beneficial rather than neutralizing activity of the antibodies (Brenner *et al.*, 2001).

Using a murine model of demyelinating disease, Ure and Rodriguez (2002) could demonstrate that remyelination of spinal cord axons is promoted by antibodies to GA. These results support the hypothesis that the antibody response in GA-treated patients may be beneficial by facilitating repair of demyelinated lesions. However, no clinical studies to date have shown either a neutralizing or a beneficial effect of anti-GA antibodies in patients with MS.

XII. Effects of GA on the T-Cell Level

In contrast to the lack of effect of GA on immune cells isolated from untreated animals, GA induces vigorous polyclonal proliferation of peripheral blood lymphocytes from untreated (unprimed) human donors (Brosnan *et al.*, 1985; Burns *et al.*, 1986; Duda *et al.*, 2000a,b; Farina *et al.*, 2001; Qin *et al.*, 2000). Like conventional peptide antigens, GA binds MHC molecules and is recognized by T cells via their antigen-specific TCR. Because GA is a pool of antigenic peptides, the strong primary T-cell proliferation it induces, even in healthy nontreated people, is not unexpected (Duda *et al.*, 2000b; Farina *et al.*, 2001). In principle, these GA-reactive T cells might be CD4+ or CD8+ (Karandikar *et al.*, 2002). In untreated patients, GA-reactive T cells have phenotypic properties of naive (antigen-inexperienced) T cells (Wiesemann *et al.*, 2001).

GA treatment induces several changes at the T-cell level *in vivo*: it modifies the GA-reactive T-cell repertoire—rendering it more diverse (Duda *et al.*, 2000a). It gradually reduces the proliferative reactivity of GA-specific CD4+ T cells (Duda *et al.*, 2000a; Farina *et al.*, 2001; Karandikar *et al.*, 2002; Ragheb *et al.*, 2001; Wiesemann *et al.*, 2003), probably by inducing apoptosis in some T cells (Rieks *et al.*, 2003; Ruggieri *et al.*, 2006). GA treatment restores GA-induced proliferation of CD8+ T cells to rates seen in healthy individuals (Karandikar *et al.*, 2002; Tennakoon *et al.*, 2006); it generates a new population of Th2 CD4+

T cells (Duda *et al.*, 2000a; Farina *et al.*, 2001; Ragheb *et al.*, 2001; Wiesemann *et al.*, 2003) and CD8+ cells (Farina *et al.*, 2001; Tennakoon *et al.*, 2006).

Overall, in our understanding, the immunologic changes are specific to GA, as it is not observed with recall antigens like tetanus toxoid and tuberculin (Farina *et al.*, 2001; Karandikar *et al.*, 2002; Tennakoon *et al.*, 2006). Theoretically, the observed decrease in GA-reactive T cells could be due to anergy induction or activation-induced cell death of GA-specific T cells.

A. Th1–Th2 Shift

GA treatment induces an *in vivo* change of the cytokine secretion pattern and the effector function of GA-reactive CD4+ Th cells, a so called Th1–Th2 shift (Aharoni *et al.*, 1997, 1998; Dabbert *et al.*, 2000; Duda *et al.*, 2000b; Farina *et al.*, 2001; Miller *et al.*, 1998; Neuhaus *et al.*, 2000). Th cells can be divided into several types based on their characteristics (Allen and Maizels, 1997). Th1 cells produce proinflammatory cytokines such as IL-2, IL-12, IFN-γ, and TNF-α. In contrast, the Th2/Th3 cells produce anti-inflammatory cytokines such as IL-4, IL-5, IL-6, IL-10, transforming growth factor (TGF)-β, and IL-13. Different lines of evidence suggest that GA treatment changes the properties of the GA-reactive T cells in such a way that they increasingly become Th2-like with time. Using intracellular double-immunofluorescence flow cytometry, Neuhaus *et al.* (2000) demonstrated that long-term GA-specific TCLs from untreated MS patients and healthy controls predominantly produce IFN-γ and are to be classified as Th1 cells, whereas GA-reactive TCLs from GA-treated MS patients predominantly produce IL-4, that is, behave like Th2 cells.

In addition, the study of Farina *et al.* (2001) demonstrated that an automated ELISPOT assay, which is able to detect cytokine production of individual PBL, allows the correct identification of GA-treated and untreated donors in most cases. GA-treated MS patients show (a) a significant reduction of GA-induced proliferation of peripheral blood mononuclear cells, (b) a positive IL-4 ELISPOT response mediated predominantly by CD4+ T cells after *in vitro* stimulation with a wide range of GA concentrations, and (c) an elevated IFN-γ response partially mediated by CD8+ T cells after stimulation with high GA concentrations. GA-reactive T cells seem to be not physically deleted, but rather they are modified in such a way that they respond to *in vitro* challenge with GA by secretion of cytokines but not by proliferation.

Chen *et al.* (2002) reported that the described immunologic changes (e.g., the Th1–Th2 shift) are sustained over long treatment periods. During GA treatment up to 9 years, there was a sustained Th2-biased GA response which is in part cross-reactive with MBP. This seems to be consistent with the long-term anti-inflammatory effects of the drug and bystander suppression.

B. GA as Altered Peptide Ligand

On the other hand, GA could act in the periphery as an "altered peptide ligand" relative to MBP (Neuhaus *et al.*, 2000; Nishimura *et al.*, 1998; Teitelbaum *et al.*, 1997b; Webb *et al.*, 1973). As a consequence, some of the circulating myelin-specific, potentially pathogenic T cells might become "anergic" or be otherwise changed in their properties, that is, in their migratory potential. This effect would be relatively antigen specific and presumably occur in the periphery at the injection sites or in the corresponding draining lymph nodes where MBP-specific T cells might be confronted with GA. Although some *in vitro* findings support this mechanism, it is not yet known whether the functional properties of MBP-specific T cells are altered in GA-treated patients. The proliferative activity and cytokine profile of MBP-specific TCLs isolated from GA-treated patients were unchanged (Duda *et al.*, 2000a; Qin *et al.*, 2000). In addition, it seems very unlikely that sufficient amounts of GA can reach the CNS to compete efficiently with locally degraded myelin antigens.

Whether GA-specific T cells migrate into the CNS in patients with MS who have been treated with this agent is not known. There are preliminary experimental data that GA-reactive Th2 cells can be cultured from CSF of GA-treated MS patients (Ziemssen *et al.*, unpublished data). In EAE, GA-reactive Th2 cells were found in the CNS (Aharoni *et al.*, 2000, 2003). Furthermore, human Th1 and Th2 GA-reactive T cells could migrate across an artificial blood–brain barrier *in vitro* (Kim *et al.*, 2004a).

C. Secretion of Neurotrophic Factors

As human immune cells can produce BDNF (Kerschensteiner *et al.*, 1999) and moreover, the receptor for BDNF, gp145Trkb, is expressed in neurons and astrocytes in MS brain lesions (Stadelmann *et al.*, 2002), we asked the following questions: (a) Can GA-reactive T lymphocytes produce BDNF, and if so, (b) do Th1- and Th2/Th3-type GA-specific T cells differ in their capacities to secrete BDNF?

To answer these questions, we had to overcome two major technical obstacles: (a) adapting our culture system to prevent added GA from affecting the BDNF enzyme-linked immunosorbent assay (ELISA) and (b) optimizing the intracellular detection of BDNF in individual T lymphocytes (Ziemssen *et al.*, 2002a). That ladder enabled us formally to demonstrate that GA-specific Th1, Th2, and Th0 cells all have the capacity to produce BDNF. We clearly show that GA-specific, activated Th0, Th1, and Th2 cells produce the neurotrophic factor BDNF—not only at the transcriptional (mRNA) level as detected by RT-PCR, but also at the protein level—using the newly developed assays for BDNF secretion and synthesis of BDNF in individual T cells. The results from all assays consistently showed a low level of basal secretion of BDNF by GA-specific T cells and an increase of BDNF production after stimulation (Fig. 4). For our analysis, we selected four well-characterized prototypic GA-specific TCLs. Similar results were obtained with >20 additional GA-specific TCLs (Ziemssen *et al.*, 2005). Chen *et al.* (2003) reproduced our results, but differing in the fact that they demonstrated a higher BDNF secretion by GA-specific Th2 versus Th1 cells.

The topographical relation between BDNF-producing immune cells and receptor-bearing neurons supports the general notion that there is a neurotrophic cross talk between the infiltrating immune cells and the autochthonous neurons. One might speculate that GA-reactive T cells, when they reach the CNS and are locally restimulated by cross-reactive myelin antigens, produce not only Th2-like immunomodulatory cytokines, but also neurotrophic factors. Therefore, these T cells might confer neuroprotection in addition to bystander suppression. These findings indicate that GA might have neuroprotective potential, which has been demonstrated in first preliminary studies using MRI.

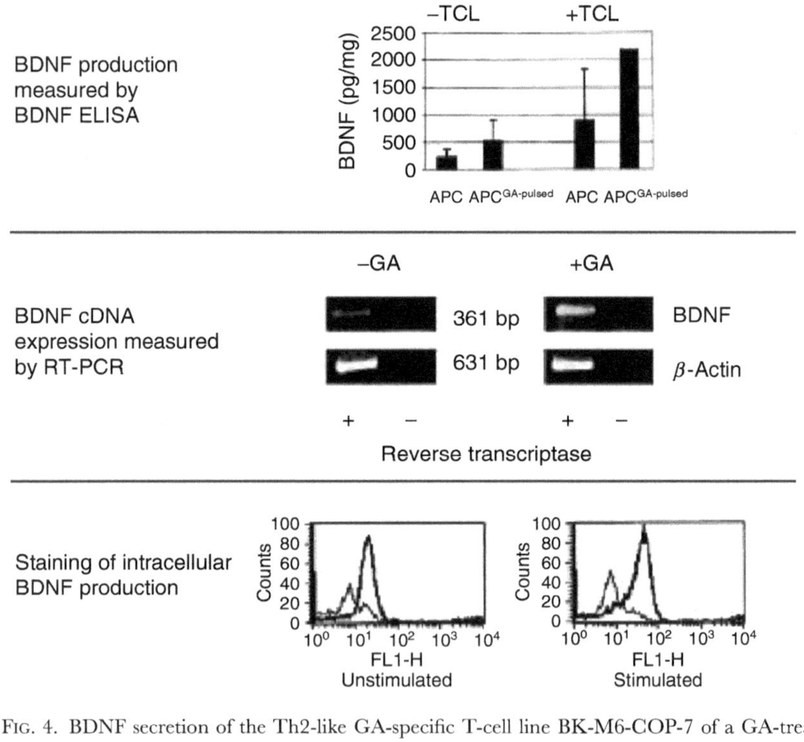

FIG. 4. BDNF secretion of the Th2-like GA-specific T-cell line BK-M6-COP-7 of a GA-treated patient, measured by BDNF ELISA, RT-PCR, and intracellular BDNF staining. There is increased BDNF production of GA-specific T cells after activation (taken with permission from Ziemssen *et al.* (2002a) by Oxford University Press).

D. Induction of GA-Specific CD8+ Suppressor T Cells

Karandikar *et al.* (2002) showed that the proliferative responses to GA are different in distinct subsets of T cells. Whereas GA-induced CD4+ T-cell responses were comparable in healthy individuals and MS patients, CD8+ T-cell responses were significantly lower in untreated MS patients. These GA-induced CD8+ T cells were regulatory/suppressor in nature. GA therapy significantly enhanced their suppressive ability in a contact-dependent manner (Tennakoon *et al.*, 2006). CD8+ T cells from GA-treated patients and healthy subjects, but not those from untreated patients with MS, exhibited potent, HLA class I-restricted, GA-specific cytotoxicity. These GA-induced cytotoxic CD8+ T cells could directly kill CD4+ T cells in a GA-specific manner, which are thought to be housed in the CD8+CD28− subset. Chronic exposure to antigen like daily GA treatment can result in the CD8+CD28− phenotype, similar to the CD4+CD25+ phenotype of acquired regulatory T cells. So GA therapy resulted in higher numbers of CD28−CD57+ cells among the GA-reactive CD8+ T cells.

E. Induction of Regulatory CD4+CD25+ T Cells

Regulatory T cells have been recognized recently as an important regulatory component that keeps autoreactive T cells in check. Significant deficiencies in the number or function of these regulatory T cells have been found to correlate with several autoimmune conditions, including MS (Viglietta *et al.*, 2004). In a study of Hong *et al.* (2005), GA treatment induced a significant increase in Foxp3 expression in CD4+ T cells in MS patients whose Foxp3 expression was reduced at baseline. CD4+CD25+ TCLs generated by GA expressed high levels of Foxp3 that correlated with an increased regulatory potential. Furthermore, it could be demonstrated that the induction of Foxp3 in T cells by GA was mediated through its ability to produce IFN-γ and, to a lesser degree, TGF-β1, as shown by antibody blocking and direct cytokine induction of Foxp3 expression in T cells.

In this regard, GA may compensate a functional deficit in this important regulatory mechanism in MS by its ability to induce or repair regulatory CD8+ or CD4+ T-cell response. The observation is also likely to offer a reasonable explanation for, or it may reconcile with, some previously described regulatory functions of GA of unknown mechanism, including so-called "bystander" inhibitory effect of GA on T-cell activation. The antigen nonspecific inhibitory effect of CD4+CD25+ regulatory T-cell response induced by GA is consistent with the spectrum of inhibition induced by GA that is often not limited to one antigen (i.e., MBP) and includes various other myelin antigens.

XIII. Conclusions

The proposed indication for the use of GA is to reduce the relapse rate in patients with RRMS, which has been linked adversely with long-term outcome. Several clinical trials have shown a consistent effect of a reduction of the relapse rate by approximately 30% and of accumulation of disability in relapsing-remitting MS patients treated with GA. The clinical benefit was supported by positive MRI effects. Clinical experience shows that GA is usually well tolerated and suitable for self-administration by patients with MS (Ziemssen *et al.*, 2001).

Concerning the specific mechanism of action, the following scenario would accommodate for both the findings in EAE and the observations in human MS (Fig. 5) (Farina *et al.*, 2005; Ziemssen, 2004):

1. Treatment with GA leads to the induction of GA-specific T-cell populations in the periphery. This may be the case because of immunomodulation of APC function by GA treatment.
2. In addition, regulatory CD8+ and CD4+CD25+ T cells are induced by GA therapy counteracting the disturbed immunoregulatory balance in MS patients.
3. Their constant activation by daily immunization should enable them to enter the CNS. The daily immunization seems to be important for the

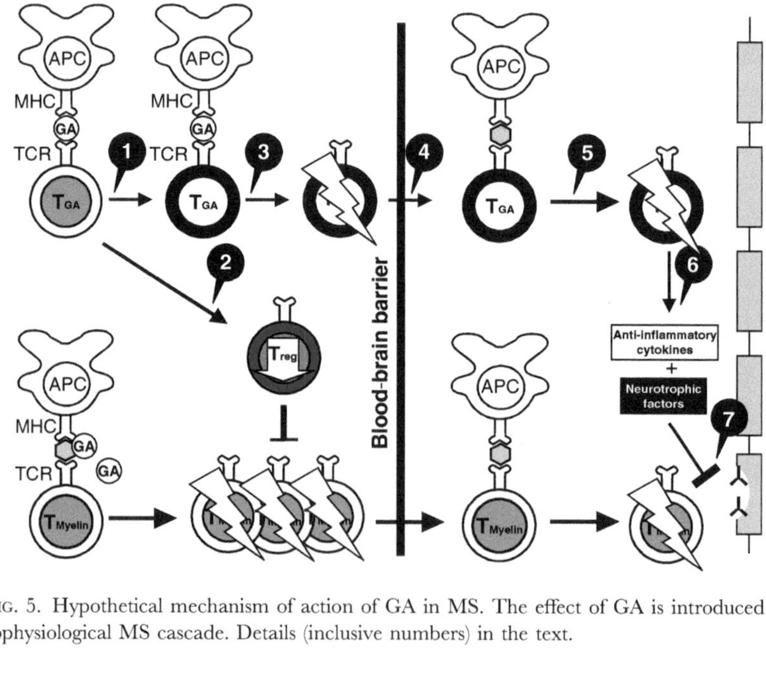

FIG. 5. Hypothetical mechanism of action of GA in MS. The effect of GA is introduced in the pathophysiological MS cascade. Details (inclusive numbers) in the text.

induction and maintenance of the regulatory/suppressive immune cell populations.

4. The migration via blood–brain barrier was demonstrated for transferred GA-reactive T cells which have been directly shown in the CNS of recipient mice in EAE models.
5. Inside the CNS, some GA-specific T cells cross-react with products of local myelin turnover presented by local APCs. In studies of Neuhaus *et al.* (2000), 8 ± 15% of GA-reactive TCLs cross-reacted with MBP as indicated by cytokine secretion, but not by proliferation. Recognition of cross-reactive antigens at the lesion site seems to be necessary for reactivation of the protective T cells and for their neuroprotective effect.
6. In response they secrete anti-inflammatory cytokines, which dampen the local inflammatory process (bystander suppression). Furthermore, they produce neurotrophic factors.
7. This might favor remyelination and axonal protection.

The ability of GA to downregulate inflammation and act directly at the lesion site facilitates axonal preservation. This effect may be related to the anti-inflammatory effect (bystander suppression) of GA and the ability of activated GA-reactive T cells to release neurotrophic factors. Because axonal loss is an early and irreversible event in MS, early treatment with drugs that mediate anti-inflammatory effects and offer neuroprotection can be an important factor in preventing disease progression and disability.

As with other immunomodulatory treatments, some GA-treated patients do not respond to this type of therapy. Early identification of such nonresponders would be very useful. The presently approved immunomodulatory treatments need to be given for at least 6–12 months before any clinical response becomes apparent. Obviously, early assessment of response would be much better. For this reason, there has been much interest in the development of appropriate biomarkers as a result of the response to treatment with IFN-β and GA. Better knowledge of the mechanisms of action of these agents should be helpful in this regard. After a first pilot study by Farina *et al.* (2002b), we just started a prospective study using the described immunologic parameters of GA treatment to identify responders and nonresponders immunologically and compare these findings with the clinical outcome of these patients (CopImmunoNet trial).

References

Abramsky, O., Teitelbaum, D., and Arnon, R. (1977). Effect of a synthetic polypeptide (COP 1) on patients with multiple sclerosis and with acute disseminated encephalomeylitis. Preliminary report. *J. Neurol. Sci.* **31,** 433–438.

Aharoni, R., Teitelbaum, D., and Arnon, R. (1993). T suppressor hybridomas and interleukin-2 dependent lines induced by copolymer 1 or by spinal cord homogenate down-regulate experimental allergic encephalomyelitis. *Eur. J. Immunol.* **23,** 17–25.

Aharoni, R., Teitelbaum, D., Sela, M., and Arnon, R. (1997). Copolymer 1 induces T cells of the T helper type 2 that crossreact with myelin basic protein and suppress experimental autoimmune encephalomyelitis. *Proc. Natl. Acad. Sci. USA* **94,** 10821–10826.

Aharoni, R., Teitelbaum, D., Sela, M., and Arnon, R. (1998). Bystander suppression of experimental autoimmune encephalomyelitis by T cell lines and clones of the Th2 type induced by copolymer 1. *J. Neuroimmunol.* **91,** 135–146.

Aharoni, R., Teitelbaum, D., Arnon, R., and Sela, M. (1999). Copolymer 1 acts against the immunodominant epitope 82–100 of myelin basic protein by T cell receptor antagonism in addition to major histocompatibility complex blocking. *Proc. Natl. Acad. Sci. USA* **96,** 634–639.

Aharoni, R., Teitelbaum, D., Leitner, O., Meshorer, A., Sela, M., and Arnon, R. (2000). Specific TH2 cells are present in the central nervous system of mice protected against EAE by copolymer 1. *Proc. Natl. Acad. Sci. USA* **97,** 11472–11477.

Aharoni, R., Kayhan, B., Eilam, R., Sela, M., and Arnon, R. (2003). Glatiramer acetate-specific T cells in the brain express T helper 2/3 cytokines and brain-derived neurotrophic factor *in situ*. *Proc. Natl. Acad. Sci. USA* **100,** 14157–14162.

Aharoni, R., Eilam, R., Domev, H., Labunskay, G., Sela, M., and Arnon, R. (2005). The immunomodulator glatiramer acetate augments the expression of neurotrophic factors in brains of experimental autoimmune encephalomyelitis mice. *Proc. Natl. Acad. Sci. USA* **102**(52), 19045–19050.

Allen, J. E., and Maizels, R. M. (1997). Th1-Th2: Reliable paradigm or dangerous dogma? *Immunol. Today* **18,** 387–392.

Angelov, D. N., Waibel, S., Guntinas-Lichius, O., Lenzen, M., Neiss, W. F., Tomov, T. L., Yoles, E., Kipnis, J., Schori, H., Reuter, A., Ludolph, A., and Schwartz, M. (2003). Therapeutic vaccine for acute and chronic motor neuron diseases: Implications for amyotrophic lateral sclerosis. *Proc. Natl. Acad. Sci. USA* **100**(8), 4790–4795.

Arnon, R., and Aharoni, R. (2004). Mechanism of action of glatiramer acetate in multiple sclerosis and its potential for the development of new applications. *Proc. Natl. Acad. Sci. USA* **101,** 14593–14598.

Arnon, R., Sela, M., and Teitelbaum, D. (1996). New insights into the mechanism of action of copolymer 1 in experimental allergic encephalomyelitis and multiple sclerosis. *J. Neurol.* **243** (Suppl. 1), S8–S13.

Basile, E., Gibbs, E., Aziz, T., and Oger, J. (2006). During 3 years treatment of primary progressive multiple sclerosis with glatiramer acetate, specific antibodies switch from IgG1 to IgG4. *J. Neuroimmunol.* **177**(1–2), 161–166.

Batchelor, P. E., Liberatore, G. T., Wong, J. Y., Porritt, M. J., Frerichs, F., Donnan, G. A., and Howells, D. W. (1999). Activated macrophages and microglia induce dopaminergic sprouting in the injured striatum and express brain-derived neurotrophic factor and glial cell line-derived neurotrophic factor. *J. Neurosci.* **19,** 1708–1716.

Besser, M., and Wank, R. (1999). Cutting edge: Clonally restricted production of the neurotrophins brain-derived neurotrophic factor and neurotrophin-3 mRNA by human immune cells and Th1/Th2-polarized expression of their receptors. *J. Immunol.* **162,** 6303–6306.

Bitsch, A., Schuchardt, J., Bunkowski, S., Kuhlmann, T., and Bruck, W. (2000). Acute axonal injury in multiple sclerosis. Correlation with demyelination and inflammation. *Brain* **123,** 1174–1183.

Bo, L., Vedeler, C. A., Nyland, H., Trapp, B. D., and Mork, S. J. (2003). Intracortical multiple sclerosis lesions are not associated with increased lymphocyte infiltration. *Mult. Scler.* **9,** 323–331.

Bornstein, M. B., Miller, A. I., Teitelbaum, D., Arnon, R., and Sela, M. (1982). Multiple sclerosis: Trial of a synthetic polypeptide. *Ann. Neurol.* **11,** 317–319.

Bornstein, M. B., Miller, A. I., Slagle, S., Arnon, R., Sela, M., and Teitelbaum, D. (1984). Clinical trials of copolymer I in multiple sclerosis. *Ann. N. Y. Acad. Sci.* **436,** 366–372.

Bornstein, M. B., Miller, A., Slagle, S., Weitzman, M., Crystal, H., Drexler, E., Keilson, M., Merriam, A., Wassertheil-Smoller, S., Spada, V., Weiss, W., Arnov, R., *et al.* (1987). A pilot trial of Cop 1 in exacerbating-remitting multiple sclerosis. *N. Engl. J. Med.* **317,** 408–414.

Brenner, T., Arnon, R., Sela, M., Abramsky, O., Meiner, Z., Riven-Krietman, R., Tarcik, N., and Teitelbaum, D. (2001). Humoral and cellular immune responses to Copolymer 1 in multiple sclerosis patients treated with Copaxone. *J. Neuroimmunol.* **115,** 152–160.

Brod, S. A., Lindsey, J. W., and Wolinsky, J. S. (2000). Combination therapy with glatiramer acetate (copolymer-1) and a type I interferon (IFN-alpha) does not improve experimental autoimmune encephalomyelitis. *Ann. Neurol.* **47,** 127–131.

Brosnan, C. F., Litwak, M., Neighbour, P. A., Lyman, W. D., Carter, T. H., Bornstein, M. B., and Bloom, B. R. (1985). Immunogenic potentials of copolymer I in normal human lymphocytes. *Neurology* **35,** 1754–1759.

Burns, J., Krasner, L. J., and Guerrero, F. (1986). Human cellular immune response to copolymer I and myelin basic protein. *Neurology* **136,** 92–94.

Butovsky, O., Koronyo-Hamaoui, M., Kunis, G., Ophir, E., Landa, G., Cohen, H., and Schwartz, M. (2006). From the cover: Glatiramer acetate fights against Alzheimer's disease by inducing dendritic-like microglia expressing insulin-like growth factor 1. *Proc. Natl. Acad. Sci. USA* **103**(31), 11784–11789.

Charcot, J. M. (1877). Lectures on the Diseases of the Nervous System (translated by R. M. May). New Sydenham Society, London.

Chen, M., Conway, K., Johnson, K. P., Martin, R., and Dhib-Jalbut, S. (2002). Sustained immunological effects of Glatiramer acetate in patients with multiple sclerosis treated for over 6 years. *J. Neurol. Sci.* **201,** 71–77.

Chen, M., Valenzuela, R. M., and Dhib-Jalbut, S. (2003). Glatiramer acetate-reactive T cells produce brain-derived neurotrophic factor. *J. Neurol. Sci.* **215,** 37–44.

Chitnis, T., Imitola, J., and Khoury, S. J. (2005). Therapeutic strategies to prevent neurodegeneration and promote regeneration in multiple sclerosis. *Curr. Drug Targets Immune Endocr. Metabol. Disord.* **5,** 11–26.

Cohen, I. R. (1992). The cognitive paradigm and the immunological homunculus. *Immunol. Today* **13,** 490–494.

Coles, A. J., Wing, M. G., Molyneux, P., Paolillo, A., Davie, C. M., Hale, G., Miller, D., Waldmann, H., and Compston, A. (1999). Monoclonal antibody treatment exposes three mechanisms underlying the clinical course of multiple sclerosis. *Ann. Neurol.* **46,** 296–304.

Comi, G., Filippi, M., and Wolinsky, J. S. (2001). European/Canadian multicenter, double-blind, randomized, placebo-controlled study of the effects of glatiramer acetate on magnetic resonance imaging—measured disease activity and burden in patients with relapsing multiple sclerosis. European/Canadian Glatiramer Acetate Study Group. *Ann. Neurol.* **49,** 290–297.

Dabbert, D., Rosner, S., Kramer, M., Scholl, U., Tumani, H., Mader, M., and Weber, F. (2000). Glatiramer acetate (copolymer-1)-specific, human T cell lines: Cytokine profile and suppression of T cell lines reactive against myelin basic protein. *Neurosci. Lett.* **289,** 205–208.

Duda, P. W., Schmied, M. C., Cook, S. L., Krieger, J. I., and Hafler, D. A. (2000a). Glatiramer acetate (Copaxone) induces degenerate, Th2-polarized immune responses in patients with multiple sclerosis. *J. Clin. Invest.* **105,** 967–976.

Duda, P. W., Krieger, J. I., Schmied, M. C., Balentine, C., and Hafler, D. A. (2000b). Human and murine CD4 T cell reactivity to a complex antigen: Recognition of the synthetic random polypeptide glatiramer acetate. *J. Immunol.* **165,** 7300–7307.

Farina, C., Then Bergh, F., Albrecht, H., Meinl, E., Yassouridis, A., Neuhaus, O., and Hohlfeld, R. (2001). Treatment of multiple sclerosis with Copaxone (COP): Elispot assay detects COP-induced interleukin-4 and interferon-gamma response in blood cells. *Brain* **124,** 705–719.

Farina, C., Vargas, V., Heydari, N., Kümpfel, T., Meinl, E., and Hohlfeld, R. (2002a). Treatment with glatiramer acetate induces specific IgG4 antibodies in multiple sclerosis patients. *J. Neuroimmunol.* **123,** 188–192.

Farina, C., Wagenpfeil, P., and Hohlfeld, R. (2002b). Immunological assay for assessing the efficacy of glatiramer acetate (Copaxone) in MS. A pilot study. *J. Neurol.* **249,** 1587–1592.

Farina, C., Weber, M. S., Meinl, E., Wekerle, H., and Hohlfeld, R. (2005). Glatiramer acetate in multiple sclerosis: Update on potential mechanisms of action. *Lancet Neurol.* **4,** 567–575.

Filippi, M., Rovaris, M., Rocca, M. A., Sormani, M. P., Wolinsky, J. S., Comi, G., and European/Canadian Glatiramer Acetate Study Group (2001). Glatiramer acetate reduces the proportion of new MS lesions evolving into "black holes." *Neurology* **57,** 731–733.

Filippi, M., Bozzali, M., Rovaris, M., Gonen, O., Kesavadas, C., Ghezzi, A., Martinelli, V., Grossman, R. I., Scotti, G., Comi, G., and Falini, A. (2003). Evidence for widespread axonal damage at the earliest clinical stage of multiple sclerosis. *Brain* **126,** 433–437.

Filippi, M., Wolinsky, J. S., Comi, G., and CORAL StudyGroup (2006). Effects of oral glatiramer acetate on clinical and MRI-monitored disease activity in patients with relapsing multiple sclerosis: A multicentre, double-blind, randomised, placebo-controlled study. *Lancet Neurol.* **5**(3), 213–220.

Ford, C. C., Johnson, K. P., Lisak, R. P., Panitch, H. S., Shifronis, G., Wolinsky, J. S., and Copaxone Study Group (2006). A prospective open-label study of glatiramer acetate: Over a decade of continuous use in multiple sclerosis patients. *Mult. Scler.* **12**(3), 309–320.

Fridkis-Hareli, M., and Strominger, J. L. (1998). Promiscuous binding of synthetic copolymer 1 to purified HLA-DR molecules. *J. Immunol.* **160,** 4386–4397.

Fridkis-Hareli, M., Teitelbaum, D., Gurevich, E., Pecht, I., Brautbar, C., Kwon, O. J., Brenner, T., Arnon, R., and Sela, M. (1994). Direct binding of myelin basic protein and synthetic copolymer 1 to class II major histocompatibility complex molecules on living antigen-presenting cells—specificity and promiscuity. *Proc. Natl. Acad. Sci. USA* **91,** 4872–4876.

Ge, Y., Grossman, R. I., Udupa, J. K., Fulton, J., Constantinescu, C. S., Gonzales-Scarano, F., Babb, J. S., Mannon, L. J., Kolson, D. L., and Cohen, J. A. (2000). Glatiramer acetate (Copaxone) treatment in relapsing-remitting MS: Quantitative MR assessment. *Neurology* **54,** 813–817.

Gilgun-Sherki, Y., Panet, H., Holdengreber, V., Mosberg-Galili, R., and Offen, D. (2003). Axonal damage is reduced following glatiramer acetate treatment in C57/bl mice with chronic-induced experimental autoimmune encephalomyelitis. *Neurosci. Res.* **47,** 201–207.

Giuliani, F., Metz, L. M., Wilson, T., Fan, Y., Bar-Or, A., and Yong, V. W. (2005). Additive effect of the combination of glatiramer acetate and minocycline in a model of MS. *J. Neuroimmunol.* **158** (1–2), 213–221.

Gold, R., Linington, C., and Lassman, H. (2006). Understanding pathogenesis and therapy of multiple sclerosis via animal models: 70 years of merits and culprits in experimental autoimmune encephalomyelitis research. *Brain* **129**(Pt. 8), 1953–1971.

Grigoriadis, N., Ben-Hur, T., Karussis, D., and Milonas, I. (2004). Axonal damage in multiple sclerosis: A complex issue in a complex disease. *Clin. Neurol. Neurosurg.* **106,** 211–217.

Hafler, D. A. (2004). Multiple sclerosis. *J. Clin. Invest.* **113,** 788–794.

Hohlfeld, R., Kerschensteiner, M., Stadelmann, C., Lassmann, H., and Wekerle, H. (2006). The neuroprotective effect of inflammation: Implications for the therapy of multiple sclerosis. *Neurol. Sci.* **27**(Suppl. 1), S1–S7.

Hong, J., Li, N., Zhang, X., Zheng, B., and Zhang, J. Z. (2005). Induction of CD4+CD25+ regulatory T cells by copolymer-I through activation of transcription factor Foxp3. *Proc. Natl. Acad. Sci. USA* **102,** 6449–6454.

Hussien, Y., Sanna, A., Soderstrom, M., Link, H., and Huang, Y. (2001). Glatiramer acetate and IFN-β act on dendritic cells in multiple sclerosis. *J. Neuroimmunol.* **121,** 102–110.

Jee, Y., Liu, R., Bai, X. F., Campagnolo, D. I., Shi, F. D., and Vollmer, T. L. (2006). Do TH2 cells mediate the effects of GA in EAE? *Int. Immunol.* **18**(4), 537–544.

Johnson, K. P. (1996). A review of the clinical efficacy profile of copolymer 1: New U.S. phase III trial data. *J. Neurol.* **243**(Suppl. 1), S3–S7.

Johnson, K. P., Brooks, B. R., Cohen, J. A., Ford, C. C., Goldstein, J., Lisak, R. P., Myers, L. W., Panitch, H. S., Rose, J. W., Schiffer, R. B., Vollmer, T., Weiner, L. P., *et al.* (1995). Copolymer 1 reduces relapse rate and improves disability in relapsing-remitting multiple sclerosis: Results of a phase III multicenter, double-blind placebo-controlled trial. The Copolymer 1 Multiple Sclerosis Study Group. *Neurology* **45,** 1268–1276.

Johnson, K. P., Brooks, B. R., Cohen, J. A., Ford, C. C., Goldstein, J., Lisak, R. P., Myers, L. W., Panitch, H. S., Rose, J. W., Schiffer, R. B., Vollmer, T., Weiner, L. P., *et al.* (1998). Extended use of glatiramer acetate (Copaxone) is well tolerated and maintains its clinical effect on multiple sclerosis relapse rate and degree of disability. Copolymer 1 Multiple Sclerosis Study Group. *Neurology* **50,** 701–708.

Jung, S., Siglienti, I., Grauer, O., Magnus, T., Scarlato, G., and Toyka, K. V. (2004). Induction of IL-10 in rat peritoneal macrophages and dendritic cells by glatiramer acetate. *J. Neuroimmunol.* **148,** 63–73.

Karandikar, N. J., Crawford, M. P., Yan, X., Ratts, R. B., Brenchley, J. M., Ambrozak, D. R., Lovett-Racke, A. E., Frohman, E. M., Stastny, P., Douek, D. C., Koup, R. A., and Racke, M. K. (2002). Glatiramer acetate (Copaxone) therapy induces CD8(+) T cell responses in patients with multiple sclerosis. *J. Clin. Invest.* **109,** 641–649.

Kayhan, B., Aharoni, R., and Arnon, R. (2003). Glatiramer acetate regulates nitric oxide and related cytokine secretion in experimental autoimmune encephalomyelitis. *Immunol. Lett.* **88,** 185–192.

Kerschensteiner, M., Gallmeier, E., Behrens, L., Leal, V. V., Misgeld, T., Klinkert, W. E., Kolbeck, R., Hoppe, E., Oropeza-Wekerle, R. L., Bartke, I., Stadelmann, C., Lassmann, H., *et al.* (1999). Activated human T cells, B cells, and monocytes produce brain-derived neurotrophic factor *in vitro* and in inflammatory brain lesions: A neuroprotective role of inflammation? *J. Exp. Med.* **189,** 865–870.

Kerschensteiner, M., Stadelmann, C., Dechant, G., Wekerle, H., and Hohlfeld, R. (2003). Neurotrophic cross-talk between the nervous system and immune system: Implications for inflammatory and degenerative neurological diseases. *Ann. Neurol.* **53**(3), 292–304.

Khan, O., Shen, Y., Caon, C., Bao, F., Ching, W., Reznar, M., Buccheister, A., Hu, J., Latif, Z., Tselis, A., and Lisak, R. (2005). Axonal metabolic recovery and potential neuroprotective effect of glatiramer acetate in relapsing-remitting multiple sclerosis. *Mult. Scler.* **11**(6), 646–651.

Kim, H. J., Biernacki, K., Prat, A., Antel, J. P., and Bar-Or, A. (2004a). Inflammatory potential and migratory capacities across human brain endothelial cells of distinct glatiramer acetate-reactive T cells generated in treated multiple sclerosis patients. *Clin. Immunol.* **111,** 38–46.

Kim, H. J., Ifergan, I., Antel, J. P., Seguin, R., Duddy, M., Lapierre, Y., Jalili, F., and Bar-Or, A. (2004b). Type 2 monocyte and microglia differentiation mediated by glatiramer acetate therapy in patients with multiple sclerosis. *J. Immunol.* **172,** 7144–7153.

Kipnis, J., Yoles, E., Porat, Z., Cohen, A., Mor, F., Sela, M., and Schwartz, M. (2000). T cell immunity to copolymer 1 confers neuroprotection on the damaged optic nerve: Possible therapy for optic neuropathies. *Proc. Natl. Acad. Sci. USA* **97,** 7446–7451.

Klein, R., Nanduri, V., Jing, S., Lamballe, F., Tapley, P., Bryant, S., Cordon-Cardo, C., Jones, K. R., Reichardt, L. F., and Barbacid, M. (1991). The trkB tyrosine protein kinase is a receptor for brain-derived neurotrophic factor and neurotrophin-3. *Cell* **66,** 395–403.

Lewin, G. R., and Barde, Y. A. (1996). Physiology of the neurotrophins. *Annu. Rev. Neurosci.* **19,** 289–317.

Li, Q. D., Milo, R., Panitch, H., Swoveland, P., and Bever, C. T. (1998). Glatiramer acetate blocks the activation of THP-1 cells by interferon-gamma. *Eur. J. Pharmacol.* **342,** 303–310.

Mancardi, G. L., Sardanelli, F., Parodi, R. C., Melani, E., Capello, E., Inglese, M., Ferrari, A., Sormani, M. P., Ottonello, C., Levrero, F., Uccelli, A., and Bruzzi, P. (1998). Effect of copolymer-1 on serial gadolinium-enhanced MRI in relapsing remitting multiple sclerosis. *Neurology* **50,** 1127–1133.

Miller, A., Shapiro, S., Gershtein, R., Kinarty, A., Rawashdeh, H., Honigman, S., and Lahat, N. (1998). Treatment of multiple sclerosis with copolymer-1 (Copaxone): Implicating mechanisms of Th1 to Th2/Th3 immune-deviation. *J. Neuroimmunol.* **92,** 113–121.

Moalem, G., Leibowitz-Amit, R., Yoles, E., Mor, F., Cohen, I., and Schwartz, M. (1999). Autoimmune T cells protect neurons from secondary degeneration after central nervous system axotomy. *Nat. Med.* **5,** 49–55.

Neuhaus, O., Farina, C., Yassouridis, A., Wiendl, H., Then Bergh, F., Dose, T., Wekerle, H., and Hohlfeld, R. (2000). Multiple sclerosis: Comparison of copolymer-1-reactive T cell lines from treated and untreated subjects reveals cytokine shift from T helper 1 to T helper 2 cells. *Proc. Natl. Acad. Sci. USA* **97,** 7452–7457.

Neuhaus, O., Farina, C., Wekerle, H., and Hohlfeld, R. (2001). Mechanisms of action of glatiramer acetate in multiple sclerosis. *Neurology* **56,** 702–708.

Nishimura, Y., Chen, Y. Z., Kanai, T., Yokomizo, H., Matsuoka, T., and Matsushita, S. (1998). Modification of human T-cell responses by altered peptide ligands: A new approach to antigen-specific modification. *Intern. Med.* **37,** 804–817.

Perry, V. H., and Anthony, D. C. (1999). Axon damage and repair in multiple sclerosis. *Phil. Trans. R. Soc. London B* **354,** 1641–1647.

Peterson, J. W., Bo, L., Mork, S., Chang, A., and Trapp, B. D. (2001). Transected neurites, apoptotic neurons, and reduced inflammation in cortical multiple sclerosis lesions. *Ann. Neurol.* **50,** 389–400.

Qin, Y., Zhang, D. Q., Prat, A., Pouly, S., and Antel, J. (2000). Characterization of T cell lines derived from glatiramer-acetate-treated multiple sclerosis patients. *J. Neuroimmunol.* **108,** 201–206.

Ragheb, S., Abramczyk, S., Lisak, D., and Lisak, R. (2001). Long-term therapy with glatiramer acetate in multiple sclerosis: Effect on T cells. *Mult. Scler.* **7,** 43–47.

Rapalino, O., Lazarov-Spiegler, O., Agranov, E., Velan, G. J., Yoles, E., Fraidakis, M., Solomon, A., Gepstein, R., Katz, A., Belkin, M., Hadani, M., and Schwartz, M. (1998). Implantation of stimulated homologous macrophages results in partial recovery of paraplegic rats. *Nat. Med.* **4,** 814–821.

Rieks, M., Hoffmann, V., Aktas, O., Juschka, M., Spitzer, I., Brune, N., Schimrigk, S., Przuntek, H., and Pöhlau, D. (2003). Induction of apoptosis of CD4+ T cells by immunomodulatory therapy of multiple sclerosis with glatiramer acetate. *Eur. Neurol.* **50,** 200–206.

Ruggieri, M., Avolio, C., Scacco, S., Pica, C., Lia, A., Zimatore, G. B., Papa, S., Livrea, P., and Trojano, M. (2006). Glatiramer acetate induces pro-apoptotic mechanisms involving Bcl-2, Bax and Cyt-c in peripheral lymphocytes from multiple sclerosis patients. *J. Neurol.* **253,** 231–236.

Salama, H. H., Hong, J., Zang, Y. C. Q., El Mongui, A., and Zhang, J. W. (2003). Blocking effects of serum reactive antibodies induced by glatiramer acetate treatment in multiple sclerosis. *Brain* **126,** 2638–2647.

Sanna, A., Fois, M. L., Arru, G., Huang, Y. M., Link, H., Pugliatti, M., Rosati, G., and Sotgiu, S. (2006). Glatiramer acetate reduces lymphocyte proliferation and enhances IL-5 and IL-13 production through modulation of monocyte-derived dendritic cells in multiple sclerosis. *Clin. Exp. Immunol.* **143**(2), 357–362.

Schori, H., Kipnis, J., Yoles, E., WoldeMussie, E., Ruiz, G., Wheeler, L. A., and Schwartz, M. (2001). Vaccination for protection of retinal ganglion cells against death from glutamate cytotoxicity and ocular hypertension: Implications for glaucoma. *Proc. Natl. Acad. Sci. USA* **98,** 3398–3403.

Schwartz, M., Moalem, G., Leibowitz-Amit, R., and Cohen, I. R. (1999). Innate and adaptive immune responses can be beneficial for CNS repair. *Trends Neurosci.* **22,** 295–299.

Serpe, C. J., Kohm, A. P., Huppenbauer, C. B., Sanders, V. J., and Jones, K. J. (1999). Exacerbation of facial motor-neuron loss after facial nerve transection in severe combined immunodeficient (SCID) mice. *J. Neurosci.* **RC7,** 1–5.

Sormani, M. P., Rovaris, M., Valsasina, P., Wolinsky, J. S., Comi, G., and Filippi, M. (2004). Measurement error of two different techniques for brain atrophy assessment in multiple sclerosis. *Neurology* **62**(8), 1432–1434.

Stadelmann, C., Kerschensteiner, M., Misgeld, T., Bruck, W., Hohlfeld, R., and Lassmann, H. (2002). BDNF and gp145trkB in multiple sclerosis brain lesions: Neuroprotective interactions between immune and neuronal cells? *Brain* **125,** 75–85.

Stasiolek, M., Bayas, A., Kruse, N., Wieczarkowiecz, A., Toyka, K. V., Gold, R., and Selmaj, K. (2006). Impaired maturation and altered regulatory function of plasmacytoid dendritic cells in multiple sclerosis. *Brain* **129**(Pt. 5), 1293–1305.

Steinman, L. (2001). Multiple sclerosis: A two-stage disease. *Nat. Immunol.* **2,** 762–764.

Stuve, O., Youssef, S., Weber, M. S., Nessler, S., von Budingen, H. C., Hemmer, B., Prod'homme, T., Sobel, R. A., Steinman, L., and Zamvil, S. S. (2006). Immunomodulatory synergy by combination of atorvastatin and glatiramer acetate in treatment of CNS autoimmunity. *J. Clin. Invest.* **116**(4), 1037–1044.

Teitelbaum, D., Meshorer, A., Hirshfeld, T., Arnon, R., and Sela, M. (1971). Suppression of experimental allergic encephalomyelitis by a synthetic polypeptide. *Eur. J. Immunol.* **1,** 242–248.

Teitelbaum, D., Webb, C., Meshorer, A., Arnon, R., and Sela, M. (1973). Suppression by several synthetic polypeptides of experimental allergic encephalomyelitis induced in guinea pigs and rabbits with bovine and human basic encephalitogen. *Eur. J. Immunol.* **3,** 273–279.

Teitelbaum, D., Webb, C., Meshorer, A., Arnon, R., and Sela, M. (1974a). Protection against experimental allergic encephalomyelitis. *Nature* **240,** 564–566.

Teitelbaum, D., Webb, C., Bree, M., Meshorer, A., Arnon, R., and Sela, M. (1974b). Suppression of experimental allergic encephalomyelitis in Rhesus monkeys by a synthetic basic copolymer. *Clin. Immunol. Immunopathol.* **3,** 256–262.

Teitelbaum, D., Sela, M., and Arnon, R. (1997a). Copolymer 1 from the laboratory to FDA. *Isr. J. Med. Sci.* **33,** 280–284.

Teitelbaum, D., Arnon, R., and Sela, M. (1997b). Copolymer 1, from basic research to clinical application. *Cell. Mol. Life Sci.* **53,** 24–28.

Teitelbaum, D., Brenner, T., Abramsky, O., Aharoni, R., Sela, M., and Arnon, R. (2003). Antibodies to glatiramer acetate do not interfere with its biological functions and therapeutic efficacy. *Mult. Scler.* **9,** 592–599.

Tennakoon, D. K., Mehta, R. S., Ortega, S. B., Bhoj, V., Racke, M. K., and Karandikar, N. J. (2006). Therapeutic induction of regulatory, CD8+ T cells in multiple sclerosis. *J. Immunol.* **176,** 7119–7129.

Torcia, M., Bracci-Laudiero, L., Lucibello, M., Nencioni, L., Labardi, D., Rubartelli, A., Cozzolino, F., Aloe, L., and Garaci, E. (1996). Nerve growth factor is an autocrine survival factor for memory B lymphocytes. *Cell* **85,** 345–356.

Trapp, B. D., Peterson, J., Ransohoff, R. M., Rudick, R., Mork, S., and Bo, L. (1998). Axonal transection in the lesions of multiple sclerosis. *N. Engl. J. Med.* **338,** 278–285.

Ure, D. R., and Rodriguez, M. (2002). Polyreactive antibodies to glatiramer acetate promote myelin repair in murine model of demyelinating disease. *FASEB J.* **16,** 1260–1262.

Vieira, P. L., Heystek, H. C., Wormmeester, J., Wierenga, E. A., and Kapsenberg, M. L. (2003). Glatiramer acetate (copolymer-1, Copaxone) promotes Th2 cell development and increased IL-10 production through modulation of dendritic cells. *J. Immunol.* **170,** 4483–4488.

Viglietta, V., Baecher-Allan, C., Weiner, H. L., and Hafler, D. A. (2004). Loss of functional suppression by CD4+CD25+ regulatory T cells in patients with multiple sclerosis. *J. Exp. Med.* **199,** 971–979.

Webb, C., Teitelbaum, D., Arnon, R., and Sela, M. (1973). *In vivo* and *in vitro* immunological cross-reactions between basic encephalitogen and synthetic basic polypeptides capable of suppressing experimental allergic encephalomyelitis. *Eur. J. Immunol.* **3,** 279–286.

Webb, C., Teitelbaum, D., Herz, A., Arnon, R., and Sela, M. (1976). Molecular requirements involved in suppression of EAE by synthetic basic copolymers of amino acids. *Immunochemistry* **13,** 333–337.

Weber, M. S., Starck, M., Wagenpfeil, S., Meinl, E., Hohlfeld, R., and Farina, C. (2004). Multiple sclerosis: Glatiramer acetate inhibits monocyte reactivity *in vitro* and *in vivo*. *Brain* **127,** 1370–1378.

Wiesemann, E., Klatt, J., Sonmez, D., Blasczyk, R., Heidenreich, F., and Windhagen, A. (2001). Glatiramer acetate (GA) induces IL-13/Il-5 secretion in naive T cells. *J. Neuroimmunol.* **119,** 137–144.

Wiesemann, E., Klatt, J., Wenzel, C., Heidenreich, F., and Windhagen, A. (2003). Correlation of serum IL-13 and IL-5 levels with clinical response to glatiramer acetate in patients with multiple sclerosis. *Clin. Exp. Immunol.* **133,** 454–460.

Wolinsky, J. S., and PROMiSe Trial Study Group (2004). The PROMiSe trial: Baseline data review and progress report. *Mult. Scler.* **10**(Suppl. 1), S65–S72.

Wolinsky, J. S., Nurayana, P. A., Johnson, K. P., Multiple, Sclerosis Study Group, and the, MRI Analysis Center (2001). United States open-label glatiramer acetate extension trial for relapsing multiple sclerosis: MRI and clinical correlates. *Mult. Scler.* **7,** 33–41.

Ziemssen, T. (2004). Neuroprotection and glatiramer acetate: The possible role in the treatment of multiple sclerosis. *Adv. Exp. Med. Biol.* **541,** 111–134.

Ziemssen, T. (2005). Modulating processes within the central nervous system is central to therapeutic control of multiple sclerosis. *J. Neurol.* **252**(Suppl. 5), v38–v45.

Ziemssen, T., and Ziemssen, F. (2005). The role of the humoral immune system in multiple sclerosis (MS) and its animal model experimental autoimmune encephalomyelitis (EAE). *Autoimmun. Rev.* **4**(7), 460–467.

Ziemssen, T., Neuhaus, O., and Hohlfeld, R. (2001). Risk-benefit assessment of glatiramer acetate in multiple sclerosis. *Drug Saf.* **24,** 979–990.

Ziemssen, T., Kümpfel, T., Klinkert, W. E., Neuhaus, O., and Hohlfeld, R. (2002a). Glatiramer acetate-specific T-helper 1- and 2-type cell lines produce BDNF: Implications for multiple sclerosis therapy. Brain-derived neurotrophic factor. *Brain* **125,** 2381–2391.

Ziemssen, T., Neuhaus, O., Farina, C., Hartung, H. P., and Hohlfeld, R. (2002b). Treatment of multiple sclerosis with glatiramer acetate—new information about Its mechanisms of action, pharmacokinetics, adverse effects and clinical studies. *Nervenarzt* **73,** 321–331.

Ziemssen, T., Kümpfel, T., Schneider, H., Klinkert, W. E., Neuhaus, O., and Hohlfeld, R. (2005). Secretion of brain-derived neurotrophic factor by glatiramer acetate-specific T-helper cell lines: Implications for multiple sclerosis therapy. *J. Neurol. Sci.* **233,** 109–112.

EVOLVING THERAPIES FOR MULTIPLE SCLEROSIS

Elena Korniychuk,* John M. Dempster,* Eileen O'Connor,†
J. Steven Alexander,‡ Roger E. Kelley,* Meghan Kenner,* Uma Menon,*
Vivek Misra,* Romy Hoque,* Eduardo Gonzalez-Toledo,§
Robert N. Schwendimann,* Stacy Smith,* and Alireza Minagar*

*Department of Neurology, Louisiana State
University Health Sciences Center, Shreveport
Louisiana 71103, USA
†Biogen-Idec, Boston, Massachusetts 02142, USA
‡Department of Cellular and Molecular Physiology
Louisiana State University Health Sciences Center, Shreveport
Louisiana 71103, USA
§Department of Radiology, Louisiana State University
Health Sciences Center, Shreveport
Louisiana 71103, USA

The introduction of immunomodulatory and immunosuppressive agents for treatment of multiple sclerosis (MS) has forever altered the natural course of this incurable and disabling neurodegenerative disorder. Despite early diagnosis of relapsing-remitting MS and early initiation of therapy, patients still experience breakthrough relapses and progression of their underlying MS pathology. The imperfect effectiveness, side effects, and toxicity of these agents, emphasize the necessity for development of more effective medications with less adverse events. This chapter presents readers with the most current information on the

INTERNATIONAL REVIEW OF
NEUROBIOLOGY, VOL. 79
DOI: 10.1016/S0074-7742(07)79025-6

nature, mechanism(s) of action, and side effects of the most promising experimental agents currently under clinical trials. Some of the agents now at different stages of clinical trial have emerged as both safe and promising. The understanding of MS etiology will lead to the development of increasingly specific, safer, and effective treatments for MS by neuroscientists and neurologists.

I. Introduction

Multiple sclerosis (MS) is an immune-mediated neurodegenerative disorder of the human central nervous system (CNS), which is clinically characterized by relapses followed by periods of clinical stability (Frohman *et al.*, 2006; Noseworthy *et al.*, 2000).

Neuropathogenesis of MS consists of activation of peripheral leukocytes against certain CNS antigen(s) such as members of the myelin basic protein family, disruption of the blood–brain barrier (BBB), ongoing destructive immune reaction within the CNS, loss of myelin/oligodendrocyte complex, and axonal/neuronal degeneration.

Of pertinence, what is clinically known as MS is a conglomerate of various disorders with different pathogenic mechanisms and potentially different responses to available treatments. On the basis of the clinical course, four types of MS have been described: (1) relapsing-remitting (RR), (2) secondary progressive (SP), (3) primary progressive (PP), and (4) relapsing-progressive (RP). Within a few years from the onset of the disease, a large number of patients with RRMS, with or without treatment, will develop SPMS which is characterized by fewer clinical relapses, a decline in the inflammation, loss of therapeutic response to immunomodulatory agents, and progressive neurodegeneration. Patients with PPMS exhibit a steady progression of neurological dysfunction, from the onset of their disease, without any clear relapses. Patients with RPMS do also demonstrate steady deterioration of their neurological status from the onset which is superimposed with frequent and devastating relapses.

Currently, the causes and cure of MS remain unknown and the majority of patients eventually develop some degree of disability. The FDA approved agents for management of MS include interferon (IFN)-β1a (Avonex and Rebif), IFN-β1b (Betaseron), glatiramer acetate (GA), natalizumab (Tysabri), and mitoxantrone (Novantrone). This chapter presents an overview of evolving treatments of MS. More effective therapies need to be designed so that patients experience less adverse effects.

II. Monoclonal Antibodies

A. Rituximab

Rituximab (Rituxan) is the first monoclonal antibody which was approved in 1997 by the FDA for treatment of relapsed, or refractory, low grade or follicular, CD20 antigen positive, B-cell non-Hodgkin's lymphoma. Rituximab is a genetically engineered, chimeric human/mouse monoclonal antibody which possesses IgG1 heavy-chain and kappa light-chain constant region sequences and mouse variable region sequences (Reff *et al.*, 1994). CD20 antigen, a 35-kDa transmembrane protein expressed by majority of B-cell non-Hodgkin's lymphomas (Anderson *et al.*, 1984), participates in cell cycle differentiation (Einfeld *et al.*, 1988; Valentine *et al.*, 1989). Interestingly, CD20 antigen is also expressed on normal B cells and its precursors. However, plasma cells, T cells, and hematopoietic stem cells do not express CD20 antigen. Rituximab is a B-cell depleting agent and causes fast elimination of CD20 antigen-carrying B cells in peripheral circulation. Since plasma cells do not express any CD20 antigen, they persist in the peripheral circulation and continue to generate antibodies.

In most patients who have been treated with rituximab, normal peripheral B cells will be generated from hematopoietic stem cells within 3–12 months of therapy (Maloney *et al.*, 1997). Proposed mechanisms of action of rituximab include complement-mediated cell lysis, antibody-dependent cellular cytotoxicity, apoptosis of B cells, suppression of cell growth, and sensitization of cells to chemotherapy (Byrd *et al.*, 2002; Clynes *et al.*, 2000; Flieger *et al.*, 2000; Pedersen *et al.*, 2002; Reff *et al.*, 1994; Shan *et al.*, 2000; Taji *et al.*, 1998). Due to its selective immunosuppressive effect on the humoral arm of the immune system, rituximab has been used in MS (Stuve *et al.*, 2005) and neuromyelitis optica (Cree *et al.*, 2005). Currently, two multicenter clinical trials of rituximab for treatment of subjects with RRMS (ClinicalTrials.gov Identifier: NCT00097188) and PPMS (ClinicalTrials.gov Identifier: NCT00087529) are being conducted in the United States and further information can be obtained at www.clinicaltrials.gov. Adverse effects of rituximab include fever, chills, nausea, vomiting, headache, and fatal infusion reactions.

B. Natalizumab

Natalizumab, also known as Tysabri, is a humanized anti-integrin monoclonal antibody which has been used for treatment of MS (Miller *et al.*, 2003; Sheremata *et al.*, 2005) and ulcerative colitis (Ghosh *et al.*, 2003). Natalizumab is an anti-adhesion therapy which targets the $\alpha4$ component of α-integrins. Integrins are one group of adhesion molecules which are involved in adhesion of activated leukocytes to underlying inflamed endothelium and consist of leukocyte function-associated integrin type-1 (LFA-1, $\alpha L\beta 2$), macrophage antigen-1 ($\alpha M\beta 2$), the $\alpha 9$ integrin ($\alpha 9\beta 7$), and two $\alpha 4$ integrins ($\alpha 4\beta 1$ and $\alpha 4\beta 7$) (also known as very late antigen-4).

During the inflammatory cascade of MS and on exposure to inflammatory cytokines such as TNF-α, the expression of adhesion molecules such as vascular cell adhesion molecule-1 (VCAM-1), E-selectin, and intercellular adhesion molecule-1 (ICAM-1) is upregulated on the surface of cerebral endothelial cells (Frohman *et al.*, 2006; Noseworthy *et al.*, 2000). The ligands for these adhesion molecules, which exist on the surface of activated leukocytes, are VLA-4 and ICAM-1/ICAM-2, respectively. Interaction between these adhesion molecules, and their ligands, plays a crucial role in transendothelial migration of activated leukocytes. Indeed, blocking the adhesion of a leukocyte to underlying endothelium may be able to suppress the inflammatory cascade significantly since without adhesion, migration will not occur. The concept of using monoclonal antibody against VLA-4 to block its interaction with VCAM-1 originated from the seminal experiment of Yednock and Steinman (Yednock *et al.*, 1992) on mice with experimental allergic encephalomyelitis (EAE). These investigators treated their EAE mouse model with anti-VLA-4 monoclonal antibody and observed that such treatment was associated with a dramatic decrease in accumulation of activated leukocytes within the CNS.

Various anti-$\alpha 4$-integrin monoclonal antibodies have been tested in animals with various immune-mediated inflammatory disorders (Jackson, 2002) and, of these, only two have been used in human clinical trials: anti-$\alpha 4\beta 1$ monoclonal antibody for MS and anti-$\alpha 7\beta 1$ antibody for ulcerative colitis. Natalizumab, which binds the associated β chain, has been tested with success in clinical trial of MS and ulcerative colitis. Natalizumab is infused intravenously (IV) once every 28 days in MS patients.

The two most significant clinical trials of natalizumab on human subjects with MS were AFFIRM and SENTINEL. The AFFIRM (Polman *et al.*, 2006) and SENTINEL (Rudick *et al.*, 2006) trials were the basis of the phase III clinical program examining natalizumab's treatment effect in patients with RRMS. These are two of the largest pivotal MS clinical trials to date and were designed as randomized, double-blind, placebo-controlled, and parallel group trials. In the AFFIRM trial, 99 centers in Europe, North America, Australia, and New Zealand began enrolling patients in November 2001, ultimately enrolling 942 patients. In the SENTINEL trial, 124 centers in Europe and the United States enrolled 1196 patients beginning in January 2002. The AFFIRM and SENTINEL trials are identical in design and endpoints examined.

Both study protocols were developed by the investigator advisory committee and the sponsors, and then were approved by the central and local ethics committees. Additionally, the AFFIRM and SENTINEL trials were overseen by an independent safety-monitoring committee, which met at least monthly to discuss study progress.

In the natalizumab phase III clinical program, three cases of progressive multifocal leukoencephalopathy (PML) occurred, two in the MS trials and one in the Crohn's trials (See package insert for Tysabri). PML is an opportunistic viral infection of the brain that usually leads to death or severe disability. As a result of this risk of roughly 1 in 1000 treated MS patient, natalizumab is indicated for monotherapy only, in the treatment of relapsing forms of MS to delay accumulation of physical disability and reduce the frequency of clinical exacerbations. Patients who were eligible to participate in the AFFIRM (monotherapy) and SENTINEL (add-on therapy) trials were between the ages of 18 and 50 years in AFFIRM and 18 and 55 years in SENTINEL and had a diagnosis of RRMS; had a score of 0–5.0 on the Expanded Disability Status Scale (EDSS); had undergone magnetic resonance imaging (MRI), showing lesions consistent with MS; and had experienced at least one medically documented relapse within the 12 months prior to study entry.

Patients in the SENTINEL trial, additionally, had to have received treatment with IFN-β1a for at least 12 months during which they experienced the above-mentioned medically documented relapse, indicative of continued disease activity despite treatment with an approved MS therapy. Patients were excluded from either study if they had a diagnosis of PPMS, SPMS, or RPMS; if they had had a relapse within 50 days before randomization; or had been treated with cyclophosphamide or mitoxantrone within the previous 12 months or treatment with GA, cyclosporine, azathioprine, methotrexate, mycophenolate, mofetil, or intravenous immune globulin within the previous 6 months. Patients participating in the AFFIRM trial, additionally, could not have received treatment with IFN-β in the year prior to randomization. Patients participating in the SENTINEL trial could not have been treated with any interferon product other than Avonex within the year prior to randomization. The AFFIRM and SENTINEL trials not only examined two different treatment protocols but also had two different patient populations as well. Patients in the SENTINEL trial were older (39 years vs 36 years), had their disease longer (8 years vs 6 years), and a higher baseline EDSS (2.5 vs 2.3).

Patients in the AFFIRM trial were randomly assigned in a 2:1 ratio to receive either natalizumab or placebo. Patients in the SENTINEL trial were randomly assigned in a 1:1 ratio to receive either natalizumab plus IFN-β1a or IFN-β1a plus placebo. Natalizumab was dosed as 300-mg IV every 4 weeks for up to 116 weeks, in both the AFFIRM and SENTINEL trials. In the SENTINEL trial, patients received 30 μg of IFN-β1a IM once weekly for up to 116 weeks.

If a relapse was suspected, the patient was evaluated by the examining neurologist. Relapses were defined as the development of new or recurrent neurological symptoms not associated with fever or infection, lasting at least 24 h, and accompanied by new, objective neurological findings. At the discretion of the treating neurologist, relapses were treated with intravenous methylprednisolone at a dose of 1000 mg/day for 3 or 5 days. Patients whose disability progression was sustained for 12 weeks were allowed to continue participation in the study but were given the option of adding an available treatment of MS as a rescue medication per protocol while continuing to receive the study drug.

Proton density-weighted or T2-weighted and gadolinium-enhanced T1-weighted MRI scans of the brain were obtained at baseline, and at weeks 52 and 104. Contiguous, 3-mm thick axial slices through the whole brain were acquired. MRI analyses were performed at the Central Reading Centers (AFFIRM—Institute of Neurology, University College London, London; SENTINEL—MS-MRI Evaluation Center, Basel, Switzerland) by blinded raters.

The AFFIRM and SENTINEL studies defined primary and secondary endpoints at two prespecified time points. At year 1, the inflammatory characteristic of the disease was assessed, utilizing the primary endpoint of rate of clinical relapse. At year 2, the progression of the irreversible destructive process was assessed with the primary endpoint defined as cumulative probability of sustained progression of disability, an increase of 1.0 or more on the EDSS from a baseline score of 1.0 or more or an increase of 1.5 or more from a baseline score of 0 that was sustained for 12 weeks. Unlike phase III trials of the other immunomodulatory therapies, disability progression was not confirmed during a relapse.

Prespecified sensitivity analyses were utilized to evaluate treatment effects according to previously used definitions (24-week confirmation, confirmation during a relapse, and so on). Secondary efficacy endpoints, at year 1, included the number of new or enlarging hyperintense lesions detected by T2-weighted MRI, the number of lesions detected by gadolinium-enhanced MRI, and the proportion of relapse-free patients. Secondary efficacy endpoints, assessed at year 2, included rate of clinical relapse, the volume of lesions as detected by T2-weighted MRI, the number of new hypointense lesions as detected by unenhanced T1-weighted MRI, and the progression of disability as measured by the multiple sclerosis functional composite (MSFC). Secondary endpoints in both AFFIRM and SENTINEL were rank-ordered and a closed testing procedure was used so that if statistical significance was not achieved for an endpoint, endpoints of a lower rank were not considered to be statistically significant. Binding antibodies against natalizumab were assessed with the use of an enzyme-linked immunosorbent assay. Positive samples (0.5 μg/ml) were further tested in a flow-cytometry assay to determine whether these antibodies interfered with the binding of natalizumab to α4-integrin. Results of the functional assay were highly correlated with the results obtained with the ELISA assay.

C. Daclizumab

Daclizumab, also known as Zenapax, is a mouse–human chimeric monoclonal antibody against interleukin-2 (IL-2R_α) receptor, which targets IL-2 receptor α chain on activated T cells. IL-2, which is also known as T-cell growth factor, plays a major role in initiation of proliferation or clonal expansion of antigen-stimulated T cells. The process of binding of IL-2 to its receptor sets the stage for clonal expansion of antigen-activated T cells and removal of the insulting antigen. This in turn translates into a decline in IL-2 receptor expression, which in turn restricts clonal expansion of a particular T cell clone. In addition, IL-2 remarkably increases the cytolytic activity of a population of natural killer cells against cancerous cells (Li *et al.*, 2004).

Daclizumab is used along with other immunosuppressors to prevent renal graft rejection (Sandrini, 2005). In an open-label study, Rose *et al.* (2004) assessed the safety and efficacy of daclizumab in 21 patients with RRMS and SPMS. During this study, 19 ambulatory MS patients with clinically active disease were treated with daclizumab for 5–25 months with two dropouts. Study subjects initially were treated with daclizumab 1 mg/kg IV followed by infusion of the same dose 14 days later, with subsequent treatments every 28 days. As treatment with daclizumab continued, the dosage of medication was adjusted based on clinical response. Of the treated MS subjects converted to monotherapy with daclizumab, two subjects continued treatment with IFN-β for the first 6 months of the study, and one subject received monthly methylprednisolone infusions once a month.

The investigators reported that five subjects with RRMS and five subjects with SPMS showed improvement and one subject with RRMS and eight subjects with SPMS remained stable. The results of this study indicated that treatment of MS subjects with daclizumab was safe, well tolerated, and was associated with improvement of EDSS score and decrease in brain MRI activity. Currently, a multicenter, double-blind, placebo-controlled, dose-ranging study of daclizumab, as a monotherapy, is being conducted to determine safety and efficacy of this monoclonal antibody subjects with RRMS (ClinicalTrials.gov Identifier: NCT00390221).

D. Alemtuzumab

Alemtuzumab, also known as Campath-1H (the word campath derives from *Cam*bridge *Path*ology), is a humanized monoclonal antibody, which is directed against the cell surface molecule CD52. CD52 is expressed on the surface of various cell populations such thymocytes, T and B cells, monocytes, and macrophages. However, stem hematopoietic cells and plasma cells do not express CD52 molecule. Alemtuzumab, which is currently being used for treatment of B-cell chronic lymphocytic leukemia, destroys cells carrying CD52 molecule via a number of mechanisms: complement-mediated lysis, antibody-dependent cell toxicity, and apoptosis.

Alemtuzumab is currently under clinical investigation for treatment of MS. The hypothesis for use of alemtuzumab in MS is based on the concept that following severe lymphocyte depletion by this monoclonal antibody, the new reconstructed pool of T cells will be devoid of autoreactive CD4 T cells which promote the inflammatory cascade in MS (Moreau *et al.*, 1994). Alemtuzumab has also been used in a cohort of 22 patients with RRMS and baseline mean EDSS score of 4.8 (Coles *et al.*, 2004). Of the 22 MS subjects, 17 were treatment-naive and the others had failed other available therapies. The study subjects had an annualized relapse rate (ARR) of 2.2 and in order to participate in this clinical trial they were required to have one-point increase in their EDSS score during the 1 year prior to study enrollment. Treatment of these subjects with alemtuzumab was associated with a significant decrease from 2.2 to 0.14 (94% decline) and decline of EDSS score from 4.8 to 2.1 during the first 2 years of follow-up.

Alemtuzumab has also been used in a pilot clinical trial involving 36 MS patients with SPMS who had an EDSS of 5.8 and had experienced one-point increase of their EDSS during the year prior to study enrollment (Coles *et al.*, 1999). Most of the study subjects were pretreated with 1 g of methylprednisolone IV prior to study drug infusion. All study subjects were treated with five consecutive daily doses of 20 mg (total of 100 mg) of alemtuzumab. This experimental treatment caused significant reduction of CD4 T cells for a median of 61 months and CD8 T cells for 30 months. Interestingly, B cells increased to 124% of pretreatment level at 27 months posttreatment. During the first posttreatment year, treatment of patients with SPMS with alemtuzumab was associated with a dramatic decline in the relapse rate and 90% reduction in the number of contrast-enhancing lesions (Paolillo *et al.*, 1999). The majority of patients (33/36) demonstrated stable EDSS scores. However, despite such promising initial results, due to pronounced immunosuppression, most of patients deteriorated clinically and after a mean 7.5 years of follow-up, only 4 MS patients remained stable and the other 32 subjects demonstrated disease progression. This clinical progression, despite profound immunosuppression, in patients with SPMS further lends support to the notion that SPMS is more of a neurodegenerative process than an inflammatory disease.

Adverse effects of alemtuzumab include infusion reaction identified by fever, rash, nausea, rigors, hypotension, and increased susceptibility to infections due to sharp drop of T lymphocytes (which reaches its lowest point in week 4 postinfusion). In addition, some of the MS patients developed Grave's disease (immune-mediated thyroiditis) following treatment with alemtuzumab. What is interesting is that cancer patients who undergo treatment with alemtuzumab do not develop immune-mediated thyroiditis, and this finding along with the fact that MS patients occasionally develop concurrent immune-mediated thyroiditis indicates that MS patients, due to the dysregulation of their immune system, possess a unique tendency to develop such complication.

A new phase II, randomized, open-label, three-arm study has compared low-dose and high-dose Campath and high-dose Rebif (IFN-β1a 44 μg subcutaneously three times weekly) in patients with early, active RRMS who have not been previously treated with immunotherapies other than steroids. This study which is no longer recruiting MS subjects has been sponsored by Genzyme (ClinicalTrials.gov Identifier: NCT00050778). For more information, the readers are referred to the website: www.clinialtrials.gov.

In conclusion, alemtuzumab is a potent immunosuppressive agent with remarkable effects on the inflammatory arm of MS pathogenesis and significant decrease in both ARR and brain MR activity. However, like other immunosuppressive agents, it comes with the price of certain potential adverse effects such as immune-mediated thyroiditis.

III. Statins

Statins, 3-hydroxy-3-methylglutaryl coenzyme A reductase inhibitors, are lipid-lowering drugs which have been demonstrated to improve disease activity in experimental colitis (Sasaki *et al.*, 2003), and in experimental MS (EAE, Youssef *et al.*, 2002). During the last few years, various immunomodulatory effects of statins have been observed. Neuhaus *et al.* (Neuhaus *et al.*, 2002) obtained peripheral blood mononuclear cells from untreated or IFN-β-treated patients with RRMS and from healthy control subjects and exposed them to various statins *in vitro*. The investigators observed that the proliferative activity of T and B cells was inhibited and the expression of activation-induced adhesion molecules on T cells was decreased. In addition, statins reduce the expression of activation-induced adhesion molecules on T cells, modifying the balance of Th1 and Th2 cytokines, reduce the secretion of matrix metalloproteinase-9, and downregulate expression of chemokine receptors on B and T cells (Mach, 2002).

Statins have been shown to reduce human BBB permeability and restrict leukocyte migration (Ifergan *et al.*, 2006). In an *in vitro* model, Ifergan *et al.* (2006) demonstrated that lovastatin and simvastatin induce a 50–60% reduction in the diffusion rates of bovine serum albumin and [(14)C]-sucrose across human BBB-ECs through blunting of isoprenylation processes. However, this was independent of the expression of the tight junction molecules occludin, VE-cadherin, JAM-1, zonula occluden-1, and zonula occluden-2. Simvastatin and lovastatin demonstrated equal potency in decreasing BBB permeability *in vitro*, with median effective concentration (EC_{50}) of $9.5 \times 10(-8)$ M and $1.0 \times 10(-7)$ M, respectively. The investigators also showed that lovastatin and simvastatin treatment of BBB-ECs significantly restricts the migration of clinically isolated syndrome (CIS)-derived and MS-derived monocytes and lymphocytes across the human BBB *in vitro*. This is through a specific reduction in the secretion of the chemokine monocyte chemotactic protein-1/CCL2 and IFN-γ-inducible protein-10/CXCL10 by BBB-ECs.

The primary mechanism for the immunomodulatory effects of statins is attributed to inhibition of posttranslational protein prenylation of small GTP-binding proteins and is largely independent of its lipid-lowering effects (Weber *et al.*, 2006). Statins have been used in clinical trials for treatment of MS. Vollmer *et al.* (2004) investigated the use of oral simvastatin (80 mg) in 30 individuals with RRMS. The mean number of contrast-enhancing lesions at months 4, 5, and 6 of treatment was compared with the mean number of lesions present on pretreatment brain MRI scans. The investigators observed that the number and volume of contrast-enhancing lesions declined by 44% ($p < 0.0001$) and 41% ($p = 0.0018$), respectively. No adverse events were reported and treatment with oral simvastatin was well tolerated. The investigators concluded that oral simvastatin may inhibit inflammatory components of MS.

A small clinical trial is currently evaluating the combination of high-dose IFN-β1a (Rebif®, 44 μg three times per week subcutaneously) with atorvastatin in patients with RRMS (Birnbaum and Altafullah, 2005). Combination therapy with intermediate (40 mg/day) and high-dose atorvastatin (80 mg/day) was generally well tolerated. However, serial MRIs revealed an increased number of T2- and Gd-enhancing lesions in two of four patients compared to baseline. This data raises concern that the combination of atorvastatin and high-dose IFN-β might not be of benefit in patients with RRMS (Birnbaum and Altafullah 2005).

Currently, two clinical trials are being conducted in the United States to assess the efficacy and safety of atorvastatin (Lipitor) in patients with CIS at risk for MS (ClinicalTrials.gov Identifier: NCT00094172) and use of early IFN-β1b and atorvastatin therapy in patients with CIS suggestive of MS (ClinicalTrials.gov Identifier: NCT00137176). The details of these clinical trials can be reached at www.clinicaltrials.gov.

Potential adverse effects of statins include elevation of hepatic transaminases and inflammatory myositis. Rhabdomyolysis has been reported in some patients treated with statins, with most cases being associated with the use of cerivastatin (Thompson *et al.*, 2003). No cases of hepatic failure have been reported and all adverse effects are reversible by reduction in dose or discontinuation of medication.

IV. Minocycline and Doxycycline

Recently, members of the tetracycline family of antibiotics, minocycline and doxycycline, have been the subject of intense debate and clinical trials for management of MS. A number of clinical studies have demonstrated that minocycline and doxycycline possess anti-inflammatory properties which are separate from their antimicrobial activity (Fan *et al.*, 2005; Giuliani *et al.*, 2005a; Yong, 2004). Minocycline, an oral semisynthetic analogue of tetracycline, is a small (495 kDa), highly lipophilic molecule with good penetration properties into the CNS. It is most commonly used for treatment of acne and rosacea. Minocycline possesses other nonantimicrobial properties which have attracted the attention of neuroscientist and immunologists. In experimental studies, minocycline has been demonstrated to reduce the neurological injuries of animal models of stroke (Yrjanheikki *et al.*, 1998, 1999), Huntington's disease (Chen *et al.*, 2000), Parkinson's disease (Du *et al.*, 2001; Wu *et al.*, 2002), amyotrophic lateral sclerosis (Kriz *et al.*, 2002), Down's syndrome (Hunter *et al.*, 2004), and spinal cord injury (Stirling *et al.*, 2004; Wells *et al.*, 2003).

Studies on animals with mild and severe EAE have demonstrated the beneficial effect of minocycline administration (Brundula *et al.*, 2002; Fan *et al.*, 2005; Guliani, 2005b; Popovic *et al.*, 2002; Yong, 2004). These studies provided more evidence that minocycline possesses multiple immunomodulatory activities, including the suppression of microglial activation (Fan *et al.*, 2005; Giuliani *et al.*, 2005c; Yong, 2004) as well as neuroprotective and antiapoptotic properties (Giuliani *et al.*, 2005b; Yong, 2004).

In an *in vitro* study, minocycline suppressed matrix metalloproteinase (MMP) activity, decreased the production of MMP-9, and reduced the transmigration of T lymphocytes across a fibronectin matrix barrier (Brundula *et al.*, 2002; Yong, 2004). These actions may be beneficial in patients with MS because MMPs are involved in transendothelial migration of the autoreactive CD4+ T cells through the BBB, and they also participate in breakdown of the myelin sheath.

Both minocycline and doxycycline have been considered for combination therapy with IFN-β (Giuliani *et al.*, 2005c; Minagar *et al.*, 2005) and with GA in a model of MS (Giuliani *et al.*, 2005b). Minagar *et al.* (2005) assessed the safety, tolerability, and efficacy of the combination of oral doxycycline (100 mg/day) and IFN-β1a (Avonex 30 μg intramuscular once weekly) in 15 patients with RRMS with active disease with breakthrough disease activity. The investigators reported that the combination of oral doxycycline and IFN-β1a was safe and well tolerated. Patients who were treated with the above combination had a lower annual relapse rate and showed a significant decrease in the number of contrast-enhancing lesions on brain MR scans.

Giuliani *et al.* (2005b) assessed the combination therapy of minocycline and GA at suboptimal doses in mice C57/BL6 with EAE. GA (250 μg per mouse) given once daily and minocycline at 25 mg/kg daily were administered. The investigators reported that treatment with a suboptimal dose of minocycline or GA by themselves as well as placebo normal saline did not affect EAE outcome. However, combination of the suboptimal doses of GA and minocycline resulted in a significant reduction of EAE severity and disease burden with attenuation of the inflammation, axonal loss, and demyelination.

Overall, it appears that minocycline and doxycycline, given in combination with IFN-β or GA, may protect study subjects against neuroinflammation and neuronal loss.

V. Vitamin D

Vitamin D, also known as cholecalciferol, is a lipid-soluble vitamin which is synthesized by the conversion of 7-dehydrocholesterol to vitamin D in the skin under the effect of ultraviolet radiation. Traditionally, vitamin D has been viewed as the main physiological regulator of the calcium homeostasis in the intestine, kidney, and bone. However, further research into the effects of vitamin D has revealed its other physiological functions, including protection against certain immune-mediated diseases such as diabetes mellitus and MS (Mark and Carson, 2006). It has been shown that in the nonobese diabetic mouse, treatment with vitamin D3 prevents development of immune-mediated insulin-dependent diabetes mellitus (Zella *et al.*, 2003).

The possible protective role of vitamin D against MS is further supported by another immune-mediated disorder of human subjects, juvenile-onset diabetes mellitus (JODM). It has been shown that children at high risk of JODM, if treated with vitamin D during infancy, are less likely to develop JODM (Hypponen *et al.*, 2001). The effects of 1,25-dihydroxyvitamin D3 (the active metabolite of vitamin D) have been studied in an animal model of MS. Cantorna *et al.* (1996) demonstrated that mice treated with dietary 1,25-dihydroxyvitamin D3 before induction of experimental autoimmune EAE did not develop symptoms of EAE, compared with a 100% incidence in the control group (Cantorna *et al.*, 1996). In addition, in mice with EAE, parenteral administration of 1,25-dihydroxyvitamin D3, followed by dietary supplementation, stopped further disease progression. These important observations provide scientific evidence that vitamin D have a therapeutic effect in MS.

In human subjects, it has been observed that level of sun exposure and vitamin D supplementation are associated with lower risk of developing MS (Hernan *et al.*, 1999; Soilu-Hanninen *et al.*, 2005). The protective mechanisms of vitamin D against disease process in MS include involvement of IL-10 signaling, which is an essential pathway for 1,25-dihydroxyvitamin D3-mediated inhibition of EAE (Spach *et al.*, 2006). Other mechanisms for immunomodulatory function of vitamin D include suppression of generated proinflammatory cytokines (Type 1) such as IFN-γ, IL-2, and IL-12 (Lemire, 1995); enhancement of the presence and function of the T regulatory cells (Hayes *et al.*, 2003); and possible downregulation of expression of HLA class II molecules by antigen-presenting cells (Tokuda and Levy, 1996).

VI. Fingolimod

Fingolimod (FTY720) is a new oral immunomodulator agent under clinical investigation for treatment of MS. Fingolimod, after phosphorylation by sphingosine kinase 2, turns into fingolimod-P which acts as a superagonist for the sphingosine 1-phosphate (S1P) receptor expressed on the surface of thymocytes and lymphocytes. Interaction between fingolimod and S1P receptor causes aberrant internalization of this receptor, which in turn causes suppression of S1P/S1P-dependent lymphocyte egress from secondary lymphoid tissue and thymus. Sequestration of lymphocytes within lymphoid organs reduces peripheral lymphocyte counts and decreases the recirculation of lymphocytes to the CNS (Brinkmann *et al.*, 2002, 2004; Mandala *et al.*, 2002).

In animal models, fingolimod does not impair activation of memory T cells or T-cell expansion in reaction to systemic viral infections (Pinschewer *et al.*, 2000). Fingolimod has been assessed in MS. Kappos *et al.* (2006) presented the results of their multicenter, placebo-controlled trial of fingolimod in RRMS. A total of 225 MS subjects were randomized to receive oral fingolimod, at a dose of 1.25 or 5.0 mg versus placebo once daily. The study subjects were followed for 6 months with brain MRI and clinical examinations (months 0 and 6). The primary endpoint of this clinical trial consisted of the total number of contrast-enhancing lesion on T2-weighted images at monthly intervals for 6 months. After this initial phase, the clinical trial was extended for another 6 months and again all subjects in the placebo arm were randomly assigned to receive one of the fingolimod doses.

The results of this clinical trial indicated that the medial total number of contrast-enhancing lesion was lower in those subjects who received 1.25 and 5.0 mg of fingolimod daily than those subjects who received placebo. The ARR was also decreased significantly in treated subjects compared to placebo. Of the 225 MS subjects who completed the clinical trial, the number of contrast-enhancing lesions and ARR remained low in those subjects who were treated with fingolimod continuously and both endpoints decreased in subjects who were switched over from placebo to active therapy with fingolimod. Potential adverse effects of fingolimod included nasopharyngitis, dyspnea, headache, diarrhea, and nausea.

VII. T-Cell Vaccination

On the basis of the concept that MS is primarily a T-cell-mediated disease, autoreactive CD4+ or CD8+ T cells may be viewed as pathogens and may be used, after attenuation, as a vaccine to stimulate the host's regulatory immune responses against these cells (Ben-Nun, 1981; Hong *et al.*, 2006). This experimental therapy may be applied to MS patients to eliminate or suppress selected clones of autoreactive T cells. Clonal expansion of MBP-reactive T cells and their accumulation within the CNS of MS patients have been considered as the major stimulus for ongoing inflammation within the CNS. The beneficial mechanisms of T-cell vaccination in MS include generation of CD4+ regulatory T cells lines which block inflammation by expression of transcription factor Foxp3 and IL-10

(Hong *et al.*, 2006), Th2 immune deviation, the peripheral idiotype–anti-idiotype network in regulating specific T cells, and anti-idiotype antibody reactivity (Hong *et al.*, 2006). Currently, there are two active clinical trials in the United States for assessment of efficacy of T-cell vaccination in MS and the details of these studies are available at www.clinicaltrials.gov.

VIII. Stem Cell Therapy

Application of stem cells to progressive neurodegenerative disorders such as MS has generated new hope and enthusiasm among neurobiologists and neuroimmunologists. Stem cells are viewed as endless source of neurons and glial cells, which can either replace or protect lost cells. One line of progenitor cells is myelin-producing oligodendrocyte progenitor cells, which reside in the adult human brain (Windrem *et al.*, 2004) and are present in chronic MS lesions (Back *et al.*, 2005). Myelin-producing oligodendrocyte progenitor cells can generate new myelin. However, in MS, astrocyte-derived hyaluronan accumulates in the plaques and prevents maturation of these cells (Back *et al.*, 2005). Another therapeutic approach is to transplant either human adult or embryonic stem cell-derived progenitor cells to subjects with progressive MS. Theoretically, these stem cells should be able to mature to lost neurons and oligodendrocytes and generate fresh myelin sheath. However, delivery of these stem cells to the sites of ongoing inflammation, due to the scattered nature of MS lesions, and maintaining their survival poses a formidable challenge to neuroscientists.

IX. Conclusions

In summary, multiple new agents are currently being assessed for treatment of MS. Our approach is intervention against the immunopathogenesis of MS through either immunosuppression or blockage of transendothelial migration of activated leukocytes. There is also the potential for desensitization of the inflammatory response via vaccination. In addition, progenitor stem cells may have potential in neurorestoration of previously affected CNS tissue in MS.

References

Anderson, K. C., Bates, M. P., Slaughenhoupt, B. L., Pinkus, G. S., Schlossman, S. F., and Nadler, L. M. (1984). Expression of human B cell-associated antigens on leukemias and lymphomas: A model of human B cell differentiation. *Blood* **63,** 1424–1433.

Back, S. A., Tuohy, T. M., Chen, H., Wallingford, N., Craig, A., Struve, J., Luo, N. L., Banine, F., Liu, Y., Chang, A., Trapp, B. D., Bebo, B. F., Jr., *et al.* (2005). Hyaluronan accumulates in demyelinated lesions and inhibits oligodendrocyte progenitor maturation. *Nat. Med.* **11,** 966–972.

Birnbaum, G., and Altafullah, I. (2005). A double blind, placebo controlled combination trial of interferon beta1a (Rebif) and atorvastatin (Lipitor) in patients with relapsing remitting multiple sclerosis [abstract# P06.158]. *In Proceedings of the 57th Meeting of the American Academy of Neurology: 2005 April 9–16; Miami Beach.* St Paul: American Academy of Neurology Press.

Brinkmann, V., Davis, M. D., Heise, C. E., Albert, R., Cottens, S., Hof, R., Bruns, C., Prieschl, E., Baumruker, T., Hiestand, P., Foster, C. A., Zollinger, M., *et al.* (2002). The immune modulator FTY720 targets sphingosine 1-phosphate receptors. *J. Biol. Chem.* **277,** 21453–21457.

Brinkmann, V., Cyster, J. G., and Hla, T. (2004). FTY720: Sphingosine 1-phosphate receptor-1 in the control of lymphocyte egress and endothelial barrier function. *Am. J. Transplant.* **4,** 1019–1025.

Brundula, V., Rewcastle, N. B., Metz, L. M., Bernard, C. C., and Yong, V. W. (2002). Targeting leukocyte MMPs and transmigration: Minocycline as a potential therapy for multiple sclerosis. *Brain* **125,** 1297–1308.

Byrd, J. C., Kitada, S., Flinn, I. W., Aron, J. L., Pearson, M., Lucas, D., and Reed, J. C. (2002). The mechanism of tumor cell clearance by rituximab *in vivo* in patients with B-cell chronic lymphocytic leukemia: Evidence of caspase activation and apoptosis induction. *Blood* **99,** 1038–1043.

Cantorna, M. T., Hayes, C. E., and DeLuca, H. F. (1996). 1,25-Dihydroxyvitamin D_3 reversibly blocks the progression of relapsing encephalomyelitis, a model of multiple sclerosis. *Proc. Natl. Acad. Sci. USA* **93,** 7861–7864.

Chen, M., Ona, V. O., Li, M., Ferrante, R. J., Fink, K. B., Zhu, S., Bian, J., Guo, L., Farrell, L. A., Hersch, S. M., Hobbs, W., Vonsattel, J. P., *et al.* (2000). Minocycline inhibits caspase-1 and caspase-3 expression and delays mortality in a transgenic mouse model of Huntington disease. *Nat. Med.* **6,** 797–801.

Clynes, R. A., Towers, T. L., Presta, L. G., and Ravetch, J. V. (2000). Inhibitory Fc receptors modulate *in vivo* cytoxicity against tumor targets. *Nat. Med.* **6,** 443–446.

Coles, A., Deans, J., and Compston, A. (2004). Campath-1H treatment of multiple sclerosis: Lessons from the bedside for the bench. *Clin. Neurol. Neurosurg.* **106,** 270–274.

Coles, A. J., Wing, M. G., Molyneux, P., Paolillo, A., Davie, C. M., Hale, G., Miller, D., Waldmann, H., and Compston, A. (1999). Monoclonal antibody treatment exposes three mechanisms underlying the clinical course of multiple sclerosis. *Ann. Neurol.* **46,** 296–304.

Cree, B. A., Lamb, S., Morgan, K., Chen, A., Waubant, E., and Genain, C. (2005). An open label study of the effects of rituximab in neuromyelitis optica. *Neurology* **64,** 1270–1272.

Du, Y., Ma, Z., Lin, S., Dodel, R. C., Gao, F., Bales, K. R., Triarhou, L. C., Chernet, E., Perry, K. W., Nelson, D. L., Luecke, S., Phebus, L. A., *et al.* (2001). Minocycline prevents nigrostriatal dopaminergic neurodegeneration in the MPTP model of Parkinson's disease. *Proc. Natl. Acad. Sci. USA* **98,** 14669–14674.

Einfeld, D. A., Brown, J. P., Valentine, M. A., Clark, E. A., and Ledbetter, J. A. (1988). Molecular cloning of the human B cell CD20 receptor predicts a hydrophobic protein with multiple transmembrane domains. *EMBO J.* **7,** 711–717.

Fan, L. W., Pang, Y., Lin, S., Rhodes, P. G., and Cai, Z. (2005). Minocycline attenuates liposaccharide-induced white matter injury in neonatal rat brain. *Neuroscience* **133,** 159–168.

Flieger, D., Renoth, S., Beier, I., Sauerbruch, T., and Schmidt-Wolf, I. (2000). Mechanism of cytotoxicity induced by chimeric mouse human monoclonal antibody IDEC-C2B8 in CD20-expressing lymphoma cell lines. *Cell Immunol.* **204,** 55–63.

Frohman, E. M., Racke, M. K., and Raine, C. S. (2006). Multiple sclerosis: The plaque and its pathogenesis. *N. Engl. J. Med.* **354,** 942–955.

Ghosh, S., Goldin, E., Gordon, F. H., Malchow, H. A., Rask-Madsen, J., Rutgeerts, P., Vyhnalek, P., Zadorova, Z., Palmer, T., and Donoghue, S., and Natalizumab Pan-European Study Group (2003). Natalizumab for active Crohn's disease. *N. Engl. J. Med.* **348,** 24–32.

Giuliani, F., Hader, W., and Yong, V. W. (2005a). Minocycline attenuates T cell and microglia activity to impair cytokine production in T cell-microglia interaction. *J. Leukocyte Biol.* **78,** 135–143.

Giuliani, F., Metz, L. M., Wilson, T., Fan, Y., Bar-Or, A., and Yong, V. W. (2005b). Additive effect of the combination of glatiramer acetate and minocycline in a model of MS. *J. Neuroimmunol.* **158,** 213–221.

Giuliani, F., Fu, S. A., Metz, L. M., and Yong, V. W. (2005c). Effective combination of minocycline and interferon-beta in a model of multiple sclerosis. *J. Neuroimmunol.* **165,** 83–91.

Hayes, C. E., Nashold, F. E., Spach, K. M., and Pedersen, L. B. (2003). The immunological functions of the vitamin D endocrine system. *Cell. Mol. Biol. (Noisy-le-grand)* **49,** 277–300.

Hernan, M. A., Olek, M. J., and Ascherio, A. (1999). Geographic variation of MS incidence in two prospective studies of US woman. *Neurology* **531,** 1711–1718.

Hong, J., Zang, Y. C., Nie, H., and Zhang, J. Z. (2006). CD4+ regulatory T cell responses induced by T cell vaccination in patients with multiple sclerosis. *Proc. Natl. Acad. Sci. USA* **103,** 5024–5029.

Hunter, C. L., Bachman, D., and Granholm, A. C. (2004). Minocycline prevents cholinergic loss in a mouse model of Down's syndrome. *Ann. Neurol.* **56,** 675–688.

Hypponen, E., Laara, E., Reunanen, A., Jarvelin, M. R., and Virtanen, S. M. (2001). Intake of vitamin D and risk of type 1 diabetes: A birth-cohort study. *Lancet* **358,** 1500–1503.

Ifergan, I., Wosik, K., Cayrol, R., Kebir, H., Auger, C., Bernard, M., Bouthillier, A., Moumdjian, R., Duquette, P., and Prat, A. (2006). Statins reduce human blood-brain barrier permeability and restrict leukocyte migration: Relevance to multiple sclerosis. *Ann. Neurol.* **60,** 45–55.

Jackson, D. Y. (2002). Alpha 4 integrin antagonists. *Curr. Pharm. Des.* **8,** 1229–1253.

Kappos, L., Antel, J., Comi, G., Montalban, X., O'Connor, P., Polman, C. H., Haas, T., Korn, A. A., Karlsson, G., and Radue, E. W., and FTY720 D2201 Study Group (2006). Oral fingolimod (FTY720) for relapsing multiple sclerosis. *N. Engl. J. Med.* **355,** 1124–1140.

Kriz, J., Nguyen, M. D., and Julien, J. P. (2002). Minocycline slows disease progression in a mouse model of amyotrophic lateral sclerosis. *Neurobiol. Dis.* **10,** 268–278.

Lemire, J. M. (1995). Immunomodulatory actions of 1,25-dihydroxyvitamin D3. *J. Steroid. Biochem. Mol. Biol.* **53,** 599–602.

Li, S., Xu, J., Makarenkova, V. P., Tjandrawan, T., Vakkila, J., Reichert, T., Gooding, W., Lagenaur, C. F., Achim, C. L., Chambers, W. H., Herberman, R. B., Whiteside, T. L., *et al.* (2004). A novel epitope of N-CAM defines precursors of human adherent NK cells. *J. Leukoc. Biol.* **76,** 1187–1199.

Mach, F. (2002). Toward a role for statins in immunomodulation. *Mol. Interv.* **2,** 478–480.

Maloney, D. G., Grillo-Lopez, A. J., White, C. A., Bodkin, D., Schilder, R. J., Neidhart, J. A., Janakiraman, N., Foon, K. A., Liles, T. M., Dallaire, B. K., Wey, K., Royston, I., *et al.* (1997). IDEC-C2B8 (Rituximab) anti-CD20 monoclonal antibody therapy in patients with relapsed low-grade non-Hodgkin's lymphoma. *Blood* **90,** 2188–2195.

Mandala, S., Hajdu, R., Bergstrom, J., Quackenbush, E., Xie, J., Milligan, J., Thornton, R., Shei, G. J., Card, D., Keohane, C., Rosenbach, M., Hale, J., *et al.* (2002). Alteration of lymphocyte trafficking by sphingosine-1-phosphate receptor agonists. *Science* **296,** 346–349.

Mark, B. L., and Carson, J. A. (2006). Vitamin D and autoimmune disease—implications for practice from the multiple sclerosis literature. *J. Am. Dietetic. Assoc.* **106,** 418–424.

Miller, D. H., Khan, O. A., Sheremata, W. A., Blumhardt, L. D., Rice, G. P., Libonati, M. A., Willmer-Hulme, A. J., Dalton, C. M., Miszkiel, K. A., and O'Connor, P. W., and International Natalizumab Multiple Sclerosis Trial Group (2003). A controlled trial of natalizumab for relapsing multiple sclerosis. *N. Engl. J. Med.* **348,** 15–23.

Minagar, A., Alexander, J. S., Brooks, R., Kelley, R. E., Schwendimann, R. N., Gonzalez-Toledo, E., and Merill-Bryanm, D. (2005). Combination therapy with intramuscular interferon-beta1a and oral doxycycline in patients with multiple slerosis: An open label trial. *Neurology* **A34**(Suppl. 2), P01.087.

Moreau, T., Thorpe, J., Miller, D., Moseley, I., Hale, G., Waldmann, H., Clayton, D., Wing, M., Scolding, N., and Compston, A. (1994). Preliminary evidence from magnetic resonance imaging for reduction in disease activity after lymphocyte depletion in multiple sclerosis. *Lancet* **344,** 298–301.

Natalizumab package insert (2006). Biogen Idec/Elan Pharmaceuticals, Cambridge, MA; June 2006.

Neuhaus, O., Strasser-Fuchs, S., Fazekas, F., Kieseier, B. C., Niederwieser, G., Hartung, H. P., and Archelos, J. J. (2002). Statins as immunomodulators: Comparsion with interferon-beta 1b in MS. *Neurology.* **59,** 990–997.

Noseworthy, J. H., Lucchinetti, C., Rodriguez, M., and Weinshenker, B. G. (2000). Multiple sclerosis. *N. Engl. J. Med.* **343,** 938–952.

Paolillo, A., Coles, A. J., Molyneux, P. D., Gawne-Cain, M., MacManus, D., Barker, G. J., Compston, D. A., and Miller, D. H. (1999). Quantitative MRI in patients with secondary progressive MS treated with monoclonal antibody Campath 1H. *Neurology* **53,** 751–757.

Pedersen, I. M., Buhl, A. M., Klausen, P., Geisler, C. H., and Jurlander, J. (2002). The chimeric anti-CD20 antibody rituximab induces apoptosis in B-cell chronic lymphocytic leukemia cells through a p38 mitogen activated protein-kinase-dependent mechanism. *Blood* **99,** 1314–1319.

Pinschewer, D. D., Ochsenbein, A. F., Odermatt, B., Brinkmann, V., Hengartner, H., and Zinkernagel, R. M. (2000). FTY720 immunosuppression impairs effector T cell peripheral homing without affecting induction, expansion, and memory. *J. Immunol.* **164,** 5761–5770.

Polman, C. H., O'Connor, P. W., Havrdova, E., Hutchinson, M., Kappos, L., Miller, D. H., Phillips, J. T., Lublin, F. D., Giovannoni, G., Wajgt, A., Toal, M., Lynn, F., *et al.* (2006). A randomized, placebo-controlled trial of natalizumab for relapsing multiple sclerosis. *N. Engl. J. Med.* **354,** 899–910.

Popovic, N., Schubart, A., Goetz, B. D., Zhang, S. C., Linington, C., and Duncan, I. D. (2002). Inhibition of autoimmune encephalomyelitis by a tetracycline. *Ann. Neurol.* **51,** 215–223.

Reff, M. E., Carner, K., Chambers, K. S., Chinn, P. C., Leonard, J. E., Raab, R., Newman, R. A., Hanna, N., and Anderson, D. R. (1994). Depletion of B cells in vivo by a chimeric mouse human monoclonal antibody to CD20. *Blood* **83,** 435–445.

Rose, J. W., Watt, H. E., White, A. T., and Carlson, N. G. (2004). Treatment of multiple sclerosis with an anti-interleukin-2 receptor monoclonal antibody. *Ann. Neurol.* **56,** 864–867.

Rudick, R. A., Stuart, W. H., Calabresi, P. A., Confavreux, C., Galetta, S. L., Radue, E. W., Lublin, F. D., Weinstock-Guttman, B., Wynn, D. L., Lynn, F., Panzara, M. A., Sandrock, A. W., *et al.* (2006). Natalizumab plus interferon beta-1a for relapsing multiple sclerosis. *N. Engl. J. Med.* **354,** 911–923.

Sandrini, S. (2005). Use of IL-2 receptor antagonists to reduce delayed graft function following renal transplantation: A review. *Clin. Transplant.* **19,** 705–710.

Sasaki, M., Bharwani, S., Jordan, P., Joh, T., Manas, K., Warren, A., Harada, H., Carter, P., Elrod, J. W., Wolcott, M., Grisham, M. B., and Alexander, J. S. (2003). The 3-hydroxy-3-methylglutaryl-CoA reductase inhibitor pravastatin reduces disease activity and inflammation in dextran-sulfate induced colitis. *J. Pharmacol. Exp. Ther.* **305,** 78–85.

Shan, D., Ledbetter, J. A., and Press, O. W. (2000). Signaling events involved in anti-CD20-induced apoptosis of malignant human B cells. *Cancer Immunol. Immunother.* **48,** 673–683.

Sheremata, W. A., Minagar, A., Alexander, J. S., and Vollmer, T. (2005). The role of alpha-4 integrin in the aetiology of multiple sclerosis: Current knowledge and therapeutic implications. *CNS Drugs* **19,** 909–922.

Soilu-Hanninen, M., Airas, L., Mononen, I., Heikkila, A., Vijanen, M., and Hanninen, A. (2005). 25-Hydroxyvitamin D levels in serum at the onset of multiple sclerosis. *Mult. Scler.* **11,** 266–271.

Spach, K. M., Nashold, F. E., Dittel, B. N., and Hayes, C. E. (2006). IL-10 signaling is essential for 1,25-dihydroxyvitamin D3-mediated inhibition of experimental autoimmune encephalomyelitis. *J. Immunol.* **177,** 6030–6037.

Stirling, D. P., Khodarahmi, K. J., Liu, L., McPhail, T., McBride, C. B., Steeves, J. D., Ramer, M. S., and Tetzlaff, W. (2004). Minocycline treatment reduces delayed oligodendrocyte death, attenuates axonal dieback, and improves functional outcome after spinal cord injury. *J. Neurosci.* **24,** 2182–2190.

Stuve, O., Cepok, S., Elias, B., Saleh, A., Hartung, H. P., Hemmer, B., and Kieseier, B. C. (2005). Clinical stabilization and effective B-lymphocyte depletion in the cerebrospinal fluid and peripheral blood of a patient with fulminant relapsing-remitting multiple sclerosis. *Arch. Neurol.* **62,** 1620–1623.

Taji, H., Kagami, Y., Okada, Y., Andou, M., Nishi, Y., Saito, H., Seto, M., and Morishima, Y. (1998). Growth inhibition of CD20-positive B lymphoma cell lines by IDEC-C2B8 anti-CD20 monoclonal antibody. *Jpn. J. Cancer Res.* **89,** 748–756.

Thompson, P. D., Clarkson, P., and Karas, R. H. (2003). Statin-associated myopathy. *JAMA* **289,** 1681–1690.

Tokuda, N., and Levy, R. B. (1996). 1,25-Dihydroxyvitamin D3 stimulates phagocytosis but suppresses HLA-DR and CD13 antigen expression in human mononuclear phagocytes. *Proc. Soc. Exp. Biol. Med.* **211,** 244–250.

Valentine, M. A., Meier, K. E., Rossie, S., and Clark, E. A. (1989). Phosphorylation of the CD20 phosphoprotein in resting B lymphocytes. Regulation by protein kinase. *C. J. Biol. Chem.* **264,** 11282–11287.

Vollmer, T., Key, L., Durkalski, V., Tyor, W., Corboy, J., Markovic-Plese, S., Preiningerova, J., Rizzo, M., and Singh, I. (2004). Oral simvastatin treatment in relapsing-remitting multiple sclerosis. *Lancet* **363,** 1607–1608.

Weber, M. S., Youssef, S., Dunn, S. E., Prod'homme, T., Neuhaus, O., Stuve, O., Greenwood, J., Steinman, L., and Zamvil, S. S. (2006). Statins in the treatment of central nervous system autoimmune disease. *J. Neuroimmunol.* **178,** 140–148.

Wells, J. E., Hurlbert, R. J., Fehlings, M. G., and Yong, V. W. (2003). Neuroprotection by minocycline facilitates significant recovery from spinal cord injury in mice. *Brain* **126,** 1628–1637.

Windrem, M. S., Nunes, M. C., Rashbaum, W. K., Schwartz, T. H., Goodman, R. A., McKhann, G., II, Roy, N. S., and Goldman, S. A. (2004). Fetal and adult human oligodendrocyte progenitor cell isolates myelinate the congenitally dysmyelinated brain. *Nat. Med.* **10,** 93–97.

Wu, D. C., Jackson-Lewis, V., Vila, M., Tieu, K., Teismann, P., Vadseth, C., Choi, D. K., Ischiropoulos, H., and Przedborski, S. (2002). Blockade of microglial activation is neuroprotective in the 1-methyl-4-phenyl-1,2,3,6-tetrahydropyridine mouse model of Parkinson disease. *J. Neurosci.* **22,** 1763–1771.

Yednock, T. A., Cannon, C., Fritz, L. C., Sanchez-Madrid, F., Steinman, L., and Karin, N. (1992). Prevention of experimental autoimmune encephalomyelitis by antibodies against alpha 4 beta 1 integrin. *Nature* **356,** 63–66.

Yong, V. W. (2004). Prospects for neuroprotection in multiple sclerosis. *J. Neuroimmunol.* **9,** 864–872.

Yong, V. W., Wellsm, J., Giuliani, F., Casha, S., Power, C., and Metz, L. M. (2004). The promise of minocycline in neurology. *Lancet Neurol.* **3,** 744–751.

Youssef, S., Stuve, O., Patarroyo, J. C., Ruiz, P. J., Radosevich, J. L., Hur, E. M., Bravo, M., Mitchell, D. J., Sobel, R. A., Steinman, L., and Zamvil, S. S. (2002). The HMG-CoA reductase inhibitor, atorvastatin, promotes a Th2 bias and reverses paralysis in central nervous system autoimmune disease. *Nature* **420,** 78–84.

Yrjanheikki, J., Keinanen, R., Pellikka, M., Hokfelt, T., and Koistinaho, J. (1998). Tetracyclines inhibit microglial activation and are neuroprotective in global brain ischemia. *Proc. Natl. Acad. Sci. USA* **95,** 15769–15774.

Yrjanheikki, J., Tikka, T., Keinanen, R., Goldsteins, G., Chan, P. H., and Koistinaho, J. (1999). A tetracycline derivative, minocycline, reduces inflammation and protects against focal cerebral ischemia with a wide therapeutic window. *Proc. Natl. Acad. Sci. USA* **96,** 13496–13500.

Zella, J. B., McCary, L. C., and DeLuca, H. F. (2003). Oral administration of 1,25-dihydroxyvitamin D3 completely protects NOD mice from insulin-dependent diabetes mellitus. *Arch. Biochem. Biophys.* **417,** 77–80.

REMYELINATION IN MULTIPLE SCLEROSIS

Divya M. Chari

Cambridge Centre for Brain Repair and Veterinary Medicine
University of Cambridge, Cambridge CB3OES, United Kingdom; and
Institute for Science and Technology in Medicine, Keele University
Staffordshire ST55BG, United Kingdom

Remyelination is the phenomenon by which new myelin sheaths are generated around axons in the adult central nervous system (CNS). This follows the pathological loss of myelin in diseases like multiple sclerosis (MS). Remyelination can restore conduction properties to axons (thereby restoring neurological function) and is increasingly believed to exert a neuroprotective role on axons. Remyelination occurs in many MS lesions but becomes increasingly incomplete/inadequate and eventually fails in the majority of lesions and patients. Efforts to understand the causes for this failure of regeneration have fueled research into the biology of remyelination and the complex, interdependent cellular and molecular factors that regulate this process. Examination of the mechanisms of repair of experimental lesions has demonstrated that remyelination occurs in two major phases. The first consists of colonization of lesions by oligodendrocyte progenitor cells (OPCs), the second the differentiation of OPCs into myelinating oligodendrocytes that contact demyelinated

INTERNATIONAL REVIEW OF
NEUROBIOLOGY, VOL. 79
DOI: 10.1016/S0074-7742(07)79026-8

axons to generate functional myelin sheaths. Several intracellular and extracellular molecules have been identified that mediate these two phases of repair. Theoretically, the repair of demyelinating lesions can be promoted by enhancing the intrinsic repair process (by providing one or more remyelination-enhancing factors or via immunoglobulin therapy). Alternatively, endogenous repair can be bypassed by introducing myelinogenic cells into demyelinated areas; several cellular candidates have been identified that can mediate repair of experimental demyelinating lesions. Future challenges confronting therapeutic strategies to enhance remyelination will involve the translation of findings from basic science to clinical demyelinating disease.

I. What is Remyelination?

Remyelination is an example of spontaneous repair in the adult central nervous system (CNS), where new myelin sheaths are generated around demyelinated axons. This follows the pathological loss of myelin [as occurs in the demyelinating diseases, notably multiple sclerosis (MS)]. Remyelination provides an exception to a primary tenet of neuroscience, namely, that the adult CNS has a poor intrinsic capacity to repair following trauma, which is summarized in Cajal's famous postulate— "everything may die, nothing may regenerate" in the CNS (Cajal, 1928). The phenomenon of remyelination was documented in early morphological descriptions of brains from MS patients (Marburg, 1906). Plaques in freshly isolated CNS tissue are easily identifiable as distinct, discolored areas of myelin loss. In addition to such classical plaques, sharply delineated areas of reduced myelin density are observed, recognizable by evenly distributed, pale myelin staining that represent the *Markschattenherde* or shadow plaques (Schlesinger, 1909) (Fig. 1). While such observations were attributed initially to incomplete demyelination (due to ongoing evolution of lesions), there had been earlier suggestions that a process of repair could follow damage to myelin (Marburg, 1906). Despite this, the verification that spontaneous remyelination could follow myelin loss was not provided till 1961, when a pioneering ultrastructural study [using the method of cerebrospinal fluid (CSF) barbotage to induce demyelination] clearly demonstrated that demyelinated axons in the adult cat spinal cord can acquire new myelin sheaths (Bunge *et al.*, 1961). This followed their earlier observations from light microscopic studies that myelin reappears at a time when animals show recovery from partial paralysis (Bunge, 1958; Bunge and Settlage, 1957). The suggestion that the thin myelin lamellae observed in demyelinated lesions could represent areas of spontaneous remyelination came from an ultrastructural study of MS plaques (Perier and Gregoire, 1965). Furthermore, other studies proved that remyelination induced not only structural recovery but also allowed axons to regain

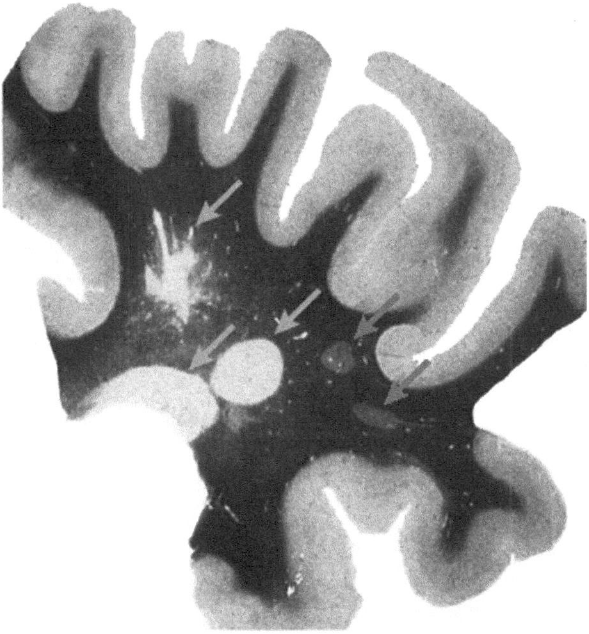

FIG. 1. Section from the brain of an MS patient stained with Luxol fast blue to identify subcortical white matter (WM). Green arrows indicate three areas of myelin loss which represent foci of chronic demyelination. Red arrows indicate areas of pale myelin staining which represent the shadow plaques where remyelination has occurred. This illustrates the key point that spontaneous remyelination can occur efficiently in MS but that this process fails on many occasions, leaving axons chronically demyelinated. Reproduced with permission from Adams (1989). (See Color Insert.)

conduction properties (Smith *et al.*, 1979) and neurological function lost as a consequence of demyelination (Jeffery and Blakemore, 1997).

Remyelination of large and medium diameter CNS axons is easily identifiable histologically, although unequivocal identification of remyelination of small diameter axons is generally more challenging or even impossible. Remyelinated axons display some distinguishing features, chief among which is a shortened length of the new internodes. Newly generated myelin sheaths are also thinner than normal; therefore, the relationship between myelin thickness and internodal length to axon diameter (that applies in normally myelinated axons) does not pertain in remyelinated axons, either in experimental lesions or MS shadow plaques. Despite this, remyelination allows axons to regain adequately secure conduction properties such that few conduction abnormalities can be recorded from remyelinated axons (this includes a restoration of latent periods of axons to normal) (Smith *et al.*, 1979, 1981).

A. Remyelination in MS

Data on the extent and distribution of remyelination in MS are mainly obtained from postmortem tissue; an obvious problem with interpreting such data is that they represent a static picture of MS lesions and therefore provide few clues about the degree and pattern of repair of dynamic, evolving lesions that exist in the clinical disease. Clinically, magnetic resonance imaging (MRI) is generally considered to be the most sensitive method to detect alterations in the white matter (WM) of MS patients. While the identification of remyelination in such lesions has so far been largely unreliable, advances in multicontrast MRI will increasingly allow for the characterization and monitoring of remyelination in MS patients (Merkler *et al.*, 2005).

The extent of remyelination observed between MS lesions is highly variable. This accompanies significant variations in plaque size with differences in patterns of demyelination and oligodendrocyte damage/loss. This may depend to some degree on the stage of progression of lesions but may also be related to variation in underlying pathogenic mechanisms (Lucchinetti *et al.*, 2000). Generally, the degree of remyelination of MS lesions correlates directly with both numbers of oligodendrocytes (expressing myelin proteins or their mRNAs) and macrophages in lesions (Lucchinetti *et al.*, 1999). Remyelination can occur rapidly and thin, short internodes can be detected in acute lesions within a few weeks of lesion genesis; remyelination can occur concomitant with tissue infiltration by macrophages (Ghatak *et al.*, 1989; Lassmann, 1983; Prineas *et al.*, 1984, 1993) and occasionally myelin destruction (Lassmann *et al.*, 1997; Prineas *et al.*, 1993; Raine and Wu, 1993). Conversely, some remyelinated lesions can undergo active demyelination (Prineas *et al.*, 1993), resulting in new lesions overlapping remyelinated areas. Shadow plaques are more commonly observed in young patients with an active relapsing-remitting course than in older patients with an inactive/chronic one (Prineas and Connell, 1979); chronic lesions from patients with long-standing disease tend to show sparse remyelination, restricted to a thin rim-bordering periplaque white matter (PPWM). Additionally, remyelination may be sparse in cases of primary progressive MS (Lassmann *et al.*, 1997).

Remyelination of central axons can sometimes be accompanied by Schwann cell invasion of the CNS (Itoyama *et al.*, 1983), which is more common near peripheral nerve root entry or exit zones in the spinal cord; this was found to be a prominent feature in some Japanese patients (Itoyama *et al.*, 1985). It has conventionally been assumed that remyelinating Schwann cells originate from Schwann cells associated with spinal roots or nerves supplying the vasculature or meninges (Franklin and Blakemore, 1993). Schwann cell remyelination is generally prominent in areas of both astrocyte and oligodendrocyte loss, leading to the suggestion that death of astrocytes constituting the glia limitans allows Schwann cell entry into the CNS; reorganization of this structure would inhibit further influx. Challenging these views, a report suggests instead that remyelinating Schwann cells may be derived from CNS precursors; this awaits formal proof (Blakemore, 2005).

The extent to which remyelination is associated with differences in clinical outcome, disease severity, and recovery in MS is currently unclear. It is suggested that remyelination is unlikely to play a significant role in early remission (as conduction deficits persist during the early stages of remission), but may be important for later stages of functional recovery (Jones, 1993). Remyelination fails in most lesions in MS and while it is likely that numerous causal factors underlie progressive clinical deterioration in the disease, the inadequacy or failure of remyelination is likely to play a significant contributory role.

B. Why is Remyelination Important?

Permanent neurological dysfunction observed in MS has traditionally been attributed to demyelination. Recent clinical studies have questioned the validity of this assumption with the growing belief that permanent axon loss is the more likely pathological correlate of irreversible functional deficit (Fig. 2). Demyelinated axons may be vulnerable to damage (during the acute phase) and possibly damage plus atrophy (during the chronic phase) of MS (Ferguson *et al.*, 1997; Lovas *et al.*, 2000; Trapp *et al.*, 1998). While the causes for such axonal atrophy are unclear, demyelination may predispose axons to secondary injury by (1) multiple molecular cytotoxicity mechanisms, including ion channel dysfunction and mitochondrial damage, or (2) failure of local target-derived neurotrophic support for myelinating oligodendrocytes. This implies that one of the principal functions of myelin is to protect axons from secondary insult (Rodriguez, 2003; Waxman, 2006). To date, there is no formal proof that remyelination does prevent axonal atrophy and loss. However, axon loss in experimental allergic encephalomyelitis (EAE) lesions is greater in demyelinated areas than remyelinated ones (Kornek *et al.*, 2000). In addition, absence of proteolipid protein (PLP), a major myelin protein, results in progressive axonal degeneration in rodents and humans (Garbern *et al.*, 2002; Griffiths *et al.*, 1998; Edgar *et al.*, 2004). Mice lacking cyclic nucleotide phosphodiesterase (CNP), an enzyme expressed by oligodendrocytes, also show severe axonal degeneration with accumulation of amyloid precursor protein (APP). This is associated with a significant motor deficit that becomes manifest by about 4 months (Lappe-Siefke *et al.*, 2003). It is not clear how oligodendrocytes maintain axonal integrity. Oligodendroglia express NGF and BDNF, key neurotrophic factors that play a role in the maintenance and survival of neurons. OPCs and differentiated oligodendrocytes can also support neuronal survival through contact-mediated/soluble mechanisms, providing potential mechanisms by which remyelination could prevent neuronal atrophy (Byravan *et al.*, 1994; Wilkins *et al.*, 2001).

Secure electrical conduction in axons is made possible by a defined organization of cellular processes and domain-specific ion channels at the nodes of Ranvier (reviewed by Sherman and Brophy, 2005). Remyelination supports the development and maturation of new nodes on central axons, which display an apparently normal ion channel distribution (Black *et al.*, 2006; Dugandzija-Novakovic *et al.*, 1995; Sasaki *et al.*, 2006). These findings add support to the known role of remyelination in restoring conduction properties to axons. Strategies aimed at promoting CNS remyelination can therefore be predicted to both restore secure

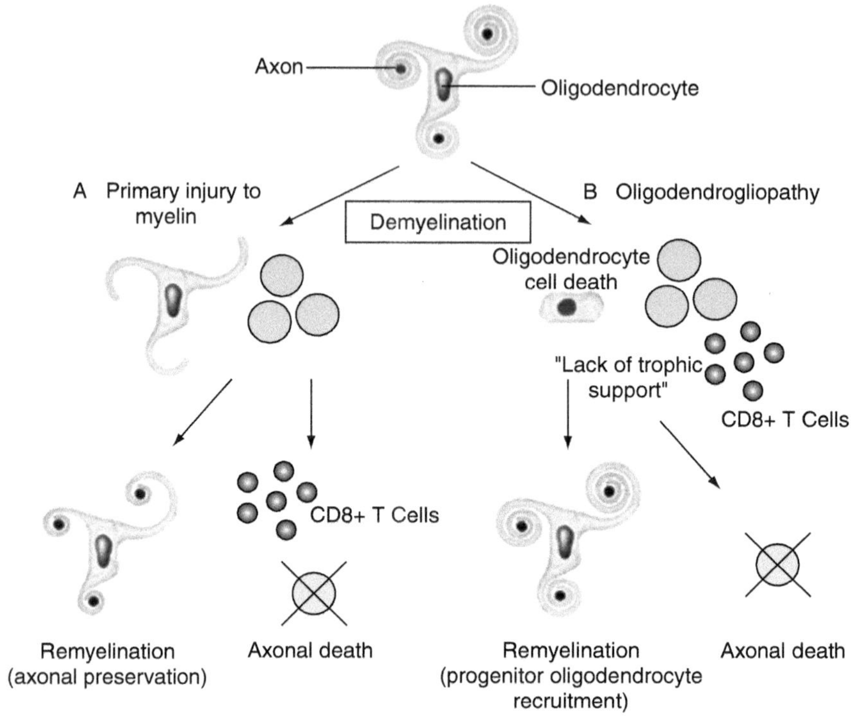

FIG. 2. There are two options following demyelination in multiple sclerosis: either (A) remyelination can take place with axonal preservation or (B) axonal death may ensue, either due to a range of cytotoxicity mechanisms or due to failure of local neurotrophic support due to glial cell injury. Remyelination most likely occurs by the recruitment of OPCs into demyelinated areas that subsequently mature into myelinating cells. Modified with permission from Rodriguez (2003).

conduction (thereby promoting functional recovery) and protect axons in inflammatory lesions from secondary injury.

II. Experimental Remyelination

Several models of experimental demyelination are currently in use.

A. Autoimmune Models

Autoimmune models of demyelination are jointly referred to as EAE models. There are many variants but most models consist of immunizing genetically susceptible animals by direct injection of myelin antigens such as myelin basic protein (MBP), myelin-associated glycoprotein (MAG), PLP, or myelin oligodendrocyte glycoprotein (MOG). Alternatively, T cells sensitized to myelin antigens can be transferred passively to animals (adoptive transfer). EAE lesions display many pathological features of MS lesions in terms of dissemination, demyelination, and infiltration by macrophages and T cells with significant similarities observed between the clinical course of EAE and MS (Kornek *et al.*, 2000; Storch *et al.*, 1998). The extent of axon loss in EAE lesions also correlates with chronic neurological deficit (Papadopoulos *et al.*, 2006). Both acute and chronic forms of the disease can be induced in EAE models, as well as relapsing-remitting and progressive forms (reviewed in Ercolini and Miller, 2006).

A targeted model of EAE has been described (Kerschensteiner *et al.*, 2004), where subthreshold immunization of Lewis rats with myelin antigens is combined with injection of cytokines (like IFN-γ and TNF-α) at focal CNS sites. This creates single prototypic, inflammatory lesions at defined CNS foci. Such lesions share some pathological features of MS lesions and have the additional advantage of ease of lesion localization. Refined behavioral testing can also be carried out in such animals. In another model, antibody-mediated demyelination can be produced by focally injecting anti-galactocerebroside (GalC) antibodies plus complement into specific CNS sites (Keirstead *et al.*, 1998).

B. Viral Demyelination

Some viruses cause CNS demyelination (reviewed in Ercolini and Miller, 2006); a widely studied model of demyelination is based on infection of mice with Theiler's murine encephalomyelitis virus (TMEV). When injected into the CNS of certain mouse strains, the virus induces demyelinating lesions associated with virus replication in oligodendrocytes. Low neurovirulent strains like DA and BeAn cause a biphasic disease in mice. There is some irrelevant matter which is not in, Disk file. The early phase consists of sublethal encephalitis while the later phase consists of viral infection of inflammatory cells, astrocytes and oligodendrocytes, multifocal demyelination, and axonal damage, depending on the mouse strain used (reviewed by Oleszak *et al.*, 2004). Experimental demyelination at multiple foci can be induced by the A59 and JHM strains of mouse hepatitis virus (MHV), a coronavirus of mice, which is followed by extensive remyelination. The Semliki Forest virus (SMV) can also induce CNS demyelination in mice.

C. Toxin-Mediated Demyelination

Many toxin-mediated models of demyelination are used currently. One approach consists of systemically administering agents like cuprizone (a copper-chelating agent) in the diets of laboratory mice. While this model does not have the disadvantages

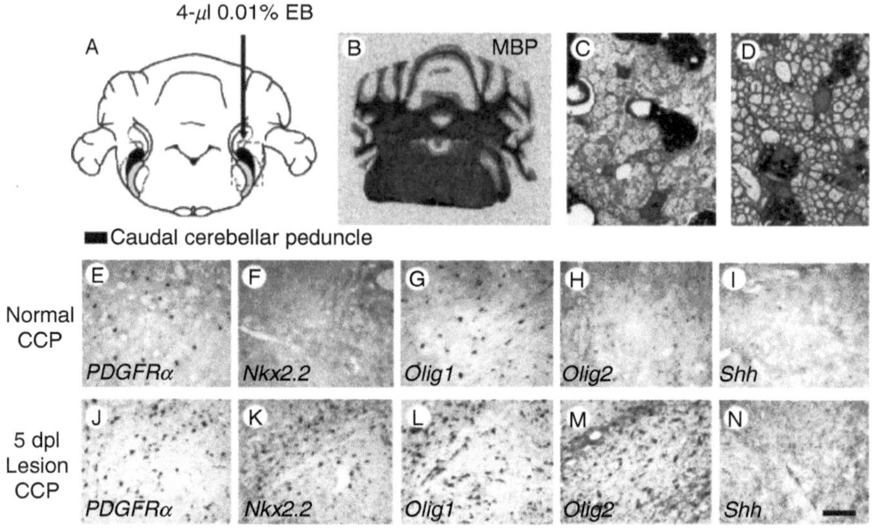

FIG. 3. OPCs increase expression of Nkx2.2 mRNA and Olig2 mRNA in response to CNS demyelination. (A) Diagram to show focal demyelinating lesions in the caudal cerebellar peduncle (CCP), a large WM tract in adult rats, created by stereotaxic injection of ethidium bromide (EB). (B) This creates focally demyelinated areas, identifiable by absence of myelin basic protein (MBP) mRNA following *in situ* hybridization. (C) Toluidine blue-stained resin section at 5 days postlesion showing demyelinated axons in transverse section and debris-filled macrophages. (D) All demyelinated axons have been reinvested with thin myelin sheaths that are characteristic of CNS remyelination, at 4 weeks postlesion. (E–H) PDGFRα, Nkx2.2, Olig1, and Olig2 mRNA expression in normal WM detected by *in situ hybridization.* (J–M) The expression of the same mRNAs in areas of demyelination 5 days postlesion. Scale bar = 1.5 mm (A), 25 μm (C–D), and 100 μm (E–N). Modified with permission from Zhao *et al.* (2005).

associated with direct toxin injection into the CNS (like CNS trauma, disruption of the blood–brain barrier, and an inflammatory response), the occurrence of demyelinating lesions may not be entirely reproducible with the doses used currently (Stidworthy *et al.*, 2003). Other models consist of focal microinjections of toxins like ethidium bromide (EB) or lysolecithin at defined CNS foci (Fig. 3A–D). Injection of EB results in death of all nucleated cells at lesion sites, whereas lysolecithin-induced lesions contain demyelinated axons present within an astrocytic environment. Toxin models have the advantage that highly reproducible demyelinating lesions can be created at defined CNS foci (Woodruff and Franklin, 1999).

D. Genetic Models of Demyelination

A number of myelin mutants exist where myelin is absent or appreciably reduced, with accompanying changes in myelin structure and oligodendrocyte number (Nave, 2001). Such mutants have proved invaluable for examining the biology of myelination and benefits of cell transplantation therapies. Myelin mutants can be divided into two categories: (1) the X-linked mutants such as the myelin-deficient (*md*) rat, shaking pup, and jimpy (*jp*) rat and (2) the autosomal recessive mutants such as the shiverer (*shi*) mouse, quaking (*qk*) mouse, and TAIEP rat. Of these, the X-linked mutants form the biggest and most well-characterized group.

Newer models to induce oligodendrocyte death include injection of viral proteins like the human endogenous retrovirus (HERV) glycoprotein–syncitin (such retroviral proteins are increasingly thought to play a role in the pathogenesis of MS). Alternatively, genetic strategies can be used to induce anatomical and lineage-targeted cell depletion using Cre-mediated gene recombination (Antony *et al.*, 2004; Brockschnieder *et al.*, 2004; Buch *et al.*, 2005; Christensen, 2005). Each of these models has particular advantages and can mimic specific aspects of MS lesions, but no model is sufficient to mimic the range of pathological presentations of the disease. Immune models are generally considered to mimic the disease process and pathogenesis in MS better than the toxin models (although there is scarce evidence that the triggers for MS are similar to those inducing demyelination in these models). A drawback of immune-mediated demyelination is the difficulty in separating cellular events associated with demyelination from repair mechanisms, as both can occur in parallel. In contrast, demyelination occurs as a single acute episode following toxin injection into the CNS, allowing demyelination to be temporally separated from regenerative events that occur in response to demyelination making toxin models particularly attractive for the study of remyelination biology. Oligodendrocyte apoptosis may be a major cause of pathology in some MS subgroups (Barnett and Prineas, 2004); toxin models causing oligodendrocyte death, and targeted oligodendrocyte depletion models do mimic this aspect of the disease.

A major factor to be considered when extrapolating experimental findings to the study of MS is that most models create small areas of demyelination which generally repair with high efficiency. This does not pertain in MS lesions which can be severalfold larger than experimental lesions; therefore, the temporal progression and dynamics of repair of experimental versus clinical lesions can be predicted to be significantly different, which may have implications for remyelination failure, as discussed later.

III. The Biology of Remyelination is Complex

A. Remyelination Occurs in Distinct Phases

Remyelination requires cell proliferation, which is highlighted by the finding that X-irradiation of demyelinating lesions (a procedure that eliminates dividing cells, notably the oligodendrocyte progenitor cells or OPCs) is sufficient to abolish remyelination (Blakemore and Patterson, 1978; Chari *et al.*, 2006b). The origins of oligodendrocytes that undertake remyelination have been controversial in the past, with two cells proposed as putative sources (1) surviving oligodendrocytes in demyelinating lesions and (2) the OPCs. The OPCs are a group of ubiquitous, slowly dividing (potentially multipotent) neural progenitors that make up roughly 3–9% of the adult CNS cell population (reviewed in Polito and Reynolds, 2005). Most evidence suggests that mature oligodendrocytes are postmitotic and incapable of generating new oligodendrocytes (Crang *et al.*, 2004; Keirstead and Blakemore, 1997). On the basis of a number of other studies, however, it is now widely accepted that new myelin sheaths are made by new oligodendrocytes derived from OPCs.

A scheme for remyelination has been proposed [based on observations on the repair of experimental (toxin-induced) lesions], whereby remyelination occurs in two major phases. The first is a *recruitment phase*, where OPCs proliferate in order to populate demyelinated areas. OPCs subsequently differentiate into premyelinating oligodendrocytes which contact demyelinated axons and differentiate into mature, myelinating oligodendrocytes that form functional myelin sheaths—this constitutes the *differentiation phase*. The recruitment and differentiation phases are associated with distinct and timed patterns of expression of molecules, notable among which are a range of growth factors. While newly generated myelin (formed as a result of remyelination) is distinct from normal myelin (formed during development), experimental analyses show some striking similarities in both processes (reviewed in Franklin, 2002). Therefore, remyelination could be considered to be a recapitulation of the developmental myelination process (Franklin and Hinks, 1999), in keeping with the widely prevailing hypothesis that regenerative mechanisms recapitulate developmental events. Despite such demonstrable parallels, it should be kept in mind that developmental myelination occurs in a normal, ordered microenvironment. By contrast, remyelination occurs within a disrupted and pathological microenvironment dominated by macrophages and T cells that contains demyelinated axons within areas of astrocytosis, none of which pertain in development (Franklin, 2002).

For remyelination in be successful, it does appear critical for the molecular signaling environment in lesions to shift from one that initially promotes OPC recruitment/proliferation to one that favors differentiation (Franklin, 2002). Therefore, perturbations to the temporal progression of the individual stages of repair could significantly alter the outcome of repair processes, as discussed in Section IV.

B. Factors Regulating OPC Activation and Proliferation

OPCs are the major proliferative population of the adult CNS [accounting for 70% and 74% of bromodeoxyuridine (BrdU)-incorporating cells in the spinal cord and cortex respectively]. Despite this, the OPCs are relatively quiescent cells that divide infrequently and have slow rates of division (Polito and Reynolds, 2005). In response to demyelination (and other CNS trauma), OPCs undergo a process of activation followed by rapid proliferation. OPC activation is associated with distinctive changes, including morphological alterations, increase in size, and upregulation of the cell surface chondroitin sulfate proteoglycan, NG2. They also demonstrate increased gene expression for transcription factors associated with developmental myelination. OPCs in WM normally express low levels of the bHLH transcription factor Olig2 and the homeodomain transcription factor Nkx2.2. However, striking increases in the expression of both Olig2 and Nkx2.2 (that is restricted to oligodendrocyte lineage cells) occur within acute demyelinating lesions (Fig. 3E–M) (Fancy *et al.*, 2004). Coexpression of Olig2 and Nkx2.2 in OPCs appears necessary for these cells to differentiate into myelinating oligodendrocytes, so it has been hypothesized that expression of these two molecules constitutes a genetic switch that converts relatively quiescent OPCs into myelinating oligodendrocytes. Nkx2.2 levels show a rapid subsequent decrease during differentiation into oligodendrocytes. Therefore, the authors hypothesize that expression patterns of Nkx2.2 and Olig2 could enable a distinction to be made between OPCs actively participating in repair from OPCs that are quiescent. The authors further speculate that such a distinction could enable actively repairing lesions to be distinguished from the ones where remyelination failure has occurred.

Following activation, OPCs undergo rapid proliferation in order to populate demyelinated areas. Several factors have been identified, which alone or in combination can promote OPC proliferation *in vivo* and *in vitro*. These include platelet-derived growth factor (PDGF), fibroblast growth factor 2 (FGF2), and glial growth factor 2 (GGF2). PDGF and FGF2 receptors are expressed on OPCs during their proliferative response to demyelination (Redwine and Armstrong, 1998). This accompanies increased expression of PDGF and FGF2 in areas of demyelination (Messersmith *et al.*, 2000; Woodruff *et al.*, 2004). *PDGF-A* knockout mice show significantly reduced numbers of OPCs and oligodendrocytes accompanied by impaired genesis of myelin (Fruttiger *et al.*, 1999). Conversely, *PDGF-A* overexpression results in significant increases in OPC densities in demyelinating lesions (Woodruff *et al.*, 2004). This study also examined if the increased availability of an OPC mitogen was associated with enhanced remyelination rates in old rats; no significant differences were observed between transgenic and control animals (despite increased OPC numbers in the former). This suggests that the OPC supply within lesions may not be rate limiting during remyelination. A recent study examining roles of individual signaling pathways has demonstrated distinctive roles *in vivo* for PDGF and FGF2 during remyelination. The authors suggest that PDGFRα signaling is likely to mainly regulate OPC proliferation (in response to demyelination), while FGF2 is a predominant inhibitor of OPC differentiation (Murtie *et al.*, 2005). Systemically administered GGF2, a neuregulin expressed by neurons, has also been shown to promote OPC expansion, prevent OPC differentiation, and enhance remyelination of experimental EAE lesions (Marchionni *et al.*, 1999). By contrast, direct administration of GGF into areas of toxin-induced demyelination did not alter the extent of remyelination, suggesting that systemic GGF2 delivery may affect remyelination by altering the immune response rather than by direct effects on oligodendrocyte lineage cells (Penderis *et al.*, 2003b).

C. Factors Regulating OPC Differentiation

Numerous extracellular and intracellular factors modulate the extent of OPC differentiation and myelin formation, but no coherent scheme has emerged so far on how the intricate process of progenitor differentiation into myelinating oligodendrocytes is regulated. Numerous transcription factors (and their regulatory molecules) mediate developmental myelination including Sox10, bHLH proteins, MyT1, Mash1, and Gtx among others; their roles in remyelination need to be elucidated (reviewed in Franklin, 2002). A key, nonredundant role for the transcription factor Olig1 has been reported in mediating remyelination (Arnett *et al.*, 2004). Repair of toxin-induced demyelinating lesions in both brain and spinal cord was significantly reduced in *Olig1* knockout mice compared with controls. Recruitment of OPCs into lesions from knockout mice appeared normal, but these cells failed to differentiate into remyelinating oligodendrocytes. The subcellular localization of Olig1 could be a critical factor mediating differentiation; nuclear Olig1, expressed by OPCs at the margins of MS lesions, was suggested to be associated with cells actively engaged in repair.

As with OPC proliferation, the involvement of growth factors in regulating OPC differentiation has been extensively documented. Insulin-like growth factor-1 (IGF-1) is suggested to play a critical role in this process based on studies in transgenic mice overexpressing IGF-1 where increases in myelin sheath thickness and numbers of myelinated axons (with increased numbers of oligodendrocytes) are observed; the expression of IGF in MS lesions is also consistent with the notion that this molecule plays a major role in the differentiation phase (Carson *et al.*, 1993; Gveric *et al.*, 1999; Luzi *et al.*, 2004; Ye *et al.*, 1995). Transforming growth factor $\beta 1$ (TGF$\beta 1$) inhibits PDGF- and FGF2-induced OPC proliferation and is suggested to be an important regulator of oligodendrocyte differentiation (McKinnon *et al.*, 1993). The observed pattern and timing of expression of both IGF-1 and TGF$\beta 1$ in demyelinating lesions supports the idea that these molecules regulate OPC differentiation during repair (Hinks and Franklin, 1999; Sim *et al.*, 2002). However, directly increasing levels of IGF-1 mRNA using adenoviral-mediated gene delivery was not sufficient to increase the extent of oligodendrocyte-mediated remyelination of toxin-induced lesions in old rats compared with young ones (O'Leary *et al.*, 2002).

It is becoming increasingly clear that electrical activity in axons, axon-derived signals, and neurotransmitter release all play important roles in the genesis, survival, and differentiation of oligodendrocytes and induction of developmental myelination (reviewed in Coman *et al.*, 2005). Blocking or increasing the firing rates of neurons can respectively inhibit or enhance myelination (Demerens *et al.*, 1996). Oligodendrocytes and OPCs express a range of ion channels and neurotransmitter receptors, the activation or inhibition of which can alter proliferation and differentiation of these cells and myelination (Gallo *et al.*, 2001). OPCs are also thought to make close contacts with the nodes of Ranvier and synapses (Bergles *et al.*, 2000; Butt *et al.*, 1999) and are therefore well placed anatomically to respond to changes in axonal function in normal and pathological situations. These observations point to a clear link between electrical activity and myelination, but the potential effects of electrical activity and conduction block in demyelinated axons on the proliferation and differentiation of adult OPCs in surrounding tissue have yet to be established. Inhibitors of oligodendrocyte differentiation (like PSA-NCAM, vitronectin, tenascins-R, GGF2, and jagged) are all expressed by axons (reviewed in Coman *et al.*, 2005). This lends support to the idea that axons could play key roles in the regulation of properties of oligodendrocyte lineage cells and myelination.

Several intracellular OPC differentiation factors have also been identified. The cyclin-dependent kinase inhibitor p27Kip-1 is found to accumulate in OPCs and on reaching a threshold appears to inhibit OPC proliferation and increased differentiation (Durand and Raff, 2000). p27Kip-1 also appears to regulate both OPC responses to mitogens and the numbers of OPCs recruited to demyelinated areas—OPC numbers were significantly increased in p27Kip-1 null mice compared with wild-type controls (Crockett *et al.*, 2005). Another recent study has demonstrated that a transmembrane CNS protein Lingo-1 is expressed by oligodendrocytes and could play a critical role in their differentiation and myelination. The authors further suggest that this molecule could represent an important therapeutic target in demyelinating and dysmyelinating disorders (Mi *et al.*, 2005).

D. Inflammation: A Double-Edged Sword?

Inflammation undoubtedly contributes to the pathogenesis of demyelinating lesions in MS. However, there is considerable evidence to support the view that the inflammatory response provides an important stimulant for remyelination. Remyelination is frequently associated with a prominent inflammatory response in both experimental and MS lesions. Remyelination is also observed to occur most efficiently in early, active lesions where the inflammatory response is robust (Graca and Blakemore, 1986; Ludwin, 1980; Raine and Wu, 1993). One study using Affymetrix microarrays to identify changes in gene expression during remyelination indicates that alterations in inflammatory genes represent the most prominent changes, emphasizing the role of the active inflammatory response (Arnett *et al.*, 2003). Furthermore, several studies have demonstrated an impairment of remyelination in the absence of lymphocytes, MHC-II, inflammatory cytokines, and macrophages (Arnett *et al.*, 2001, 2003; Bieber *et al.*, 2003, Kotter *et al.*, 2001, 2005; Mason *et al.*, 2001). This is supported by the findings that anti-inflammatory drugs, such as minocycline and corticosteroids (such as dexamethasone, methylprednisolone, and prednisone), significantly delay remyelination of toxin-induced demyelinating lesions (Chari *et al.*, 2006a; Li *et al.*, 2005; Triarhou and Herndon, 1986) (Fig. 4). Anti-inflammatory agents appear to act mainly by inhibiting OPC differentiation, as colonization of lesions by OPCs was found to be essentially normal in corticosteroid (CS)-treated animals. Such agents do not permanently impair the repair properties of myelinogenic cells as normal levels of remyelination can be found with increased survival times (following treatment withdrawal) (Chari *et al.*, 2006a). While CS administration has numerous known benefits, these findings will have obvious implications for the use of anti-inflammatory therapies in MS and raise the possibility that prolonged anti-inflammatory treatment may predispose to enhanced axon loss in MS lesions (Chari *et al.*, 2006a).

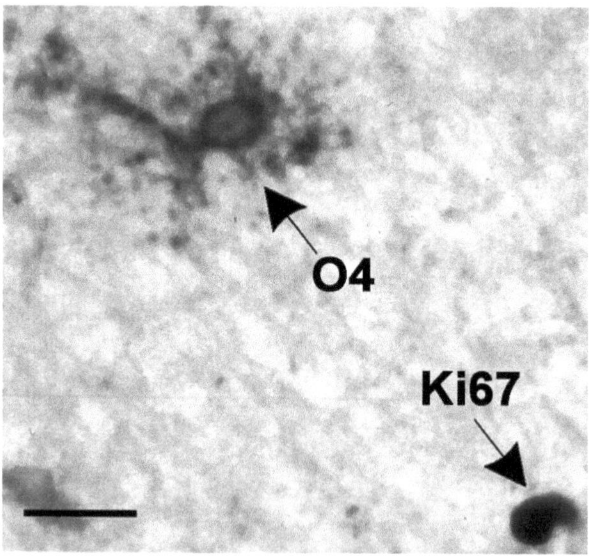

FIG. 4. Chronic MS lesions contain O4-positive OPCs that fail to bind the Ki67 antibody (a marker of cell proliferation), suggesting that these are a relatively quiescent population. Some lesions do contain some Ki67-immunoreactive nuclei, such as the one indicated toward the lower right of the figure. Reproduced with permission from Wolswijk (1998).

How could inflammation influence remyelination? The exact mechanisms are unclear at present. Macrophages/microglia are suggested to produce molecules such as growth factors that mediate repair either directly or by activation of astrocytes to produce remyelination-associated growth factors. The addition of peritoneal macrophages, *in vitro*, to fetal neuron glia cocultures increases the synthesis of myelin proteins like MBP, and increases myelin deposition (Loughlin *et al.*, 1997; Nathan, 1987). Inflammation can also enhance the survival and migration of transplanted OPCs, in part, due to upregulation of multiple factors implicated in oligodendrocyte development (Setzu *et al.*, 2006; Tourbah *et al.*, 1997). A similar, remyelination-promoting role has been suggested for T cells in demyelinating lesions (Bieber *et al.*, 2003). Macrophages play a major role in myelin debris clearance with the suggestion that OPC maturation can be inhibited by contact with myelin (Miller, 1999). Indeed, introduction of purified myelin into demyelinating lesions in the brainstem causes a significant impairment of remyelination which is attributable to arrested OPC differentiation rather than impaired recruitment (Kotter *et al.*, 2005). This impairment may be due to a physical restriction of contact between myelinating oligodendrocytes and demyelinated axons, or alternatively due to the presence of inhibitory molecules on myelin debris.

IV. What Causes Remyelination Failure in MS

Remyelination is regulated by complex mechanisms and the individual phases of repair are associated with distinct, timed patterns of expression of numerous molecules. Given the complexity of such regulatory processes and the prominent heterogeneity in pathogenesis and extent of cell survival in MS lesions, it is clear that the success of repair will be influenced by several interdependent variables. These include (1) OPC and oligodendrocyte availability in lesions, (2) astrocyte and macrophage presence in demyelinated areas, (3) the timed and coordinated expression of regulatory molecules, and (4) the ability of axons and OPCs/remyelinating oligodendrocytes to interact in a manner appropriate for successful remyelination to occur. These parameters are further influenced by age (Chari *et al.*, 2003a; Gilson and Blakemore, 1993; Sim *et al.*, 2002), sex (Li *et al.*, 2006), and possibly lesion size (Chari and Blakemore, 2002b). Therefore, a single causal factor in remyelination failure appears unlikely and the postulated causes for regenerative failure in MS remain largely speculative. Several issues must be considered.

First, given that OPCs are critical mediators of remyelination, the question of whether these cells are depleted in MS lesions is a crucial but unresolved one. Histopathological studies using markers like O4, NG2, and PDGFRα suggest that

OPCs are present in chronic MS lesions in the brain and spinal cord in equal/higher numbers than periplaque WM (Chang *et al.*, 2000; Scolding *et al.*, 1998; Wolswijk, 1998). Such OPCs were, however, suggested to be relatively quiescent based on the failure to label with a nuclear proliferation antigen (detected by the Ki-67 antibody) (Fig. 5) and lack of oligodendrocytes in the center of MS lesions. It was suggested that these OPCs may be incapable of differentiation or alternatively that the lesion environment lacked factors to stimulate regenerative mechanisms (or contained factors that inhibited it) (Wolswijk, 1998). One study examining the specific possibility that remyelination fails due to local OPC destruction along with oligodendrocytes demonstrated that chronic lesions from spinal cords of MS patients contained significant numbers of O4 positive, GalC negative OPCs (Wolswijk, 2002). While interpreting such observations, it should be kept in mind

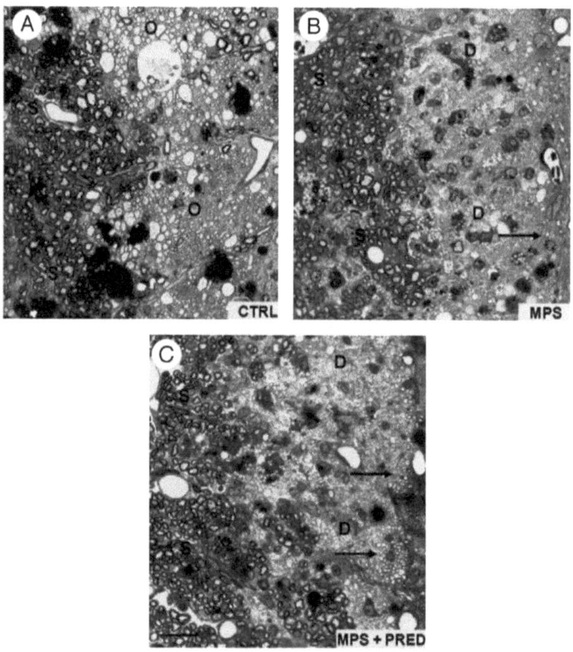

FIG. 5. Corticosteroid (CS) treatment significantly reduces oligodendrocyte-mediated remyelination of EB-induced lesions in the spinal cord at 1 month. Treatment groups consisted of injection of high-dose MPS with and without subsequent PRED taper, in order to mimic current CS treatment regimens used clinically to treat acute relapses in MS. (A) Extensive oligodendrocyte and Schwann cell-mediated remyelination in saline-injected control rats. (B, C) Reduced oligodendrocyte remyelination in CS-treated rats with Schwann cell remyelination unaffected. Oligodendrocyte remyelination is indicated with arrows. MPS, methylprednisolone, PRED, prednisone. Modified with permission from Chari *et al.* (2006a).

that it is currently impossible to distinguish OPCs that survive in lesions from OPCs that colonize lesions from surrounding normal tissue. Therefore, an observed abundance of OPCs in lesions provides few clues as to their potential destruction during early stages of lesion development. This study further reported that with increasing age of subjects and duration of clinical symptoms (1) OPC densities declined significantly, (2) genesis of new oligodendrocytes appeared increasingly impaired, and (3) lesions did not appear to be significantly repopulated by OPCs (Wolswijk, 2002). Another study has reported that NG2 positive cells could *not* be detected in acute inflammatory lesions (although small numbers of lesions were examined in this study) (Chang *et al.*, 2000). Support for the idea that OPCs could be killed in MS lesions comes from the demonstration that molecules like MOG and GalC (previously considered to be restricted to oligodendrocytes) are also present on OPCs (Li *et al.*, 2002; Shi *et al.*, 1998). Autoimmunity to MOG has been suggested to play an important role in the pathogenesis of MS lesions, so expression of such molecules on OPCs could, in theory, result in OPC destruction. The finding that patients with relapsing-remitting MS synthesize antibodies to the OPC-specific surface glycoprotein AN2 (which is absent in the CSF of patients with inactive disease) (Niehaus *et al.*, 2000) lends support to this idea.

If OPCs are significantly depleted along with oligodendrocytes in demyelinated areas, then the repopulation of lesions with repair-mediating OPCs will clearly constitute the first step toward successful remyelination. An influential hypothesis regarding the failure of remyelination in MS has been the exhaustion of the OPC pool, which suggests that OPCs become depleted around lesions following remyelination. However, one study has demonstrated that focal areas of experimental OPC depletion can be fully repopulated by OPCs from surrounding normal tissue, a process that restores OPC densities to normal (Chari and Blakemore, 2002a). This is supported by the finding that repeated episodes of demyelination do not lead to depletion of the OPC pool or an impairment of remyelination (Penderis *et al.*, 2003a). The repopulation of OPC-depleted areas (by OPCs from adjoining, normal tissue) is a slow process [approximately 0.5 mm/week, a rate that declines further with age due to intrinsic changes in OPCs (Chari *et al.*, 2003b)]. On the basis of such observations, it is clear that large areas of demyelination in MS could require several months to become completely repopulated by OPCs (assuming that tissue colonization rates for human OPCs are similar to that of rat OPCs). As acute inflammation appears to provide a major stimulus for OPC recruitment, it has been suggested that the abating of inflammation in demyelinated areas, with time, will reduce the drive for remyelination, thereby causing repair failure. In other words, remyelination failure may occur due to a "Temporal Mismatch" between the presence of a remyelination-promoting environment and the arrival of OPCs in demyelinated areas (Chari and Blakemore, 2002a,b). This would imply that a critical time window exists for the interaction of OPCs with demyelinated axons, outside of which the likelihood of successful remyelination will be significantly reduced. This hypothesis could provide an explanation for the finding that remyelination can be prominent at the margins of MS lesions but not the core (i.e., with increasing distance from lesion boundaries). It may also account for the finding that a relatively quiescent OPC population is present in chronic lesions, which does not appear to contribute to remyelination. Another study has suggested that the presence of dysfunctional OPCs may inhibit further colonization of tissue by OPCs from normal tissue (Chari *et al.*, 2003b), which could potentially explain the failure of repopulation of lesions by OPCs from normal WM with disease progression in MS (Wolswijk, 2002).

In contrast, a study examining the distribution and morphologies of oligodendrocytes in chronic lesions found that most lesions contained oligodendrocytes that extended multiple processes and contacted axons, but failed to remyelinate them. Such axons were found to be dystrophic with multiple swellings. The authors concluded that remyelination failure could not be accounted for either by the absence of OPCs or their failure to generate oligodendrocytes, but instead suggested that demyelinated axons were themselves no longer receptive to remyelination (Chang *et al.*, 2002). A recent review suggests that demyelination that induces major alterations in axonal and oligodendroglial signaling events could contribute to remyelination deficits. This may be linked to the reexpression of cell adhesion molecules like PSA-NCAM on demyelinated axons in MS patients (which has an inhibitory effect on remyelination). The authors also speculate that modification of domain-specific aggregations of nodal and perinodal proteins during demyelination (or disruption of axoglial interactions in these domains) could contribute to repair failure (Coman *et al.*, 2005).

Reactive astrocytosis is usually evaluated by increased GFAP immunoreactivity and can be an early response to demyelination (Ayers *et al.*, 2004). This accompanies inflammatory changes in MS lesions (Hemmer *et al.*, 2002) which are characterized by a dense matrix of astrocytic processes termed a "glial scar." Scarring astrocytes are widely believed to inhibit remyelination by ensheathing axons and restricting access of repair-mediating cells to demyelinated axons (Groves *et al.*, 1993; Rosen *et al.*, 1989). Some studies do indicate that astrocytes may restrict migration of OPCs and inhibit remyelination (Blakemore *et al.*, 2002, 2003). However, in the TAIEP mutant rat, the presence of chronically demyelinated axons and astrocytosis does not appear to represent a significant barrier to tissue colonization by OPCs (Foote and Blakemore, 2005). Other reports suggest that reactive astrocytes could be a major source of remyelination mediators in demyelinated areas (Franklin *et al.*, 1991; Frost *et al.*, 2003; Hinks and Franklin, 1999; Redwine and Armstrong, 1998). Therefore, the exact roles played by reactive astrocytes during demyelination and repair are currently far from clear.

Notch-jagged signaling pathways (consisting of Notch-1 receptor and its ligand Jagged-1) are powerful inhibitors of OPC differentiation in development.

Lesions from brains of MS patients contain elevated levels of Notch-1 and Jagged-1 (John *et al.*, 2002), suggesting the Notch pathway could potentially underlie the limited degree of remyelination observed in MS. More recently, it has been demonstrated that γ -secretase-mediated inhibition of Notch signaling in oligodendrocytes in mice with EAE could significantly promote remyelination and reduce axonal damage in the CNS. This was accompanied by an enhancement of clinical recovery (Jurynczyk *et al.*, 2005). Other molecules like the neuregulins (a family of ligands that exerts trophic effects on neurons and glia) appear necessary for normal oligodendrocyte development. Astrocyte-derived neuregulin is dramatically reduced in both chronic and active lesions; this reduced expression is also suggested to contribute to the paucity of remyelination in MS lesions (Viehover *et al.*, 2001). It has been demonstrated recently that the glycosaminoglycan hyaluronan accumulates in MS lesions and mice with induced EAE. Hyaluronan was found to inhibit remyelination of experimental demyelinating lesions; interestingly, OPCs were found to colonize such lesions but did not appear to mature into myelinating oligodendrocytes in areas of hyaluronan expression (Back *et al.*, 2005). The identification of such molecules in demyelinated areas will aid in defining key therapeutic targets to enhance remyelination clinically.

Integrating some of the current hypotheses, the "Dysregulation Hypothesis" suggests that remyelination may fail because the complex signaling environment within lesions becomes inappropriately regulated or "dysregulated." Therefore remyelination may fail because the precise coordination between the various cellular events necessary for successful remyelination is lost in demyelinated areas (Fig. 6) (Franklin, 2002). Any or all of the above factors can be predicted to contribute to such dysregulation.

V. MS Therapy: Strategies to Enhance Remyelination

With increasing evidence for a link between absence of myelin and the development of irreversible neurological deficits in MS, it is clear that achieving widespread remyelination of axons will constitute a major component of future therapies that aim to limit neurological disability and restore function in demyelinating disease (along with disease-modifying and neuroprotective strategies). The strategies to enhance remyelination fall into two broad categories.

A. Strategies to Stimulate or Enhance Endogenous Remyelination

As remyelination in MS lesions appears to fail (or be incomplete), a potential strategy to repair MS lesions is to enhance the naturally occurring regenerative mechanisms. As detailed in previous sections, a number of molecules have been

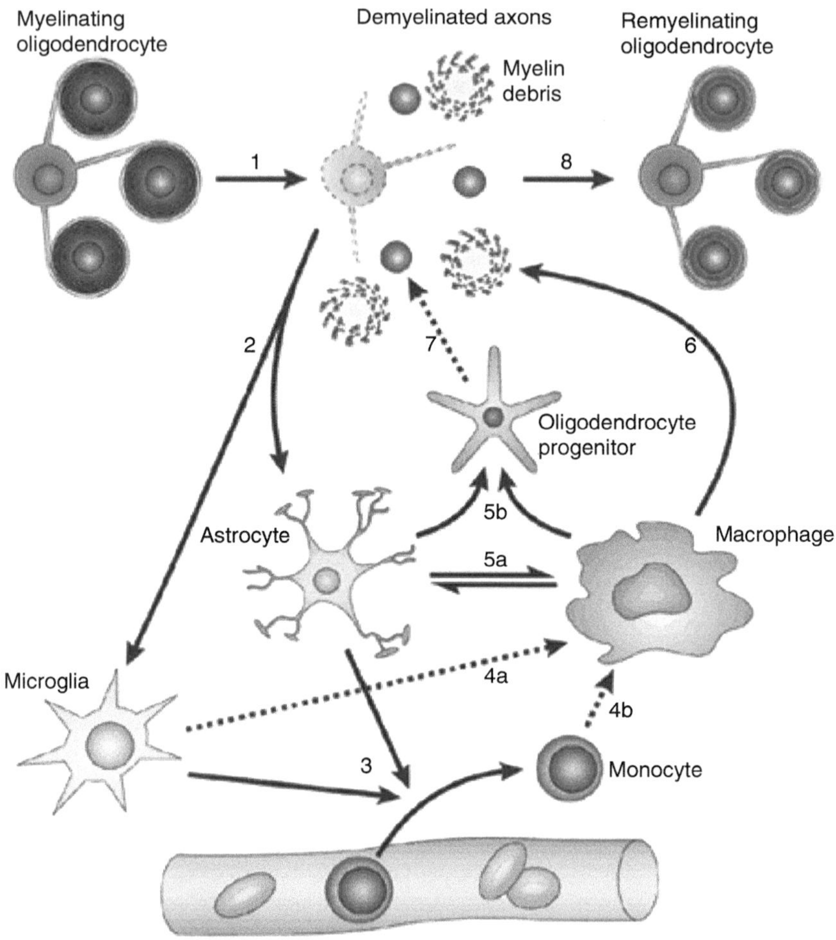

FIG. 6. Diagram illustrating that remyelination involves a complex sequence of tightly coordinated events, the "dysregulation" of which will result in remyelination impairment. In response to a demyelinating insult, the myelinated axons undergo demyelination (1). This generates myelin debris. Demyelination causes activation of resident astrocytes and microglia (2). Activated astrocytes and microglia produce factors that act to recruit monocytes from the vasculature (3). Microglia (4a) and recruited monocytes (4b) differentiate into macrophages. Activated astrocytes and macrophages produce factors that can mutually activate each other (5a). As a result of such activation, both produce growth factors that act on OPCs to alter their proliferation and differentiation (5b). Macrophages remove myelin debris (6), a function that is beneficial for remyelination. Under the influence of factors produced by astrocytes and macrophages, recruited OPCs engage with demyelinated axons (7) and differentiate into remyelinating oligodendrocytes (8). Modified with permission from Franklin (2002). (See Color Insert.)

identified (including growth factors, transcription factors, and cell adhesion molecules), which represent potential therapeutic targets to enhance remyelination. Although this is attractive in principle, there are several challenges confronting such a strategy. (1) Given that the regulation of remyelination is a complex, multifactorial process, the identification of relevant molecules to enhance repair will be critical. This is complicated by the considerable redundancy in the regulatory mechanisms that mediate remyelination. (2) The extensive variability observed to occur in MS lesions raises important issues in relation to the nature, timing, and mode of delivery of therapeutic agents for individual patients. (3) Effective, targeted delivery of therapeutic agents to multifocal sites of demyelination may be essential in order to avoid the nonspecific side effects of systemic delivery—the use of new drug/biomolecule-targeting strategies/technologies could prove very useful in this regard.

Antibody therapy: Interestingly, studies show that immunoglobulins (generally preparations of intact IgG molecules) can promote remyelination (Warrington *et al.*, 2000 and reviewed in Trebst and Stangel, 2006). The mechanisms by which immunoglobulins enhance repair are not clear but the current postulated mechanisms can be divided into two categories—(1) *Direct* via an effect on OPCs (by modifying their proliferation, migration, and differentiation) and oligodendrocytes (by rescuing them from cell death) and (2) *Indirect* via an immunomodulatory role (by decreasing inflammatory damage, modifying cytokine release, and phagocytosis) in lesions.

B. Transplantation of Myelinogenic Cells

The alternative strategy to enhance remyelination is to introduce myelin-forming cells into areas of demyelination. Several cell types (immature and adult) have been identified that achieve extensive remyelination (with restoration of saltatory conduction) in experimental demyelinating lesions and in myelin mutants. These include OPCs, Schwann cells, olfactory ensheathing cells (OECs) as well as various categories of stem cells [like neural stem cells (NSCs), embryonic stem cells (ESCs), and bone marrow-derived stem cells (BMSCs)] and their derivatives—these have been the subject of a number of excellent reviews (Dubois-Dalcq *et al.*, 2005; Duncan, 2005; Ibrahim *et al.*, 2006; Keirstead, 2005; Pluchino and Martino, 2005; Pluchino *et al.*, 2005; Tepavcevic and Blakemore, 2005) (Fig. 7). Numerous issues need to be considered when evaluating the clinical feasibility and benefits of cell transplantation-based strategies in MS. These include (1) the possibility that transplanted cells may, in turn, become targets of the demyelinating pathology, (2) the ability of the transplanted population to proliferate and differentiate in sufficient numbers to mediate remyelination (myelination potential, in general, appears to be dependent on the degree of enrichment of

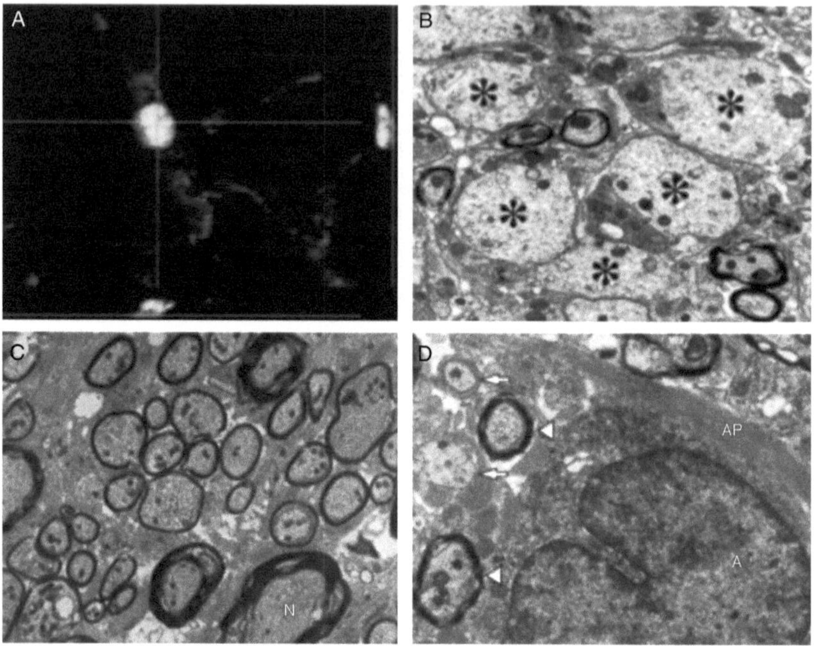

FIG. 7. Transplantation of human ESC-derived predifferentiated oligodendrocyte progenitors into a region of demyelination results in differentiation of ESCs into oligodendrocytes and remyelination. (A) Assessment of cell fate and differentiation following transplantation using double immunohistochemical staining for human nuclei (green) and the oligodendrocyte marker APC-CC1 (red) confirms that transplanted cells survive and differentiate into mature oligodendrocytes. (B) Electron micrograph (EM) of the transplant environment 7 days postinjury. Asterisks indicate demyelinated axons in an astrogliosis-free microenvironment. (C) EM of a rat spinal cord that was transplanted with human ESC-derived oligodendrocyte progenitors 7 days postinjury. Oligodendrocyte-remyelinated axons (R; with characteristically thin myelin sheaths) are evident among few normally myelinated axons (N). (D) EM of a spinal cord injury site 10 months postinjury, illustrating an astrocyte (A) with a large intermediate filament-rich process (AP) extending to demyelinated axons (arrows) and myelinated axons (arrowheads), both surrounded by intermediate filament-rich astrocytic processes. This illustrates the point that demyelinated axons in areas of chronic injury are ensheathed by astrocytic processes, which may prevent remyelination by endogenous or transplanted cells. Reproduced with permission from Keirstead (2005). (See Color Insert.)

immature cells in the cell preparation). ESCs are particularly attractive in this regard due to their capacity for unlimited expansion *in vitro*, broad differentiation potential (including differentiation into multipotent neural precursors and OPCs), and their amenability to genetic engineering (Keirstead, 2005; Nistor *et al.*, 2005). Other studies have shown that supernumerary transplanted OPCs do not survive, proliferate, or migrate in the adult CNS (Franklin *et al.*, 1996; O'Leary and Blakemore, 1997) and indicate that considerable depletion of the resident OPC population (>50%) is prerequisite to introducing significant numbers of OPCs into adult CNS tissue (Chari *et al.*, 2006b). This may not represent a problem in lesions where OPC depletion occurs along with oligodendrocyte loss. However, as dysfunctional OPCs appear to inhibit tissue repopulation by new OPCs (Chari *et al.*, 2003b), it remains to be established if transplant-derived OPCs can, in fact, mediate repair of lesions that contain significant numbers of endogenous, dysfunctional OPCs. (3) Astrogliosis is a pathological hallmark of chronic MS lesions, so the ability of transplanted cells to survive within astrogliosed areas will be a key. OECs but not Schwann cells have been shown to survive in areas of astrocytosis (Barnett and Riddell, 2004), but it is not clear to what extent OECs can be expanded indefinitely in culture. OPCs, on the other hand, possess the ability to colonize areas of astrocytosis (with the caveat that prior depletion of the endogenous OPC population or the presence of an acute inflammatory environment is critical for transplant survival) (Foote and Blakemore, 2005). (4) As multifocal demyelination is another key feature of MS, the ability of transplanted cells to migrate within the CNS may be important unless it is feasible to introduce cells directly into lesions. Multipotent NSCs injected intravenously or intraventricularly can enter areas of demyelination, generate myelinating oligodendrocytes, and mediate functional recovery (possibly because these cells express a range of inflammatory molecules like adhesion molecules, cytokines, and chemokines). This "homing" behavior makes them attractive candidates to mediate targeted repair of multifocal lesions. In addition, such cells have been shown to exert significant anti-inflammatory effects and reduce astrocytosis in demyelinated areas. Multipotent NSCs have been isolated from human subcortical WM; these cells can be maintained *in vitro* prior to transplantation and give rise to neurons and glia, when transplanted into the rodent CNS (Nunes *et al.*, 2003). (5) Given that inflammation appears to be beneficial for remyelination and transplant survival, it will be important to establish the effects of common anti-inflammatory and immunomodulatory treatments (used in MS treatment) on transplant survival and repair. (6) It is critical that the genetic stability and safety of the transplanted cells (including the elimination of pathogens and tumorigenic potential) be established prior to transplantation, (7) potential rejection of transplanted cells and the need for tissue-type matching must be taken into account, and (8) a number of ongoing ethical and political issues relating to the derivation of cells for transplantation will have a serious impact on which therapies can be advanced to clinical trials. Additionally, there are many practical concerns when surgically introducing cells into neural tissue. These include (1) the consequences of disruption of the blood–brain barrier, (2) the possibility of inducing hemorrhage at the transplant site, (3) the potential expense associated with multiple surgical injections to introduce transplanted cells into multifocal lesions, (iv) the significant challenge associated with accurately targeting transplanted cells to inaccessible CNS lesions sites, and (5) the risk of inducing CSF occlusion or emboli when introducing cells intraventricularly or intravenously. It is clear, therefore, that the benefits of a range of therapeutic strategies can be demonstrated in experimental conditions. The translation of such approaches to the clinical setting will constitute a major objective for future studies.

Acknowledgments

D.M.C. is supported by a Junior Research Fellowship from the Multiple Sclerosis Society of the United Kingdom and Northern Ireland. I am grateful to Professor Bill Blakemore and Professor Robin Franklin for their helpful comments on this chapter.

References

Adams, C. W. M. (1989). "A Colour Atlas of Multiple Sclerosis and Other Myelin Disorders." Wolfe Medical Publications, London.

Antony, J. M., van Marle, G., Opii, W., Butterfield, D. A., Mallet, F., Yong, V. W., Wallace, J. L., Deacon, R. M., Warren, K., and Power, C. (2004). Human endogenous retrovirus glycoprotein-mediated induction of redox reactants causes oligodendrocyte death and demyelination. *Nat. Neurosci.* **7,** 1088–1095.

Arnett, H. A., Mason, J., Marino, M., Suzuki, K., Matsushima, G. K., and Ting, J. P. (2001). TNF alpha promotes proliferation of oligodendrocyte progenitors and remyelination. *Nat. Neurosci.* **4,** 1116–1122.

Arnett, H. A., Wang, Y., Matsushima, G. K., Suzuki, K., and Ting, J. P. (2003). Functional genomic analysis of remyelination reveals importance of inflammation in oligodendrocyte regeneration. *J. Neurosci.* **23,** 9824–9832.

Arnett, H. A., Fancy, S. P., Alberta, J. A., Zhao, C., Plant, S. R., Kaing, S., Raine, C. S., Rowitch, D. H., Franklin, R. J., and Stiles, C. D. (2004). bHLH transcription factor Olig1 is required to repair demyelinated lesions in the CNS. *Science* **17,** 2111–2115.

Ayers, M. M., Hazelwood, L. J., Catmull, D. V., Wang, D., McKormack, Q., Bernard, C. C., and Orian, J. M. (2004). Early glial responses in murine models of multiple sclerosis. *Neurochem. Int.* **45,** 409–419.

Back, S. A., Tuohy, T. M., Chen, H., Wallingford, N., Craig, A., Struve, J., Luo, N. L., Banine, F., Liu, Y., Chang, A., Trapp, B. D., Bebo, B. F., *et al.* (2005). Hyaluronan accumulates in demyelinated lesions and inhibits oligodendrocyte progenitor maturation. *Nat. Med.* **11,** 966–972.

Barnett, M. H., and Prineas, J. W. (2004). Relapsing and remitting multiple sclerosis: Pathology of the newly forming lesion. *Ann. Neurol.* **55,** 458–468.

Barnett, S. C., and Riddell, J. S. (2004). Olfactory ensheathing cells (OECs) and the treatment of CNS injury: Advantages and possible caveats. *J. Anat.* **204,** 57–67. Review.

Bergles, D. E., Bergles, D. E., Roberts, J. D., Somogyi, P., and Jahr, C. E. (2000). Glutamatergic synapses on oligodendrocyte precursor cells in the hippocampus. *Nature* **405,** 187–191.

Bieber, A. J., Kerr, S., and Rodriguez, M. (2003). Efficient central nervous system remyelination requires T cells. *Ann. Neurol.* **53,** 680–684.

Black, J. A., Waxman, S. G., and Smith, K. J. (2006). Remyelination of dorsal column axons by endogenous Schwann cells restores the normal pattern of Nav1. 6 and Kv1. 2 at nodes of Ranvier. *Brain* **129,** 1319–1329.

Blakemore, W. F. (2005). The case for a central nervous system (CNS) origin for the Schwann cells that remyelinate CNS axons following concurrent loss of oligodendrocytes and astrocytes. *Neuropathol. Appl. Neurobiol.* **31,** 1–10.

Blakemore, W. F., and Patterson, R. C. (1978). Suppression of remyelination in the CNS by X-irradiation. *Acta Neuropathol. Berl.* **42,** 105–113.

Blakemore, W. F., Chari, D. M., Gilson, J. M., and Crang, A. J. (2002). Modelling large areas of demyelination in the rat reveals the potential and possible limitations of transplanted glial cells for remyelination in the CNS. *Glia* **15,** 155–168.

Blakemore., W. F., Gilson, J. M., and Crang, A. J. (2003). The presence of astrocytes in areas of demyelination influences remyelination following transplantation of oligodendrocyte progenitors. *Exp. Neurol.* **184,** 955–963.

Brockschnieder, D., Lappe-Siefke, C., Goebbels, S., Boesl, M. R., Nave, K. A., and Riethmacher, D. (2004). Cell depletion due to diphtheria toxin fragment A after Cre-mediated recombination. *Mol. Cell. Biol.* **24,** 7636–7642.

Buch, T., Heppner, F. L., Tertilt, C., Heinen, T. J., Kremer, M., Wunderlich, F. T., Jung, S., and Waisman, A. (2005). A Cre-inducible diphtheria toxin receptor mediates cell lineage ablation after toxin administration. *Nat. Methods* **2,** 419–426.

Bunge, M. B., Bunge, R. P., and Ris, H. (1961). Ultrastructural study of remyelination in an experimental lesion in adult cat spinal cord. *J. Biophys. Biochem. Cytol.* **10,** 67–94.

Bunge, R. P. (1958). Regenerative phenomena in the feline spinal cord in late stages of lesions inflicted by cerebrospinal fluid manipulation. *Anat. Rec.* **130,** 279.

Bunge, R. P., and Settlage, P. H. (1957). Neurological lesions in cats following cerebrospinal fluid manipulation. *J. Neuropathol. Exp. Neurol.* **16,** 471–491.

Butt, A. M., Duncan, A., Hornby, M. F., Kirvell, S. L., Hunter, A., Levine, J. M., and Berry, M. (1999). Cells expressing the NG2 antigen contact nodes of Ranvier in adult CNS white matter. *Glia* **26,** 84–91.

Byravan, S., Foster, L. M., Phan, T., Verity, A. N., and Campagnoni, A. T. (1994). Murine oligodendroglial cells express nerve growth factor. *Proc. Natl. Acad. Sci. USA* **91,** 8812–8816.

Cajal, S. R. (1928). "Degeneration and Regeneration of the Central Nervous System," pp. 734–760. Humphrey Milford/Oxford University Press, London.

Carson, M. J., Behringer, R. R., Brinster, R. L., and McMorris, F. A. (1993). Insulin-like growth factor I increases brain growth and central nervous system myelination in transgenic mice. *Neuron* **10,** 729–740.

Chang, A., Nishiyama, A., Peterson, J., Prineas, J., and Trapp, B. D. (2000). NG2-positive oligodendrocyte progenitor cells in adult human brain and multiple sclerosis lesions. *J. Neurosci.* **20,** 6404–6412.

Chang, A., Tourtellotte, W. W., Rudick, R., and Trapp, B. D. (2002). Premyelinating oligodendrocytes in chronic lesions of multiple sclerosis. *N. Engl. J. Med.* **346,** 165–173.

Chari, D. M., and Blakemore, W. F. (2002a). Efficient recolonisation of progenitor-depleted areas of the CNS by adult oligodendrocyte progenitor cells. *Glia* **37,** 307–313.

Chari, D. M., and Blakemore, W. F. (2002b). New insights into remyelination failure in multiple sclerosis: Implications for glial cell transplantation. *Mult. Scler.* **8,** 271–277. Review.

Chari, D. M., Crang, A. J., and Blakemore, W. F. (2003a). Decline in rate of colonization of oligodendrocyte progenitor cell (OPC)-depleted tissue by adult OPCs with age. *J. Neuropathol. Exp. Neurol.* **62,** 908–916.

Chari, D. M., Huang, W. L., and Blakemore, W. F. (2003b). Dysfunctional oligodendrocyte progenitor cell (OPC) populations may inhibit repopulation of OPC depleted tissue. *J. Neurosci. Res.* **73,** 787–793.

Chari, D. M., Zhao, C., Kotter, M. R., Blakemore, W. F., and Franklin, R. J. (2006a). Corticosteroids delay remyelination of experimental demyelination in the rodent central nervous system. *J. Neurosci. Res.* **83,** 594–605.

Chari, D. M., Gilson, J. M., Franklin, R. J., and Blakemore, W. F. (2006b). Oligodendrocyte progenitor cell (OPC) transplantation is unlikely to offer a means of preventing X-irradiation induced damage in the CNS. *Exp. Neurol.* **198,** 145–153.

Christensen, T. (2005). Association of human endogenous retroviruses with multiple sclerosis and possible interactions with herpes viruses. *Rev. Med. Virol.* **15,** 179–211. Review.

Coman, I., Barbin, G., Charles, P., Zalc, B., and Lubetzki, C. (2005). Axonal signals in central nervous system myelination, demyelination and remyelination. *J. Neurol. Sci.* **233,** 67–71. Review.

Crang, A. J., Gilson, J. M., Li, W. W., and Blakemore, W. F. (2004). The remyelinating potential and *in vitro* differentiation of MOG-expressing oligodendrocyte precursors isolated from the adult rat CNS. *Eur. J. Neurosci.* **20,** 1445–1460.

Crockett, D. P., Burshteyn, M., Garcia, C., Muggironi, M., and Casaccia-Bonnefil, P. (2005). Number of oligodendrocyte progenitors recruited to the lesioned spinal cord is modulated by the levels of the cell cycle regulatory protein p27Kip-1. *Glia* **49,** 301–308.

Demerens, C., Stankoff, B., Logak, M., Anglade, P., Allinquant, B., Couraud, F., Zalc, B., and Lubetzki, C. (1996). Induction of myelination in the central nervous system by electrical activity. *Proc. Natl. Acad. Sci. USA* **93,** 9887–9892.

Dubois-Dalcq, M., Ffrench-Constant, C., and Franklin, R. J. (2005). Enhancing central nervous system remyelination in multiple sclerosis. *Neuron* **48,** 9–12. Review.

Dugandzija-Novakovic, S., Koszowski, A. G., Levinson, S. R., and Shrager, P. (1995). Clustering of Na+ channels and node of Ranvier formation in remyelinating axons. *J. Neurosci.* **15,** 492–503.

Duncan, I. D. (2005). Remyelination and restoration of axonal function by glial cell transplantation. *Ernst Schering Res. Found. Workshop* **53,** 115–132. Review.

Durand, B., and Raff, M. (2000). A cell-intrinsic timer that operates during oligodendrocyte development. *Bioessays* **22,** 64–71. Review.

Edgar, J. M., McLaughlin, M., Yool, D., Zhang, S. C., Fowler, J. H., Montague, P., Barrie, J. A., McCulloch, M. C., Duncan, I. D., Garbern, J., Nave, K. A., and Griffiths, I. R. (2004). Oligodendroglial modulation of fast axonal transport in a mouse model of hereditary spastic paraplegia. *J. Cell Biol.* **166,** 121–131.

Ercolini, A. M., and Miller, S. D. (2006). Mechanisms of immunopathology in murine models of central nervous system demyelinating disease. *J. Immunol.* **176,** 3293–3298. Review.

Fancy, S. P., Zhao, C., and Franklin, R. J. (2004). Increased expression of Nkx2. 2 and Olig2 identifies reactive oligodendrocyte progenitor cells responding to demyelination in the adult CNS. *Mol. Cell. Neurosci.* **27,** 247–254.

Ferguson, B., Matyszak, M. K., Esiri, M. M., and Perry, V. H. (1997). Axonal damage in multiple sclerosis lesions. *Brain* **120,** 393–399.

Foote, A. K., and Blakemore, W. F. (2005). Inflammation stimulates remyelination in areas of chronic demyelination. *Brain* **128,** 528–539.

Franklin, R. J. (2002). Why does remyelination fail in multiple sclerosis? *Nat. Rev. Neurosci.* **3,** 705–714. Review.

Franklin, R. J., and Hinks, G. L. (1999). Understanding CNS remyelination: Clues from developmental and regeneration biology. *J. Neurosci. Res* **58,** 207–213. Review.

Franklin, R. J., Bayley, S. A., and Blakemore, W. F. (1996). Transplanted CG4 cells (an oligodendrocyte progenitor cell line) survive, migrate, and contribute to repair of areas of demyelination in X-irradiated and damaged spinal cord but not in normal spinal cord. *Exp. Neurol.* **137,** 263–276.

Franklin, R. J. M., and Blakemore, W. F. (1993). Requirements for Schwann cell migration within CNS environments: A viewpoint. *Int. J. Dev. Neurosci.* **11,** 641–649.

Franklin, R. J. M., Crang, A. J., and Blakemore, W. F. (1991). Transplanted type-1 astrocytes facilitate repair of demyelinating lesions by host oligodendrocytes in adult rat spinal cord. *J. Neurocytol.* **20,** 420–430.

Frost, E. E., Nielsen, J. A., Le, T. Q., and Armstrong, R. C. (2003). PDGF and FGF2 regulate oligodendrocyte progenitor responses to demyelination. *J. Neurobiol.* **54,** 457–472.

Fruttiger, M., Karlsson, L., Hall, A. C., Abramsson, A., Calver, A. R., Bostrom, H., Willetts, K., Bertold, C. H., Heath, J. K., Betsholtz, C., and Richardson, W. D. (1999). Defective oligodendrocyte development and severe hypomyelination in PDGF-A knockout mice. *Development* **126,** 457–467.

Gallo, V., Ghiani, C. A., and Yuan, X. (2001). The role of ion channels and neurotransmitter receptors in glial cell development. *In* "Glial Cell Development" (K. R. Jessen and W. D. Richardson, Eds.), 109–130. Oxford University Press, New York.

Garbern, J. Y., Yool, D. A., Moore, G. J., Wilds, I. B., Faulk, M. W., Klugmann, M., Nave, K. A., Sistermans, E. A., van der Knaap, M. S., Bird, T. D., Shy, M. E., Kamholz, J. A., *et al.* (2002). Patients lacking the major CNS myelin protein, proteolipid protein 1, develop length-dependent axonal degeneration in the absence of demyelination and inflammation. *Brain* **125,** 551–561.

Ghatak, N. R., Leshner, R. T., Price, A. C., and Felton, W. L. (1989). Remyelination in the human central nervous system. *J. Neuropathol. Exp. Neurol.* **48,** 507–518.

Gilson, J., and Blakemore, W. F. (1993). Failure of remyelination in areas of demyelination produced in the spinal cord of old rats. *Neuropathol. Appl. Neurobiol.* **19,** 173–181.

Graca, D. L., and Blakemore, W. F. (1986). Delayed remyelination in the rat spinal cord following ethidium bromide injection. *Neuropathol. Appl. Neurobiol.* **12,** 593–605.

Griffiths, I. R., Klugmann, M., Anderson, T., Yool, D., Thomson, C., Schwab, M. H., Schneider, A., Zimmermann, F., McCulloch, M., Nadon, N., and Nave, K. A. (1998). Axonal swellings and degeneration in mice lacking the major proteolipid of myelin. *Science* **280,** 1610–1613.

Groves, A. K., Barnett, S. C., Franklin, R. J., Crang, A. J., Mayer, M., Blakemore, W. F., and Noble, M. (1993). Repair of demyelinated lesions by transplantation of purified O-2A progenitor cells. *Nature* **362,** 453–455.

Gveric, D., Cuzner, M. L., and Newcombe, J. (1999). Insulin-like growth factors and binding proteins in multiple sclerosis plaques. *Neuropathol. Appl. Neurobiol.* **25,** 215–225.

Hemmer, B., Archelos, J. J., and Hartung, H. P. (2002). New concepts in the immunopathogenesis of multiple sclerosis. *Nat. Rev. Neurosci.* **3,** 291–301. Review.

Hinks, G. L., and Franklin, R. J. M. (1999). Distinctive patterns of PDGF-A, FGF-2, IGF-I, and TGF-beta1 gene expression during remyelination of experimentally-induced spinal cord demyelination. *Mol. Cell. Neurosci.* **14,** 153–168.

Ibrahim, A., Li, Y., Li, D., Raisman, G., and El Masry, W. S. (2006). Olfactory ensheathing cells: Ripples of an incoming tide? *Lancet Neurol.* **5,** 453–457. Review.

Itoyama, Y., Webster, H. D., Richardson, E. P., and Trapp, B. D. (1983). Schwann cell remyelination of demyelinated axons in spinal cord multiple sclerosis lesions. *Ann. Neurol.* **14,** 339–346.

Itoyama, Y., Ohnishi, A., Tateishi, J., Kuroiwa, Y., and Webster, H. D. (1985). Spinal cord multiple sclerosis lesions in Japanese patients: Schwann cell remyelination occurs in areas that lack glial fibrillary acidic protein (GFAP). *Acta Neuropathol. (Berl.)* **65,** 217–223.

Jeffery, N. D., and Blakemore, W. F. (1997). Locomotor deficits induced by experimental spinal cord demyelination are abolished by spontaneous remyelination. *Brain* **120,** 27–37.

John, G. R., Shankar, S. L., Shafit-Zagardo, B., Massimi, A., Lee, S. C., Raine, C. S., and Brosnan, C. F. (2002). Multiple sclerosis: Re-expression of a developmental pathway that restricts oligodendrocyte maturation. *Nat. Med.* **8,** 1115–1121.

Jones, S. J. (1993). Visual evoked potentials after optic neuritis. Effect of time interval, age and disease dissemination. *J. Neurol.* **240,** 489–494.

Jurynczyk, M., Jurewicz, A., Bielecki, B., Raine, C. S., and Selmaj, K. (2005). Inhibition of Notch signaling enhances tissue repair in an animal model of multiple sclerosis. *J. Neuroimmunol.* **170,** 3–10.

Keirstead, H. S. (2005). Stem cells for the treatment of myelin loss. *Trends Neurosci.* **28,** 677–683.

Keirstead, H. S., and Blakemore, W. F. (1997). Identification of post-mitotic oligodendrocytes incapable of remyelination within the demyelinated adult spinal cord. *J. Neuropathol. Exp. Neurol.* **56,** 1191–1201.

Keirstead, H. S., Levine, J. M., and Blakemore, W. F. (1998). Response of the oligodendrocyte progenitor cell population (defined by NG2 labelling) to demyelination of the adult spinal cord. *Glia* **22,** 161–170.

Kerschensteiner, M., Stadelmann, C., Buddeberg, B. S., Merkler, D., Bareyre, F. M., Anthony, D. C., Linington, C., Bruck, W., and Schwab, M. E. (2004). Targeting experimental autoimmune encephalomyelitis lesions to a predetermined axonal tract system allows for refined behavioral testing in an animal model of multiple sclerosis. *Am. J. Pathol.* **164,** 1455–1469.

Kornek, B., Storch., M. K., Weissert, R., Wallstroem, E., Stefferl, A., Olsson, T., Linington, C., Schmidbauer, M., and Lassmann, H. (2000). Multiple sclerosis and chronic autoimmune encephalomyelitis: A comparative quantitative study of axonal injury in active, inactive, and remyelinated lesions. *Am. J. Pathol.* **157,** 267–276.

Kotter, M. R., Setzu, A., Sim, F. J., Van Rooijen, N., and Franklin, R. J. M. (2001). Macrophage depletion impairs oligodendrocyte remyelination following lysolecithin-induced demyelination. *Glia* **35,** 204–212.

Kotter, M. R., Zhao, C., van Rooijen, N., and Franklin, R. J. (2005). Macrophage-depletion induced impairment of experimental CNS remyelination is associated with a reduced oligodendrocyte progenitor cell response and altered growth factor expression. *Neurobiol. Dis.* **18,** 166–175.

Lappe-Siefke, C., Goebbels, S., Gravel, M., Nicksch, E., Lee, J., Braun, P. E., Griffiths, I. R., and Nave, K. A. (2003). Disruption of Cnp1 uncouples oligodendroglial functions in axonal support and myelination. *Nat. Genet.* **33,** 366–374.

Lassmann, H. (1983). "Comparative Neuropathology of Chronic Experimental Allergic Encephalomyelitis and Multiple Sclerosis." Springer, Berlin.

Lassmann, H., Bruck, W., Lucchinetti, C., and Rodriguez, M. (1997). Remyelination in multiple sclerosis. *Mult. Scler.* **3,** 133–136. Review.

Li, G., Crang, A. J., Rundle, J. L., and Blakemore, W. F. (2002). Oligodendrocyte progenitor cells in the adult rat CNS express myelin oligodendrocyte glycoprotein (MOG). *Brain pathol.* **12,** 463–471.

Li, W. W., Penderis, J., Zhao, C., Schumacher, M., and Franklin, R. J. (2006). Females remyelinate more efficiently than males following demyelination in the aged but not young adult CNS. *Exp. Neurol.* **202,** 250–254.

Li, W. W., Setzu, A., Zhao, C., and Franklin, R. J. M. (2005). Minocycline-mediated inhibition of microglia activation impairs oligodendrocyte progenitor cell responses and remyelination in a non-immune model of demyelination. *J. Neuroimmunol.* **158,** 58–66.

Loughlin, A. J., Copelman, C. A., Hall, A., Armer, T., Young, B. C., Landon, D. N., and Cuzner, M. L. (1997). Myelination and remyelination of aggregate rat brain cell cultures enriched with macrophages. *J. Neurosci. Res.* **47,** 384–392.

Lovas, G., Szilagyi, N., Majtenyi, K., Palkovits, M., and Komoly, S. (2000). Axonal changes if chronically demyelinated cervical spinal cord plaques. *Brain* **123,** 308–317.

Lucchinetti, C., Bruck, W., Parisi, J., Scheithauer, B., Rodriguez, M., and Lassmann, H. (1999). A quantitative analysis of oligodendrocytes in multiple sclerosis lesions. A study of 113 cases. *Brain* **122,** 2279–2295.

Lucchinetti, C., Bruck, W., Parisi, J., Scheithauer, B., Rodriguez, M., and Lassmann, H. (2000). Heterogeneity of multiple sclerosis lesions: Implications for the pathogenesis of demyelination. *Ann. Neurol.* **47,** 707–717.

Ludwin, S. K. (1980). Chronic demyelination inhibits remyelination in the central nervous system. *Lab. Invest.* **43,** 382–387.

Luzi, P., Zaka, M., Rao, H. Z., Curtis, M., Rafi, M. A., and Wenger, D. A. (2004). Generation of transgenic mice expressing insulin-like growth factor-1 under the control of the myelin basic protein promoter: Increased myelination and potential for studies on the effects of increased IGF-1 on experimentally and genetically induced demyelination. *Neurochem. Res.* **29,** 881–889.

Marburg, O. (1906). Die sogenannte 'akute multiple Sklerose' (Encephalomyelitis periaxialis scleroticans). *J. Neurol. Psychiatry* **27,** 213–312.

Marchionni, M. A., Cannella, B., Hoban, C., Gao, Y. L., Garcia-Arenas, R., Lawson, D., Happel, E., Noel, F., Tofilon, P., Gwynne, D., and Raine, C. S. (1999). Neuregulin in neuron/glial interactions in the central nervous system. GGF2 diminishes autoimmune demyelination, promotes oligodendrocyte progenitor expansion, and enhances remyelination. *Adv. Exp. Med. Biol.* **468,** 283–295.

Mason, J. L., Suzuki, K., Chaplin, D. D., and Matsushima, G. K. (2001). Interleukin-1beta promotes repair of the CNS. *J. Neurosci.* **21,** 7046–7052.

McKinnon, R. D., Piras, G., Ida, J. A., and Dubois-Dalcq, M. (1993). A role for TGF-beta in oligodendrocyte differentiation. *J. Cell Biol.* **121,** 1397–1407.

Merkler, D., Boretius, S., Stadelmann, C., Ernsting, T., Michaelis, T., Frahm, J., and Bruck, W. (2005). Multicontrast MRI of remyelination in the central nervous system. *NMR Biomed.* **18,** 395–403.

Messersmith, D. J., Murtie, J. C., Le, T. Q., Frost, E. E., and Armstrong, R. C. (2000). Fibroblast growth factor 2 (FGF2) and FGF receptor expression in an experimental demyelinating disease with extensive remyelination. *J. Neurosci. Res.* **62,** 241–256.

Mi, S., Miller, R. H., Lee, X., Scott, M. L., Shulag-Morskaya, S., Shao, Z., Chang, J., Thill, G., Levesque, M., Zhang, M., Hession, C., Sah, D., *et al.* (2005). LINGO-1 negatively regulates myelination by oligodendrocytes. *Nat. Neurosci.* **8,** 745–751.

Miller, R. H. (1999). Contact with central nervous system myelin inhibits oligodendrocyte progenitor maturation. *Dev. Biol.* **216,** 359–368.

Murtie, J. C., Zhou, Y. X., Le, T. Q., Vana, A. C., and Armstrong, R. C. (2005). PDGF and FGF2 pathways regulate distinct oligodendrocyte lineage responses in experimental demyelination with spontaneous remyelination. *Neurobiol. Dis.* **19,** 171–182.

Nathan, C. F. (1987). Secretory products of macrophages. *J. Clin. Invest.* **79,** 319–326. Review.

Nave, K. A. (2001). Myelin specific genes and their mutations in the mouse. *In* "Glial Cell Development" (K. R. Jessen and W. D. Richardson, Eds.), 177–208. Oxford University Press, Oxford, UK.

Niehaus, A., Shi, J., Grzenkowski, M., Diers-Fenger, M., Archelos, J., Hartung, H. P., Toyka, K., Bruck, W., and Trotter, J. (2000). Patients with active relapsing-remitting multiple sclerosis synthesize antibodies recognizing oligodendrocyte progenitor cell surface protein: Implications for remyelination. *Ann. Neurol.* **48,** 362–371.

Nistor, G. I., Totoiu, M. O., Haque, N., Carpenter, M. K., and Keirstead, H. S. (2005). Human embryonic stem cells differentiate into oligodendrocytes in high purity and myelinate after spinal cord transplantation. *Glia* **49,** 385–396.

Nunes, M. C., Roy, N. S., Keyoung, H. M., Goodman, R. R., McKhann, G., Jiang, L., Kang, J., Nedergaard, M., and Goldman, S. A. (2003). Identification and isolation of multipotential neural progenitor cells from the subcortical white matter of the adult human brain. *Nat. Med.* **9,** 439–447.

O'Leary, M. T., and Blakemore, W. F. (1997). Oligodendrocyte precursors survive poorly and do not migrate following transplantation into the normal adult central nervous system. *J. Neurosci. Res.* **48,** 159–167.

O'Leary, M. T., Hinks, G. L., Charlton, H. M., and Franklin, R. J. (2002). Increasing local levels of IGF-I mRNA expression using adenoviral vectors does not alter oligodendrocyte remyelination in the CNS of aged rats. *Mol. Cell. Neurosci.* **19,** 32–42.

Oleszak, E. L., Chang, J. R., Friedman, H., Katset, C. D., and Platsoucas, C. D. (2004). Theiler's virus infection: A model for multiple sclerosis. *Clin. Microbiol. Rev.* **17,** 174–207. Review.

Papadopoulos, D., Pham-Dinh, D., and Reynolds, R. (2006). Axon loss is responsible for chronic neurological deficit following inflammatory demyelination in the rat. *Exp. Neurol.* **197,** 373–385.

Penderis, J., Shields, S. A., and Franklin, R. J. M. (2003b). Impaired remyelination and depletion of oligodendrocyte progenitors does not occur following repeated episodes of focal demyelination in the rat central nervous system. *Brain* **126,** 1382–1391.

Penderis, J., Woodruff, R. H., Lakatos., A., Li, W. W., Dunning, M. D., Zhao, C., Marchionni, M., and Franklin, R. J. (2003a). Increasing local levels of neuregulin (glial growth factor-2) by direct infusion into areas of demyelination does not alter remyelination in the rat CNS. *Eur. J. Neurosci.* **18,** 2253–2264.

Perier, O., and Gregoire, A. (1965). Electron microscopic features of multiple sclerosis lesions. *Brain* **88,** 937–952.

Pluchino, S., and Martino, G. (2005). The therapeutic use of stem cells for myelin repair in autoimmune demyelinating disorders. *J. Neurol. Sci.* **233,** 117–119. Review.

Pluchino, S., Zanotti, L., Deleidi, M., and Martino, G. (2005). Neural stem cells and their use as therapeutic tool in neurological disorders. *Brain Res. Brain Res. Rev.* **48,** 211–219.

Polito, A., and Reynolds, R. (2005). NG2-expressing cells as oligodendrocyte progenitors in the normal and demyelinated adult central nervous system. *J. Anat.* **207,** 707–716. Review.

Prineas, J. W., and Connell, F. (1979). Remyelination in multiple sclerosis. *Ann. Neurol.* **5,** 22–31.

Prineas, J. W., Kwon, E. E., Cho, E. S., and Sharer, L. R. (1984). Continual breakdown and regeneration of myelin in progressive multiple sclerosis plaques. *Ann. N. Y. Acad. Sci.* **436,** 11–32.

Prineas, J. W., Barnard, R. O., Revesz, T., Kwon, E. E., Sharer, L., and Cho, E. S. (1993). Multiple sclerosis. Pathology of recurrent lesions. *Brain* **116,** 681–693.

Raine, C. S., and Wu, E. (1993). Multiple sclerosis: Remyelination in acute lesions. *J. Neuropathol. Exp. Neurol.* **52,** 199–204.

Redwine, J. M., and Armstrong, R. C. (1998). *In vivo* proliferation of oligodendrocyte progenitors expressing PDGFalphaR during early remyelination. *J. Neurobiol.* **37,** 413–428.

Rodriguez, M. (2003). A function of myelin is to protect axons from subsequent injury: Implications for deficits in multiple sclerosis. *Brain* **126,** 751–752.

Rosen, C. L., Bunge, R. P., Ard, M. D., and Wood, P. M. (1989). Type 1 astrocytes inhibit myelination by adult rat oligodendrocytes *in vitro*. *J. Neurosci.* **9,** 3371–3379.

Sasaki, M., Black, J. A., Lankford, K. L., Tokuno, H. A., Waxman, S. G., and Kocsis, J. D. (2006). Molecular reconstruction of nodes of Ranvier after remyelination by transplanted olfactory ensheathing cells in the demyelinated spinal cord. *J. Neurosci.* **26,** 1803–1812.

Schlesinger, H. (1909). Zur Frage der akuten multiplen Sklerose und der encephalomyelitis disseminata im Kindesalter. *Arb. Neurol. Inst. (Wien.)* **17,** 410–432.

Scolding, N., Franklin, R., Stevens, S., Heldin, C. H., Compston, A., and Newcombe, J. (1998). Oligodendrocyte progenitors are present in the normal adult human CNS and in the lesions of multiple sclerosis. *Brain* **121,** 2221–2228.

Setzu, A., Lathia, J. D., Zhao, C., Wells, K., Rao, M. S., Ffrench-Constant, C., and Franklin, R. J. M. (2006). Inflammation stimulates myelination by transplanted oligodendrocyte precursor cells. *Glia.* **54,** 297–303.

Sherman, D. L., and Brophy, P. J. (2005). Mechanisms of axon ensheathment and myelin growth. *Nat. Rev. Neurosci.* **6,** 683–690.

Shi, J., Marinovich, A., and Barres, B. A. (1998). Purification and characterization of adult oligodendrocyte precursor cells from the rat optic nerve. *J. Neurosci.* **18,** 4627–4636.

Sim, F. J., Zhao, C., Penderis, J., and Franklin, R. J. M. (2002). The age-related decrease in CNS remyelination efficiency is attributable to an impairment of both oligodendrocyte progenitor recruitment and differentiation. *J. Neurosci.* **22,** 2451–2459.

Smith, E. J., Blakemore, W. F., and McDonald, W. I. (1979). Central remyelination restores secure conduction. *Nature* **280,** 395–396.

Smith, K. J., Blakemore, W. F., and McDonald, W. I. (1981). The restoration of conduction by central remyelination. *Brain* **104,** 383–404.

Stidworthy, M. F., Genoud, S., Suter, U., Mantei, N., and Franklin, R. J. M. (2003). Quantifying the early stages of remyelination following cuprizone-induced demyelination. *Brain Pathol.* **13,** 329–339.

Storch, M. K., Stefferl, A., Brehm, U., Weissert, R., Wallstrom, E., Kerschensteiner, M., Olsson, T., Linington, C., and Lassmann, H. (1998). Autoimmunity to myelin oligodendrocyte glycoprotein in rats mimics the spectrum of multiple sclerosis pathology. *Brain Pathol.* **8,** 681–694.

Tepavcevic, V., and Blakemore, W. F. (2005). Glial grafting for demyelinating disease. *Philos. Trans. R. Soc. Lond. B Biol. Sci.* **360,** 1775–1795.

Tourbah, A., Linnington, C., Bachelin, C., Avellana-Adalid, V., Wekerle, H., and Baron-Van Evercooren, A. (1997). Inflammation promotes survival and migration of the CG4 oligodendrocyte progenitors transplanted in the spinal cord of both inflammatory and demyelinated EAE rats. *J. Neurosci. Res.* **50,** 853–861.

Trapp, B. D., Peterson, J., Ransohoff, R. M., Rudick, R., Mork, S., and Bo, L. (1998). Axonal transection in the lesions of multiple sclerosis. *N. Engl. J. Med.* **338,** 278–285.

Trebst, C., and Stangel, M. (2006). Promotion of remyelination by immunoglobulins: Implications for the treatment of multiple sclerosis. *Curr. Pharm. Des.* **12,** 241–249.

Triarhou, L. C., and Herndon, R. M. (1986). The effect of dexamethasone on L-alpha-lysophosphatidyl choline (lysolecithin)-induced demyelination of the rat spinal cord. *Arch. Neurol.* **43,** 121–125.

Viehover, A., Miller, R. H., Park, S. K., Fischbach, G., and Vartanian, T. (2001). Neuregulin: An oligodendrocyte growth factor absent in active multiple sclerosis lesions. *Dev. Neurosci.* **23,** 377–386.

Warrington, A. E., Asakura, K., Bieber, A. J., Ciric, B., Van Keulen, V., Kaveri, S. V., Kyle, R. A., Pease, L. R., and Rodriguez, M. (2000). Human monoclonal antibodies reactive to oligodendrocytes promote remyelination in a model of multiple sclerosis. *Proc. Natl. Acad. Sci. USA* **97,** 6820–6825.

Waxman, S. G. (2006). Ions, energy and axonal injury: Towards a molecular neurology of multiple sclerosis. *Trends Mol. Med.* **12,** 192–195.

Wilkins, A., Chandran, S., and Compston, A. (2001). A role for oligodendrocyte-derived IGF-1 in trophic support of cortical neurons. *Glia* **36,** 48–57.

Wolswijk, G. (1998). Chronic stage multiple sclerosis lesions contain a relatively quiescent population of oligodendrocyte precursor cells. *J. Neurosci.* **18,** 601–609.

Wolswijk, G. (2002). Oligodendrocyte precursor cells in the demyelinated multiple sclerosis spinal cord. *Brain* **125,** 338–349.

Woodruff, R. H., and Franklin, R. J. M. (1999). Demyelination and remyelination of the caudal cerebellar peduncle of adult rats following stereotaxic injections of lysolecithin, ethidium bromide, and complement/anti-galactocerebroside: A comparative study. *Glia* **25,** 216–228.

Woodruff, R. H., Fruttiger, M., Richardson, W. D., and Franklin, R. J. M. (2004). Platelet-derived growth factor regulates oligodendrocyte progenitor numbers in adult CNS and their response following CNS demyelination. *Mol. Cell. Neurosci.* **25,** 252–262.

Ye, P., Carson, J., and D'Ercole, A. J. (1995). *In vivo* actions of insulin-like growth factor-I (IGF-I) on brain myelination: Studies of IGF-I and IGF binding protein-1 (IGFBP-1) transgenic mice. *J. Neurosci.* **15,** 7344–7356.

Zhao, C., Fancy, S. P., Kotter, M. R., Li, W. W., and Franklin, R. J. (2005). Mechanisms of CNS remyelination—the key to therapeutic advances. *J. Neurol. Sci.* **233,** 87–91.

TRIGEMINAL NEURALGIA: A MODERN-DAY REVIEW

Kelly Hunt and Ravish Patwardhan

The Interactive Neuroscience Center, LLC, Shreveport, Louisiana 71115, USA

Objective: To review the disorder of trigeminal neuralgia (TN) in the context of previously known data and recent advancements.

Methods: A PubMed search (November 25, 2006) was conducted, with review of existing literature related to this search. Review articles were accessed along with new advancements in diagnosis, pathophysiology, treatment, and prognosis.

Results: The diagnosis of TN is primarily based on clinical suspicion; ancillary tests can help rule out other associated pathology. Treatment includes medical and surgical treatment as previously described; recent radiosurgical treatment offers some success as well. Insights into pathophysiology have also been afforded by advances in imaging.

Conclusions: TN is a complex but important disorder to diagnose; in the modern-day context, appropriately diagnosed TN can be treated with a generally good prognosis.

I. Introduction

Trigeminal neuralgia (TN) is a disorder characterized by lancinating, paroxysmal facial pain which usually lasts several seconds. Its typical presentation, diagnosis, pathophysiology, treatment, and prognosis are reviewed here. Although diagnosed initially and described well over a century ago, recent technology has

INTERNATIONAL REVIEW OF
NEUROBIOLOGY, VOL. 79
DOI: 10.1016/S0074-7742(07)79027-X

0074-7742/07 $35.00

allowed further insight into pathophysiology, associated disorders, and treatment, with improved control rates.

The purpose of this chapter is to briefly introduce the basics of TN, along with recent advances which have allowed greater elucidation.

II. Methods

A PubMed search (November, 2006) was conducted, with review of existing literature related to this search. Review articles were accessed along with new advancements in diagnosis, pathophysiology, treatment, and prognosis.

Articles with abstracts were reviewed initially, with pertinent articles investigated in greater detail for relevant information. Particular focus on more recent data was considered, to allow this chapter to be more up-to-date. Finally, although this chapter focuses on human TN, animal models have been briefly investigated as well, as these provide valuable insight into human TN pathophysiology.

III. Results and Discussion

A. Clinical Diagnosis

As previously noted, the fundamental diagnosis of TN is clinical. In 1688, Drs. Johannes Michael Fehr and Elias Schmidt described the first case in the following manner: *a sharp, shooting pain in the maxilla which prevented eating solid food, varying in time—the person died of malnutrition* (Dewhurst, 1957; Stookey and Ransohoff, 1959). Further characterization in a succinct manner was done by Dr. John Fothergill, a British physician, who described the lancinating nature while not being able to be affected by direct pressure to the area.

Present clinical diagnosis essentially reflects Dr. Fothergill's account in 1773:

> The affection seems to be peculiar to persons advancing in years, and to women more than to men .... The pain comes suddenly and is excruciating; it lasts but a short time, perhaps a quarter or half a minute, and then goes off; it returns at irregular intervals, sometimes in half an hour, sometimes there are two or three repetitions in a few minutes .... Eating will bring it on some persons. Talking, or the least motion of the muscles of the face affects others; the gentlest touch of a hand or a handkerchief will sometimes bring on the pain, whilst a strong pressure on the part has no effect. (Fothergill, 1773)

The key components of the above paragraph are that the pain onset is sudden and is excruciating [which differentiates TN from atypical facial pain (AFP)]; the pain can be triggered by certain movements or touch. As characterized in greater detail later, the pain often involves the lower two-third distribution of the trigeminal nerve (branches V2 and V3). Also, as noted above, typically elderly people are afflicted, and there appears to be a greater affliction of women versus men (Ashkenazi and Levin, 2004).

B. Imaging Diagnosis

Imaging serves two important roles in diagnosis/association of TGN: (1) it may help identify a structural compression of the trigeminal nerve or adjacent brainstem (e.g., a vascular loop, tumor, and so on) or (2) it may identify demyelination, suggestive of multiple sclerosis (MS). In rare other cases, it may identify a stroke or other lesion associated with TGN.

Identification of a vascular loop may be better identified using specialized techniques versus conventional MRI. For example, using conventional MRI, Benes *et al.* identified a vascular loop to be present in 52.4% of imaging studies later confirmed in the operating room. Using a specialized (3D-FIESTA) sequence, neurovascular compression was found to be in accordance with the intraoperative finding in 71.4%. While admittedly imaging is not yet capable of identifying vascular compression definitively, it is useful if depicted; absence of imaging findings of vascular compression does not implicate its absence intraoperatively or pathophysiologically (Benes *et al.*, 2005).

A study investigating MR-angiography (MRA) revealed sensitivity of 88.5% but specificity of only 50%, based on surgical findings as reference (Boecher-Schwarz *et al.*, 1998).

Findings of MS were found to correlate reasonably well according to a study by Gass *et al.* (1997), as all patients with T2-sequence MRI scans showed brainstem involvement of trigeminal fibers when suffering from TGN. In no cases, of note, was there any vascular ectasia demonstrated, suggestive of a compressive syndrome according to this study. Demyelination was implicated as the underlying pathophysiology, possibly leading to ephaptic transmission (as discussed under Section III.C).

C. Pathophysiology and Classification of TN

The relationship of TN's clinical manifestations, known collectively as "Fothergill's disease," was not made to the trigeminal nerve until the work of Charles Bell anatomically characterized the trigeminal nerve (Wilkins and Brody, 1969). On the basis of this, the trigeminal nerve was directly implicated in the pathophysiology, hence terming the disease "trigeminal neuralgia."

An association of TN and relief of pain was made by Dandy (1925), thus implicating that the trigeminal nerve sectioning led to loss of pain. However, it was an interesting observation quite consistently made by Dandy in over 500 operations implicating a vascular loop compressing the nerve, which led to further insight into pathophysiology (Dandy, 1963).

On the basis of compression as a possible cause as well as the association with MS (a demyelinating diseases), a theory of ephaptic transmission (or more simplistically, "short-circuiting") has been proposed. This has since been supported by "focal loss of myelin and close apposition of the demyelinated axons with few intervening astrocytic processes," as reported by Hilton *et al.* (1994).

A classification scheme based on specific, objective, and reproducible criteria for TN and related facial pain syndromes is proposed. Such a classification scheme is based on information provided in the patient's history and incorporates seven diagnostic criteria as follows: (1 and 2) TN types 1 and 2 (TN1 and TN2) refer to idiopathic, spontaneous facial pain that is either predominantly episodic (as in TN1) or constant (as in TN2) in nature. (3) Trigeminal neuropathic pain results from unintentional injury to the trigeminal nerve from trauma or surgery. (4) Trigeminal deafferentation pain results from intentional injury to the nerve by peripheral nerve ablation, gangliolysis, or rhizotomy in an attempt to treat either TN or other related facial pain. (5) Symptomatic TN results from MS. (6) Postherpetic TN follows a cutaneous herpes zoster outbreak in the trigeminal distribution. (7) The category of AFP is reserved for facial pain secondary to a somatoform pain disorder and requires psychological testing for diagnostic confirmation. The purpose of a classification scheme like this is to advocate a more rigorous, standardized natural history and outcome studies for TN and related facial pain syndromes (Eller *et al.*, 2005). Although not universally adopted, this scheme is useful in consideration of TGN with its various characteristics, in choice of treatment algorithm.

D. Differential Diagnosis

While the aforementioned criteria are used to make a clinical diagnosis of TN, it is critical to differentiate TN from similar disorders. For our consideration here, these include tooth problems (often implicating wisdom teeth erroneously, which are then extracted), temporomandibular joint (TMJ) problems, AFP, MS, herpes zoster infection, and other types of scalp/juxtafacial pain (e.g., occipital neuralgia, glossopharyngeal neuralgia, and so on). The diagnosis of each, with distinguishing criteria in brief, is mentioned here.

1. *Tooth Problems and Other Oral Pain*

Many patients have been misdiagnosed in either manner (either as TN being misdiagnosed as a tooth disease, resulting in multiple procedures of tooth extraction and others, or a primary tooth/bone cavity problem being misdiagnosed as TN (Shankland, 1993). There are also some reports implicating a tooth problem as a basis of TN (Roberts *et al.*, 1984); however, in our experience, this is rare, and we have seen more commonly patients who had dental procedures were not diagnosed properly for TN. Although sometimes difficult to distinguish, lancinating pain lasting a few seconds and then disappearing is more suggestive of TN than an underlying tooth problem, which is typically more constant.

In a review, Israel *et al.* noted the following percentages of facial pain encountered at an academic center: myofascial pain (50%), atypical facial neuralgia (40%), depression (30%), TMJ synovitis (14%), TMJ osteoarthritis (12%), TN (10%), and TMJ fibrosis (2%) (Israel *et al.*, 2003).

2. *Atypical Facial Pain*

This category is a catch-all for facial pain which does not clearly follow the set criteria for TN. As per Quail (2005):

> Although in the vast majority of cases the cause is dental disease or tempero-mandibular joint dysfunction, the remaining patients are often difficult to diagnose and treat." While "a detailed history, clinical examination, imaging and laboratory investigations may be required to establish the cause of atypical facial pain, an assessment of the patient's mental state is mandatory, as depression or psychological overlay is common. In a small number of cases, the diagnosis remains unclear despite detailed investigation. These patients may have their symptoms ameliorated by empirical pharmacological therapy. (Quail, 2005)

3. *Multiple Sclerosis*

MS is technically not in the differential diagnosis of TN, but rather TN may be a symptomatic manifestation of MS. As such, on diagnosis of TN the clinician must be vigilant to exclude other associated findings (such as varying neurological subsystems being affected in varying time, which is characteristic of MS). As discussed later, this pathophysiology also has further implications in treatment, since MS involves demyelination resulting in ephaptic transmission, whereas in other cases a vascular compression syndrome may exist. In a recent study, 4.9% of patients with MS had TGN (Osterberg *et al.*, 2005).

4. *Postherpetic Neuralgia*

While both postherpetic neuralgia (PHN) and TN may present with lancinating pain, key attributes may help differentiate the two (Ashkenazi and Levin, 2004). For example, PHN involves primarily the supraorbital branch, or V1 distribution, whereas TN involves V2 and/or V3 typically. In addition, a prior history of varicella infection can be helpful, with prodromic characteristics noted. The intermittently remitting pain of TGN may also be a valuable characteristic in differentiation.

5. *Occipital Neuralgia*

This syndrome is differentiated from TGN primarily due to its distribution, which involves the posterior scalp, as the greater occipital nerve ascends from its cervical origin to innervate that region. Aside of distribution, sufferers can often obtain relief from occipital nerve blocks, which may serve as both diagnostic and therapeutic in this instance (Hecht, 2004).

6. *Glossopharyngeal Neuralgia*

This disorder is characterized by paroxysmal, generally unilateral, pain with radiation from the oropharynx and base of the tongue to the ear. It is precipitated by swallowing. In rare cases, it can be accompanied by bradycardia or syncope (Ratis *et al.*, 1990). The distribution of pain and associated cardiovascular findings help differentiate this disorder from TGN in most circumstances.

E. Treatment and Prognosis

1. *Early Treatments*

The treatment of TGN has evolved over time, in part paralleling understanding of the pathophysiology of the disease. For example, after its characterization but before understanding its pathophysiology, treatments including "purging" via laxatives were used. However, as the fifth cranial nerve's role became clearer, therapies aimed at this nerve were then developed. Initial experience with agents including quinine, ether, opium, mercury, arsenic, and many other ingredients was variable (Cole *et al.*, 2005; Fothergill, 1773; Hutchinson, 1820; Stookey and Ransohoff, 1959).

2. *Antiepileptic and Muscle Relaxant Agents*

In 1942, a derivative of phenytoin was used to successfully treat TGN (Bergouignan, 1942). A precursor, diphenylhydantoin, was used successfully initially. Development of phenytoin subsequently led to its use for the same problem until discovery of carbamazepine (Blom,1962), which has since been used as a first-line medication. Efficacy rates of up to 90% have been reported using carbamazepine (Cole *et al.*, 2005), and as per recent recommendations phenytoin may be substituted if carbamazepine is not an option (although control rate approaches 50% with this medication).

Other medications have been used as alternatives as well. For example, gabapentin (Mack, 2003), baclofen, and valproic acid, as well as oxycarbamazepine (which is thought to achieve similar efficacy with fewer side effects) have been used. Few studies support use of such recent drugs. For example, a small double-blind crossover study of the effects of baclofen was conducted, with baclofen significantly decreasing the number of painful paroxysms in 7 of the 10 patients. An open trial in another 50 patients with TN refractory to or unable to tolerate carbamazepine showed that 37 (74%) were relieved of their attacks by baclofen, either alone or in combination with previously ineffective doses of carbamazepine or phenytoin. On long-term follow-up (1–5 years), 30% continued pain-free.

Other antiepileptic agents have been used to varying success rates. Unfortunately, there is no clear evidence for some such drugs such as gabapentin being warranted, and it is recommended that off-label use be considered cautiously (Mack, 2003).

3. *Percutaneous Lysis/Lesioning Procedures*

In 1876, Bartholow described the use of chloroform in the vicinity of the trigeminal trunks resulting in some relief of TGN (Cole *et al.*, 2005; Stookey and Ransohoff, 1959). This was subsequently followed by use of alcohol, which was somewhat less desirable since it had the potential to disseminate and affect other adjacent neural structures as well. In addition, the mechanical disruption using these and other agents resulted in dysesthetic pain and other side effects related to the compound itself. Recurrent pain was also noted, and hence injection of a substance as a treatment was largely considered temporary (as the nerve regenerated to some extent).

Search of alternative methods, including a destructive substance, yielded similar findings of dysesthetic side effects. In attempting to direct gamma rays toward the Gasserian ganglion, however, a serendipitous finding of efficacy using glycerol (intended originally as a radiation marker, but found to be an agent causing relief by itself) was noted (Hakanson, 1981). Although it is suggested that "all injection therapy is temporary," some have found a persistent pain control rate of approximately 70% at 3-year follow-up (Cole *et al.*, 2005). Of note, recent modification of this technique involves an extraoral approach, directed at the foramen ovale; subsequent filling of the trigeminal cistern is visualized using the radiopaque dye.

Radiofrequency lesioning of the trigeminal nerve fibers has also been described, with several modifications to technique attempting to target the sensory fibers specifically. Although originally described with significant complication rate, this technique has since been modified to allow more precise localization and selective lesioning (Sweet and Wepsic, 1974; Tew and Taha, 1996).

In order to effect "gentle trauma" to the trigeminal nerve's Gasserian ganglion, another technique involved introducing a Fogarty balloon and gently compressing the nerve (Mullan and Lichtor, 1983). Although this technique was met with significant success (~92% initially), it has significant associated side effects including bradycardia and brief hypotension (in two-thirds), and postoperative numbness in over 57% (Brown *et al.*, 1996). As such, this technique has recently largely been abandoned and substituted by either glycerol injection (as previously mentioned), microvascular (open surgical) decompression, or radiosurgical (stereotactic) lesioning, discussed in the following sections.

4. *Open Surgical Procedures: Nerve Sectioning and Microvascular Decompression*

Walter Dandy, in his initial experience using a "cerebellar approach," showed reasonably good results with nerve root sectioning. Interestingly, he made an observation in several of his accounts of a blood vessel compressing the nerve (Dandy, 1963); however, he did not make an attempt to transpose the vessel. Later work by Dr. Peter Jannetta (Jannetta, 1967), which involved separating the vessel and placing Teflon padding between the blood vessel (usually a superior cerebellar artery branch or a vein) and the nerve, showed significant success (over 90% initial success) in terms of pain resolution, with relatively minor side effects. Nonetheless, this did involve an open operation, and as such had associated risks of cerebrospinal fluid leak, meningitis, and possible injury to surrounding structures. In a comprehensive study of TGN treated via microvascular decompression, the following information was noted by Barker *et al.* (1996):

> Most postoperative recurrences of TGN occurred within the first 2 years post-operatively; 30% had recurrences within 20 years, with 11% undergoing re-operation. At 10-year follow-up, 70% were pain-free without medication. At 10 years post-operatively, the recurrence rate was less than 1%. Recurrence was predicted by female sex, duration of symptoms over 8 years, venous compression (vs. arterial), and lack of immediate post-operative pain cessation. Major complications included two deaths shortly after the operation (0.2%) and one brain-stem infarction (0.1%); 1% had ipsilateral hearing loss.

In a separate analysis, Bederson and Wilson (1989) reported data that following initial microvascular decompression, 83% had initial good relief; however, for those who did not improve substantially, reoperation primarily involved rhizotomy (or sectioning of some portion of the nerve), leading to excellent or good results in 85%. It is important to note here that in cases of MS, the general consensus is that the pathophysiology results from demyelination and resultant ephaptic transmission rather than vascular compression; hence, partial sectioning (preserving the corneal reflex by sectioning the inferior two-thirds of the nerve) is the preferred treatment modality.

Overall, it appears that open surgical decompression or rhizotomy has good initial success rate with lesser recurrence rate than injections/balloon compression; however, it may not be appropriate for poor surgical candidates (who suffer from multiple medical problems). As such, options include either the previously described injections or the balloon compression, or a more recently refined technique of stereotactic radiosurgery as discussed in the following section.

5. *Stereotactic Radiosurgery for TN Treatment*

Recent advances in radiosurgery have allowed precise targeting of the nerve prior to its entry into the brainstem. Two of the major techniques in widespread use include the Leksell Gamma Knife and the linear accelerator. The basic technique is to focus multiple radiation particles on a defined focus, hence directing maximal energy toward creating a lesion, while sparing such effect to the brain and surrounding structures. A significant body of data has been accumulated for the Gamma Knife, which has shown generally good efficacy with minimal side effects. An initial study by Kondziolka *et al.* (2002) showed, using a median dose of 80 Gy, that complete or partial relief was achieved in 86% of patients at 1-year follow-up. Seventy percent had complete relief at 1 year, this number fell to 56% at 5-year follow-up. Over 61% had surgery prior to Gamma Knife radiosurgery, and prior surgery was found to be a negative predictor versus no previous procedure. Ten percent developed new paresthesias.

While most of the studies examining radiosurgery for TN have used a Gamma Knife, a recent study described use of a linear accelerator on 20 patients, with dose 82.3–100 Gy (Kubicek *et al.*, 2004). With a 56.5-month follow-up (and 70% with prior history of surgery), 78% had significant pain improvement. Sixty-three percent of those achieving initial pain relief had recurrence at a mean time of 21.5 months, according to that study. The overall results in terms of initial efficacy appear similar to Gamma Knife radiosurgery as noted, and hence this is considered a viable option as well by some.

IV. Conclusions

TN is a very specific disorder, with primary diagnosis derived from clinical symptomatology. Evolution in treatment has largely advanced concomitantly with observations regarding pathophysiology, aided by imaging and surgical techniques. Minimally invasive techniques, in closed/open surgical techniques, as well as radiosurgery, have allowed myriad options which can be specifically suited to the patient to obtain the best outcome.

References

Ashkenazi, A., and Levin, M. (2004). Three common neuralgias. How to manage trigeminal, occipital, and postherpetic pain. *Postgrad. Med.* **116,** 16–18, 21–24, 31–32 passim.

Barker, F. G., II, Jannetta, P. J., Bissonette, D. J., Larkins, M. V., and Jho, H. D. (1996). The long-term outcome of microvascular decompression for trigeminal neuralgia. *N. Engl. J. Med.* **334,** 1077–1083.

Bederson, J. B., and Wilson, C. B. (1989). Evaluation of microvascular decompression and partial sensory rhizotomy in 252 cases of trigeminal neuralgia. *J. Neurosurg.* **71,** 359–367.

Benes, L., Shiratori, K., Gurschi, M., Sure, U., Tirakotai, W., Krischek, B., and Bertalanffy, H. (2005). Is preoperative high-resolution magnetic resonance imaging accurate in predicting neurovascular compression in patients with trigeminal neuralgia? A single-blind study. *Neurosurg. Rev.* **28,** 131–136.

Bergouignan, M. (1942). Cures heureuses de neurologies essentielles par le dephenyl hydantoinate de sounde. *Rev. Laryngol. Otol. Rhinol.* **63,** 34–41.

Blom, S. (1962). Trigeminal neuralgia: Its treatment with a new anticonvulsant drug (G-32883). *Lancet* **1,** 839–840.

Boecher-Schwarz, H. G., Bruehl, K., Kessel, G., Guenthner, M., Perneczky, A., and Stoeter, P. (1998). Sensitivity and specificity of MRA in the diagnosis of neurovascular compression in patients with trigeminal neuralgia. A correlation of MRA and surgical findings. *Neuroradiology* **40,** 88–95.

Brown, J. A., Chittum, C. J., Sabol, D., and Gouda, J. J. (1996). Percutaneous balloon compression of the trigeminal nerve for treatment of trigeminal neuralgia. *Neurosurg. Focus* **1**(2), e4; discussion 1 p. following e4.

Cole, C. D., Liu, J. K., and Apfelbaum, R. I. (2005). Historical perspectives on the diagnosis and treatment of trigeminal neuralgia. *Neurosurg. Focus* **18**(5), E4.

Dandy, W. E. (1925). Section of the sensory root of the trigeminal nerve at the pons: Preliminary report of the operative procedure. *Bull. Johns Hopkins Hosp.* **36,** 105–106.

Dandy, W. E. (1963). The brain. *In* "Lewis-Waltersá; Practice of Surgery" (W. Walters and F. H. J. Ellis, Eds.), Vol. XII, pp. 1–671.

Dewhurst, K. (1957). A symposium on trigeminal neuralgia: With contributions by Locke, Syndenham, and other eminent seventeenth century physicians. *J. Hist. Med. Allied Sci.* **12,** 21–36.

Eller, J. L., Raslan, A. M., and Burchiel, K. J. (2005). Trigeminal neuralgia: Definition and classification. *Neurosurg. Focus* **18**(5), E3.

Fothergill, J. (1773). Of a painful affection of the face. *In* "Society of Physicians in London: Medical Observations and Inquiries," Vol. 5, pp. 129–142. T Cadell, London.

Gass, A., Kitchen, N., MacManus, D. G., Moseley, I. F., Hennerici, M. G., and Miller, D. H. (1997). Trigeminal neuralgia in patients with multiple sclerosis: Lesion localization with magnetic resonance imaging. *Neurology* **49,** 1142–1144.

Hakanson, S. (1981). Trigeminal neuralgia treated by the injection of glycerol into the trigeminal cistern. *Neurosurgery* **9,** 638–646.

Hecht, J. S. (2004). Occipital nerve blocks in postconcussive headaches: A retrospective review and report of ten patients. *J. Head Trauma Rehabil.* **19**(1), 58–71.

Hilton, D. A., Love, S., Gradidge, T., and Akham, H. B. (1994). Pathological findings associated with trigeminal neuralgia caused by vascular compression. *Neurosurgery* **35,** 299–303; discussion 303.

Hutchinson, B. (1820). "Cases of Tic Douloureux Successfully Treated." Longmans, London.

Israel, H. A., Ward, J. D., Horrell, B., and Scrivani, S. J. (2003). Oral and maxillofacial surgery in patients with chronic orofacial pain. *J. Oral Maxillofac. Surg.* **61,** 662–667.

Jannetta, P. J. (1967). Arterial compression of the trigeminal nerve at the pons in patients with trigeminal neuralgia. *J. Neurosurg.* **26**(Suppl.), 159–162.

Kondziolka, D., Lunsford, L. D., and Flickinger, J. C. (2002). Stereotactic radiosurgery for the treatment of trigeminal neuralgia. *Clin. J. Pain* **18,** 42–47.

Kubicek, G. J., Hall, W. A., Orner, J. B., Gerbi, B. J., and Dusenbery, K. E. (2004). Long-term follow-up of trigeminal neuralgia treatment using a linear accelerator. *Stereotact Funct. Neurosurg.* **82,** 244–249.

Mack, A. (2003). Examination of the evidence for off-label use of gabapentin. *J. Manag. Care Pharm.* **9,** 559–568.

Mullan, S., and Lichtor, T. (1983). Percutaneous microcompression of the trigeminal ganglion for trigeminal neuralgia. *J. Neurosurg.* **59,** 1007–1012.

Osterberg, A., Boivie, J., and Thuomas, K. A. (2005). Central pain in multiple sclerosis—prevalence and clinical characteristics. *Eur. J. Pain* **9,** 531–542.

Quail, G. (2005). Atypical facial pain—a diagnostic challenge. *Aust. Fam. Physician* **34,** 641–645.

Ratis, G., Sicuro, L., Giordano, M., and Pomatto, E. (1990). Glossopharyngeal neuralgia. *Minerva Stomatol.* **39,** 755–758.

Roberts, A. M., Person, P., Chandran, N. B., and Hori, J. M. (1984). Further observations on dental parameters of trigeminal and atypical facial neuralgias. *Oral Surg. Oral Med. Oral Pathol.* **58,** 121–129.

Shankland, W. E., II (1993). Osteocavitation lesions (Ratner bone cavities): Frequently misdiagnosed as trigeminal neuralgia—a case report. *Cranio* **11,** 232–326.

Stookey, B., and Ransohoff, J. (1959). "Trigeminal Neuralgia: Its History and Treatment." Charles C. Thomas, Springfield, IL.

Sweet, W. H., and Wepsic, J. G. (1974). Controlled thermocoagulation of trigeminal ganglion and rootlets for differential destruction of pain fibers. 1. Trigeminal neuralgia. *J. Neurosurg.* **40,** 143–156.

Tew, J. M., Jr., and Taha, J. M. (1996). Percutaneous rhizotomy in the treatment of intractable facial pain (trigeminal, glossopharyngeal, and vagal nerves). *In* "Operative Neurosurgical Techniques: Indications, Methods and Results" (H. H. Schmidek and W. H. Sweet, Eds.), 3rd ed. WB Saunders, Philadelphia.

Wilkins, R. H., and Brody, I. A. (1969). Bellá;s palsy and Bellá;s phenomenon. *Arch. Neurol* **21,** 661–662.

OPTIC NEURITIS AND THE NEURO-OPHTHALMOLOGY OF MULTIPLE SCLEROSIS

Paramjit Kaur* and Jeffrey L. Bennett*,†

*Department of Neurology, University of Colorado at Denver and Health Sciences Center
Denver, Colorado 80262, USA

†Department of Ophthalmology, University of Colorado at Denver and Health Sciences Center
Denver, Colorado 80262, USA

Multiple sclerosis (MS) is the most common cause of neurological disability in young adults. Since approximately 40% of the brain is devoted to vision, demyelination commonly affects visual function, resulting in a myriad of neuro-ophthalmic symptoms. In this chapter, we examine the seminal afferent and efferent neuro-ophthalmological manifestations of MS, highlighting those history and examination findings critical for the diagnosis and treatment of various visual and ocular motor disorders. Among the topics, a special emphasis will be placed on optic neuritis, the most common clinically isolated demyelinating syndrome. This chapter focuses on the evaluation and treatment of visual sensory and oculomotor disorders in MS. The objective is to provide the reader with a working model for enhancing their care of patients with demyelinating disease.

INTERNATIONAL REVIEW OF
NEUROBIOLOGY, VOL. 79
DOI: 10.1016/S0074-7742(07)79028-1

0074-7742/07 $35.00

I. Disorders of the Afferent Visual System

Central nervous system (CNS) demyelination frequently results in afferent visual complaints. While loss of visual acuity predominates, visual field loss and cognitive visual impairment may also occur. Since the management and prognosis of prechiasmal and postchiasmal visual problems are quite distinct, it is important for the clinician to recognize these distinct problems and treat them accordingly.

A. Optic Neuritis

Optic neuritis (ON) is the most common cause of vision loss in multiple sclerosis (MS). The annual incidence is estimated to be 5:100,000 with a prevalence of 115:100,000 (Rodriguez *et al.*, 1995). ON is the presenting symptom of MS in approximately 25% of patients, and 65% of MS patients suffer ON at sometime during their illness (Sorensen *et al.*, 1999).

1. *Clinical Course*

While the clinical presentation of ON has been documented for well over 100 years, only recently has organized clinical efforts codified the relevant clinical symptoms. ON typically presents with acute painful vision loss. It is typically monocular, but may be bilateral in 15% of cases. Sixty-five percent of cases, however, are subclinically bilateral as evidenced by visual field analysis (Keltner *et al.*, 1994). Vision loss ensues over a few hours to days, but rarely extends beyond 2 weeks. Pain occurs in 90% of affected individuals, and the discomfort typically increases with eye movement. Overall, the clinical signs and symptoms of ON are identical to any other optic neuropathy: reduced central acuity, diminished color vision, visual field loss, and a relative afferent pupillary defect. In two-third of cases, the optic nerve appears normal; however, one-third of the time there is optic disc swelling. Typically, the swelling is mild (Fig. 1A), but in 5% of cases there is severe swelling with hemorrhages (Fig. 1B). The presence of disc edema in patients with monosymptomatic ON may modify the risk of developing MS (see below); therefore, it is important for every patient to receive a careful funduscopic exam.

Visual recovery following ON is generally good. Recovery characteristically begins within 2 weeks and is complete by 6 months. The mean level of recovery is 20/15, and only 5% of individuals recover to less than 20/200 (Beck *et al.*, 1994). The severity of vision loss does not appear to influence the chance for good visual recovery. Although there were more individuals with severe vision loss (<20/200) in the cohort with poor visual recovery in the ONTT, the difference was not

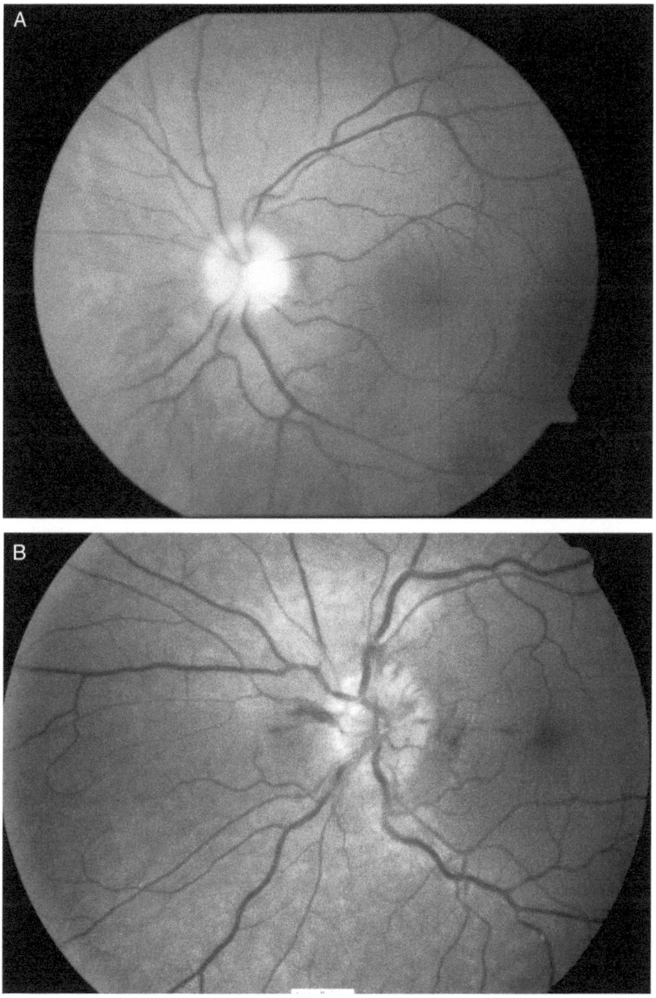

FIG. 1. Optic nerve head edema in ON. (A) Mild optic nerve head papillitis evidenced by peripapillary nerve fiber layer edema. (B) Severe optic disc edema characterized by nerve fiber layer swelling and hemorrhages. (See Color Insert.)

statistically significant (Beck and Cleary, 1993). Visual worsening for more than 2 weeks or lack of recovery by 6–8 weeks should raise concern for the diagnosis and/or treatment.

Despite their excellent prognosis, many patients with ON have persistent symptoms of visual impairment including diminished contrast acuity, impaired color perception, visual field loss, and abnormal motion perception (Cleary *et al.*, 1997).

Clinical recovery likely reflects the combined effects of axonal injury, gliosis, and compensatory neuronal recruitment. A study using ocular coherence tomography (OCT) has demonstrated that axonal injury is common in ON (Costello *et al.*, 2006). Retinal nerve fiber layer (RNFL) thinning was observed in 74% of individuals with acute ON and typically manifested 3 months following injury. In this study and a complementary cross-sectional analysis of MS patients with ON (Fisher *et al.*, 2006), RNFL thickness was significantly reduced in the affected eye when compared to fellow eyes or disease-free controls. In both studies, the increased RNFL thinning correlated with diminished visual function (visual field mean deviation, low contrast vision, and high contrast vision). In a cohort of patients with poorly recovered ON, a similar correlation was observed between RNFL thickness and color vision and visual evoked potential (VEP) amplitude (Trip *et al.*, 2005). The modest correlation coefficients ($r = 0.26$–0.33) observed between RNFL axonal loss and low- and high-contrast acuity are likely a reflection of the complex interplay between axonal loss, gliosis, and secondary neuronal recruitment in determining functional visual recovery. Indeed, the magnitude of VEP latency delay following ON correlates with the severity of demyelination as measured by magnetization transfer ratio (Thorpe *et al.*, 1995), and functional magnetic resonance imaging (MRI) imaging demonstrates adaptive cortical reorganization following acute injury (Russ *et al.*, 2002; Toosy *et al.*, 2002, 2005). Consistent with this notion, the average loss in RNFL thickness associated with a one-line loss in low-contrast acuity is higher than that observed for a one-line loss in high-contrast acuity (3.8 vs 2.9 μm) despite the increased sensitivity of low-contrast vision to optic nerve injury (Fisher *et al.*, 2006). This paradox may be explained by low-contrast acuity being exquisitely sensitive to demyelination or relatively resistant to cortical compensation.

2. *Ancillary Testing*

While the diagnosis of ON can often be made by clinical history and funduscopic examination; occasionally, cases of ON require diagnostic confirmation. Two tools that have proven useful for the diagnosis of ON are MRI imaging of the orbits and VEP testing.

a. MRI Orbits. MRI imaging of the optic nerves provides the most sensitive tool for the diagnosis of clinical and subclinical ON. A protocol that includes 3-mm contiguous fat-suppressed contrast-enhanced images and fat-suppressed T2-weighted (STIR) images has a high sensitivity (>90%) for detecting ON lesions within 4 weeks of onset (Gass *et al.*, 1996; Kupersmith *et al.*, 2002; Miller *et al.*, 1993). STIR sequences demonstrate characteristic high-signal intensity foci that characteristically extend into the perineural tissue. These acute ON lesions frequently enhance following the administration of intravenous contrast. The sensitivity is increased by double-dose contrast but such methodology is not used in typical clinical practice. MRI of optic nerves, however, cannot distinguish demyelinating ON from inflammatory neuropathy due to other causes such as syphilis or sarcoidosis (Carmody *et al.*, 1994). While there is no correlation between the nature of the optic nerve lesion and the future development of MS, some studies have observed that the length and location of the lesion may be associated with a worse visual prognosis (Hickman *et al.*, 2004; Kupersmith *et al.*, 2002). The data, however, remains limited.

b. Visual Evoked Potential. The VEP is a low-voltage change in the occipital electroencephalographic potential that occurs in response to visual stimulation. Any pathology that damages the visual pathways will affect the VEP (Asselman *et al.*, 1975). Factors influencing the VEP include both the nature of the pathology (demyelination, compression, infection, and ischemia) and test conditions (stimulus size and frequency, background illumination, patient age, fatigue, attention and concentration). Alterations in the VEP, particularly prolonged latency, are almost invariably present in subjects affected by ON. Subsequent normalization of the VEP latency following recovery is rare, occurring in only 5–29% of cases (Bynke *et al.*, 1980; Shahrokhi *et al.*, 1978). As a result, the VEP offers a sensitive but nonspecific tool for the diagnosis of ON in cases where important clinical findings are absent or equivocal.

c. Pathology. An understanding of the nature of the immunopathology in ON is critical to the development of effective therapies that facilitate the rate and extent of visual recovery and ameliorate axonal injury. In general, the composition of the inflammatory infiltrate in ON has been inferred from studies of analogous acute lesions in MS brains. Tsoi *et al.* (2006) reported that a 10-month-old ON lesion demonstrated similar findings to those observed in chronic active MS plaques: loss of myelin protein, the presence of myelin breakdown products, and the peripheral infiltration of macrophages and activated microglia. The composition of the lymphocytic infiltrate was not characterized. Therefore, the absolute contribution of cellular and humoral immunity in acute and chronic optic nerve lesions remains unclear.

Pathological analyses of demyelinating lesions in MS suggest that acute demyelination is initiated by CD4 T-cell-mediated immunity, perhaps triggered by prior exposure to an infectious agent. This hypothesis is based on: (1) the disproportionate number of CD4 T cells relative to other lymphocytes found at the margins of active plaques (Traugott *et al.*, 1983); (2) increased expression of HLA class II antigens on resident CNS cells (Traugott *et al.*, 1985); (3) restricted expression of T cell transcripts within MS lesions (Oksenberg *et al.*, 1991); and (4) the established role of T cells in experimental allergic encephalomyelitis (EAE), a noninfectious prototype CNS demyelinative disorder (Zamvil and Steinman, 1992). Detailed evaluation of MS plaques, however, has failed to reveal temporally defined stages of T-cell-mediated inflammation (Lassman *et al.*, 1998). For example, both Th1 and Th2 cytokines are expressed in MS lesions and ON cerebrospinal fluid (CSF) (Baranzini *et al.*, 2000; Soderstrom, 2001).

The dysregulation of self-reactive T cells may also play a role in ON and MS lesions. Although there is an increase in the frequency of MBP- and other myelin-specific T cells in blood and CSF of some ON and MS patients (Olsson *et al.*, 1992; Soderstrom, 2001; Sun *et al.*, 1991), similar autoreactive T cells are observed in the CSF and blood of normal individuals and inflammatory controls (Diaz-Villaslada *et al.*, 1999; Lindert *et al.*, 1999; Vu *et al.*, 2001). These autoreactive T cells may be activated in postinfectious encephalomyelitis (Johnson *et al.*, 1984) or by molecular mimicry (Wucherpfennig and Strominger, 1995).

In addition to CD4 T-cell-mediated demyelination, there is evidence implicating humoral immunopathology in MS and ON patients (Archelos *et al.*, 2000; Lucchinetti *et al.*, 2000). First, B lymphocytes and plasma cells are present in active and late stage MS lesions (Esiri, 1977; Lucchinetti *et al.*, 2000; Prineas, 1985). Second, CSF from ON and MS patients contain clonal populations of B lymphocytes and plasma cells (Haubold *et al.*, 2004; Qin *et al.*, 2003; Ritchie *et al.*, 2004). Third, histological analyses of MS plaques have revealed evidence of several Ig-mediated effector functions: (1) the capping of surface IgG on macrophages involved in myelin breakdown (Prineas and Graham, 1981), (2) the codeposition of IgG and complement proteins at plaque borders (Gay *et al.*, 1997; Lucchinetti *et al.*, 2000; Storch *et al.*, 1998), and (3) the presence of complement-enriched membrane attack vesicles in the CSF (Scolding *et al.*, 1989). Finally, acute ON responds to therapies that diminish humoral-mediated immunity (Beck *et al.*, 1992; Ruprecht *et al.*, 2004).

Although the mechanism of injury in ON is inflammatory demyelination, the etiology remains uncertain. Possible causes include infection, parainfectious inflammation (molecular mimicry), and isolated autoimmunity. Two lines of evidence suggest some role for infection in the initiation of ON. First, ON occurs at an uneven frequency across the seasons with highest incidence in spring and lowest in winter (Jin *et al.*, 1999, 2000). This uneven frequency implies that some causative factor(s) may be more prevalent or virulent during different seasons. Similarly, seasonal-varying viral infections are the most frequent event preceding relapses in MS patients (Panitch, 1994; Sibley *et al.*, 1985). Second, ON CSF frequently demonstrates intrathecal IgG synthesis and oligoclonal bands (OCBs) (Rolak *et al.*, 1996; Soderstrom *et al.*, 1998). OCBs are seen almost exclusively in infectious CNS disorders such as acute viral meningitis and encephalitis, neurosyphilis, tuberculous and fungal meningitis, and SSPE. Finally, OCBs are not observed in postrabies vaccination-induced ON (Gupta *et al.*, 2004). Indeed, ON observed after inoculation with spinal cord extract (rabies vaccine) or viral proteins (measles vaccine) fails to generate OCBs and relapsing-remitting disease.

Genetic linkage supporting a primary autoimmune mechanism in ON is lacking. The most consistently observed genetic influence on ON arises from genes encoded by or linked to the major histocompatibility complex (MHC). There is a higher prevalence of DR2 haplotypes in both ON and MS cohorts

(Francis *et al.*, 1997; Hely *et al.*, 1986; Soderstrom *et al.*, 1998). While there are no studies suggesting a genetic susceptibility to ON among DR2-positive individuals, studies are mixed regarding the relationship between DR2, DR3, and other HLA alleles in ON patients and the future risk of MS (Francis *et al.*, 1997; Hauser *et al.*, 2000; Hely *et al.*, 1986; Nilsson *et al.*, 2005; Soderstrom *et al.*, 1998). This risk appears to be heightened among individuals with an abnormal baseline MRI (Hauser *et al.*, 2000). Therefore, while genetic factors, such as HLA haplotype, may influence the risk of MS following ON, there is no data to suggest they confer susceptibility to ON.

4. *Treatment Of ON*

a. Corticosteroids. While corticosteroids do not improve visual acuity following acute ON, treatment with high-dose intravenous methylprednisolone provides faster visual recovery, improved contrast sensitivity, and improved color vision. In the Optic Neuritis Treatment Trial (ONTT), patients were randomized to one of the three treatment groups: (1) oral prednisone (1 mg/kg/day) for 14 days, (2) intravenous methylprednisolone (250-mg qid for 3 days) followed by an oral prednisone (1 mg/kg/day) for 11 days, and (3) oral placebo for 14 days. After 6 months, all three treatment groups showed a median visual acuity of 20/16, and less than 10% of the patients in each group had a visual acuity of 20/50 or worse. Despite a lack of effect on visual acuity, at 6 months, contrast vision and color vision were significantly improved in the methylprednisolone treatment arm. After 1 year, however, there was no significant difference in the mean visual acuity, color vision, contrast sensitivity, or visual fields in the treated and untreated patients. Oral prednisone did not speed recovery of vision and resulted in an increased risk of recurrent ON 2, 5, and 10 years following the sentinel event.

In the ONTT, treatment with high-dose methylprednisolone significantly reduced the development of clinically definite MS (CDMS) for 2 years following the initial sentinel event (Beck *et al.*, 1993a). A similar effect, however, was not observed for all individuals enrolled in the ONTT trial (Beck *et al.*, 1992), and additional studies investigating the effects of high-dose steroid therapy on relapse rates in MS patients have failed to observe a similar salutatory effect. Therefore, the statistical significance observed in the ONTT may be result of subgroup analysis.

b. Intravenous Immunoglobulin. Intravenous immunoglobulin (IVIg) has been studied in patients with acute and chronic vision loss from ON. In a double-blind, randomized trial of IVIg treatment in acute ON, there was no effect on long-term visual function (Roed *et al.*, 2005). At 6 months, there was no effect of IVIG treatment on visual acuity, contrast sensitivity, color vision, or VEP latency. Noseworthy *et al.* (2001) have examined the ability of IVIg to reverse chronic visual impairment in MS patients with ON. In their study, IVIg administration did not significantly reverse persistent vision loss in affected individuals. Similarly,

IVIg administration did not facilitate recovery from acute MS relapses when added to intravenous methylprednisolone (Sorensen *et al.*, 2004).

c. Plasmapheresis. Plasma exchange has demonstrated efficacy in the treatment fulminate demyelinating attacks in MS. In a retrospective analysis of plasma exchange in the treatment of refractory demyelination in MS patients, only individuals with Type II (mixed T- and B-cell infiltrate) demonstrated therapeutic efficacy (Keegan *et al.*, 2005). The authors hypothesized that the benefit of plasma exchange in this subgroup of patients was secondary to the prominent role of the humoral immune response in their active lesions. A similar benefit has been observed for plasma exchange in a group of refractory ON patients (Ruprecht *et al.*, 2004). By analogy, the success of plasma exchange in restoring vision in refractory cases of ON suggests a prominent role for humoral immune activation in acute ON lesions. The results will need to be evaluated in a larger cohort of patients since the number of individuals examined in the study was small and the duration of disease was short. In the ONTT, only 50% of individuals had recovered spontaneously from ON 3 weeks after the onset of vision loss (Beck *et al.*, 1992); therefore, a significant fraction of these refractory ON patients may have spontaneously recovered.

5. *Risk of Developing MS in Isolated ON*

MRI and CSF are powerful tools for predicting the future development of MS in patients with acute ON. In the ONTT, patients with two or more white matter lesions had an increased risk of developing CDMS over the subsequent 2 years when compared to patients with one or fewer lesions (35.9% vs 5.4%). At 10 years, the risk of CDMS was 56% in ON patients with one or more lesions on MRI compared to 22% for those with no baseline MRI lesions (Beck *et al.*, 2003). Higher number of lesions on MRI did not further increase the risk. In the ONTT, the risk of CDMS was also increased at 2 and 5 years in individuals with oligoclonal banding in the CSF. At 2 and 3 years following isolated ON, oligoclonal banding increases the risk of CDMS in patients with a positive MRI scan (Jin *et al.*, 2003; Rolak *et al.*, 1996). A similar phenomenon is not observed 5 years following ON (Cole *et al.*, 1998; Soderstrom *et al.*, 1998), suggesting that MRI and oligoclonal banding may be reporting different information on the short-term risk of CDMS in ON patients.

In addition to MRI and CSF, important features of the clinical exam may modify the future risk of CDMS in ON patients. In ON patients without MRI lesions, the future risk of CDMS is reduced twofold by the presence of ocular inflammation such as mild disc edema, uveitis, or retinal periphlebitis (Fig. 2A) (Beck *et al.*, 2003). In fact, in the ONTT, no patient with a normal MRI scan and severe disc edema with retinal hemorrhages, retinal exudates, or lack of pain developed CDMS 5 years or 10 years following their event of vision loss (Beck *et al.*, 2003).

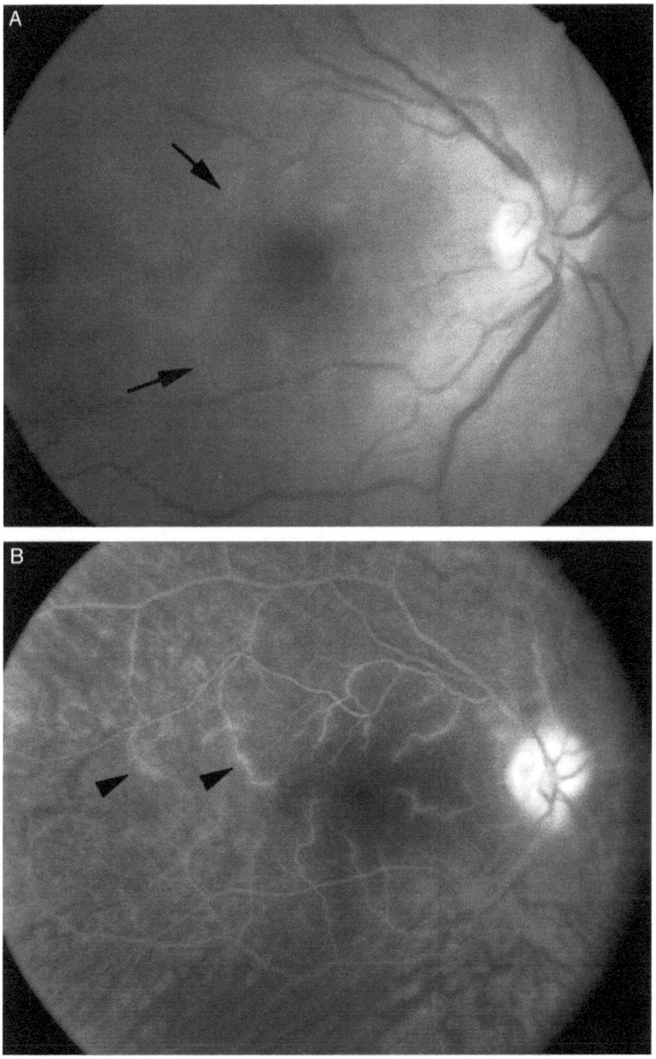

FIG. 2. Retinal periphlebitis associated with ON. (A). Funduscopy demonstrates mild disk edema with perivascular sheathing of the retinal veins (arrows). (B) Fluorescein angiography of the same eye shows corresponding dye leakage from the inflamed retinal veins (arrowheads). (See Color Insert.)

6. *Early Treatment of High-Risk ON Patients*

Three major clinical trials have investigated the utility of early immunomodulatory following a clinically isolated demyelinating syndrome (CIS) such as ON: Early Treatment of Multiple Sclerosis (ETOMS), Controlled High Risk Avonex Multiple

Sclerosis Prevention Study (CHAMPS), and Betaseron in Newly Emerging MS for Initial Treatment (BENEFIT). Each trial demonstrated that treatment with β-interferon significantly reduced the chance of developing CDMS in high-risk CIS patients. The design, entry criteria, clinical endpoints, and results of these trials are presented in Table I. When compared to the treatment effects observed using identical doses of β-interferon in established MS populations, the results of these early treatment trials demonstrate a robust treatment effect on clinical and MRI activity. Whether this will translate into long-term clinical benefit remains unclear. In the (CHAMPIONS) trial, a 3-year open-label follow-up to the CHAMPS trial, early treatment with β-interferon significantly reduced the probability of a second clinical attack and significantly delayed the time to a second clinical attack (Kinkel *et al.*, 2006). Endpoints, such as MRI lesion burden, disability, and relapse-rate (years 3–5), however, were not significantly different between the early and delayed treatment groups. While the lack of a significant effect in these areas may be due to inadequate powering, future studies will be necessary to clarify whether early treatment of high-risk CIS patients provides added benefit over waiting until a diagnosis can be made by the international criteria (McDonald *et al.*, 2001).

B. Chronic ON

ON may develop subacutely or subclinically in some MS patients. In many instances, subclinical optic nerve injury occurs during an episode of unilateral ON involving the fellow eye. In the ONTT, 68.8% of patients with acute unilateral ON demonstrated subclinical visual field loss in their fellow eye, and one-third of these cases remained abnormal at 1 year (Keltner *et al.*, 1994). Abnormalities in the fellow eye were also noted in visual acuity, contrast sensitivity, and color vision (Beck *et al.*, 1993b). Since only 7% of individuals reported a previous episode of ON in the fellow eye and visual function improved in a large fraction of affected eyes, ON may be bilateral but asymmetric in the majority of cases. In MS, vision loss may result from chronic demyelination and axonal injury. Similar to findings observed in secondary progressive patients, there is little evidence for chronic inflammation of the optic nerves in such affected individuals. Treatment of chronic vision loss in MS has not been systematically studied. In a small cohort of MS patients with chronic visual impairment from ON, no improvement was noted with the use of IVIg (Noseworthy *et al.*, 2001).

C. Ocular Inflammation

Several forms of ocular inflammation may be observed in MS patients: anterior and posterior uveitis, pars planitis, retinal periphlebitis, and retinal vasculitis. In MS patients, uveitis may precede or follow the onset of demyelinating disease,

TABLE I

ENHANCED EFFICACY OF EARLY IMMUNOMODULATORY THERAPY IN MULTIPLE SCLEROSIS

						Results of phase III trials[d]	
Clinical trial	N	Treatment	Enrollment criteria	Relative reduction in conversion to CDMS (%)	Change in median T2-lesion volume	Relative reduction in patients with ≥ 1 exacerbation (%)	Change in median T2-lesion volume
ETOMS	308	22-μg sq β-interferon-1a qW	One clinical event; three or more MRI lesions	24	+8.8% placebo versus -13% β-interferon-1a	0[a]	+1.5% placebo versus +2.8% β-interferon-1a[a]
CHAMPS	383	30-μg sq β-interferon-1a qW	One clinical event; two or more MRI lesions	44[b]	+16% placebo versus −1% β-interferon-1a	20.8[c]	−6.5% placebo versus −13.2% β-interferon-1a
BENEFIT	480	250-μg sq β-interferon-1b qod	One clinical event; two or more MRI lesions	50	N.A.	22.6	+11.9% placebo versus −5.6% β-interferon-1b

[a]1-year results.

[b]3-year results.

[c]18-month results.

[d]phase III trials: OWIMS (OWIMS Study Group, 1999), Multiple Sclerosis Collaborative Research Group (Jacobs *et al.*, 1996), and IFNB Multiple Sclerosis Study Group (IFNB Multiple Sclerosis Study Group, 1993).

qW, every week; qod, every other day; sq, subcutaneous.

and severe vision loss is uncommon. Within a cohort of MS patients identified from a large academic uveitis practice, pars planitis was the most frequent form of uveitis identified (Zein *et al.*, 2004). Anterior uveitis was common and frequently granulomatous in nature. Concurrent uveitis in MS patients does not appear to influence the pattern of demyelinating disease (Schmidt *et al.*, 2001), but ocular inflammation may present at times of increased CNS inflammation (Tola *et al.*, 1993).

Retinal periphlebitis is well described in MS patients. Pathologically, periphlebitis is characterized by sheathing or cuffing of the retinal veins by lymphocytes and plasma cells (Arnold *et al.*, 1984; Shaw *et al.*, 1987). Funduscopically, periphlebitis is evident as white exudates that surround the retinal veins (Fig. 2A). Fluorescein angiography demonstrates leakage of dye from inflamed veins (Fig. 2B). Retinal periphlebitis may be observed in other infectious and inflammatory disorders of eye such as sarcoidosis, syphilis, toxoplasmosis, and Eale's disease. Therefore, monosymptomatic ON associated with retinal periphlebitis should prompt a detailed evaluation for causes other than idiopathic demyelinating disease.

Retinal vasculitis may occur in up to 23% of MS patients (Engell and Andersen, 1982). Funduscopic exam typically demonstrates arteriolar sheathing, preretinal or intraretinal hemorrhage, and vitritis (Fig. 3). Patients should be

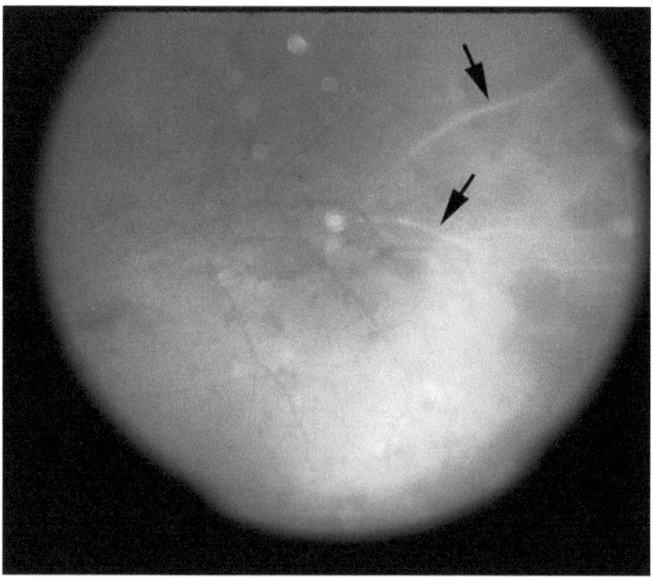

FIG. 3. Retinal vasculitis. Funduscopy reveals peripheral arteriolar occlusion (arrows) and intraretinal hemorrhages in a patient complaining of peripheral visual field loss. The vasculitis responded to high-dose oral steroids. (See Color Insert.)

monitored closely for subsequent retinal neovascularization (Friedman, 2005). As with retinal periphlebitis, affected individuals should be evaluated for alternative causes such as syphilis, Behçet disease, sarcoidosis, systemic lupus erythematosus, toxoplasmosis, and viral retinitis.

D. Demyelination of the Optic Chiasm, Optic Tracts, and Optic Radiations

Demyelination within the visual pathways outside of the optic nerves occurs infrequently in MS patients. Demyelination of the optic chiasm may produce a classic bitemporal hemianopsia or result in a junctional scotoma: central field loss in one eye and temporal field loss in the fellow eye (Fig. 4). Demyelination involving the optic tract or optic radiations may result in variable degrees of homonymous visual field loss. Involvement of the optic tract typically produces an incongruous hemianopsia and an afferent pupillary defect involving the eye with greater visual field loss. Involvement of the optic radiations may produce complete or incomplete hemianopic field loss. Incongruous hemianopic field loss may be misinterpreted as optic nerve injury (Fig. 5). The extent of field recovery following postchiasmal visual pathway demyelination has not been studied systematically. Recovery may not be as favorable as for ON.

E. Higher Cortical Visual Dysfunction

Patients with demyelinating disease affecting the frontal, parietal lobes, or thalamus may develop problems with higher order visual processing. These may include symptoms such as palinopsia (visual perseveration), visual allesthesia, and hallucinosis (Jacome, 1985; Nicolai and Lazzarino, 1995; Regan and Simpson, 1995; Seigel, 1988). Difficulties in visual cognitive function generally occur in conjunction with generalized cognitive decline but, on occasion, may occur as an isolated problem (Laatu *et al.*, 2001). In a functional MRI study of verbal working memory in MS patients, visual processing rehearsal systems appeared less active in a number of individuals (Sweet *et al.*, 2004).

Higher cortical dysfunction may manifest at any time during the course of MS, but is typically found in secondary progressive disease. Rarely, higher cortical dysfunction may be the initial presentation of MS. Similar to other forms of cognitive decline, the treatment of higher order visual dysfunction in MS is difficult. Symptom management includes treating underlying fatigue and depression, avoiding aggravating factors such as stress and heat, and cognitive evaluation and therapy. The benefit of acetylcholinesterase inhibitors for disorders of visual cognition is uncertain. Some patients with bothersome visual hallucinations, such

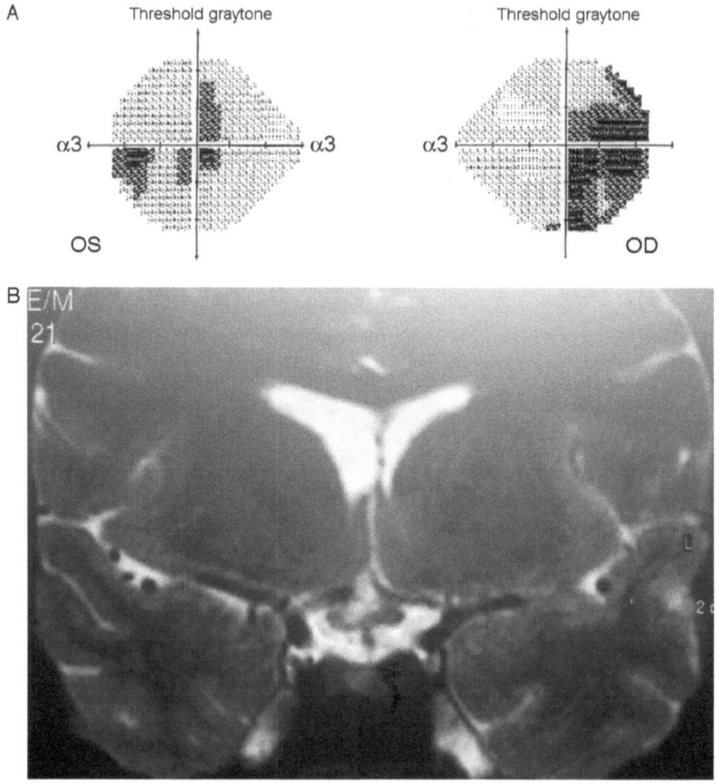

FIG. 4. Junctional scotoma secondary to chiasmal demyelination in MS. (A). Junctional scotoma defined by temporal field loss in the right eye (OD) and central field loss in the left eye (OS). (B). Coronal T2-weighted MRI image demonstrating demyelination of the body of the optic chiasm. (See Color Insert.)

as polyopsia, may benefit from low-dose neuroleptics (e.g., olanzapine). These medications should be used with caution to avoid worsening fatigue and cognitive function.

II. Disorders of the Ocular Motor System

MS patients frequently develop disorders of ocular motor function. Abnormalities of ocular fixation, ocular motility, and ocular alignment result in complaints of loss of visual clarity, double vision, oscillopsia, visual blur, and poor visual tracking. On examination, findings may include monocular cranial nerve palsies,

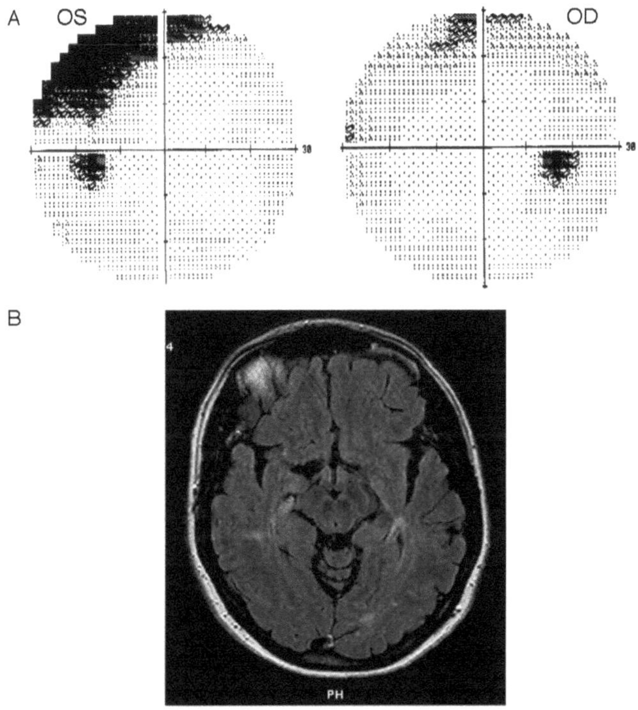

FIG. 5. Incongruous homonymous field loss in MS. (A). Automated threshold perimetry demonstrates an incongruous left homonymous field loss. (B). FLAIR imaging demonstrates demyelination involving the right anterior optic radiations.

supranuclear gaze palsies, disorders of smooth pursuit and saccades, and nystagmus. In certain instances, a careful examination is necessary to differentiate symptoms arising from ocular rather than CNS disease. Therefore, for MS patients with ocular motor dysfunction, a thorough neuro-ophthalmological examination is often required to address both diagnostic and therapeutic questions.

A. DIPLOPIA

Double vision in MS may arise from ophthalmological or neurological causes. Monocular diplopia is rarely of neurological origin. Typically, monocular diplopia is caused by refractive aberration: astigmatism, corneal pathology, cataract, or an epiretinal membrane. On examination, monocular diplopia from ocular pathology usually resolves with pinhole correction allowing it to be differentiated from rare cases of cerebral monocular diplopia.

While ocular dysmotility may result from demyelination to supratentorial or infratentorial white matter, binocular diplopia usually arises from injury to ocular motor pathways in the brainstem or cerebellum. A detailed neurological history can often yield important preliminary information regarding lesion localization. The nature of the diplopia (horizontal, vertical, or oblique), the severity with near or distance vision, the magnitude in various directions of gaze, and the presence of associated neurological symptoms (ptosis, oscillopsia, head tilt, and so on) provide detailed information regarding the position of the lesion in the posterior fossa. Horizontal diplopia that is worse at distance and in left gaze suggests a left sixth nerve palsy, while vertical diplopia that is similar in all directions of gaze and accompanied by a head tilt suggests a skew deviation.

Ocular dysmotility in MS may be the result of injury to supranuclear, nuclear, infranuclear, and internuclear pathways that control ocular motility. While defining the nature and location of the injury is often unnecessary for initiating acute therapy, a detailed understanding of the mechanism may be important for documenting therapeutic response.

1. *Supranuclear Lesions*

Demyelination in the white matter of the cerebral hemispheres or brainstem may produce supranuclear gaze impairment. When the lesion resides in the cerebral hemispheres, ocular saccades and pursuits are impaired but oculocephalic function is preserved (Barnes and McDonald, 1992). Lesions in the frontal white matter involving descending fibers from the frontal eye fields result in a supranuclear horizontal gaze palsy to the side contralateral to the lesion, whereas lesions involving white matter tracts descending from the parieto-occipital eye fields result in supranuclear horizontal gaze impairment to the side ipsilateral to injury. Demyelination in the pons may result in unilateral or bilateral supranuclear horizontal gaze palsies through involvement of the paramedian pontine reticular formation (PPRF) (Tan and Kansu, 1990). Unilateral lesions result in ipsilateral gaze impairment. Dorsal PPRF lesions may spare oculocephalic eye movements (Deleu *et al.*, 1988).

While bilateral cerebral white matter injury may impair vertical gaze, supranuclear vertical gaze impairment typically results from demyelination in the midbrain. Lesions involving pathways descending from the rostral interstitial nucleus of the medial longitudinal fasciculus, nucleus of the posterior commissure, or mesencephalic reticular formation may produce supranuclear impairment of upgaze or downgaze (Fig. 6). Pupillary and eyelid abnormalities may be present as a result of injury to white matter in the posterior commissure and pretectum. Rarely, a dorsal midbrain syndrome (supranuclear vertical gaze palsy, pupillary light-near dissociation, lid retraction, and convergence-retraction nystagmus) may be evident (Quint *et al.*, 1993).

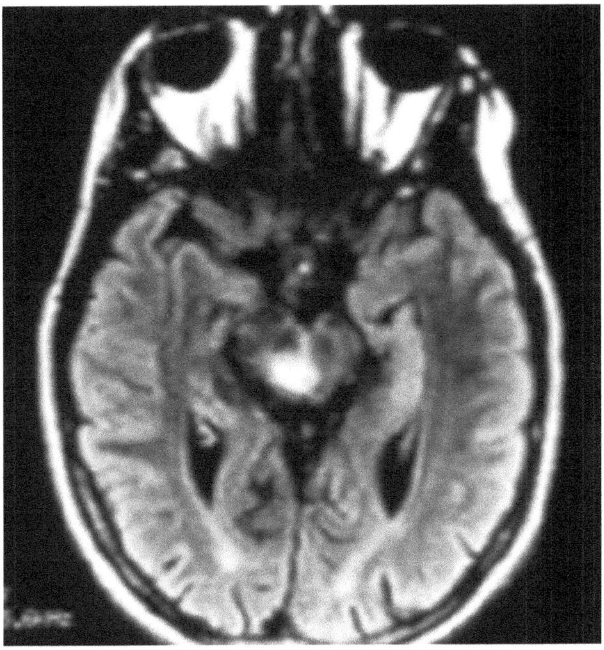

FIG. 6. Midbrain demyelination resulting in supranuclear upgaze and downgaze impairment. Flair imaging reveals a large demyelinating lesion involving the rostral midbrain.

Skew deviation is a vertical misalignment of the eyes secondary to interruption of supranuclear inputs from the vestibular nuclei onto the third and fourth cranial nuclei. Because the fibers connecting these nuclei travel within the medial longitudinal fasciculus (MLF), skew deviation may be accompanied by internuclear ophthalmoplegia (INO) (see Section II.A.4). Rostral pontine or midbrain lesions result in an ipsilateral hyperdeviation, while caudal pontine and medullary lesions result in a contralateral hyperdeviation. Skew deviation is frequently accompanied by a subjective tilt of the visual vertical; less often there is a head tilt toward the side of the lower eye (Brandt and Dieterich, 1993). Together, skew deviation accompanied by subjective tilt of the visual vertical and a head tilt constitutes the ocular tilt reaction.

2. *Nuclear Lesions*

Similar to gray matter lesions in the cerebral cortex, demyelinating lesions of the brainstem may selectively injure ocular motor nuclei. Nuclear third nerve lesions characteristically result in bilateral ptosis and upgaze impairment (de Seze *et al.*, 2006); the pupil may or may not be affected. In rare instances, an isolated muscle may be involved simulating an orbital process (Lee *et al.*, 2000). Nuclear fourth nerve lesions are unusual and indistinguishable from infranuclear injury. Due to the decussation of the fourth nerve fascicles in the inferior colliculus, the fourth nerve palsy is contralateral to the side of the lesion. Nuclear sixth nerve lesions produce an ipsilateral gaze palsy to the side of the lesion. If the lesion involves the ipsilateral MLF, then a one-and-a-half syndrome will result (see Section II.A.5).

3. *Infranuclear Lesions*

Ocular motor palsies may result from infranuclear injury to the third, fourth, or sixth nerves. Sixth nerve palsies are the most common, while fourth nerve palsies are the least. Occasionally, ocular motor nerve paresis may be the presenting sign of demyelinating disease (Chou *et al.*, 2004; Jacobson *et al.*, 1999; Lee *et al.*, 2000). Third nerve lesions may be complete or incomplete, pupil-involving or pupil-sparing (Ksiazek *et al.*, 1989; Lee *et al.*, 2000). MRI typically shows demyelination involving the brainstem fascicles of the relevant ocular motor nerve; however, in an unusual case of an MS patient with an isolated painful third nerve palsy, MRI showed isolated enhancement of the cisternal portion of the third nerve (Bhatti *et al.*, 2003).

4. *Internuclear Lesions*

Demyelination within a segment of the MLF results in an INO: paresis of the adducting eye during a horizontal saccade (Keane, 2005). In addition to the impairment of the adducting saccade, there is a horizontal dissociated nystagmus that is worse in the abducting eye. The mechanism of this nystagmus remains unclear, but potential explanations include dissociated gaze-evoked nystagmus and adaptive compensation (Baloh *et al.*, 1978). An INO may also be accompanied by vertical eye movement abnormalities. These include deficiencies in vertical gaze-holding, vertical pursuits, and vertical vestibuloocular and optokinetic responses (Ranalli and Sharpe, 1988). If otolithic pathways in the MLF are disrupted, then the INO will be accompanied by a skew deviation. In these instances, the ipsilateral eye is typically hyperdeviated.

Within the MLF, the demyelinating lesion interrupts axons from abducens interneurons as they travel to innervate the contralateral medial rectus subnucleus. Most lesions occur in the dorsal pons or midbrain. The clinical presentation may vary depending on the location of the lesion along the rostral–caudal axis. Rostral lesions of the MLF may injure vergence pathways resulting in impaired convergence. If the medial rectus motor neurons are injured, then the patient may present with a bilateral INO and a severe exotropia in primary gaze (wall-eyed bilateral INO). In affected individuals, MRI commonly demonstrates the offending lesion. Proton density and T2-weighted imaging are the most sensitive techniques, successfully identifying the causative lesion in 100% and 88% of affected individuals, respectively, in a case series (Frohman *et al.*, 2001).

5. *Combined Lesions*

Because of the disseminated nature of demyelinating attacks, MS patients may present with complex patterns of ocular motor dysfunction secondary to combined supranuclear, nuclear, internuclear, and infranuclear injury. INO has been reported in combination with both infranuclear and supranuclear motility disturbances. INO has been observed in conjunction with trochlear (Vanooteghem *et al.*, 1992), abducens (Frohman and Frohman, 2003), and oculomotor nerve palsies (Gnanaraj and Rao, 2000). INO combined with supranuclear horizontal gaze paresis is termed the one-and-a-half syndrome (Wall and Wray, 1983). The one-and-a-half syndrome consists of a horizontal gaze paresis in one direction and an INO in the opposite direction. Patients may be aligned or exotropic (paralytic pontine exotropia) in primary gaze (Sharpe *et al.*, 1974). The disorder results from a lesion or lesions that involve the MLF and the adjacent PPRF or abducens nucleus. Due to coincident injury to the central tegmental tract, patients with the one-and-a-half syndrome may be predisposed to develop oculopalatal myoclonus (Wolin *et al.*, 1996).

6. *Treatment of Ocular Motor Dysfunction in MS Patients*

MS patients with disturbances in ocular motility may note a wide variety of difficulties: double vision, loss of depth perception, visual fatigue, transient oscillopsia (dissociated nystagmus), and visual confusion. Since most symptoms will resolve spontaneously with resolution of the demyelinating event, acute therapy is best limited to symptomatic care. Patching or monocular occlusion will relieve double vision and visual fatigue at the expense of depth perception. Typically, the patient would occlude the eye with limited motility, but in some MS patients, impaired acuity secondary to prior ON may necessitate that the opposite eye is covered. If recovery is incomplete, then prism therapy may be effective for monocular nerve palsies and concomitant skew deviations. Patients and physicians should be cautioned to avoid early prismatic correction since it is often unnecessary and may produce paradoxical worsening as motility recovers. If recovery is incomplete and the magnitude of the ocular malalignment remains significant, then strabismus surgery is an option. Strabismus surgery should only be considered after the measured deviation is stable for at least 6 months. This will insure that surgical correction is not performed prematurely.

B. Nystagmus

Nystagmus occurs frequently in MS patients as a result of brainstem or cerebellar demyelination. The nystagmus may be "jerk" or pendular. Jerk nystagmus is characterized by a slow drift of gaze from its fixed position followed by a resetting, corrective saccade. Pendular nystagmus is a back-and-forth slow oscillation of the eyes that may have horizontal, vertical, and torsional components. Jerk nystagmus results from disruption of gaze holding networks originating from the vestibular nuclei, cerebellar flocculus, and nuclei prepositus hypoglossi. Pendular nystagmus results from interruption of feedback circuits that connect critical brainstem and cerebellar structures (Averbuch-Heller *et al.*, 1995). In MS patients, pendular nystagmus may be aggravated by prior or concurrent optic nerve demyelination. In a group of MS patients with dissociated pendular nystagmus, the eye with larger oscillations demonstrated evidence of more severe optic nerve impairment (Barton and Cox, 1993). A direct investigation of the effect of altering visual fixation in acquired pendular nystagmus, however, failed to demonstrate any effect on the amplitude of the ocular oscillations (Averbuch-Heller *et al.*, 1995).

Multiple forms of jerk nystagmus have been reported in MS patients: upbeat (Fisher *et al.*, 1983; Ohkoshi *et al.*, 1998), downbeat (Baloh and Spooner, 1981; Masucci and Kurtzke, 1988), rebound (Hood *et al.*, 1973; Solingen *et al.*, 1977), and periodic alternating nystagmus (PAN) (Keane, 1974; Matsumoto *et al.*, 2001). Upbeat or downbeat nystagmus that is present in central position results from demyelinating injury to central vestibular pathways. Demyelinating lesions causing downbeat nystagmus localize to the vestibulocerebellum or craniocervical junction; lesions causing upbeat nystagmus vary in position from the medulla to the midbrain. Clinically, both upbeat and downbeat nystagmus tend to respect Alexander's Law (the nystagmus increases with gaze into the direction of the quick phase), but in rare instances upbeat nystagmus may worsen in downgaze and downbeat nystagmus may worsen in upgaze. On examination, downbeat nystagmus typically worsens in lateral gaze. While upbeat nystagmus does not increase with lateral gaze, it may vary in intensity or convert to downbeat nystagmus with convergence. Upbeat and downbeat nystagmuses are often associated with disturbances of vertical vestibular and smooth-pursuit function.

Rebound nystagmus is a pathological jerk nystagmus that occurs when the eyes are returned to primary position from eccentric gaze (slow phase in the direction of prior gaze). On examination, patients with rebound nystagmus may demonstrate a reversal in the direction of their gaze-evoked nystagmus with sustained eccentric gaze (centripetal nystagmus) (Leech *et al.*, 1977). In MS patients, rebound nystagmus emerges following cerebellar demyelination. It is thought to develop as an adaptive mechanism to compensate for the centripetal drift of gaze-evoked nystagmus.

PAN is a spontaneous horizontal, direction-reversing nystagmus. Each phase lasts approximately 2 min, and the brief null period between phases may be interrupted by downbeating or upbeating nystagmus. PAN may be associated with periodic alternating head turns in the direction of the quick phase of the nystagmus. PAN results from injury to the cerebellar flocculus and uvula, resulting in a loss of inhibitory control over vestibular rotational responses. While loss of visual fixation may elicit PAN (Jay *et al.*, 1985), PAN has not been described in an MS patient following ON.

Acquired pendular nystagmus in MS patients may be horizontal, vertical, or elliptical. When bilateral, the ocular oscillations may be highly asymmetric and out of phase. If the offending lesion involves Mollaret's triangle (the pathways connecting the inferior olive, dentate nucleus, and red nucleus), then pendular nystagmus is accompanied by palatal myoclonus (Revol *et al.*, 1990). Additional forms of pendular nystagmus described in MS patients include seesaw nystagmus (Samkoff and Smith, 1994; Sandramouli *et al.*, 2005) and convergence-evoked nystagmus (Barton *et al.*, 1999). Seesaw nystagmus is a vertical–torsional nystagmus consisting of alternating phases of elevation and intorsion of one globe with simultaneous depression and extorsion of the other globe. Pendular seesaw nystagmus is most frequently reported in association with parasellar, chiasmal, or midbrain lesions. Convergence-evoked nystagmus is an asymmetric, horizontal pendular nystagmus induced by convergence. Opposite to the nystagmus observed with dorsal midbrain injury, the two eyes move in a conjugate fashion with convergence-evoked nystagmus. Both optic nerve and cerebellar lesions have been documented in affected individuals (Barton *et al.*, 1999).

C. Treatment of Nystagmus

Several medications may prove beneficial in the treatment of acquired nystagmus in MS patients (Table II) (Rucker, 2005; Strupp and Brandt, 2006). Gamma-aminobutyric acid (GABA) agonists have been reported to provide symptomatic improvement in patients suffering from upbeat and downbeat nystagmus. Clonazepam, a $GABA_A$ agonist, has been demonstrated to be effective in reducing the intensity, slow phase velocity, and clinical symptomatology of downbeat nystagmus (Currie and Matsuo, 1986; Young and Huang, 2001). Baclofen, a $GABA_B$ agonist, has been reported to reduce the slow phase velocity and clinical symptoms of upbeat and downbeat nystagmus (Dieterich *et al.*, 1991). Baclofen is also uniquely effective in the treatment of PAN. The potassium channel antagonists 3,4-diaminopyridine and 4-aminopyridine have shown success in reducing the slow phase velocity and oscillopsia of patients with downbeat nystagmus (Kalla *et al.*, 2004; Strupp *et al.*, 2003). 4-Aminopyridine may also prove beneficial in the treatment of upbeat nystagmus (Glasauer *et al.*, 2005).

TABLE II
Pharmacological Therapy of Nystagmus and Saccadic Oscillations

Disorder	Treatment
Downbeat nystagmus	Clonazepam; baclofen; 3,4-diaminopyridine; 4- diaminopyridine; scopolamine
Upbeat nystagmus	Clonazepam; baclofen; 4-diaminopyridine
Pendular nystagmus	Gabapentin; valproic acid; memantine; pregabalin (?)
Periodic alternating nystagmus	Baclofen
Square wave jerks	Clonazepam; baclofen; anticonvulsants
Macrosaccadic oscillations	Baclofen; clonazepam
Ocular flutter and opsoclonus	Unknown

Intravenous scopolamine has been reported to reduce downbeat nystagmus (Barton *et al.*, 1994), but transdermal delivery has not demonstrated similar success (Kim *et al.*, 2001).

Acquired pendular nystagmus may be treated with a spectrum of medications that act on a diverse array of ion channels. GABA agonists (clonazepam, valproate, and isoniazid), anticholinergics (scopolamine) (Rucker, 2005; Traccis *et al.*, 1990), NMDA antagonists (memantine) (Starck *et al.*, 1997), and P/Q-type voltage-dependent calcium channel antagonists (gabapentin) (Averbuch-Heller *et al.*, 1997) have demonstrated benefit in the treatment of acquired pendular nystagmus. For the treatment of acquired pendular nystagmus in MS patients, gabapentin is the first-line choice (Averbuch-Heller, 1999). Seesaw nystagmus may respond to clonazepam (Cochin *et al.*, 1995).

D. Disorders of Saccades

MS patients may be troubled by inappropriate or inaccurate saccades. Saccadic intrusions are extraneous, inappropriate saccades that interrupt fixation. The most common form in MS is square wave jerks: small-amplitude eye movements that move away from and back to fixation and are punctuated by a period of intersaccadic latency. Similar but larger amplitude movements are termed macro-square wave jerks. Ocular flutter is characterized by back-to-back horizontal saccades that are not interrupted by a period of intersaccadic latency. Opsoclonus is random back-to-back saccades in multiple directions (horizontal, vertical, and oblique). Saccadic intrusions typically arise from injury to the caudal brainstem and cerebellum that disrupt connections from pause cell neurons in the pontine raphe.

Inaccurate saccades are common in MS patients. Cerebellar lesions affecting the dorsal vermis or posterior fastigial nuclei result in saccadic dysmetria. Injury to the deep cerebellar nuclei results in hypermetric saccades. Injury solely to the dorsal vermis results in hypometric saccades. Demyelinating lesions involving the cerebellar peduncles may produce varying effects of saccadic accuracy. Lesions involving the inferior cerebellar peduncle result in hypometric saccades with gaze toward the side of the lesion, whereas lesions involving the superior cerebellar peduncle result in hypometric saccades with gaze away from the lesion.

In general, saccadic intrusions have proven difficult to treat. In an isolated report, clonazepam, diazepam, and barbiturates have been reported to suppress square wave jerks and macrosaccadic oscillations (Fukazawa *et al.*, 1986). Additional anecdotal success has been reported for propranolol and verapamil (Pranzatelli, 1992).

III. Summary

The afferent and efferent visual systems are commonly affected in MS patients. ON is the most common manifestation of demyelination in the afferent visual system. ON typically manifests as acute painful vision loss, but chronic vision loss may occur as a manifestation of secondary progressive disease. The degree of axonal loss from optic nerve demyelination appears to correlate with the level of residual visual acuity, but additional studies will be required to determine whether current anti-inflammatory and immunomodulatory modify the degree of axonal injury following ON. Additional manifestations of afferent visual impairment in MS patients include uveitis, visual field loss, and visual cognitive dysfunction.

Double vision, visual blurring, oscillopsia, and visual fatigue are common symptoms of demyelination involving ocular motor pathways. Symptomatic therapy is the mainstay of treatment for ocular motor impairment in MS. Patching is often sufficient to provide acute symptomatic relief of diplopia prior to spontaneous recovery. Prism therapy and strabismus surgery allow for the correction of recalcitrant diplopia. Increasing options are becoming available for the pharmacological therapy of nystagmus and saccadic intrusions. Clonazepam and baclofen may be useful in some individuals for the treatment of upbeat and downbeat nystagmus, and novel agents such as 3,4-diaminopyridine and 4-aminopyridine offer new promise for effective therapy in symptomatic patients. Gabapentin and valproate are the primary medications available for the treatment of acquired pendular nystagmus. Saccadic intrusions, such as square wave jerks, ocular flutter, and opsoclonus, remain difficult to treat. Further research into the mechanisms underlying various forms of nystagmus and saccadic oscillations should allow for the development of more effective therapies.

Acknowledgments

J.L.B. is supported by grants from the National Eye Institute (EY14573) and the Multiple Sclerosis Society (RG3908).

References

Archelos, J., Storch, M., and Hartnung, H. (2000). The role of B cells and autoantibodies in multiple sclerosis. *Ann. Neurol.* **47,** 694–706.

Arnold, A. C., Pepose, J. S., Hepler, R. S., and Foos, R. Y. (1984). Retinal periphlebitis and retinitis in multiple sclerosis. I. Pathologic characteristics. *Ophthalmology* **91,** 255–262.

Asselman, P., Chadwick, D. W., and Marsden, D. C. (1975). Visual evoked responses in the diagnosis and management of patients suspected of multiple sclerosis. *Brain* **98,** 261–282.

Averbuch-Heller, L. (1999). Acquired Nystagmus. *Curr. Treat. Options Neurol.* **1,** 68–73.

Averbuch-Heller, L., Zivotofsky, A. Z., Das, V. E., DiScenna, A. O., and Leigh, R. J. (1995). Investigations of the pathogenesis of acquired pendular nystagmus. *Brain* **118**(Pt. 2), 369–378.

Averbuch-Heller, L., Tusa, R. J., Fuhry, L., Rottach, K. G., Ganser, G. L., Heide, W., Buttner, U., and Leigh, R. J. (1997). A double-blind controlled study of gabapentin and baclofen as treatment for acquired nystagmus. *Ann. Neurol.* **41,** 818–825.

Baloh, R. W., and Spooner, J. W. (1981). Downbeat nystagmus: A type of central vestibular nystagmus. *Neurology* **31,** 304–310.

Baloh, R. W., Yee, R. D., and Honrubia, V. (1978). Internuclear ophthalmoplegia. I. Saccades and dissociated nystagmus. *Arch. Neurol.* **35,** 484–489.

Baranzini, S., Elfstrom, C., Chang, S., Butonoi, C., Murray, R., Higuchi, R., and Oksenberg, J. (2000). Transcriptional analysis of multiple sclerosis brain lesions reveals a complex pattern of cytokine expression. *J. Immunol.* **165,** 6576–6582.

Barnes, D., and McDonald, W. I. (1992). The ocular manifestations of multiple sclerosis. 2. Abnormalities of eye movements. *J. Neurol. Neurosurg. Psychiatry* **55,** 863–868.

Barton, J. J., and Cox, T. A. (1993). Acquired pendular nystagmus in multiple sclerosis: Clinical observations and the role of optic neuropathy. *J. Neurol. Neurosurg. Psychiatry* **56,** 262–267.

Barton, J. J., Huaman, A. G., and Sharpe, J. A. (1994). Muscarinic antagonists in the treatment of acquired pendular and downbeat nystagmus: A double-blind, randomized trial of three intravenous drugs. *Ann. Neurol.* **35,** 319–325.

Barton, J. J., Cox, T. A., and Digre, K. B. (1999). Acquired convergence-evoked pendular nystagmus in multiple sclerosis. *J. Neuroophthalmol.* **19,** 34–38.

Beck, R. W., and Cleary, P. A. (1993). Recovery from severe visual loss in optic neuritis. *Arch. Ophthalmol.* **111,** 300.

Beck, R. W., Cleary, P. A., Anderson, M. M., Jr., Keltner, J. L., Shults, W. T., Kaufman, D. I., Buckley, E. G., Corbett, J. J., Kupersmith, M. J., Miller, N. R., *et al.* (1992). A randomized, controlled trial of corticosteroids in the treatment of acute optic neuritis. The Optic Neuritis Study Group. *N. Engl. J. Med.* **326,** 581–588.

Beck, R. W., Cleary, P. A., Trobe, J. D., Kaufman, D. I., Kupersmith, M. J., Paty, D. W., and Brown, C. H. (1993a). The effect of corticosteroids for acute optic neuritis on the subsequent development of multiple sclerosis. The Optic Neuritis Study Group. *N. Engl. J. Med.* **329,** 1764–1769.

Beck, R. W., Kupersmith, M. J., Cleary, P. A., and Katz, B. (1993b). Fellow eye abnormalities in acute unilateral optic neuritis. Experience of the optic neuritis treatment trial.. *Ophthalmology* **100,** 691–697; discussion 697–698.

Beck, R. W., Cleary, P. A., and Backlund, J. C. (1994). The course of visual recovery after optic neuritis. Experience of the Optic Neuritis Treatment Trial. *Ophthalmology* **101,** 1771–1778.

Beck, R. W., Trobe, J. D., Moke, P. S., Gal, R. L., Xing, D., Bhatti, M. T., Brodsky, M. C., Buckley, E. G., Chrousos, G. A., Corbett, J., Eggenberger, E., Goodwin, J. A., *et al.* (2003). High- and low-risk profiles for the development of multiple sclerosis within 10 years after optic neuritis: Experience of the optic neuritis treatment trial. *Arch. Ophthalmol.* **121,** 944–949.

Bhatti, M. T., Schmalfuss, I. M., Williams, L. S., and Quisling, R. G. (2003). Peripheral third cranial nerve enhancement in multiple sclerosis. *AJNR Am. J. Neuroradiol.* **24,** 1390–1395.

Brandt, T., and Dieterich, M. (1993). Skew deviation with ocular torsion: A vestibular brainstem sign of topographic diagnostic value. *Ann. Neurol.* **33,** 528–534.

Bynke, H., Rosen, I., and Sandberg-Wollheim, M. (1980). Correlation of visual evoked potentials, ophthalmological and neurological findings after unilateral optic neuritis. *Acta Ophthalmol. (Copenh.)* **58,** 673–687.

Carmody, R. F., Mafee, M. F., Goodwin, J. A., Small, K., and Haery, C. (1994). Orbital and optic pathway sarcoidosis: MR findings. *AJNR Am. J. Neuroradiol.* **15,** 775–783.

Chou, K. L., Galetta, S. L., Liu, G. T., Volpe, N. J., Bennett, J. L., Asbury, A. K., and Balcer, L. J. (2004). Acute ocular motor mononeuropathies: Prospective study of the roles of neuroimaging and clinical assessment. *J. Neurol. Sci.* **219,** 35–39.

Cleary, P. A., Beck, R. W., Bourque, L. B., Backlund, J. C., and Miskala, P. H. (1997). Visual symptoms after optic neuritis. Results from the Optic Neuritis Treatment Trial. *J. Neuroophthalmol.* **17,** 18–23; quiz 24–28.

Cochin, J. P., Hannequin, D., Do Marcolino, C., Didier, T., and Augustin, P. (1995). Intermittent sea-saw nystagmus successfully treated with clonazepam. *Rev. Neurol. (Paris)* **151,** 60–62.

Cole, S. R., Beck, R. W., Moke, P. S., Kaufman, D. I., and Tourtellotte, W. W. (1998). The predictive value of CSF oligoclonal banding for MS 5 years after optic neuritis. Optic Neuritis Study Group. *Neurology* **51,** 885–887.

Costello, F., Coupland, S., Hodge, W., Lorello, G. R., Koroluk, J., Pan, Y. I., Freedman, M. S., Zackon, D. H., and Kardon, R. H. (2006). Quantifying axonal loss after optic neuritis with optical coherence tomography. *Ann. Neurol.* **59,** 963–969.

Currie, J. N., and Matsuo, V. (1986). The use of clonazepam in the treatment of nystagmus-induced oscillopsia. *Ophthalmology* **93,** 924–932.

de Seze, J., Vukusic, S., Viallet-Marcel, M., Tilikete, C., Zephir, H., Delalande, S., Stojkovic, T., Defoort-Dhellemmes, S., Confavreux, C., and Vermersch, P. (2006). Unusual ocular motor findings in multiple sclerosis. *J. Neurol. Sci.* **243,** 91–95.

Deleu, D., Solheid, C., Michotte, A., and Ebinger, G. (1988). Dissociated ipsilateral horizontal gaze palsy in one-and-a-half syndrome: A clinicopathologic study. *Neurology* **38,** 1278–1280.

Diaz-Villaslada, P., Shih, A., Shao, L., Genain, C., and Hauser, S. (1999). Autoreactivity to myelin antigens: Myelin/oligodendrocyte glycoprotein is a prevalent autoantigen. *J. Neuroimmunol.* **99,** 36–43.

Dieterich, M., Straube, A., Brandt, T., Paulus, W., and Buttner, U. (1991). The effects of baclofen and cholinergic drugs on upbeat and downbeat nystagmus. *J. Neurol. Neurosurg. Psychiatry* **54,** 627–632.

Engell, T., and Andersen, P. K. (1982). The frequency of periphlebitis retinae in multiple sclerosis. *Acta Neurol. Scand.* **65,** 601–608.

Esiri, M. (1977). Immunoglobulin-containing cells in multiple sclerosis plaques. *Lancet* **2,** 478–480.

Fisher, A., Gresty, M., Chambers, B., and Rudge, P. (1983). Primary position upbeating nystagmus. A variety of central positional nystagmus. *Brain* **106**(Pt. 4), 949–964.

Fisher, J. B., Jacobs, D. A., Markowitz, C. E., Galetta, S. L., Volpe, N. J., Nano-Schiavi, M. L., Baier, M. L., Frohman, E. M., Winslow, H., Frohman, T. C., Calabresi, P. A., Maguire, M. G., *et al.* (2006). Relation of visual function to retinal nerve fiber layer thickness in multiple sclerosis. *Ophthalmology* **113,** 324–332.

Francis, D., Compston, D., Batchelor, J., and McDonald, I. (1997). A reassessment of the risk of multiple sclerosis developing in patients with optic neuritis after extended follow up. *J. Neurol. Neurosurg. Psych.* **50,** 758–765.

Friedman, S. M. (2005). Retinal vasculitis as the initial presentation of multiple sclerosis. *Retina* **25,** 218–219.

Frohman, E. M., and Frohman, T. C. (2003). Horizontal monocular saccadic failure: An unusual clinically isolated syndrome progressing to multiple sclerosis. *Mult. Scler.* **9,** 55–58.

Frohman, E. M., Zhang, H., Kramer, P. D., Fleckenstein, J., Hawker, K., Racke, M. K., and Frohman, T. C. (2001). MRI characteristics of the MLF in MS patients with chronic internuclear ophthalmoparesis. *Neurology* **57,** 762–768.

Fukazawa, T., Tashiro, K., Hamada, T., and Kase, M. (1986). Multisystem degeneration: Drugs and square wave jerks. *Neurology* **36,** 1230–1233.

Gass, A., Moseley, I. F., Barker, G. J., Jones, S., MacManus, D., McDonald, W. I., and Miller, D. H. (1996). Lesion discrimination in optic neuritis using high-resolution fat-suppressed fast spin-echo MRI. *Neuroradiology* **38,** 317–321.

Gay, F., Drye, T., Dick, G., and Esiri, M. (1997). The application of multifactorial clustering analysis in the staging of plaques in early multiple sclerosis: Identification and characterization of the primary demyelinating lesion. *Brain* **120,** 1461–1483.

Glasauer, S., Strupp, M., Kalla, R., Buttner, U., and Brandt, T. (2005). Effect of 4-aminopyridine on upbeat and downbeat nystagmus elucidates the mechanism of downbeat nystagmus. *Ann. N. Y. Acad. Sci.* **1039,** 528–531.

Gnanaraj, L., and Rao, V. J. (2000). Partial unilateral third nerve palsy and bilateral internuclear ophthalmoplegia: An unusual presentation of multiple sclerosis. *Eye* **14**(Pt. 4), 673–675.

Gupta, V., Bandyopadhyay, S., Bapuraj, J. R., and Gupta, A. (2004). Bilateral optic neuritis complicating rabies vaccination. *Retina* **24,** 179–181.

Haubold, K., Owens, G. P., Kaur, P., Ritchie, A. M., Gilden, D. H., and Bennett, J. L. (2004). B-lymphocyte and plasma cell clonal expansion in monosymptomatic optic neuritis cerebrospinal fluid. *Ann. Neurol.* **56,** 97–107.

Hauser, S., Oksenberg, J., Lincoln, R., Garovoy, J., Beck, R., Cole, S., Moke, P., Kip, K., Gal, R., and Long, D. (2000). Interaction between HLA-DR2 and abnormal brain MRI in optic neuritis and early multiple sclerosis. *Neurology* **54,** 1859–1861.

Hely, M. A., McManis, P. G., Doran, T. J., Walsh, J. C., and McLeod, J. G. (1986). Acute optic neuritis: A prospective study of risk factors for multiple sclerosis. *J. Neurol. Neurosurg. Psychiatry* **49,** 1125–1130.

Hickman, S. J., Toosy, A. T., Miszkiel, K. A., Jones, S. J., Altmann, D. R., MacManus, D. G., Plant, G. T., Thompson, A. J., and Miller, D. H. (2004). Visual recovery following acute optic neuritis—a clinical, electrophysiological and magnetic resonance imaging study. *J. Neurol.* **251,** 996–1005.

Hood, J. D., Kayan, A., and Leech, J. (1973). Rebound nystagmus. *Brain* **96,** 507–526.

IFNB Multiple Sclerosis Study Group (1993). Interferon beta-1B is effective in relapsing-remitting multiple sclerosis. *Neurology* **43,** 655–661.

Jacobs, L. D., Cookfair, D. L., Rudick, R. A., Herndon, R. M., Richert, J. R., Salazar, A. M., Fischer, J. S., Goodkin, D. E., Granger, C. V., Simon, J. H., Alam, J. J., Bartoszak, D. M., *et al.* (1996).

Intramuscular interferon beta-1a for disease progression in relapsing multiple sclerosis. The Multiple Sclerosis Collaborative Research Group (MSCRG). *Ann. Neurol.* **39,** 285–294.

Jacobson, D. M., Moster, M. L., Eggenberger, E. R., Galetta, S. L., and Liu, G. T. (1999). Isolated trochlear nerve palsy in patients with multiple sclerosis. *Neurology* **53,** 877–879.

Jacome, D. E. (1985). Palinopsia and bitemporal visual extinction on fixation. *Ann. Ophthalmol.* **17,** 251–252, 257.

Jay, W. M., Williams, B. B., and De Chicchis, A. (1985). Periodic alternating nystagmus clearing after cataract surgery. *J. Clin. Neuroophthalmol.* **5,** 149–152.

Jin, Y. P., de Pedro-Cuesta, J., Soderstrom, M., and Link, H. (1999). Incidence of optic neuritis in Stockholm, Sweden, 1990–1995: II. Time and space patterns. *Arch. Neurol.* **56,** 975–980.

Jin, Y. P., de Pedro-Cuesta, J., Soderstrom, M., Stawiarz, L., and Link, H. (2000). Seasonal patterns in optic neuritis and multiple sclerosis: A meta-analysis. *J. Neurol. Sci.* **181,** 56–64.

Jin, Y. P., de Pedro-Cuesta, J., Huang, Y. H., and Soderstrom, M. (2003). Predicting multiple sclerosis at optic neuritis onset. *Mult. Scler.* **9,** 135–141.

Johnson, R., Griffin, D., Hirsch, J., Wolinsky, J., Rodenbeck, S., Lindo de Soriano, I., and Vaisberg, A. (1984). Measles encephalomyelitis: clinical and immunological studies. *NEJM* **310,** 137–141.

Kalla, R., Glasauer, S., Schautzer, F., Lehnen, N., Buttner, U., Strupp, M., and Brandt, T. (2004). 4-aminopyridine improves downbeat nystagmus, smooth pursuit, and VOR gain. *Neurology* **62,** 1228–1229.

Keane, J. R. (1974). Periodic alternating nystagmus with downward beating nystagmus. A clinicoanatomical case study of multiple sclerosis. *Arch. Neurol.* **30,** 399–402.

Keane, J. R. (2005). Internuclear ophthalmoplegia: Unusual causes in 114 of 410 patients. *Arch. Neurol.* **62,** 714–717.

Keegan, M., Konig, F., McClelland, R., Bruck, W., Morales, Y., Bitsch, A., Panitch, H., Lassmann, H., Weinshenker, B., Rodriguez, M., Parisi, J., and Lucchinetti, C. F. (2005). Relation between humoral pathological changes in multiple sclerosis and response to therapeutic plasma exchange. *Lancet* **366,** 579–582.

Keltner, J. L., Johnson, C. A., Spurr, J. O., and Beck, R. W. (1994). Visual field profile of optic neuritis. One-year follow-up in the Optic Neuritis Treatment Trial. *Arch. Ophthalmol.* **112,** 946–953.

Kim, J. I., Averbuch-Heller, L., and Leigh, R. J. (2001). Evaluation of transdermal scopolamine as treatment for acquired nystagmus. *J. Neuroophthalmol.* **21,** 188–192.

Kinkel, R. P., Kollman, C., O'Connor, P., Murray, T. J., Simon, J., Arnold, D., Bakshi, R., Weinstock-Gutman, B., Brod, S., Cooper, J., Duquette, P., Eggenberger, E., *et al.* (2006). IM interferon beta-1a delays definite multiple sclerosis 5 years after a first demyelinating event. *Neurology* **66,** 678–684.

Ksiazek, S. M., Repka, M. X., Maguire, A., Harbour, R. C., Savino, P. J., Miller, N. R., Sergott, R. C., and Bosley, T. M. (1989). Divisional oculomotor nerve paresis caused by intrinsic brainstem disease. *Ann. Neurol.* **26,** 714–718.

Kupersmith, M. J., Alban, T., Zeiffer, B., and Lefton, D. (2002). Contrast-enhanced MRI in acute optic neuritis: Relationship to visual performance. *Brain* **125,** 812–822.

Laatu, S., Revonsuo, A., Hamalainen, P., Ojanen, V., and Ruutiainen, J. (2001). Visual object recognition in multiple sclerosis. *J. Neurol. Sci.* **185,** 77–88.

Lassman, H., Raine, C., Antel, J., and Prineas, J. (1998). Immunopathology of multiple sclerosis: Report on an international meeting held at the Institute of Neurology of the University of Vienna. *J. Neuroimmunol.* **86,** 213–217.

Lee, A. G., Tang, R. A., Wong, G. G., Schiffman, J. S., and Singh, S. (2000). Isolated inferior rectus muscle palsy resulting from a nuclear third nerve lesion as the initial manifestation of multiple sclerosis. *J. Neuroophthalmol.* **20,** 246–247.

Leech, J., Gresty, M., Hess, K., and Rudge, P. (1977). Gaze failure, drifting eye movements, and centripetal nystagmus in cerebellar disease. *Br. J. Ophthalmol.* **61,** 774–781.

Lindert, R., Haase, C., Brehm, U., Linington, C., Wekerle, H., and Hohfeld, R. (1999). Multiple sclerosis B- and T-cell responses to the extracellular domain of th emyelin oligodendrocyte glycoprotein. *Brain* **122,** 2089–2099.

Lucchinetti, C., Bruck, W., Parisis, J., Schiethauer, B., Rodriguez, M., and Lassmann, H. (2000). Heterogeneity of multiple sclerosis lesions: Implications for the pathogenesis of demyelination. *Ann. Neurol.* **47,** 707–717.

Masucci, E. F., and Kurtzke, J. F. (1988). Downbeat nystagmus secondary to multiple sclerosis. *Ann. Ophthalmol.* **20,** 347–348.

Matsumoto, S., Ohyagi, Y., Inoue, I., Oishi, A., Goto, H., Nakagawa, T., Yamada, T., and Kira, J. (2001). Periodic alternating nystagmus in a patient with MS. *Neurology* **56,** 276–277.

McDonald, W. I., Compston, A., Edan, G., Goodkin, D., Hartung, H. P., Lublin, F. D., McFarland, H. F., Paty, D. W., Polman, C. H., Reingold, S. C., Sandberg-Wollheim, M., Sibley, W., *et al.* (2001). Recommended diagnostic criteria for multiple sclerosis: Guidelines from the International Panel on the diagnosis of multiple sclerosis. *Ann. Neurol.* **50,** 121–127.

Miller, D. H., Mac Manus, D. G., Bartlett, P. A., Kapoor, R., Morrissey, S. P., and Moseley, I. F. (1993). Detection of optic nerve lesions in optic neuritis using frequency-selective fat-saturation sequences. *Neuroradiology* **35,** 156–158.

Nicolai, A., and Lazzarino, L. G. (1995). Peduncular hallucinosis as the first manifestation of multiple sclerosis. *Eur. Neurol.* **35,** 241–242.

Nilsson, P., Larsson, E. M., Maly-Sundgren, P., Perfekt, R., and Sandberg-Wollheim, M. (2005). Predicting the Outcome of Optic Neuritis Evaluation of risk factors after 30 years of follow-up. *J. Neurol.* **252,** 396–402.

Noseworthy, J. H., O'Brien, P. C., Petterson, T. M., Weis, J., Stevens, L., Peterson, W. K., Sneve, D., Cross, S. A., Leavitt, J. A., Auger, R. G., Weinshenker, B. G., Dodick, D. W., *et al.* (2001). A randomized trial of intravenous immunoglobulin in inflammatory demyelinating optic neuritis. *Neurology* **56,** 1514–1522.

Ohkoshi, N., Komatsu, Y., Mizusawa, H., and Kanazawa, I. (1998). Primary position upbeat nystagmus increased on downward gaze: Clinicopathologic study of a patient with multiple sclerosis. *Neurology* **50,** 551–553.

Oksenberg, J., Stuart, S., Begovich, A., Bell, R., Erlich, H., Steinman, L., and Bernard, C. (1991). Limited heterogeneity of rearranged T-cell receptor Vα transcripts in brains of multiple sclerosis patients. *Nature* **345,** 344–347.

Olsson, T., Sun, J., Hillert, J., Hojeberg, B., Andersson, G., Olerup, O., and Link, H. (1992). Increased numbers of T cells recognizing multiple myelin basic protein epitopes in multiple sclerosis. *Eur. J. Immunol.* **22,** 1083–1087.

OWIMS Study Group (1999). Evidence of interferon beta-1a dose response in relapsing-remitting MS: The OWIMS Study. The Once Weekly Interferon for MS Study Group. *Neurology* **53,** 679–686.

Panitch, H. (1994). Influence of infection on exacerbations of multiple sclerosis. *Ann. Neurol.* **36**(Suppl.), S25–S28.

Pranzatelli, M. R. (1992). The neurobiology of the opsoclonus-myoclonus syndrome. *Clin. Neuropharmacol.* **15,** 186–228.

Prineas, J. (1985). The neuropathology of multiple sclerosis. *In* "Handbook of Clinical Neurology" (P. Vinken, G. Bruyn, H. Klawans, and J. Koetsier, Eds.), Vol. 3, pp. 213–257. Elsevier Science Publishers, Amsterdam, NY.

Prineas, J., and Graham, J. (1981). Capping of surface immunoglobulin G on macrophages engaged in myelin breakdown. *Ann. Neurol.* **10,** 149–158.

Qin, Y., Duquette, P., Zhang, Y., Olek, M., Da, R. R., Richardson, J., Antel, J. P., Talbot, P., Cashman, N. R., Tourtellotte, W. W., Wekerle, H., and Van Den Noort, S. (2003). Intrathecal B-cell clonal expansion, an early sign of humoral immunity, in the cerebrospinal fluid of patients with clinically isolated syndrome suggestive of multiple sclerosis. *Lab. Invest.* **83,** 1081–1088.

Quint, D. J., Cornblath, W. T., and Trobe, J. D. (1993). Multiple sclerosis presenting as Parinaud syndrome. *AJNR Am. J. Neuroradiol.* **14,** 1200–1202.

Ranalli, P. J., and Sharpe, J. A. (1988). Vertical vestibulo-ocular reflex, smooth pursuit and eye-head tracking dysfunction in internuclear ophthalmoplegia. *Brain* **111**(Pt. 6), 1299–1317.

Regan, D., and Simpson, T. (1995). Multiple sclerosis can cause visual processing deficits specific to texture-defined form. *Neurology* **45,** 809–815.

Revol, A., Vighetto, A., Confavreux, C., Trillet, M., and Aimard, G. (1990). Oculo-palatal myoclonus and multiple sclerosis. *Rev. Neurol. (Paris)* **146,** 518–521.

Ritchie, A. M., Gilden, D. H., Williamson, R. A., Burgoon, M. P., Yu, X., Helm, K., Corboy, J. R., and Owens, G. P. (2004). Comparative analysis of the CD19+ and CD138+ cell antibody repertoires in the cerebrospinal fluid of patients with multiple sclerosis. *J. Immunol.* **173,** 649–656.

Rodriguez, M., Siva, A., Cross, S. A., O'Brien, P. C., and Kurland, L. T. (1995). Optic neuritis: A population-based study in Olmsted County, Minnesota. *Neurology* **45,** 244–250.

Roed, H. G., Langkilde, A., Sellebjerg, F., Lauritzen, M., Bang, P., Morup, A., and Frederiksen, J. L. (2005). A double-blind, randomized trial of IV immunoglobulin treatment in acute optic neuritis. *Neurology* **64,** 804–810.

Rolak, L. A., Beck, R. W., Paty, D. W., Tourtellotte, W. W., Whitaker, J. N., and Rudick, R. A. (1996). Cerebrospinal fluid in acute optic neuritis: Experience of the optic neuritis treatment trial. *Neurology* **46,** 368–372.

Rucker, J. C. (2005). Current treatment of nystagmus. *Curr. Treat. Options Neurol.* **7,** 69–77.

Ruprecht, K., Klinker, E., Dintelmann, T., Rieckmann, P., and Gold, R. (2004). Plasma exchange for severe optic neuritis: Treatment of 10 patients. *Neurology* **63,** 1081–1083.

Russ, M. O., Cleff, U., Lanfermann, H., Schalnus, R., Enzensberger, W., and Kleinschmidt, A. (2002). Functional magnetic resonance imaging in acute unilateral optic neuritis. *J. Neuroimaging* **12,** 339–350.

Samkoff, L. M., and Smith, C. R. (1994). See-saw nystagmus in a patient with clinically definite MS. *Eur. Neurol.* **34,** 228–229.

Sandramouli, S., Benamer, H. T., Mantle, M., and Chavan, R. (2005). See-saw nystagmus as the presenting sign in multiple sclerosis. *J. Neuroophthalmol.* **25,** 56–57.

Schmidt, S., Wessels, L., Augustin, A., and Klockgether, T. (2001). Patients with Multiple Sclerosis and concomitant uveitis/periphlebitis retinae are not distinct from those without intraocular inflammation. *J. Neurol. Sci.* **187,** 49–53.

Scolding, N., Morgan, B., Houston, W., Linington, C., Campbell, A., and Compston, D. (1989). Vesicular removal by oligodendrocytes of membrane attack complexes formed by activated complement. *Nature* **339,** 620–622.

Seigel, A. M. (1988). Inverted vision in MS. *Neurology* **38,** 1335.

Shahrokhi, F., Chiappa, K. H., and Young, R. R. (1978). Pattern shift visual evoked responses. Two hundred patients with optic neuritis and/or multiple sclerosis. *Arch. Neurol.* **35,** 65–71.

Sharpe, J. A., Rosenberg, M. A., Hoyt, W. F., and Daroff, R. B. (1974). Paralytic pontine exotropia. A sign of acute unilateral pontine gaze palsy and internuclear ophthalmoplegia. *Neurology* **24,** 1076–1081.

Shaw, P. J., Smith, N. M., Ince, P. G., and Bates, D. (1987). Chronic periphlebitis retinae in multiple sclerosis. A histopathological study. *J. Neurol. Sci.* **77,** 147–152.

Sibley, W., Bamford, C., and Clark, K. (1985). Clinical viral infections and multiple sclerosis. *Lancet* **1,** 1313–1315.

Soderstrom, M. (2001). Clues to the immunopathogenesis of multiple sclerosis by investigating untreated patients during the very early stage of disease. *Neurol. Sci.* **22,** 145–149.
Soderstrom, M., Ya-Ping, J., Hillert, J., and Link, H. (1998). Optic neuritis: Prognosis for multiple sclerosis from MRI, CSF, and HLA findings. *Neurology* **50,** 708–714.
Solingen, L. D., Baloh, R. W., Myers, L., and Ellison, G. (1977). Subclinical eye movement disorders in patients with multiple sclerosis. *Neurology* **27,** 614–619.
Sorensen, P. S., Haas, J., Sellebjerg, F., Olsson, T., and Ravnborg, M. (2004). IV immunoglobulins as add-on treatment to methylprednisolone for acute relapses in MS. *Neurology* **63,** 2028–2033.
Sorensen, T., Frederiksen, J., Bronnum-Hansen, H., and Petersen, H. (1999). Optic neuritis as onset manifestation of multiple sclerosis: A nationwide long-term survey. *Neurology* **53,** 473–478.
Starck, M., Albrecht, H., Pollmann, W., Straube, A., and Dieterich, M. (1997). Drug therapy for acquired pendular nystagmus in multiple sclerosis. *J. Neurol.* **244,** 9–16.
Storch, M. K., Piddlesden, S., Haltia, M., Iivanainen, M., Morgan, P., and Lassmann, H. (1998). Multiple sclerosis: *In situ* evidence for antibody- and complement-mediated demyelination. *Ann. Neurol.* **43,** 465–471.
Strupp, M., and Brandt, T. (2006). Pharmacological advances in the treatment of neuro-otological and eye movement disorders. *Curr. Opin. Neurol.* **19,** 33–40.
Strupp, M., Schuler, O., Krafczyk, S., Jahn, K., Schautzer, F., Buttner, U., and Brandt, T. (2003). Treatment of downbeat nystagmus with 3,4-diaminopyridine: A placebo-controlled study. *Neurology* **61,** 165–170.
Sun, J., Link, H., Olsson, T., Xiao, B., Andersson, G., Ekre, H., Linington, C., and Diener, P. (1991). T and B cell responses to myelin oligodendrocyte glycoprotein in mulitple sclerosis. *J. Immunol.* **146,** 1490–1495.
Sweet, L. H., Rao, S. M., Primeau, M., Mayer, A. R., and Cohen, R. A. (2004). Functional magnetic resonance imaging of working memory among multiple sclerosis patients. *J. Neuroimaging* **14,** 150–157.
Tan, E., and Kansu, T. (1990). Bilateral horizontal gaze palsy in multiple sclerosis. *J. Clin. Neuroophthalmol.* **10,** 124–126.
Thorpe, J. W., Barker, G. J., Jones, S. J., Moseley, I., Losseff, N., MacManus, D. G., Webb, S., Mortimer, C., Plummer, D. L., Tofts, P. S., *et al.* (1995). Magnetisation transfer ratios and transverse magnetisation decay curves in optic neuritis: Correlation with clinical findings and electrophysiology. *J. Neurol. Neurosurg. Psychiatry* **59,** 487–492.
Tola, M. R., Granieri, E., Casetta, I., Monari, P., Scorrano, R., Mazzeo, V., Paolino, E., Monetti, V. C., and Govoni, V. (1993). Retinal periphlebitis in multiple sclerosis: A marker of disease activity? *Eur. Neurol.* **33,** 93–96.
Toosy, A. T., Hickman, S. J., Miszkiel, K. A., Jones, S. J., Plant, G. T., Altmann, D. R., Barker, G. J., Miller, D. H., and Thompson, A. J. (2005). Adaptive cortical plasticity in higher visual areas after acute optic neuritis. *Ann. Neurol.* **57,** 622–633.
Toosy, A. T., Werring, D. J., Bullmore, E. T., Plant, G. T., Barker, G. J., Miller, D. H., and Thompson, A. J. (2002). Functional magnetic resonance imaging of the cortical response to photic stimulation in humans following optic neuritis recovery. *Neurosci. Lett.* **330,** 255–259.
Traccis, S., Rosati, G., Monaco, M. F., Aiello, I., and Agnetti, V. (1990). Successful treatment of acquired pendular elliptical nystagmus in multiple sclerosis with isoniazid and base-out prisms. *Neurology* **40,** 492–494.
Traugott, U., Reinherz, E., and Raine, C. (1983). Multiple sclerosis: Distribution of T cell subsets within acute chronic lesions. *Science* **219,** 308–310.
Traugott, U., Scheinberg, L., and Raine, C. (1985). On the presence of Ia-positive endothelial cells and astrocytes in multiple sclerosislesions and its relevance to antigen presentation. *J. Neuroimmunol.* **8,** 1–14.

Trip, S. A., Schlottmann, P. G., Jones, S. J., Altmann, D. R., Garway-Heath, D. F., Thompson, A. J., Plant, G. T., and Miller, D. H. (2005). Retinal nerve fiber layer axonal loss and visual dysfunction in optic neuritis. *Ann. Neurol.* **58,** 383–391.

Tsoi, V. L., Hill, K. E., Carlson, N. G., Warner, J. E., and Rose, J. W. (2006). Immunohistochemical evidence of inducible nitric oxide synthase and nitrotyrosine in a case of clinically isolated optic neuritis. *J. Neuroophthalmol.* **26,** 87–94.

Vanooteghem, P., Dehaene, I., Van Zandycke, M., and Casselman, J. (1992). Combined trochlear nerve palsy and internuclear ophthalmoplegia. *Arch. Neurol.* **49,** 108–109.

Vu, T., Myers, L., Ellison, G., Mendoza, G., and Bronstein, J. (2001). T-cell specific responses to oligodendrocye-specific protein in multiple sclerosis. *J. Neurosci. Res.* **66,** 506–509.

Wall, M., and Wray, S. H. (1983). The one-and-a-half syndrome—a unilateral disorder of the pontine tegmentum: A study of 20 cases and review of the literature. *Neurology* **33,** 971–980.

Wolin, M. J., Trent, R. G., Lavin, P. J., and Cornblath, W. T. (1996). Oculopalatal myoclonus after the one-and-a-half syndrome with facial nerve palsy. *Ophthalmology* **103,** 177–180.

Wucherpfennig, K., and Strominger, J. (1995). Molecular mimicry in T cell-mediated autoimmunity: Virla peptides activate human T cell clones specific for myelin basic protein. *Cell* **80,** 695–705.

Young, Y. H., and Huang, T. W. (2001). Role of clonazepam in the treatment of idiopathic downbeat nystagmus. *Laryngoscope* **111,** 1490–1493.

Zamvil, S., and Steinman, L. (1992). The pathogenesis of demyelinating disease in the central nervous system and the development of selective immunotherapy. *In* "Current Neurology" (S. Appel, Ed.), Vol. 12, pp. 44–82. Mosby Year Book, St. Louis.

Zein, G., Berta, A., and Foster, C. S. (2004). Multiple sclerosis-associated uveitis. *Ocul. Immunol. Inflamm.* **12,** 137–142.

NEUROMYELITIS OPTICA: NEW FINDINGS ON PATHOGENESIS

Dean M. Wingerchuk

Mayo Clinic College of Medicine, Scottsdale, Arizona 85259, USA

Neuromyelitis optica (NMO) is an idiopathic CNS demyelinating disorder that preferentially involves the optic nerve and spinal cord. Diverse sources of evidence support the hypothesis that NMO is distinct from classical multiple sclerosis (MS) and that the pathogenesis of NMO is dominated by humoral mechanisms. Such evidence includes clinical observations that systemic autoimmune diseases often coexist with NMO and that therapeutic plasmapheresis may provide meaningful rescue therapy for severe clinical attacks, immunopathologic studies that demonstrate prominent complement activation and immunoglobulin deposition, and the discovery of the serum autoantibody NMO-IgG, a potential NMO biomarker that targets aquaporin-4 (AQP4). The NMO-IgG marker is present in a majority of patients with "NMO-spectrum disorders," including isolated or recurrent longitudinally extensive transverse myelitis, recurrent optic neuritis with negative brain imaging, and the Asian optic-spinal form of MS. Preliminary experiments demonstrate that NMO-IgG can modulate AQP4 function and fix complement, characteristics that suggest it has the potential to be pathogenic in NMO. Other immunologic differences among NMO, NMO-spectrum disorders, and classical MS are reviewed.

INTERNATIONAL REVIEW OF
NEUROBIOLOGY, VOL. 79
DOI: 10.1016/S0074-7742(07)79029-3

0074-7742/07 $35.00

I. Introduction

Neuromyelitis optica (NMO; also known as Devic's syndrome or Devic's disease) is an inflammatory demyelinating disorder of the central nervous system (CNS) (Cree *et al.*, 2002; de Seze, 2003; Wingerchuk *et al.*, 1999). The burden of clinical disease in NMO falls on the optic nerves and spinal cord therefore, its clinical expression is in the form of severe attacks of inflammatory optic neuritis and acute myelitis. Optic neuritis and myelitis attacks also occur frequently in multiple sclerosis (MS). Because until recently there was no validated means for separating the disorders, it has long been controversial whether NMO is a distinct disorder or whether it simply represents a severe but topographically restricted form of MS. While MS is a putative autoimmune disease with T-cell-predominant immunologic mechanisms, mounting evidence reviewed herein strongly suggests that B cells and the humoral arm of the immune system are primarily responsible for NMO pathogenesis (Table I) (Weinshenker *et al.*, 2006b; Wingerchuk, 2006). The most important advance has been the discovery of the serum autoantibody neuromyelitis optica immunoglobulin-G (NMO-IgG), which targets the water channel aquaporin-4 (AQP4) (Lennon *et al.*, 2004, 2005). Serological–clinical association studies showed that NMO-IgG is extremely specific for NMO to the exclusion of typical MS. NMO may represent the first of a new class of autoimmune channelopathy.

II. NMO: Evolution of Disease Definition and Spectrum

A. Historical Background

Although he was not the first to recognize the clinical association of visual loss and spinal cord disease, Eugene Devic made several important observations that consolidated the association in the nineteenth century (Devic, 1894). The term "Devic's syndrome" or "Devic's disease" was thereafter typically applied to patients who experienced an acute, near-simultaneous combination of bilateral inflammatory optic neuritis and "transverse" myelitis causing complete or nearly complete loss of sensory, motor, and bowel and bladder function below the level of the cord lesion. Debate about the place of NMO in the nosological structure of CNS demyelinating diseases flared with each new case series reported during the twentieth century but lack of a definitive disease marker failed to resolve the issue (Cree *et al.*, 2002; Wingerchuk, 2004). The reports did, however, provide numerous exceptions to the conception that NMO was always monophasic. Several large contemporary series made the case for substantial clinical variability in

TABLE I

CONVERGENT DATA THAT SUPPORT HUMORAL AUTOIMMUNE MECHANISMS IN NMO

- Systemic autoimmunity often coexists with NMO
- Distinctive immunopathology
  - Eosinophilic and neutrophil inflammatory infiltrates
  - Immunoglobulin deposition
  - Complement deposition
- Optic-spinal distribution of pathology in specific animal models
- Very high specificity of NMO-IgG (anti-AQP4) autoantibody for NMO
- Clinical treatment responses
  - Relatively selective benefit of plasmapheresis for acute attacks
  - Lack of clear benefit of MS immunomodulatory therapies for attack prevention
  - Apparent benefit of immunosuppressive therapies for attack prevention, including specific B-cell depletion therapy

NMO; the disease could cause either unilateral or bilateral optic neuritis, some milder attacks occurred, and a chronic relapsing course was, in fact, the rule (O'Riordan *et al.*, 1996; Wingerchuk *et al.*, 1999). Most attacks were quite severe and patients were at risk for developing neurogenic respiratory failure and death owing to ascending cervical myelitis. Furthermore, one study demonstrated that half of patients were blind (visual acuity less than 20/200) or required ambulatory assistance within 5 years of disease onset (Wingerchuk and Weinshenker, 2003). The advent of magnetic resonance imaging (MRI) represented a major advance. Brain MRI is usually normal or reveals cerebral white matter lesions that do not meet diagnostic imaging criteria for MS. The most powerful criterion is the finding of a longitudinally extensive spinal cord MRI lesion spanning three or more vertebral segments and best detected in the setting of acute myelitis (Fig. 1). In 1999, the cumulative observational experience was consolidated into empiric diagnostic criteria that emphasized the restriction of clinical disease to the optic nerves and spinal cord and the severity of attacks and included a hierarchy of supportive MRI and cerebrospinal fluid (CSF) characteristics (Table II) (Wingerchuk *et al.*, 1999).

B. NMO-IgG DISCOVERY AND REVISED NMO DIAGNOSTIC CRITERIA

In 2004, Lennon *et al.* (2004) reported discovery of a serum autoantibody marker for NMO termed neuromyelitis optica immunoglobulin G (NMO-IgG). This marker, the characteristics of which will be discussed later, was 73% sensitive and 90% specific for differentiating NMO from classical MS in patients presenting with simultaneous or sequential optic neuritis and myelitis. The NMO-IgG marker appears to be extremely specific for NMO in that it is not found in normal controls or patients with classical MS presenting with symptoms other than optic

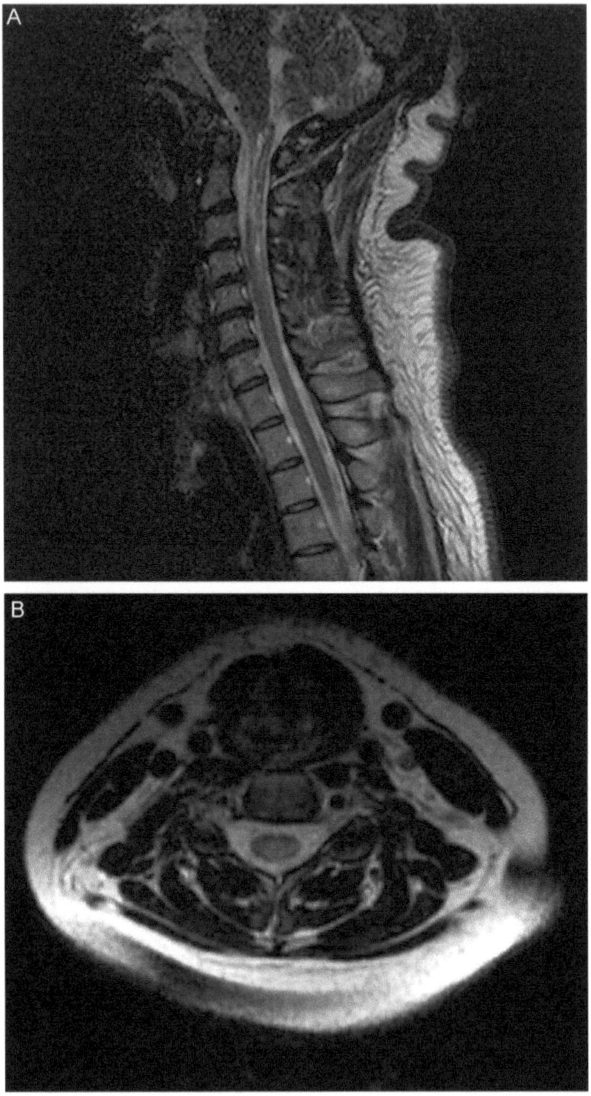

Fig. 1. Panel (A) demonstrates a characteristic longitudinally extensive and contiguous cervical spinal cord lesion with extension into the brain stem (T2-weighted sagittal section). Panel (B) reveals the centrally based abnormality that involves most of the cervical spinal cord cross-sectional area (T2-weighted axial section).

neuritis and myelitis. Extensive longitudinal experience with NMO-IgG led our group in 2006 to propose revised diagnostic criteria that incorporate the marker (Wingerchuk *et al.*, 2006). The revised criteria require optic neuritis, acute myelitis,

TABLE II
DIAGNOSTIC CRITERIA FOR NMO (1999)

All absolute criteria plus:
- 1/3 major supportive criteria

or
- 2/3 minor supportive criteria

are required to confirm diagnosis

Absolute criteria:
Optic neuritis
Acute myelitis
No clinical disease outside of the optic nerves and spinal cord

Major supportive criteria:
Negative brain MRI at disease onset (normal or not meeting radiological diagnostic criteria for MS)
Spinal cord MRI with T2 signal abnormality extending over three or more vertebral segments
CSF pleocytosis (>50 WBC/mm^3) OR > 5 neutrophils/mm^3

Minor supportive criteria:
Bilateral optic neuritis
Severe ON with fixed visual acuity worse than 20/200 in at least one eye
Severe, fixed, attack-related weakness (MRC grade 2 or less)

TABLE III
REVISED NMO DIAGNOSTIC CRITERIA (2006)

Diagnosis requires fulfillment of absolute criteria *and* at least two of three supportive criteria:
Absolute criteria:
Optic neuritis
Acute myelitis
Supportive criteria:
Negative brain MRI at disease onset
Spinal cord MRI with contiguous T2-weighted signal abnormality extending over three or more vertebral segments
NMO-IgG seropositivity

and two of three supportive elements: normal brain MRI (or nonspecific lesions only), a longitudinally extensive cord lesion visible on spinal cord MRI, or NMO-IgG seropositive status (Table III). These criteria are 99% sensitive and 90% specific for NMO and are expected to have greater ability to distinguish NMO from patients with early, mild MS who have a normal brain MRI despite experiencing at least two attacks (including optic neuritis and partial myelitis). Importantly, the criteria allow for the finding that some patients with NMO experience symptoms referable to CNS regions other than the optic nerves and spinal cord (e.g., the brain stem and cerebrum). Furthermore, the criteria acknowledge that the presence of brain MRI lesions reduces the probability that a patient has NMO but does not exclude the diagnosis completely, acknowledging the evidence that white matter lesions accumulate over time in NMO. Although the lesions usually follow no specific pattern, about 10% of patients will ultimately have lesions that meet Barkhof criteria for MS (Pittock *et al.*, 2006a). Some other patients' scans contain T2-weighted signal abnormalities that with unusual extensive subcortical or deep gray lesions (Nakashima *et al.*, 2006; Pittock *et al.*, 2006c). Further validation is required, but the simplified and highly specific criteria appear to have significant clinical utility especially since the treatment approaches for NMO and MS are now widely divergent (Wingerchuk and Weinshenker, 2005).

C. The NMO Spectrum of Disorders

The recognition that other clinical syndromes belong within a spectrum of NMO disorders is a significant advance aided by the discovery of NMO-IgG. Patients with "idiopathic" isolated or recurrent longitudinally extensive transverse myelitis (LETM; acute myelitis accompanied by MRI evidence of a lesion extending over three or more vertebral segments) have an NMO-IgG seropositivity rate of approximately 50–60% and many will experience a future event of optic neuritis that confirms NMO (Lennon *et al.*, 2004; Weinshenker *et al.*, 2006c). Some, however, experience many recurrent myelitis attacks without optic neuritis over many years or decades of follow-up. A similar scenario occurs with patients who experience recurrent bouts of unilateral or bilateral optic neuritis without clinical evidence for myelitis; such patients are less frequently NMO-IgG seropositive (less than 30%). Some patients with systemic autoimmune diseases, such as systemic lupus erythematosus (SLE) or Sjogren syndrome (SS), develop optic neuritis or LETM and are NMO-IgG seropositive; recent evidence (discussed later in greater detail) suggests that these patients have coexisting disorders rather than an optic-spinal pattern of SLE or SS-related neurological complications. Finally, about half of patients with Asian optic-spinal MS, traditionally recognized and described in Japan, are NMO-IgG seropositive (Lennon *et al.*, 2004; Nakashima *et al.*, 2006; Weinshenker *et al.*, 2006c). Most of these patients would be recognized as having NMO by current criteria. The clinical syndromes that fall under the current concept of NMO spectrum disorders are summarized in Table IV.

III. Evidence for Humoral Autoimmunity in NMO

The pathogenesis of NMO is not well understood but increasing evidence supports a primary role for antibody-mediated mechanisms. This section will summarize data from animal models, clinical association studies, immunopathologic

TABLE IV
PROPOSED NMO SPECTRUM DISORDERS

Neuromyelitis optica (2006 definition)
Limited forms of NMO
"Idiopathic" LETM, single or recurrent
Optic neuritis, recurrent or simultaneous bilateral
Asian optic-spinal MS
Optic neuritis or longitudinally extensive myelitis associated with systemic autoimmune disease
Optic neuritis or myelitis associated with NMO-specific brain lesions (hypothalamic, periventricular, brain stem)

findings, and therapeutic observations, each of which supports the fundamental importance of antibody-mediated disease mechanisms. In the final section, more detail will be presented regarding NMO-IgG, its antigen target AQP4, and its potential for being the primary pathogenic antibody in NMO.

A. Coexisting Systemic Autoimmune Disease and Serological Data

Large case series demonstrate that about 30% or more of cases of clinically definite NMO are associated with disorders such as autoimmune thyroid disease (Hashimoto's thyroiditis or Grave's disease), SS and less common diseases such as pernicious anemia, SLE, idiopathic thrombocytopenic purpura, ulcerative colitis, and primary sclerosing cholangitis (Wingerchuk *et al.*, 1999). Furthermore, at least 50% of NMO patients harbor one or more of antinuclear, anti-double stranded DNA, anti-microsomal, or antiphospholipid antibodies, extractable nuclear antigen, or rheumatoid factor. The SSA autoantibody has been associated with increased risk of relapse after a first event of transverse myelitis (Hummers *et al.*, 2004).

The fundamental NMO phenotype (simultaneous or sequential optic neuritis and myelitis) is not specific for NMO. As outlined above, it may occur in classical MS, though in much less severe form and without the specific MRI characteristics or NMO-IgG seropositivity required by revised NMO diagnostic criteria. In some situations, however, more characteristic NMO findings, such as LETM, occur in the context of a known or newly recognized systemic autoimmune disease such as SLE or SS or in patients who have serum autoantibodies associated with those disorders (including antinuclear antibody, anti-double-stranded DNA antibody and SSA or SSB) (Boumpas *et al.*, 1990; Manabe *et al.*, 2000). In some reported cases, existence of the systemic autoimmune disorder was established well in advance of the emergence of neurological symptoms and was confirmed using standard rheumatologic diagnostic criteria. In contrast, in many other instances, the systemic diagnosis was inferred by the detection of one or more serum autoantibodies during investigation of the sentinel optic neuritis or myelitis event. The obvious question is whether NMO coexists with other autoimmune diseases or whether the NMO syndrome is a common neurological phenotype that results when the CNS is involved by systemic autoimmune diseases, perhaps in the form of vasculitis. Current data that take advantage of the specificity of NMO-IgG suggest that NMO and systemic autoimmune diseases usually coexist (Pittock *et al.*, 2006b; Weinshenker *et al.*, 2006a). In a study of 38 patients independently classified by French investigators and tested for NMO-IgG in blinded fashion by Mayo Clinic, 22 had NMO ($n = 14$) or were deemed "at risk" for NMO ($n = 8$) by virtue of experiencing prior optic neuritis or myelitis. Six of the NMO patients and all of the "at risk" patients had SLE or SS. Seven of the 22 patients (32%) were NMO-IgG seropositive and there was no difference between those with concurrent SLE or SS. In contrast, a group of SLE or SS patients ($n = 16$) without NMO symptoms was uniformly NMO-IgG seronegative ($p = 0.01$ by Pearson chi-square analysis). These data strongly suggest that NMO and SLE or SS are coassociated diseases and that the pathogenesis of NMO, optic neuritis, and LETM in the setting of SLE or SS may be the same as for uncomplicated NMO.

B. Human Immunopathologic Findings in NMO

Optic nerve and spinal cord lesions in NMO and MS both contain features of demyelination, macrophage and activated microglia infiltration, and axonal loss. Lesions from NMO patients usually also reveal involvement of both white and gray matter, necrosis and cavitation owing to the destructive nature of the inflammatory process. Distinguishing characteristics of NMO lesions include a high proportion of eosinophils and neutrophils in the inflammatory lesion and the presence of hyalinized blood vessels of indeterminate significance (Lefkowitz and Angelo, 1984; Lucchinetti *et al.*, 2002; Ortiz de Zarate *et al.*, 1968). Furthermore, NMO lesions display prominent antibody deposition and complement activation that is similar to, but not precisely the same, as one of the four distinct immunopathologic patterns described in classical MS (Lefkowitz and Angelo, 1984; Lucchinetti *et al.*, 2000, 2002; Ortiz de Zarate *et al.*, 1968). Whereas "pattern 2" MS lesions reveal deposition of complement components and immunoglobulin in areas of active demyelination, in NMO lesions, these features are located around penetrating microvessels and they stain with characteristic ring and rosette patterns. Overall, the findings suggest that the perivascular region is targeted by an antibody-dependent, complement-mediated immune attack that could be amplified by eosinophil recruitment with local degranulation.

The highly selective (though not absolute) restriction of NMO pathology to optic nerve and spinal cord is not understood. Possible mechanisms include increased relative permeability of the blood–brain barrier at these sites, allowing greater penetration of small amounts of pathogenic antibody. Another potential mechanism is greater accessibility to target antigen because of its distribution, concentration, or orientation within optic nerve and spinal cord. If NMO-IgG is responsible for primary pathogenesis of NMO (see below), potential mechanisms would have to be account for the fact that its target antigen, AQP4, is widely distributed throughout the CNS (Agre and Kozono, 2003).

C. Immunologic Studies in NMO

There are relatively few studies comparing immunologic function between NMO and classical MS. An early investigation by Mandler *et al.* (2001) demonstrated that matrix metalloproteinase-9 levels were elevated in the CSF of MS patients but not NMO patients when fluid was collected during the evolution of an acute exacerbation. The authors interpreted the results as potentially reflective of a different primary immunopathogenic mechanism between the diseases.

The CSF characteristically lacks unique oligoclonal bands in NMO. This could be explained if the inciting pathogenic mechanism originates outside of the CNS and the disease activity occurs without development of intrathecal B-cell clonal expansion. An additional factor may be a relative lack of an IgG1 isotype in patients with relapsing NMO (Nakashima *et al.*, 2004); in contrast, MS patients have an elevated IgG1% and IgG1 index indicative of Th1-dominant immunologic mechanisms. Indirectly, these results support a humoral or Th2-type immune response. More specific evidence for this response pattern derived from a recent peripheral blood and CSF study showed that, compared with age- and sex-matched MS controls and healthy subjects, NMO patients demonstrated a measurable humoral response pattern exclusively in the CSF (Correale and Fiol, 2004). This response was inferred after detection of greater antimyelin oligodendrocyte glycoprotein (MOG) mononuclear cell production of interleukin (IL)-5, IL-6, and IgM in NMO patients. In addition, eotaxin-2, eotaxin-3, and eosinophil cationic protein (ECP) levels were significantly higher in NMO patients' CSF, a result compatible with the known pattern of eosinophilic infiltration of acute NMO lesions.

Ishizu *et al.* (2005) undertook a comprehensive evaluation of CSF cytokines and chemokine profiles [including IL-1β, IL-2, IL-4, IL-5, IL-6, IL-7, IL-8, IL-10, IL-12, IL-13, IL-17, interferon-γ, tumor necrosis factor-α, granulocyte colony-stimulating factor (G-CSF), monocyte chemoattractant protein-1 (MCP-1), and macrophage inflammatory protein-1β (MIP-1β)] during acute relapse periods in 20 patients with Asian optic-spinal MS and 20 patients with conventional relapsing–remitting MS. Samples from patients with spinocerebellar degenerations were used as controls. Evidence of Th1 cell upregulation was seen in both types of demyelinating disease but increased IL-17 and IL-8 levels were noted specifically in the optic-spinal MS group and the results correlated with length of spinal cord lesions on MRI. Three of six spinal cord autopsy specimens from optic-spinal MS patients showed numerous myeloperoxidase-positive neutrophils infiltrating necrotic lesions. The results suggested to the authors that intrathecal activation of the IL-17/IL-8 axis was associated with prominent neutrophil infiltration of the spinal cord, in part explaining the development of long-cord lesions.

D. Animal Models with NMO-Like Pathological Patterns

Experimental autoimmune encephalomyelitis (EAE) is a putative animal model of MS. A variety of different phenotypes and pathological patterns may be induced by using specific genetic strains of animal and different inciting antigens (Gold and Linington, 2002). Immunization of the Brown Norway rat with soluble MOG in Freund's adjuvant results eosinophil-containing infiltrates, antibody deposition, and complement activation in a pattern of optic-spinal pathology reminiscent of NMO (Storch *et al.*, 1998). This finding does not necessarily implicate MOG as the primary antigenic target for human NMO, but suggests that humoral mechanisms predominate in development of an optic-spinal pathological pattern.

A new animal model of spontaneous autoimmune CNS demyelination, termed opticospinal EAE (OSE) by one group of investigators (Krishnamoorthy *et al.*, 2006), may represent an important advance in understanding NMO-like pathological patterns (Ransohoff, 2006). Two independent laboratories reported that more than half of double-transgenic mice expressing T- and B-cell receptors that recognize MOG developed spontaneous inflammation of the optic nerves and spinal cord with involvement of both meninges and parenchyma (Bettelli *et al.*, 2006; Krishnamoorthy *et al.*, 2006). Cytokine profiles from affected tissues and cell cultures showed a mix of Th1 and Th2 cytokines, including interferon-γ, IL-5, and IL-17. Some important features of human NMO pathology that are lacking in the OSE model include the absence of lesional immunoglobulin deposition and complement, absence of longitudinally extensive spinal cord lesions (most lesions were quite small), and lack of AQP4 antibodies. Despite these important differences, the new model represents an important advance by providing new information on the ability of B cells to operative as antigen-presenting cells and the mechanisms and results of spontaneous EAE models.

E. The NMO-IgG Autoantibody Marker

1. *Discovery and Clinical -Serological Associations*

The recognition of NMO as an entity distinct from classical MS was strongly facilitated by the discovery of the serum autoantibody marker NMO-IgG by Lennon *et al.* (2004). The distinctive indirect immunofluorescence pattern, which provided early clues that the target antigen is AQP4, was first detected in a small number of patient sera sent to the Mayo Clinic Neuroimmunology Laboratory for paraneoplastic antibody testing. Retrospective examination of case histories revealed that most of the patients whose serum was associated with this staining pattern had a clinical history compatible with NMO or recurrent severe transverse myelitis. Mayo Clinic investigators then initiated a prospective evaluation of patients with NMO, recurrent LETM, recurrent optic neuritis with negative brain imaging, and patients with an optic-spinal clinical presentation in whom the final clinical diagnosis was classical MS. This study demonstrated that NMO-IgG is sensitive (73%; 95% confidence interval = 60–86%) and highly specific (91%; 95% confidence interval = 79–100%) for distinguishing NMO from classical MS using final clinical diagnosis as the reference standard. NMO-IgG seropositivity was also detected in 52% of patients with recurrent isolated transverse myelitis (fitting the definition of LETM) and 25% of patients with recurrent optic neuritis, suggesting that many patients fitting these "idiopathic" disease descriptions are at "high risk" for future development of NMO or have a *form fruste* of the disease. Furthermore, Japanese investigators contributed to the study by sending serum samples from patients with optic-spinal MS, classical MS, or stroke for NMO-IgG evaluation. These samples were tested by the Mayo Clinic laboratory investigators who were blinded to clinical information. Seven of 12 (58%) optic-spinal MS patients (one had a "high-risk" syndrome) were NMO-IgG seropositive, whereas all "Western-type" MS and stroke patients were seronegative. In sum, these data demonstrated strong specificity of NMO-IgG for NMO, suggested that Asian optic-spinal MS and NMO may be identical (at least in some cases) and that the spectrum of NMO disorders includes the "high-risk" syndromes of recurrent LETM and recurrent isolated optic neuritis.

Further experience at Mayo Clinic with NMO-IgG has confirmed its specificity; it has not been detected in patients with classical MS whose disease course does not consist of optic neuritis or myelitis. The uncommon cases in which NMO-IgG is positive in patients with a diagnosis of classical MS have proven to be the most clinically challenging. Retrospective analysis of individual cases has demonstrated that in some instances the occurrence of brain lesions or minor brain stem symptoms may have led to a diagnosis of classical MS. Recognition that NMO may at times be associated with non-optic-spinal symptoms or be accompanied by brain lesions should minimize clinical misclassification in future research studies as well as clinical practice.

We used our longitudinal experience with NMO-IgG testing and clinical follow-up of 96 Mayo Clinic patients to systematically develop more accurate NMO diagnostic criteria (Wingerchuk *et al.*, 2006). We evaluated the sensitivity and specificity of each individual component criterion of the 1999 diagnostic criteria and then various combinations of the most accurate variables in order to determine which model provided the likelihood ratios with greatest clinical utility. Final clinical diagnosis (NMO or MS) was used as the reference standard. We determined that about 15% of patients with NMO experienced neurological symptoms referable to CNS regions other than the optic nerve and spinal cord, confirming that arbitrary exclusion of an NMO diagnosis in such patients was invalid. We found that the best diagnostic model (Table III) required the occurrence of optic neuritis and acute myelitis with support from at least two of three additional elements: a longitudinally extensive spinal cord lesion, onset brain MRI normal or nondiagnostic for MS, or NMO-IgG seropositive status. These criteria are extremely sensitive (99%; 95% confidence interval 97–100%) and specific (90%; 95% confidence interval 80–100%) for NMO in patients presenting with a clinical optic-spinal syndrome. The associated likelihood ratios [for a positive test, 10.2 (95% confidence interval 3.49–30.0); for a negative test, 0.01 (95% confidence interval 0.002–0.09)] demonstrate that the proposed diagnostic model has large impact on determining the posttest probability of NMO diagnosis.

Since the 2004 report of NMO-IgG discovery, other groups have independently confirmed that NMO-IgG or anti-AQP4 antibodies have high specificity for NMO and are frequently detected in patients with LETM. Nakashima *et al.* (2006) showed that NMO-IgG was detected in 12/19 (63%) of Japanese optic-spinal MS patients and 2/13 (15%) of Japanese patients with conventional MS. A closer look at the latter group revealed that the two NMO-IgG seropositive patients had unusual brain lesions but in other respects were compatible with optic-spinal MS. NMO-IgG positive status was also associated with longitudinally extensive spinal cord lesions and more severe optic-spinal clinical disease, including complete blindness. Spanish investigators showed that 50% of 10 NMO patients and 2/2 (100%) of LETM patients were NMO-IgG seropositive, whereas 0/8 MS patients were seropositive (Zuliani *et al.*, 2006a). Investigators from the United Kingdom found identical NMO-IgG seropositive frequencies in 10 NMO and 2 LETM patients and also reported that all 10 MS patients tested seronegative (Littleton *et al.*, 2006). A French group showed that 55% of 20 NMO patients and 45% of 11 LETM patients were NMO-IgG seropositive; 9% of 52 MS patients with optic-spinal presentations were seropositive but some of these patients had longitudinally extensive cord lesions (Marignier *et al.*, 2006). Finally, a European group using a novel detection approach found that 54% of 35 NMO patients and 80% of 5 LETM patients had anti-AQP4 antibodies compared with only 3% of 140 classical MS patients (Jarius *et al.*, 2006). Overall, the results are highly congruent and suggest that the occasional cases of seropositivity in MS may owe more to the imperfect nature of clinical diagnosis than to the biological specificity of the antibody testing.

NMO-IgG also has prognostic utility for patients with LETM. A prospective study of 29 patients with LETM as a first-ever demyelinating event revealed that 37.9% were NMO-IgG seropositive (Weinshenker *et al.*, 2006c). Twenty-three patients were followed for at least a year. Nine were NMO-IgG seropositive; of those, five (55%) had a second neurological event (four experienced recurrent LETM and one had optic neuritis). In contrast, none of the 14 NMO-IgG seronegative patients relapsed ($p = 0.004$). These data provide further support for the concept that LETM is a high-risk syndrome that often leads to NMO or is a limited NMO variant.

It is not yet clear whether NMO-IgG serological status has prognostic value in patients with fully developed NMO. In the initial report describing the autoantibody, there were no differences in the clinical features or course between seropositive (73%) and seronegative (27%) patients but further studies are required. The explanation for seronegative cases is also not clear. One potential contributor might be the sensitivity of the NMO-IgG assay, which has improved over time owing to technical refinements. Another possibility is that target antigens other than AQP4 might be of primary importance in some cases of clinical NMO. In a study of five NMO patients and five healthy controls, serum microarray analysis detected three potential autoantibodies in a single NMO case (Lalive *et al.*, 2006). The autoantibodies bound to the 73-kDa subunit of cleavage and polyadenation specificity factor (CPSF-73), ring finger protein 141, and myosin light chain 2, precursor lymphocyte-specific IgG. Levels of one of these autoantibodies, anti-CPSF-73, became undetectable after B-cell depletion therapy with rituximab. The relevance and role of this finding is not clear but the approach may be useful to identify other potential autoantigens in NMO and related diseases.

2. *NMO-IgG, AQP4, and NMO Pathogenesis*

Lennon *et al.* (2005) reported that AQP4 is the target antigen of NMO-IgG. Aquaporins are membrane water channel proteins that are involved in fluid homeostasis and are especially important in times of metabolic or osmotic stress (Agre and Kozono, 2003). Most aquaporins exclusively transmit water but some allow passage of solutes such as glycerol or urea. There are now more than a dozen known members of the aquaporin family, each with a distinct distribution in body organs and cells. The CNS contains aquaporin types 1, 4, and 9. In rats, AQP4 is present on astrocytic foot processes along endothelial tight junctions, on the abluminal side of cerebral microvessels, within the cerebellar Purkinje cell layer, and in the hypothalamus (Nielsen *et al.*, 1997). It is not associated with neurons, oligodendrocytes, or microglia. It is anchored in the astrocytic foot process membrane by the dystroglycan complex (Amiry-Moghaddam and Ottersen, 2003). It has been implicated in control of mechanisms related to cytotoxic edema and is a potential therapeutic target in brain trauma, stroke, epilepsy, preeclampsia, and other disorders.

Lennon and colleagues used clues from the indirect immunofluorescence staining pattern of NMO-IgG in mouse tissues (Fig. 2). The antibody binds to the abluminal surface of microvessels, pia, subpia, and the Virchow–Robin spaces in mouse CNS tissues. It also colocalizes with laminin, consistent with the antigen being a component of the glia limitans at the blood–brain barrier. In a series of further experiments, they demonstrated that NMO-IgG did not bind to myenteric and submucosal plexi or arteriole-associated sympathetic nerves in stomach, kidney, or liver but it did bind significantly to distal collecting tubules in the renal medulla and parietal cells in the gastric crypts. This distribution suggested AQP4 as a candidate antigen. The investigators then used dual immunostaining with AQP4-specific rabbit IgG and confocal microscopy to demonstrate that NMO patients' IgG bound to an antigen that colocalized that with AQP4 in all of these tissue types (Lennon *et al.*, 2005).

The investigative group then went on to demonstrate that NMO-IgG staining is abolished in an AQP4 knockout mouse and that NMO-IgG selectively immunoprecipitates AQP4, but not other constituents of the dystroglycan protein complex that anchor AQP4 to the cell membrane. In sum, these experiments conclusively demonstrate that astrocyte-based water channel AQP4 is the target antigen of NMO-IgG. Recently, other groups have independently confirmed these findings. Zuliani *et al.* (2006b) used confocal microscopy and dual immunostaining to confirm the colocalization of NMO-IgG with anti-AQP4 IgG in two patients with NMO. Jarius *et al.* (2006) developed a quantitative method for detection of anti-AQP4 antibodies by cloning the cDNA sequences of human AQP4, expressing the protein in a eukaryotic transcription/translation system, and establishing a 35S-methionine radioimmunoprecipitate assay (RIPA). This technique was used to confirm the specificity of anti-AQP4 antibodies for NMO and LETM.

Recent publications have demonstrated interesting correlations between the distribution of AQP4 in the CNS and a rather specific MRI lesional pattern in some NMO patients. Several groups have reported a pattern of brain MRI T2-weighted signal abnormality involving the third ventricle, hypothalamus, periaqueductal and peri-fourth ventricular regions, periependymal regions surrounding the lateral ventricles, superior cerebellar peduncle, and subpial regions of the cerebellar hemispheres (Fig. 3) (Pittock *et al.*, 2006a,c; Poppe *et al.*, 2005). Pittock *et al.* (2006c) observed this distinctive pattern in 8/120 (6.7%) of NMO-IgG seropositive patients.

Preliminary work to evaluate AQP4 in pathological samples has revealed loss of AQP4 in central gray matter of spinal cord lesions, especially in perivascular

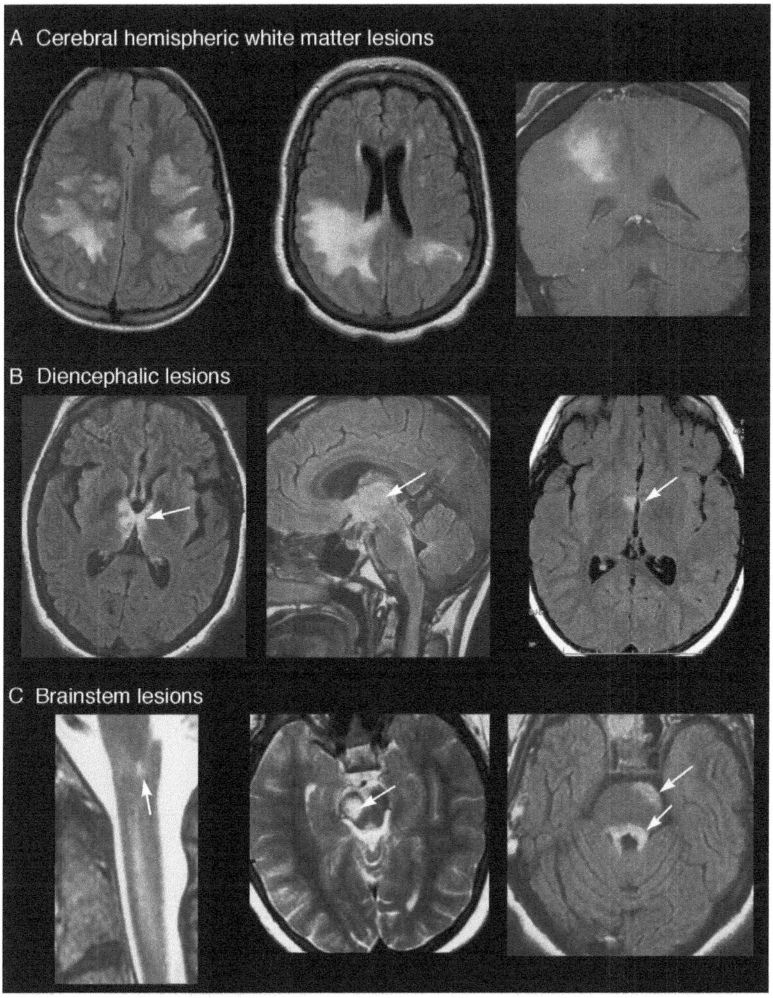

FIG. 2. Illustration of atypical brain lesions encountered in patients with NMO seen on MRI. (A) Left: This image demonstrates extensive bihemispheric subcortical nonenhancing white matter fluid-attenuated inversion recovery (FLAIR) signal abnormality. (A) Center: There is large confluent FLAIR signal abnormality in the right parietal area that demonstrates diffuse gadolinium enhancement (A, right). (B) Hypothalamic FLAIR abnormality (B right, arrow) and right cerebral peduncle (C center, arrow). (B) Center: FLAIR signal abnormality is present in the thalamus (arrow), hypothalamus, and optic chiasm extending into the superior cerebellar peduncle and the floor of the fourth ventricle. (B) Left: Confluent nonenhancing signal abnormality is seen from the anterosuperior thalamus-hypothalamus (arrow) to the optic tracts behind the chiasm to the superior surface of the mesencephalon extending to the periaqueductal area (right > left) to the superior cerebellar peduncles, and the pontine tegmentum (C right, arrows). Extension of T2-weighted MRI signal abnormality into the medulla (C left, arrow) is seen in a different patient with an otherwise normal brain MRI. Figure and revised legend from Pittock *et al.* (2006a), with permission.

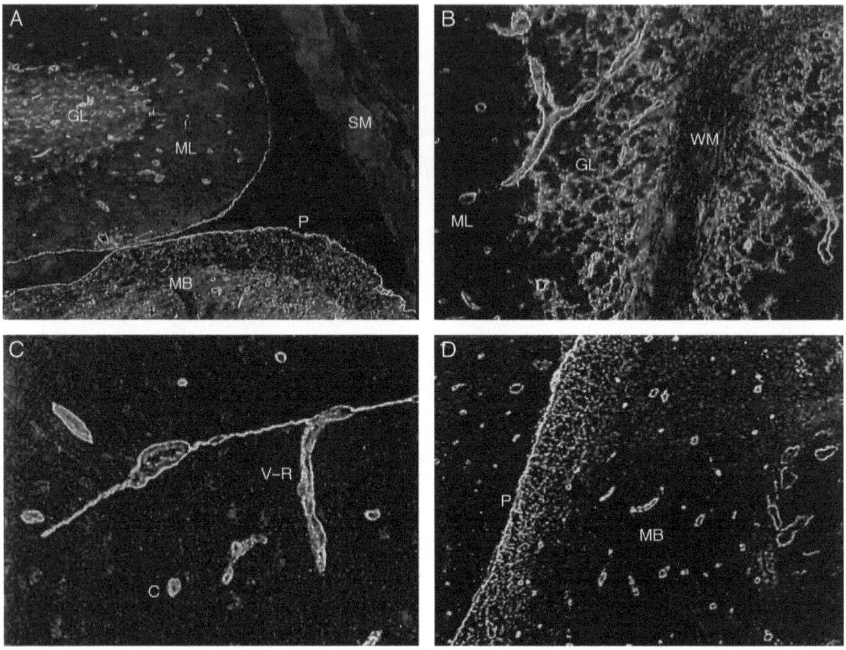

FIG. 3. Immunofluorescence pattern of bound NMO-IgG in mouse CNS (A) Linear staining of juxtaposed pial membranes (P) of cerebellar cortex and midbrain (MB) and their microvessels. Adjacent gut smooth muscle, submucosa and vessels (SM) not stained (×200). (B) Prominent microvessel staining in cerebellar molecular layer (ML), granular layer (GL), and white matter (WM) (×400). (C) Linear staining in cerebellar cortex includes pia, pial lining of Virchow–Robin (V–R) spaces, and continues along microvessels, including capillaries (C). (D) Staining of the subpia of midbrain (×400). Figure and legend from Lennon *et al.* (2004), with permission. (See Color Insert.)

regions near deposits of complement and immunoglobulin; however, it should be noted that this loss of AQP4 corresponded in some regions to areas of necrosis and cavitation. Glial fibrillary acidic protein (GFAP) staining was reduced or lost at these sites (Misu *et al.*, 2006a,b). In contrast, MS lesions showed preservation or upregulation of AQP4 in demyelinated lesions. Details concerning the stage of demyelinating activity of the NMO and MS lesions and other potential factors that might influence AQP4 expression or detection were not provided.

The discovery of NMO-IgG represents the first known specific autoantibody marker of a CNS inflammatory demyelinating disease. Although its clinical utility is now well established, it is not yet known whether the NMO-IgG is the primary effector of the disease or simply a secondary or epiphenomenon. The specificity of NMO-IgG as a marker for clinically confirmed NMO and its immunoreactive sites in the spinal cord (abluminal surface of blood vessels and astrocytic foot processes), where NMO pathology occurs, is consistent with its potential involvement in disease pathogenesis; NMO may turn out to be an autoimmune channelopathy. If NMO-IgG is the fundamental cause of the disease, it is hypothesized that peripheral antibody must somehow access the CNS—either through regions of absent blood–brain barrier (e.g., circumventricular organs), susceptible barrier (e.g., possibly the spinal cord), or at sites of barrier damage. Lennon *et al.* (2005) postulated that NMO-IgG might cause complement activation, a fundamental pathological finding in NMO, by one of two mechanisms. The homotetrameric structure of AQP4 might directly induce complement activation after NMO-IgG binding. Alternatively, NMO-IgG may interfere with aquaporin-regulated fluid homeostatic mechanisms, resulting in endothelial leakage and secondary complement activation. These potential mechanisms remain speculative and await the results of passive transfer experiments and evaluation of NMO animal models, but the discovery of this putative biomarker will allow a new dimension of hypothesis-driven investigations in CNS inflammatory demyelinating diseases.

F. Observations from Treatment Responses in NMO

Therapeutic observations provide an opportunity to make inferences about disease mechanisms. There are very few controlled therapeutic studies that include patients with NMO; however, there is growing observational experience for both treatment of acute exacerbations and use of immunotherapies to prevent future NMO exacerbations. Both areas provide evidence further supporting humoral autoimmunity in NMO.

1. *Therapy for Acute Exacerbations*

Acute NMO attacks are typically severe episodes of optic neuritis, causing blindness, or myelitis, resulting in paraplegia, quadriplegia, sensory loss, impairment of bowel and bladder function, and sometimes neurogenic respiratory failure. They are associated with the immunopathologic findings reviewed earlier in this chapter. The mainstay of initial therapy for acute NMO attacks is parenteral corticosteroids. This approach is arbitrary and based on extension of principles of treatment of MS attacks but appears to be effective for some patients. The treatment effects of corticosteroids, which include broad-ranging anti-inflammatory mechanisms, can be evident almost immediately. Unfortunately, many patients experience progression of neurological symptoms despite the use of high-dose corticosteroids over several days. In this circumstance, evidence from a randomized, blinded, crossover trial supports the use of plasmapheresis as second-line therapy. Weinshenker *et al.* (1999) studied treatment responses in patients with very severe, corticosteroid-refractory CNS inflammatory/demyelinating events owing to MS, NMO, recurrent or single-event transverse myelitis, acute disseminated encephalomyelitis, Marburg's MS variant, and focal cerebral demyelination. Participants were randomized to initially receive plasmapheresis or a sham exchange procedure. The crossover design of the protocol allowed initial nonresponders to receive the second intervention, meaning that those who were randomized to first receive sham exchange each had an opportunity to undergo the true plasmapheresis procedure. The primary outcome was moderate or greater clinical improvement in the "targeted neurological deficit" (coma, aphasia, hemiplegia, paraplegia, or quadriplegia). The results confirmed the efficacy of plasmapheresis; 8/19 patients (42.1%) responding during that treatment phase compared with 1/19 (5.9%) demonstrating improvement during sham exchange. Evaluation of the responders showed that most were patients with NMO or acute transverse myelitis. A separate retrospective evaluation of response to early plasmapheresis therapy in NMO noted that 6/10 patients with severe, corticosteroid-refractory myelitis attacks had moderate or marked improvement in neurological function within 2 weeks of treatment initiation (Keegan *et al.*, 2002).

These therapeutic observations and controlled trial results are consistent with the hypothesis that one or more humoral factors present in plasma, and therefore removed by plasmapheresis, may play important roles in the pathogenesis of NMO. The responsible factors are not known but might include pathogenic antibodies directed at CNS antigens, circulating immune complexes, complement or activated complement components, or cytokines (Weinshenker *et al.*, 1999). In MS patients, plasmapheresis specifically benefits patients with the "pattern 2" immunopathologic subtype that is associated with complement activation and immunoglobulin deposition (Keegan *et al.*, 2005). As outlined earlier, there are similar (though not identical) pathological findings in NMO. Perhaps extraction of complement components and specific immunoglobulins improves neurological function by restoring immediate neural conduction in affected optic nerve and spinal cord regions with relatively intact axons. In the controlled trial, plasmapheresis responders showed sustained clinical improvement to at least the final examination visit (208 days after treatment). One possible interpretation of this result is that plasmapheresis contributed to disease stabilization by interrupting an active humorally mediated inflammatory cascade.

2. *Therapy to Prevent Attacks*

Two aspects of long-term therapy allow inferences about the immunologic mechanisms that underlie NMO: the apparent failure of standard MS immunomodulatory therapies to prevent NMO attacks and observations that immunosuppression is associated with a more favorable disease course (Wingerchuk and Weinshenker, 2005).

Many patients with NMO are first diagnosed and treated as having atypical or severe MS; hopefully the frequency of early misclassification is shrinking with better awareness of the disease and the availability of NMO-IgG testing. Many clinicians who care for NMO patients have observed that standard MS immunotherapies, such as β-interferon preparations and glatiramer acetate, fail to impact the frequency or severity of attacks. There are no randomized controlled trials dedicated to determining whether standard MS immunomodulatory drugs are efficacious for NMO. A Japanese study compared two doses of subcutaneous interferon β-1b and confirmed superiority of the higher dose preparation (Saida *et al.*, 2005). Twenty-one percent of enrolled patients had the optic-spinal form of MS and the remainder had "Western" MS. An analysis stratified by disease type revealed a trend toward interferon β-1b benefit for relapse reduction (but not relapse freedom) for the optic-spinal group; however, the results were not statistically significant. The study was not powered to detect a difference for this subgroup; therefore, the true effect of β-interferon on Japanese optic-spinal MS is not known.

The apparent failure of β-interferon and glatiramer acetate to stabilize the course of NMO is not unexpected if disease activity is driven by humoral mechanisms. β-Interferon preparations result in immunomodulation independent of antigen, mainly by reducing the secretion of proteolytic matrix metalloproteinases that mediate the T-cell migration across the blood–brain barrier (Yong *et al.*, 1998). They may also downregulate major histocompatibility complex (MHC) class II on various antigen-presenting cells. Glatiramer acetate probably alters T-cell activation by binding to MHC class II molecules (regardless of haplotype), and may cross-react with a variety of CNS antigens (Neuhaus *et al.*, 2001). It may also function as an altered peptide ligand and induce regulatory glatiramer-reactive T cells that produce "bystander suppression" when reactivated within the CNS. Therefore, most if not all direct and indirect actions of the β-interferons and glatiramer acetate occur through T-cell-related processes. The role of T-cell-dependent immune mechanisms in NMO is not known but is probably minor in comparison to B-cell processes.

Clinicians who treat NMO on a regular basis have observed and reported a more favorable outcome with long-term immunosuppressive therapy. In a case series of seven patients, the combination of prednisone and azathioprine was associated with freedom from attacks for up to 18 months (Mandler *et al.*, 1998). Neurological function improved during the period of quiescence. This regimen is still employed as a first-line therapy by many neurologists. A recently published case series ($n = 5$) described the disease course after treatment with mitoxantrone, an anthracenedione approved for worsening relapsing–remitting or secondary progressive MS (Weinstock-Guttman *et al.*, 2006). Mitoxantrone has T- and B-cell suppressive properties. A dose of 12 mg/m^2 administered monthly for 3 or 6 consecutive months, followed by additional doses every 3 months up to a maximum cumulative dose of 100 mg/m^2, was associated with more favorable posttreatment course compared with the pretreatment exacerbation frequency. Others immunosuppressive therapies, such as mycophenolate mofetil, cyclophosphamide, and methotrexate, have been noted to induce short-term remissions. The mechanism of these nonspecific immunosuppressive drugs includes effects on B cells and antibody production in addition to alteration of T-cell responses.

A more directed evaluation of B-cell-specific immunotherapy comes from a case series of active NMO patients treated with the monoclonal antibody rituximab (Cree *et al.*, 2005). Rituximab is a chimeric murine/human anti-CD20 monoclonal antibody that specifically depletes pre-B and mature B cells. Eight patients with active, relapsing NMO that failed to respond to other immunotherapies, such as β-interferon ($n = 4$), glatiramer acetate ($n = 1$), azathioprine ($n = 3$), intravenous immunoglobulin ($n = 3$), and mitoxantrone ($n = 1$), improved after receiving rituximab. Four weekly 375-mg/m^2 intravenous infusions of rituximab resulted in sustained peripheral B-cell depletion and treatment was associated with maintenance of relapse-free status over the next 12 months (range of follow-up 6–18 months) in six of eight patients. When B cells reemerged, as assessed by flow-cytometry detection of CD19-positive cells, retreatment was offered with two consecutive infusions of 1000 mg administered 2 weeks apart. The median annualized attack rate was 2.6 attacks per patient per year in the pretreatment period and 0 attacks per patient per year after rituximab therapy ($p = 0.0078$). Seven of eight patients had significant recovery of neurological function with reduction in the median Expanded Disability Status Scale score from 7.5 before treatment to 5.5 at last posttreatment examination ($p = 0.013$). This uncontrolled observation requires confirmation in the setting of a randomized and appropriately controlled clinical trial but is promising and, based on the weight of evidence for humoral autoimmunity in NMO, represents a logical and more specific therapeutic approach compared with other general immunosuppressive agents.

IV. Conclusions

The case for NMO as a distinct disorder has now been built on the foundations of clinical, diagnostic imaging, immunology, immunopathology, treatment effects, and the high specificity of NMO-IgG. The spectrum of the disorder has been widened to include a proportion of patients with LETM, recurrent optic neuritis, Asian optic-spinal MS, and patients with coexisting connective tissue diseases. The weight of this evidence strongly implicates humoral autoimmune mechanisms in the immunopathogenesis of NMO and NMO spectrum disorders.

The identification of AQP4 as the target antigen of NMO-IgG now sets the stage for focused laboratory studies to determine whether this autoantibody is indeed pathogenic. Unequivocal demonstration that NMO is an aquaporin channelopathy would create a new class of primary CNS disease and clearly delineate, for the first time, the primary pathogenesis of one of the inflammatory demyelinating diseases of the CNS. This would, in turn, increase confidence that a similar approach integrating clinical and laboratory observations can identify other distinct diseases that can be "split" from those currently combined under the umbrella term of "MS."

References

Agre, P., and Kozono, D. (2003). Aquaporin water channels: Molecular mechanisms for human diseases. *FEBS Lett.* **555,** 72–78.

Amiry-Moghaddam, M., and Ottersen, O. P. (2003). The molecular basis of water transport in the brain. *Nat. Rev. Neurosci.* **4,** 991–1001.

Bettelli, E., Baeten, D., Jager, A., Sobel, R. A., and Kuchroo, V. K. (2006). Myelin oligodendrocyte glycoprotein-specific T and B cells cooperate to induce a Devic-like disease in mice. *J. Clin. Invest.* **116,** 2393–2402.

Boumpas, D. T., Patronas, N. J., Dalakas, M. C., Hakim, C. A., Klippel, J. H., and Balow, J. E. (1990). Acute transverse myelitis in systemic lupus erythematosus: Magnetic resonance imaging and review of the literature. *J. Rheumatol.* **17,** 89–92.

Correale, J., and Fiol, M. (2004). Activation of humoral immunity and eosinophils in neuromyelitis optica. *Neurology* **63,** 2363–2370.

Cree, B. A., Goodin, D. S., and Hauser, S. L. (2002). Neuromyelitis optica. *Semin. Neurol.* **22,** 105–122.

Cree, B. A., Lamb, S., Morgan, K., Chen, A., Waubant, E., and Genain, C. (2005). An open label study of the effects of rituximab in neuromyelitis optica. *Neurology* **64,** 1270–1272.

de Seze, J. (2003). Neuromyelitis optica. *Arch. Neurol.* **60,** 1336–1338.

Devic, E. (1894). Myelite aigue compliquee de nevrite optique. *Bull. Med. (Paris)* **8,** 1033–1034.

Gold, R., and Linington, C. (2002). Devic's disease: Bridging the gap between laboratory and clinic. *Brain* **125,** 1425–1427.

Hummers, L. K., Krishnan, C., Casciola-Rosen, L., Rosen, A., Morris, S., Mahoney, J. A., Kerr, D. A., and Wigley, F. M. (2004). Recurrent transverse myelitis associates with anti-Ro (SSA) autoantibodies. *Neurology* **62,** 147–149.

Ishizu, T., Osoegawa, M., Mei, F. J., Kikuchi, H., Tanaka, M., Takakura, Y., Minohara, M., Murai, H., Mihara, F., Taniwaki, T., and Kira, J. (2005). Intrathecal activation of the IL-17/IL-8 axis in opticospinal multiple sclerosis. *Brain* **128,** 988–1002.

Jarius, S., Friedemann, P., Aktas, O., Franciotta, D., Bergamaschi, R., Siffrin, V., Bauer, O., Bluthner, M., Hohlfeld, R., Vincent, A., Seelig, H. P., and Zipp, F. (2006). Antibodies to aquaporin-4 in the diagnosis and differential diagnosis of neuromyelitis optica. Late Breaking Abstract, 22nd Congress of the European Commitee for Treatment and Research in Multiple Sclerosis (ECTRIMS), Madrid, Spain.

Keegan, M., Pineda, A. A., McClelland, R. L., Darby, C. H., Rodriguez, M., and Weinshenker, B. G. (2002). Plasma exchange for severe attacks of CNS demyelination: Predictors of response. *Neurology* **58,** 143–146.

Keegan, M., Konig, F., McClelland, R., Bruck, W., Morales, Y., Bitsch, A., Panitch, H., Lassmann, H., Weinshenker, B., Rodriguez, M., Parisi, J., and Lucchinetti, C. F. (2005). Relation between humoral pathological changes in multiple sclerosis and response to therapeutic plasma exchange. *Lancet* **366,** 579–582.

Krishnamoorthy, G., Lassmann, H., Wekerle, H., and Holz, A. (2006). Spontaneous opticospinal encephalomyelitis in a double-transgenic mouse model of autoimmune T cell/B cell cooperation. *J. Clin. Invest.* **116,** 2385–2392.

Lalive, P. H., Menge, T., Barman, I., Cree, B. A., and Genain, C. P. (2006). Identification of new serum autoantibodies in neuromyelitis optica using protein microarrays. *Neurology* **67,** 176–177.

Lefkowitz, D., and Angelo, J. N. (1984). Neuromyelitis optica with unusual vascular changes. *Arch. Neurol.* **41,** 1103–1105.

Lennon, V. A., Wingerchuk, D. M., Kryzer, T. J., Pittock, S. J., Lucchinetti, C. F., Fujihara, K., Nakashima, I., and Weinshenker, B. G. (2004). A serum autoantibody marker of neuromyelitis optica: Distinction from multiple sclerosis. *Lancet* **364,** 2106–2112.

Lennon, V. A., Kryzer, T. J., Pittock, S. J., Verkman, A. S., and Hinson, S. R. (2005). IgG marker of optic-spinal multiple sclerosis binds to the aquaporin-4 water channel. *J. Exp. Med.* **202,** 473–477.

Littleton, E. T., Jacob, A., Boggild, M., and Palace, J. (2006). An audit of the diagnostic usefulness of the NMO-IgG assay for neuromyelitis optica. *Mult. Scler.* **12,** S156.

Lucchinetti, C., Bruck, W., Parisi, J., Scheithauer, B., Rodriguez, M., and Lassmann, H. (2000). Heterogeneity of multiple sclerosis lesions: Implications for the pathogenesis of demyelination. *Ann. Neurol.* **47,** 707–717.

Lucchinetti, C. F., Mandler, R. N., McGavern, D., Bruck, W., Gleich, G., Ransohoff, R. M., Trebst, C., Weinshenker, B., Wingerchuk, D., Parisi, J. E., and Lassmann, H. (2002). A role for humoral mechanisms in the pathogenesis of Devic's neuromyelitis optica. *Brain* **125,** 1450–1461.

Manabe, Y., Sasaki, C., Warita, H., Hayashi, T., Shiro, Y., Sakai, K., Kashihara, K., and Abe, K. (2000). Sjogren's syndrome with acute transverse myelopathy as the initial manifestation. *J. Neurol. Sci.* **176,** 158–161.

Mandler, R. N., Ahmed, W., and Dencoff, J. E. (1998). Devic's neuromyelitis optica: A prospective study of seven patients treated with prednisone and azathioprine. *Neurology* **51,** 1219–1220.

Mandler, R. N., Dencoff, J. D., Midani, F., Ford, C. C., Ahmed, W., and Rosenberg, G. A. (2001). Matrix metalloproteinases and tissue inhibitors of metalloproteinases in cerebrospinal fluid differ in multiple sclerosis and Devic's neuromyelitis optica. *Brain* **124,** 493–498.

Marignier, R., De Seze, J., Durand-Dubief, F., Vukusic, S., Zephir, H., Vermersch, P., Cabre, P., Cavillon, G., Honnorat, J., and Confavreux, C. (2006). NMO-IgG: A French experience. *Mult. Scler.* **12,** S4.

Misu, T., Fujihara, K., Kakita, A., Murakami, K., Konno, H., Nakamura, M., Endo, M., Watanabe, S., Takahashi, T., Nakashima, I., Takahashi, H., and Itoyama, Y. (2006a). Loss of aquaporin-4 in active perivascular lesions in neuromyelitis optica. 22nd Congress of the European Committee for Treatment and Research in Multiple Sclerosis, Madrid, Spain.

Misu, T., Fujihara, K., Nakamura, M., Murakami, K., Endo, M., Konno, H., and Itoyama, Y. (2006b). Loss of aquaporin-4 in active perivascular lesions in neuromyelitis optica: A case report. *Tohoku J. Exp. Med.* **209,** 269–275.

Nakashima, I., Fujihara, K., Fujimori, J., Narikawa, K., Misu, T., and Itoyama, Y. (2004). Absence of IgG1 response in the cerebrospinal fluid of relapsing neuromyelitis optica. *Neurology* **62,** 144–146.

Nakashima, I., Fujihara, K., Miyazawa, I., Misu, T., Narikawa, K., Nakamura, M., Watanabe, S., Takahashi, T., Nishiyama, S., Shiga, Y., Sato, S., Weinshenker, B. G., *et al.* (2006). Clinical and MRI features of Japanese patients with multiple sclerosis positive for NMO-IgG. *J. Neurol. Neurosurg. Psychiatr.* **77,** 1073–1075.

Neuhaus, O., Farina, C., Wekerle, H., and Hohlfeld, R. (2001). Mechanisms of action of glatiramer acetate in multiple sclerosis. *Neurology* **56,** 702–708.

Nielsen, S., Nagelhus, E. A., Amiry-Moghaddam, M., Bourque, C., Agre, P., and Ottersen, O. P. (1997). Specialized membrane domains for water transport in glial cells: High-resolution immunogold cytochemistry of aquaporin-4 in rat brain. *J. Neurosci.* **17,** 171–180.

O'Riordan, J. I., Gallagher, H. L., Thompson, A. J., Howard, R. S., Kingsley, D. P., Thompson, E. J., McDonald, W. I., and Miller, D. H. (1996). Clinical, CSF, and MRI findings in Devic's neuromyelitis optica. *J. Neurol. Neurosurg. Psychiatr.* **60,** 382–387.

Ortiz de Zarate, J. C., Tamaroff, L., Sica, R. E., and Rodriguez, J. A. (1968). Neuromyelitis optica versus subacute necrotic myelitis. II. Anatomical study of two cases. *J. Neurol. Neurosurg. Psychiatr.* **31,** 641–645.

Pittock, S. J., Lennon, V. A., Krecke, K., Wingerchuk, D. M., Lucchinetti, C. F., and Weinshenker, B. G. (2006a). Brain abnormalities in neuromyelitis optica. *Arch. Neurol.* **63,** 390–396.

Pittock, S. J., Lennon, V. A., Wingerchuk, D. M., Homberger, H. A., Lucchinetti, C. F., and Weinshenker, B. G. (2006b). The prevalence of non-organ-specific autoantibodies and NMO-IgG in neuromyelitis optica (NMO) and related disorders. *Neurology* **66,** A307.

Pittock, S. J., Weinshenker, B. G., Lucchinetti, C. F., Wingerchuk, D. M., Corboy, J. R., and Lennon, V. A. (2006c). Neuromyelitis optica brain lesions localized at sites of high aquaporin 4 expression. *Arch. Neurol.* **63,** 964–968.

Poppe, A. Y., Lapierre, Y., Melancon, D., Lowden, D., Wardell, L., Fullerton, L. M., and Bar-Or, A. (2005). Neuromyelitis optica with hypothalamic involvement. *Mult. Scler.* **11,** 617–621.

Ransohoff, R. M. (2006). A mighty mouse: Building a better model of multiple sclerosis. *J. Clin. Invest.* **116,** 2313–2316.

Saida, T., Tashiro, K., Itoyama, Y., Sato, T., Ohashi, Y., and Zhao, Z. (2005). Interferon beta-1b is effective in Japanese RRMS patients: A randomized, multicenter study. *Neurology* **64,** 621–630.

Storch, M. K., Stefferl, A., Brehm, U., Weissert, R., Wallstrom, E., Kerschensteiner, M., Olsson, T., Linington, C., and Lassmann, H. (1998). Autoimmunity to myelin oligodendrocyte glycoprotein in rats mimics the spectrum of multiple sclerosis pathology. *Brain Pathol.* **8,** 681–694.

Weinshenker, B. G., O'Brien, P. C., Petterson, T. M., Noseworthy, J. H., Lucchinetti, C. F., Dodick, D. W., Pineda, A. A., Stevens, L. N., and Rodriguez, M. (1999). A randomized trial of plasma exchange in acute central nervous system inflammatory demyelinating disease. *Ann. Neurol.* **46,** 878–886.

Weinshenker, B. G., De Seze, J., Vermersch, P., Pittock, S., and Lennon, V. (2006a). The relationship between neuromyelitis optica and systemic autoimmune disease. *Neurology* **66,** A380–A381.

Weinshenker, B. G., Wingerchuk, D. M., Nakashima, I., Fujihara, K., and Lennon, V. A. (2006b). OSMS is NMO, but not MS: Proven clinically and pathologically. *Lancet Neurol.* **5,** 110–111.

Weinshenker, B. G., Wingerchuk, D. M., Vukusic, S., Linbo, L., Pittock, S. J., Lucchinetti, C. F., and Lennon, V. A. (2006c). Neuromyelitis optica IgG predicts relapse after longitudinally extensive transverse myelitis. *Ann. Neurol.* **59,** 566–569.

Weinstock-Guttman, B., Ramanathan, M., Lincoff, N., Napoli, S. Q., Sharma, J., Feichter, J., and Bakshi, R. (2006). Study of mitoxantrone for the treatment of recurrent neuromyelitis optica (Devic disease). *Arch. Neurol.* **63,** 957–963.

Wingerchuk, D. M. (2004). Neuromyelitis optica: Current concepts. *Front. Biosci.* **9,** 834–840.

Wingerchuk, D. M. (2006). Evidence for humoral autoimmunity in neuromyelitis optica. *Neurol. Res.* **28,** 348–353.

Wingerchuk, D. M., and Weinshenker, B. G. (2003). Neuromyelitis optica: Clinical predictors of a relapsing course and survival. *Neurology* **60,** 848–853.

Wingerchuk, D. M., and Weinshenker, B. G. (2005). Neuromyelitis optica. *Curr. Treat. Options Neurol.* **7,** 173–182.

Wingerchuk, D. M., Hogancamp, W. F., O'Brien, P. C., and Weinshenker, B. G. (1999). The clinical course of neuromyelitis optica (Devic's syndrome). *Neurology* **53,** 1107–1114.

Wingerchuk, D. M., Lennon, V. A., Pittock, S. J., Lucchinetti, C. F., and Weinshenker, B. G. (2006). Revised diagnostic criteria for neuromyelitis optica. *Neurology* **66,** 1485–1489.

Yong, V. W., Chabot, S., Stuve, O., and Williams, G. (1998). Interferon beta in the treatment of multiple sclerosis: Mechanisms of action. *Neurology* **51,** 682–689.

Zuliani, L., Blanco, Y., Tavolato, B., Giometto, B., Graus, F., and Saiz, A. (2006a). Neuromyelitis optica autoantibody (NMO-IgG) in patients with suspected NMO or limited forms of NMO. *Mult. Scler.* **12,** S155–S156.

Zuliani, L., Lopez de Munain, A., Ruiz Martinez, J., Olascoaga, J., Graus, F., and Saiz, A. (2006b). [NMO-IgG antibodies in neuromyelitis optica: A report of 2 cases]. *Neurologia* **21,** 314–317.

INDEX

D

H

I

J

O

Q

R

U

V

W

X

Z

CONTENTS OF RECENT VOLUMES

Volume 39

Volume 40

Volume 41

Volume 42

Volume 43

Volume 44

Volume 45

Volume 49

Volume 50

Volume 51

Volume 52

Volume 53

Volume 54

Volume 55

Volume 56

Volume 57

Volume 58

Volume 59

Volume 60

Volume 61

Volume 62

Volume 63

Volume 64

Volume 65

Volume 66

Volume 67

Volume 68

Volume 69

Volume 70

Volume 71

Volume 72

Volume 73

Volume 74

Volume 75

Volume 76

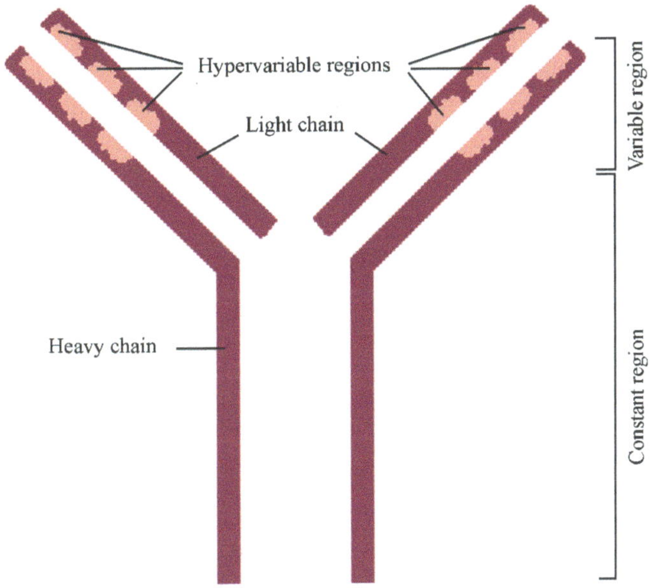

NIKBIN *ET AL.*, CHAPTER 2, FIG. 1. (See Legend in Text.)

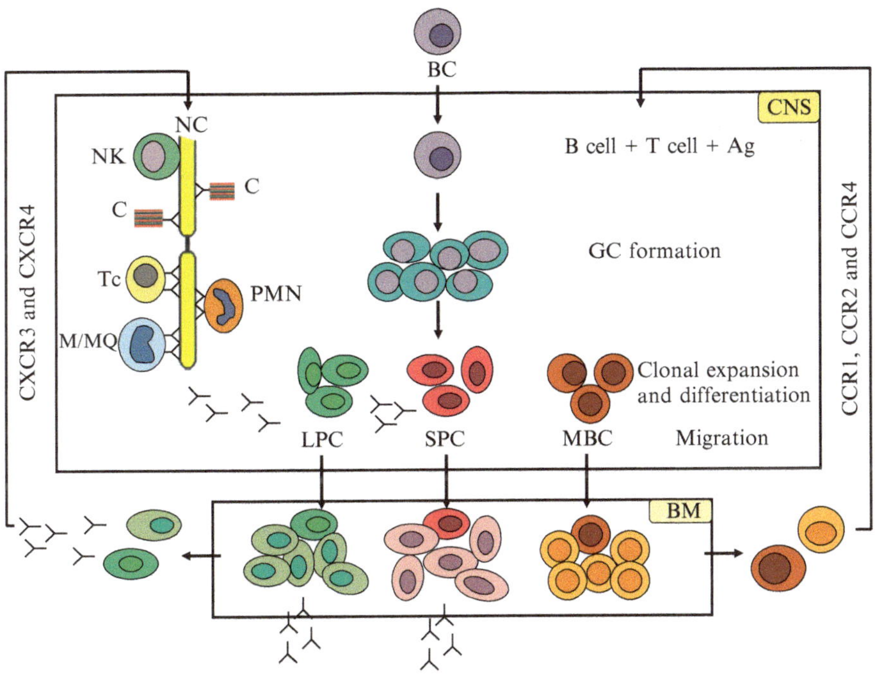

NIKBIN *ET AL.*, CHAPTER 2, FIG. 3. (See Legend in Text.)

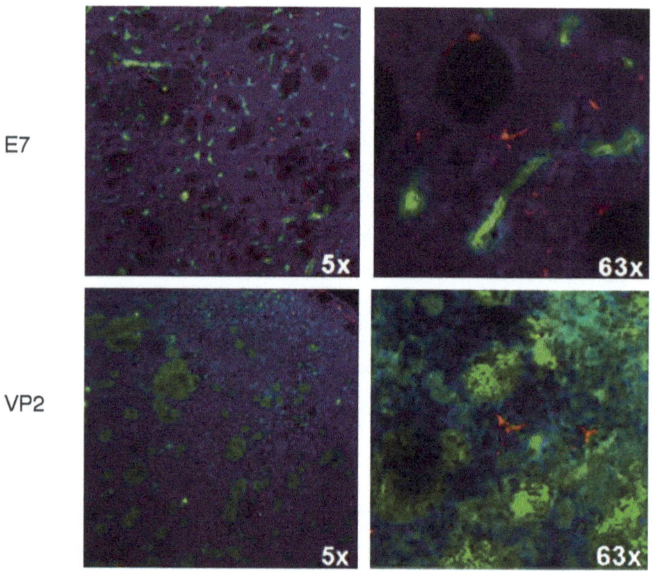

JOHNSON *ET AL.*, CHAPTER 4, FIG. 5. (See Legend in Text.)

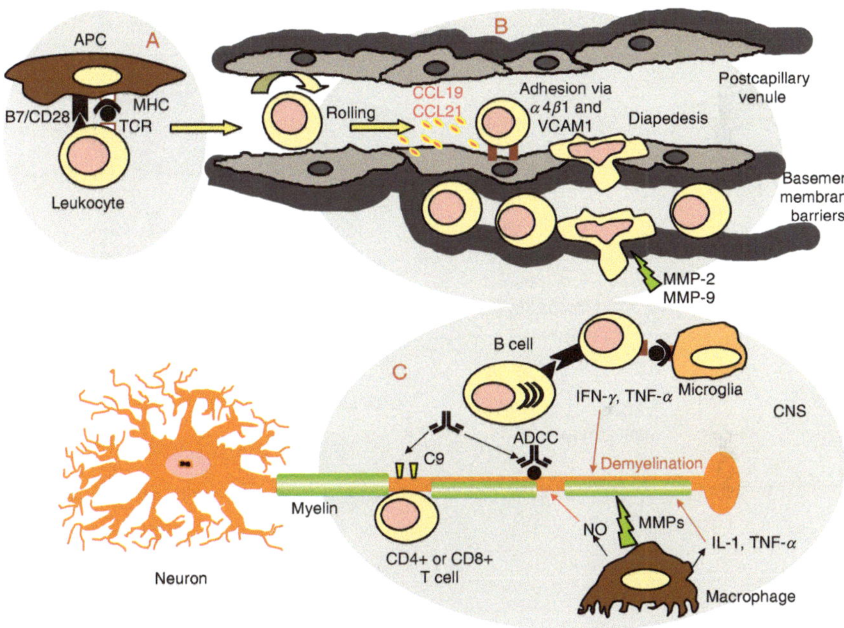

AGRAWAL AND YONG, CHAPTER 5, FIG. 1. (See Legend in Text.)

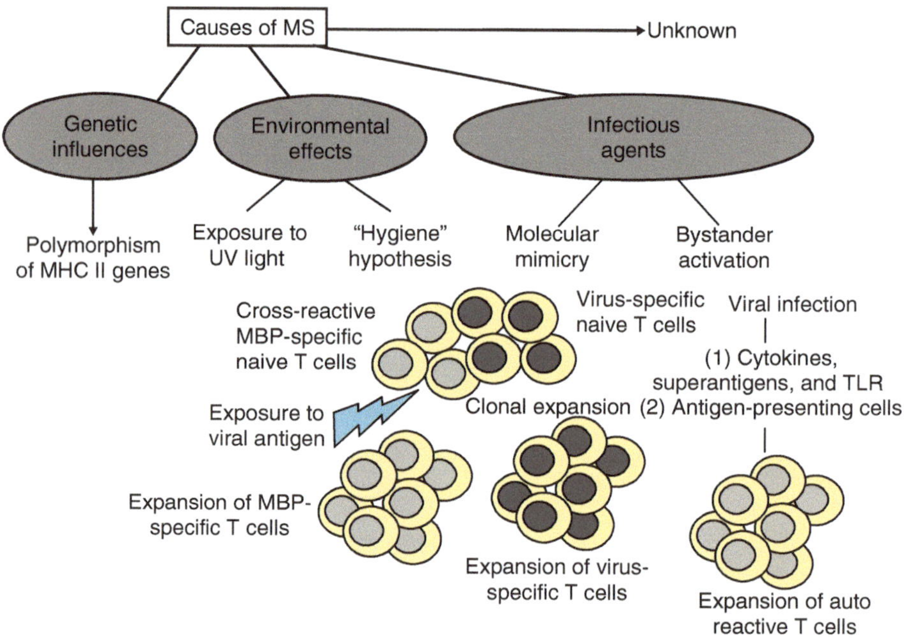

Agrawal and Yong, Chapter 5, Fig. 2. (See Legend in Text.)

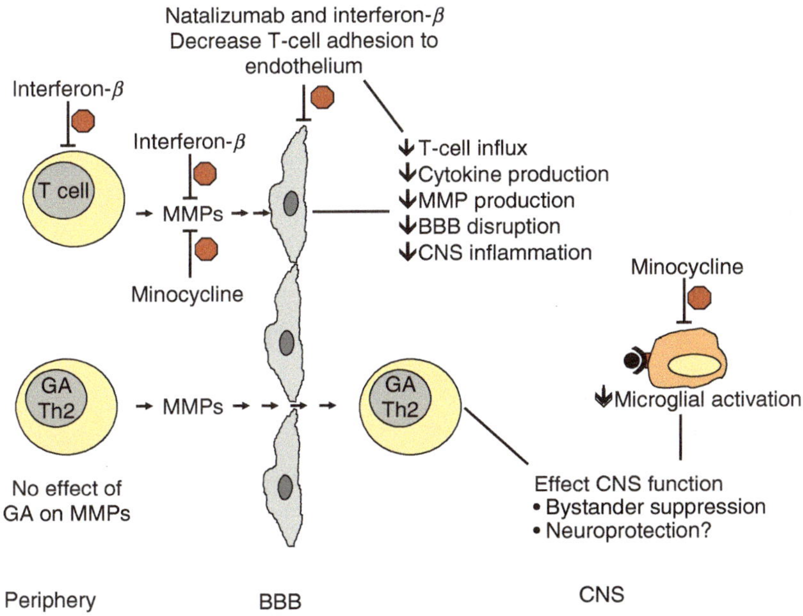

AGRAWAL AND YONG, CHAPTER 5, FIG. 4. (See Legend in Text.)

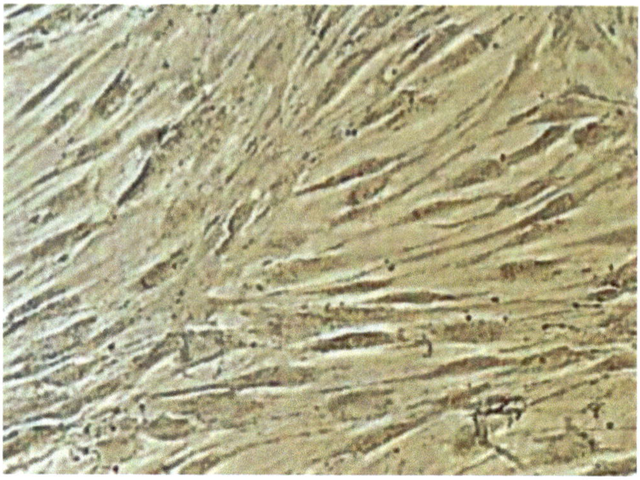

NIKBIN *ET AL.*, CHAPTER 8, FIG. 1. (See Legend in Text.)

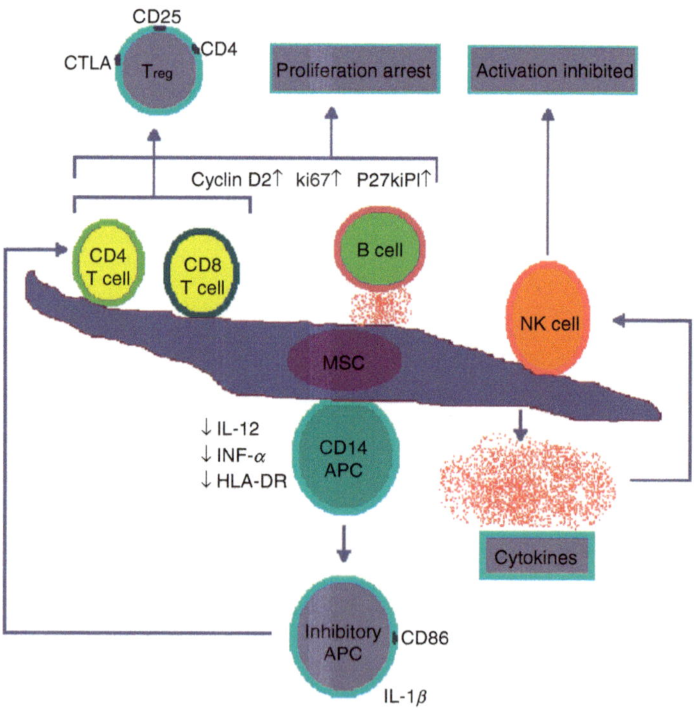

NIKBIN *ET AL.*, CHAPTER 8, FIG. 2. (See Legend in Text.)

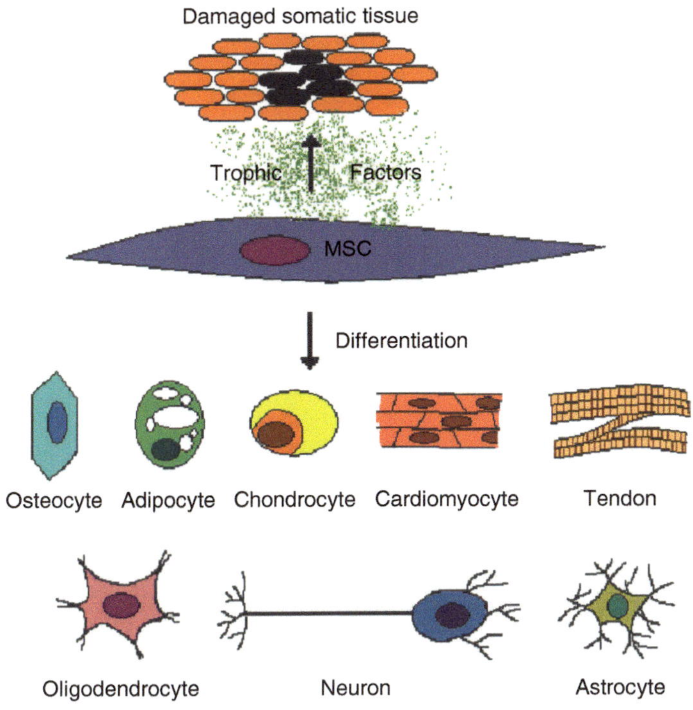

NIKBIN *ET AL.*, CHAPTER 8, FIG. 3. (See Legend in Text.)

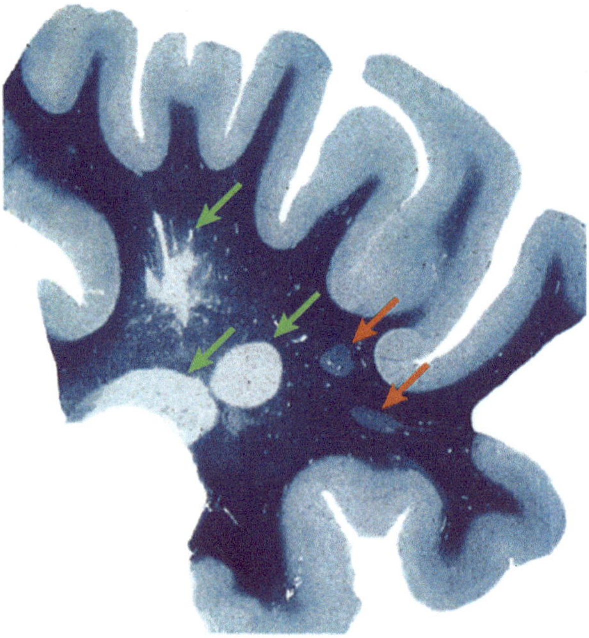

CHARI, CHAPTER 26, FIG. 1. (See Legend in Text.)

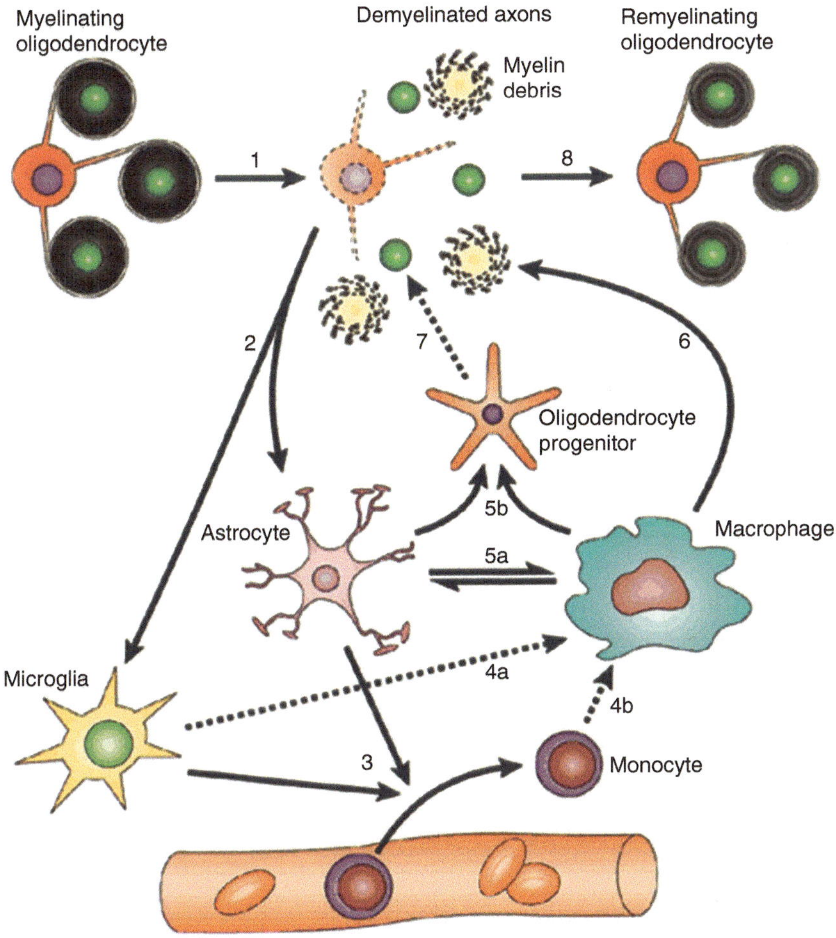

CHARI, CHAPTER 26, FIG. 6. (See Legend in Text.)

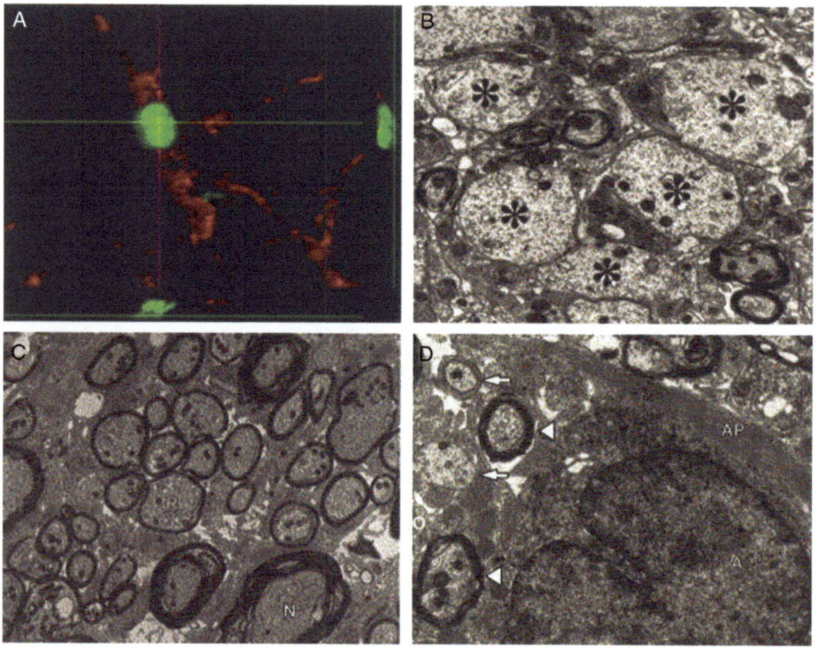

CHARI, CHAPTER 26, FIG. 7. (See Legend in Text.)

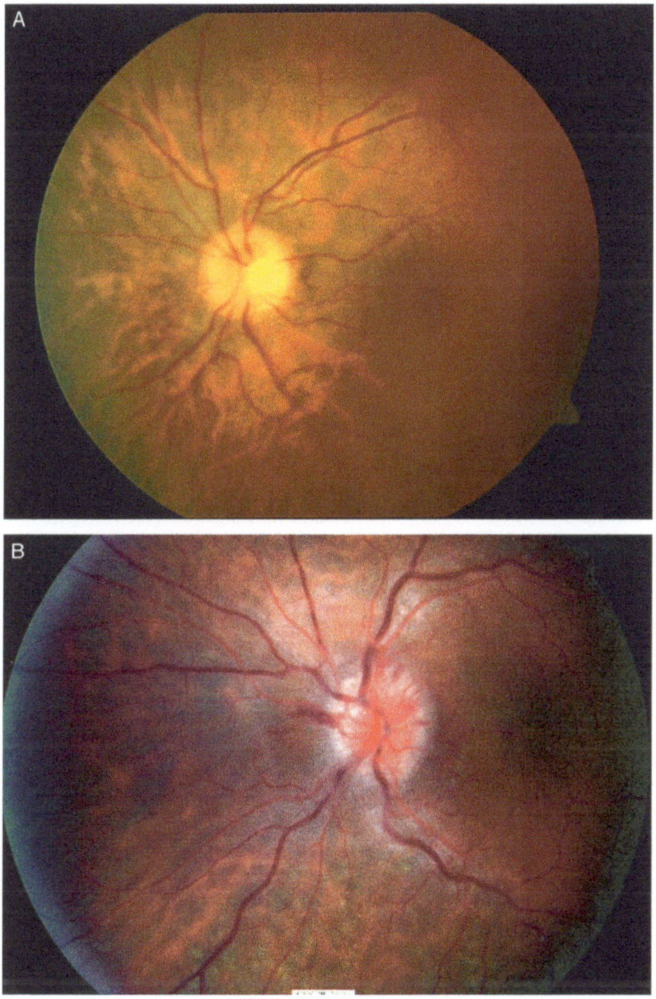

KAUR AND BENNETT, CHAPTER 28, FIG. 1. (See Legend in Text.)

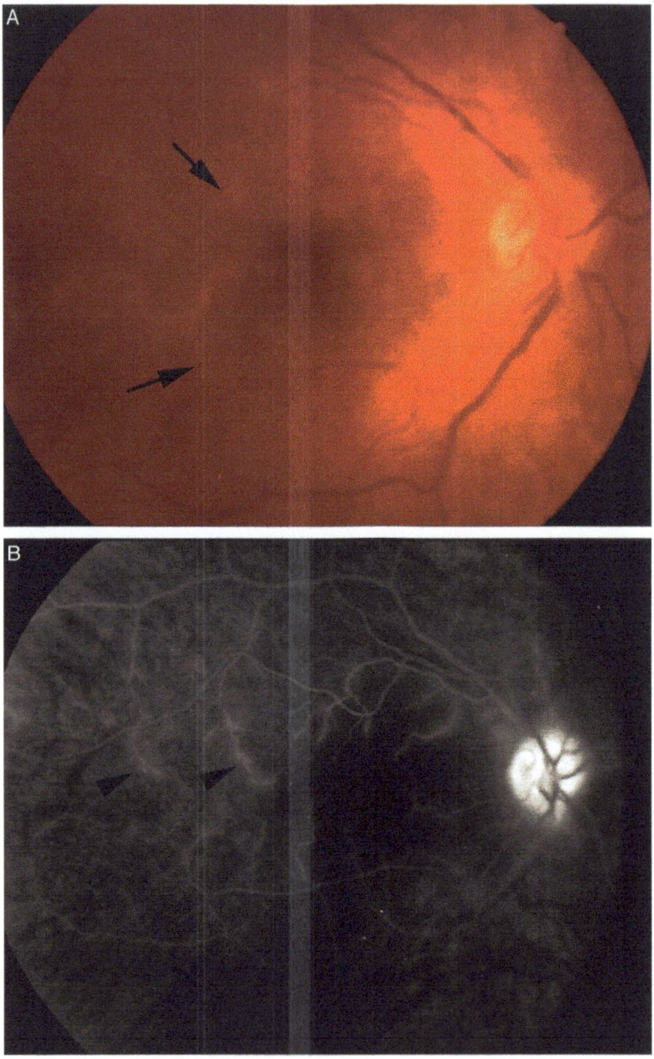

Kaur and Bennett, Chapter 28, Fig. 2. (See Legend in Text.)

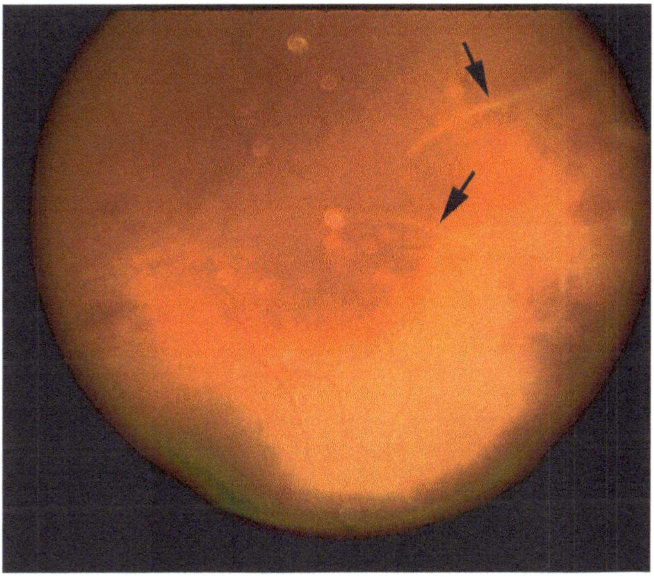

KAUR AND BENNETT, CHAPTER 28, FIG. 3. (See Legend in Text.)

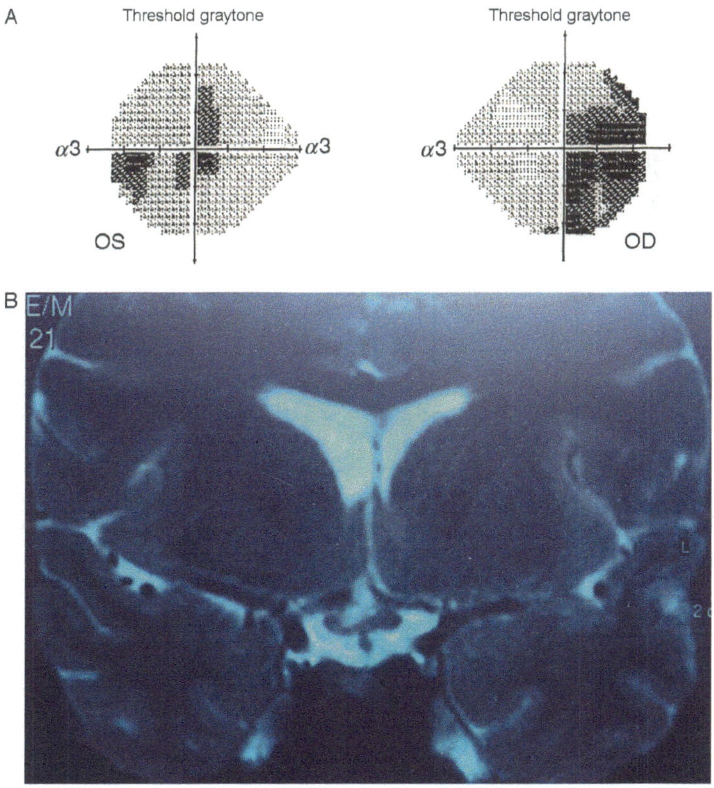

KAUR AND BENNETT, CHAPTER 28, FIG. 4. (See Legend in Text.)

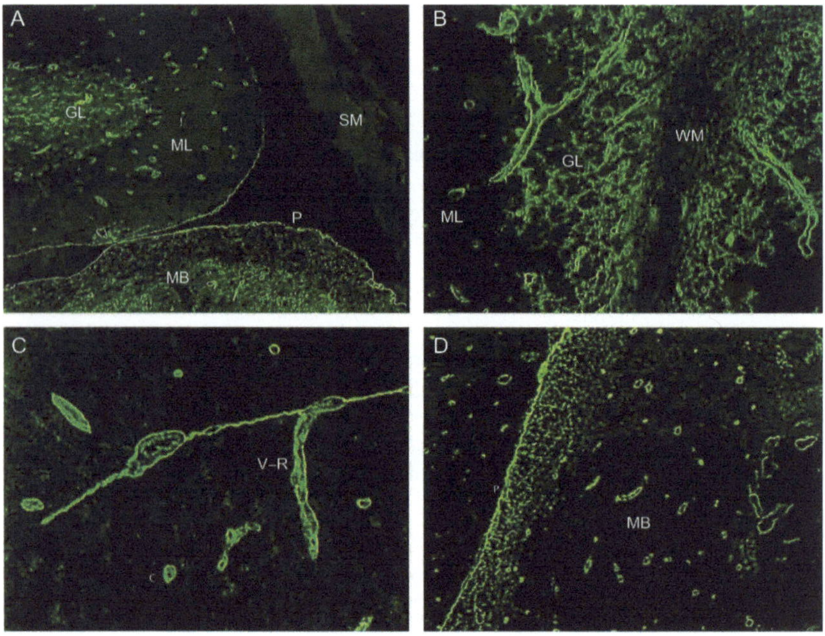

WINGERCHUK, CHAPTER 29, FIG. 3. (See Legend in Text.)

Kolibri 360 GmbH
Ferdinand-Jühlke-Straße 7
99095 Erfurt, Deutschland
produktsicherheit@kolibri360.de